Geophysical Exploration for Hydrocarbon Reservoirs, Geothermal Energy, and Carbon Storage

Geophysical Exploration for Hydrocarbon Reservoirs, Geothermal Energy, and Carbon Storage

New Technologies and AI-based Approaches

Edited by

Said Gaci
Sonatrach's Central Research and Development Directorate (DC R&D),
Boumerdes, Algeria

WILEY

This edition first published 2026

Registered Offices
John Wiley & Sons, Inc., 111 River Street, Hoboken, NJ 07030, USA
John Wiley & Sons Ltd, Oldlands Way, Bognor Regis PO22 9NQ, UK

For details of our global editorial offices, customer services, and more information about Wiley products visit us at www.wiley.com.

The manufacturer's authorized representative according to the EU General Product Safety Regulation is Wiley-VCH GmbH, Boschstr. 12, 69469 Weinheim, Germany, e-mail: Product_Safety@wiley.com.

Wiley also publishes its books in a variety of electronic formats and by print-on-demand. Some content that appears in standard print versions of this book may not be available in other formats.

Library of Congress Cataloging-in-Publication Data:

Names: Gaci, Said editor
Title: Geophysical exploration for hydrocarbon reservoirs, geothermal energy, and carbon storage : new technologies and AI-based approaches / Edited by Said Gaci, Sonatrach-Algerian Institute of Petroleum (IAP), Boumerdes, Algeria.
Description: Hoboken, NJ : John Wiley & Sons Ltd, 2026. | Includes bibliographical references and index.
Identifiers: LCCN 2025021123 | ISBN 9781394261536 hardback | ISBN 9781394261550 adobe pdf | ISBN 9781394261543 epub | ISBN 9781394261567 ebook
Subjects: LCSH: Prospecting—Geophysical methods | Geothermal resources | Hydrocarbon reservoirs | Carbon sequestration | Petroleum—Prospecting | Artificial intelligence—Geophysical applications
Classification: LCC TN269 .G3985 2026 | DDC 622/.1028563—dc23/eng/20250709
LC record available at https://lccn.loc.gov/2025021123

Cover Design: Wiley
Cover Image: © Jouni Niskakoski/Shutterstock

Set in 9.5/12.5pt STIXTwoText by Lumina Datamatics

Contents

List of Contributors

Mohamed Amrouche
slb – D&I Digital Subsurface Solutions, Algiers, Algeria

Aldjia Boualam
North Dakota University, Grand Forks, USA

Amar Boudella
Geophysics Department – FSTGAT,
University of Sciences and Technology Houari Boumediene (USTHB), Algiers, Algeria

Salah Boufenchouche
Sonatrach, Exploration Division, Exploration Operations Directorate, Ouargla, Algeria

Abdelbasset Boulassel
Geophysics Department – FSTGAT,
University of Sciences and Technology Houari Boumediene (USTHB), Algiers, Algeria
Sonatrach, Exploration Division, Exploration Operations Directorate, Ouargla, Algeria

Zinelaabidine Boumelit
Geophysics Department – FSTGAT,
University of Sciences and Technology Houari Boumediene (USTHB), Algiers, Algeria

Fethi Ali Cheddad
Geophysics Department – FSTGAT,
University of Sciences and Technology Houari Boumediene (USTHB), Algiers, Algeria
Sonatrach, Exploration Division, Exploration Operations Directorate, Ouargla, Algeria

Hichem Aymen Katib Chellal
University of North Dakota, Grand Forks, USA

Sofiane Djezzar
North Dakota University, Grand Forks, USA

Mohammed Farfour
Department of Earth Sciences, College of Science, Sultan Qaboos University, Muscat, Oman

Said Gaci
Director of Scientific and Technical Support to Research at Sonatrach's Central Research and Development Directorate (DC R&D)

William Gosnold
University of North Dakota, Grand Forks, North Dakota, USA

Ghoulem Ifrene
University of North Dakota, Grand Forks, North Dakota, USA

Elena Karnaukhova
S. Subbotin Institute of Geophysics NAS of Ukraine, Kyiv, Ukraine

Aziz Khelalef
Sonatrach – Division Exploration, Boumerdes, Algeria

Ignat Korchagin
Institute of Geophysics, NAS of Ukraine, Kyiv, Ukraine

Valeriy Korchin
S. Subbotin Institute of Geophysics NAS of Ukraine, Kyiv, Ukraine

Aimen Laalam
Colorado School of Mines, Golden, Colorado, USA

Soraya Makhlouf
Geophysics Department – FSTGAT, University of Sciences and Technology Houari Boumediene (USTHB), Algiers, Algeria

Ahmed Merzoug
University of Texas at Austin, Austin, Texas, USA

Hachay Olga Alexandrovna
Institute of Geophysics, UB RAS, Ekaterinburg, Russia

Kuldeep Singh
Earth Science Department, Kent State University, Kent, Ohio, USA

Valery Soloviev
Institute of Geophysics, NAS of Ukraine, Kyiv, Ukraine

Steven A. Tedesco
Running Foxes Energy Inc., Centennial, Colorado, USA

Mykola Yakymchuk
Institute of Applied Problems of Ecology, Geophysics and Geochemistry, Kyiv, Ukraine

Hasna Yazid
Department of Computer Science, University of M'Hamed Bougerra (UMBB), Boumerdes, Algeria

Khachay Oleg Yurievitch
Ural Federal University, Ekaterinburg, Russia

Naima Zaourar
Geophysics Department – FSTGAT, University of Sciences and Technology Houari Boumediene (USTHB), Algiers, Algeria

Badis Zegagh
Department of Geophysics, Faculty of Hydrocarbon and Chemistry, University of Boumerdes M'Hamed BOUGARA (UMBB), Independence Avenue, Boumerdes, Algeria

Preface

Nowadays, finding sustainable energy alternatives and practicing optimal resource management offers both opportunities and challenges for a viable energy future. Subsurface exploration plays an increasingly important role in hydrocarbon extraction, geothermal energy production, and carbon storage in order to decarbonize the world's energy supply. Advanced technology, data-driven decision-making, and creative solutions are needed to unlock new possibilities in the fields of petroleum exploration, to broader subsurface energy systems, and to assist the development of alternative energy sources including geothermal energy and the use of CO_2 sequestration, in addition to improving hydrocarbon recovery.

The multidisciplinary approach presented in this book, *Geophysical Exploration for Hydrocarbon Reservoirs, Geothermal Energy, and Carbon Storage: New Technologies and AI-based Approaches*, combines conventional petroleum exploration techniques with innovative advances in artificial intelligence (AI). This volume aims to offer a thorough understanding of subsurface energy systems by linking the most recent technical advancements with new geological insights, and providing solutions for geothermal energy development, petroleum exploration, reservoir characterization, and CO_2 sequestration. Every topic is supported by strong theoretical frameworks and illustrated by real-world case studies from diverse worldwide contexts.

This book aims to give professionals, scholars, and students the knowledge and skills necessary to navigate this changing field as the global energy sector evolves. It provides insightful guidance not only for those working in petroleum exploration industry but also for those involved in geothermal energy development and carbon sequestration projects, both of which are essential for lowering our carbon footprint.

Geophysical Exploration for Hydrocarbon Reservoirs, Geothermal Energy, and Carbon Storage: New Technologies and AI-based Approaches, which combines the latest technologies with well-established geological knowledge, is well-positioned to be an essential tool for anyone hoping to remain at the forefront of subsurface energy exploration in the twenty-first century.

We hope that the knowledge included in this volume will stimulate even more creativity and cross-disciplinary cooperation, fostering the quest for economical and sustainable energy solutions in the future.

Said Gaci

About the Book

by Said Gaci and Mohammed Farfour

Geophysical Exploration for Hydrocarbon Reservoirs, Geothermal Energy, and Carbon Storage: New Technologies and AI-based Approaches is an interdisciplinary book that integrates the fields of petroleum exploration, reservoir characterization, and energy transition. It also highlights the use of new geological knowledge and artificial intelligence techniques in reservoir characterization. Each topic is presented through theoretical frameworks, accompanied by recent case studies from various global contexts. This book provides practical insights into petroleum exploration, reservoir characterization, geothermal energy development, and CO_2 sequestration.

Section 1. New Technologies and Insights into Petroleum Exploration

Chapter 1. Gas Seepage in Marginal Structures as Additional Shallow and Deep Hydrocarbon Systems Indicator (Some of Recent FR Scanning Results), Valery Soloviev, Mykola Yakymchuk, and Ignat Korchagin

Chapter 2. The Role of the LVZ and of Increased Seismicity in the Localization of Abiogenic HC in the Crystalline Crust of Transcarpathia, Valeriy Korchin and Elena Karnaukhova

Chapter 3. Precambrian Mid-Continent Rift Potential for Hosting Numerous Helium and Hydrogen Accumulations, Central USA, Steven A. Tedesco

Chapter 4. Production from Desmoinesian and Atokan Age Coalbed Methane and Carbonaceous Mudstone and Their Relationship to Structure and Geologic History of the Cherokee Basin, Kansas and Oklahoma, USA, Steven A. Tedesco

Chapter 5. Geophysical Research and Monitoring Within the Framework of a Block-Layered Model with Inclusions of a Hierarchical Structure, Olga Hachay and Andrey Khachay

Chapter 1

Direct frequency-resonance (FR)-sounding technologies can identify the source and depth of gas migration in continental margins of the World Ocean (Heggland, 1998; Levashov et al., 2012; Yakymchuk et al., 2022; Yakymchuk and Korchagin, 2023). Discussed cases study taken from marginal migration centers of the Greenland shelf, Norwegian, Barents, and North Seas, and other structures of continental margins validate the impact of crust-mantle gas fluids on degassing mechanisms and support the "volcanic model" for continental margin formation in several part of the Earth. Using FR technologies, seeps can be served as supplementary indicators of hydrocarbon deposits in regions of gas release, helping in the construction of models pertaining to deep structure and abiogenic formation mechanisms. Seeps of methane and oil spills may be signs of volcanic activity and a factor in the changing worldwide climate.

References

Heggland, R. (1998). Gas Seepage As an Indicator of Deeper Prospective Reservoirs. A Study Based on Exploration 3D Seismic Data. Marine and Petroleum Geology 15 (1), pp. 1–9.

Levashov, S. P., Yakymchuk, N. A., & Korchagin, I. N. (2012). Frequency-Resonance Principle, Mobile Geoelectric Technology: A New Paradigm of Geophysical Investigations. Geofizicheskiy zhurnal 34 (4), pp. 166–176 (in Russian).

Yakymchuk, M. & Korchagin, I. (2023). About the Opportunity of Direct-Prospecting Methods Application for Detection Areas of Gas and Natural Hydrogen Migration to The Surface and in The Atmosphere. *Publisher.agency: Proceedings of the 2nd International Scientific Conference «Interdisciplinary Science Studies» (May 4–5, 2023)*. Dublin, Ireland, pp. 220–251. ISBN 978-2-6400-5011-9. https://doi.org/10.5281/zenodo.7905551

Yakymchuk, M., Korchagin, I., Levashov, S., et al. (2022). *Volcanism and Degassing Processes in the Structures of the Earth's Polar Regions (review based on the results of frequency-resonance studies)*. Dodo Books Indian Ocean Ltd. And OmniScriptum S.R.L Publishing Group, p. 276 (in Ukrainian). ISBN: 978-620-0-63606-5. https://morebooks.de/shop-ui/shop/search?q=978-620-0-63606-5&page=1

Chapter 2

The deep genesis of hydrocarbons and their vertical movement along faults are becoming more and more involved in exploration efforts based on the organic hypothesis of oil origin. This means that in order to explain both local anomalies and potential for hydrocarbon exploration and production, a certain region's deep geological structure must be meticulously modeled. The most efficient technique takes into account geothermal, seismic sections of DSS, tectonics, and the physical characteristics of rocks, namely petrophysical thermobaric modeling (PTBM) (Korchin et al., 2020; Korchin and Karnaukhova, 2021). Low velocity zones (LVZs) validate the new hypothesis of thermobaric mineral matter decompaction in the Earth's crust at depths of 5–25 km. Exploring the deep structure of this area and, in particular, determining the location and configuration of the LVZs in the future will greatly increase the potential for finding hydrocarbon resources in Transcarpathia (Ukraine).

References

Korchin, V. A. & Karnaukhova, O. Y. (2021). Peculiarities of Fluid Dynamics in the Seismically Active Zone of the Earth's Crust of the Transcarpathian Depression (Ukraine). Science and education: problems, prospects and innovations. Proceedings of the 8th International Scientific and Practical Conference. CPN Publishing Group. Kyoto, Japan. pp. 45–53. ISBN: 978-4-9783419-5-2. URL: https://sci-conf.com.ua/viii-mezhdunarodnaya-nauchno-prakticheskaya-konferentsiyascience-and-education-problems-prospects-and-innovations-28-30-aprelya-2021-godakioto-yaponiya-arhiv/

Korchin, V. A., Rusakov, O. M., Butrny, P. A., et al. (2020). The Origin of the Low Density Zones in the Crystalline Crust of the Transcarpathian Depression (Ukraine) From Petrophysical Thermobaric Modelling. Geodynamics 1 (28), pp. 81–93. https://doi.org/10.23939/jgd2020.01.081.

Chapter 3

The Mid-Continent Rift (MCR) is a 1.1-billion-year-old Precambrian feature in the central USA, known for its base metal and iron production in the northern regions. It underlies oil and gas-producing areas like as the Nemaha Ridge in Nebraska, Oklahoma, and Kansas, the Michigan Basin, Findley Arch in Ohio, and Indiana and Cincinnati Arch in Kentucky and Tennessee, though its connection to Paleozoic production is unclear. Reactivated Precambrian faults in the MCR provide structural traps and pathways for migrating helium (He), hydrogen (H), and dolomitizing fluids, enhancing Paleozoic reservoirs. Helium and hydrogen show have been reported in several wells, especially in the Kansas zone and near Babbitt, Minnesota. Helium is vital for technology and medicine, and global shortages persist, with key production centers in the USA, Canada, and other regions. Despite exploration efforts in Nebraska, Iowa, Kansas, and Michigan, significant He or H finds have yet to emerge. The scale of the MCR makes exploration challenging, with limited success so far, but the region remains a potential target for He and H resources (Tedesco, 2022; Maiga et al., 2023).

References

Tedesco, S. A. (2022). *Geology of Production of Helium and Associated Gases*. Amsterdam: Elsevier, p. 394.

Maiga, O., Deville, E., Laval, J., et al. (2023). Characterization of the Spontaneously Recharging Natural Hydrogen Reservoirs of Bourakebougou in Mali. www.nature.com/scientificreports.

Chapter 4

The Cherokee Basin, located in southeastern Kansas and northeastern Oklahoma, has been producing gas from Lower Desmoinesian and Atokan-age coalbed methane and carbonaceous mudstone (CBMCM) reservoirs since the 1920s. A major development effort in the 1990s expanded production until gas prices collapsed in 2008, halting further growth. By that time, over 6000 wells were active, with more than 1000 awaiting connection. However, as of 2023, CBMCM exploitation has not resumed, given more profitable shale gas plays across the USA. Despite low production rates, many wells continue to operate without significant decline. The basin is part of the Pennsylvanian Interior Basin, with key gas-producing coals such as the Riverton, Drywood-Rowe, Weir-Pittsburg, and Mulky coals. The most productive wells are located near the Chautauqua Arch, where heat from underlying Precambrian plutons facilitated gas generation. From 1982 to 2011, the basin produced 355 BCF of gas, about 12.6% of the USGS's estimated 2.8 TCF. Despite earlier projections of higher reserves, actual recovery per well has been lower than expected. This study proposes an exploration model for finding similar CBMCM deposits elsewhere (Tedesco, 2014; Newell et al., 2004).

References

Tedesco, S. A. (2014). Reservoir Characterization and Geology of the Coals and Carbonaceous Shale of the Cherokee Group in the Cherokee Basin, Kansas, Missouri and Oklahoma, U.S.A, PhD. Colorado School of Mines 2, p. 201.

Newell, K. D., Johnson, T. A., Brown, W.M., et al. (2004). Geological and Geochemical Factors Influencing the Emerging Coalbed Gas Play in the Cherokee and Forest City Basins in Eastern Kansas, Kansas Geological Survey Open-File Report, 2004–2017, p. 12.

Chapter 5

This review discusses the development of new models of continuous media mechanics. These models are applied to the study of events occurring in permafrost under the impact of thawing processes, as well as to composite and statistically heterogeneous media, new structural materials, difficult arrays in mine and ground settings, and studying phenomena in permafrost under thawing processes (Panin et al. 1985; Goldin, 2002). The authors focus on studying non-powerful and internal degrees of freedom, which are not described by classical elasticity theory and could potentially predict the complex, nonlinear, and synergistic effects of catastrophic processes in rock massifs. In addition, the discussed models could be of high interest during the development processes of oil and gas fields.

References

Goldin, S. V. (2002). Lithosphere Destruction and Physical Mesomechanics. Physical Mesomechanics. 5 (5), pp. 5–22.

Panin, V. E., Likhachev, V. A., & Grinjaev, Yu, V. (1985). Structural Levels of Deformation of Solids. Novosibirsk: From the USSR Academy of Science. p. 226.

Section 2. Reservoir Characterization Concepts and Workflows

Chapter 6. A Review on Shear Wave Velocity Estimation Methods, Said Gaci and Mohammed Farfour

Chapter 7. Geomechanics in Petroleum Exploration, Development, and Energy Transition, Ghoulem Ifrene and Kuldeep Singh

Chapter 8. Size Scaling and Spatial Clustering of Natural Fracture Networks Using Fractal Analysis, Sofiane Djezzar and Aldjia Boualam

Chapter 9. Application of Seismic Attributes on Digital Elevation Model: Fractures Detection and Reservoir Implication, Sofiane Djezzar and Aldjia Boualam

Chapter 10. Structural Analysis and Fracture Kinematics Using Seismic 2D and Geological Maps, Sofiane Djezzar and Aldjia Boualam

Chapter 11. A New Method for Reservoir Fracture Characterization and Modeling Using Surface Analog, Sofiane Djezzar and Aldjia Boualam

Chapter 12. An Integrated Workflow for Multiscale Fracture Analysis in Reservoir Analog, Sofiane Djezzar and Aldjia Boualam

Chapter 6

In petroleum exploration, particularly in geomechanical, rock physics, and geophysical studies of reservoirs, shear wave velocity (Vs) is an essential parameter. Using the available data (e.g. well logs, seismic data, numerical models, and findings from laboratory core analyses) determines which model should be used to estimate it. Vs predictions can be achieved using three categories of methods: empirical relationships, intelligent systems, and rock physical models (Kwak et al., 2015; Gaci 2017; Avseth et al., 2021).

References

Kwak, D. Y., Brandenberg, S. J., Mikami, A., et al. (2015). Prediction Equations for Estimating Shear-Wave Velocity from Combined Geotechnical and Geomorphic Indexes based on Japanese Data Set. Bulletin of the Seismological Society of America 105 (4), pp.179–188.

Gaci, S. (2017). A Novel Model to Estimate S-Wave Velocity Integrating Hölderian Regularity, Empirical Mode Decomposition, and Multilayer Perceptron Neural Networks, In: Oil and Gas Exploration: Methods and Application (ed. S. Gaci and O. Hachay), ch. 12, pp. 181–200. Wiley Editions-USA.

Avseth, P., Lehocki, I., Kjøsnes, Ø., et al. (2021). Data-Driven Rock Physics Analysis of North Sea Tertiary Reservoir Sands. Geophysical Prospecting 69, pp. 608–621. https://doi.org/10.1111/1365-2478.12986.

Chapter 7

This chapter highlights the vital role that geomechanics plays in promoting sustainable energy practices and improving efficiency and safety in petroleum exploration and production (E&P) phases, and supporting sustainable energies (Zoback, 2010; Zoback and Kohli, 2019). In E&P activity, geomechanics is significantly used in several aspects such as formation pressure prediction, casing program design, optimization of well trajectory, wellbore stability analysis, fractured performance evaluation, geological modeling of fractures, etc. In addition, it is of high interest for developing unconventional hydrocarbon reservoirs, as well as in geological carbon storage (GCS) (for determining the suitability of a site, injection rates, and pressure limitations), and hydrogen storage and production.

References

Zoback, M. D. (2010). Reservoir Geomechanics. Cambridge: Cambridge University Press, p. 449 pages.

Zoback, M. D. & Kohli, A. H. (2019). Unconventional Reservoir Geomechanics: Shale Gas, Tight Oil, and Induced Seismicity. Cambridge: Cambridge University Press, p. 492.

Chapter 8

Fractal geometry is a technique for investigating and measuring the recurrent geometric patterns of fractures at numerous scales (Mandelbrot, 1982). It offers insights into the scaling and clustering of complex fracture networks and aids in the comprehension of the spatial distribution of fractures. In hydrocarbon exploration, it plays a key role in forecasting the presence and properties of subseismic fractures, which can considerably impact reservoir quality.

This work focuses on the southern edge of the Mouydir Basin in the Saharan platform, where fracture distribution, connectivity, and kinematics can be better understood and characterized by using the Cambro-Ordovician analog. Major and minor fractures are identified using digital elevation models (DEMs), and the fractal dimensions of major and minor fracture networks are estimated using fractal dimension analysis (Djezzar, 2019). This thorough fracture map offers insights into the subsurface dynamics of the Cambro-Ordovician reservoir analog by assisting in the identification of major and minor fracture sets and in understanding their fractography. This information is essential for upcoming research projects in the understudied Mouydir Basin.

References

Mandelbrot, B. (1982). The Fractal Geometry of Nature Freeman (Ed.). San Francisco, p. 461.

Djezzar, S. (2019). Fracture Detection and Prediction In Unconventional Reservoirs for Finding Sweet Spot. Theses and Dissertations. 2551. https://commons.und.edu/theses/2551/

Chapter 9

The exploration and development of unconventional reservoirs have seen a significant increase in hydrocarbon discoveries, with natural fractures playing a crucial role in hydrocarbon production. To improve hydrocarbon production, geoscientists seek for zones having dense networks of interconnecting fractures. However, understanding and forecasting these complex networks is difficult and requires high-quality datasets (3D seismic and well data) (Djezzar and Boualam, 2020).

Analyzing reservoir analogs provides a more practical method with less particular data needed. An illustration of this approach is the Ordovician formation in the Mouydir Basin (Algeria), which is well-known for its modest reservoir qualities (Djezzar and Boualam, 2021). Using particular techniques on a Digital Elevation Model (DEM), this study suggests a workflow for characterizing and modeling natural fractures, and focuses on the reservoir analog of the Ordovician at the southern boundary of the Mouydir Basin. The obtained results are promising for understanding naturally fractured reservoirs.

References

Djezzar, S. & Boualam., A. (2020). Analysis and modeling of tight oil fractured reservoir. SEG 2020 Annual Meeting. TX, USA. https://doi.org/10.1190/segam2020-3425011.1.

Djezzar, S. & Boualam, A. (2021). Impact of Natural Fractures on the Productivity of Lower Devonian Reservoirs in Reggane Basin, Algeria. 55th US Rock Mechanics/Geomechanics Symposium. 21-A-1295-ARMA.

Chapter 10

The chapter presents a structural analysis and fracture kinematics in the Mouydir Basin using 2D seismic profiles, geological maps, and gravity data. It entails subsurface studies and outcrop observations, with data comparisons to establish relationships between these datasets to understand subsurface structures (Djezzar et al., 2019; Djezzar and Boualam, 2020). Afterwards, the analysis is focused on the fault network within the Mouydir Basin, where the different geometrical parameters of the fault sets (position, length and orientation) are studied.

The suggested model provides insights into the structural complexity of the two basins (ahnet and Mouydir) by tracing the formation and extension of the fault network from surface measurements to the subsurface.

References

Djezzar, S., Rasouli, V., Boualam, A., et al. (2019). Size Scaling and Spatial Clustering of Natural Fracture Networks Using Fractal Analysis. 53rd American Rock Mechanics Association Symposium, NY, USA. 19-A-375-ARMA.

Djezzar, S. & Boualam, A. (2020). Fractures Characterization and Their Impact on the Productivity of Hamra Quartzite Reservoir (Haoud-Berkaoui field, Oued-Mya basin, Algeria). 54th American Rock Mechanics Association Symposium, CO, USA. 20-A-1438-ARMA.

Chapter 11

The study provides a novel and a comprehensive workflow for the characterization and modeling of natural fractures, for the Cambro-Ordovician reservoir analog taken from the Algerian Mouydir Basin. The method is based on two approaches. The first one involves a comprehensive evaluation of fracture networks in the study area using various data sources, such as digital elevation models (DEMs), geology maps, seismic data, satellite imagery, and curvature illumination attributes from seismic data, to create a fracture map (Djezzar et al., 2020). However, the second approach focuses on the several Cambro-Ordovician stratigraphic units and the basement rocks fractures. This finer-scale approach ensures a coherent and thorough fracture characterization across various geological contexts by using the same array of input data to infer equivalent parameters and results.

For every geological unit, 3D deterministic fracture models are made in order to show the distribution and sources of fractures. By highlighting the impact of basement fractures on the overlying sedimentary cover, these models aid in the understanding of the structural evolution of basins and managing future exploration programs. By combining various data sources and methodologies, this original method improves the comprehension of fracture networks and optimizes petroleum exploration.

References

Djezzar, S., Rasouli, V., Boualam, A., et al. (2020). An Integrated Workflow for Multiscale Fracture Analysis in Reservoir Analog. Arabian Journal of Geosciences 13, 161. https://doi.org/10.1007/s12517-020-5085-6.

Chapter 12

Modeling and predicting fracture networks at different scales is a challenge when studying naturally fractured reservoirs. In contrast with different methods requiring advanced (seismic and well log) data and taking much time (Djezzar, 2019), alternative approaches like studying reservoir analogs offer more available data, and allow the fracture characterization and modeling across different scales various workflows. In this view, an integrated workflow is presented to characterize and model natural fractures in Algeria's Cambro-Ordovician reservoir analog, located in southern Mouydir Basin on the Saharan platform in Algeria (Djezzar et al., 2020).

To locate fault planes, the suggested workflow makes use of digital elevation models, geological maps, and satellite photos. Seismic characteristics are used to identify both large and small fractures. It allows the investigation of significant faults influencing the target zone and the detailed examination of fractures impacting the basement and the diverse Cambro-Ordovician stratigraphic units, and improves reservoir analog comprehension by producing graphical representations. Next, 3D deterministic fault models are built to show the different properties of fractures (influence, spatial and temporal distribution of fractures, origins, etc.) and the effect of basement faults on the overlaying layers.

References

Djezzar, S. (2019). Fracture Detection and Prediction in Unconventional Reservoirs for Finding Sweet Spot. Theses and Dissertations 2551. https://commons.und.edu/theses/2551/

Djezzar, S., Rasouli, V., Boualam, A., et al. (2020). An Integrated Workflow for Multiscale Fracture Analysis in Reservoir Analog. Arabian Journal of Geosciences 13, 161. https://doi.org/10.1007/s12517-020-5085-6.

Section 3. Artificial Intelligence Applied to Reservoir Characterization

Chapter 13. Exploring the Depths: Satellite Image Processing and Artificial Intelligence in the Oil and Gas Industry, Hasna Yazid and Said Gaci

Chapter 14. Modern AI Usage in the Oil and Gas Industry for Reservoir Characterization and Lithofacies Forecasting (Rock Typing), Hasna Yazid, Said Gaci, and Mohammed Farfour

Chapter 15. Logging-Data-Driven Fluid Prediction in Clastic Reservoir Based on Fractal Attributes and Machine Learning Methods, Abdelbasset Boulassel, Soraya Makhlouf, Fethi Ali Cheddad, Zinelaabidine Boumelit, Badis Zegagh, Salah Boufenchouche, Amar Boudella, Naima Zaourar, and Said Gaci

Chapter 16. Unlocking Deeper Insights: Using Machine Learning to Predict Dynamic Shear Wave Slowness from Well Logs, Abdelbasset Boulassel, Soraya Makhlouf, Zinelaabidine Boumelit, Badis Zegagh, Salah Boufenchouche, Fethi Ali Cheddad, Amar Boudella, Naima Zaourar, and Said Gaci

Chapter 13

Advanced image processing methods and artificial intelligence (AI) algorithms, in conjunction with satellite imagery, have enhanced operational efficiency and enabled intelligent decision-making processes by yielding valuable insights across a range of oil and gas domains (Sang et al., 2020; Wu et al., 2022; Newaz et al., 2023). This chapter explores the different AI-based algorithms that are utilized to analyze satellite images and derive important information. In addition, it showcases the main applications of satellite imagery and AI in various activities of oil and gas industry, including exploration, drilling, pipeline monitoring, and environmental monitoring, and their contribution to optimize operations, reduce expenses, and improve the sector's sustainability and efficiency.

References

Sang, X., Xue, L., Ran, X., et al. (2020). Intelligent High-Resolution Geological Mapping Based on SLIC-CNN. ISPRS International Journal of Geo-Information 9, 99.

Wu, M., Lin, N., Li, G., et al. (2022). Hyperspectral Estimation of Petroleum Hydrocarbon Content in Soil using Ensemble Learning Method and LASSO Feature Extraction. Environmental Pollutants and Bioavailability 34 (1), pp. 308–320. https://doi.org/10.1080/26395940.2022.2102543.

Newaz, S., Tahmid, M. T., Al-Aboody, N., et al. (2023). A Ubiquitous Method for Predicting Underground Petroleum Deposits based on Satellite Data. Scientific Reports, 13.

Chapter 14

Well core analysis is crucial for reservoir characterization studies. Since the spatial heterogeneity of geological formations often requires large numbers of well cores, leading to higher costs, an alternative approach is required to overcome this drawback (Yazid and Gaci, 2022; 2024). This chapter explores the use of artificial intelligence (AI) techniques for reservoir characterization and lithofacies forecasting, focusing on identifying and classifying rock type units. Machine learning methods, including support vector machine (SVM), one-class support vector machine (OCSVM), radial basis function (RBF), logistic regression (LR), K-nearest neighbors (KNN), multilayer perceptron neural networks (MLP), decision trees (DT), both unsupervised and supervised self-organizing maps (SOM), random forest (RF), gradient boost model (GBM), and adaptive boost (AB), have been applied to well log data from ten boreholes across different locations. The results show superior efficiency and improved accuracy compared to conventional methods.

References

Yazid, H. & Gaci, S. (2022). Machine Learning-Based Techniques for Reservoir Characterization using Rock-Typing from Well Logs. In: 2022 7th International Conference on Image and Signal Processing and Their Applications (ISPA). Mostaganem, Algeria, 2022, pp. 1–6. https://doi.org/10.1109/ISPA54004.2022.9786300.

Yazid, H. & Gaci, S. (2024). Integrating Hybrid Weak Learners for Lithofacies Classification using Well Log Data. In: 2024 8th International Conference on Image and Signal Processing and their Applications (ISPA). Biskra, Algeria, pp. 1–5. https://doi.org/10.1109/ISPA59904.2024.10536797.

Chapter 15

This chapter proposes a new method to combine machine learning (ML) techniques with fractal attributes inferred from well logs to increase the accuracy of fluid prediction in clastic reservoirs (Habib et al., 2017; Boulassel et al., 2021). To predict the reservoir fluid types (oil or gas), several fractal features are computed from different well logs – including gamma ray (GR), acoustic logging (AC), density logging (DEN), neutron logging (CNL), electrical resistivity logging (R), photoelectric absorption factor (PF), and natural gamma ray spectroscopy (GRS)-, and processed using ML algorithms – specifically, Logistic Regression, Random Forests (RF), K-Nearest Neighbors (KNN), Support Vector Machines (SVM), and Discriminant Analysis. The obtained results show the RF algorithm yields the highest accuracy with the lowest misclassification rate. This method based on the fractal analysis combined with the RF algorithm provides a notable enhancement in reservoir characterization.

References

Habib, M., Guangqing, Y., Xie, C., et al. (2017). Optimizing Oil and Gas Field Management through a Fractal Reservoir Study Model. Journal of Petroleum Exploration and Production Technology 7, 43–53. https://doi.org/10.1007/s13202-016-0252-8.

Boulassel, A., Zaourar, N., Gaci, S., Boudella, A. (2021). A New Multifractal Analysis-Based for Identifying the Reservoir Fluid Nature. Journal of Applied Geophysics 185, p. 104185. ISSN 0926-9851. https://doi.org/10.1016/j.jappgeo.2020.104185.

Chapter 16

Shear wave slowness is a key parameter in hydrocarbon reservoir characterization, geomechanical studies, etc. (Seyedalireza et al., 2018; Khatibi and Aghajanpour, 2020). Since its measurements are not available at wells, their prediction with high accuracy remains challenging. In this view, this chapter presents a ML-based approach to predict shear wave slowness from well logs. In this sturdy, a large dataset of well logs from different geological settings is considered. The used dataset is preprocessed and undergone four (04) machine learning models – extreme gradient boosting (XGBoost), random forest, support vector machine (SVM), and K-nearest neighbors (KNN). This chapter emphasizes how these methods – in particular, Random Forest and XGBoost – can enhance the accuracy of predicting shear wave slowness from well logs, and hence the results of integrated reservoir characterization studies.

References

Khatibi, S., & Aghajanpour, A. (2020). Machine Learning: A Useful Tool in Geomechanical Studies, a Case Study from an Offshore Gas Field. Energies 13, p. 3528. https://doi.org/10.3390/en13143528.

Seyedalireza, K., Aghajanpour, A., Ostadhassan, M., et al. (2018). Evaluating Single-Parameter Parabolic Failure Criterion in Wellbore Stability Analysis. Journal of Natural Gas Science and Engineering 50, pp. 166–180.

Section 4. Energy Transition: New Perspectives on Geothermal Energy Exploration and Development and CO_2 Sequestration

Chapter 17. Energy Transition and the Role of AI: Statistics, Trends, and Implications, Said Gaci, Hasna Yazid, Aziz Khelalef and Mohammed Farfour

Chapter 18. On the Importance of Integrating Geomodeling in Geothermal Studies, Mohamed Amrouche

Chapter 19. Advancements, Challenges, and Outlook of Geothermal Reservoir Operations, Ghoulem Ifrene, Singh Kuldeep, and William Gosnold

Chapter 20. Multiscale Reservoir Characterization of a CO_2 Storage Aquifer: Mineralogical, Geomechanical, and Petrophysical Analyses for aCCS Project in North Dakota?, Aimen Laalam, Ahmed Merzoug, and Hichem Aymen Katib Chellal

Chapter 21. Anthropogenic Carbon Sequestration into the Subsurface: Caveats and Pitfalls, Steven A. Tedesco

Chapter 17

The goal of the energy transition is to mitigate climate change and achieve sustainability by moving the world toward renewable energy sources. It supports the replacement of fossil fuel-based infrastructure with clean, renewable energy sources like solar and wind power, and promotes distributed generation (Harichandan et al., 2022). Another goal of the energy transition is to reduce energy waste and losses by implementing more efficient practices, technologies, and services. Renewable energy sources are becoming increasingly incorporated into the grid because of their erratic nature and strong dependence on weather (Sayed-Mouchaweh, 2020).

This chapter focuses on the connection between artificial intelligence (AI) and the global energy transition, emphasizing on the opportunities, trends, and difficulties of using AI to promote the shift to renewable and sustainable energy sources. By enhancing the security, sustainability, and efficiency of energy systems, the application of AI in the energy transition can have a major impact on the energy industry.

References

Harichandan, S., Kar, S. K., Bansal, R., et al. (2022). Energy Transition Research: A Bibliometric Mapping of Current Findings and Direction for Future Research. Cleaner Production Letters 3. 100026. https://doi.org/10.1016/j.clpl.2022.100026.

Sayed-Mouchaweh, M. (2020). Artificial Intelligence Techniques for a Scalable Energy Transition: Advanced Methods, Digital Technologies, Decision Support Tools, and Applications (1st edition 2020). Springer Cham, Switzerland. https://doi.org/10.1007/978-3-030-42726-9.

Chapter 18

Because it is low-carbon, renewable, and significant component of decarbonization efforts, geothermal energy is gaining an increasing interest. By combining geological and geophysical data, geomodeling builds a numerical model of geothermal reservoirs that may be used to locate and describe active provinces. In fact, exploration and production of oil and gas reservoirs currently leverage this technique, which helps to maximize production of the targeted reservoir, and to investigate their geothermal potential.

This chapter discusses the use of geocellular subsurface modeling in geothermal studies, combining non-destructive geophysical techniques (gravity and magnetic prospecting) and geomechanical investigations, in order to analyze the natural fractures of geothermal reservoirs (Saibi et al., 2022; Melouah et al., 2023).

References

Saibi, H., Amrouche, M., Bâtir, J., et al. (2022). Magnetic and Gravity Modeling and Subsurface Structure of Two Geothermal Fields in the UAE. Geothermal Energy 10, 28. https://doi.org/10.1186/s40517-022-00240-4.

Melouah, O., Ebong, E. D., Abdelrahman, K., et al. (2023). Lithospheric Structural Dynamics and Geothermal Modeling of the Western Arabian Shield. Scientific Reports 13, p.11764. https://doi.org/10.1038/s41598-023-38321-4.

Chapter 19

This chapter discusses geomechanical reservoir operations, and the importance of geomechanics, and advanced drilling techniques in the development of enhanced geothermal systems (EGS), and highlights the potential of the fishbone drilling method, which increases EGS effectiveness, compared to the conventional drilling techniques in terms of enhancements to the system's overall performance, heat extraction efficiency, and reservoir connectivity (Manshad et al., 2019; Aihar et al., 2023).

The chapter also covers the technical aspects of using fishbone drilling with EGS, including fluid circulation, heat exchange optimization, reservoir characterization, and well design. In addition, the advantages and disadvantages of using carbon dioxide (CO_2) as an injection fluid in EGS are examined. Moreover, social and environmental factors are taken into account, with an emphasis on geothermal energy integration with other renewable energy sources and sustainable energy practices.

References

Aihar, A., Bouabdallah, N., Ifrene, G., et al. (2023). Comparing Fishbone Drilling and Hydraulic Fracturing in Ultra-Low Permeability Geothermal Reservoirs. 2023 AAPG Rocky Mountain Section Meeting, Bismarck, North Dakota. June 4–6, 2023, https://doi.org/10.1306/80737Aihar2023.

Manshad, A. K., Dastgerdi, M. E., Ali, J. A., et al. (2019). Economic and Productivity Evaluation of Different Horizontal Drilling Scenarios: Middle East Oil Fields as Case Study. Journal of Petroleum Exploration and Production Technology 9, pp. 2449–2460.

Chapter 20

Carbon capture, utilization, and storage (CCUS) technology is of high interest for mitigating climate change and fulfilling the objectives of the Paris Agreement. It prevents the release of carbon dioxide emissions and promotes the development of a circular carbon economy by absorbing emissions from a variety of sources, such as industrial operations and the atmosphere (Friedlingstein et al., 2020). Using CCUS technologies can help industries including chemical manufacture, steel, and cement (Fuss et al., 2014). The International Energy Agency (IAE) and the Intergovernmental Panel on Climate Change (IPCC) emphasize how crucial CCUS is to reaching net-zero emissions by 2050 (Salvia et al., 2021). Many studies highlight the importance of CCUS technology in achieving carbon neutrality targets and moving towards a low-carbon future (Ampomah et al., 2016).

In this view, this chapter investigates the saline aquifer's potential for CO_2 storage, illustrated by CCUS Projects in North Dakota (USA), and examines the benefits of machine learning algorithms in improving CCUS technology. Moreover, a case application focusing on the storage aspects of CCS, including a detailed analysis of the saline aquifer and the geological layers that act as confining zones above and below it. The purpose of this study is to evaluate the aquifer's feasibility for efficiently and securely storing CO_2.

References

Friedlingstein, P., O'Sullivan, M., Jones, M., et al. (2020). Global Carbon Budget 2020. Earth System Science Data 12 (4), pp. 3269–3340. https://doi.org/10.5194/essd-12-3269-2020.

Fuss, S., Canadell, J., Peters, G., et al. (2014). Betting on Negative Emissions. Nature Climate Change 4 (10), pp. 850–853. https://doi.org/10.1038/nclimate2392.

Salvia, M., Reckien, D., Pietrapertosa, F., et al. (2021). Will Climate Mitigation Ambitions Lead to Carbon Neutrality? An analysis of the local-level plans of 327 cities in the EU. Renewable and Sustainable Energy Reviews 135, p. 110253. https://doi.org/10.1016/j.rser.2020.110253.

Ampomah, W., Balch, R., Cather, M., et al. (2016). Evaluation of CO_2 Storage Mechanisms in CO_2 Enhanced Oil Recovery Sites: Application to Morrow Sandstone Reservoir. Energy & Fuels 30 (10), pp. 8545–8555. https://doi.org/10.1021/acs.energyfuels.6b01888.

Chapter 21

In order to lessen global warming, carbon capture and sequestration (CCS) programs seek to extract anthropogenic carbon dioxide (CO_2) from the atmosphere. In addition to soil, ocean, and minerals, carbon dioxide can be stored in depleted oil, gas, and carbon dioxide reservoirs, carbonaceous mudstones, coal-bed methane, and saline aquifers. Any industrial, commercial, or residential source may be used to provide the anthropogenic CO_2 for sequestration. CO_2 is then moved to an injection well and buried by truck, pipeline, or both. The transportation and injection of CO_2 into the subsurface are done at a pressure of about 100 bars, where it dissolves into subsurface oil, gas, or brine or is adsorbed into organic matter.

Because CO_2 may react with the minerals in the reservoir, it is important to monitor the plume of CO_2.

Surface issues for CCS projects in the United States include site, groundwater protection, permitting requirements, CO_2 level monitoring, and surface facility quality. In addition, high porosity and moderate to high permeability are desirable qualities in reservoir rock for CCS storage. In order to address surface concerns prior to injection, a thorough geology, geophysical, and engineering analysis is necessary for CCS project evaluation.

This chapter addresses sequestering CO_2 in brine-filled aquifers or depleted oil, gas, and CO_2 reservoirs, and shows that the only differences between the two types of processes are whether or not petroleum or CO_2 are present in the reservoir and project criteria (Service, 2018; McDonnell, 2021). In contrast with the features of virgin, saline aquifer reservoirs which can only be inferred from neighboring wells and 2D/3D seismic surveys, prior to actual development, depleted reservoirs are usually well-documented.

References

Service, R. F. (2018). Cost Plunges for Capturing Carbon Dioxide from the Air: Approach Could Make It Affordable to Produce Gasoline from Air, Water, and Renewable Electricity, Science, June 7, www.science.org/content/article/cost-plunges-capturing-carbon-dioxide-air.

McDonnell, T. (2021). The World's Biggest Carbon-Sucking Machine is Switching on in Iceland, World Economic Forum. September 13, www.weforum.org/agenda/2021/09/worlds-biggest-carbon-machine-iceland/.

Section I

New Technologies and Insights into Petroleum Exploration

1

Gas Seepage in Marginal Structures as Additional Shallow and Deep Hydrocarbon Systems Indicator (Some of Recent FR Scanning Results)

Valery Soloviev[1], *Mykola Yakymchuk*[2], *Ignat Korchagin*[1]

[1] *Institute of Geophysics, NAS of Ukraine, Kyiv, Ukraine*
[2] *Institute of Applied Problems of Ecology, Geophysics and Geochemistry, Kyiv, Ukraine*

> *The indications of sustained gas seepage show that gas seepage is a periodic process in geological time. The presence of oil and gas at a deeper prospect, assumed to be the source of the migrating gas, suggests that the presence of gas seepage and associated features may be indicative of deeper prospective reservoirs.*
>
> (Heggland, 1998)

1.1 Introduction

The idea of degassing processes in the deep zones of the Earth is an important factor in the development and evolution of the planet (Shestopalov et al., 2018). It allows us to reconsider the interconnected processes of global volcanism and the intrusion of deep fluids, mud volcanism, and local and long-lived gas seeps – zones of the anomalous release of carbon dioxide, methane, hydrogen, helium, nitrogen, and other gases due to the functioning of gas pipes (point channels) and linear faults (Bogoyavlensky et al., 2020; Yakymchuk et al., 2019, 2022; WMO, 2022). The interpretation of the remote sensing data yields new independent information that can be useful at any stage of seeps, pockmarks, and sleeks prospecting.

The formation of oil and gas deposits, accumulations of gas hydrates and sub-hydrate gas, as well as the presence of areas of mud volcanic activity and centers of gas emission, can be considered the result of long-term active influence of localized flows of hydrocarbon fluids (mantle degassing products) on the rocks of deep and near-surface structures of various ages and genesis. An important role in the inflow of deep fluids into the upper horizons of the earth's crust is played by the presence of a developed network of faults, as well as processes of tectonic activation, due to which degassing and concentration of gas emissions in emission centers occur. These assumptions do not exclude the possible mixed (abiogenic and secondary biogenic) nature of methane in separate

Geophysical Exploration for Hydrocarbon Reservoirs, Geothermal Energy, and Carbon Storage: New Technologies and AI-based Approaches, First Edition. Edited by Said Gaci.

structures, as well as the formation (if necessary and under sufficient conditions) of hydrocarbon accumulations in a wide range of lithosphere depths. This report briefly describes the possible directions for using mobile direct-prospecting methods of frequency-resonance (FR) processing of satellite images and photographs in a new model of prospecting for oil and gas in sea and ocean structures.

Some results of the use of the technology of FR research on methane seeps at gas emission sites of various genesis and modern gas releases in various structures of the shelves and continental margins are considered below.

1.2 General Principles and Methods

The express technologies used in the work include modified methods of FR processing and decoding of ERS data and photographs. They are based on the experimentally proven assumption that a large amount of a homogeneous substance creates an electromagnetic field characteristic of a given substance, the radiation power of which is proportional to the concentration of the substance (Levashov et al., 2012; Yakymchuk et al., 2019, 2022, 2023; Yakymchuk and Korchagin, 2019, 2023). According to the new paradigm of conducting geophysical research, the focus is not on the selection of certain structural elements in the section and the determination of their physical properties, but on the detection of the specific substances in the section.

Databases of chemical elements, minerals, oil, and condensate samples, as well as sedimentary, igneous, and metamorphic rocks, classified and divided into separate types and groups, the resonance frequencies of which are used in the processing satellite images and photographs, became important components in the modifications of direct search methods. Photographs of used sets of samples of sedimentary, metamorphic, and igneous rocks are borrowed from the electronic document on the website http://rockref.vsegei.ru/petro/ (Yakymchuk et al., 2022).

For the experiment, areas of seep activity were selected in some structures of the continental margins of the Arctic region, the North and Norwegian Seas shelf, and other Regions (Figure 1.1). The results obtained should be considered the first approximation of quantitative calculations of possible deep sources of Earth degassing processes (Levashov et al., 2012; Soloviev et al., 2023; Yakymchuk et al., 2019, 2022, 2023; Yakymchuk and Korchagin, 2019, 2023).

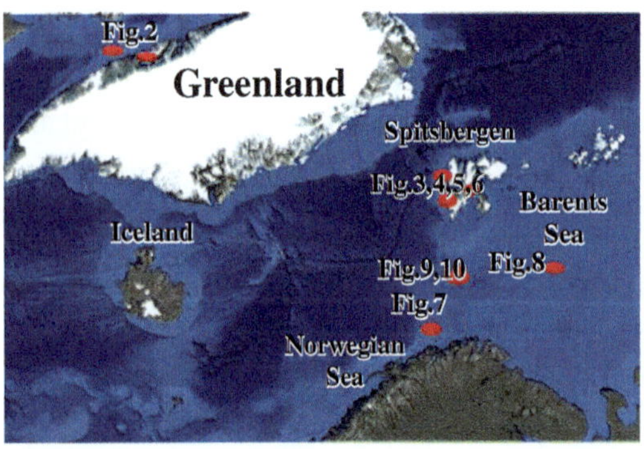

Figure 1.1 Position of some FR - scanning points (SP) in the Polar region. Google Earth.

1.3 Gas Fluids as Additional Hydrocarbon Processes Indicator in Some Continental Margin Structures

AI tells us that there is a connection between the fields of gas emissions in the structures of the polar continental margins and oil and gas content, but it is quite difficult to accurately and unambiguously assess it due to the many factors influencing the oil and gas content of this region. However, existing gas emissions may not be a direct indicator of the presence of oil and gas in commercial quantities. Off-gassing may be caused by other causes, such as natural decomposition of microbial processes (Judd and Hovland, 2007).

In addition, it is necessary to take into account the geological structure of the region and the presence of geological traps that can hold oil and gas in commercially significant volumes. In general, there is a connection between the gas emission fields and oil and gas content, but it requires more detailed research and analysis. Additional geological and geophysical research could help better understand this relationship and identify potential oil and gas deposits in the Polar region.

Below we will consider examples of studying the possible connection between areas of seep fields (and pockmarks) and the presence of hydrocarbon accumulations, the characteristics of these gas eruptions, and their connection with known data on the deep geology and tectonics of these areas.

1.3.1 Seepage Areas in the Central West Greenland Margin

Greenland Island covers an area of approximately 2.2 million square kilometers and has over 44,000 km of coastline. Despite its huge size, only 14 wells have been drilled offshore Greenland in the past 40 years. The U.S. Geological Survey estimates that there could be 17.5 billion undiscovered barrels of oil and 148 trillion cubic feet of natural gas off Greenland, although the island's remote location and harsh weather allow limited exploration (Schumann et al., 2012; Yakymchuk et al., 2023).

There should be no expectation that oil and gas exploration and production will take off any time soon in Greenland; oil and gas production is, at best, 15–20 years away. Although the timing is uncertain, it seems inevitable that as technologies improve, environmental concerns abate, and market demand increases, Greenland will become a significant hydrocarbon province of the future (http://www.oxfordenergy.org/wpcms/wp-content/uploads/2014/11/WPM-56.pdf) (Hogan et al., 2012).

In the central part of the Western Greenland margin (Disko Bay region), two FR scanning points (PG2011-12 and PG2012-03, Figures 1.1 and 1.2), located within fields of pockmarks and seeps, were selected for research (Christiansen et al., 2020; Cook, 2019; Greinert et al., 2010; Nielsen et al., 2014; Yakymchuk et al., 2023).

Sounding point PG2011-12 (Ilulissat area) is located at the south-western flank of the Nuussuaq Basin, and near the mouth of the Isfjord (Figure 1.2a), at a depth of 433 m, inside the largest pockmark, the width of which reaches 200 m, and the depth is 23 m. No acoustic indications for hydrocarbons within the Mesozoic strata have been reported for this distal part of the Nuussuaq Basin (Nielsen et al., 2014).

Seismic data showed that the pockmarks resulted due to expulsing fluids, neo-tectonic faulting, and crustal uplift (Hogan et al., 2012). Active modern gas migration in pockmarks was not observed, and it can be supposed that the pockmarks are the result only of neo-tectonic activity and gas hydrate was not involved in their formation in this area (Nielsen et al., 2014).

Our FR-sounding data of the **PG2011-12** satellite image confirmed the modern active migration of gas, but there are no signals from gas hydrates. Salt signals are recorded in the range of 435–882 m. The joint signal of gas and salt is fixed in the interval of 468–742 m. FR data fixed the salt volcano presence with the root at 99 km. A hydrocarbon synthesis signal is recorded at a depth of 57 km (Yakymchuk et al., 2023).

Scanning point site **PG2012-03** is located in the deepest part of a pockmark-like structure with a large diapir in the upper sediment unit where gas/fluid escape features lie close to the seabed

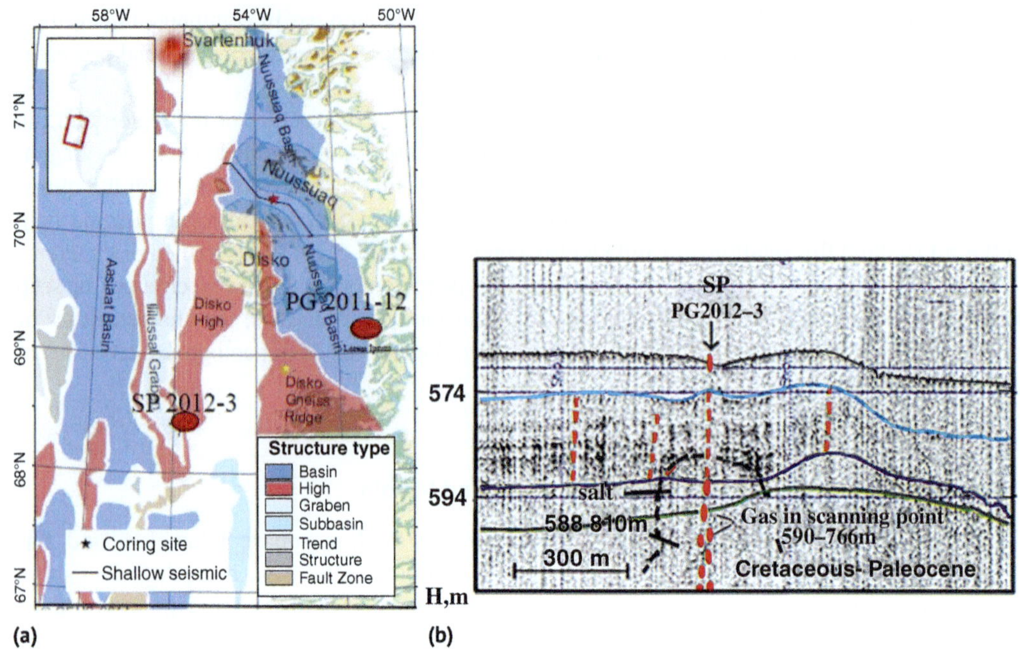

Figure 1.2 Structure types (Nielsen et al., 2014) and scanning point position in central West Greenland continental margin (a); (b) high-resolution seismic profile structure across PG 2012-03 point with FR results data (Nielsen et al., 2014; Yakymchuk et al., 2023). In (b), the light blue, dark blue, and green lines are the units in 40 m thick sediment sequence; the orange lines – gas-fluid escape features, by Nielsen et al. (2014) / Springer Nature.

(Nielsen et al., 2014). The seismic data show that in an elongated depression (20 × 35 km large, 575 m deep) on the inner shelf west of Disko Bay, younger sediments (Figure 1.2b) cover the Cretaceous–Paleocene strata with gas-hydrates and a bottom simulating reflector (BSR) below seabed (Nielsen et al., 2014).

The presence of gas hydrates is significant for future petroleum exploration offshore Greenland but gas hydrates there are mainly in sectors with a thin young sediment cover. The new geochemical data point to methane migration from a deeper-lying petroleum system (Nielsen et al., 2014).

These data are confirmed by the FR-sounding results at point PG 2012-03, where the signals of synthesis of oil, gas condensate, and gas at a depth of 57 km are recorded, and there are no signals of gas hydrates (Yakymchuk et al., 2023).

The salt signal is fixed in the interval of 588–810 m, and the joint signal of salt and gas is fixed in the interval of 590–766 m (Figure 1.2). The signals of dolomites (0.81–1.3 km, 2.0–99 km) and the ultramafic magma group (1.3–2.0 km, 99–470 km) are recorded. A degassing process with gas migration from the surface to the atmosphere is fixed though no sign of present-day gas seepage was found in the area (Nielsen et al., 2014).

These results support previous predictions that Mesozoic sedimentary basins in the central West Greenland margins could hold high quantities of oil and gas, but the processes of mature source rocks in the subsurface, their transformation, and the vertical and lateral migration of hydrocarbons into traps are enough debatable.

The Cretaceous and Palaeogene structures together with the oil seeps and offshore reservoir quality sandstones known onshore indicate that the offshore area west and south of Disko could potentially be prospective.

1.3.2 Spitsbergen, Isfjorden Area

The degassing processes of investigation in this area contributed to a detailed study of the formation and evolution of gas hydrate and free gas accumulations, as well as the nature of their sources (Yakymchuk et al., 2022). Particular emphasis in these works was placed on the study of the dynamics of the dissociation of gas hydrates and gas emissions associated with paleoclimatic changes in the region. Their important result was the identification of numerous fields of pockmarks, as well as associated "degassing pipes" and systems of tectonic faults extending to considerable (Andreassen et al., 2017; Betlem et al., 2021; Blinova et al., 2013; Bouriak et al., 2000; Forwick et al., 2009; Roy et al., 2015, 2019; Sarkar et al., 2012; Soloviev et al., 2023; Veloso et al., 2015; Vogt et al., 1999; Yakymchuk et al., 2022; Wallmann et al., 2018; Westbrook et al., 2009).

The distribution of pockmarks in Isfjorden has a mosaic character (Blinova et al., 2013; Römer et al., 2021; Roy et al., 2015, 2019; Sarkar et al., 2012), and their most actively expressed fields are spatially associated with known active fault zones of the Svalbard fold-thrust belt and other faults (Billefiorden, Isfiorden Fault Zones).

More than 1300 pockmarks (diameter from 14 to 265 m, depth from 1 to 11 m, and frequency of distribution up to 20 per km^2) were found in Isfjorden, and 535 pockmarks – in Nordfjorden (Roy et al., 2015, 2019; Vogt et al., 1999). Their genesis can be determined by the influx of hydrocarbon fluids (of biogenic or thermogenic origin), decomposition of gas hydrates, thawing of permafrost, and many other factors operating in polar latitudes. Geochemical analysis of hydrocarbon anomalies in the near-surface marine deposits of the fjords in the west and around Svalbard indicates a mixed (thermogenic and biogenic) nature of gases (Blinova et al., 2013; Roy et al., 2015, 2019; Yakymchuk et al., 2023).

In the central part of Isfjord (Figure 1.3), the flow of thermogenic gas along fault zones and large tectonic lineaments may be the main geological factor in the formation of seeps and pockmarks, and other factors (the presence of free gas accumulations, filtration processes, seismic activity, etc.) may be important.

Seismical work results and geological observations onshore demonstrate a continuation of lithological boundaries and tectonic structures across the floor of Isfjorden (Blinova et al., 2013).

In the central part of Spitsbergen, the flow of thermogenic gas along fault zones and large tectonic lineaments may be the main geological factor in the formation of seeps and pockmarks, and other factors (the presence of free gas accumulations, filtration processes, seismic activity, etc.) may be also important.

In SP1, FR signals from oil, gas, gas condensate, carbon dioxide, methane-oxidizing bacteria, yellow phosphorus, gas hydrates, and anthracite have been registered. The sounding was carried out to a depth of 4789 m. Responses at the gas frequency were obtained at depths: 1335–1932 m, 2459–3321 m, and 3933–4467 m (Figure 1.3).

Geochemical analysis of hydrocarbon anomalies in the near-surface marine deposits of the fjords in the west Spitsbergen indicates a mixed (thermogenic and biogenic) nature of gases, and faults, high permeability layers, heavily fractured units, and igneous intrusions channel the gas seepage into the water column (Yakymchuk et al., 2023).

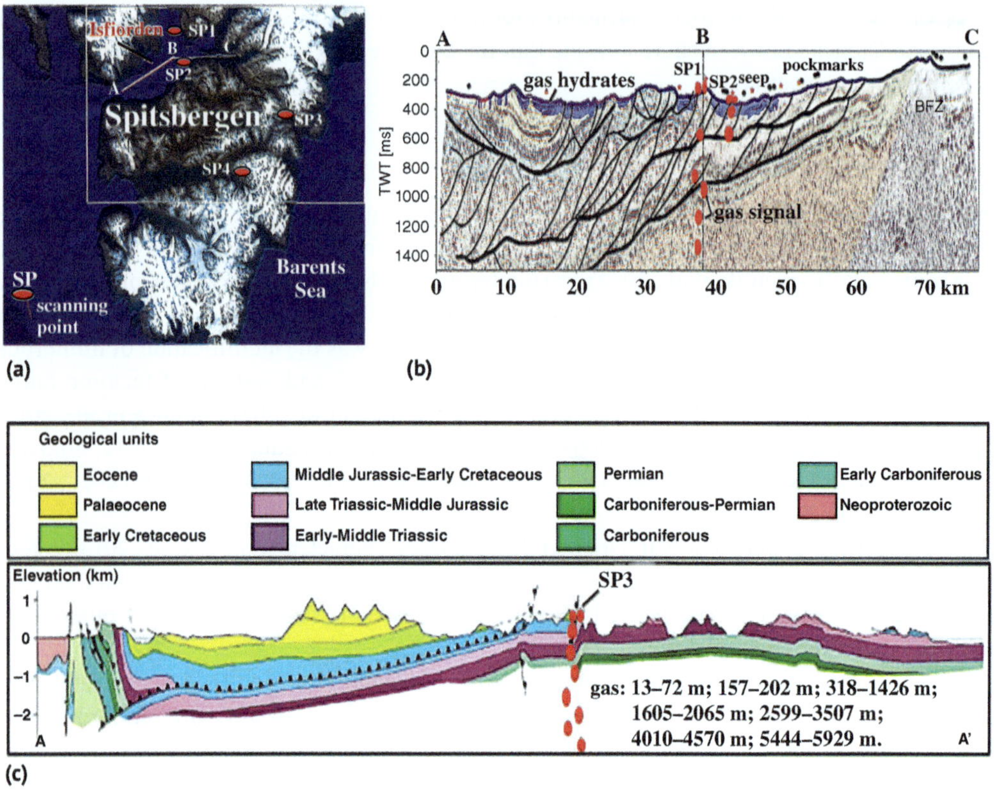

Figure 1.3 SP1-SP4 in the Central Spitsbergen area (a) Google Earth; (b) geological cross-section with GHSZ extents of A-B-C profile (Blinova et al., 2013; Roy et al., 2015, 2019). Pockmarks and flares within 250 m of the transect have been projected onto the profile. BFZ: Billefjorden Fault Zone; (c) profile cross-section with FR-sounding point (SP3) results.

In SP2, the signal of gas and yellow phosphorus and their degassing from the surface is fixed, the seventh sediment, the gas signal is fixed in the intervals: 290–615 m, 996–1159 m, and 1198–1314 m (Figure 1.3). The degassing processes and gas migration into the atmosphere at sounding point 2 (**SP2**) was fixed too.

Scanning points SP3 and SP4 are located in areas where the supersaturated with methane groundwater springs with isotopic evidence of a thermogenic source have formed, and annual methane emissions from proglacial ground waters to be up to 2.31 kt across the Spitsbergen archipelago (Vogt et al., 1999).

In **SP3** (Figure 1.3), the signals of oil and gas condensate are fixed and gas degassing was recorded. The gas signal was recorded in the depth range of 12.7–72 m, 152–202 m, 318–1426 m, 1605–2065 m, 2599–3507 m, 4010–4570 m, and 5444–5929 m, further from probing was not conducted. The depth ranges of 2599–3507 m, 4010–4570 m, and 5444–5929 m are the most perspectives.

In **SP4** (Figure 1.3), the signals of oil and gas condensate are fixed and gas degassing is recorded. The gas signal was recorded in the depth range 17.1–70 m, 161–215 m, step 10 cm 563–916 m, 1139–1523 m, 2012–2638 m, 3148–4282 m, 4412–5105 m, and 5465–5963 m, further from probing was not conducted. The depth ranges of 2012–2638 m, 3148–4282 m, 4412–5105 m, and 5465–5963 m are the most perspectives (Yakymchuk et al., 2023).

The processes of active gases entering the water column and the atmosphere can change over time; they depend on many parameters, including the thickness of the water column, temperature, and wave dynamics (Kleber et al., 2023; Veloso et al., 2015).

1.3.3 Area of Prince Karls Forland Island

FR sounding was carried out on the shelf of the continental margin of Svalbard (Figure 1.4) in the area of Prince Karls Forland (PKF) island. A study of the seasonal dynamics of dune fields was previously conducted here, which confirmed the increased emission of methane into the water column in the summer period but did not reveal a significant release of methane into the atmosphere.

FR-probing showed (Yakymchuk et al., 2022) that the accumulation of gases was isolated in the sedimentary stratum at depths of 700–1171 m and 2209–2976 m (deeper probing was not carried out).

Emissions of gases into the atmosphere were also detected at the research site (Yakymchuk et al., 2022). Its presence is confirmed by satellite measurements of methane near the western shores of Svalbard (for the period 2010–2014), which show the presence of two local maxima of methane emissions to the north and south of the island (Myhre et al., 2014).

The data obtained from the studies of alluvial fields and pockmarks in various structures located west of Spitsbergen confirm the complex genesis of cold alluviums in the structures of the Polar Arctic, which are formed due to the dissociation of methane hydrates of the sedimentary stratum, as well as the influx of deep gases.

An analysis of multiple SAR satellite images from 2014 revealed persistent oil slicks on the sea surface, 10 km offshore of PKF, located at approximately 78°29.6′ N, 10°31.5′ E (Panieri et al., 2024).

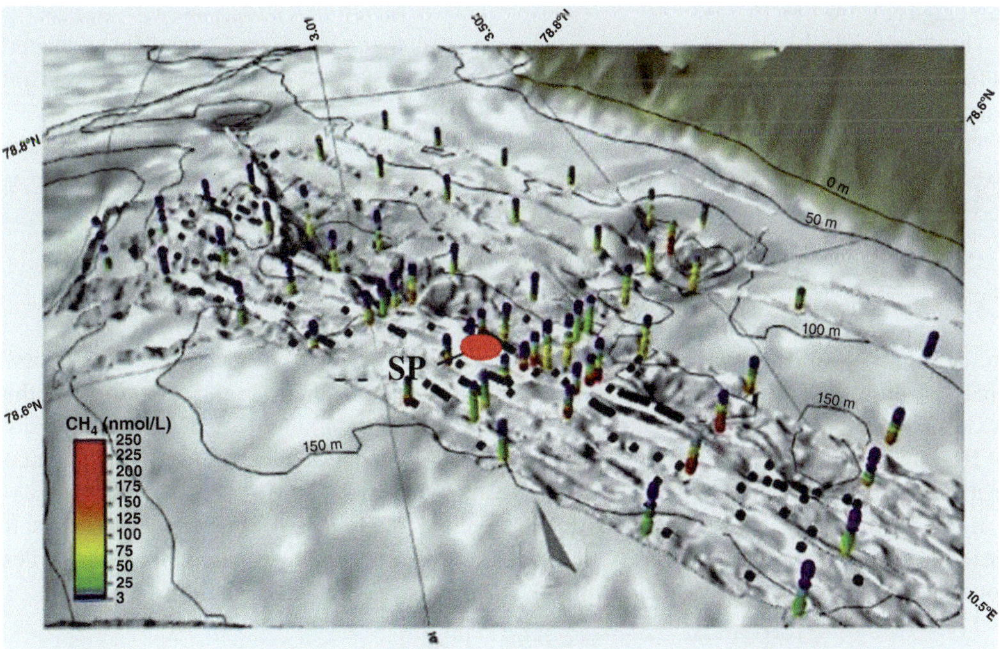

Figure 1.4 CH_4 concentrations from a hydrocast survey offshore of Prins Karls Forland. CH_4 concentrations in the ocean are illustrated by colored dots (scale on the bottom left in nmol L^{-1}). Black dots indicate the location of the gas flares Myhre C., et.al (2014) / with permission of John Wiley & Sons.

Interestingly, an oil seep site offshore Prins Karls Forland (80–100 m water depth) based on multisource data from remote sensing, geochemistry, and biology was discovered (Panieri et al., 2024).

The oil slick at PKF originates from the seep beneath and consists of severely biodegraded oil from an unknown source rock at depth, with a Tertiary age or younger, also confirmed by the dominant biodegraded thermogenic methane. Authors suggest emissions from fractures along the Hornsund fault system (Panieri et al., 2024).

Perhaps, the depth of these methane sources in this region has partially coincided with our results of FR studies.

1.3.4 Barents Sea areas

Between Norway and the Svalbard archipelago, there are large seeps and pockmarks areas, gas chimneys, and leaking fault systems that show hydrocarbon leakage and fluid migration in structures of this region. This region may become after 20–30 years an important hydrocarbon production area (Andreassen et al., 2017; Yakymchuk et al., 2023).

Here, we will repeat only some of the studies that clearly show the quantitative connection of seep fields with deep hydrocarbons of deep horizons, as well as the results of those measurements in which hydrocarbon FR signals are recorded.

1.3.5 Northern Norwegian Barents Sea, Sentralbanken High Area

As shown in Andreassen et al. (2017) and Serov et al. (2023), methane fluxes from >7000 seeps in the Arctic significantly deplete seawater, but reach the sea surface and may transfer to the atmosphere. A total of 4137 acoustic "flares" diagnostic of bubble emission sites (seeps) were identified at the Sentralbanken area with the highest gas flare density in the central part of this structural high (Figure 1.5).

The signals of oil, gas, and gas condensate are recorded at a depth of 57 km. The gas signal is fixed in depth intervals: 158–501 m, 864–1158 m, and 1458–1835 m, further probing was not carried out.

The gas signals are recorded in the depth range of 1835–10 km. The signal of gas hydrates is fixed in the interval 0–400 m. The degassing process from the water surface into the atmosphere is recorded (Yakymchuk et al., 2023).

1.3.6 Borealis and Håkon Mosby Mud Volcanoes

In the summer of 2023, a new mud volcano in the Southwestern Barents Sea waters was discovered, the Borealis Mud Volcano (Figure 1.6). The discovery was made in the framework of the AKMA 3 project (Advancing Knowledge of Methane in the Arctic) of UiT, the Arctic University of Norway, in partnership with REV Ocean, whose goal is to better understand the Arctic natural methane emissions (Foucher et al., 2010; REV Ocean, 2023; Bludd, 2023).

It was only the second (except Håkon Mosby) mud volcano found within Norwegian waters. It lies at the outer part of Bjørnøyrenna (Outer Bear Island Trough), approximately 70 nautical miles south of Bear Island (REV Ocean, 2023; Bludd, 2023) and at 400 m deep (Figure 1.6).

1.3.6.1 Borealis Mud Volcano

This volcano lies inside a crater that is approximately 300 m wide and 25 m deep and is most likely the result of a catastrophic, natural blowout that abruptly released massive methane just after the

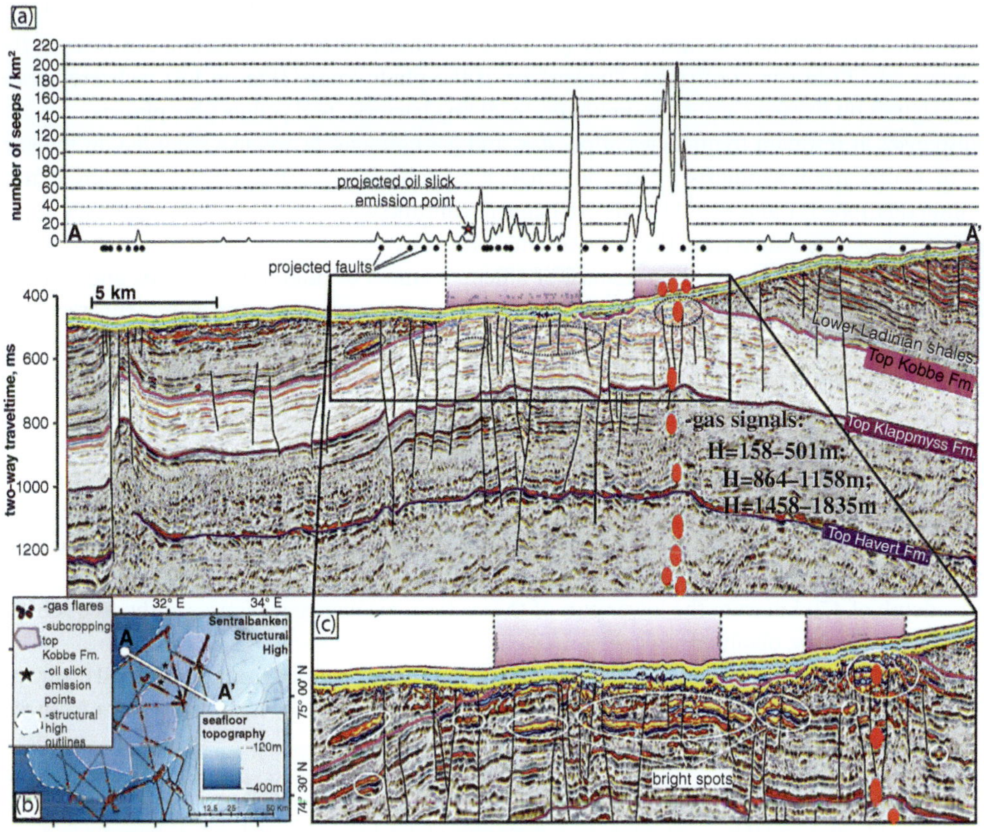

Figure 1.5 Gas flares of the Sentralbanken high (Serov et al., 2023 / Springer Nture / CC BY 4.0) are correlated with gas release and sub-cropping reservoir Kobbe Formation and faults. (a) Seepage density, faults, and sub-cropping Kobbe Formation along transect A–A'; (b) transect A–A' location; (c) bright spots distribution within the apex of Sentralbanken structural high (Serov et al., 2023 / Springer Nture / CC BY 4.0). There is also a possible depth of gas signals by FR scanning results.

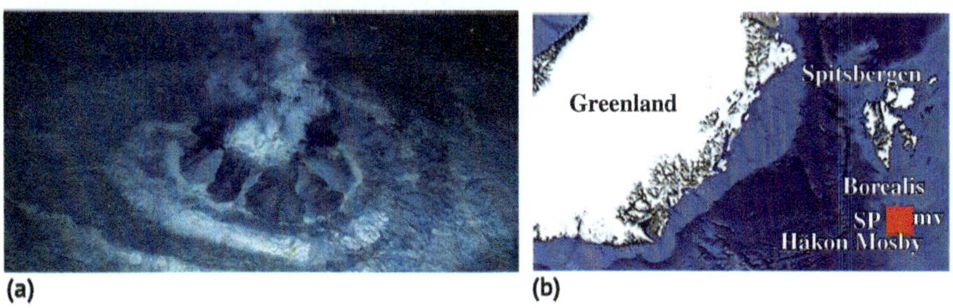

Figure 1.6 Photography of a new Borealis mud volcano (a) seabed morphology UiT The Arctic University of Norway / https://www.revocean.org/mud-volcano/ / last accessed Aug 05, 2025 (b) the Borealis and the Håkon Mosby mud volcanoes position.

last glaciation period, 18,000 years ago. Currently, the Borealis Mud Volcano, which is 7 m in diameter and 2.5 m high, continuously emits fluids rich in methane (Foucher et al., 2010; REV Ocean, 2023; Bludd, 2023; Shilov et al., 1999).

FR data discovered that the Borealis volcano location, a volcanic complex with a root at a depth of 470 km, filled with limestones, and the presence of conditions for hydrocarbon synthesis at the boundary of 57 km was confirmed. The surface of the photo image (Figure 1.6) registered resonance signals on frequencies of gas, methane-oxidizing bacteria, and yellow phosphorus.

Responses at resonant frequencies of oil, gas, and gas condensate are also recorded on the surface of hydrocarbon synthesis 57 km. Resonance responses at gas frequencies registered from the intervals 166–511 and 882–1131 m. No deeper scanning was carried out, but responses at gas frequencies were recorded from the interval 1.131–10.0 km.

These signals may indicate the presence of a deep channel of migration of fluids and mineral matter to the upper horizons of the section in the study area. Instrumental measurements recorded the migration of gas and yellow phosphorus into the atmosphere. This may indicate the activity of the volcanic structure and the synthesis of hydrocarbons in it at present, which is confirmed by the modern processes of gas eruption in the crater of the Borealis mud volcano.

In this regard, as well as taking into account previous surveys of regions of gas migration into the atmosphere (Bogoyavlensky et al., 2020), it can be stated that in active volcanic complexes, the synthesis of hydrocarbons and other gases is currently carried out regularly, and, therefore, their migration into the atmosphere. In this regard, large volumes of harmful gases that pollute the environment in the areas of active volcanic complexes are also possible.

At a depth of 470 km from the lower part of the section, resonance responses at the frequencies of granites (old) were obtained.

Resonance responses at the frequencies of gas hydrates were obtained from the interval 0–400 m. Instrumental measurements confirmed the facts of gas migration into water and further from the surface into the atmosphere.

1.3.6.2 The Håkon Mosby Mud Volcano

Seafloor mud volcano Håkon Mosby with a diameter of 1.5 km is located in SW Barents Sea at a depth of 1250 m in the area between the coast of Norway and Bear Island (Figure 1.7).

It was discovered in 1995. The main products of its eruptions are mud and gases, of which more than 99% is methane. Under the mud volcano, there is a 1- to 2-km-wide disturbed zone with free gas which extends to a depth of >3 km (Shilov et al., 1999).

In the process of FR processing of a fragment of a photograph with a volcano from the surface, resonant signals were registered at the frequencies of gas, oil, bacteria methane oxidizers, and yellow phosphorus. Repeated acoustic surveys performed in 2002 and 2003 suggest continuous methane emissions to the ocean. From seafloor visual observations, we estimated a gas flux of 0.2 (0.08–0.36) mol s^{-1} which translates to several hundred tons per year under the assumption of a steady discharge (Foucher et al., 2010).

Resonance responses (signals) at frequencies of the seventh group (limestones) of sedimentary rocks were recorded only up to a depth of 2 km. From the intervals of 2–57–99 and 99–723 km, resonant responses (signals) were obtained at the frequencies of the ninth group (marl) of sedimentary rocks and the eleventh (kimberlite) group of igneous rocks. Instrumental measurements confirmed the facts of gas and yellow gas migration to the ocean surface and into the atmosphere.

On the site of the Håkon Mosby volcanic complex, there are no conditions for the synthesis of hydrocarbons at the 57 km limit. The FR results in the location of the Håkon Mosby volcanic complex were of a reconnaissance nature, so they have limited accuracy in determining the depth characteristics of this mud volcano.

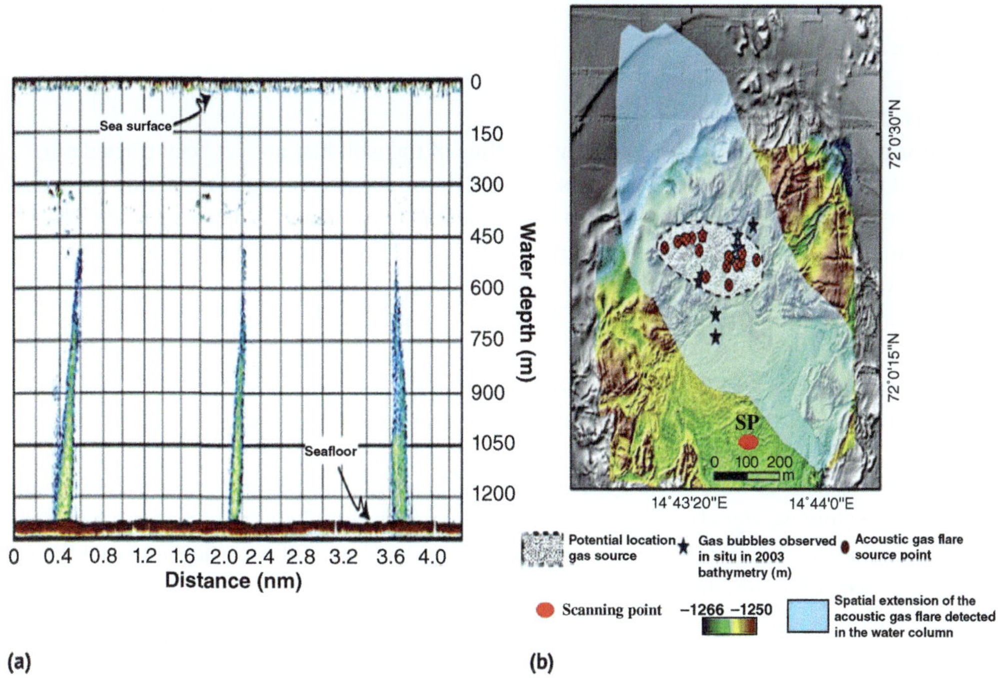

Figure 1.7 Acoustic gas flares (a) and gas bubbles (Foucher et al., (2010) / with permission of Springer Nature) were detected in the water column above the Håkon Mosby mud volcano (b).

1.3.7 The North Sea Area

Examples of FR technologies used to obtain depth characteristics of some discovered seepage areas in the North Sea and the central part of the Norwegian Sea (Yakymchuk et al., 2022, 2023; Soloviev et al., 2023) are considered below.

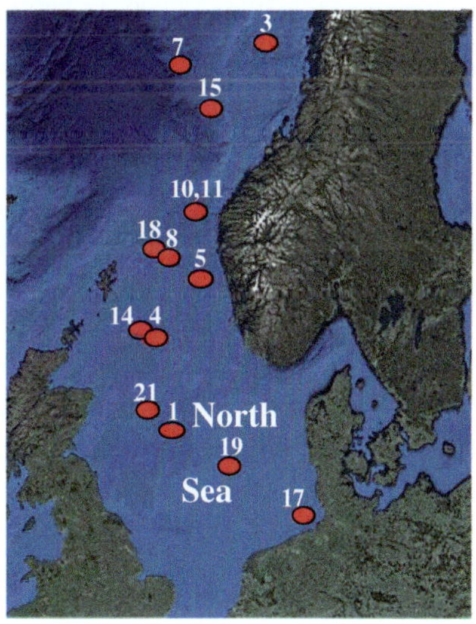

Figure 1.8 Bathymetry map of the North Sea and Norwegian Sea with the location of the SP points. Google Earth.

1.3.7.1 Well 6508/1-3 (SP3, Figure 1.8)

FR processing was carried out on a small fragment of this image in the area of the marker. During the processing of a local fragment of the image, responses from oil, condensate, and gas were not recorded. Signals were received from the eighth group of sedimentary rocks (dolomites). The root of a deep channel (volcano) filled with these rocks was determined at a depth of 194 km (layer of the plastic state of rocks).

1.3.7.2 Well 16/5-7 (SP4, Figure 1.8)

During FR processing of a fragment of this image in the well area, responses from oil, condensate, and gas **were not recorded**. As in the previous section, in the area of the well, it was established the presence of a volcano with a root at 194 km, filled with sedimentary rocks of the eighth group (dolomites). Responses from water on the surface of 69 km were also obtained here.

1.3.7.3 Well 32/4-3 S (SP5, Figure 1.8)

At the time of processing the image fragment, drilling of the well had not yet been carried out.

In the FR processing of this well, responses from oil, condensate, and gas **were not recorded**. Signals were received from the ninth group of sedimentary rocks (marls). The root of the volcano filled with these rocks was recorded at a depth of 194 km. In this area, responses from water at the surface of 69 km were also obtained. As follows from the results of FR processing of a satellite image of the well location, it turned **out to be dry.**

1.3.7.4 Local Glengorm Well Drilling Site (SP21, Figure 1.8)

The depth of the drilled well is 5056 m. The well-location coordinates were taken from messages on websites.

Anomalous responses were obtained at the frequencies of oil, and gas condensate (intense signal) and gas (intense signal) using the resonant frequencies of oil and gas condensate samples from the local area. We established the presence in this local area of a vertical channel for the migration of deep fluids and mineral matter.

Determination of condensate depth (by scanning from 86 m) established signals in intervals: (i) 300–475 m; (ii) 2200–2477 m; (iii) 3626–3738 m: (iv) 4855–5224 m; (v) 10,000–14,400 m (strong); (vi) 19,880–26,900 m; and (vii) 28,600–56,000 km.

Gas depth determination (by scanning from 80 m) established signals in intervals: (i) 286–470 m; (ii) 1789–2040 m; (iii) 3700–4082 m; (iv) 4873–5100 m; and (v) 5210–5380 m.

Determination of oil depth by scanning established signals in intervals) 5240–5765 m up to 10 km traced; further 13,900–14,880 m. The predicted oil reservoirs are located deeper than the bottom of the drilled well.

1.3.7.5 Nyegga Pockmarks (SP15, Figure 1.8)

With the help of FR technologies, sounding was carried out at a point on the Nyegga pockmark field, located on the shelf of the continental slope of Middle Norway in the southern part of the Voring plateau. More than 230 pockmarks were discovered here, the diameter of which reaches 600 m, and the depth is 15 m (Jedari-Eyvazi et al., 2023; Plaza-Faverola et al., 2010, 2011).

The peculiarities of the structure and the possible age of pockmarks of the Nyegga site are closely related to "gas pipes" structures and were interpreted in detail as a result of the long-term evolution of upward deep gaseous fluids through the sedimentary thickness (Hogan et al., 2012; Judd and Hovland, 2007; Yakymchuk et al., 2022, 2023; Plaza-Faverola et al., 2010, 2011; Soloviev et al., 2023).

In the Nyegge area, pockmarks are often expressed in relief by local structures, although not all "pipes" reach the surface of the bottom. Note the fact that in the presented models of "gas pipes," the main source of the formation of gas fluids is gas accumulations lying below the zone of gas hydrates (Figure 1.9).

The free gas system is mainly microbial methane from *in situ* production (Judd and Hovland, 2007). However, the contribution of thermogenic gas through deeper polygonal faults to the free gas system seems evident.

Conditions for thermogenic methane formation may exist below the bottom surface at depths of 1500 m and more. We suppose that there were pulsed inflows of fluids from gas accumulations rich in methane and do not exclude the possible influence of deep crustal (and mantle) gas fluids flow on pockmarks and seeps formation.

Our experience studying "gas pipes" has shown that many gas fluid flows and degassing processes in continental margin structures are of primary crustal origin.

During reconnaissance processing of the entire area of Nyegga, signals were recorded at the frequencies of hydrogen, red phosphorus, hydrogen bacteria, living water, marls, and siliceous sedimentary rocks, as well as signals from basalts and ultramafic rocks (Yakymchuk et al., 2022).

When probing the pockmark at the sounding point feedback was obtained for several groups of sedimentary rocks, gas hydrates, hydrocarbons, carbon dioxide, and yellow phosphorus. Gas hydrates are fixed in the section to a depth of 555 m.

Degassing processes were recorded on the water surface, as well as for depths of 340–586 m, 1669–1933 m, 3547–3999 m, and 4156–4819 m (deeper probing was not carried out).

The obtained results show that gas fluids of a sedimentary thickness of about 5 km are involved there in the formation of pockmarks. It is not excluded that degassing processes may exist in deeper horizons too. The depth of gas hydrates at the sounding point exceeds the model values of the accepted depth of the zone of stability of gas hydrates in this area (Figure 1.9).

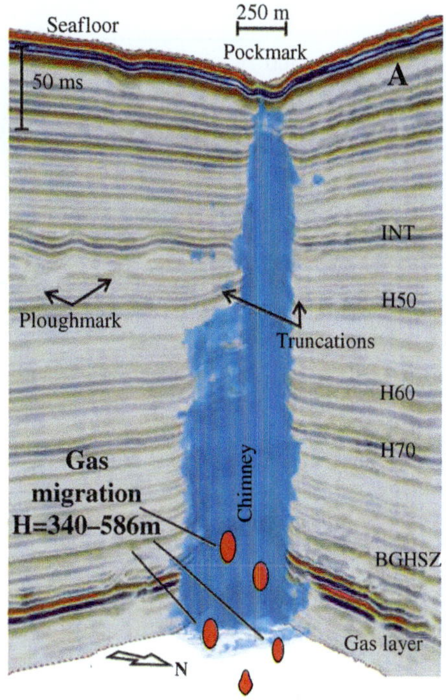

Figure 1.9 Pockmark and "gas pipe" in the Nyegga area (Plaza-Faverola et al., 2010, 2011) with schematic FR soundings results of pockmark's deep gas migration.

Thus, the results of the application of FR technologies showed that the sediments of the Nyegga area are an important and characteristic part of the system of geofluids and gas hydrates on the continental slope of Middle Norway, in the formation of which deep methane flows play a real role (Yakymchuk et al., 2022).

The obtained sounding results confirm the presence of deep accumulations of hydrocarbons in the Nyegga area. This area is known to be located in a gas-prospective zone, and the estimated gas reserves from gas hydrates, sub-hydrate "free" gas, and the "chimney zone" are roughly comparable to other recoverable gas reserves.

These obtained data indicate the existence of additional sources and processes of deep degassing that can explain the magnitude of the formation and manifestation of seep and pockmark fields in various structures of the Barents and other Seas.

1.3.7.6 The Area of the Emergency Well 2/4-14 (SP1, Figure 1.8)

In January 1989, exploration well 2/4-14, drilled in the southern part of the Norwegian sector of the North Sea (Figure 1.10), exploded at a depth of 4.7 km due to extremely high reservoir pressure (Landrø et al., 2019; Leifer and Judd, 2015).

The root of the volcano's sedimentary rocks was determined to be at a depth of 723 km. On the surface of 4300 m from the upper part of the section, there are signals from the seventh group of sedimentary rocks; at a depth of 4000 m, signals from these formations have already been daily. At the surface of 4500 m, gas signals were recorded from the lower and upper parts of the section (Figure 1.10).

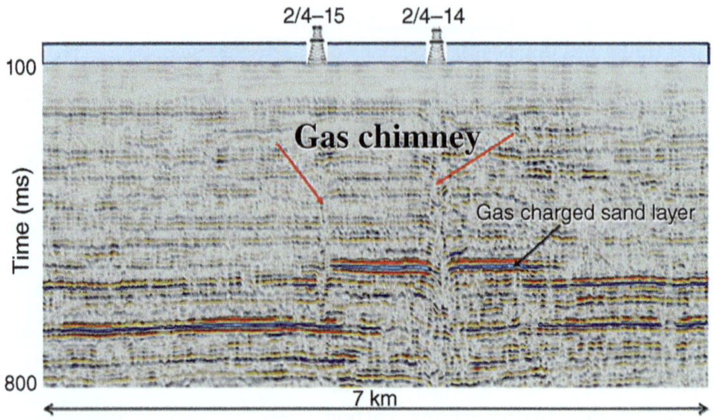

Figure 1.10 Schematic vertical profile from 3D seismic (1991) through 2/4-14 and 2/4-15 wells (Landrø et al., 2019 / with permission of Elsevier).

This indicates the migration of gas through water into the atmosphere, which confirms the genetic connection of sulfur fields with clay accumulations of carbohydrates, which is manifested in rich structures of the Polar (and other) regions.

1.3.7.7 Seepage Fields at the "Berta" Area (SP19, Figure 1.8)

The studied area is located within the Central Graben-Mesozoic rift structure of the northwestern part of the German sector of the North Sea, which belongs to one of the large provinces promising for hydrocarbons. In this area (Figure 1.11), about 170 places of seepage of gases into the water column were discovered and studied (Müller, 2023).

Salt volcanoes (diapirs) of the Late Jurassic age are widely developed in this part of the North Sea. Their growth led to the appearance of sub-vertical faults, disruption of sediment integrity, and the formation of gas migration paths in the process of deep degassing.

Well B11-4 is located away from a possible "gas pipe," and the values of methane concentration in bottom waters obtained in this area do not increase toward the well, although they differ in sharply increased values. Shallow (30–50 m) bottom depths contributed to the methane emission into the upper layers of the water column, as well as the emission of gases into the atmosphere (Müller, 2023).

The main part of the mounds is not manifested in the topography of the bottom and is connected with the salt volcano (diapir) "Berta" (Figure 1.11), on the eastern edge of which the most active manifestation of methane flares is observed.

All registered flares are located near the salt structure or above its upper part, which is located at a depth of about 2.0 km. The main concentration of methane flares (Figure 1.11) is located approximately 150–450 m to the northeast of this structure (Müller, 2023).

The results of the application of FR technologies showed (sounding point 1) that the sources of gas inflow are not only near the bottom but also its deep accumulations (gas pockets), located at depths from 1094 m to 6000 m. The thermogenic gas from deep horizons may be only a part of the gases at this site. The total thickness of a series of gas-bearing horizons may indicate the presence of a large reservoir at the eastern slope of the Berta salt volcano (diapir), located within the "gas pipe" with a localized area of seeps.

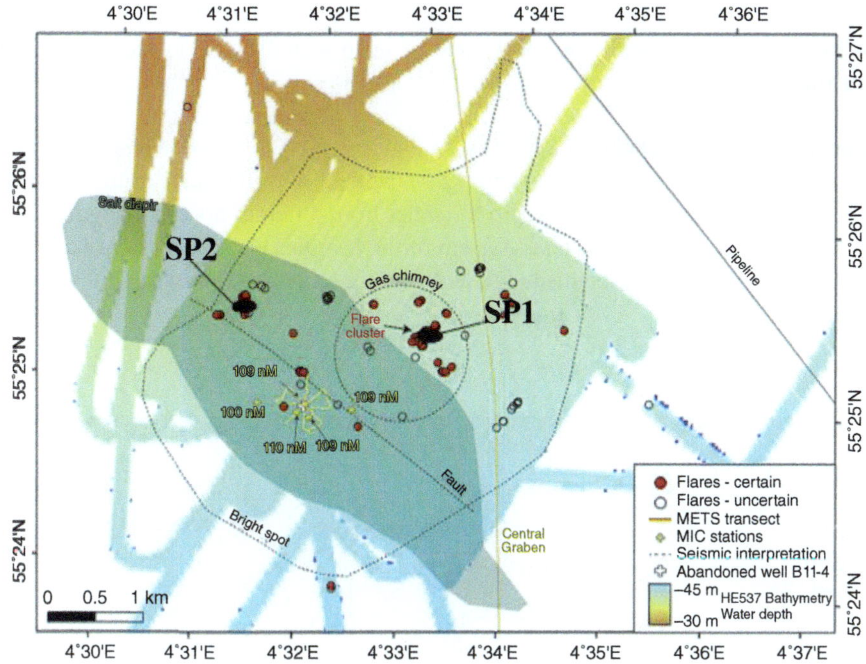

(a)

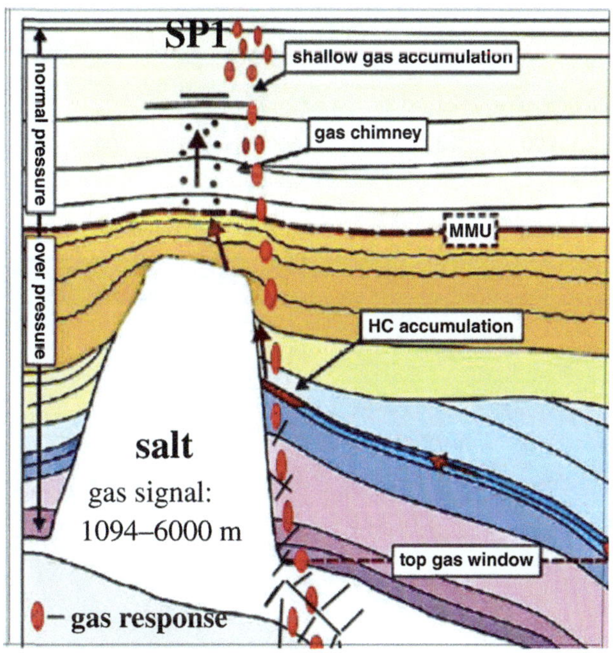

(b)

Figure 1.11 Map (a) of salt structure "Berta" in the northwestern part of the German sector of the North Sea, and the location of the FR sounding point (Römer et al., 2021 / Frontiers Media S.A. / CC BY 4.0); schematic map (b) of the seepage field, and gas accumulation in the German Central Graben (Adapted from Müller, 2023, modified).

The absence of gas signals at sounding point 2 (Figure 1.11) confirms the connection of the migration paths of deep gases with the local zone of deep-seated faults located outside the salt-dome structure.

1.3.7.8 Scanner Structure (SP16, Figure 1.8)

The results of studies of some seeps and pockmarks ("Scanner," "Challenger") in the North Sea showed that methane emission through "gas pipes" comes from different horizons (Figure 1.12).

An important component of the "gas pipe" system of the Scanner pockmark is the fault zone, which consists of many vertical cracks filled with near-surface and deep-seated gas (Callow et al., 2021; Kutas, 2020).

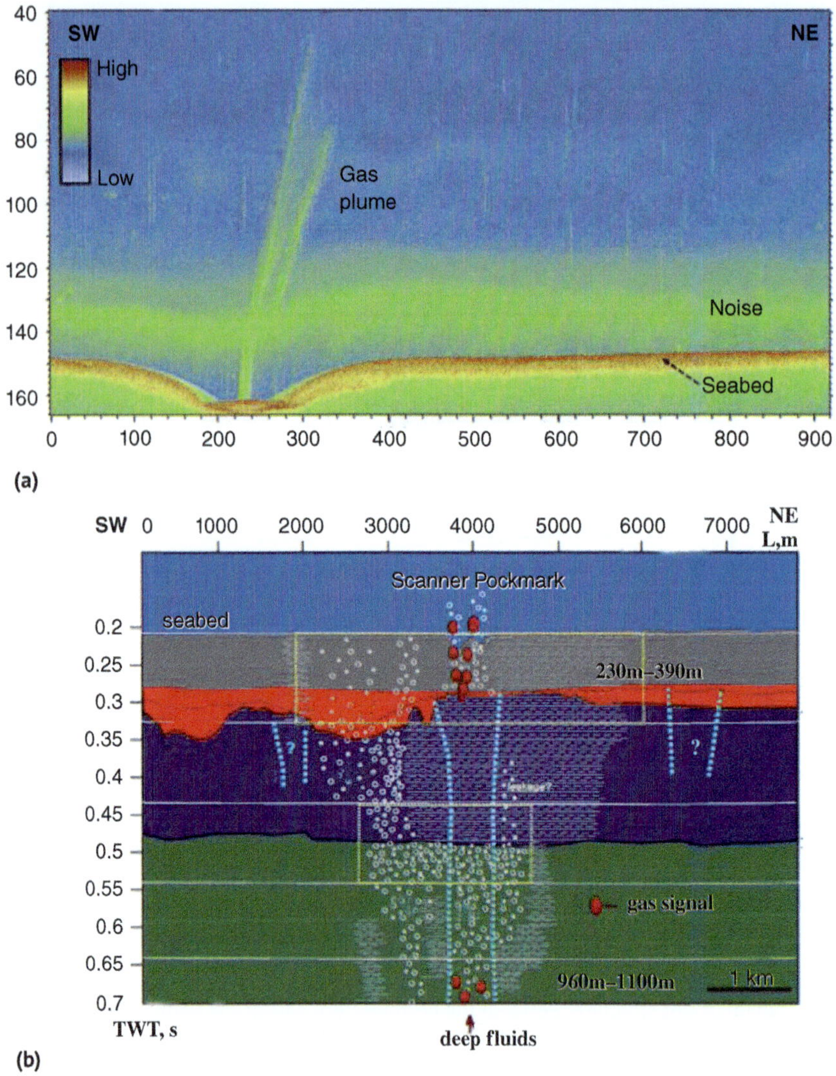

Figure 1.12 Scanner pockmark (a) with gas plume (Adapted from Robinson et al., 2021); (b) structure's section (Jedari-Eyvazi et al., 2023, modified / Oxford University Press / CC BY 4.0) with part of FR results.

A. Judd (Hogan et al., 2012) assumed that in the area where the Scanner and Challenger active pockmarks were located, the degassing process covered not only bottom sediments at depths less than 100 m below the bottom.

We have investigated the Scanner pockmark complex and obtained the FR data to determine the characteristics of the active area of focus.

Gas accumulations were found here at depths of 230–390 m and 960–1100 m, which indicated deep sources of degassing in the sediments.

The FR data have also shown that the gas accumulations are located in the upper part of sediments with a total thickness of 4800 m (Figure 1.12).

Gas signals at the other pockmark (the "Challenger") were recorded at the bottom level and a depth of 839–1635 m, i.e. the degassing process covered a part of the sedimentary sequence (Paleogene), below which the presence of a salt volcano was assumed. The results of studies of some seeps and pockmarks ("Scanner," "Challenger") in the North Sea showed that methane emission through "gas pipes" comes from different horizons (Figure 1.12).

1.3.7.9 Pockmark Field in the Gulf of Patras, Ionian Sea

It is known that the Patras Gulf Pockmark field (PGPF) is located in shallow waters of the Ionian Sea offshore (Figure 1.13) Patras City (Greece). Patras Gulf is the western part of the Patras-Corinth rift zone, and the bathymetry of the gulf is controlled by linear structures (extensive active faults) trending WNW–ESE. Six normal faults were recorded in the PGPF area. They appear in many local seismic profiles, show a sufficient displacement of the Holocene/Pleistocene boundary, and can be correlated to bathymetric features (Christodoulou et al., 2023).

The field has been under investigation since 1996 and is considered one of the most spectacular and best-documented fluid seepage activities. The PGP field consists of numerous (72) craters of variable shapes and extents, limited to an area of about 2.5 km^2.

There are many acoustic anomalies related to the presence of gas (mainly methane of microbial origin) in sediments (Christodoulou et al., 2023) and in the water column (gas flare recordings were revealed).

The main mechanisms for the development of the pockmark field in the Gulf are the local tectonism and the internal characteristics of the sedimentary layers. The spatial distribution of the earthquake epicenters for the time interval 2011–2022 shows medium seismicity in the eastern part of Patras Gulf and low in the central and it's western part (Figure 1.13). Authors suppose that the main earthquakes (1993 and 2008) were strong enough to induce cyclic loading in the uppermost sediments of the area and consequently trigger upward gas migration and emissions from the entire field (Christodoulou et al., 2023).

The scanning point (SP) is located south of the large ATF fault, and also away from the small fault (Christodoulou et al., 2023), in which the displacement of sediments is 12 m (Figure 1.13).

It can be assumed that in the Patras Gulf region, there are less-expressive fault zones, which are associated with minor disturbances of the bottom topography. The seismic studies carried out were local, therefore the deep structure of this PGPF area remains unstudied, as does the nature of methane in the sediments of the Gulf of Patras, although the authors classify it as a near-surface gas of typical microbial origin (Christodoulou et al., 2023).

Our preliminary investigations of FR scanning show that the joint signal of gas and salt is fixed in the interval 228 (step 1 cm)–1219 (step 50 cm), 1520–3153 m (step 50 cm). The salt volcano is fixed to a depth of 99 km. Further sounding was not carried out. Synthesis of hydrocarbons at a depth of 57 km is in a salt volcano.

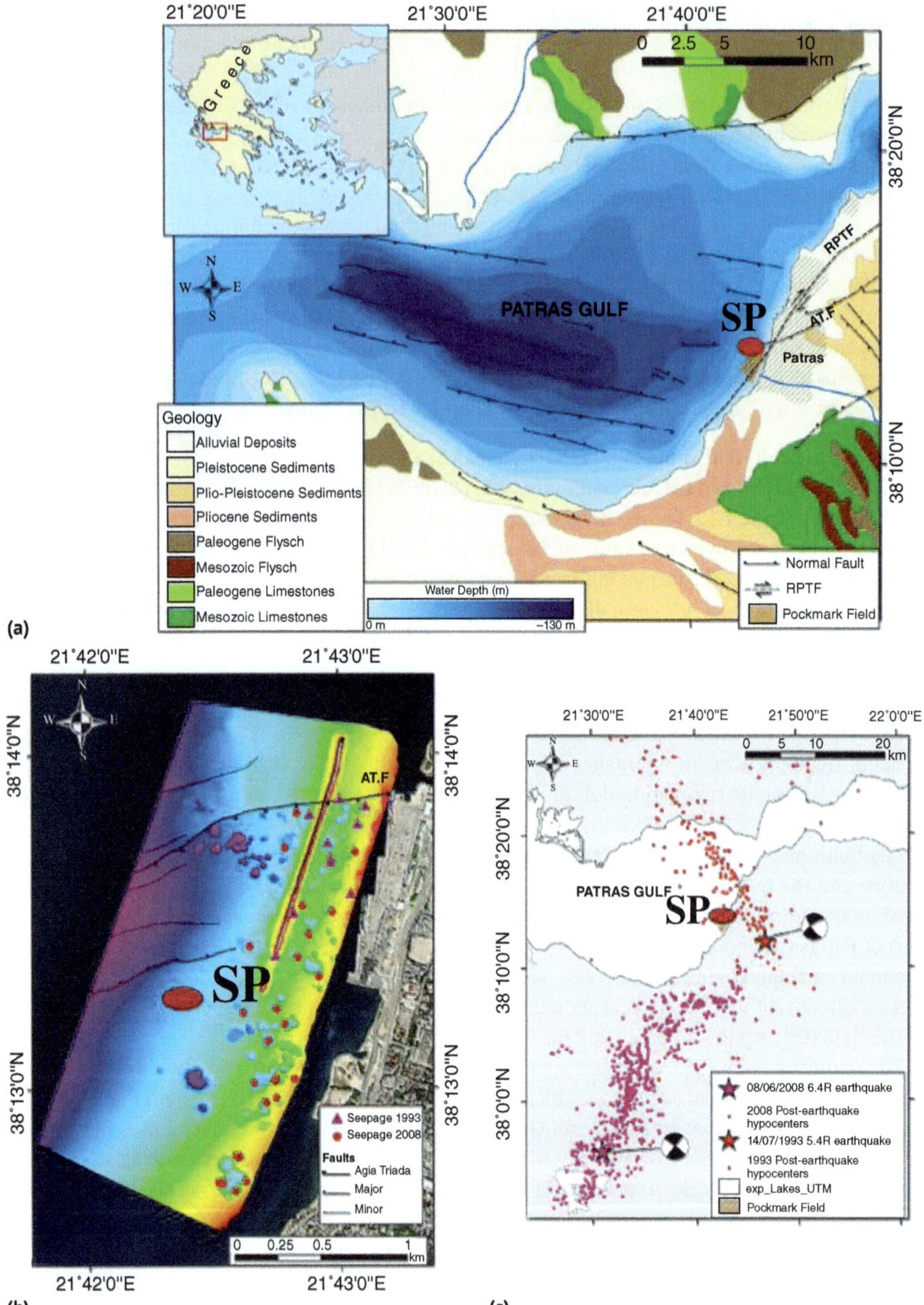

Figure 1.13 Simplified geological map of Patras Gulf (a), with the major faults, the pockmark field, and SP location (Christodoulou et al., 2023 / MDPI / CC BY 4.0); (b) area bathymetry with seepage location; (c) map with the location of the epicenters of the two major (1993 and 2008) earthquakes (Christodoulou et al., 2023 / MDPI / CC BY 4.0), their focal mechanism and the distribution of the post-earthquake events, and SP location. ATF, Agia Triada Fault.

However, we think the appearance of new arguments in favor of their deep genesis does not exclude the possible microbial nature of the pockmark of the local field with gas manifestations in the PGPF area.

Our experience of researching the sites of accumulations of sepals and pockmarks in many structures of the continental margins of the World Ocean indicates the presence of mainly deep (crustal and mantle) sources of the formation of a wide class of hydrocarbons, genetically united by degassing processes.

1.3.7.10 Some SP Scanning Results in the Black Sea

Below are some results of FR studies for scanning points (SP1–SP3) of gas emission centers in the western part of the Black Sea, and in the deltaic structures of paleo-rivers (Kalanchak, Dnipro, Dniester, and Danube).

SP1 ("degassing pipe") is located near the Pradniprovska-2 parametric well (Ludmann et al., 2004; Naudts et al., 2009; Shnyukov and Yanko-Hombach, 2020). During the process of scanning the image of the entire area, responses from oil, condensate, gas, oil shale, gas hydrates, anthracite, hydrogen, etc. were recorded.

Signals at gas frequencies were obtained (Figure 1.15) from the intervals: (i) 865–1100 m, (ii) 1400–1780 m, (iii) 2170–(2400-intense) – (2650-very intense – 2700 m) – (2780–2800-very intense – 2900) –2940 m, (iv) 3500–3690 m, (v) 4185–(5040-very intense) –5100 m, (v) 5445–(5560-intense)–5650 m.

Volcanic complexes were found within the entire area, in the contours in which the synthesis of hydrocarbons is carried out, as well as a basalt volcano, through which hydrogen migrates into the atmosphere.

The results of the scans indicate the prospects of certain local areas within the Dniester zone (west of the Pradniprovska-2 well), which confirms the predictive assessment given in Ludmann et al. (2004) and Vogt et al. (1999).

The scanning point **SP2** is located in the area of the paleo-delta of the Dnieper, and near several promising oil and gas-bearing structures (Figures 1.14 and 1.15). This zone of the shelf is characterized by a developed system of tectonic disturbances of various ages and extents, which control the position of the main blocks, as well as the paths of the ascent of coro-mantle fluids (Kutas, 2020; Kutas et al., 2004; Shnyukov and Yanko-Hombach, 2020; Shnyukov et al., 2013; Shnyukov and Kobolev, 2020). The results of the scan were carried out to a depth of 1320 m, recorded signals from oil, gas condensate, and gas (intervals: 565–702 m; 1130–1315 m), and confirmed the active emission of gases into the atmosphere. Signals from gas hydrates were not received in SP2 (in contrast to point SP1).

The scanning point **SP3** is located over the MSU mud volcano in the deep-sea basin of the Black Sea (Figures 1.14 and 1.16).

MSU MV is characterized by a regular cone-shaped structure (about 2.5 km in diameter at the base) and a height of about 60 m (Shnyukov et al., 2013) under the sea bottom (2100 m).

The common signals at the frequencies of gas and the seventh group of limestones were recorded in the intervals of 2495–4496 m, 5998–7249 m, and 7716–8086 m (Figure 1.16). The shape of the crater and FR data reflect the multi-stage (two or three) nature of the strong eruptions with different mudflows on the flanks.

The facts of gas and yellow phosphorus migration into the water column and the atmosphere were confirmed by instrumental measurements.

Gas flares in the water column have been recorded in over 5000 locations in the Black Sea and range from seeps related to deep-seated mud volcanoes (Shnyukov and Yanko-Hombach, 2020; Shnyukov et al., 2013).

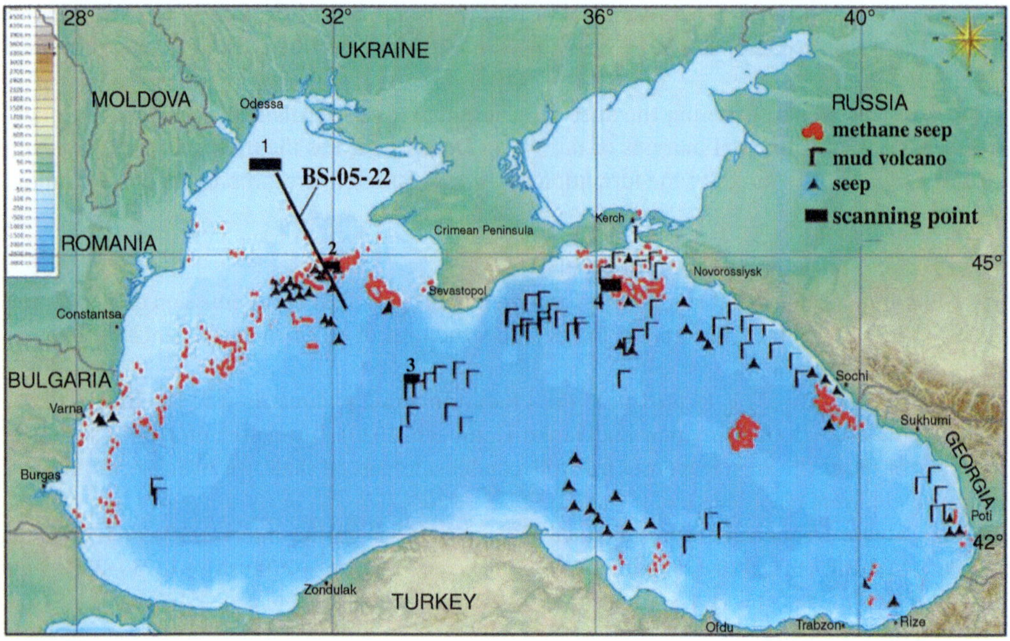

Figure 1.14 Schematic map of the distribution of gas seeps and mud volcanoes in the Black Sea (Shnyukov and Yanko-Hombach, 2020; Shnyukov et al., 2013) with locations of FR-scanning points (SP1–SP3) and seismical profile BS-05-22 (Vogt et al., 1999) in the western part of the Black Sea.

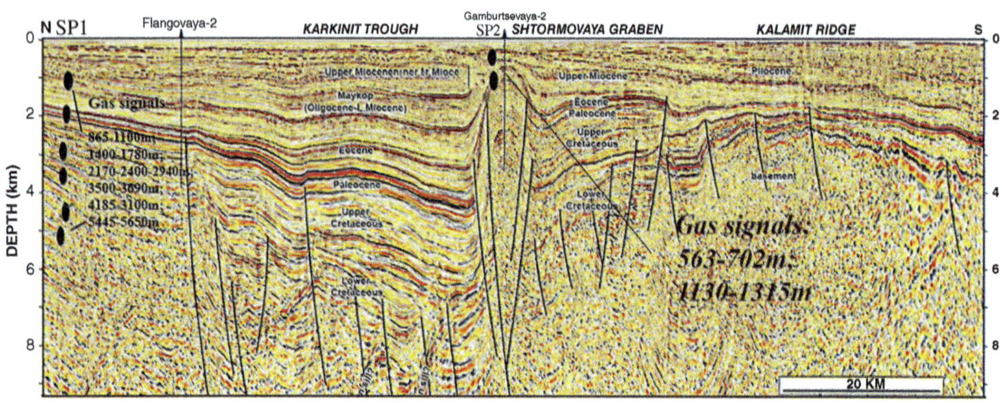

Figure 1.15 Schematic seismic profile BS-05-22 (Adapted from Vogt et al., 1999) across the Odesa Shelf (Black Sea) with the location of SP1 and SP2 and the FR scanning results.

It is known that there is evidence of both thermogenic and biogenic sources of gas (mostly methane) origin in the Black Sea region. Seismic data confirm the migration of deep gases (Popescu et al., 2007, Figure 4).

Mud volcano of Mitin. The depth of the sea in the area with the volcano is 680 m (Figure 1.17). This mud volcano, like some of the other mud volcanoes of the Black Sea, does not have a clearly defined connection with diapiric structures (Figure 1.17) but is confined to a nearby fault system,

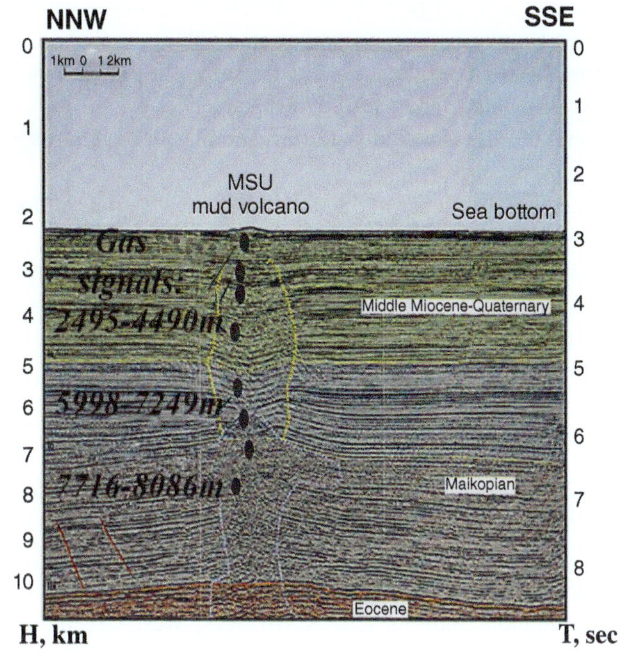

Figure 1.16 The MSU MV in the western part of the Black Sea (Adapted from Shnyukov and Yanko-Hombach, 2020) with FR gas signal results.

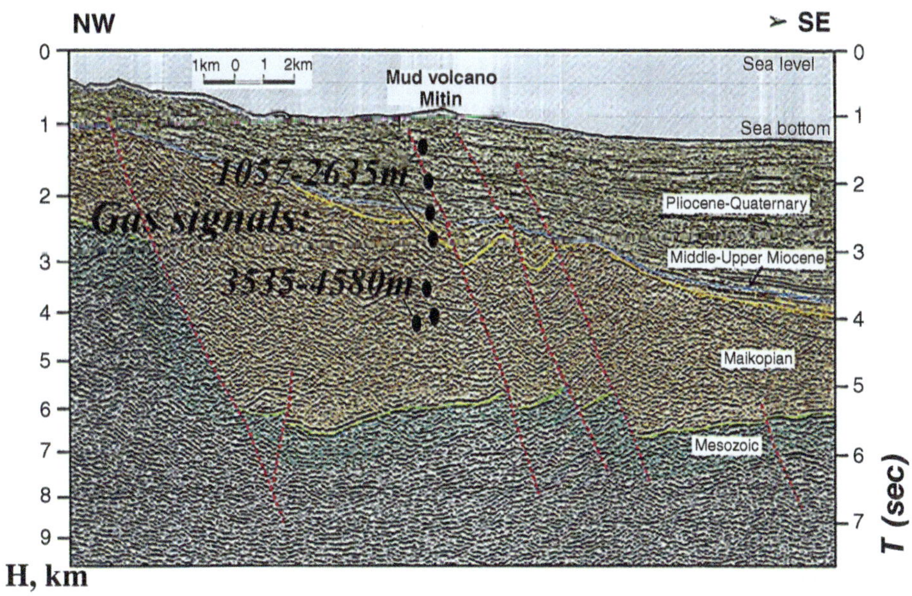

Figure 1.17 Structure of the Mitin MV and FR scanning point data (Adapted from Shnyukov and Yanko-Hombach, 2020, simplified).

consisting of several faults in the upper part of the sedimentary sequence with a thickness of about 2 km (Shnyukov and Yanko-Hombach, 2020; Shnyukov et al., 2013).

Responses were obtained at the frequencies of the seventh group of igneous (ultramafic) rocks and the ninth group of sedimentary rocks (marls), respectively.

Signals at the frequencies of oil, gas condensate, gas, and black phosphorus were recorded from the surface.

When scanning cross-section to a depth of 5300 m, the general signals at the frequencies of gas and ultramafic rocks were in the intervals of 1057–2635 m and 3536–4580 m.

Instrumental measurements confirmed the existence of conditions for the synthesis of oil, gas condensate, gas, and black phosphorus at the 57 km depth, as well as the facts of gas and black phosphorus migration into the water column and the atmosphere.

1.4 Conclusions

This chapter considered some new results of FR research using methane seeps at gas emission sites of various genesis and modern gas releases in various structures of the shelf and continental margins.

The materials of experimental reconnaissance studies in various regions of the globe (degassing areas of the Arctic Region, Barents and North Seas, West Greenland and Spitsbergen, Mediterranean, and Black Sea) are important arguments in favor of the "volcanic" model of formation of various structural elements of Earth.

Methane seeps and oil slicks can serve as indicators of the activity of volcanoes in which hydrocarbons may synthesized. In the contours of such volcanoes, there are deep channels (signals at the frequencies of oil, condensate, and gas were also recorded) through which oil, condensate, and gas may migrate to the upper horizons of cross-section and can replenish the already formed deposits on hydrocarbon fields.

FR scanning data for different seeps and pockmarks areas showed that the intensity and dynamics of their formation partly depended on the inflow of gas fluids from deep sources because of active degassing processes. The FR-sounding technologies make it possible to determine the probable depth of geological sources at active migration centers. The interpretation of the remote sensing data yields new additional (and independent) information, and their study has confirmed that these structures (seeps and pockmarks) can be a possible indicator of shallow and deep hydrocarbon accumulations.

R. Yoshida's statement "Science is truthful, but not truth itself" emphasizes the need for further development of science and the search for objective evidence of the reality of facts (especially unexpected ones) obtained when applying new technologies to study the earth's interior. FR studies, some results of which are given above, need development and improvement, but the already obtained results allow us to hope that they can be actively (additionally) used in conducting regional and local geophysical research.

Author Contributions

M.Y. and I.K. carried out the experiments and developed the concepts of this research. V.S. collected the data and wrote this manuscript with contributions on discussion from M.Y. and I.K. All authors contributed to reading, manuscript revision, read, and approved the submitted version.

Conflict of Interests

The authors declare no conflict of interest.

References

Andreassen, K., Hubbard, A., Winsborrow, M., et al. (2017). Massive blow-out craters formed by hydrate-controlled methane expulsion from the Arctic seafloor. Science 356 (6341), pp. 948–953.

Betlem, P., Roy, S., Birchall, T., et al. (2021). Modeling of the gas hydrate potential in Svalbard's fjords. Journal of Natural Gas Science and Engineering 94, p. 104127. https://doi.org/10.1016/j.jngse.2021.104127.

Blinova, M., Faleide, J. I., Gabrielsen, R. H., et al. (2013). Analysis of structural trends of sub-sea-floor strata in the Isfjorden area of the West Spitsbergen Fold-and-Thrust Belt based on multichannel seismic data. Journal of the Geological Society, London 170, pp. 657–668.

Bludd, E. K. (2023). New volcano discovered in the Barents Sea – UiT – 10 May 2023.

Bogoyavlensky, V. I., Sizov, O. S., Nikonov, R. A., et al. (2020). Earth degassing in the Arctic: the genesis of natural and anthropogenic methane emissions. Arctica: Ecology and Economy 39 (3), pp. 6–22. https://doi.org/10.25283/2223-4594-2020-3-6-22.

Bouriak, S., Vanneste, M., & Saoutkine, A. (2000). Inferred gas hydrates and clay diapirs near the Storegga Slide on the southern edge of the Voring Plateau, offshore Norway. Marine Geology 163 (1–4), pp. 125–148.

Callow, B., Bull, J. M., Provenzano, G., et al. (2021). Seismic chimney characterization in the North Sea – implications for pockmark formation and shallow gas migration. Marine and Petroleum Geology, 133, p. 105301.

Christiansen, F. G., Bojesen-Koefoed, J., Dam, G., et al. (2020). A review of oil and gas seepage in the Nuussuaq Basin, West Greenland – implications for petroleum exploration. GEUS Bulletin 44 (4567), pp. 1–21. https://doi.org/10.34194/geusb.v44.4567.

Christodoulou, D., Papatheodorou, G., Geraga, M., et al. (2023). Geophysical and geochemical exploration of the pockmark field in the Gulf of Patras: New insights on formation, growth and activity. Applied Sciences 13, p. 10449. https://doi.org/10.3390/app131810449.

Cook, L. (2019). Documenting cold seeps on the west coast of Greenland using videos of the seabed. UCL MRes Biodiversity, Evolution and Conservation, BIOS0013_18-19, pp. 1–48.

Forwick, M., Baeten, N. J., & Vorren, T. O. (2009). Pockmarks in Spitsbergen fjords. Norwegian Journal of Geology 89, pp. 65–77.

Foucher, J. -P., Dupré, S., Scalabrin, C. et al. (2010). Changes in seabed morphology, mud temperature, and free gas venting at the Håkon Mosby mud volcano, offshore northern Norway, over the time period 2003–2006. Geo-Marine Letters 30 (3–4), pp. 157–167. http://dx.doi.org/10.1007/s00367-010-0193.

Greinert, J., McGinnis, D. F., Naudts, L., et al. (2010). Atmospheric methane flux from bubbling seeps: Spatially extrapolated quantification from a Black Sea shelf area. Journal of Geophysical Research 115, p. C01002. https://doi.org/10.1029/2009JC005381.

Heggland, R. (1998). Gas seepage as an indicator of deeper prospective reservoirs. A study based on exploration 3D seismic data. Marine and Petroleum Geology 15 (1), pp. 1–9.

Hogan, K. A., Dowdeswell, J. A., & Ó Cofaigh, C. (2012). Glacimarine sedimentary processes and depositional environments in an embayment fed by West Greenland ice streams. Marine Geology 311–314, pp. 1–16. http://www.oxfordenergy.org/wpcms/wp-content/uploads/2014/11/WPM-56.pdf.

Jedari-Eyvazi, F., Bayrakci, G., Minshull, T. A., et al. (2023). Seismic characterization of a fluid escape structure in the North Sea: the Scanner Pockmark complex area. Geophysical Journal International 234 (1), pp. 597–619. https://doi.org/10.1093/gji/ggad078.

Judd, A. & Hovland, M. (2007). *Seabed Fluid Flow: The Impact on Geology, Biology, and the Marine Environment.* Cambridge. p. 475.

Kleber, G. E., Hodson, A. J., Mager, L., et al. (2023). Groundwater springs formed during glacial retreat are a large source of methane in the high Arctic. Nature Geoscience 16, pp. 597–604. https://doi.org/10.1038/s41561-023-01210-6.

Kutas, R. I. (2020). Geotectonic and geothermal conditions of fluid and gas discharge in the Black Sea. Geofizicheskiy Journal 42 (5), pp. 16–52.

Kutas, R. I., Paliy, S. I., & Rusakov, O. M. (2004). Deep faults, heat flow and gas leakage in the northern Black Sea. Geo Marine Letters 24, pp. 163–168. https://doi.org/10.1007/S00367-004-0172-3.

Landrø, M., Wehner, D., Vedvik, N., et al. (2019). Gas flow through shallow sediments – a case study using passive and active seismic field data. International Journal of Greenhouse Gas Control 87, pp. 121–133.

Leifer, I. & Judd, A. (2015). The UK22/4b blowout 20 years on: Investigations of continuing methane emissions from sub-seabed to the atmosphere in a North Sea context. Marine and Petroleum Geology 68, pp. 706–717. https://doi.org/10.1016/j.marpetgeo.2015.11.012.

Levashov, S. P., Yakymchuk, N. A., & Korchagin, I. N. (2012). Frequency-resonance principle, mobile geoelectric technology: A new paradigm of geophysical investigations. Geofizicheskiy Zhurnal 34 (4), pp. 166–176 (in Russian).

Ludmann, T., Wong, H. K., Konerding, P., et al. (2004). Heat flow and quantity of methane deduced from a gas hydrate field in the vicinity of the Dnieper Canyon, northwestern Black Sea. Geo-Marine Letters 24, pp. 182–193.

Müller, S. M. (2023). The German Central Graben (North Sea): Tectonostratigraphic evolution and hydrocarbon systems. Hannover: Gottfried Wilhelm Leibniz Universität. Diss. IX, p. 206. https://doi.org/10.15488/13228.

Naudts, L., Batist, M., Greinert, J., et al. (2009). Geo and hydro-acoustic manifestations of shallow gas and gas seeps in the Dnepr paleodelta, northwestern Black Sea. The Leading Edge. https://doi.org/10.1190/1.3236372.

Nielsen, T., Laier, T., Kuijpers, A., et al. (2014). Fluid flow and methane occurrences in the Disko Bugt area offshore West Greenland: Indications for gas hydrates? Geo-Marine Letters 34, pp. 511–523. https://doi.org/10.1007/s00367-014-0382-2.

Panieri, G., Argentino, C., Ramalho, S. P., et al. (2024). An Arctic natural oil seep investigated from space to the seafloor. Science of the Total Environment 907, 167788, pp. 1–13.

Plaza-Faverola, A., Bünz, S., & Mienert, J. (2010). Fluid distributions inferred from P-wave velocity and reflection seismic amplitude anomalies beneath the Nyegga pockmark field of the mid-Norwegian margin. Marine and Petroleum Geology 27 (1), pp. 46–60.

Plaza-Faverola, A., Bünz, S., & Mienert, J. (2011). Repeated fluid expulsion through sub-seabed chimneys offshore Norway in response to glacial cycle. Earth and Planetary Science Letters 305, pp. 297–308.

Popescu, I., Lericolais, G., Panin, N., et al. (2007). Seismic expression of gas and gas hydrates across the western Black Sea. Geo-Marine Letters 27 (2–4), pp. 173–183.

REV Ocean (2023). New mud volcano discovered in the Barents Sea. https://www.revocean.org/mud-volcano/.

Robinson, A. H., Callow, B., Bottner, C., et al. (2021). Multiscale characterization of chimneys/pipes: Fluid escape structures within sedimentary basins. International Journal of Greenhouse Gas Control 106, p. 103245. https://doi.org/10.1016/j.ijggc.2020.103245.

Römer, M., Blumenberg, M., Heeschen, K., et al. (2021). Seafloor methane seepage related to salt diapirism in the northwestern part of the German North Sea. Frontiers in Earth Science 9, p. 556329. https://doi.org/10.3389/feart.2021.556329.

Roy, S., Hovland, M., Noormets, R., et al. (2015). Seepage in Isfjorden and its tributary fjords, West Spitsbergen. Marine Geology 363, pp. 146–159. http://dx.doi.org/10.1016/j.margeo.2015.02.003.

Roy, S., Senger, K., Hovland, M., et al. (2019). Geological controls on shallow gas distribution and seafloor seepage in an Arctic fjord of Spitsbergen, Norway. Marine and Petroleum Geology 107, pp. 237–254. http://dx.doi.org/10.1016/j.marpetge.

Sarkar, S., Berndt, C., Minshull, T. A., et al. (2012). Seismic evidence for shallow gas-escape features associated with a retreating gas hydrate zone offshore west Svalbard. Journal of Geophysical Research Solid Earth 117, 09102, pp. 1–18.

Schumann, K., Völker, D., & Weinrebe, W. R. (2012). Acoustic mapping of the Illulissat Ice Fjord mouth, West Greenland. Quaternary Science Reviews 40, pp. 78–88.

Serov, P., Mattingsdal, R., Winsborrow, M., et al. (2023). Widespread natural methane and oil leakage from submarine Arctic reservoirs. Nature Communications 14, p. 1782. https://doi.org/10.1038/s41467-023-37514-9.

Shestopalov, V. M., Lukin, A. E., Zgonik, V. A., et al. (2018). *Essays on Earth's Degassing*. Kyiv: BADATA-Intek Service (in Russian).

Shilov, V., Druzhinina, N., Vasilenko, L., et al. (1999). Stratigraphy of sediments from the Håkon Mosby mud volcano area. Geo-Marine Letters 19, pp. 48–56.

Shnyukov, E. & Yanko-Hombach, V. (2020). *Mud Volcanoes of the Black Sea Region and Their Environmental Significance*. Springer Nature Switzerland AG. https://doi.org/10.1007/978-3-030-40316-4.

Shnyukov, E. F. & Kobolev, V. P. (2020). Blind mud volcanoes of the Black Sea. Geology and Drawings of the Light Ocean 16 (2), pp. 49–65. https://doi.org/10.15407/gpimo2020.02.049.

Shnyukov, E. F., Kobolev, V. P., & Pasynkov, A. A. (2013). *Gas Volcanism of the Black Sea*. Kyiv. p. 383 (in Russian).

Soloviev, V., Yakymchuk, N., & Korchagin, I. (2023). Pockmarks, seep sources, and degassing processes in the polar region structures. New Concepts in Global Tectonics Journal 11 (1), pp. 35–47. ISSN 2202-0039. http://www.ncgtjournal.co11.m/journals.html.

Veloso, M., Greinert, J., Mienert, J., et al. (2015). A new methodology for quantifying bubble flow rates in deep water using split-beam echo-sounders: Examples from the Arctic offshore NW-Svalbard. Limnology and Oceanography: Methods 13, pp. 267–287.

Vogt, P., Gardner, J., & Crane, K. (1999). The Norwegian-Barents Svalbard (NBS) continental margin: Introducing a natural laboratory of mass wasting, hydrates, and the ascent of sediment, pore water, and methane. Geo-Marine Letters 19 (1–2), pp. 2–21.

Wallmann, K., Riedel, M., Hong, W. L., et al. (2018). Gas hydrate dissociation off Svalbard induced by isostatic rebound rather than global warming. Nature Communications 9 (1), p. 83.

Westbrook, G. K., Thatcher, K. E., Rohling, E. J., et al. (2009). Escape of methane gas from the seabed along the West Spitsbergen continental margin. Geophysical Research Letters 36 (15), p. L15608. https://doi.org/10.1029/2009GL039191.

WMO (2022). WMO Greenhouse Gas Bulletin (GHG Bulletin): The State of Greenhouse Gases in the Atmosphere Based on Global Observations through 2021 (No. 18 | 26 October 2022).

Yakymchuk, M. & Korchagin, I. (2023). About the opportunity of direct-prospecting methods application for detection areas of gas and natural hydrogen migration to the surface and in the atmosphere. *Publisher.agency: Proceedings of the 2nd International Scientific Conference «Interdisciplinary Science Studies» (May 4–5, 2023)*. Dublin, Ireland, pp. 220–251. ISBN 978-2-6400-5011-9. https://doi.org/10.5281/zenodo.7905551.

Yakymchuk, M., Korchagin, I., Levashov, S., et al. (2022). *Volcanism and Degassing Processes in the Structures of the Earth's Polar Regions (review based on the results of frequency-resonance studies)*. Dodo Books Indian Ocean Ltd. And OmniScriptum S.R.L Publishing Group, p. 276 (in Ukrainian). ISBN: 978-620-0-63606-5. https://morebooks.de/shop-ui/shop/search?q = 978-620-0-63606-5&page=1.

Yakymchuk, M., Korchagin, I., & Soloviev, V. (2023). Some results of direct FR technology applied to study methane seepage areas in the Arctic region. Advances in Geological and Geotechnical Engineering Research 5 (3), pp. 25–38. https://ojs.bilpublishing.com/index.php/agger.

Yakymchuk, N. A., & Korchagin, I. N. (2019). Technology of frequency-resonance processing of remote sensing data: results of practical approbation during mineral searching in various regions of the globe. Part I. Geoinformatika, 2019, no. 3, pp. 29–51; Part II. Geoinformatika. 2019. no. 4, pp. 30–58; Part III. Geoinformatika. 2020. no. 1, pp. 19–41; Part IV. Geoinformatika. 2020. no. 3, pp. 29–62; Part V. Geoinformatika. 2021. no. 3–4, pp. 51–88 (in Russian).

Yakymchuk, N. A., Korchagin, I. N., Bakhmutov, V. G., et al. (2019). Geophysical investigation in the Ukrainian marine Antarctic expedition of 2018: mobile measuring equipment, innovative direct-prospecting methods, new results. Geoinformatika 1, pp. 5–27 (in Russian).

2

The Role of the LVZ and of Increased Seismicity in the Localization of Abiogenic HC in the Crystalline Crust of Transcarpathia

Valeriy Korchin, Elena Karnaukhova

S. Subbotin Institute of Geophysics NAS of Ukraine, Kyiv, Ukraine

2.1 Introduction

Hydrocarbons formed in the Earth's interior are not of sufficient interest if they are inaccessible to humans in the upper horizons. Therefore, the study of the path of their movement from the depths and their synthesis deserves special attention. The migration and localization of abiogenic hydrocarbons require detailed study since they provide the necessary information for their successful search and production. Summarizing the world experience in studying the territories of large hydrocarbon (HC) deposits confined to the crystalline crust, and also taking into account their inexhaustible reserves, the main provisions of their possible deep, abiogenic origin have been developed. Accumulations of hydrocarbons of abiogenic genesis, as a rule, are associated with fault zones (rifting), intersection of fault structures (rift node), plumetectonic objects (deep-seated diapirs), disjunctive dislocation zones of metamorphic and igneous rocks (the presence of decompaction zones), a thick layer of sedimentary cover and high-temperature fluid flows (Korchin, 2013, 2017; Korchin and Rusakov, 2021; Korchin et al., 2021; Krayushkin and Shevchenko, 2018).

Analyzing materials (over more than a hundred years of study history) of geology, geophysics and geodynamics of the Carpathian region, we can assume the predominant role of the abiogenic component of hydrocarbon deposits in the region. In the process of research, the features of the geological construction, structure and nature of elastic and density anomalies of the Earth's crust with the activation of seismotectonic processes and localization of hydrocarbon deposits in Transcarpathia were studied in detail. Poor knowledge of the deep structure of Transcarpathia, especially the upper part of the Earth's crust in the depth range of 3–10 km, did not allow solving a number of fundamental issues of tectonics and seismicity, zoning of minerals in the study area. Therefore, a detailed model of the distribution of rock types with depth was developed and the features of the structure of the Earth's crust of the Transcarpathian trough along the RP-17 profile (Chop-Velykyi Bychkiv) were clarified, taking into account the modern geodynamics of the area using petrophysical thermobaric modeling (PTBM) techniques (Burtnyi et al., 2013;

Geophysical Exploration for Hydrocarbon Reservoirs, Geothermal Energy, and Carbon Storage: New Technologies and AI-based Approaches, First Edition. Edited by Said Gaci.

Chekunov et al., 1969; Korchin, 2017, 2021; Korchin et al., 2012, 2013, 2018b, 2019, 2020b, 2022; Korchin and Karnaukhova, 2021; Korchin and Rusakov, 2021). This made it possible to some extent to explain the physical nature of some local velocity anomalies along the profile. A more detailed study of the deep structure of this region, and especially the position and structure of the LVZs, in the future will significantly expand the possibilities of searching for hydrocarbon deposits in Transcarpathia.

2.2 Basic Principles of Petrophysical Thermobaric Modeling

To model the petrophysical structural characteristics of the deep horizons of the Transcarpathian depression (TD), we proposed petrophysical thermobaric modeling techniques using experimental data on changes in the physical characteristics of rocks in various *PT* conditions of experiments corresponding to those for specific geological regions. To understand the modeling process itself, in Sections 2.2–2.5, we used our previously published works (Burtnyi et al., 2013, 2023; Korchin, 2010, 2013, 2015a, 2015b, 2017; Korchin et al., 2009, 2012, 2013, 2018a, 2018b, 2020a, 2020b, 2022, and others). A sample of the rock in the experiment is like being "submerged" to the definite depth. This technique made it possible to obtain a large amount of information about the change in elastic-density parameters of the mineral substance with depth for specific regions in a short period of time. Below is an optimal block diagram of small-scale petrophysical deep structural modeling (Figure 2.1).

2.3 Influence of *PT*-Regimes on the Elastic Characteristics and Density of Rocks

We conducted long-term investigations of the experimental changes in velocities and densities with depth for rocks under *PT* conditions relevant to specific regions and at specific depths. The study established the occurrence of the velocity inversion of elastic longitudinal waves (V_P) and density (ρ) at different depths (H). With increasing depth (i.e., with increasing pressure (P) and temperature (T) affecting the specimen of the rock), after some initial increase in V_P and ρ, their decrease is observed. Then the velocities and densities increase again (Figure 2.2a, b) (Burtnyi et al., 2013, 2023; Korchin, 2013; Korchin et al., 2013, 2018a, 2020a, 2020b, 2021, 2022).

Thus, a low velocity zones are manifested on the $V_P = f(PT) = f(H)$ curves whose configuration and location correlate well with elastic anomalies registered by deep seismic sounding (DSS) information from the Earth's crust at depths of 3–25 km. The changes of $V_P = f(H)$ can also be obtained from measuring the isobars of velocities ($V_P = f(T)$ at $P = const$) and their isotherms ($V_P = f(P)$ at $T = const$) (Figure 2.2d). Detailed studies showed that both methods of determining $V_P = f(H)$ plots give identical results (Burtnyi et al., 2023; Korchin, 2010, 2013; Korchin et al., 2013, 2020b, 2021, 2022).

Regardless of the methods of determining velocities, using several *PT* programs, or calculating the isobars and isotherms, there is a threshold value of changes of the temperature with depth ($\partial T/\partial H$) when the anomalous elastic state of mineral matter arises – LVZs. The changes of velocity of elastic waves with depth (V_p) in the rock of constant mineral composition can be calculated by the ratio of $\frac{\partial V}{\partial H} = \left(\frac{\partial V}{\partial P}\right)_T \cdot \frac{\partial P}{\partial H} + \left(\frac{\partial V}{\partial T}\right)_P \cdot \frac{\partial T}{\partial H}$. The LVZs in the Earth's crust are determined by the

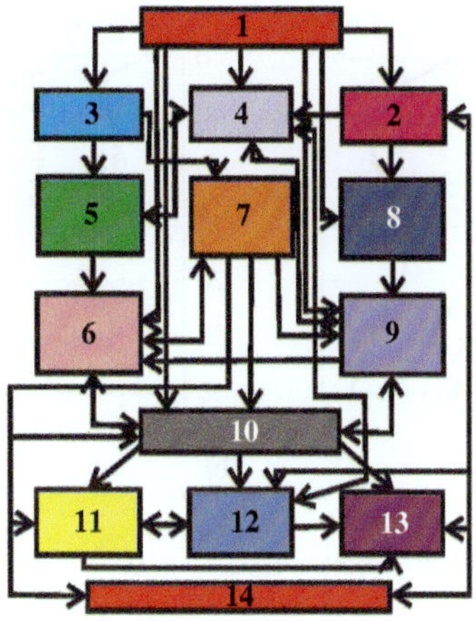

Figure 2.1 Block diagram of small-scale petrovelocity deep structural modeling (Korchin et al., 2009). (1) The study of structural and tectonic features of the region; selection of the main groups of rocks forming the geological environment; its further dismemberment into separate blocks; (2) the analysis of a priori geological and geophysical information, taking into account the data of the deep thermodynamic situation; (3) selection of a collection of rock samples of the studied region, including possible deep analogues (based on the information of blocks 1, 2); (4) the drawing up program variants of experimental study based on the prediction of the distribution with depth of *P* and *T* (based on geothermy and gravimetry materials) and their subsequent use in experimental studies of rock samples in high-pressure-temperature apparatus; (5) the study of elastic and density parameters in different thermobaric conditions based on rock samples collections; (6) statistical processing and analysis of the results of laboratory experiments; (7) complex processing of petrological information and experimental data; the search for correlational dependencies between relevant parameters; drawing up work sheets or tables; (8) the analysis of materials of field geophysical observations (primarily DSS); the construction of high-speed cuts in blocks and layers along the corresponding profile; (9) comparing the results of laboratory petro-velocity *PT* studies with seismic information; complex primary interpretation; (10) construction of a lithological model of the possible distribution with depth of surface analogues of deep rocks based on a systematic analysis of relevant information (blocks 1, 2, 6, 7 and 9); (11) reconstruction based on the lithological model (block 10) of petrovelocity sections (V_P, V_S) according to experimental data (blocks 6, 7); (12) construction of a density cross-section with subsequent cyclic correlation of anomalies of the gravity field with velocity and physical cross-sections; (13) creation and analysis of models of distribution with depth of elastic-density (Young's and shear modules, Poisson's ratio, compressibility, etc.) of mineral matter by individual blocks; comparison with primary geophysical information; and (14) the construction of complex petrovelocity sections, which are based on the data of experimental *PT* studies and information accumulated in 10–13 blocks (Burtnyi et al., 2013; Korchin, 2013; Korchin et al., 2009, 2013, 2020b, 2022).

assumption of $(\partial V/\partial H) < 0$. As far as $(\partial V/\partial P)_T\,(\partial P/\partial H)_P\,(\partial T/\partial H)$ are positive, $(\partial V/\partial T)_P < 0$, then to form a zone it is necessary to fulfil the assumption for absolute values:

$$\left|\left(\frac{\partial V}{\partial P}\right)_T \cdot \frac{\partial P}{\partial H}\right| < \left|\left(\frac{\partial V}{\partial T}\right)_P \cdot \frac{\partial T}{\partial H}\right| \tag{2.1}$$

In most cases, the change of the lithostatic pressure with the depth can be considered a constant number. Therefore, $(\partial P/\partial H)$ is $0.24 \div 0.32$ kbar/km at depths from 3 to 40 km for ancient shields. The temperature gradient at these depths varies within the wide range (from 5 to 25 °C/km) (Korchin, 2013, 2017; Korchin et al., 2013, 2020b). The relative increase of velocities with pressure under room temperature is characterized by two intervals: $P = (0 \div 2)$ kbar is the interval of maximum increase of velocity, $P > 2$ kbar where the minimum velocity gradient is observed. As a rule, the velocity change with temperature under atmospheric pressure has three intervals: $T < 80 \div 100\,°C$ (minimal changes), $T \approx 80 \div 250\,°C$ (maximal changes).

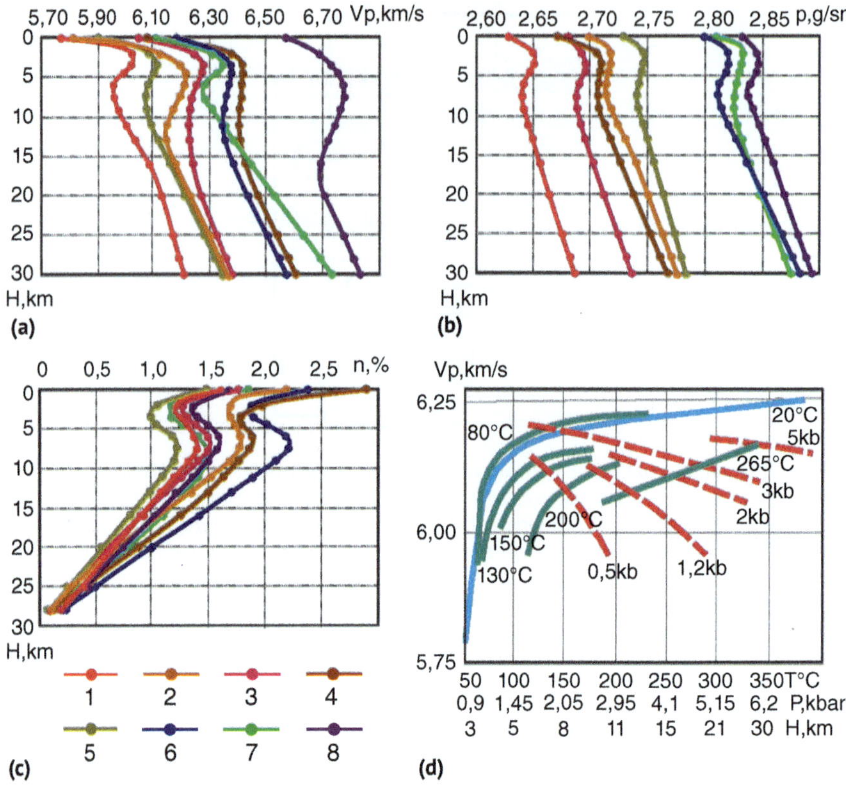

Figure 2.2 Changes of elastic longitudinal waves (a), density (b) and porosity (c) relationships with depth; the isobars ($V_P = f(T)$ at $P = const$) and isotherms ($V_P = f(P)$ at $T = const$) of velocities (d) (Korchin, 2010; Korchin et al., 2021). 1 – evenly granular granites, 2 – porphyroid granites, 3 – trachytoid granites, 4 – rapakivi granites, 5 – plagiogranites, 6 – charnockitoides, 7 – diorites, 8 – anortosites.

Further heating within the interval of $T = 250 \div 600\,°C$ results in the small decrease of the velocity. The relative changes of velocity under compensating constant pressure (isobars) and constant external temperature (isotherms) are different in absolute values. Within the interval of $20 \div 70\,°C$ under $P < 0.7$ kbar, the changes of velocity with temperature are negligible, i.e., up to a depth of $2 \div 3$ km the velocities always increase intensively. It is caused by the growth of V_P with pressure due to induration of rocks. The interval $T = 100 \div 250\,°C$ is the interval of the most intensive changes of $V_P = f(T)$. Here decrease of velocity by two times is possible because of temperature influence under atmospheric pressure and by about 10–20% under compensating pressure of $P \approx 1 \div 4$ kbar. It is in this interval of pressures and temperatures ($P \approx 1.2 \div 3.5$ kbar; $T \approx 110 \div 250\,°C$) that the greatest negative changes of velocity of the elastic waves are observed and zones of low velocities (LVZs) are revealed. The experiments found $(\partial V/\partial T)_P = -2.7 \pm 0.5$ ($P \approx 0.5$ kbar); -0.7 ± 0.3 ($P = 2$ kbar); -0.33 ± 0.1 m/s · °C ($P = 5$ kbar) and V_P from pressure under different constant temperatures $(\partial V/\partial P)_T = 0.8 \pm 0.3$ (pressure interval $0 \div 2$ kbar, temperature $20 \div 80\,°C$); 0.01 ± 0.005 (under $P \approx 2 \div 5$ kbar, $T \approx 20 \div 80\,°C$); 0.04 ± 0.01 (under $P \approx 2 \div 5$ kbar, $T \approx 265\,°C$). Based on these data and experiments under the low and high-temperature regimes and calculations, it was revealed that under low temperature ($(\partial T/\partial H) < 9 - 11\,°C/km$), the zones of the velocity inversion are not reflected on the curves of $V_P = f(PT) = f(H)$. In this case, the condition of (1) is not executed. If the temperature gradient is equal to $(\partial T/\partial H)<15-20\,°C/km$ in pressures interval of $1.8 \div 3.5$ kbar, the relationships of $V_P = f(PT)$ clearly revealed LVZs. The decrease in the velocities of different samples varies from −10 to −250 m/s in these zones.

Similar calculations were also performed according to experimental studies of rocks in various thermodynamic conditions of experiments performed by other authors (Christensen, 1989; Christensen and Mooney, 1995). However, these data can be used for modeling only at depths more than 10–15 km.

Information on changes density vs. depth was obtained from studies of the volume decrement under *PT*-experiments and the ultrasonic determinations of the compressibility of the rocks (Figure 2.2b).

The density characteristics of the rocks nonlinearly change with depth similarly to the elastic parameters. The $\rho = f(PT) = f(H)$ relationships demonstrate maximums and minimums. It implies that deep simultaneous influence of *P* and *T* on mineral matter results in the formation of zones of density inversion. This mechanism for the origin of the LVZ is corroborated by studies of rock density under programming influence of *P* and *T*. As it was expected, under *PT* relevant to the LVZs, density also decreases. The values of $\partial\rho/\partial H$ sometimes are negative reflecting a considerable density decrease of rocks that creates zones of low density in the crust. Under thermodynamics conditions at 5–15 km depth, the gradient of the density increment falls to zero or becomes negative (Figure 2.2b). The additional experiments show that the LVZs depend weakly on a mineral composition of the crust matter. Obviously, the structural transformations and chemical composition play a key role in the variations of rock density (Burtnyi et al., 2023; Korchin et al., 2018a, 2020a, 2022).

The low density is very sensitive to the temperature conditions of the Earth's crust similarly to the LVZs. Increasing deep heat flow (HF) decreases rock density and activates capability to decrease the rock density and increase their permeability and hygroscopicity (activates a process of fluid movement) that, for example, causes metamorphic transformation of rocks. In other words, these LVZs are to be the most active horizons of present-day geological-geophysical transformations of the crystalline crust mineral environment (Burtnyi et al., 2023; Korchin, 2013, 2017; Korchin et al., 2018a, 2020a, 2020b). A multidisciplinary analysis of DSS data, geothermal and petro-structural modeling showed that the crystal domains of higher temperature gradients are distinguished by more complex character of variations in seismic velocity with depth. Here intensive LVZs occur. In "colder" crystal domains, the thickness of the LVZs is negligible or zones are absent at all. In addition, we confirmed that the variations in the mineral composition of the rocks at depths of 5–20 km influence negligibly on position of LVZs and their intensity (Korchin, 2013, 2017; Korchin et al., 2022).

The comprehensive experiments on rocks and minerals under different *PT*-conditions disclosed an increase in relative deformation of grains and their twinning in the LVZs (5–15 km depth). The density of rocks increases inside the blocks of dislocations and decreases on intergrain boundaries. The defects of minerals packing increase. The intergrain boundaries expanse due to their mylonitization, the amount of main micro fissures increases. There is depressurizing or opening of the gas-liquid inclusions in minerals in heating because of excess inner pressure. Moreover, the rocks are characterized the lowered density (Korchin, 2013, 2017; Korchin et al., 2022).

Previously performed the petrophysical thermobaric modelling of the crystalline crust on the Ukrainian Shield (USh) demonstrated that characteristics of low-velocity zones depend slightly on mineral composition of rocks at relevant depths. They are mainly related to the geothermal regimes of the crystalline crust in study areas. Being large-scale systems of transcrustal deep distortions, zones of low elastic properties and densities are the most permeable for ascending hydrothermal solutions and fluids of the mantle origin during the activation of tectonomagmatic processes accompanied with intensive HF. Different processes of metasomatosis, formation and localization of mineral resources (e.g. hydrocarbons) most likely occur in these zones.

A satisfactory coincidence was obtained between the position of LVZs (decompaction zones) and mineral deposits in the Kirovograd ore region of the USh, on the NW Black Sea shelf, in certain areas of the Caspian Sea (Burtnyi et al., 2013, 2023; Korchin, 2017; Korchin et al., 2013, 2020b; Korchin and Rusakov, 2021). It follows that LVZs of the thermobaric nature serve as the diagnostic indicator of deep mineral deposits. The LVZs are also domains of intensive relaxation of stresses of tectonically active processes due to weakened elastic properties of their rocks. Here is the action of mechanisms for dilatancy destruction of the mineral environment at certain depths of the crystalline crust. Thermobaric LVZs (area of the lowered elasticity, density and increased porosity of rocks) are the trigger mechanism for the origin of earthquakes and faults of different orientation in the Earth's crust that is confirmed by the greatest number of focuses of crustal earthquakes in the LVZs (Korchin, 2013; Korchin et al., 2020b).

2.4 The LVZs in the Crystalline Crust as Zones of Increased Porosity of Mineral Matter

The physical parameters of different types of crystalline (igneous and metamorphic) rocks are mainly controlled by the composition and structural-textural peculiarities of mineral material although a jointing and porosity (n) of the rocks, the state of intergrain boundaries also influence on these characteristics. The occurrence of different types of pores and fissures is governed by the conditions of their formation and subsequent processes of transformations. Porosity in the crystalline rocks became the subject of the intensive studies due to examining migration of gas-liquid flows, particularly hydrocarbons, at different depths in the lithosphere (Burtnyi et al., 2023; Geguzin and Krivoglaz, 1971; Korchin, 2013, 2015a, 2017; Korchin et al., 2022; Tripolsky and Sharov, 2004). Based on the studies of the relationship between porosity and velocity under high PT in the granite samples from the USh, it became possible to characterize more clearly changing porosity and jointing vs. PT environments at different depths. The study also explains some crustal decompaction anomalies of rocks, which, in turn, can act as migration pathway for hydrocarbons of the mantle origin. Under atmospheric pressure, the clear nonlinear relationship is revealed between the velocity and porosity for three groups of the granites, which differ in a grain size of rock-forming minerals. The gradients of changing $V_P/f(n)$ increase with increasing the size of grains (Figure 2.3a) (Burtnyi et al., 2023; Korchin, 2017; Korchin et al., 2020b).

Velocity values in almost "poreless" granites agree well with the extrapolated linear sections of the dependence $V_P = f(P)$ to the intersection with the ordinate axis (V_E) (Figure 2.3b). The curvilinear sections of the $V_P = f(P)$ graphs characterize the increase in speed, mainly due to the closing of various types of pores and microcracks (Vpn), as well as due to the change in the elastic properties of rock-forming minerals (Vpn = 0). The average pressure values, at which there is practically no influence of the pore space on the change in velocity, for fine-, medium-, and coarse-grained granites are equal to 2.6 (2.0–2.8), 2.2 (1.8–2.6) and 1.9 (1.8–2.4) kbar (Figure 2.3c) (Korchin, 2015a).

The $\frac{n_0 - n_P}{n_0} = \frac{\Delta n}{n_0} = f(P)$ plots characterize the relative change in the pore space (in %) vs. pressure in the granites, where $n = n_0\left(1 - \sqrt{\frac{V_P - DP - V_0}{V_E - V_0}}\right)$, D is the relative change in the velocity vs. pressure plot on the linear segment due to the elastic deformation of the rock-forming minerals (Figure 2.3c). The largest gradients of decrease in the volume of the pore space with increasing pressure are observed in coarse-grained granites. The smallest – in fine-grained.

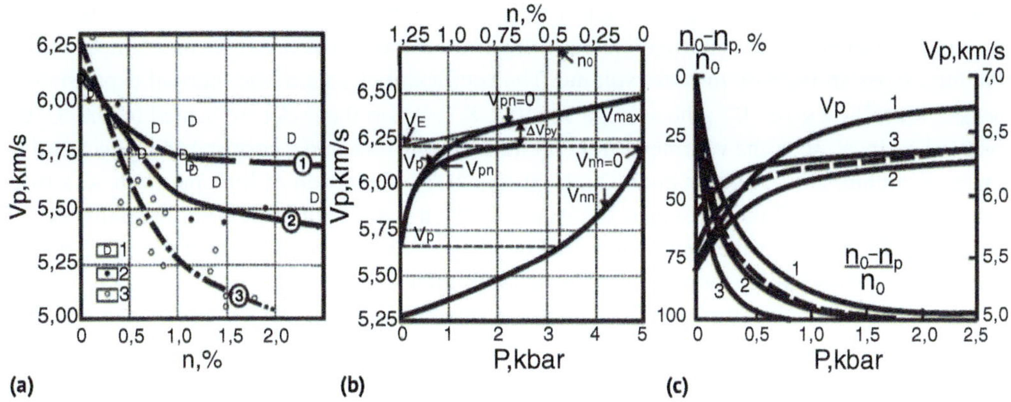

Figure 2.3 Changes in velocity and pore space with pressure in granites. (a) Change of $V_P = f(n)$ for granites (Korchin, 2015a): 1 – fine-grained, 2 – medium-grained, 3 – coarse-grained; (b) change of mean values velocity of longitudinal waves for the medium-grained granites vs. pressure ($V_p = f(P)$) and porosity ($V_{p_n} = f(n)$); (c) relative change of pore space under the increase of pressure in granites: 1 – fine-grained, 2 – medium-grained, 3 – coarse-grained.

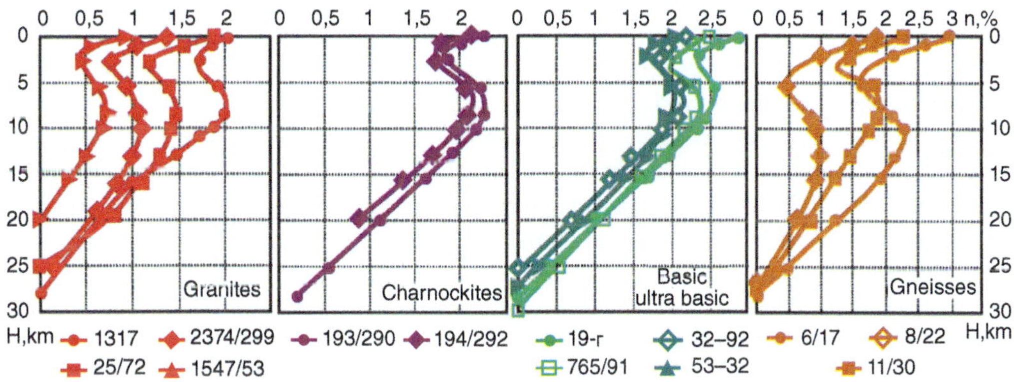

Figure 2.4 Change in porosity of the crystalline rocks under thermodynamic conditions at the different depths (Korchin, 2015b; Korchin et al., 2016).

It is assumed that with an increase in pressure, first of all, microcracks and slit-like pores close. Therefore, in coarse-grained granites, a predominance of fissure porosity is possible, while fine-grained granites are characterized by volumetric porosity.

The given ratio makes it possible to estimate changes in the size of the pore space of rocks depending on the applied pressure (depending on the depth) and temperature, based on the data of the measurement of the speed of the propagation of elastic waves in rocks at different depths together with the change in the density of rocks in different *PT* conditions (Figure 2.3c).

Graphical presentations and theoretical calculations demonstrate that the rock porosity decreases by 30–50% at a 3–5 km depth from the initial one, measured under atmospheric conditions. In the low-velocities zone (4–15 km) (the zone of decompaction of rocks, loosening of intergrain boundaries), an increase in porosity is observed (Figures 2.2c and 2.4).

In the crystalline crust, rocks are subject to compression, heating up, and various structural and sometimes material transformations resulting from the action of pressures and temperatures. Heating causes an increase in rocks volume. The coefficient of volumetric thermal expansion of rocks: $2 \times 10^{-6} - 4 \times 10^{-4} K^{-1}$; the average $3 \times 10^{-5}\ K^{-1}$. When the rocks are heated to 300 °C, we observed an increase in the volume of mineral matter by 1–1.5% and a corresponding decrease in density (Korchin, 2013; Korchin et al., 2013, 2018b, 2022). The action of high pressure, as a rule, leads to compression of the substance, the value of which is determined by compressibility or the coefficient of comprehensive contraction. At a depth of 5–10 km, the pressure of $P = 2.5$–3.0 kbar can increase the crystalline rock densities by no more than ~0.5%.

Thus, thermodynamic forces that facilitate the development of local high microstresses will predominate at depths from 3 to 10 km, causing volumetric destruction of structural integrity of the rocks, that results in the formation of deconsolidation of Earth's crust mineral matter at these depths. In this *PT* interval in the crystalline rocks, a substance decomposes due to the internal multi-oriented stresses, which in local mineral contacts reach values exceeding the strength of individual minerals. This results in a fragile micro-destruction of the environment. The number of main microcracks increases (loosening). At the same time, the migration of free liquid and gas through micro-disruption of minerals and inter-grain contacts activates in the intergranular space, destroying the rock and forming secondary pores and cavities in it. This is a key process for the movement of deep fluids through rocks in the LVZs under natural conditions. In the LVZs, an increase in porosity is observed by 10–25% compared with values at depths of overlying layers (Figure 2.4) (Korchin, 2013, 2015a; Korchin et al., 2016, 2022; Kissin, 1996; Reider, 1987).

At the same time, these microstructural disturbances in the rock provide conditions for its significant transformations with a further increase in pressure and temperature. With an increase in the software *PT* parameters acting on the sample ($H > 25$ km, $P > 5.5$ kbar, $T > 300\,°C$), an intensive growth of the elastic characteristics of the rock is observed. After the rock is loosened with depth ($H > 10$–15 km), the mechanisms characteristic of plastic deformations of the environment is included. Due to material and structural (at the level of elementary defects) reconstructions, the rock is improved, which improves its elastic characteristics, increases the density and the velocity inversion of elastic waves in them and reduces porosity and cracking (Burtnyi et al., 2023; Korchin, 2013, 2017; Korchin et al., 2013, 2022).

2.5 A Comparison of Experimental Data and Geophysical Observations

The DSS studies revealed anomalies of elastic behavior of mineral matter in the form of LVZs (Figures 2.2 and 2.4). They were also registered during the laboratory experiments on the rock samples simulating thermobaric parameters relevant to 3–25 km depths. A comparison of interdisciplinary experimental and field data allows us to put forward several ideas on the physical nature of elastic zoning mineral matter in the upper layers of the crystalline crust (Korchin, 2013, 2017; Korchin et al., 2013, 2020b).

Model and natural processes or phenomena are considered to be similar if their determining criteria are numerically equal. The theory of dimension and similarity makes it possible to use effective methods of selecting necessary relationships (Korchin, 2013; Korchin et al., 2013, 2020b, 2022).

The criteria of similarity (C) are dimensionless quantities, which characterize the given phenomenon or process. The similarity analysis was applied to select the dimensionless quantities which mostly control the velocity of the elastic wave in mineral matter. As the velocity mostly reflects the physical state of matter under different thermobaric conditions, it functionally depends on pressure (P), temperature (T), density (ρ), volume (U), specific heat capacity (C_{ou}) and the Poisson coefficient (σ):

$$V = f(P,T,\rho,U,C_{ou},\sigma) \tag{2.2}$$

As objects under study have finite dimensions and are subjected to changing PT environments, it is reasonable to consider depending V on changing PT-regimes with a depth. Substituting T and P by their gradients (*grad T* and *grad P*), we shall get the following dimensions: $[V] = \text{m} \cdot \text{s}^{-1}$; $[grad\ P] = \text{kg} \cdot \text{m}^{-2}\text{s}^{-2}$; $[grad\ T] = {}^\circ\text{C} \cdot \text{m}^{-1}$; $[\rho] = \text{kg} \cdot \text{m}^{-3}$; $[\text{U}] = \text{m}^3$; $[\text{z}] = \text{m}$; $[C_{ou}] = \text{m}^2 \cdot \text{s}^{-2}\ ({}^\circ\text{C})^{-1}$; $[\sigma]$= dimensionless.

Solving the matrix of their dimensions, we shall obtain the followings criteria of similarity:

$$C_1 = \frac{gradP \cdot z}{V^2 \cdot \rho}, \tag{2.3}$$

$$C_2 = \frac{gradT \cdot z \cdot C_{ou}}{V^2} \tag{2.4}$$

The first (C_1) characterizes the influence of pressure on the velocity of the elastic wave of the object of study within a boundary thickness of z, the second (C_2) characterizes the influence of temperature. Consequently, if a model (m) and its natural (n) affinity consist of the same matter and $\rho_m = \rho_n$ and $C_{oum} = C_{oun}$, equality of the velocity in the model and natural settings will occur in satisfying the following condition:

$$gradP \cdot z = inv, \qquad gradT \cdot z = inv, \tag{2.5}$$

where inv is invariant.

Based on these criteria of similarity, equality of mean values of P and T in the model and natural environments is enough to compare the velocity in these both settings. It implies that the V_P/V_s ratio (C_3) must be constant. The mean value of this ratio for the Earth's crust is 1.77 while its laboratory mean value is 1.8 ± 0.1 with 0.95 confidence interval. It follows that the C_3 criterion is the invariant number, e.g., it is numerically equal in nature and models. Therefore, the velocity of elastic wave in the relevant model must not differ from that in natural conditions. Accordingly, the scale coefficients can be taken to be equal to 1. The equations of the relationships between V_P, X_1, P, and T from experiments are invariant and can be applied without changes in simulating natural environments at different depths. Generally, the essence of the petrovelocity thermobaric modeling is the comparability of elastic wave velocity in the lithosphere from DSS method and experimental measurements of this parameter under high PT regimes pressures. From it follows that the LVZs at 3–25 km depth are most likely to be of thermobaric origin (Burtnyi et al., 2013; Korchin, 2013; Korchin et al., 2013, 2022).

2.6 Geological Interpretation of the PTBM Results

A comprehensive analysis of seismicity, geodynamics, volcanic activity and mineral deposits in the Ukrainian Transcarpathian depression (TD) is mainly based on new information on geological, geophysical, geodesic, structural, and geomorphological studies (Chekunov et al., 1969, 1994; Gordienko et al., 2011; Kutas, 2014, 2016; Nazarevich and Nazarevich, 2002, 2004; Tretyak et al., 2015).

The TD is the part of the Carpathian orogeny where the significant restructuring of the tectonic pattern took place at the turn of the Paleogene and Neogene periods that resulted in a significant

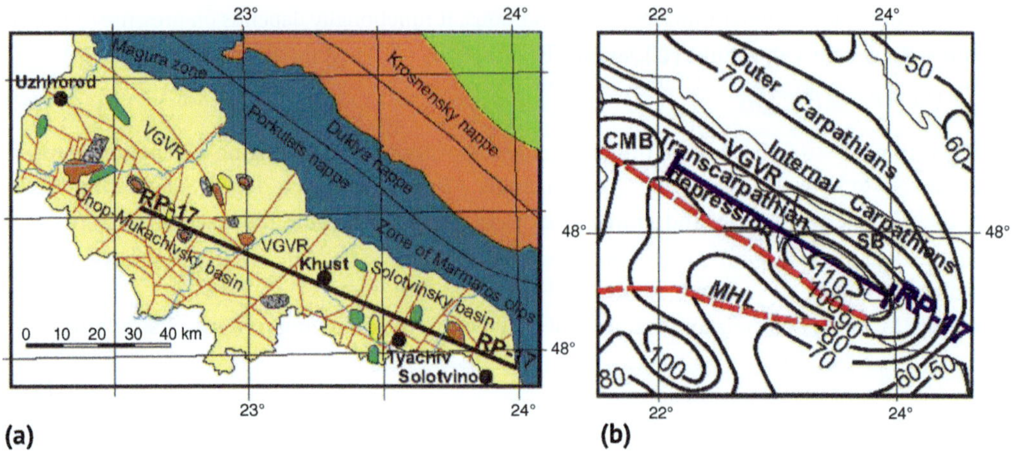

Figure 2.5 Schematic maps of the regional tectonic units (a, b), location of the RP-17 seismic profile (a, b) and isolines of heat flow (b) (Kutas, 2014, 2016; Korchin et al., 2022; Nazarevich and Nazarevich, 2002, 2004; Tretyak et al., 2015). CMB – Chop-Mukachivsky basin, VGVR – Vygorlat-Guta volcanic ridge, SB – Solotvinsky basin, MHL – Mid Hungarian line.

difference in the Earth's crust structure of the region. Active post-Alpine tectonic processes caused the formation of both the depression itself and young volcanic features. The depression extends from the NW to the SE and consists of two major units (Figure 2.5). They are the NW Chop-Mukachivsky basin on the Central massifs of the Internal Carpathians and SE Solotvinsky basin superimposed on the flysch zone of the Outer Carpathians. These units are separated from each other by the Vygorlat-Guta volcanic ridge. The Chop-Mukachivsky basin descends, while elevation, horizontal movement and high seismicity are characteristics of the Solotvinsky basin (Korchin et al., 2020b, 2022; Tretyak et al., 2015).

Drilling data show that the basement of the Transcarpathian depression consists of the Paleozoic metamorphic schists, the Triassic dolomites and dolomitized limestones, the Upper Cretaceous marls and mudstones with sandstones and the Palaeogene sandy-clay formations (Korchin et al., 2020b; Nazarevich and Nazarevich, 2002, 2004; Tretyak et al., 2015).

The Paleozoic sediments are exposed in the NW depression. The Cretaceous and Paleogene sediments crop out mostly in the south-east, in the Solotvinsky basin. Thus, from the NW to the SE, the younger sediments gradually comprise the basement of the Transcarpathian depression where numerous faults, clearly registered by both geophysical and geological methods give rise to a complicated block nature of its structure (Kutas, 2014, 2016; Korchin et al., 2020b; Nazarevich and Nazarevich, 2002, 2004).

The block tectonics is a specific characteristic of the Chop-Mukachivsky basin, while it is much less expressed in the Solotvinsky depression. Active vertical and horizontal movements and high seismicity can occur here (Chekunov et al., 1969; Gordienko et al., 2011; Korchin et al., 2020b; Kutas, 2014, 2016; Nazarevich and Nazarevich, 2002, 2004; Tretyak et al., 2015). Along the TD, the RP-17 profile (Chekunov et al., 1969) revealed the following horizons: (1) the sedimentary cover (V_P = 1.7 to 4.8 km/s), (2) the Mesozoic basement of the depression (V_P = 5.3 to 6.0 km/s), and (3) the Paleozoic basement (V_P = 6.0 to 6.4 km/s). In the Solotvinsky basin, Precambrian rocks (V_P = 6.2–6.5 km/s) and the Conrad boundary (V_P = 6.6 to 6.8 km/s) are also delineated (Figure 2.6) (Korchin et al., 2018a, 2020b).

The velocities of V_P = 7.5–8.5 km/s are registered on the Moho discontinuity. The minimal thickness of the folded basement of the depression is 3 km in the SE Chop-Mukachivsky basin.

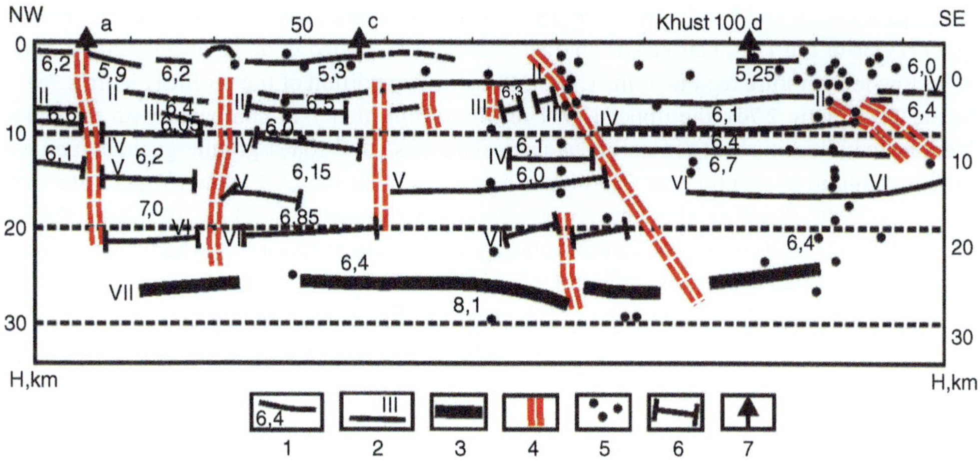

Figure 2.6 Seismic section of the Earth's crust and local earthquake *hypocentres* along the TD (Chekunov et al., 1969; Korchin et al., 2018a; Nazarevich and Nazarevich, 2002, 2004; Tretyak et al., 2015). 1 – refraction layers and velocities along them; 2 – reflection horizons and their numbers; 3 – Moho discontinuity; 4 – faults; 5 – earthquake focuses; 6 – seismic boundaries; 7 – hydrocarbon exploitation wells.

It increases to 4.0–4.5 km in the Solotvinsky basin. The layer of "granitoids" is determined by the velocity of 6.1–6.4 km/s for the Horizon II at depths from 4.5 to 8 km. The LVZs (Horizon III-LVZ) with the velocity of 6.0–6.1 km/s are situated below.

The seismic horizon IV (6.2–6.5 km/s) at depths of 6–14 km was earlier believed to be the Conrad boundary (V, K_2). However, it is now interpreted to be LVZ of the thermobaric origin located at depths of 6–17.5 km (Korchin, 2017). The Horizon VI is the second LVZ with a velocity of about 6.4 km/s in the crystalline crust of the TD. This layer is located above the Moho discontinuity (Figures 2.5 and 2.6). Its velocity values are small for basic rocks. The explanation of the nature of the upper and lower LVZs is possible in analyzing HF in the TD. There are its stationary component and non-stationary anomalies due to geodynamic processes and the distribution of radiogenic heat sources and thermal parameters of the environment (Gordienko et al., 2011; Korchin et al., 2020b; Kutas, 2014, 2016). In the depressions of the TD, HF varies from 90 to 120 mVt/m^2. The intensive anomalies (110 mVt/m^2) occur along the Central TD. Temperatures are very heterogeneous laterally and vertically: 5 km–85–150 °C; 10 km–180–300, 20 km–300–470, 30 km–450–650, at the M–600–800 °C (Figures 2.5 and 2.6).

In order to model the petrophysical structural characteristics of Transcarpathian Depression deep horizons, we proposed petrophysical modeling techniques using experimental data on changes in the physical characteristics of rocks in various *PT* conditions of experiments corresponding to those for specific geological regions (Burtnyi et al., 2013; Korchin et al., 2019, 2013, 2022).

2.7 The Nature of LVZ along the DSS Profile (RP-17) Using the PTBM Methodology

The PTBM allows us to explain the origin of the LVZ along the profile (Chekunov et al., 1969; Korchin et al., 2018b, 2019, 2020b, 2021, 2022; Korchin and Karnaukhova, 2021). For this purpose, the samples are selected of slates, quartzites, metaconglomerates, mylonites, granite-gneisses, diorites, gabbroides, and basalts from the USh and Transcarpathia because they seem to form the deep horizons of the TD. Their elastic characteristics were studied in detail at relevant *PT* conditions at low-temperature regime of the Chop-Mukachivsky portion of the profile (Figure 2.7a).

In the TD (Figure 2.7d), the upper LVZ-I resulted from high *T* in compared with its average values due to radiogenic heat generation of clays, siltstones, sandstones, granite-gneisses, and with some contribution of mantle heat (Gordienko et al., 2011). In the LVZ-I, the temperature varies from 160 to 260 °C, therefore, this zone is of the thermobaric nature (Korchin, 2013, 2017; Korchin et al., 2020b, 2022) and consists of decompaction granite-gneiss with schists, which are vulnerable to destructive external influences.

In constructing models of other parts of the profile with high-temperature regime (abnormally high for the territory of Ukraine), temperature corrections were made for the values of velocities

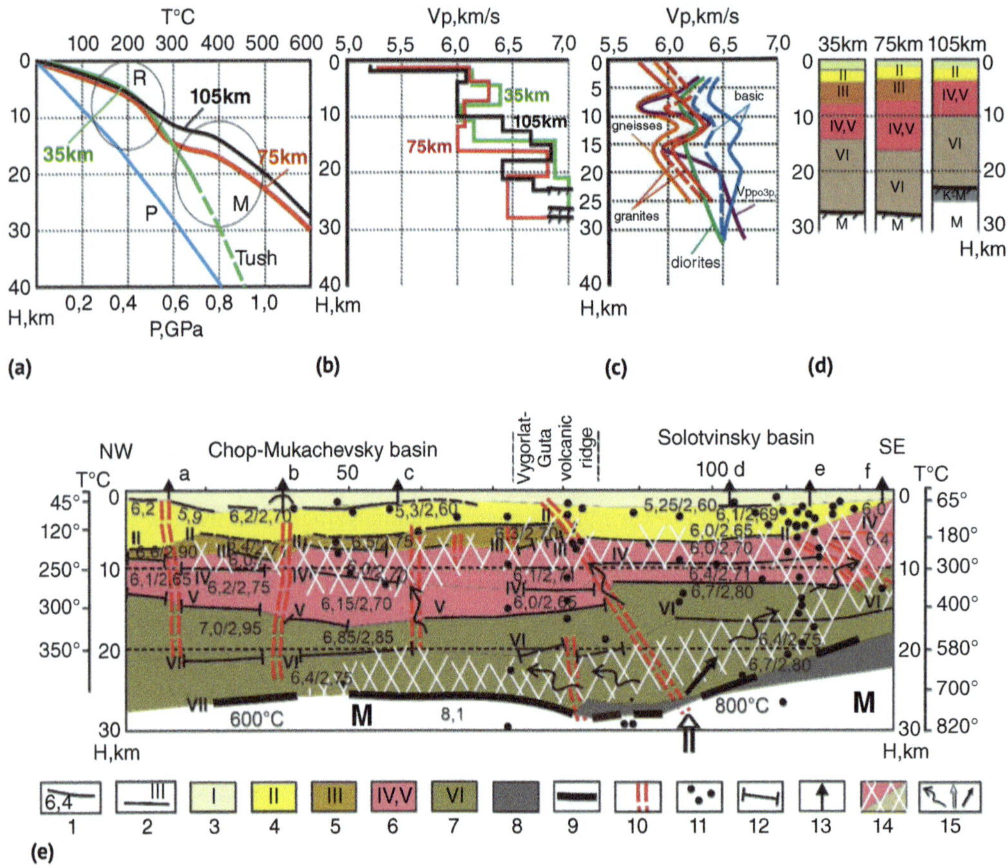

Figure 2.7 Elements of PTBM for areas of 35, 75, 105 km of the RP-17 DSS profile (a–d) and the model of rocks distribution with a depth along TD (e). This study and data from (Chekunov et al., 1969; Korchin et al., 2020b, 2022; Nazarevich and Nazarevich, 2004; Tretyak et al., 2015). (a) – temperature regimes; (b) – seismic velocity; (c) – calculated $V_P = f(PT) = f(H)$ for thermal regime of 105 km; (d) – distribution models for mineral materials vs. depth according to these velocities. 1 – refraction layers and velocities along them; 2 – reflection horizons and their numbers; 3 – Neogene layer; 4 – folded basement; 5 – Paleozoic amphibolite facies (gneisses); 6 – Proterozoic (Precambrian) gneisses with LVZs; 7 – basic rocks; 8 – crust-mantle mixture; 9 – Moho discontinuity; 10 – faults; 11 – earthquake focuses; 12 – seismic boundaries; 13 – hydrocarbon exploitation wells; 14 – LVZs; 15 – fluid flows.

of the relevant rocks. For depths of 10–35 km, relationships of $V_p = f(PT) = f(H)$ were obtained applying isobars and isotherms of velocities obtained by the authors of this study, as well as data on rocks forming the deep horizons of the Earth's crust (Christensen, 1989; Christensen and Mooney, 1995). As an example of such a modeling, for the areas around 35, 75, 105 km of the RP-17 DSS profile, Figure 2.7a–d demonstrates the data on temperature regimes, seismic velocities, calculated $V_p = f(PT) = f(H)$. A model is also presented for the distribution of mineral environments with depth corresponding to these velocities. Having analyzed these materials, we made an assumption about the nature of the LVZs zones in the TD.

Below the LVZ-I, the temperatures are low because of the decrease in radiogenic heat. Besides, the mantle component of HF is controlled by the physical state of the underlying rocks that causes the formation of a layer with high velocity because of the dominance of *P* over *T*. This horizon consists of basic rocks. At the depth of 18–26 km, temperature regimes produced a new LVZ-II. The input of heat from the mantle warms up the rocks up to 600 °C at 18–26 km depth (Figure 2.7a, e). The zone is composed of diorites, gabbroides, basaltoides, and other products of volcanic activity. The ruptured structure of Moho allows us to suppose the occurrence of crust-mantle mixture here (Korchin, 2013, 2017; Korchin et al., 2018b, 2019, 2020b).

There are two voluminous domains in the LVZs along the profile. The first occurs in the Vygorlat-Guta volcanic ridge area, which is apparently related to an intense inflow of mantle heat along the fault. The second one is located at a 90–120 km depth. The Chop-Mukachivsky basin is localized within the Transcarpathian mantle fault zone where volcanism was widespread. It now descends due to the powerful cooling of volcanogenic formations of low elastic parameters in the lower crust (Gordienko et al., 2011; Korchin et al., 2020b; Tretyak et al., 2015). Additional decompaction of basified formations above the Moho discontinuity and the velocity decrease arc associated with shear stresses directed along the TD (Korchin et al., 2020b; Tretyak et al., 2015), which causes the dilatant decompaction of rocks (Nikolaevsky, 1996).

Higher pressure ceases the growth of the zone due to the compensating effect of structural distortions under pressure. Thus, the zone modifies the configuration: its thickness can be increased due to the increase of intensity of deep HF or the zone can disappear in case of lowering it. Such an instability of the thermodynamics of the LVZs causes their episodic occurrence in the Earth's crust as well as their vertical and horizontal migration depending on of the temperature variations in the deep horizons of Earth (Korchin, 2013, 2017; Korchin et al., 2013, 2020b, 2022; Tripolsky and Sharov, 2004).

The seismic K_2 boundary seems to be thermobaric in origin which results from the intensive increase in V_P above the LVZ. However, it is not excluded that the intensive increase in the velocity above the LVZ can also be associated with occurring of the high velocity rocks instead if low velocity ones. In this case the deep thermobaric conditions facilitate a higher velocity jump.

An intense warming up of the rocks zone is a characteristic of in Solotvinsky basin, which formed a large LVZs above the M discontinuity (in turn, here the latter ascends to 18–21 km). The constant heat inflow from the mantle heats the rocks at depths of 13–18 km and a new LVZ-II zone is formed. Some contribution to the increase in *T* in the zone is made by the rocks of low thermal conductivity of the LVZ-I above the zone LVZ-II that reduces the outflow of the mantle heat from the LVZ-II (Korchin et al., 2022).

This is the region of transition from acidic rocks to rocks of intermediate composition, and gabbroids and diorites seem to underlain it. There is the development of LVZs in the Earth's crust in active zones of quite intense stresses. In the process of destruction of the mineral matter at certain depths, the development of LVZs can be associated with the action of the dilatancy effect (Korchin et al., 2022; Nikolaevsky, 1996; Milanovsky and Nicolaevsky, 2010). Theoretically, it means the linear proportionality of inelastic increases in volume and shear, or in other words, the volumetric and shear inelastic deformations are functions of pressure and tangent stress. In crystalline rocks, the effect of the development of microcracks is manifested when the stresses satisfy the criterion for the onset of dilatant pre-destruction or, more simply, the elastic limit. It is assumed that the characteristic changes of types of brittle fractures observed in laboratory experiments on rocks occur within the crust. On this basis, the relevant models are developed for defects of the deep horizons of the crust. A certain disadvantage of our model is the requirement of the presence in the Earth's crust of shear stresses that reach the strength limits of the rocks. However, these phenomena are possible for the territory of Transcarpathia under study. Additional decompaction of basified formations above the M discontinuity and reduction of the velocity are associated with dilatant decompaction of the rocks due to shear stresses directed along the TD and the presence of the mantle Transcarpathian fault (Chekunov et al., 1994; Tretyak et al., 2015). Moreover, earthquakes are concentrated at a shallow depth of the TD (Figure 2.7e), which, in our opinion, is closely associated with the thermobaric zones of the upper crust decompaction. Here, the seismic activity contributes to the widening of the channels for motion of fluid from the asthenosphere lens below the Carpathian region, which is described in Chekunov et al. (1994); Gordienko and Gordienko (2019); Korchin et al. (2020b).

2.8 Elastic Characteristics of the Mineral Substance along the DSS Profile (RP-17)

The elastic characteristics of a mineral substance in *PT* conditions are a source of important information about the action of internal stresses in the crust and qualitatively determine the intensity of the development of decompaction and fragile and plastic deformations in it (Burtnyi et al., 2023; Directory of Physical Constants of Rocks, 1969; Korchin et al., 2022). At depths of 5–15 km of the Earth's crust with increased HFs, which are not provided with the necessary pressure compensation, destructive temperature stresses occur in the sample. This leads to the decompaction of rocks, a decrease in V_P and, as a result, a low seismic velocity zone (registered by DSS) and anomalous behavior of rock modules appears at these depths. For isotropic media, modules can be calculated from $V_{P,S}$ and their density. In stationary isotropic conditions, solids can be roughly divided into plastic and fragile. When P and T influence them, their transition from one state to another is possible. Poisson coefficient is an indicator of these processes. Conventionally, for most fragile materials, $0.10 < \sigma < 0.25$, and for ductile materials, $0.25 < \sigma < 0.47$. If it is deformed and heated, inelastic phenomena develop in the substance, then the values can go beyond these limits up to negative values. Inelasticity, as a rule, appear itself in the form of fragile fracture or flow. From these positions, we qualitatively characterized the fragile and plastic state of the rocks of the upper horizons of the Earth's crust on the basis of thermobaric petrophysical modeling, which actually exists in the depths of the TD.

The depression extends from the NW to the SE and consists of two major units (Figure 2.8a) (Korchin et al., 2022).

They are the NW Chop-Mukachivsky basin on the Central massifs of the Internal Carpathians and SE Solotvinsky basin superimposed on the flysch zone of the Outer Carpathians. The Vygorlat-Guta volcanic ridge separates these units from each other. The Chop-Mukachivsky basin descends while elevation, horizontal movement, and high seismicity are characteristics of the Solotvinsky basin.

According to DSS data (RP-17), PTBM was carried out along the TD. The simulation results are presented for the individual structures of the above-mentioned TD (pickets by profile P-35 CMB, P-75 VGVR, P-105 SB). Figure 2.8b–d shows the average calculated values according to the V_P,V_S

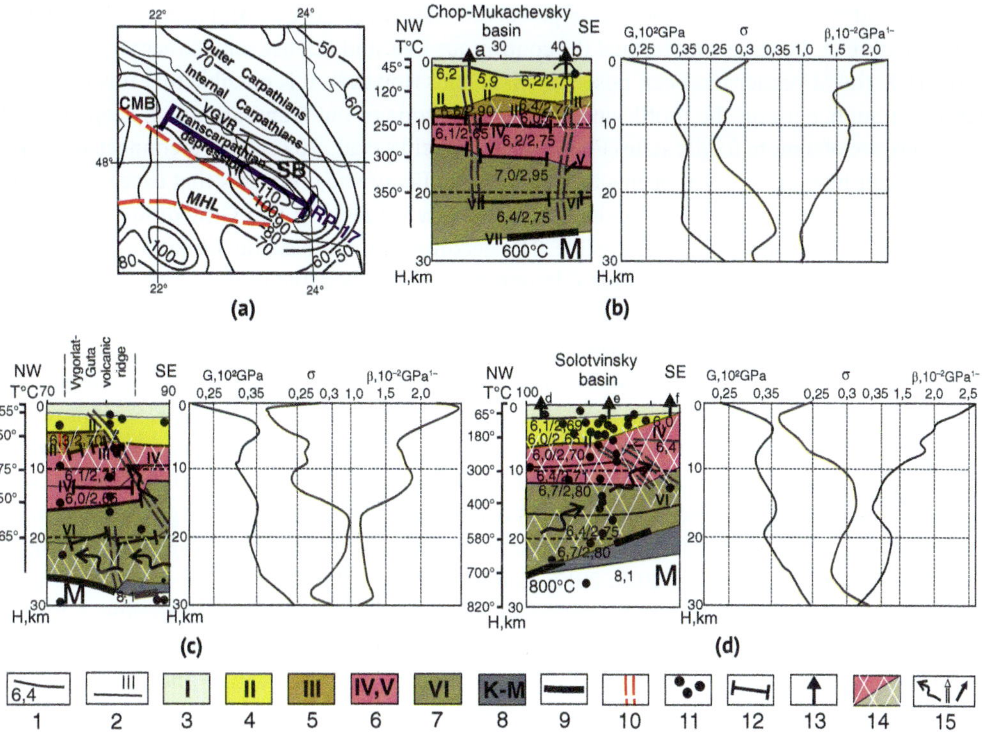

Figure 2.8 (a) The layout of the seismic profile (RP-17) and heat flows in the Transcarpathian depression; (b–d) RP-17 sections (35, 75, and 105) – horizons and structures according to the data of DSS and PTBM and corresponding modules (notational designations – see Figure 2.7) (Korchin et al., 2022).

experimental data, density and values of the corresponding modules of rocks (G – shear modulus, σ – Poisson coefficient, β – compressibility).

In the LVZ, the upper horizons are composed of unconsolidated rocks and are homogeneous in terms of elastic parameters. With depth, they become fragile, with high compressibility (Figure 2.8) (Korchin et al., 2022). Below them, at depths from 4.5 to 8 km, there are layers of "granite-gneiss." Here, crustal of low seismic velocities zones were discovered the temperature varies from 160 to 260 °C. Therefore, the zone has a thermobaric nature (Korchin, 2013, 2017; Korchin et al., 2013, 2022) and it is composed of thermally decompacted granito-gneisses and slates, which are exposed to destructive external influences. The calculated modules (Figure 2.8b–d) indicate this. At depths of 6–12 km, E, σ are decrease, G, β are increase slightly, and rocks acquire quasi-fragile properties. The seismic horizon at depths from 6 to 17.5 km is composed of rocks of medium composition, and as the depth changes, σ and β acquire increased values in all three sections of the TD. This horizon is composed of rocks in a quasi-plastic state.

The second layer is the LVZ in the crust of the TD with a velocity of about 6.4 km/s, located above the M boundary at a depth below 18–26 km, and caused by temperature conditions (Figure 2.8c, d) (Korchin et al., 2022). The influx of heat from the mantle heats rocks H = 18–26 km to 600 °C (Figures 2.6 and 2.7). The zone is composed of diorites and basaltoides. LVZ-2 along the profile has two volumetric areas – the first in the VGVZ area, associated with an intense influx of mantle heat along the fault (Christensen, 1989).

The second area is located in the area of pickets 90–120 km (Figure 2.8d). The SB is characterized by the heating of the rocks that formed the continuous LVZ and intense seismic activity. According to the DSS, the M border has been raised to 15–21 km. However, it is possible that it is not clearly defined enough, perhaps the profile entered the large zone of the Transcarpathian deep fault. The rocks here are of a more fragile state (Figure 2.8d) (Korchin et al., 2022). This is connected with the development here of a system of macrocracks at significant shear stresses that exceed the elastic limit of rocks at these depths, as evidenced by the low values of σ and G modulus. The appearance of auxetics at these depths is possible. The decompaction of basaltic formations above the M boundary is provoked by the dilatational loosening of the rocks, due to the action of shear stress directed along and across the depression and mantle Transcarpathian deep fault (Figures 2.6 and 2.7) (Korchin et al., 2018a, 2020b). Therefore, a mass of earthquake epicenters is concentrated at a shallow depth of the TD (Figure 2.7), which, in our opinion, is closely related to the thermobaric zones of decompaction of the upper crust, their low elastic characteristics, and σ values. Seismic activity contributes to the effect of expanding the channels of fluid movement, under the action of which the disintegration of the mineral substance of the TD is activated. Under thermodynamic conditions, which correspond to the LVZs at depths from 3–5 km to 12–20 km, changes are observed in the rocks, when their cataclastic transformation takes place with an increase in the cracking and porosity of the rocks (Korchin, 2013, 2017; Korchin et al., 2018b, 2019, 2020b, 2022). The migration of free water and gas through the microcracks of the rock is activated, destroying it, forming secondary cracks and caverns. At the same time, a so-called degassing pipe is formed, which is a source for the accumulation of abiogenic hydrocarbons in the crust, which under certain conditions (high pressure and decompression) penetrate into the near-surface horizons of the sedimentary cover, where explosive deposits are formed.

2.9 Conclusions

The LVZs are the most active horizons of the present-day transformation of the mineral environment and changes in the structural features of the Earth's crust. The LVZs in the crystalline crust are its integral part stabilizing the lithospheric state. In our opinion, active heating of the Earth's interior, as a result of mantle cataclysms and the localization of radioactive elements, inevitably facilitates the formation of the LVZs, which are the horizons of thermobaric decompression of the rocks. Such zones under the influence of fluid migration, stress, multi-directional deformations, and vibrations become strongly dislocated media, provoking intense relaxation of tectonic stresses in the form, for example, of earthquakes. In turn, seismic activity causes widening migration channels for the mantle fluids producing the so-called degassing tubes. Such tubes are the essential elements of the formation of hydrocarbon accumulations. This changes outdated ideas on the geological and structural features of the regions under study. Therefore, when interpreting the materials of geophysical studies, it is necessary to take into account the presence of crustal decompaction zones (LVZs).

Based on geological-geophysical data, geodynamic researches and PTBM, the TD is considered to be a region of the present-day active geological processes where mineral deposits are formed and earthquakes occur. Among known deep anomalies of the geophysical fields, LVZs are the most accessible to study using different geological-geophysical methods, including superdeep drilling. Taking into consideration the present analysis of the nature and origin of LVZs in the crust of Transcarpathia, it is possible to fulfill the most detailed and perspective study aiming at searching for mineral resources and clarifying the deep structure of the Earth, as well as explaining and predicting crustal earthquakes. Detailed further study of the deep structure of the TD, and especially the position of the LVZs, can significantly expand the possibilities for hydrocarbons searching. Moreover, legends in Figure 2.8 show hydrocarbon fields, most likely of abiogenic origin.

The LVZs are also domains of intensive relaxation of stresses of tectonically-active processes due to weakened elastic properties of their rocks. Here, act mechanisms for dilatancy destruction of the mineral environment at certain depths of the crystalline crust. Thermobaric LVZs (area of the lowered elasticity, density and increased porosity of rocks) are the trigger mechanism for the origin of earthquake and faults of different orientation in the Earth's crust that is confirmed by the greatest number of focuses of crustal earthquake in the LVZs.

The conducted research makes possible to clarify the geological and structural features of the crust of the region, to adequately interpret distribution of geophysical fields and decipher the peculiarities of local geodynamics and seismic-tectonic process, to clarify the level and nature of geo-ecological hazards, and to more effectively predict and investigate deep spatial distribution of mineral resources.

Some of the materials presented in Chapter 2 were previously published by us in other works, links to which are presented in the list of references below.

References

Burtnyi, P. A., Korchin, V. A., & Karnaukhova, O. Y. (2013). *Modeling Material Composition of Deep Horizons of the Crust: A New Conception of the Interpretation of Geophysical Data*. Saarbrücken, Deutschland: LAP Lambert Academic Publishing, p. 188. ISBN: 978-3-659-38626-8 (in Russian).

Burtnyi, P. O., Karnaukhova, O. Y., Kobolev, V. P., et al. (2023). *Directory (Cadastre) of Physical Parameters of Rocks of Ukrainian Shield at High Pressures and Temperatures*. Kiev: Naukova Dumka Publishing House, p. 281. https://doi.org/10.15407/978-966-00-1878-5 (in Ukrainian).

Chekunov, A. V., Livanova, L. P., & Geyko, V. S. (1969). Deep structure and certain peculiarities of the tectonics of the Transcarpathian depression. Soviet Geology 10, pp. 57–68 (in Russian).

Chekunov, A. V., Sologub, N. V., Starostenko, V. I., et al. (1994). The deep structure and geodynamics of the Carpathians. *Lithosphere of Central and Eastern Europe: Young Platforms and Alpine Fold Complex*. Edited by Chekunov, A.V. Kiev: Naukova Dumka Publishin House, pp. 121–174 (in Russian).

Christensen, N. (1989). Reflectivity and seismic properties of the deep continental crust. Journal of Geophysical Research 94 (17), pp. 793–804. https://doi.org/10.1029/JB094iB12p17793.

Christensen, N. & Mooney, W. (1995). Seismic velocity structure and composition of the continental crust: A global view. Journal of Geophysical Research 100 (B7), pp. 9761–9788. https://doi.org/10.1029/95JB00259.

Directory of Physical Constants of Rocks. (1969). Editor S. Clark Jr. Moscow: Mir, p. 544.

Geguzin, Y. E. & Krivoglaz, M. A. (1971). *Migration of Macroscopic Inclusions in Solid Bodies*. Moscow: Metallurgia, p. 344 (in Russian).

Gordienko, V. V., Gordienko, I. V., Zavgorodnyaya, O. V., et al. (2011). *The Ukrainian Carpathians (Geophysics, Deep Processes)*. Kiev: Logos Publishing House, p. 128 (in Russian).

Gordienko, V. V. & Gordienko, L. Y. (2019). Asthenospheric lenses in the mantle of oil and gas regions. Geology of Mineral Resources and the World Ocean. 2, pp. 35–51. https://doi.org/10.15407/gpimo2019.02.035 (in Russian).

Kissin, I. G. (1996). Fluid saturation of the Earth's crust, electrical conductivity, seismicity. Earth Physics 4, pp. 30–40 (in Russian).

Korchin, V. A. (2010). Thermobaric elastic heterogeneity of the earth's crust and the dynamics of its change. Vestnik ONZ RAS, 2, p. NZ6019. https://doi.org/10.2205/2010NZ000037 (in Russian).
Korchin, V. A. (2013). *Thermodynamics of Crustal Low Velocity Zones (A New Scientific Hypothesis)*. Saarbrücken, Deutschland: LAP Lambert Academic Publishing, p. 280 (in Russian).
Korchin, V. A. (2015a). Crust thermobaric anomalies of rocks lowered density and enhanceable porosity. 14th EAGE International Conference on Geoinformatics – Theoretical and Applied Aspects, May 2015, Volume 2015, pp. 1–5. https://doi.org/10.3997/2214-4609.201412383 (in Russian).
Korchin, V. A. (2015b). Low velocity zones of thermobaric origin in the crystalline crust. Geophysical Journal 37 (5), pp. 46–65. https://doi.org/10.24028/gzh.0203-3100.v37i5.2015.111145 (in Russian).
Korchin, V. A. (2017). Anomalies of low density in the crystalline crust of thermobaric origin: A new insight into migration and localization of hydrocarbons. *Oil and Gas Exploration: Methods and Application*. Edited by Gasi, S. and Hachay, O. Monograph Number 72. Wiley, pp. 237–257. https://doi.org/10.1002/9781119227519.ch15.
Korchin, V. A. (2021). Brittle and plastic deformations of crustal rocks of the Transcarpathian depression. *Geophysics and Geodynamics: Forecasting and Monitoring of the Geological Medium*. Edited by Maksimchuk, V., Lviv: Rastr-7, pp. 96–99 (in Ukrainian).
Korchin, V. A., Burtnyi, P. A., & Karnaukhova, E. E. (2009). Prognosis of geological medium of the Earth crust on materials of DSS and petrophysical thermobaric research of the mineral matter. Geodynamics 1 (8), pp. 67–75. https://doi.org/10.23939/jgd2009.01.067 (in Russian).
Korchin, V., Burtnyi, P., & Karnaukhova, E. (2012). Deep modeling of the mineral composition of the lithosphere according to PT-petrophysics data and DSS materials. 7th EUROGEO: European on Regional Geoscientific Cartography and Information Systems, June 12–15, 2012, Bologna, Italy, pp. 231–232.
Korchin, V., Burtnyi, P., & Karnaukhova, O. (2016). Thermobaric decompaction of crustal rocks – areas of increased porosity and localization of minerals. «Geophysical Technologies of Forecasting and Monitoring of the Geological Environment»: Proceedings of the VI International Scientific Conference, Lviv, 20–23 September, 2016. Lviv: SPOLOM, pp. 108–110 (in Ukrainian).
Korchin, V. A., Burtnyy, P. A., & Karnaukhova, E. E. (2018a). Decompaction of metamorphic rocks under thermodynamic conditions of the Earth's crust (experimental data). Geophysical Journal 40 (4), pp. 107–130. https://doi.org/10.24028/gzh.0203-3100.v40i4.2018.140612 (in Russian).
Korchin, V. A., Burtnyi, P. A., & Karnaukhova, E. E. (2018b). Thermobaric decompression zones of the crustal rocks are natural satellites of seismo-tectonically active regions. Modern Problems of Mechanics 33 (3), pp. 399–409. ISSN 1694-6065 (in Russian).
Korchin, V. A., Burtnyi, P. A., & Karnaukhova, E. E. (2019). The prognosis Transcarpathia petrophysical thermobaric model along the RP-17 profile. *Geophysics and Geodynamics: Forecasting and Monitoring of the Geological Medium*. Edited by Maksimchuk, V., Lviv: Rastr-7, pp. 68–70 (in Ukrainian).
Korchin, V. A., Burtnyi, P. A., & Karnaukhova, E. Y. (2020a). The density of magmatic, ultrametamorphic rocks of the Ukrainian shield in model deep conditions of the Earth's crust (experimental data). Geophysical Journal 42 (5), pp. 148–171. https://doi.org/10.24028/gzh.0203-3100.v42i5.2020.215078 (in Russian).
Korchin, V. A., Burtnyi, P. A., & Kobolev, V. P. (2013). *Thermobaric Petrophysical Modelling in Geophysics*. Kiev: Naukova Dumka Publishin House, p. 312. ISBN: 978-966-00-1360-5 (in Russian).
Korchin, V. A. & Karnaukhova, O. Y. (2021). Peculiarities of fluid dynamics in the seismically active zone of the Earth's crust of the Transcarpathian depression (Ukraine). Science and Education: Problems, Prospects and Innovations. Proceedings of the 8th International Scientific and Practical

Conference. CPN Publishing Group. Kyoto, Japan. pp. 45–53. ISBN: 978-4-9783419-5-2. URL: https://sci-conf.com.ua/viii-mezhdunarodnaya-nauchno-prakticheskaya-konferentsiyascience-and-education-problems-prospects-and-innovations-28-30-aprelya-2021-godakioto-yaponiya-arhiv/.

Korchin, V. & Rusakov, O. M. (2021). The low-velocity zone in the crystalline crust of the NW Black Sea shelf as a potential regional methane trap of the thermobaric type. *Methods and Applications in Petroleum and Mineral Exploration and Engineering Geology*. Edited by Gaci, S. et al. Springer Series in Earth Sciences/Geology, pp. 169–182. https://doi.org/10.1016/B978-0-323-85617-1.00025-4.

Korchin, V. A., Rusakov, O. M., Butrny, P. A., et al. (2020b). The origin of the low density zones in the crystalline crust of the Transcarpathian Depression (Ukraine) from petrophysical thermobaric modelling. Geodynamics 1 (28), pp. 81–93. https://doi.org/10.23939/jgd2020.01.081.

Korchin, V. A., Rusakov, O. M., & Karnaukhova, O. Y. (2021). Low velocity zone in the upper crystalline crust and their relevance to accumulation of abiogenic hydrocarbonces. Worldscience: Problems, Prospects and Innovations. *Abstracts of the 7th International Scientific and Practical Conference*. Toronto, Canada: Perfect Publishing, pp. 91–99. https://sci-conf.com.ua/vii-mezhdunarodnaya-nauchno-prakticheskaya-konferentsiya-world-science-problems-prospects-and-innovations-24-26-marta-2021-goda-toronto-kanada-arhiv/.

Korchin, V., Rusakov, O., & Karnaukhova, O. (2022). The construction of the model of the regional trap of abiogenic hydrocarbons in the crystalline crust of the Transcarpatian depression (Ukraine). Norwegian Journal of development of the International Science 92, pp. 3–14. https://doi.org/10.5281/zenodo.7078322/.

Krayushkin, V. A. & Shevchenko, N. B. (2018). To the problem of non-biogenic nature of petroleum. Geology and Minerals of the World Ocean 2, pp. 66–85. ISSN 1999-7566. https://doi.org/10.15407/gpimo2018.02.065 (in Russian).

Kutas, R. I. (2014). Heat flow and geothermal models for the Earth's crust of the Ukrainian Carpathians. Geophysical Journal 36 (6), pp. 3 27. https://doi.org/10.24028/gzh.0203-3100.v36i6.2014.111016 (in Russian).

Kutas, R. I. (2016). Geothermal conditions and the Mesozoic-Cenozoic evolution of the Carpathian-Pannonian region. Geophysical Journal 38 (5), pp. 75–107. https://doi.org/10.24028/gzh.0203-3100.v38i5.2016.107823 (in Russian).

Milanovsky, S. Y. & Nicolaevsky, V. N. (2010). Processes of transfer (migration) in the system of the Earth's crust. Modern Methods of Seismic Survey in Searching for Oil and Gas under Conditions of Complex Structures (Seismo 2012), 16–22 September 2010, Kurortnoe, AR Crimea, Ukraine. The International Scientific and Practical Conference, pp. 37–44 (in Russian).

Nazarevich, A. V. & Nazarevich, L. E. (2002). Deep trap-collector tectonic structures in the lithosphere of the Carpathian region of Ukraine: nature, origin and perspective resources. Scientific Bulletin Ivano-Frankivsk National Technical University of Oil and Gas 3 (4), pp. 10–21 (in Ukrainian).

Nazarevich, L. E. & Nazarevich, A. V. (2004). Technique of refining the parameters of the hypocenters of the Carpathian earthquakes. Geodynamics 1 (4), pp. 53–62 (in Ukrainian).

Nikolaevsky, V. N. (1966). *Geomechanics and Fluid Dynamics*. Moscow: Nedra, p. 447 (in Russian).

Reider, E. (1987). *Fluid Inclusions in Minerals*. Moscow: Mir, p. 632 (in Russian).

Tretyak, K. P., Maksimchuk, V. Y., & Kutas, R. I. (Eds.). (2015). *Modern Geodynamics and Geophysical Fields of Carpathians and Contiguous Territories*. Lviv: Polytechnic Publishing House, p. 420. ISBN 978-617-607-763-3 (in Russian).

Tripolsky, A. A. & Sharov, N. V. (2004). *The Lithosphere of the Precambrian Shields of the Northern Hemisphere Derived from Seismic Data*. Petrozavodsk: Karelian Scientific Centre RAS, p. 159 (in Russian).

3

Precambrian Mid-Continent Rift Potential for Hosting Numerous Helium and Hydrogen Accumulations, Central USA

Steven A. Tedesco

Running Foxes Energy Inc., Centennial, Colorado

The Middle Proterozoic Precambrian (1.1 Ga) aged curvilinear Mid-Continent Rift (MCR) in the Central part of the USA is a world-class producer of base metals and iron at its northern end. The MCR underlies several oil and gas producing areas such as the Nemaha Ridge in Nebraska, Oklahoma and Kansas, the Michigan Basin, Findley and Cincinnati arches in Ohio, Indiana, Kentucky, and Tennessee, USA. The MCR's relationship with shallower Paleozoic oil and gas production has not been defined. In many areas, reactivation of Precambrian fault systems related to the MCR does provide structural traps and pathways for migrating, helium (He) and hydrogen (H), low-temperature hydrothermal fluids that dolomitize reservoirs providing a reservoir in the Paleozoic strata. Light elements such as He and H have been reported in natural gas and nitrogen gas flows from several wells along the Kansas zone of the MCR and He in one localized area near Babbitt, Minnesota, in the MCR. Helium is critical to western civilization for growing silicon and germanium chips for computers, MRIs and other medical equipment, welding, and as an inert gas for a variety of applications. In 2025, a worldwide shortage for helium continues. Helium is typically extracted from natural gas in certain areas in the world but predominantly in the USA, Canada, Qatar, Poland, Algeria, South Africa, and Russia. Many areas of the world are devoid of He resources even though significant volumes of natural gas are present such as in China, North Sea, Gulf of Mexico, Permian Basin, Appalachia Basin, etc. Recent discoveries in the USA and Canada are found associated with carbon dioxide (CO_2) and nitrogen (N_2) as the host gas rather than methane (CH_4). The southwestern end of the MCR in central Kansas has had numerous He shows of over 1% in post-Mississippian rocks. Hydrogen is critical for civilization in certain industrial applications as well and there is a growing demand for it to be a primary energy source for society. The MCR in the 1980s and 1990s in Central Iowa and in northern Wisconsin, Upper Peninsula of Michigan, was subject to oil and gas exploration by major oil companies, exploration was defined by 5000+ miles of seismic which identified several sedimentary sag basins along with the rift. There have been several deep tests along the entire MCR that have penetrated Precambrian clastic and defined sediments that are interbedded, overlain, or underlain by granitic and mafic rocks. Recently, there has been increase drilling specifically for He and H in Nebraska, Iowa, and Kansas with no significant results to date. Michigan saw a deep MCR test, the Sparks-Eckelbarger-Whightsil No. 1-8 in Gratiot County, which did not encounter any petroleum but encountered several thousands of feet/meters of Precambrian sediments. Deep tests

Geophysical Exploration for Hydrocarbon Reservoirs, Geothermal Energy, and Carbon Storage: New Technologies and AI-based Approaches, First Edition. Edited by Said Gaci.

along the southeastern part of the rift have been limited to areas associated with the Rome Trough that found 0.5–2.0% He hosted in N_2 gas associated with Cambrian age sandstones. The potential of finding significant He and H accumulations in the MCR are unknown and in the early stages of exploration. The scale of the MCR is such the question has to be asked where does one start? This chapter is generally summarizing an overall view of the MCR geology, geophysics, and H and He in the MCR at present.

3.1 Introduction

The MCR System for discussion in this chapter includes the typical area as discussed in numerous publications plus the Fort Wayne Rift (FWR) system located in southwestern Michigan, northeastern Indiana and west-central Ohio; and the East Continent Gravity High (ECGH) located in west-central Ohio, central Kentucky and Tennessee, and northwestern Alabama (Figure 3.1). The reason to include these segments is that several wells that penetrated the Precambrian strata in the FWR and ECGH encountered mafic and sedimentary rocks similar to those found in Lake Superior (Stein et al., 2018). MCR is typical of many rift systems in that it is a series of linear depressions formed by extension and filled with igneous and sedimentary rock. The MCR was formed approximately 1.1 Ga years ago during the formation of a plate boundary that eventually failed. What is unusual is that the rift system is relatively narrow and deep and contains significant volumes of flood basalts. Rifts typically are shallow and wide, associated with sedimentary and low-density igneous rocks that eventually form into plate boundaries such as the East Africa Rift System (EARS). The mechanism of the formation of the MCR is not well understood and various tectonic processes occurring 1.1 Ga years ago maybe different from today due unknown factors affecting the Earth in the past. Flood basalts, which have a high density, are typically associated with broad regions of volcanism formed by sublithospheric processes and not with crustal extension and faulting. In addition, the majority of rifts are associated with negative Bouguer anomalies and positive ones as with the MCR. The majority of rifts are filled with low-density sediments, but in the case of the MCR, the flood basalts raise the density significantly forming a positive Bouguer anomaly. The MCR has been divided into ten zones and they are the Oklahoma, Kansas, Iowa, Superior, Mackinaw, Maumee, Fort Wayne Rift, ECGH (North and South), and Alabama (Stein et al., 2018) (Figure 3.1).

The MCR also has a major iron (Fe), copper (Cu), and nickel (Ni) mining district in the Superior and Mackinaw zones. Petroleum is not found in rift sediments but typically in the Paleozoic strata above. The MCR faulting in some areas was reactivated through time and help to create structural traps in the overlying Paleozoic rocks that ultimately became petroleum productive.

Helium (He) and hydrogen (H) are the focus of this chapter and at present, there is no production of either along the rift (Tedesco, 2022). There are numerous shows of He and H in and around the Kansas zone of the MCR. There recently has been a He show in test well in the Superior zone. There is anomalous He in shallow Mississippian rocks, which overlie the Mackinaw zone part of the MCR. There are He shows in several wells along the southern ECGH in Kentucky where it is in contact with the Rome Trough (Figure 3.1).

The MCR presents a rather large He and H target but its exploration is in its early stages. The main problem is where does one start. For He exploration, the Kansas, Mackinaw, and Kentucky ECGH zones are the most likely areas of interest. Only the Kansas zone for H at present. As the need for He and H for society increases, the exploration of the MCR for these elements will intensify.

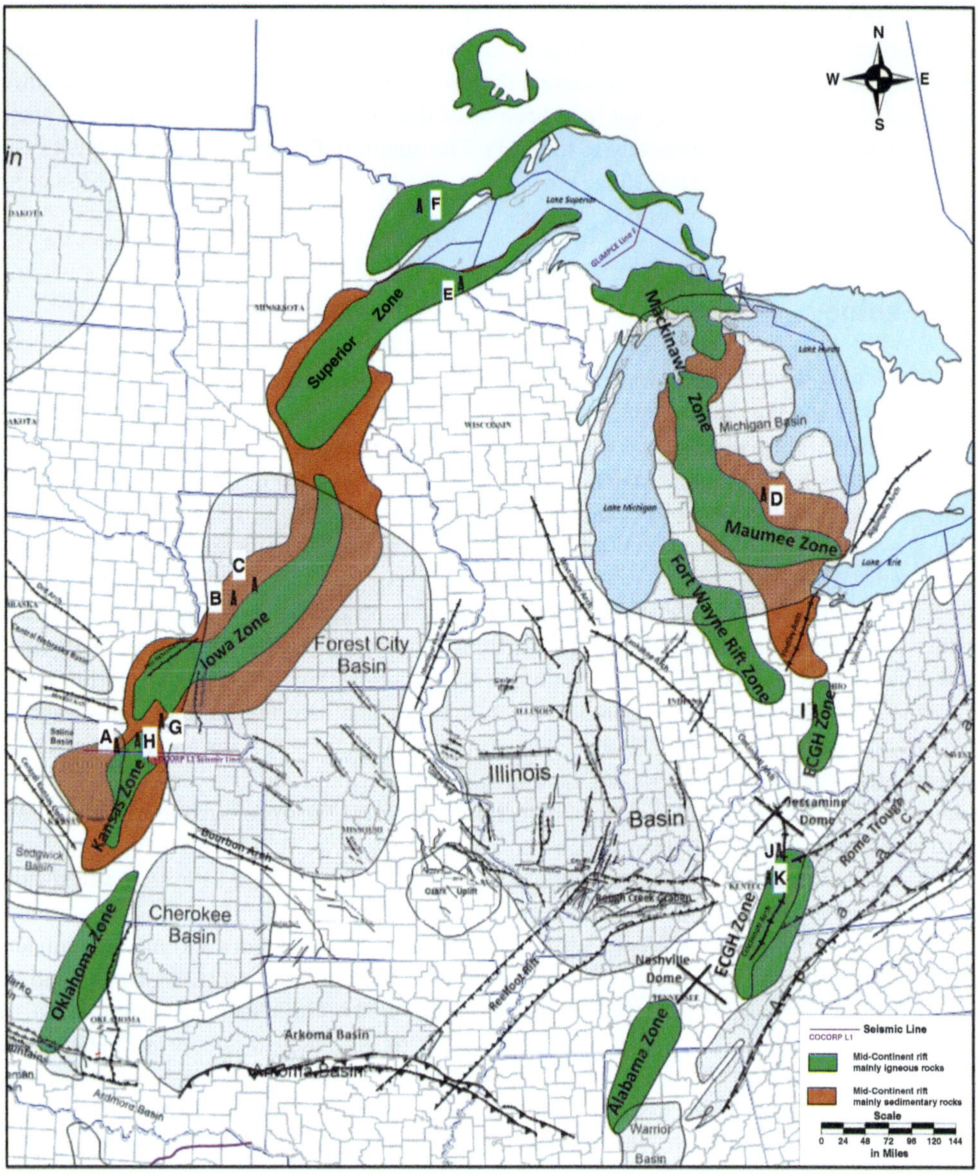

Figure 3.1 Location of the Mid-Continent Rift System (MCR). Wells mentioned in this text are labeled as follows: A – Poersch #1; B – Eischeid $1; C – Van Diest 14-3; D – Sparks-Eckelbarger-Whightsil No. 1-8; E – Terra-Patrick 7-22; F – Jetstream #1; G – Finn –4; H – Friedrich 1-8; and I – Friend No. 1 Mattinson. ECGH – Eastern Continental Gravity High. Basin and structural features are indicated.

3.2 The Formation of the Mid-Continent Rift System

The MCR lies underneath significant Phanerozoic cover through most of its extent which prevents detail mapping except through geophysics and leads to uncertainties about its true stratigraphy, architecture, geometry, and history. The rift is a failed triple junction structural feature with the intensity of rifting varying away from its center caused when the North American craton split apart in Keweenawan times (Azmi and Goldberg, 2024). The rift may have been the result of extensional forces that occurred with the Grenville Orogeny to the east. When the forces later became compressive during the Glenville Orogeny, this caused the rift to fail. There are indications that the rifting and deposition of sediments may not have been all at the same time across the MCR. The length of time of the rifting is approximately 15 to 20 MY. The amount of igneous rock in the MCR is significantly greater than the EARS (Hinze et al., 1997). The MCR sits above plate boundaries and as such provides a unusual record of this process.

3.3 Geology

The geology of the MCR is fairly well defined in the Superior Zone because of all the intense mining but elsewhere along the features extent each zone has limited geologic information. Discussed below is general information on the stratigraphy of Precambrian Keweenawan Supergroup rocks of approximately 1.1 Ga old and starting from the southwestern side of the MCR and extending northward and then turning southward concluding in the ECGH in Kentucky. Many of the ages of the Keweenawan Supergroup are based on dating of the igneous rocks as fossils are almost non-existent.

In Kansas and Nebraska, the rift sedimentary rocks associated with the MCR are defined as the Rice Fm which is a thick sequence of feldspathic sandstone, arkose, and red and sandy micaceous shale located in a half graben rift basin that is overlain by the Cambrian Reagan (Mt. Simon Sandstone elsewhere) Sandstone (Berendsen, 1994, 1997). The Rice Fm represents alluvial fans, braided fluvial systems, and possibly aeolian environments (Figure 3.2). Azmi and Goldberg (2024) did detail analysis of the Poersch #1 Keweenawan Supergroup rocks and determined that the lower sedimentary section is dominated by aeolian sediments whereas the upper section is dominated by fluvial and lake environments (Figure 3.2).

The Keweenawan Precambrian stratigraphy (Figure 3.3) in the Eischeid #1 in Iowa was divided into Lower and Upper Red Clastic Sequences (Witzke, 1990). Unit A in Figure is the granite as defined by Schmus et al. (1990). The Lower Red Clastic Sequence (Figure 3.3) has been divided into Units B through D and is equivalent to the Oronto Group in Minnesota, Wisconsin and the Upper Michigan Peninsular (Daniels, 1997). The Upper Red Clastic Sequence (Figure 3.3) has been divided into Units E, F, and G and is equivalent to the Bayfield Group in Michigan and Wisconsin (Daniels, 1997). Unit H, as defined by Witzke (1990), is potentially a transition zone from Keweenawan rocks to the overlying Mt. Simon sandstone of Cambrian age.

In the Superior Zone of the MCR in Minnesota, the Upper Peninsula of Michigan and Wisconsin, the stratigraphy is well-defined due to extensive mining over the past 100+ years. Figure 3.4 is a general stratigraphic column for Keweenawan Supergroup in the Upper Peninsula of Michigan and Wisconsin (Baumann, 2023). The Oronto Group is a member of the Keweenawan, composed of thick arkoses and shale underlying the Bayfield Group, and is subdivided in ascending order, the Copper Harbor Conglomerate, Nonesuch Shale, and Freda Sandstone in Upper Michigan Peninsular, Minnesota, and Wisconsin, USA (Hinze et al., 1997; Witthuhn-Rolf, 1997; Baumann, 2023). These rocks in Iowa are known as the Lower Red Clastic Sequence. Similar rocks are found in Kansas and Nebraska but have not received a formal name. The group is devoid of fossils but based on its stratigraphic position is thought to be around 1.1 GA. The basal unit is the alluvial Copper Harbor Conglomerate Formation (Hinze et al., 1997; Baumann, 2023). The overlying Nonesuch shale is a lacustrine deposit that is mature, organic rich shale in some areas, 500 to 700 ft (150 to 210 m) thick, and yields oil in the White Pine copper mine in the Upper Peninsula of

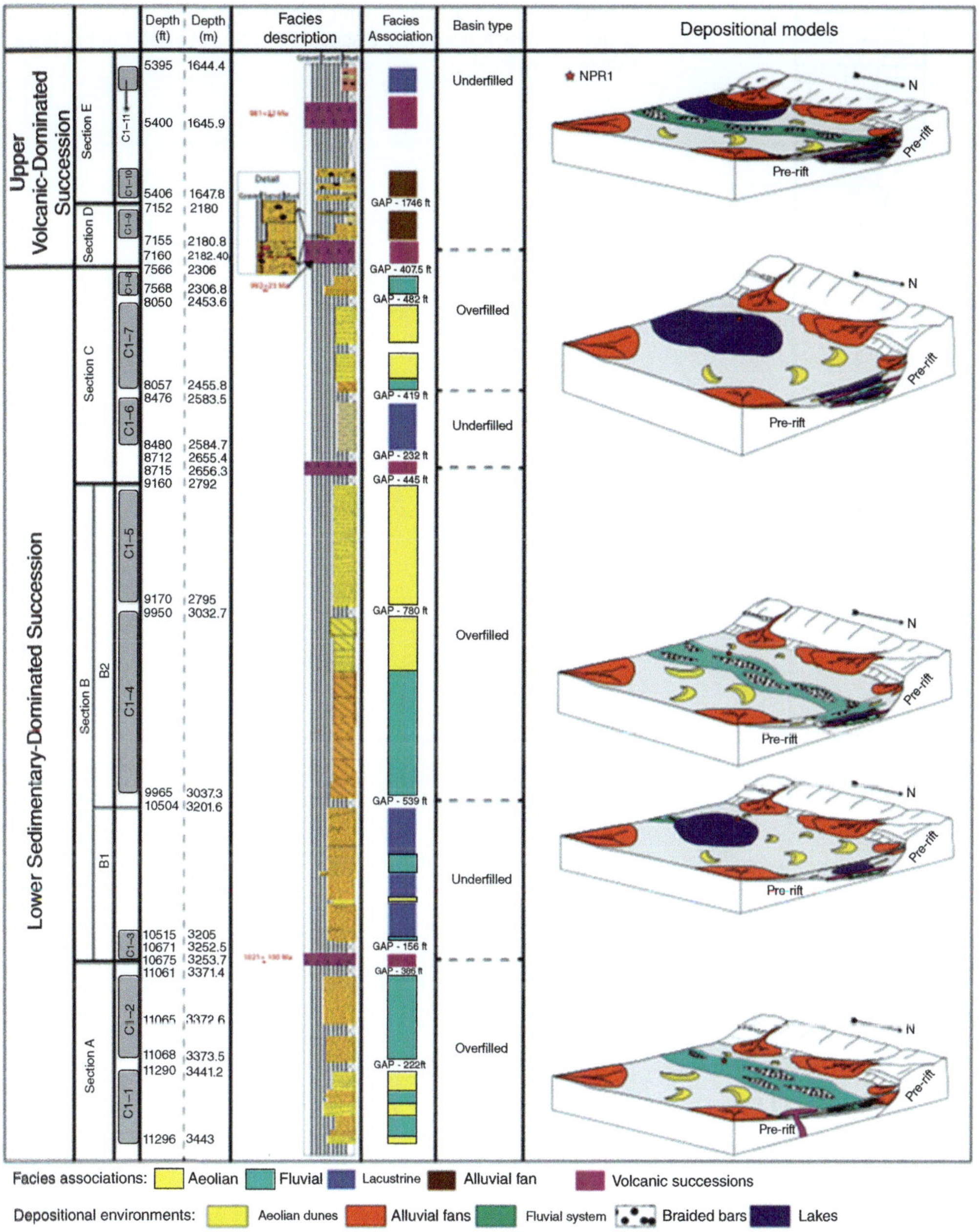

Figure 3.2 Depositional model for different sections of NP#1, based on the facies associations identified from the description of the available core intervals (CI) (Azmi and Goldberg, 2024).

Michigan (Hinze et al., 1997; Baumann, 2023). The Nonesuch Shale was deposited in lakes across the Superior Zone of the MCR (IGS, 2006). The Nonesuch Shale is a major source of copper and based on present data from several wells in other states seems to not likely be a potential petroleum source rock. The Nonesuch Shale is conformably overlain by the fluvial Freda Sandstone which is consists of fine grained to conglomeratic sandstones and arkoses (Hinze et al., 1997; Baumann, 2023). The Lake Shore Trap is a localized rock unit composed of mafic lavas (White, 1972).

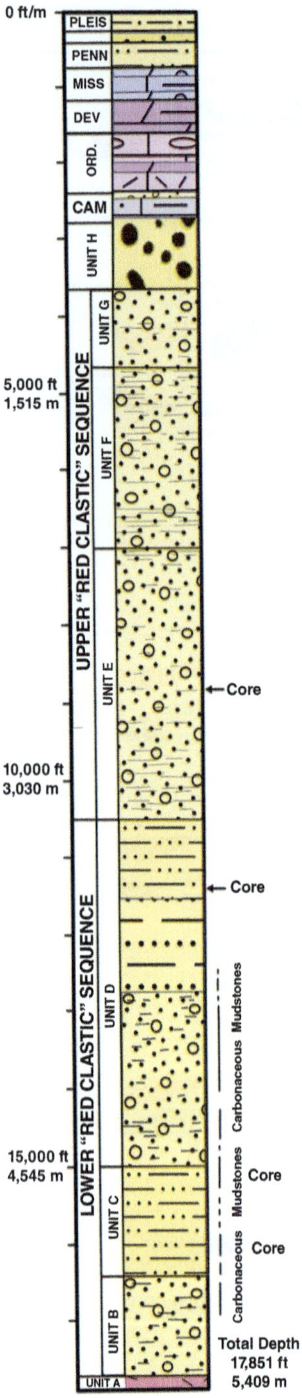

Figure 3.3 General stratigraphic sequence in the Precambrian strata from the Amoco M. G. Eischeid #1 well in Carroll County, Iowa (Hinze et al., 1997).

The Jacobsville or Bayfield Group has five members and they are: L'Anse, Orienta, Devil's Island, Jackson Creek, and Chequamegon formations (Figure 3.4). The Jacobsville (Bayfield) Group is primarily sandstones that increase upward in textural and mineralogical maturity (Hinze et al., 1997; Baumann, 2023). The environment of deposition are alluvial plains that are dominated by meandering river systems, with more localized eolian and lacustrine. The Jacobsville (Bayfield) Group sediments are believed to have been deposited in broad sag basins that are centered over the rift but extended well beyond the margins of the rift trough (Hinze et al., 1997).

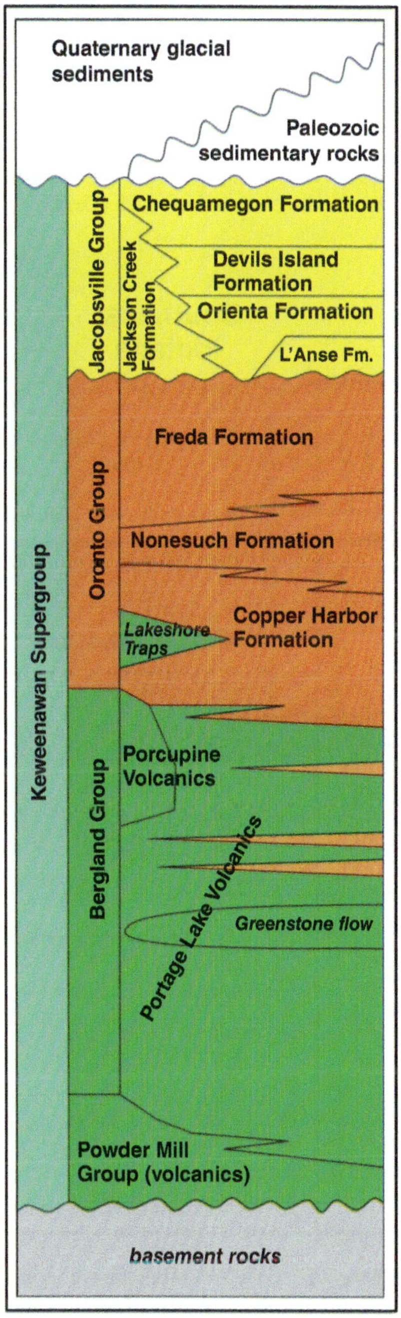

Figure 3.4 Stratigraphic column of the Superior Zone of the MCR (Baumann, 2023).

The Mackinaw and Maumee zones in Michigan have only one well, the Sparks-Eckelbarger-Whightsil No. 1-8 located in Gratiot County, Michigan (Figure 3.1) (discussed below). The Precambrian rocks in the well are thought to be equivalent to the Bayfield and Oronto groups of Wisconsin, Upper Peninsula of Michigan and Wisconsin, and the Upper and Lower Red Clastic Sequences defined in the Eischeid well in Iowa.

The Keweenawan Middle Run sandstone of Ohio and Kentucky (Figure 3.5) developed in the ECGH areas and is dominated thick to massively bedded, red (caused by hematite coatings and rhyolite grains) to gray, fine-to-medium-grained sandstones (lithic arenites) with minor red siltstones and shales (Drahovzal et al., 1992). The sandstones are non-porous because of compaction and hematite, quartz, feldspar, and clay cementation. Interpretation of some of the cores from the ECGH zone is that a distal alluvial fans or braided plain deposit which is consistent with a rift basin setting (Drahovzal et al., 1992). Extrusive felsic flows (rhyolites and trachytes) occur in parts of the basin and rock fragments found in the Middle Run Formation were derived from intrabasinal and extrabasinal extrusive rocks (Drahovzal et al., 1992).

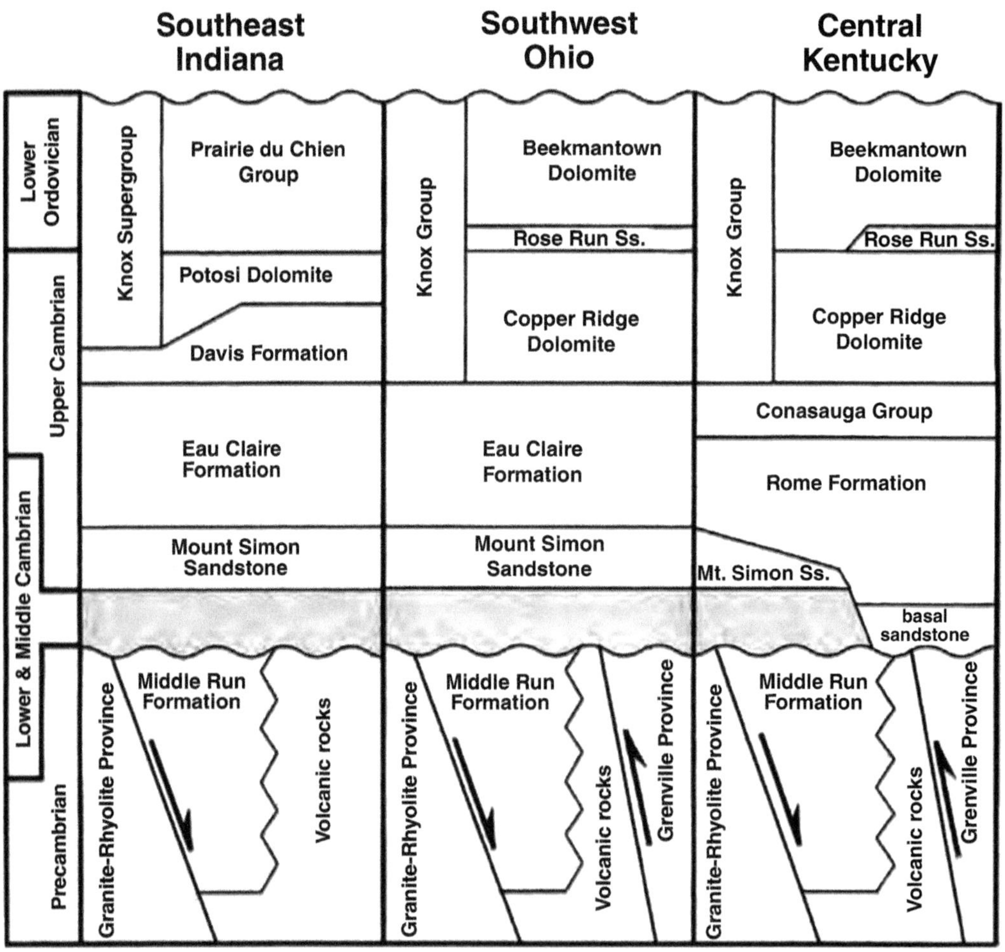

Figure 3.5 General stratigraphy for Pre-Knox geology for the ECGH in Ohio, Kentucky, and Tennessee (Drahovzal et al., 1992). No vertical or horizontal scale implied.

3.4 Wells of Interest

There are a number of tests drilled by the petroleum industry in the 1980s and 1990s that were looking for potential hydrocarbons in Precambrian sediments. Recent tests searching for He and H have only recently been drilled and many remain on a confidentiality basis at the time of this article. The early tests raised the flowing questions:

1. No hydrocarbon shows encountered while drilling,
2. Concerns about if the source rock was viable to expel hydrocarbons,
3. Timing of oil petroleum expulsion confiding with formation of trap,
4. Reservoir quality (permeability and porosity) and lithologic characteristics (clean fine to coarse sandstones versus alluvial or fluvial conglomerates and poorly sorted sandstone),
5. If any effective seals are present.

This collapse of petroleum oil prices in the 1980s and 1990s, abandoned exploration along the MCR. The recent phase of exploration starting in 2022 is targeting both the rift basins and mafic/volcanic rocks for He and H.

There are numerous wells along the MCR that penetrate a limited amount of its section. In the case of the Superior Zone (Figure 3.1), there are thousands of shallow mining test bores. Many of these did not penetrate the entire clastic or volcanic section. Presented here is a few of the significant tests drilled over the years.

The Poersch #1 well (Figure 3.1 – labeled A) located in section 31, Township 5 South, Range 5 east, Washington County, Kansas, USA that drilled to 11,300 ft (3444 m), did not encounter any petroleum shows. The upper section encountered the initial Precambrian unit of a 300 ft (91 m) thick gabbro at 2901 ft (879 m) followed by basalts to 7429 ft (2251 m) with minor pegmatites, red oxidized siltstone, and arkose occur in this interval (Berendsen et al., 1988). Below this basalt interval to total depth, the strata are mainly of fine-grained to conglomeratic red arkose and subarkose, together with minor amounts of red oxidized siltstone and shale and is called the Rice formation (Berendsen et al., 1988). The source for the arkoses is terrestrial in origin and derived locally and probably from weathering of adjacent Precambrian igneous rocks. No organic matter was observed in the cuttings from the well (Berendsen et al., 1988). Within this volcanic and mafic units, sulfide mineralization is present. The basalts are likely the only viable seal unit for trapping any hydrocarbons, helium or hydrogen.

The Amoco Eischeid well #1 (Figures 3.1 – labeled B and 3.2) is located in Carroll County, Iowa, USA was drilled by AMOCO Production for petroleum which drilled the sedimentary section of the rift and encountered a gabbro dike or sill at 5374 to 5376 m (17,733 to 17,742 ft) with an age of 1250 to 1270 MA (Anderson, 1990a, 1990b, 1990c; Van Schmus et al., 1990). The well was had no petroleum shows (Figure 3.2).

The Van Diest 14-3A was drilled specifically for H in 2023 and is located section 3, Township 90 North, Range 27 West, Webster County, Iowa, USA (Figure 3.1 – labeled C). The well was drilled to 6100 ft (1848 m), is shut-in, and data is presently confidential.

The Sparks-Eckelbarger-Whightsil No. 1-8 is located in section 8 Township 10 North Range 2 West, Gratiot County, Michigan, USA, and was drilled in 1976 to 17,466 ft (5 293 m) and penetrated Precambrian rift clastic sediments from 12,176 ft (3690 m) to total depth (Figure 3.1 – labeled D). The Precambrian lithologies were highly variable between red to maroon shale, quartzites, poorly to well sorted sandstones, siltstones with basalt, fine-to-medium coarsely crystalline diorite and clay-like serpentine near the bottom of the well (McClure Oil, 1976). The well purpose was to find hydrocarbons, but there was no indication of any type of inert gas or petroleum in the Precambrian clastic rocks. Within the area there are several fields that produce natural gas with He from the Mississippian stray sandstone (yellow on the map). This sandstone is an under-pressured reservoir that exists in sheets and lenses, traps are a combination of structure and stratigraphic configuration, compartmentalized due to fracture and diagenetic controls (Nowaczewski, 1994), and contains anomalous He (Figure 3.6). The anomalous He is 0.3% or greater and based on present prices that would be economic. Many of the natural gas productive Mississippian stray sandstones contain anomalous He and lie directly above the Mid-continent rift system in the area.

The Terra-Patrick 7-22 located in section 22, Township 47 North, Range 6 West, Bayfield County, Wisconsin, USA (Figure 3.1 – labeled E). The well reached total depth 4966 ft (1504 m) in the Copper Harbor Formation of the Oronto Group. The Terra-Patrick 7-22 well penetrated in descending order 2877 ft (872 m) of the Freida Fm, 873 ft (264 m) of the Copper Harbor Formation, 403 ft (122 m) of the Nonesuch Shale, 361 ft (109 m) of the Freida-Nonesuch Transition Zone, and 162 ft (49 m) of the Nonesuch Shale – Copper Harbor Transition Zone. Daniels (1977) suggested that the Oronto Group in the well was equivalent to the Lower Red Clastic Sequence in the Eischeid #1 well in Iowa, USA.

The Jetstream well was drilled near Babbitt, Minnesota, USA, and offsets a shallow mineral test well that encountered He rich gas (Figure 3.1 – labeled F). The well is discussed later in the text.

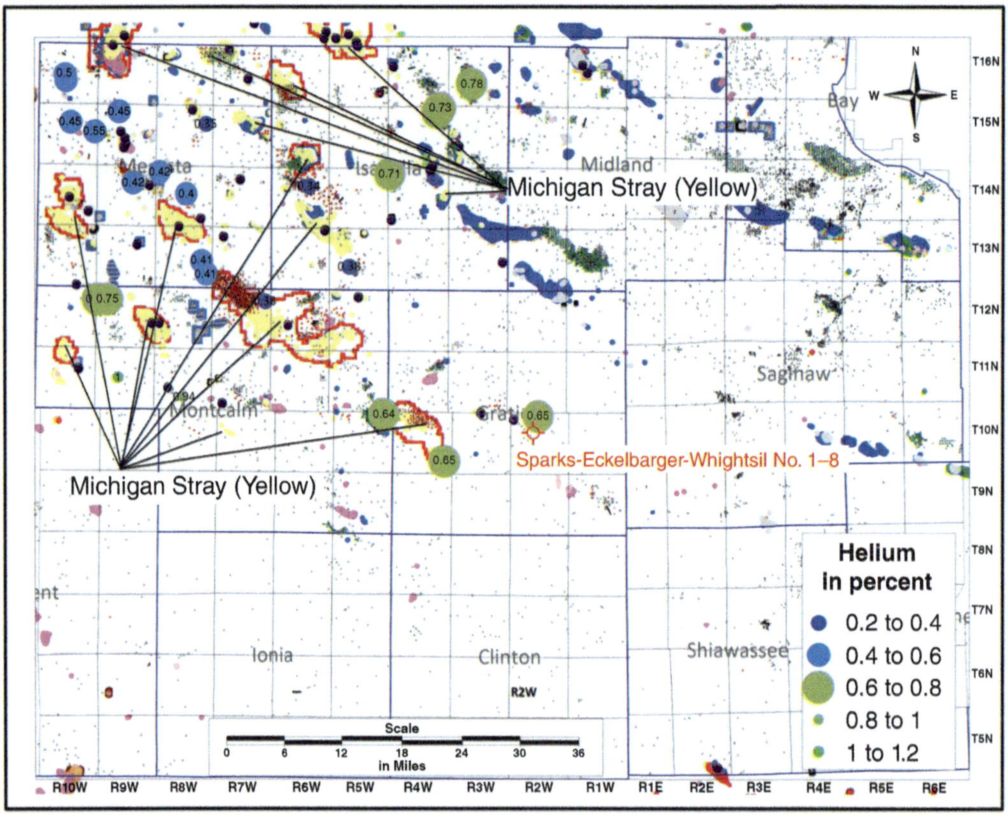

Figure 3.6 Location of the Sparks-Eckelbarger-Whightsil No. 1-8 in Gratiot County, Michigan. USA. The yellow areas are Michigan Stray sandstones (Mississippian) that have anomalous He and would be considered economic based on their percentage. The other colors are Paleozoic oil and gas production that is stratigraphically deeper.

The Finn #1-4 well located in section 4, Township 4 South, Range 7 East, Marshall County, Kansas, USA encountered in the upper part of the well several hundred feet of medium-to-dark-gray occasionally calcareous siltstone and shale with minor gray limestones (Figure 3.1 – labeled G). Below this interval to total depth in the well is composed of red oxidized arkoses and siltstones, and minor mafic rocks at the bottom of the well (Cullers and Berendsen, 1993).

The Friederich #1-8 well located in section 35, Township 7 South, Range 4 east, Riley Co., Kansas, USA which encountered a thick gabbro sill and overlie basaltic rocks that are partially oxidized, overlaying varicolored rocks that are predominantly volcaniclastics (Cullers and Berendsen, 1993) (Figure 3.1 – labeled H).

The Friend No 1 Mattinson (Figure 3.1 – labeled I) located in Clark County, Ohio, USA which drilled 1281 ft (388 m) of Keweenawan Middle Run sandstone, basalts, and rhyolite (Drahovzal et al., 1992).

The Sherrer 1 is located in Jessamine County, Kentucky, USA and is found near the border of the Rome Trough and the ECGH Zone in Kentucky (Figure 3.1 – labeled J). The well penetrated 2008 ft (608 m) of red-colored sandstones, siltstones, and minor shales known as the Middle Run sandstone along with basalts (Drahovzal et al., 1992).

The Spears 1 is located in Lincoln County, Kentucky, USA and is found near the border of the Rome Trough and the ECGH Zone in Kentucky (Figure 3.1 – labeled K). The Spears 1 encountered 357 ft (108 m) of thin sandstones and trachyte (felsic) flows but no Keweenawan Middle Run sandstone (Drahovzal et al., 1992). The trachyte flows may be equivalent to the North Shore Volcanic Group in Minnesota which lie below the Oronto Group in age (Drahovzal et al., 1992).

3.5 Trap and Seal

Petroleum requires a trap and seal in order to develop an economic accumulation of oil and/or gas. The optimal trap would be an anticlinal closure with a dense shale or anhydrite or salt or combination of these lithologies to facilitate accumulation. In some cases, volcanic rocks could act as a seal. Unlike He and H, petroleum molecules tend to be large and trapped easily even with a marginal seal. Within the MCR, the zone that can act like a seal that would be the Nonesuch shale. Also, the timing of petroleum generation has to be post formation of the trap being present. Based on the present understanding of the MCR system the timing of petroleum generation, migration and accumulation in the trap is generally speculative. However, based on present knowledge, the presence of an effective seal and an associated trap within the MCR is low.

Recent work has indicated that He is almost always accumulated in areas of structural closure with associated basement faulting and fracturing defined by aeromagnetics and gravity (Tedesco, 2022). The seal for a He or H trap is problematic as it typically, unlike for the large molecule size for CH_4 and other hydrocarbons, is more like a baffle that delays the elements migration to the surface (Tedesco, 2022). He or H can be held in the trap by changes in capillary pressure in the seal. Typically, the higher the He % the reservoir will be significantly under-pressured preventing migration through the seal, which is typically higher pressured (Tedesco, 2022). Within the MCR rift system, the seal is likely to be the thick Nonesuch shale and/or flood basalts. There are numerous sandstone reservoirs within the Nonesuch and the underlying Copper Harbor clastic sediments. The reservoirs vary from conglomerates to fine grained sandstones, presence of "old pore water" (critical to a He accumulation) and are highly lenticular as they were deposited in an alluvial, marine or lacustrine setting. Other potential reservoirs are open fractures or voids filled with He or H and other gases trapped within mafic and igneous rocks. Presently H and He has been trapped, based on data from existing wells, in the shallower Pennsylvanian and Permian rocks which have a significant amount of shale, salt, and anhydrites that act as a seal.

3.5.1 Source Rocks

Source rocks for the Precambrian rocks in the Mid-Continent Rift to data is limited from a few tests along the MCR. The carbonaceous shales within the Precambrian sediments of the MCR typically have low organic matter that is mature to over mature. The Nonesuch shale has been expelling oil at the White Pine Copper Mine in the Upper Michigan Peninsular for decades (Dickas and Mudrey, 1997).

The Upper Red Clastic Zone of carbonaceous mudstones in the Eischeid #1 well typically had <0.1 to 0.05% total organic carbon (TOC) and have hydrocarbon potential (Anderson, 1990a, 1990c). The Lower Red Clastic Zone is divided into Unit C equivalent to the Nonesuch Shale and Copper Harbor Fm and Unit D to the Freida Fm of the Oronto Group. Unit C can be divided in the equivalent the Copper Harbor Fm overlain by the Nonesuch Shale. The Copper Harbor has 0.04 to 0.21% TOC and has no potential for petroleum generation (Palacas et al., 1990). The Nonesuch Shale varies from 0.06 to 1.40% with a Hydrogen Index of 21 to 77, a T_{max} of 498 to 508 °C and no potential for petroleum generation. Units E through H have <0.01 to 0.08 TOC and has no potential as a source rock (Palacas et al., 1990). The Nonesuch shale was evaluated by Burruss and Palacas (1999) at the Terra-Patrick 7-22 well in Wisconsin and found that the average T_{max} was 438 °C, Hydrogen Index of 190, in drill cutting from 0.1 to 044% TOC, and as such the rocks are considered marginally immature.

Helium is sourced from uranium (Ur) and thorium (Th) rich minerals (Tedesco, 2022). These minerals need to be present in the crust and upper mantle rocks, below the MCR to generate sufficient He through decay of Ur and Th to be accumulated later in economic quantities. The process of decay from Ur and Th typically takes 1 BY+ to generate one He molecule (Tedesco, 2022). Some percentage, it would be impossible to determine if it is 1% or 99%, of the He migrating from the Pre-Paleozoic rocks were also trapped during the formation of the planet (Tedesco, 2022).

Hydrogen exploration is relatively new and so is the understanding of what is generating the element in the geologic subsurface. Sources of H are according to Wang et al. (2023) and Jackson et al., 2024:

1. Magma degassing of deep H from the crust and mantle,
2. reaction of H_2O with ultrabasic rocks that is called serpentinization,
3. water in contact with reducing agents with the crust and mantle will release H,
4. decomposition of hydroxyl ions in the structure of minerals to release H,
5. natural H_2O radiolysis,
6. decomposition of organic matter,
7. biological activity.

Most of the processes to generate H are slow and their effectiveness to produce an economic accumulation of H is uncertain. Hydrogen is also a very reactive element; it will typically recombine with other elements and compounds forming new gases, fluids, and minerals. Microbes will consume H as part of a food source. This inhibits H ability to form large accumulations. There have been suggestions of doing serpentinization of rocks in the ground and extract H using that process. However, as serpentinization process proceeds, the porosity and permeability decreases as H is removed from the minerals. The MCR has significant volumes of igneous rocks that could be a viable target for serpentinization and extraction of H.

3.6 Gravity/Magnetics

MCR presents a distinct positive Bouguer gravity (Figure 3.7) as a result of the high density of igneous and sedimentary rocks deposited in a narrow rift system during formation. The aeromagnetic signature of the MCR (Figure 3.8) is less definitive but still shows a similar pattern to the Bouguer gravity anomaly in many areas of the rift. Detail analysis of both magnetic and gravity data can assist in determining location of Precambrian faults that can act as conduits for migrating fluids and gases (Hydrocarbons, CO_2, N_2, He, and H). Basement structural changes may also lead to finding closures that could potentially act as traps for various economic elements or compounds (Gay, 1995).

The majority of analysis of the MCR has come from modeling of the magnetic and gravity data, especially in areas where there is no drilling. Confirmation of many of the interpretations of the magnetic and gravity modeling is not likely by drilling unless there is a resource boom of some type.

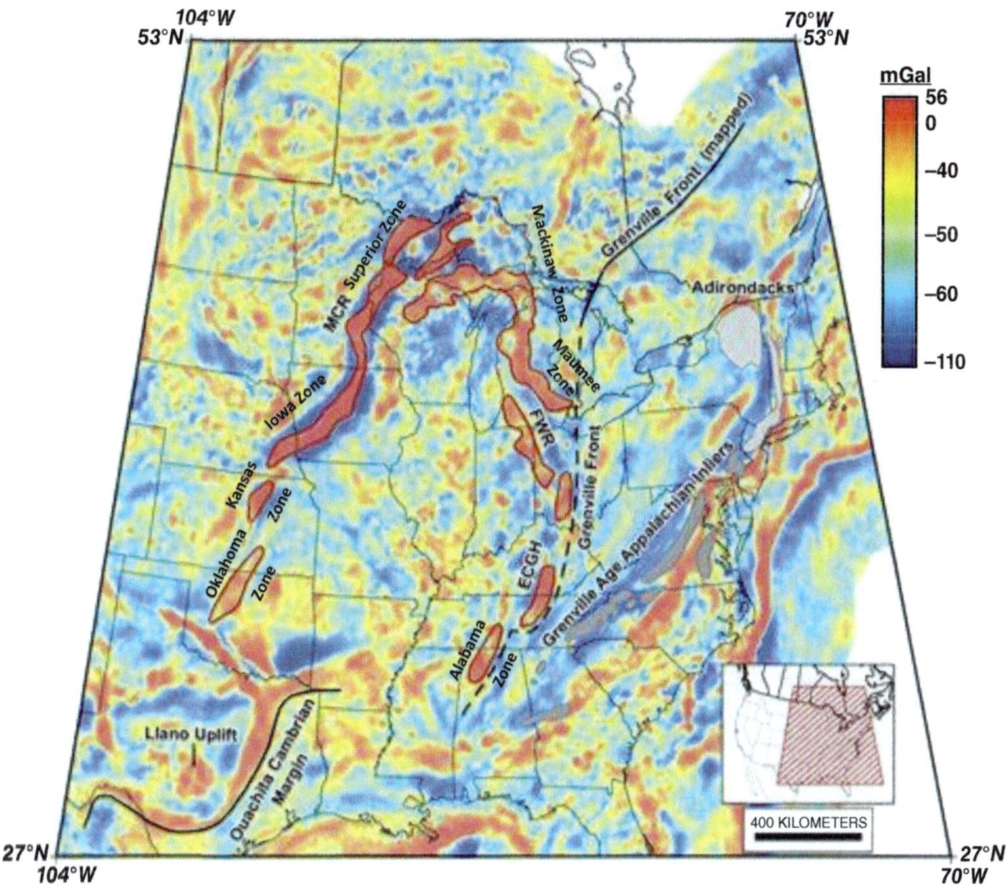

Figure 3.7 Gravity Map indicating the MCR (modified from Stein et al., 2018). The west arm extends southward from Lake Superior to at least to Oklahoma. The eastern arm extends through Michigan, USA along the Fort Wayne rift, through the ECGH to Alabama, USA. The Grenville Front is shown by a solid line where observed and dashed line where inferred.

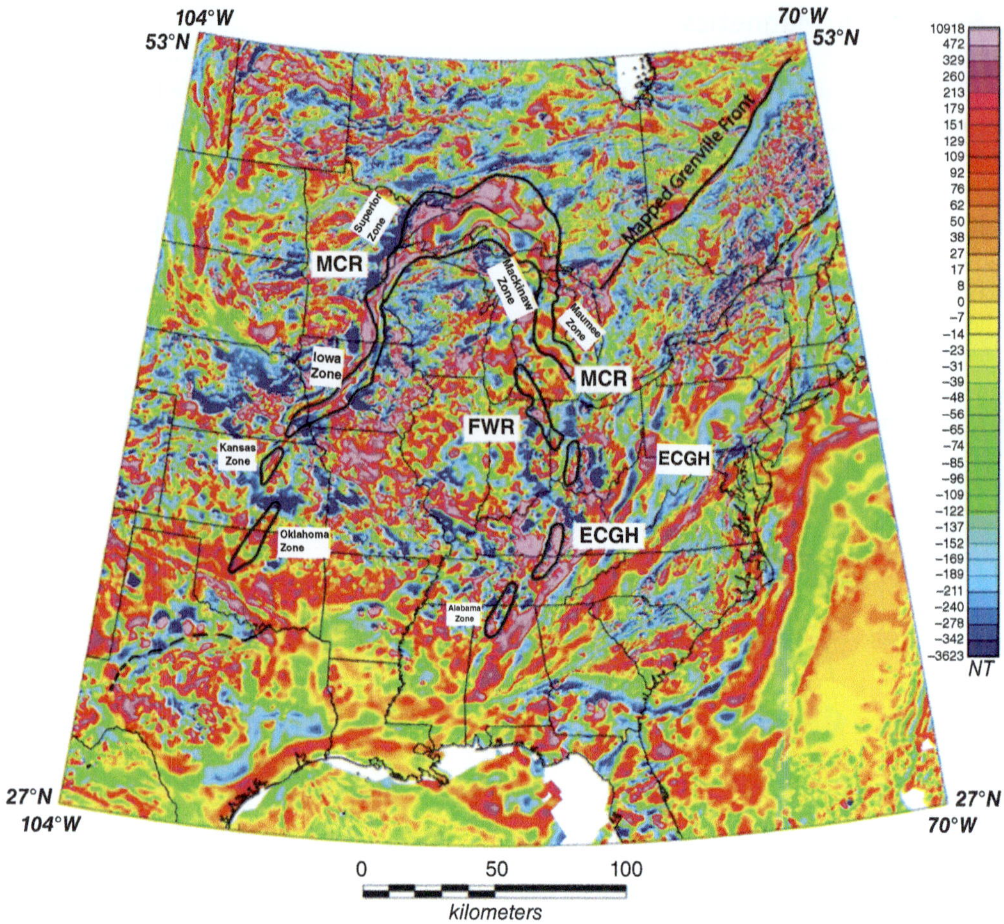

Figure 3.8 Aeromagnetic map of the MCR (modified from Stein et al., 2018).

3.7 Seismic

There has been historically a limited amount of seismic data that has been available across the MCR. The majority of data until recently was from the 1970s and 1980s when several companies carried out major exploration programs looking for hydrocarbons along the MCR. Seismic programs over the Superior Lake region of the MCR were carried out by the USGS plus other operators and Figures 3.1 and 3.9 is GLIMPCE Line F. GLIMPCE Line F indicates a Felsic Unit underlying the Oronto Group and overlying basaltic volcanic flows. Magnetic and gravity profile are indicated above the GLIMPCE Line F that has been interpreted as a magnetic, low-density felsic volcanic unit that was extruded near the end of volcanic activity (Hinze et al., 1997).

The COCORP L1 line in Kansas (Figures 3.1 and 3.10) indicates the rift with several clastic units overlying the basalts (Woelk and Hinze, 1995; Hinze et al., 1997). The magnetic and gravity data observed and modeled are presented above the COCORP L1. Deeper in the section is a number of mafic intrusions that could have provided heat to shallow carbonaceous

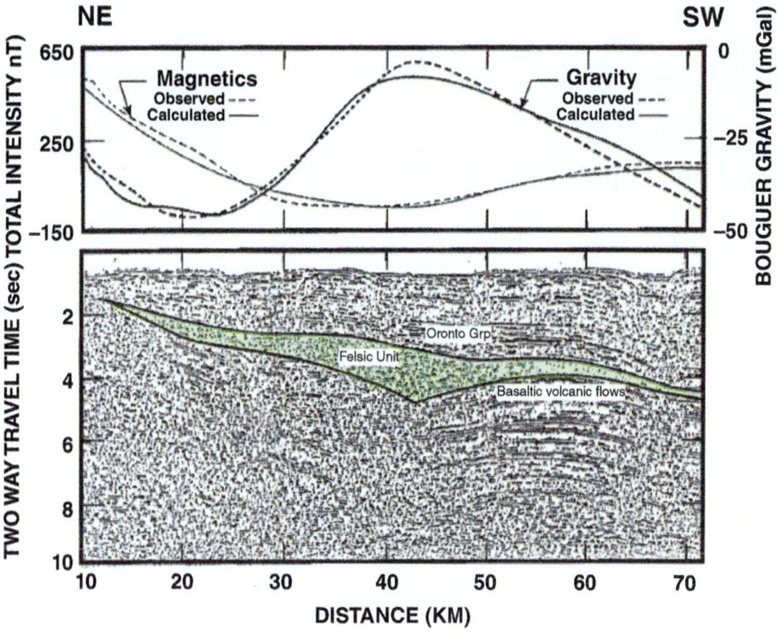

Figure 3.9 Final stack seismic reflection record section detail of GLIMPCE Line F (Figure 3.1) in east-central Lake Superior showing the outline of the interpreted felsic flow used in gravity and magnetic models. The observed and calculated Bouguer gravity and total intensity magnetic anomalies (Hinze et al., 1997).

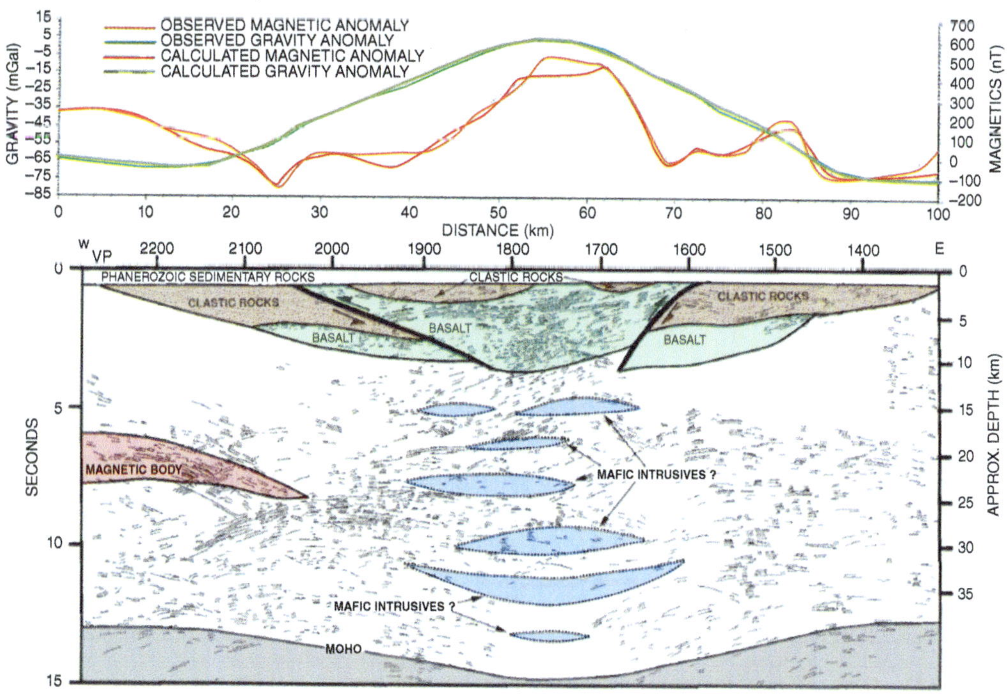

Figure 3.10 Crustal cross section (west-east) of the Midcontinent Rift in northeastern Kansas using a COCORP L1 seismic line (Figure 3.1) with associated based gravity modeling that is constrained by magnetic anomaly modeling (modified from Woelk and Hinze, 1995; Hinze et al., 1997).

mudstones that may have generated oil if they were emplaced later after deposition of the Nonesuch Shale. The mafic intrusions could also be a source of base and precious metals, or be the source of He or H.

3.8 Oil and Gas Exploration and Production

The MCR has several areas where there is petroleum production from overlying Paleozoic strata but to date no petroleum production from Keweenawan age rift sediments. The southern end of the western limb in Kansas and Oklahoma has prolific petroleum production from Ordovician to Permian age rocks whose trap development was related to minor tectonic movements in MCR basement faulting (Gay, 1995; Berendsen and Blair, 1996). Berendsen and Blair (1996) mapped the Precambrian surface and identified multiple strike-slip fault systems along the MCR's Nemaha Ridge that indicated a strong relationship to the location of petroleum production (Figure 3.11). The Mackinaw and Maumee zones extend through Michigan, which has prolific petroleum production. Several Paleozoic age petroleum fields overlying the MCR are producing from Ordovician reservoirs in the Mackinaw, Maumee, and ECGH zones that are associated with strike-slip faulting, brecciation, and formation of low-temperature dolomites that are related to basement faulting. Two examples of this are the Albion-Scipio (175MMBO+) and Stoney Point fields in Hillsdale County, Michigan, that is linear, reservoir is a dolomitized breccia. This breccia type of petroleum reservoir and several economic zinc ore bodies production are found in the ECGH zone in south-central Kentucky and north-central Tennessee.

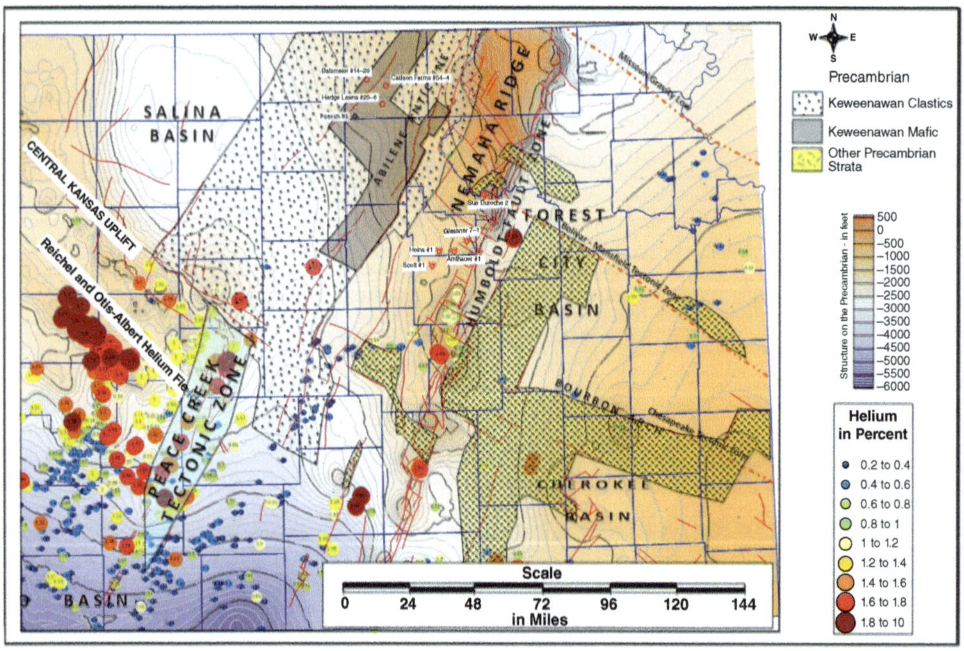

Figure 3.11 Kansas Zone of the MCR. He numbers indicated from tested wells. The clastic and mafic rocks of the MCR are listed. Other Precambrian strata that are not associated with the MCR. The Central Kansas Uplift which contains one of the historically large He fields in the US and is part of the Central Kansas Uplift.

The potential of the MCR for producing economic petroleum is considered remote due to problems with: (i) maturity of any source rocks to expel hydrocarbons, (ii) presence of a trap/seal for petroleum accumulation, (iii) heterogeneous reservoirs in terms of permeability and porosity, and (iv) timing of expulsion of hydrocarbons from organic matter in relation to when the traps were formed.

3.9 Iron and Base Metals

Iron (Fe) ore has or is being mined from several ranges or elongated ore bodies or trends in Southwestern Ontario, Northeastern Minnesota, Western Michigan, and Northern Wisconsin, USA along the northern part of the Superior and Mackinaw zones (Figure 3.1) of the MCR. Only the Mesabi and Marquette iron ranges are still active. The Mesabi Iron Range, named by the local Indian Ojibwe tribe, is a mining district primarily in Itasca and Saint Louis counties of northeastern Minnesota along the northwestern part of the Mid-Continent Ridge (Trendall, 1968; Jirsa et al., 2004; Tieberg, 2023). The ore deposits are found in the Animikie Group of the Middle Precambrian of 2.5 to 1.8 Ga and were formed by precipitation of Fe and Si during the middle Precambrian. Erosion caused the release of Fe and Si in to the local seas (Trendall, 1968; Jirsa et al., 2004; Tieberg, 2023). Marine algae present in the local sea raised the level of atmospheric O_2 through biologic processes. This O_2 increase caused the Fe to precipitate into the banded iron formations found in northern Minnesota and other members of the Animikie Group (Trendall, 1968; Jirsa et al., 2004; Tieberg, 2023). The Animikie Group underlies the MCR but could provide potential reservoirs for He and H.

Copper and Nickel are found in the Duluth Group, which is age equivalent to Copper Harbor Conglomerate of the Oronto Group and underlying North Shore Volcanic Group, all of Keweenawan age (Cooper, 1978). The Duluth Group is a long arcuate mafic intrusion that has been subdivided into three intervals in ascending order: (i) anorthositic and (ii) troctolitic and felsic series (Cooper, 1978). Copper is also mined from the Nonesuch Shale (Butler and Burbank, 1929). The MCR has extensive mineralization for Fe and base metals and will likely remain a major mining district for some time to come.

3.10 Impact Craters

The Mason crater located in Pocahontas, Humboldt, Webster, and Calhoun counties; Iowa, USA impacted the Iowa zone of the MCR 73.8 MY ago (Anderson, 1999). The impact affected Paleozoic rocks, MCR sediments and groundwater resources to present day. Several Unit C of the Lower Re Clastic Sequence of Keweenawan age have been encountered in several shallow wells drilled in the crater (IGS, 2006). There are no known impact craters that were emplaced during the formation of the MCR.

3.11 Helium

He for use by civilization presently relies almost entirely on extraction from natural gas in specific areas around the world. More than 80% of the He in the world is produced from five areas: Texas Panhandle-Hugoton, Big Piney-La Barge Field in Wyoming, South Pars in Qatar, and Amur in

Central Russia and Algeria (Tedesco, 2022). The remainder is produced from small fields generally located in the USA and Canada, specifically along the northern flank of the Williston Basin, the Colorado Plateau, Central Kansas Uplift and the Four Corners area in New Mexico and Arizona. Other areas in the world that produce He are Poland and South Africa (Tedesco, 2022).

Helium, as well as H, was not typically analyzed for from natural gas samples consistently until the last decade. In the last five years, He analysis has become standard in natural gas sampling because of its high price of over $400 (US) an MCF. Gas analysis from a reservoir that has 2.0% or higher He historically tend to generally be associated with significantly under pressured reservoirs pressure and volume. Reservoirs that have 0.3 to 2.0% He concentrations are more economically viable relative because of higher flow rates, more reasonable reservoir pressures and greater areas of accumulation. The accuracy of analytical equipment has increased through time in that He ppm data collected pre-1980 has to be accepted with caution.

Significant He has been found in non-petroleum gases associated with rift systems such as the EARS in Tanzania and Kenya. In the 1950s, the British government actively pursued He, and He is presently the primary target by at least two companies (Tedesco, 2022). The EARS has yielded gas contents from local hot springs that typically are greater than 5% He in the bubbling gas streams but to date no economic accumulations have been discovered. Presently, the exploration for economic He in the Tanzania part of EARS is being pursued by Helium One (www.helium-one.com) and Noble Helium (www.noblehelium.com.au). The difficulty in the EARS is that it is an active volcanic region, the rift system is actually starting to split the continent, and the concern is how can reservoirs be preserved without repeatedly being breached by tectonic activity. The age of EARS is fairly recent which typically would preclude any He accumulation being developed (Tedesco, 2022).

The MCR has He shows on the southwestern end in Kansas, USA from numerous Paleozoic oil and gas tests, from a mining test well in the northern top of the western limb (Minnesota), and southern part of the eastern limb where it butts up to the Rome Trough (Kentucky) in Cambrian age sandstones (Figure 3.1). The MCR crosses several states such as Iowa, Wisconsin, Michigan, Indiana, Ohio, Tennessee, and Alabama, USA; there is no gas analysis indicating He present. Of course, for many decades, He and H were not analyzed for on gas samples from either test wells or production in these areas.

The area of the MCR in Kansas, USA that has He in the natural gas stream is located in Chase, Marion, Wabaunsee, and Sumner counties (Figure 3.11). The area of the rift that is productive is along a structural high of the Nemaha Ridge with the southern portion where most of the Middle Ordovician, Hunton, Mississippian, and Pennsylvanian Cherokee Group has been removed. Over 18 BCF of natural gas have been produced from the area but no He even though it is present from 0.03 to 1.45% He (Bhattacharya et al., 2008). The area has multiple natural gas reservoirs and the He is found only in the Pennsylvanian Lansing Kansas City to the Permian Admire Group. The higher He contents are related to where the Arbuckle under lies the Pennsylvanian Lansing-Kansas City directly. The Council Grove Group section is over 500 ft (151 m) thick and has anhydrites and salts that are laterally extensive and are interbedded with the multiple sandstone reservoirs (Bhattacharya et al., 2008).

There has been a moderate show of He in two wells drilled in Lake County near Babbitt, Minnesota, USA drilled in 2024 to 2200 ft (671 m) (Pulsar, 2024a, 2024b, 2024c) (Figure 3.1 – labeled F). The first well was drilled by a mining company searching for nickel and encountered a gas reservoir. The gas flow was not significant as it caused no serious issues to the workers or the area. The gas from that well was reported to have 10.5% He, 13.2% N_2, 73.8% CO_2 and

2.4% CH_4 (Barry et al., 2022). The well was plugged without significant effort despite flowing gas, for four days. A second well, Jetstream #1, a direct offset, 50 ft (165 m) from the first well, was drilled in 2024. Analysis of the gas from the Jetstream 1 contained up to 71% N_2, 24% CO_2, and 8.7 to 14.5% He, no water, and with a bottomhole pressure of 162 psi. The reservoir is fractures encased in the Bald Eagle intrusion which is composed of interlayered troctolite and anorthosite rocks (Pulsar, 2024a, 2024b, 2024c). Logging of the well was inhibited by the lack of fluid in the hole minimizing the effectiveness of the mechanical logs to define the reservoir. The operator seems to have perforated from 1605 to 2200 ft (503 to 671 m). The well after reaching total depth had to be cleaned out with high pressure air in order to evaluate the well for gas flow, gas composition and flow rate (Pulsar, 2024a, 2024b, 2024c). This implies the reservoir was under pressured and could not flow naturally without difficulty to surface. A normally pressured reservoir at 2200 ft (671 m), using a 0.45 psi gradient per foot would be 990 psi. The maximum flow rate achieved was 821 MCFGPD using artificial compression. Under natural conditions, the well flowed 150 MCFGPD on a six-hour test (Pulsar, 2024a, 2024b, 2024c). The Jetstream 1 has characteristics of many high percentage He wells in that the reservoir is significantly under pressured. The area has been historically subject to base metal and iron mining for over 100 years but no previous gas flow rates have been reported. Similar significantly under pressured He plays (based on He percentage alone) are found in the Colorado Plateau in Utah, New Mexico, and Arizona; and on the Apishapa Arch, Las Animas, Colorado, all in the USA (Tedesco, 2022). The well is awaiting installation of process facility to extract He and potentially CO_2 as well as drilling of additional wells.

3.12 Hydrogen

Hydrogen is a perceived potential energy source that could potentially replace fossil fuels. Hydrogen presently is generated by steam reformatting of natural gas, which results in generation of carbon dioxide and H. Historically, natural gas exploration gas samples are analyzed for a variety of hydrocarbons, inert elements, and compounds, but only recently H. Models for natural hydrogen generation, migration, and economic accumulation are to date speculation. There is only one actual H producing reservoir located in the southern portion of the Taoudeni Mega Basin, Bourakebougou, and Mali (Maiga et al., 2023). That reservoir contains gas that is 98% hydrogen, 1% methane, and 1% nitrogen. The reservoirs are carbonates and sandstones with alternating dolerite sill layers. There have been 25 wells drilled that have defined a large are of H gas but only one well produces at 5 to 8 MCFGPD providing power to a local village (Maiga et al., 2023). The company involved in the field has as a subsequent reserve report claiming several BCF of recoverable H. Further development of the field is unlikely anytime soon due to the political instable of Mali and lack of a market for the remaining H.

Hydrogen is present in three wells in Riley and Geary counties, Kansas, USA (Figure 3.11) located on the Kansas zone (Guelard et al., 2017). All three wells demonstrated high amplitude variation attributed to different well conditions during sampling and flow rates of less than 10 MCFGPD (Figure 3.12). Recently, a series of wells have been drilled to the north of these wells in Washington and Marshall counties, Kansas, USA but no H has been reported from the wells (Figure 3.11). Those wells are the Hedge Lawns #25-6, Carlson #54-4, and the Balsmeier #14-29 (Figure 3.1). The Iowa zone also had at least one test in 2023, Van Diest 14-3A, and to date there is no indication of success or failure (Figure 3.1).

Sample	Sampling Date	Sampling Hours	Gas Type	Noble Gas (ppm)							σ							
				^{4}He	^{20}Ne	^{36}Ar	^{84}Kr	$^{40}Ar/^{36}Ar$	$^{3}He/^{4}He$	$^{4}He/^{40}Ar^{*}$	^{4}He	^{20}Ne	^{36}Ar	^{84}Kr	$^{40}Ar/^{36}Ar$	$^{3}He/^{4}He$	$^{4}He/^{40}Ar^{*}$
Atm	-	-	-	5.24	16.45	31.57	0.65	295.5	1.39E − 06								
ASW	-	-	-	0.05	0.17	1.07	0.04	0.24	2.01								
D#2	2008	-	(1)	n.a.	n.a.	n.a.	n.a.	n.a.	n.a.	n.a.	n.a.	n.a.	n.a.	n.a.	n.a.	n.a.	n.a.
D#2	2011	-	(2)	n.a.	n.a.	n.a.	n.a.	n.a.	n.a.	n.a.	n.a.	n.a.	n.a.	n.a.	n.a.	n.a.	n.a.
D#2-1	16 Mar 2012	-	(2)	10,978	3.45	8.53	0.19	419.2	2.1E − 07	10.40	1238.8	0.4	0.3	1.1E − 02	1.29	2.6E − 08	0.09
D#2-2	17 Mar 2012	-	(2)	8199	2.43	8.20	0.20	413.4	2.0E − 07	8.48	924.8	0.3	0.3	1.2E − 02	1.69	9.4E − 08	0.09
D#2-Vacu 3	17 Mar 2012	-	(2)	n.a.	n.a.	n.a.	n.a.	n.a.	n.a.	n.a.	n.a.	n.a.	n.a.	n.a.	n.a.	n.a.	n.a.
D#2-1	12 Nov 2013	10:25 a.m.	Air	1050	16.01	28.83	0.62	306.8	1.4E − 07	3.24	76.0	2.4	0.9	2.2E − 02	3.62	1.5E − 08	0.77
D#2-2	12 Nov 2013	10:30 a.m.	(3)	12,786	4.78	10.24	0.25	376.9	1.5E − 07	15.33	921.7	0.7	0.3	9.3E − 03	4.42	2.6E − 08	0.11
D#2-3	12 Nov 2013	11:00 a.m.	(3)	16,495	3.07	6.25	0.16	453.9	1.4E − 07	16.65	1189.1	0.5	0.2	6.3E − 03	5.60	1.8E − 08	0.06
D#2-4	12 Nov 2013	05:00 p.m.	(3)	15,676	2.16	6.63	0.18	459.3	1.4E − 07	14.43	1131.0	0.3	0.2	7.5E − 03	5.66	1.5E − 08	0.06
D#2-5	13 Nov 2013	09:45 a.m.	(3)	20,458	2.23	6.66	0.17	457.6	1.5E − 07	18.94	1484.3	0.3	0.1	5.7E − 03	2.02	1.5E − 08	0.03
D#2 Vacu7	13 Nov 2013	10:00 a.m.	(3)	n.a.	n.a.	n.a.	n.a.	n.a.	n.a.	n.a.	n.a.	n.a.	n.a.	n.a.	n.a.	n.a.	n.a.
D#2-1	16 May 2014	06:00 p.m.	(3)	14,413	3.19	6.57	0.17	417.8	1.3E − 07	17.95	668.3	0.2	0.1	4.5E − 03	6.59	1.8E − 08	0.03
D#2-2	17 May 2014	02:30 p.m.	(2)+(3)	31,076	4.93	9.72	0.22	436.5	1.5E − 07	22.68	1440.3	0.3	0.1	5.5E − 03	6.64	1.3E − 08	0.03
D#2-Vacu1	17 May 2014	02:30 p.m.	(2)+(3)	n.a.	n.a.	n.a.	n.a.	n.a.	n.a.	n.a.	n.a.	n.a.	n.a.	n.a.	n.a.	n.a.	n.a.

D#2-3	18 May 2014	11:30 a.m.	(2)	28,089	4.30	9.45	0.22	442.4	1.4E − 07	20.24	1301.5	0.4	0.1	5.4E − 03	6.76	1.3E − 08	0.03
D#2-Vacu2	18 May 2014	11:30 a.m.	(2)	n.a.	n.a.	n.a.	n.a.	n.a.	n.a.	n.a.	n.a.	n.a.	n.a.	n.a.	n.a.	n.a.	n.a.
D#2-4	18 May 2014	01:00 p.m.	(2)	31,159	4.91	10.36	0.24	441.4	1.3E − 07	20.61	1443.5	0.3	0.1	6.0E − 03	6.67	1.3E − 08	0.03
D#2-5	18 May 2014	01:45 p.m.	(2)	31,870	4.59	10.32	0.25	440.7	1.3E − 07	21.26	1477.4	0.4	0.1	6.6E − 03	6.85	1.7E − 08	0.03
D#2-1	29 Aug 2014	00:20 p.m.	(2)	28,343	6.01	10.68	0.26	429.0	1.3E − 07	19.88	675.0	0.5	0.2	5.2E − 03	1.47	1.5E − 08	0.04
D#2-2	29 Aug 2014	03:20 p.m.	(2)	25,523	5.33	9.65	0.24	428.0	1.5E − 07	19.96	626.8	0.5	0.2	4.9E − 03	1.46	1.8E − 08	0.04
D#2-3	29 Aug 2014	04:45 p.m.	(2)	25,145	4.76	9.85	0.24	431.7	1.4E − 07	18.74	605.2	0.4	0.2	5.1E − 03	1.58	4.4E − 08	0.04
D#2-4	30 Aug 2014	02:00 p.m.	(2)	25,859	5.56	10.90	0.32	415.3	1.5E − 07	19.79	617.4	0.5	0.2	6.3E − 03	1.34	2.4E − 08	0.04
D#2-5	30 Aug 2014	05:40 p.m.	(2)	21,817	6.65	11.88	0.30	397.5	1.4E − 07	18.01	520.8	0.6	0.2	6.3E − 03	1.18	1.5E − 08	0.05
D#2-6	31 Aug 2014	11:45 a.m.	(2)	28,860	6.19	11.51	0.27	418.4	1.7E − 07	20.40	687.8	0.6	0.2	5.3E − 03	1.30	2.5E − 08	0.04
D#2-7	31 Aug 2014	02:30 p.m.	(2)	27,467	5.21	9.95	0.25	432.5	1.4E − 07	20.15	655.2	0.5	0.2	5.0E − 03	1.40	1.4E − 08	0.04
D#2-8	2 Sep 2014	11:45 a.m.	(2)	26,757	5.61	11.24	0.28	421.1	1.4E − 07	18.94	638.5	0.5	0.2	5.6E − 03	1.26	1.6E − 08	0.04

Note: (-) not available. (n.a.) not analyzed.

Figure 3.12 Noble well gas composition and isotopic ratios for the Sue Duroche #2 well (Guelard et al., 2017).

3.13 Summary

The MCR is Keweenawan-age Precambrian feature that formed 1.1 Ga years ago that presently only produces base metals and Fe in the Superior Zone of the feature. There is shallow Paleozoic oil and gas production gas along the southeastern and southwestern limbs in Paleozoic rocks but none from any of the MCR sediments. There are numerous H and He shows in the Kansas Zone, a He show in the Superior zone, and He shows in the Kentucky portion of the ECGH Zone. The sedimentary rocks of the MCR are the Oronto Group (Lower Red Clastic Zone) and the Jacobsville/Bayfield (Upper Red Clastic Zone) are correlate effectively across the feature. Helium, which is the second smallest molecule, in the 0.3 to 2.0% typically requires a very effective seal such as a thick shale column and/or a shale sequence that contains salts and anhydrite. Higher He percentages in a reservoir generally require significantly low reservoir pressure relative to the capillary pressure in the seal in order not migrate through the seal and out of the trap.

Hydrogen is very similar to He except it is the smallest element on the periodic table. In addition, H is highly reactive with many other elements and compounds making it difficult to accumulate in a reservoir. The reservoir lithology would have to be relatively inert and like He, the accumulation has to be significantly under pressured relative to the seal.

The Keweenawan rocks that have been analyzed to date based on oil and gas drilling their effective permeability and porosity would not be conducive to a petroleum accumulation. Petroleum fields have developed in the overlying Paleozoic rocks by reactivation of Precambrian faults that create structural settings to create a trap. The faults also provide conduits for migrating fluids and gases that can facilitate both petroleum, H, and He migration to a trap. The Nonesuch of the Oronto Group is the source oil in the White Pine Copper Mine in the Upper Michigan Peninsular. This is not a petroleum accumulation and is more of small geologic anomaly. The Nonesuch Shale does not seem to be a major source rock for petroleum.

Exploration of the MCR will probably continue for a few years until either no H and/or He economic accumulations are found. The drilling may actually encounter base and precious metals. Regardless the stratigraphy and stratigraphy of the MCR will be advanced by this new data.

References

Anderson, R. R. (1990a). Review of the Precambrian geological history of the central United States and the Midcontinent Rift System. Iowa Geological Survey, Special Report Series 2, pp. 1–25.

Anderson, R. R. (1990b). Interpretation of geophysical data over the Midcontinent Rift System in the area of the M.G. Eischeid #1 petroleum test, Carroll County. Iowa, Iowa Geological Survey, Special Report Series 2, pp. 27–38.

Anderson, R. R. (1990c). Review of current status of Proterozoic rocks in the Amoco M.G. Eischeid #1 petroleum test, Carroll County, Iowa. Iowa Geological Survey, Special Report Series 2, pp. 175–184.

Anderson, R. R. (1999). Iowa's Manson Impact Structure. Iowa City: Iowa Geological Survey. http://www.igsb.uiowa.edu/Browse/manson99/manson.htm.

Azmi, I. & Goldberg, K. (2024). Facies analysis and depositional model of the Midcontinent Rift System in Kansas, USA. The Depositional Record 10 (2), pp. 282–297.

Barry, P., Sherwood-Lollar, B., & Ballentine, C., et al. (2022). High helium concentrations in well gases from the Duluth Complex, Minnesota, USA. GSA, Annual Meeting October 11, Abstract. 54 (5).

Baumann, S. D. J. (2023). Written by: Ages of select units from the Keweenawan Supergroup in Wisconsin and Upper Peninsula of Michigan. Midwest Institute of Geosciences and Engineering. www.researchgate.net/publication/369020128_Written_by_Ages_of_select_units_from_the_Keweenawan_Supergroup_in_Wisconsin_and_Upper_Peninsula_of_Michigan.

Bhattacharya, S., Watney, L., & Newell, D. (2008). *Kansas low-BTU gas compositions and production potential at Elmdale Field.* New York: Chase. Kansas, KGS, OFR 2008-34, p. 67.

Berendsen, P. (1994). Review of Precambrian rift stratigraphy. In: *Revision of Stratigraphic Nomenclature in Kansas* (ed. D. L. Baars), pp. 1–4. Lawrence: Kansas Geological Survey. Bull 230. ww.kgs.ku.edu/Publications/Bulletins/230/01_rift.html.

Berendsen, P. (1997). Tectonic evolution of the midcontinent rift system in Kansas. Geological Society of America Special Papers 312, pp. 235–241.

Berendsen, P. & Blair, K. (1996). Structural configuration of the Precambrian Basement. Lawrence: Kansas Geological Survey. Map M-45:1.

Berendsen, P., Borcherding, R. M., Doveton, J., et al. (1988). Texaco Poersch #1, Washington County, Kansas – Preliminary geologic report of the pre-Phanerozoic rocks. Lawrence: Kansas Geological Survey. OFR 88-22, www.kgs.ku.edu/Publications/OFR/1988/OFR88_22/index.html.

Burruss, R .C. & Palacas, J. G. (1999). Chapter 8: Evaluation of the petroleum source-rock potential of the Middle Proterozoic Nonesuch Formation, Terra-Patrick #7-22 Borehole, Bayfield, Wisconsin. In: Terra-Patrick 7-22 Deep Hydrocarbon Test, Bayfield County, Wisconsin, Investigation and Final Report (ed. A. B. Dickas & M. G. Mudrey). Madison: Wisconsin Geological and Natural History Survey, 97-1, pp. 87–93.

Butler, B. S. & Burbank, W. S. (1929). The Copper deposits of Michigan. Reston, Virginia: USGS. Prof. Paper 144, p. 235.

Cooper, R. W. (1978). Geology of the regional copper-nickel study area. St Paul, MN: Minnesota Environmental Quality Board, p. 86.

Cullers, R. L. & Berendsen, P. (1993). Composition of rift-related igneous and sedimentary rocks of the Keweenawan Supergroup in the Poersch no. 1, OZ-1, Finn, and Friederich wells, northeastern Kansas. Geological Survey Bulletin 235, pp. 55–72.

Daniels, P. A. (1997). Chapter 7. Stratigraphy and lithology of the Keweenawan sedimentary rocks penetrated by the Terra-Patrick #7-22 Borehole, Bayfield County, Michigan. In: *Terra-Patrick 7-22 Deep Hydrocarbon Test, Bayfield County, Wisconsin, Investigation and Final Report* (ed. A. B. Dickas & M. G. Mudrey). Madison: Wisconsin Geological and Natural History Survey, 97-1, pp. 75–85.

Dickas, A. B. & Mudrey, M. G. (1997). Summary. In: *Terra-Patrick 7-22 Deep Hydrocarbon Test, Bayfield County, Wisconsin, Investigation and Final Report* (ed. A. B. Dickas & M. G. Mudrey). Madison: Wisconsin Geological and Natural History Survey, 97-1, p. 126.

Drahovzal, J. A., Harris, D. C., Wickstrom, L. H., et al. (1992). The East Continent Rift Basin: a new discovery. Cincinnati Arch Consortium, Special Publication 18, Series XI, p. 31.

Gay, S. P. (1995). Basement control of selected oil and gas fields in Kansas as determined by detailed residual aeromagnetic data. Lawrence: Kansas Geological Survey. Bull 237, p. 7. www.kgs.ku.edu/Publications/Bulletins/237/Gay1/basement.pdf.

Guelard, J., Beaumont, V. Rouchon, V., et al. (2017). Natural H_2 in Kansas: Deep or shallow origin, AGU Publications. Geochemistry, Geophysics, Geosystems, Special Section, Research Article. 18 (5), pp. 1841–1865. https://doi.org/10.1002/2016GC006544.

Hinze, W. J., Allen, D. J. Braile, L. W., et al. (1997). The Midcontinent Rift System: A major Proterozoic continental rift. In: *Middle Proterozoic to Cambrian Rifting* (ed. R. W. Ojakangas, A. B. Dickas & J. C. Green). Boulder, Colorado: Central North America. Geological Society of America. Special Paper 312, pp. 7–36.

IGS (2006). Exploring the Midcontinent Rift. Iowa City: Iowa Geological Survey. Our Common Ground, p. 4.

Jackson, O., Lawrence, S. R., Hutchinson, I. P., et al. (2024). Natural hydrogen: sources, systems and exploration plays. Geoenergy, May 7. www.pubs.geoscienceworld.org/gsl/geoenergy/article/2/1/geoenergy2024-002/638133/Natural-hydrogen-sources-systems-and-exploration.

Jirsa, M. A., Boerboom, T. J., Green, J. C., et al. (2004). Institute on Lake Superior Geology, Field Trip 5 – Classic Outcrops of Northeastern Minnesota. Proceedings of the 50th Annual Meeting – Part 2, Field Trip Guidebook. pp. 129–169.

Maiga, O., Deville, E., Laval, J., et al. (2023). Characterization of the spontaneously recharging natural hydrogen reservoirs of Bourakebougou in Mali. www.nature.com/scientificreports.

McClure Oil (1976). Completion for the Sparks-Eckelbarger-Whightsil No. 1-8 is located in the Ne Sw Nw Section 8 Township 10 North Range 2 West, Gratiot County, Michigan. Michigan Department of Environment, Great Lakes and Energy. www.egle.state.mi.us/GeoWebface/GeoWebface/DL/057/29739_DL.pdf.

Nowaczewski, S. F. (1994). Relevance of trapping mechanisms in certain Michigan formation Stray sandstone gas reservoirs in gas storage operations. AAPG, Annual Convention, September 18–20, Abstract.

Palacas, J. G., Schmoker, J. W., Daws, T. A., et al. (1990). Petroleum source-rock assessment of middle Proterozoic (Keweenawan) sedimentary rocks, Eischeid #1 Well, Carrol County, Iowa. IGS, Special Report Series 2, pp. 119–134.

Pulsar (2024a). Pulsar receives laboratory results for the laboratory results for the Jetstream #1 Appraisal well at the Topaz Project, March 14, 2024. www.pulsarhelium.com.

Pulsar (2024b). Pulsar receives positive results from wireline logs for the wireline logs for the Jetstream #1 appraisal well at the Topaz helium project, March 25, 2024. www.pulsarhelium.com.

Pulsar (2024c). Pulsar announces helium flowing to surface at Jetstream #1 in Minnesota, June 6, 2024. www.pulsarhelium.com.

Stein, S., Stein, C. A. Elling, R., et al. (2018). Insights form North America's failed Midcontinent Rift into the evolution of continental rifts and passive continental margins. Tectonophysics 744, pp. 403–421.

Tedesco, S. A. (2022). *Geology of Production of Helium and Associated Gases*. Amsterdam: Elsevier, p. 394.

Tieberg, A. (2023). Mesabi Iron Range. MNOPEDIA. www.mnopedia.org/place/mesabi-iron-range.

Trendall, A. F. (1968). Three great basins of Precambrian Banded Iron Formation Deposition: A Systematic Comparison. GSA 79 (11), p. 1527.

Van Schmus, W. R., Willin, E. T. & Shearer, C. K. (1990). Age relationships for gabbro from the Amoco M. G. Eischeid #1 well, Iowa Geological Survey. Special Report Series 2, pp. 67–76.

Wang, L., Jin, Z., Chen, X., et al. (2023). The origin and occurrence of natural hydrogen. Energies 16, p. 18.

White, W. S. (1972). The base of the Upper Keweenawan, Michigan and Wisconsin, USGS. Geological Survey Bull 1354-F, p. 36.

Witthuhn-Rolf, K. M. (1997). A structural analysis of the Midcontinent rift in Michigan and Minnesota. In: *Middle Proterozoic to Cambrian Rifting* (ed. R. W. Ojakangas, A. B. Dickas & J. C. Green), Boulder, Colorado: Central North America. Geological Society of America. pp. 97–114.

Witzke, B. J. (1990). General stratigraphy of the Phanerozoic and Keweenawan Sequence, M. G. Eischeid #1 Drillhole, Carrol County, Iowa. Iowa Geological Survey, Special Report Series 2, pp. 39–57.

Woelk, T. S. & Hinze, W. J. (1995). Midcontinent rift system in northeastern Kansas. Kansas Geological Survey. Bull 237, p. 6. www.kgs.ku.edu/Publications/Bulletins/237/Woelk1/index.html.

4

Production from Desmoinesian and Atokan Age Coalbed Methane and Carbonaceous Mudstone and Their Relationship to Structure and Geologic History of the Cherokee Basin, Kansas and Oklahoma, USA

Steven A. Tedesco

Running Foxes Energy Inc., Centennial, Colorado, USA

4.1 Introduction

Southeastern Kansas and northeastern Oklahoma comprise the majority of the Cherokee Basin (Figure 4.1), in the 1920s to the 1930s, was the largest natural gas producing region in the country for and almost exclusively from CBMCM (Stoeckinger, 1990; Tedesco, 2014). The Cherokee Basin lies along the Ouachita – Marathon Orogeny, which is also associated with several other productive unconventional basins such as the Appalachian (Marcellus and coals), Warrior (coals), Arkoma (Fayetteville and Hartshorne coal), the Anadarko (Woodford and other Pennsylvanian shales), and the Fort Worth (Barnet) and the Permian basins (Tedesco, 2014). Historically, mine explosions in underground mines have been reported both in northeastern Oklahoma, southwestern Missouri, and southeastern Kansas prior to the existence of the oil and gas industry (Bartow, 1906; Tedesco, 2014). The shale gas industry, which included coals, started in the 1920s and 1930s in eastern Kansas, northeastern Oklahoma, and Missouri as a result of depletion of conventional sandstone reservoirs primarily from the Bartlesville, Warner, and Squirrel sandstone reservoirs (Charles and Page, 1929; Green, 1933; Deason, 1984; Deason and Bohm, 1988). The CBMCM was productive from Kansas City, Kansas, to Tulsa Oklahoma through the 1920s (Tedesco, 2014). Upon the discovery of higher volume wells in the Hugoton Field in western Kansas and Texas Panhandle Field in Texas and Oklahoma, production from the CBMCM in the Cherokee Basin rapidly declined.

The establishment of federal tax credits by the US Congress in the 1980s and early 1990s for drilling coal-bed methane caused a boom in the drilling of these unconventional reservoir (Stoeckinger, 1990, 2000). Initially in the 1990s, the area of focused CBMCM activity was in northeast Oklahoma (Tedesco, 2014). In 2001, Devon Energy Corp. bought two projects in northeast Oklahoma and southeastern Kansas, which caused a land rush by a number of other companies (Tedesco, 2014). The CBMCM activity in the Cherokee Basin spilled over into the adjacent Arkoma and Forest City basins (Tedesco, 1992; Bostic et al., 1993; Smith, 2001; Cardott, 1999, 2000; Tedesco, 2014). The exploitation of the CBMCM continued but started to wane from 2005 to 2008 not only in the

Geophysical Exploration for Hydrocarbon Reservoirs, Geothermal Energy, and Carbon Storage: New Technologies and AI-based Approaches, First Edition. Edited by Said Gaci.

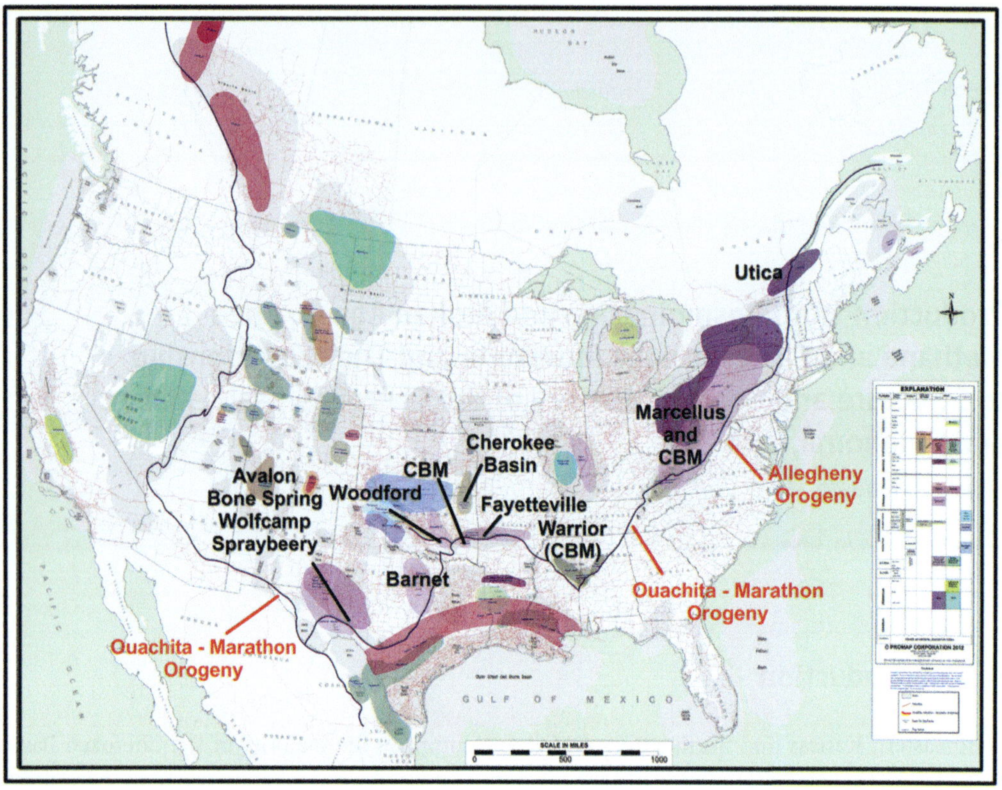

Figure 4.1 Location of the Ouachita-Marathon Orogeny and associated US and Canadian orogenies represented by a black line. The present shale and coal bed methane plays in the USA and Canada are indicated. The unconventional resources associated with the Ouachita-Marathon and Allegheny (northern extension) orogenies. (Reproduced with permission from Promap Corp. / Steven A. Tedesco).

Cherokee Basin but in other areas such as the Arkoma, Appalachian, Alberta, San Juan, Warrior, and Powder River basins due to more prolific shale plays elsewhere. In 2008 to 2009 with gas prices declining from $8 to $10(US) an MCF to less than $3 (US) an MCF and coinciding with a substantial increase in shale gas production, the CBMCM play in the Cherokee Basin stopped. The Oklahoma portion of the basin gas production peaked in 2006 at 18.9 BCF (Brady et al., 2011). The peak year for Kansas production was in 2008 at 49.1 BCF (Newell and Birdie, 2021). The peak production for the basin as a whole peaked in 2008 (Brady et al., 2011; Tedesco, 2014).

Drilling and completion techniques are critical in achieving economic and optimal production from conventional and unconventional reservoirs (Tedesco, 2014). CBMCM reservoirs in the Cherokee Basin require no unique methods to protect the reservoirs. Completion methods for coal-bed methane reservoirs started out using cross-link gel on individual seams with good results. This transitioned to slick water fracture stimulation at low barrels per minute rate to present the stimulation leaving the reservoir. Carbonaceous mudstones reservoirs require a small skin fracture stimulation to get by cement damage around the wellbore (Tedesco, 2014). Evaluating production data in its relation to drilling and completion techniques for CBMCM reservoirs in many basins can be a long-term event as these reservoirs typically take time to dewater and eventually produce gas. The CBMCM reservoirs in the Cherokee Basin on the other hand almost always produce gas immediate and the production profile is similar to a conventional gas well (Tedesco, 2014).

4.2 Geology

The Cherokee Basin is bounded to the north by the Bourbon Arch, the west by the Nemaha Ridge, the east by the Ozark Uplift and to the south by the Arkoma Basin (Rascoe and Alder, 1983) (Figure 4.1). The Cherokee Basin is a shallow inter-cratonic depression that has had limited sediment accommodation and formed in the Late Devonian over the paleo-Chautauqua Arch (Figure 4.2). The basin accumulated a maximum 4000 ft (1200 m) of sediment, based on present evidence, in Pennsylvanian-Permian times (Tedesco, 2014). An additional 4000 to 6000 ft (1200 to 1820 m) of sediment was deposited in the Permian to Tertiary times but has since been removed. The Atokan and Desmoinesian CBMCM were deposited across a broad area that included the adjacent Forest City and Arkoma basins. The petroleum presently found in the Ordovician to Pennsylvanian reservoirs migrated and accumulated in various reservoirs in late Pennsylvanian to Early Permian time.

The Chautauqua Arch, which lies under the Cherokee Basin, became a positive feature in Devonian time that resulted in the removal of all of the Middle Ordovician to Late Devonian sediments across it (Merriam, 1963). The Chautauqua Arch was beveled down to an erosional surface on top of the Lower Ordovician Arbuckle Group (Figure 4.3). The arch ceased to be a positive feature in Early Mississippian and the highly rich organic Chattanooga shale across the Chautauqua Arch (Merriam, 1963). Mississippian carbonates were deposited over the Chattanooga Shale and in Pre-Pennsylvanian times the Cherokee Basin was subject to erosion and karsting. The petroleum in the Cherokee Basin is either derived from the Ordovician Simpson and Maquoketa shales and the Mississippian age Chattanooga Shale (Tedesco, 2014). The gas from the CBMCM reservoirs was generated from internal from these lithologies (Tedesco, 2014).

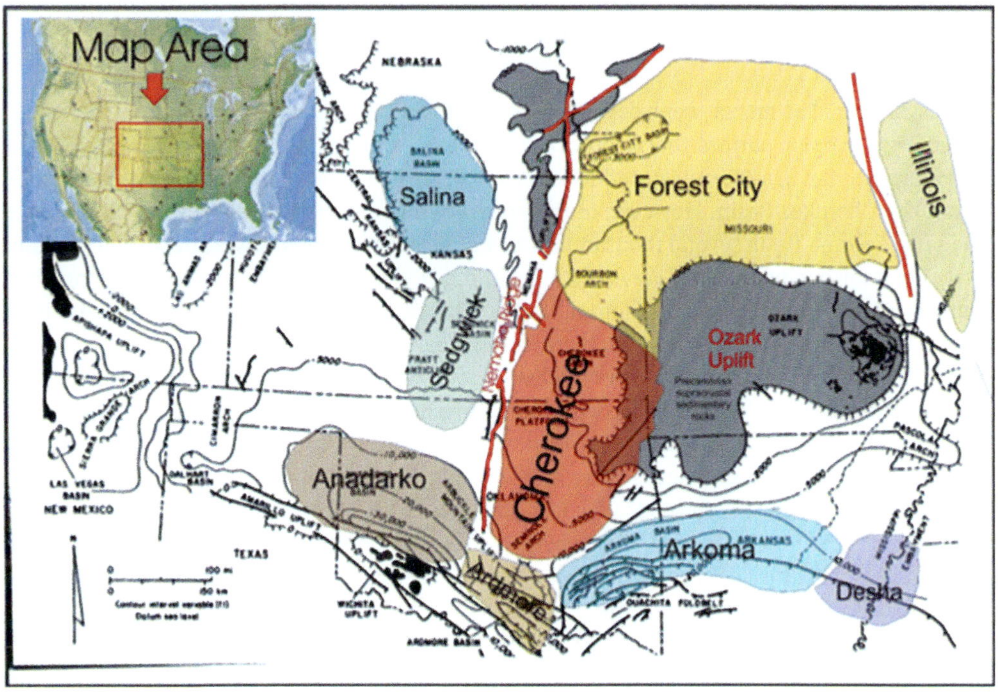

Figure 4.2 Present-day structure and tectonic features surrounding the Cherokee Basin. Modified from Rascoe and Adler (1983). The Cherokee, Forest City, Arkoma, Salina, Anadarko, and Desha basins are indicated. The Western Interior Basin includes the Forest City, Cherokee, and Arkoma basins.

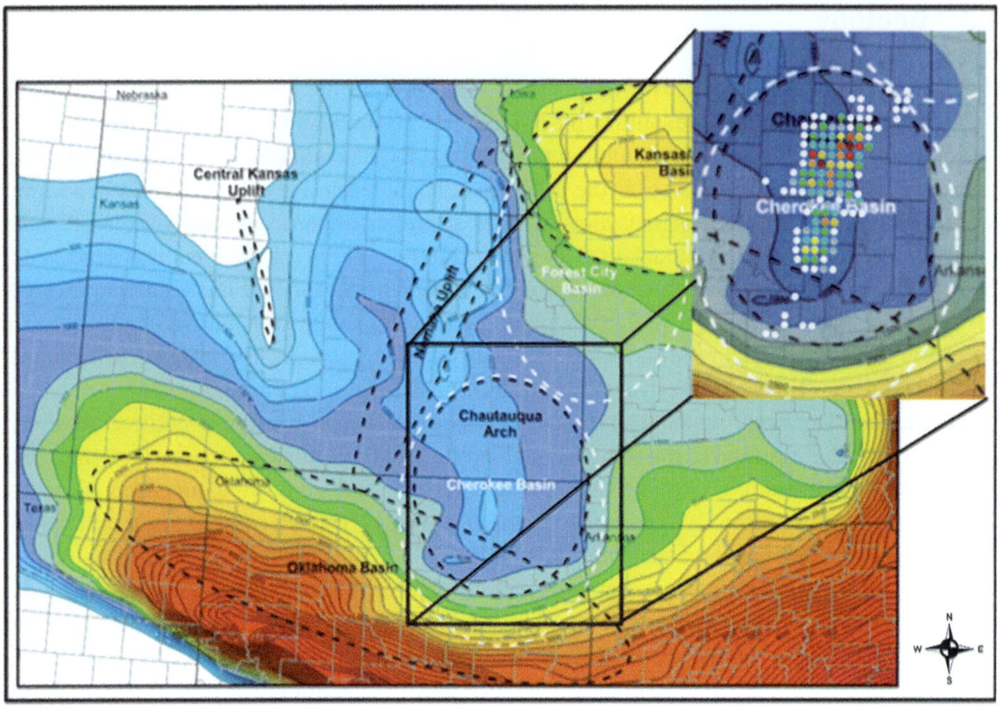

Figure 4.3 Overlay of the production on a township basis over the pre-Pennsylvanian isopach of Paleozoic rocks. The majority of robust production represented by blue, green, yellow, orange, and red dots lie generally in the maximum thin area of the isopach in the Cherokee Basin.

Pennsylvanian sediments were deposited across the Paleozoic Western Interior Basin on a karsted Mississippian surface that had been subaerially exposed (Wanless, 1964; Stanton, 1984). The Pennsylvanian age Morrowan sediments were deposited across much of the Mid-Continent but not the Cherokee Basin. However, to the north in the Forest City Basin several hundred feet and in the Arkoma Basin several thousands of feet of Morrow sediments were deposited (Tedesco, 2014). The Morrow delta in the Forest City Basin did not cross the Bourbon Arch, the northern boundary of the Cherokee Basin (Stanton, 1984). The earliest sediments in the Cherokee Basin were of Atokan age but were limited to the Riverton coal and associated shale.

CBMCM reservoirs represent two very different rock types and were deposited in considerably different environments, and each has unique compositional characteristics. The CBMCM reservoirs in the Cherokee Basin are found in the Atokan and Desmoinesian Cherokee Groups of Pennsylvanian Desmoinesian age (Figure 4.4). The productive coals, in ascending order, are the Riverton, Bw, Aw, Rowe, Drywood, Dbj, Cbj, Bbj, Abj, Weir-Pittsburg, Tebo B, Tebo, Scammon B, Scammon A, Mineral, Fleming, Croweberg, Bevier, Iron Post, Mulky, Summit, and Mulberry (Tedesco, 2014) (Figure 4.4). The carbonaceous mudstones in ascending order are the Bbj, Tebo B, Tebo, V, Excello, Little Osage, and Anna (Figure 4.4). The coals that actually produced consistently are in ascending order: Riverton, Rowe, Drywood, Weir-Pittsburg, and Mulky (Tedesco, 2014). The primary carbonaceous mudstone is the Excello.

These thin coals and carbonaceous mudstones have lateral continuity across the basin and were deposited across a very stable and relatively flat platform that extended from the Cherokee Basin to

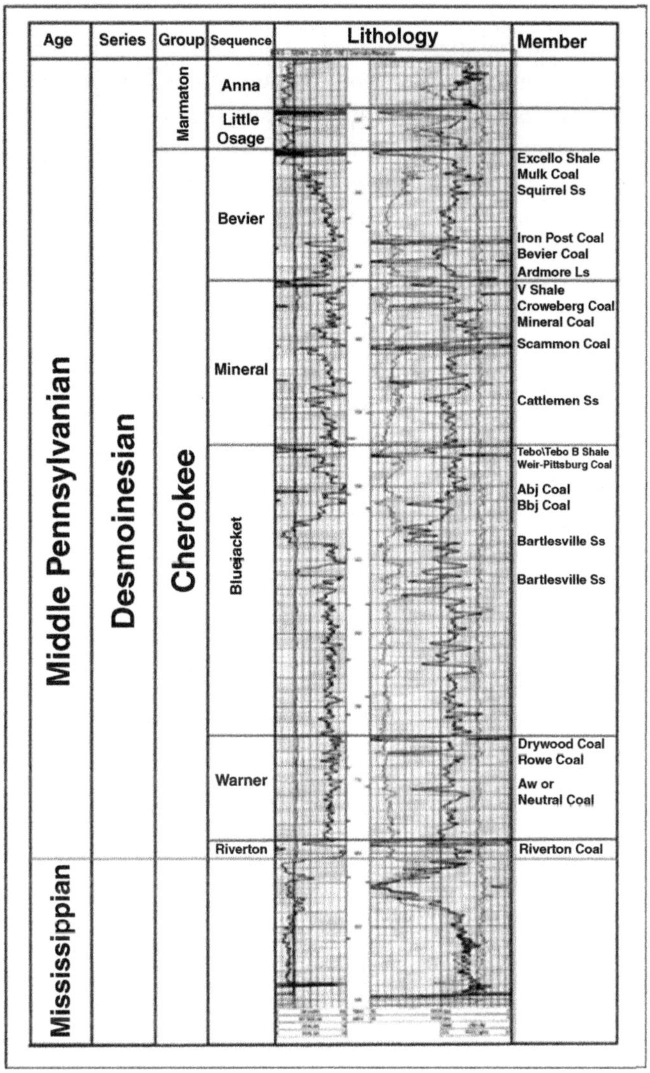

Figure 4.4 General stratigraphic column for the Atoka and Cherokee age strata for the Cherokee Basin (Tedesco, 2014).

the eastern portion of the Illinois Basin (Wanless, 1964). The source of clastic material for the basin is thought to be the Precambrian Shield to the north and the Nemaha Ridge to the west (Merriam, 1963; Wanless, 1964); The Ozark Uplift to the east was a subtle feature and did not seem to be undergoing extensive erosion (Merriam, 1963; Wanless, 1964).

The Desmoinesian strata is divided into the Cherokee and Marmaton groups. The Cherokee Group in the basin is dominated by transgressive and highstand system tracts (Lange, 2003; Johnson, 2004). Walton (1995) completed a general study relating sequence stratigraphy to sandstone reservoirs in the Oklahoma part of the Cherokee Basin that indicated that numerous sequences exist, based on paleosols and basal contacts of down-cutting channel systems (Tedesco, 2014). Walton (1995) defined the sandstone reservoirs in the Cherokee Group as a series of valley-fill fluvial or estuarine deposits. Deposition of deltaic and near-shore marine sediment in the Cherokee Basin has been interpreted to be the product of accommodation space created by relative sea-level rises rather than significant subsidence (Tedesco, 2014). The coals, paleosols, carbonaceous shale, or thin marine limestones are types of markers that are widespread across the basin as such the correlation of individual sandstones became simpler and more realistic (Walton, 1995; Tedesco, 2014).

Desmoinesian (Pennsylvanian) strata of the Cherokee Basin coals and carbonaceous mudstones were deposited in restricted near-shore, coastal, or lacustrine environments (Stanton, 1984; Tedesco, 2014). The shale-dominated facies were considered prodeltaic, near-shore marine, transgressive system tract deposits, or final deposits in a valley-fill sequence (Tedesco, 2014). The Cherokee Group lacks any significant carbonate deposits except the thin Ardmore limestone. The overlying Marmaton Group is dominated by shale and marine carbonates and significantly less sandstone. Many of the CBMCM in the Desmoinesian strata are thin and laterally extensive that suggest a relatively flat coastal or lower delta plain which was easily subjected to marine or tidal inundations over a large area (Tedesco, 2014). The thin coals suggest short periods of mire development or a lack of accommodation space for organic matter accumulation. The coals have a high ash content indicating an unfavorable and rapidly changing conditions for organic sediment accumulation during the time of mire development (Tedesco, 2014). Tedesco (2014) suggested that because of the macerals present, implied forest fires were common. A tidal-influenced coast prograded into the basin westward, which resulted in subaerial exposure and pedogenic processes.

The coals in the Cherokee Basin are high-volatile A. The shallow depth of burial of the basin would preclude this high rank for the coals and therefore another mechanism has to be determined (Coveney, 1992; Coveney and Sangster, 1995; Wojcik et al., 1995; Newell, 1997; Newell et al., 2004). The presence of the Chautauqua Arch and the removal of several thousand feet of strata during its formation placed the Precambrian basement closer to the CBMCM reservoirs closer to greater hotter thermal source. Numerous plutons in the Precambrian basement-based aeromagnetic data are thought to have been emplaced under the Chautauqua Arch sometime in the past, which would provide heat and drive low temperature hydrothermal fluids through the basin (Coveney, 1992; Coveney and Sangster, 1995; Wojcik et al., 1995; Newell, 1997; Newell et al., 2004; Tedesco, 2014; Newell and Birdie, 2021). The drive for the LTHF would be the adjacent Ouachita-Marathon Orogeny. The orogenic event also borders several unconventional gas productive basins such as the Appalachian, Warrior, Arkoma, Anadarko, Barnett, and Permian. The Ouachita-Marathon event would have been the main driver of LTHF during the Late Pennsylvanian-Permian when it underwent compression. The coals contain sphalerite and galena in their cleats (Tedesco, 2014). Several MVT deposits exist across the basin with the largest being the Tri-State district located on the east side (Coveney, 1992; Coveney and Sangster, 1995; Wojcik et al., 1995; Newell, 1997; Newell et al., 2004; Tedesco, 2014). Some of the gas within the CBMCM reservoirs is reported to be partially microbial (Newell et al., 2004). The microbes would benefit from a warmer environment in terms of easting organic matter and expelling more CH_4.

4.2.1 Desorption, Adsorption, and Proximate Analysis

Desorption, adsorption, and proximate analysis data for the Cherokee Basin is presented in Table 4.1. The Cherokee coals are of high-volatile A rank based on proximate analysis. The coals are low in fixed carbon and rich in inertinite and fusinite suggesting that the mires were inundated by tidal and fluvial processes that deposited clay in the seams and there were numerous forest fires (Tedesco, 2014). The coals in the basin are typically thin and generally less than 2 ft (0.65 m) thick. The coals have moderate sulfur. The desorption and adsorption gas contents are typically marginal compared to other basins with Pennsylvanian coals (Tedesco, 2014). The main coals are the Riverton, Rowe and Drywood, and Weir-Pittsburg, which have good to poor desorption and adsorption data.

Table 4.1 Desorption, adsorption, and proximate analysis of the coals and carbonaceous mudstones from the Cherokee Basin (Tedesco, 2014).

	Unit	Desorbed gas content SCF per ton DAF median	Absorbed gas content SCF per ton DAF median	Fixed carbon	Moisture	Volatile matter	Ash	Sulfur	BTU DAF
Carbonaceous Mudstones	Little Osage	118.6	144.17	7.9	3.6	14.7	75.1	1.49	10,876
	Excello	97.9	165.95	10.3	3.1	12.3	74.5	1.97	11,831
	V	102.9	124.94	8.5	3.2	18.5	72.8	1.41	11,541
	Tebo & Tebo B	56.4	Not available	22.7	3.8	20.1	48.4	3.37	13,170
Coal	Mulky	138.9	Not available	43.8	1.9	36.4	15.8	4.24	14,932
	Bevier	130.5	212.08	32.4	2.6	27.1	32.7	1.97	14,298
	Croweberg	129.9	Not available	41.2	2.7	35.0	22.9	2.74	15,241
	Mineral	172.7	120.14	33.4	3.6	27.8	32.3	3.13	14,458
	Scammon	107.0	133.91	28.9	5.0	26.0	34.6	5.53	14,829
	Weir-Pittsburg	98.3	207.92	42.6	2.1	35.8	18.1	4.76	15,143
	Drywood	149.7	138.24	41.1	3.4	32.0	23.3	4.63	15,086
	Rowe	123.9	157.62	44.5	2.9	33.7	17.7	5.31	14,929
	Aw	121.7	116.61	43.7	4.2	34.1	18.8	5.64	14,985
	Riverton	121.8	157.62	44.7	2.6	34.1	17.2	6.24	14,994

The Mineral coal is typically not productive but has one of the best desorption and adsorption results. The other coals are generally non-productive across the basin (Tedesco, 2014). Gas is typically produced in every well from day 1 of production. It can be assumed that much of the gas produced in first years is from the fractures and pores in CBMCM reservoirs.

The carbonaceous mudstones have good to poor desorbed and adsorbed gas content (Table 4.1). The Excello is the primary productive shale, whereas the Little Osage, V and Tebo, and Tebo B rarely produce gas (Tedesco, 2014). The carbonaceous mudstones are of marine origin. The Excello shale extends from the Denver Basin in Colorado, across the Cherokee and Forest City basins in the central USA into the Illinois Basin to the east (Tedesco, 2014). The Excello represents a major marine event in Pennsylvanian times.

4.3 Production

Production data in the basin was derived from the Oklahoma and Kansas geological surveys. Approximately 6119 wells as of 2011 have been designated coal-bed methane or shale gas wells by both surveys. There are probably an additional 1000+ wells that would have produce gas from coals and carbonaceous shales but were never completed (Tedesco, 2014). Some gas wells produced into a common flow meter, which makes evaluation challenging. A number of wells the actual date of first production may be several months to a few cases years after the date of completion. Reasons is primarily due to the lack of pipeline access, price of gas, cost of using third-party gathering services, or for reasons unknown. Therefore, for this chapter, the data was normalized by evaluating the wells from year two in terms of production (Tedesco, 2014). Production could only be evaluated on a yearly basis as for many wells daily or monthly production was not available.

The wells used for evaluation of well production are indicated by the blue dots in Figure 4.5. These are producing wells as of 2011 and CBMCM producers (Tedesco, 2014). Many of these wells have been actually plugged recently under the orphan well program. There are at least 200 CBMCM wells in Osage County, Oklahoma, but this area is managed by the Osage Tribe and BIA and the data is not available. Figure indicates that there is generally a productive area in the central part of the basin extending in a north-south direction (Tedesco, 2014). The total production on a township basis (a township is six miles by six miles) as of December, 2011 (Figure 4.6). The better production to date is located in portions of Neosho, Wilson, Montgomery, and Labette

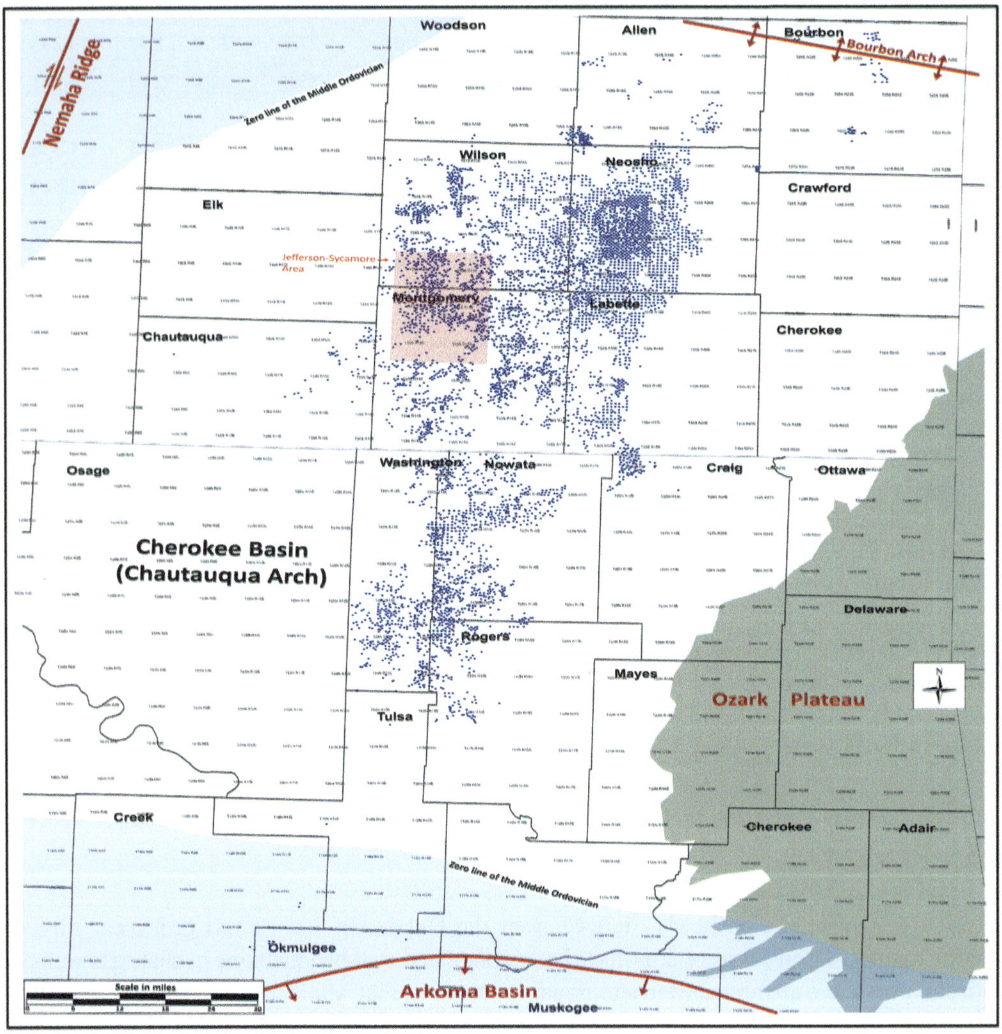

Figure 4.5 The location of wells (blue dots) CBMCM producing wells (blue dots) in the Cherokee Basin (modified from Tedesco, 2014). The red hatched area is the detail CBMCM area discussed in the text.

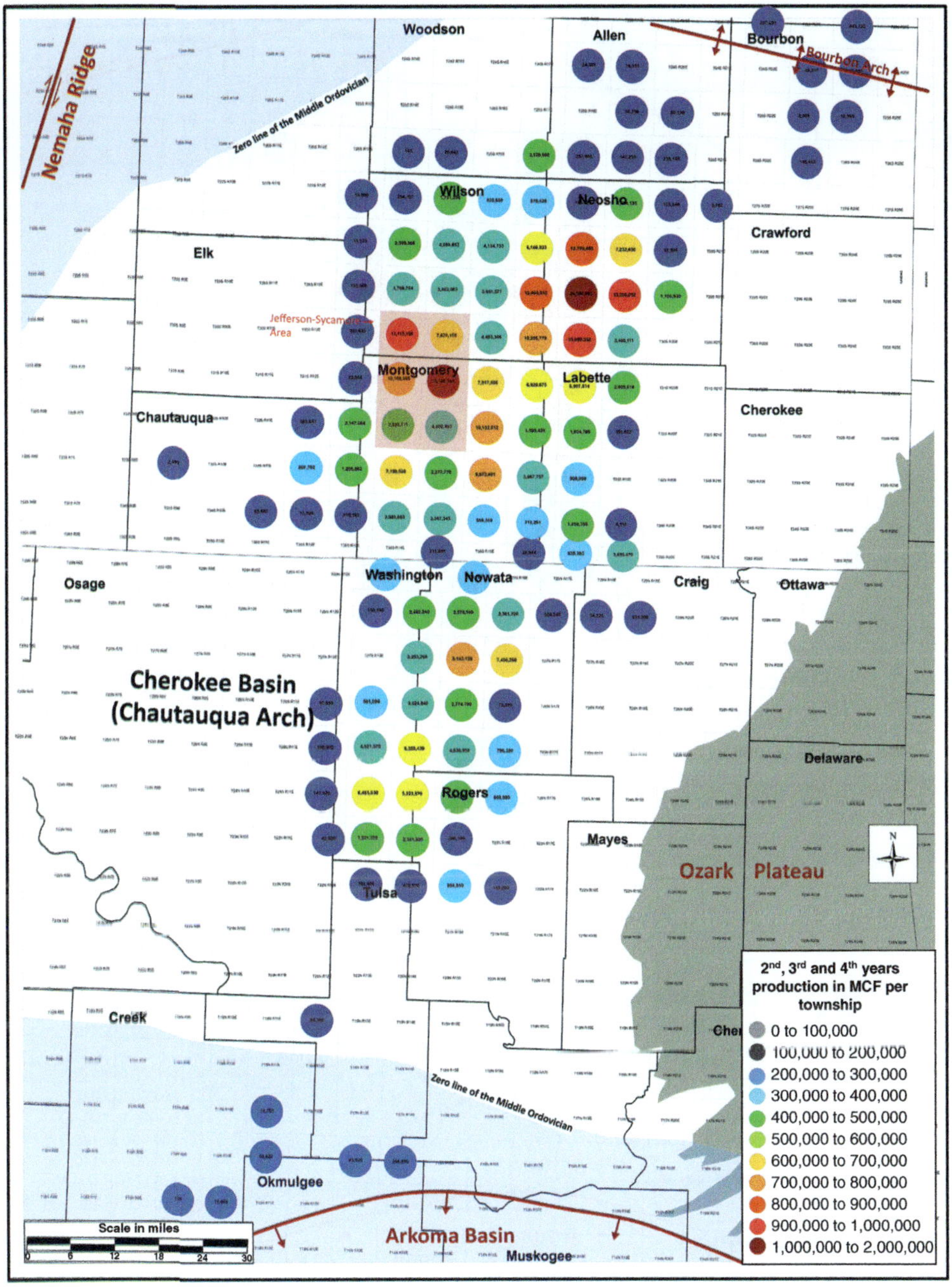

Figure 4.6 Total production in the Cherokee Basin from CBMCM wells on a township basis. The number in the circle corresponds to the actual gas produced as of December, 2011 (modified from Tedesco, 2014).

counties, Kansas and Washington, Rogers, and Nowata counties, Oklahoma. There has been a considerable amount of testing around the basin and the absence of production in many areas reflect the lack of gas, poor reservoir quality, lack of pipelines, or lack of reservoir in certain areas (Tedesco, 2014). Figure 4.7 is production on a per well basis. There is no particular pattern or trend in terms of areas that uniformly are productive (Tedesco, 2014). The Cherokee Basin

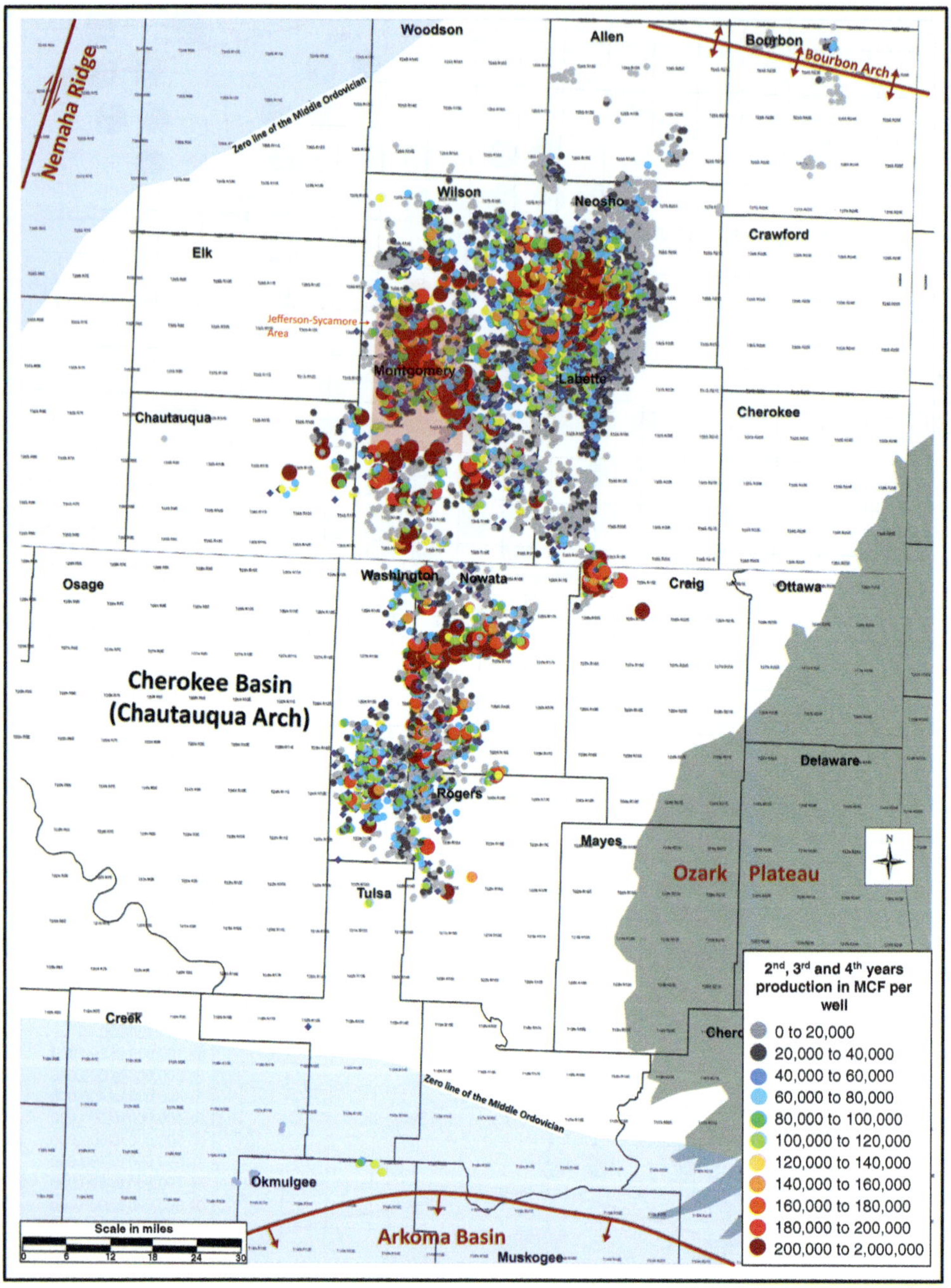

Figure 4.7 The total CBMCM production for each well in the Cherokee Basin. The total production has not been normalized and is not related to when the wells started production (modified from Tedesco, 2014).

CBMCM play is typical of many unconventional plays in that there is a percentage of the wells that are exception in terms of production whereas the majority are marginal to subeconomic. The difference in good to exceptional production (greater than 140,000 MCF) to some degree is related to completion techniques as will be discussed below.

Figure 4.8 indicates the either a well had multiple completions or a single zone was completed. Oklahoma area of the CBMCM Play is dominated by single zone completions. This reflects a decrease in coals deposited or were removed by later fluvial erosion across the delta (Tedesco, 2014). As a result, the Oklahoma part of the play is typically dominated by single zone completion. In some areas, such as southwestern Allen County, Kansas multiple Warner, Bartlesville, Weir, Peru, Chelsea, Squirrel, etc., sandstone deposits are stacked on top of each and have removed many of the seams (Tedesco, 2014).

The average gas profile for a CBMCM is presented in Figure 4.9. The production curve is more typical of a conventional well decline. This suggests the gas is not being desorbed from the organic

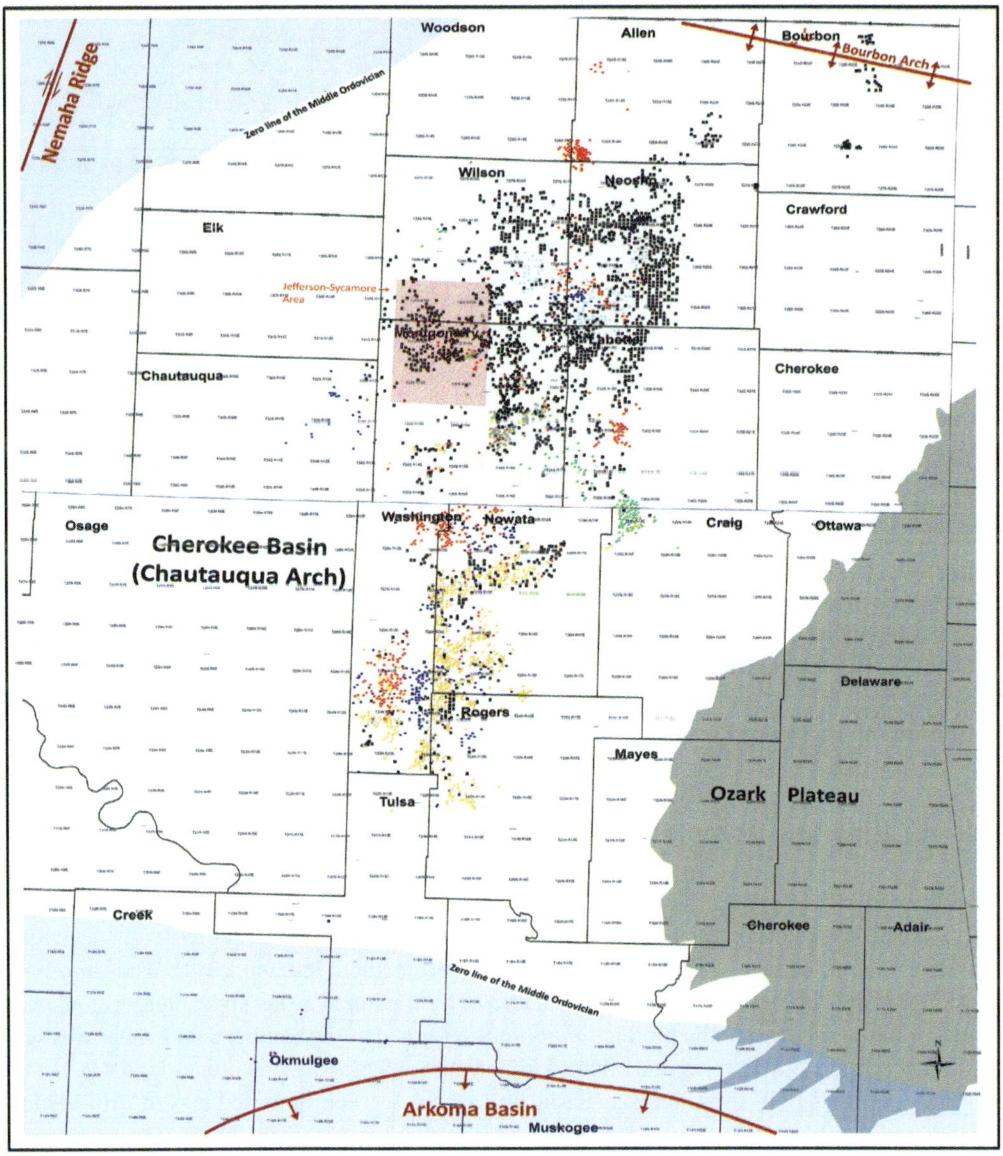

Figure 4.8 The productive zone for each well as based on data submitted by an operator to the State of Kansas and Oklahoma (modified from Tedesco, 2014).

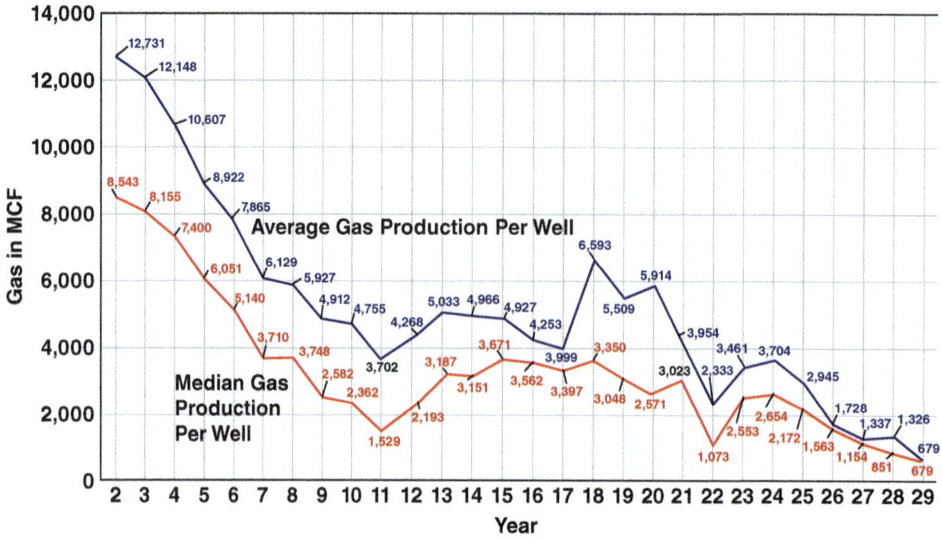

Figure 4.9 This is the average production decline curve for all the wells within the basin (Tedesco, 2014). The average of the data is indicated in blue and the median in red. Many of the wells saw some sort of restimulation of existing zones and/or opening of new zones. The production curve is more typical of a conventional well than a typical CBMCM well that requires dewatering before gas starts desorbing and flowing (Tedesco, 2014).

matter during water removal but actually is present in the pore and fracture spaces of the reservoir. The CBMCM reservoirs produce very little water compared to other CBMCM plays. A typical coal-bed methane well usually has a period of dewatering then gas begins desorbing and flowing to the wellbore. The later peak in Figure 4.9 in many wells was either restimulation of existing zones or adding of new zones in the borehole (Tedesco, 2014).

The total production by township is overlain on top of the pre-Pennsylvanian thickness map as indicated in Figure 4.3. The production generally matches a thinning of Pre-Pennsylvanian strata indicating a possible relationship between the location of production and a geologic anomaly. There has been no economic or significant gas production from CBMCM reservoirs outside the present area of production (Tedesco, 2014).

4.4 Drilling and Completion Methods

Prior to 1960, drilling and completion practices in the Cherokee Basin for producing gas from coal and carbonaceous mudstone reservoirs included drill to the top of the target reservoir, set casing with only minimal cement, and then cable tool a few feet further (Tedesco, 2014). Wells may have been stimulated with nitroglycerine or allowed to produce naturally. Gas and water were allowed to flow to the surface inside the production tubing, separated at the wellhead and water was sent down the backside of the tubing and allowed to flow into any porous zone (Tedesco, 2014).

Drilling technology in the Cherokee Basin utilizes small truck mounted rigs to drill vertical wells to total depth with air or an air mist as a drilling medium (Tedesco, 2014). Horizontal wells have proven unsuccessful in increasing production. Reservoir damage is minimal drilling methods with air or air mist as drilling medium and well bore stability is generally good as the rocks are

Pennsylvanian in age (Tedesco, 2014). Many drill rigs in the area cannot run pipe and a well may be left open for several weeks before casing is cemented. While this might not be optimal, there is anecdotal data that suggests this may affect production since services were in short supply from 2000 to 2008 (Tedesco, 2014).

Acid is used prior to the stimulation to clean-up cement damage and connect to the wellbore to the CBMCM zone. Completion methods begin in the basin in 1990 with the use of cross-link gel with 20,000 to over 40,000 pounds of sand per two-foot coal (Tedesco, 2014). Cross-link gel fracture stimulations is able to carry higher rates of proppant into the zone whereas slick water has difficulty doing that. There was no attempt to keep the stimulation in the zone. As the industry developed completion methods for CBMCM in the later part of 1990s into 2000s, the trend was to use slick water fracture stimulations with rates of 8 to 12 barrels of fluid per minute with 1000 to 2000 pounds of sand per foot (Tedesco, 2014). The Cherokee Basin is an under-pressured structural feature and this presents a challenge when fracture stimulating a reservoir. Using low fluid injection rates during stimulation with slick water attempts restrain the stimulation to remain in the CBMCM zone. Acid stimulations, usually hydrochloric (HCL), were rare in the basin as the concern is that the fluids will dissolve non-organic matter within the CBMCM zone. The acid would dissolve carbonates and release them into the gas and fluid flow causing problems with production (Tedesco, 2014).

Prior to 2000, most CBMCM wells were completed in a single zone. Post 2000, a number of operators begin completing multiple zones in new wells and/or recompleting additional zones in existing wells (Tedesco, 2014). The incremental gas that was added due to recompletions was highly variable. Many of the minor coals did not contain substantial gas (Tedesco, 2014).

4.5 Jefferson-Sycamore Area

The Jefferson-Sycamore area is located in Townships 30, 31 and 32 South, Ranges 14 and 15 East, Kansas (Figures 4.5–4.8 and 4.10). The Jefferson-Sycamore area is one of the oldest areas where the most recent exploitation phase started in Kansas in the 1990s and one of the most prolific CBMCM areas in the Cherokee Basin (Tedesco, 2014). There are approximately 610 wells in the Jefferson Sycamore area with completion data. Unless otherwise indicated the median is used for discussion purposes rather than the average which may be skewed by a few anomalous data points. Part of the Jefferson-Sycamore area was not developed for CBMCM and this may be due to topography, lack of interest by landowners, geologic reasons, perceived lack of gas in CBMCM reservoirs, decline in gas prices, or a number of other unknown reasons.

There are six operators in the Jefferson-Sycamore area and their median gas decline curves are indicated Figure 4.11. Stroud Oil Properties, Jones Gas, and Great Eastern began producing in the 1990s and their completion methods were initially dominated by one to two zone completion with cross-link gel (Tedesco, 2014). Layne Operating, Dart, and Bluejay Operating began drilling and producing in approximately 1998 to 2002, completed multiple zones, and which resulted in production profiles that have a more rapid production decline curve (Tedesco, 2014). The decline curves are more typical of conventional gas wells than CBMCM wells in other geologic basins. Around year 12, Jones Gas and Great Eastern either recompleted existing zones or opened other zones in each well (Tedesco, 2014).

As will be discussed below, Stroud had the best production for completing a single zone.

The CBMCM in the area started the focus solely on the Weir-Pittsburg coal and the standard at the time was fracture stimulating with cross-link gel and 20,000 to 40,000 lbs of sand (Tedesco, 2014). As

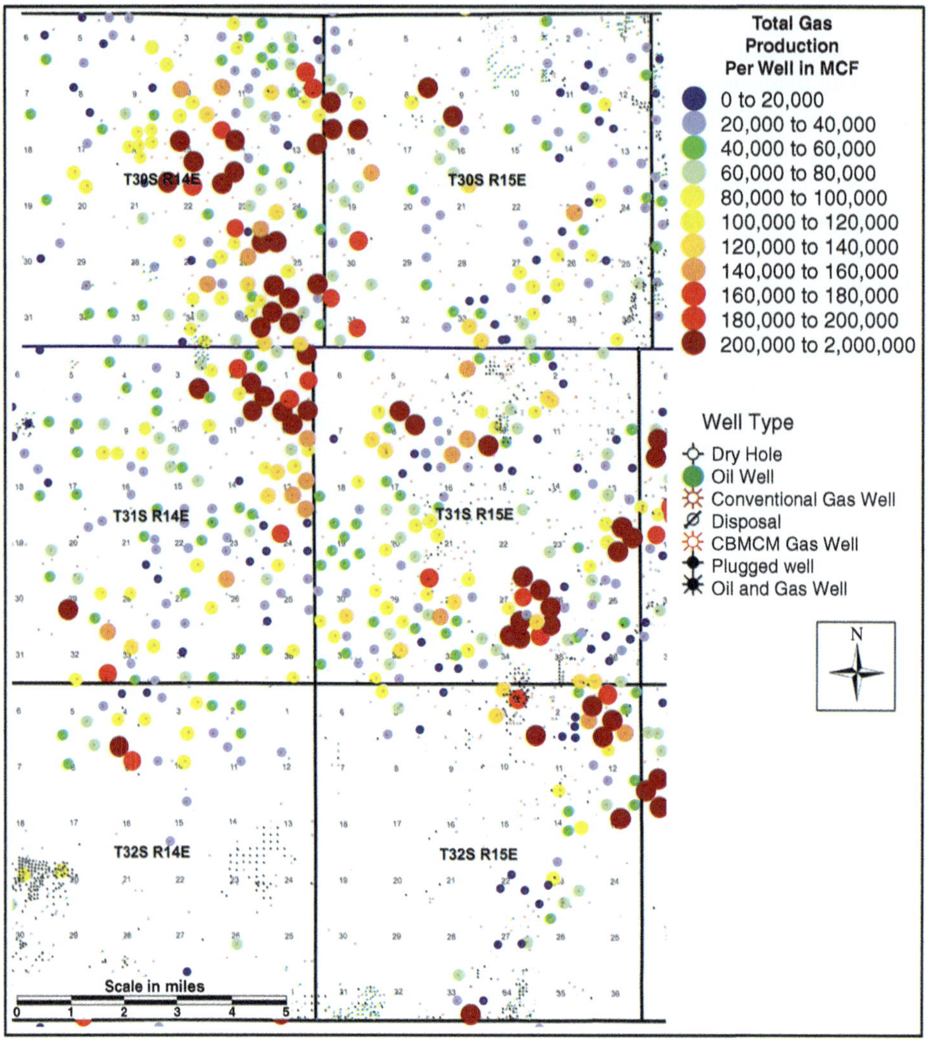

Figure 4.10 Jefferson-Sycamore area in the Cherokee Basin with CBMCM wells.

the exploitation of CBMCM in the area expanded, several operators began completing multiple coals and shales either individually or as package in a well (Tedesco, 2014). Nitrogen was attempted in a small number of fracks, had better production results but the costs were not economic to pursue as a standard completion practice (Tedesco, 2014). Geologic mapping on the top of the Mississippian and Cherokee strata with the production total for the second, third, and fourth years in the Jefferson-Sycamore area (Figures 4.12 and 4.13). There does not seem to be any correlation between structural highs or lows (Figures 4.12 and 4.13). There is a potential northeast-southwest trending fault in the northwest area in Figures 4.12 and 4.13 and could potentially enhance fracturing and thus gas rate. Based on the present production data, there is no definitive relationship between the fault and higher CBMCM production (Tedesco, 2014).

Isopach maps of the Riverton and Rowe coals (Figures 4.14 and 4.15) with the total CBMCM for production total for the second, third, and fourth years do not indicate any discernable relationship with thicks or thins of either seam. The Riverton and Rowe zones were typically completed in wells

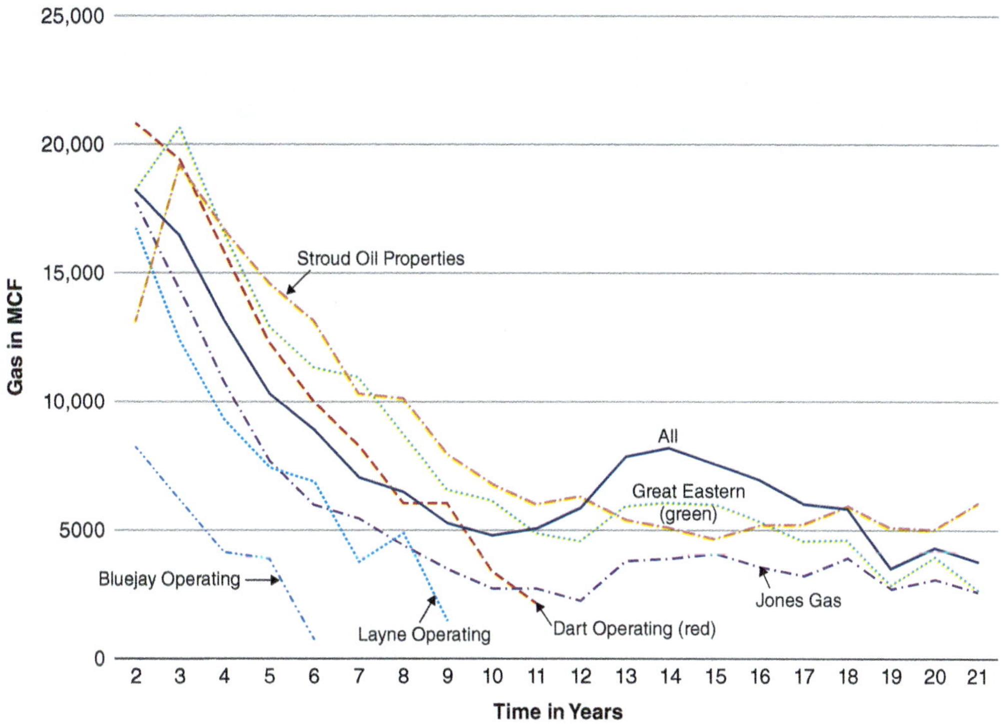

Figure 4.11 Gas decline curves from year 2 to year 21 in the Jefferson-Sycamore area (Tedesco, 2014 / Steven A. Tedesco).

where multiple reservoirs were completed (Tedesco, 2014). The isopach map of the Weir-Pittsburg coal (Figure 4.16) with the total CBMCM for the second, third, and fourth years of production overlain, and there is a stronger correlation with production than with the Riverton and Rowe coals. However, several areas of thick Weir-Pittsburg are not very productive. This suggests there may be other factors for why the gas is productive from CBMCM reservoirs. These factors can include but are not limited to: geologic factors, the number of zones that are open, the zones are completed individually or in packages, no gas in the zone, etc.

The first stimulation method used in the area was cross-link gel and 20,000 to 40,000 lbs of sand for a single zone. As the exploitation boom expanded, this changed to slick water with 2000 lbs of sand per foot with low rates of water injection (Tedesco, 2014). Table 4.2 is the summary of the fracture stimulation components used by the operators in the Jefferson-Sycamore area. The Cherokee Basin is an under-pressured basin with a 0.34 psi per foot gradient. Therefore, fracture stimulation has to attempt to stay within zone by using low rate per minute of fluid entry or the proppant will leave zone (Tedesco, 2014). This rate, based on the authors experience, is 8 to 10 barrels of fluid with proppant per minute. Nitrogen had the best results but was not economic based on the resulting production and gas price at the time. Cross-link gel is typically at rates of 30 barrels per minute or more and will leave the CBMCM reservoir and fracture stimulate some amount of overlying geologic section (Tedesco, 2014). The cross-link gel stimulation did better than slick water based on Table 4.2.

Table 4.3 is the cumulative production (median and not the average) for one to seven zones completed. The best production was seven zones completed individually. This essentially would fracture

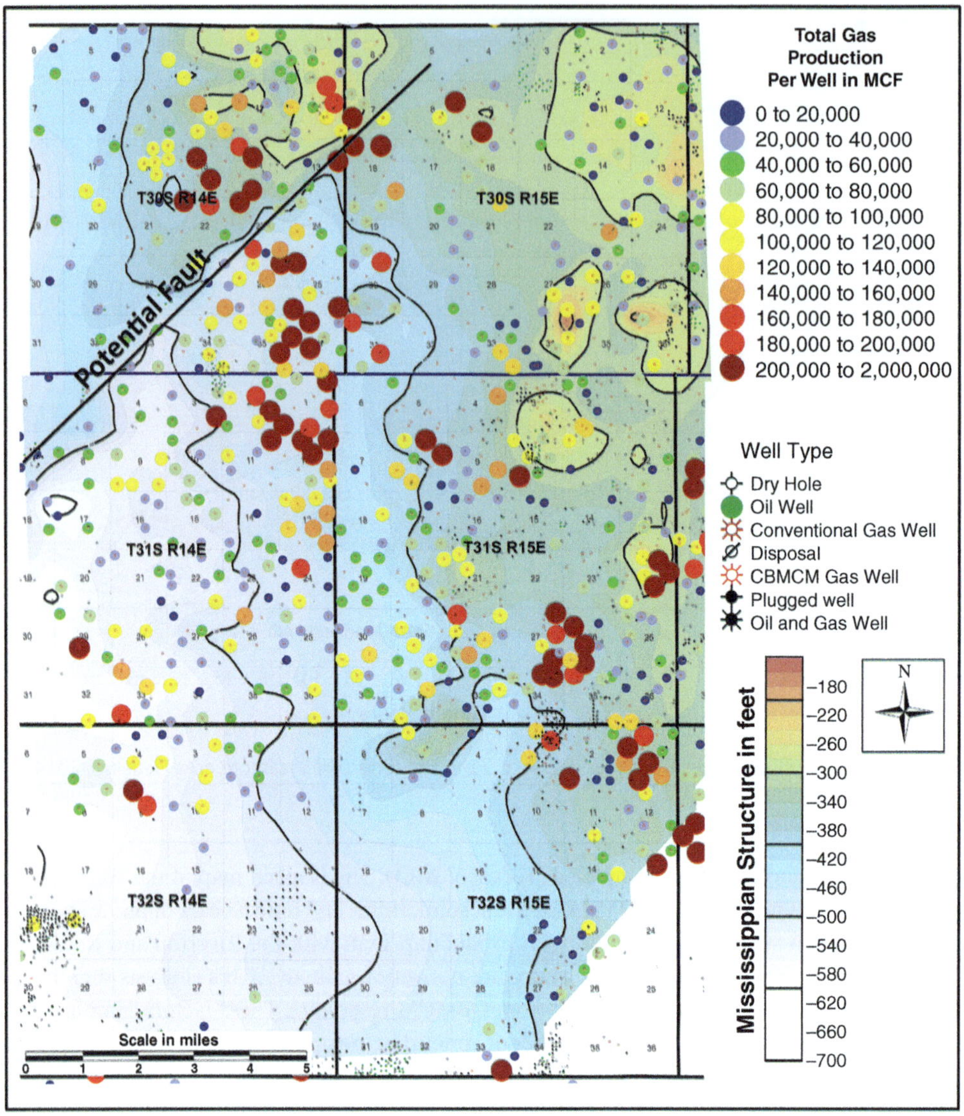

Figure 4.12 Mississippian structure map with a classed post of the total CBMCM for the second, third, and fourth years of production in the Jefferson-Sycamore area (modified from Tedesco, 2014).

stimulate the entire Cherokee and potentially the lower Marmaton Group section (Tedesco, 2014). The second optimal completion was fracture stimulation of single zone, usually the Weir-Pittsburg, which typically used cross-link gel with a higher amount of sand for completion (Tedesco, 2014). There were several wells that completed only the Riverton and Rowe, which was done by either cross-link gel or slick water that also had positive results (Tedesco, 2014). The cumulative production for the second, third, and fourth years' shows that completion of five to seven zones had better production than the one zone for the same period. However if compared to Figure 4.11, with the knowledge that Stroud Oil Properties had a more robust decline curve because it focused on single zones, multiple zone completion had higher costs and less return in terms of profit (Tedesco, 2014)

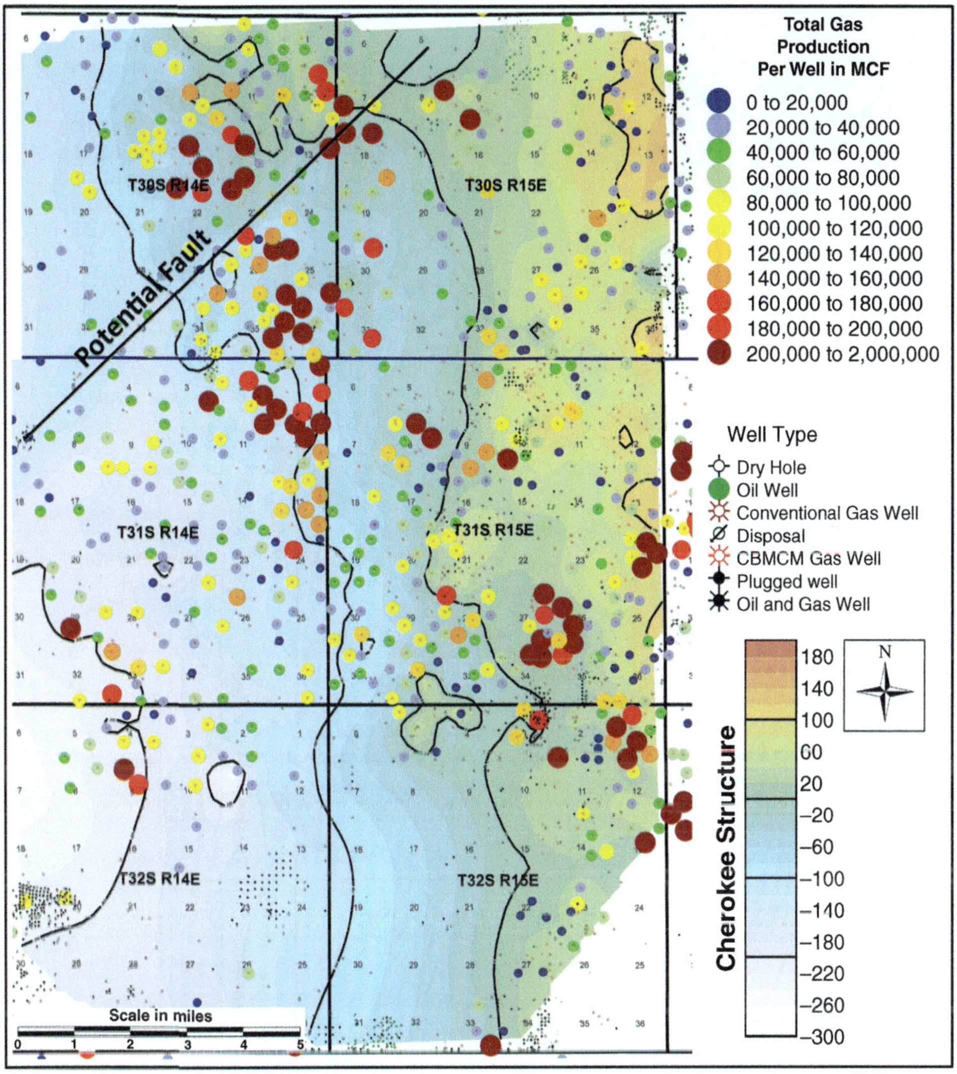

Figure 4.13 Cherokee structure map with a classed post of the total CBMCM for the second, third, and fourth years of production in the Jefferson-Sycamore area (modified from Tedesco, 2014).

Table 4.4 is a comparison of production from various zones, various components of the fracture stimulation and zones either present or not present in completion. The best zone was a single Weir-Pittsburg coal, which typically was stimulated by cross-link gel. The Weir-Pittsburg also had the best daily production during the second, third, and fourth years (Tedesco, 2014). The worst productive zone was the Excello and Osage shales in the Jefferson-Sycamore area (Tedesco, 2014). The majority of the wells were completed with slick water, which resulted in lower daily production rates. The total production makes comparison difficult because many of the wells have produced for different length of times (Tedesco, 2014).

Table 4.5 is a comparison between operators in terms of types of fracture stimulation and resulting production. The problem with using the average for statistics it tends to be biases towards anomalies or isolated wells that have really stellar or poor results but do not represent the overall sample population.

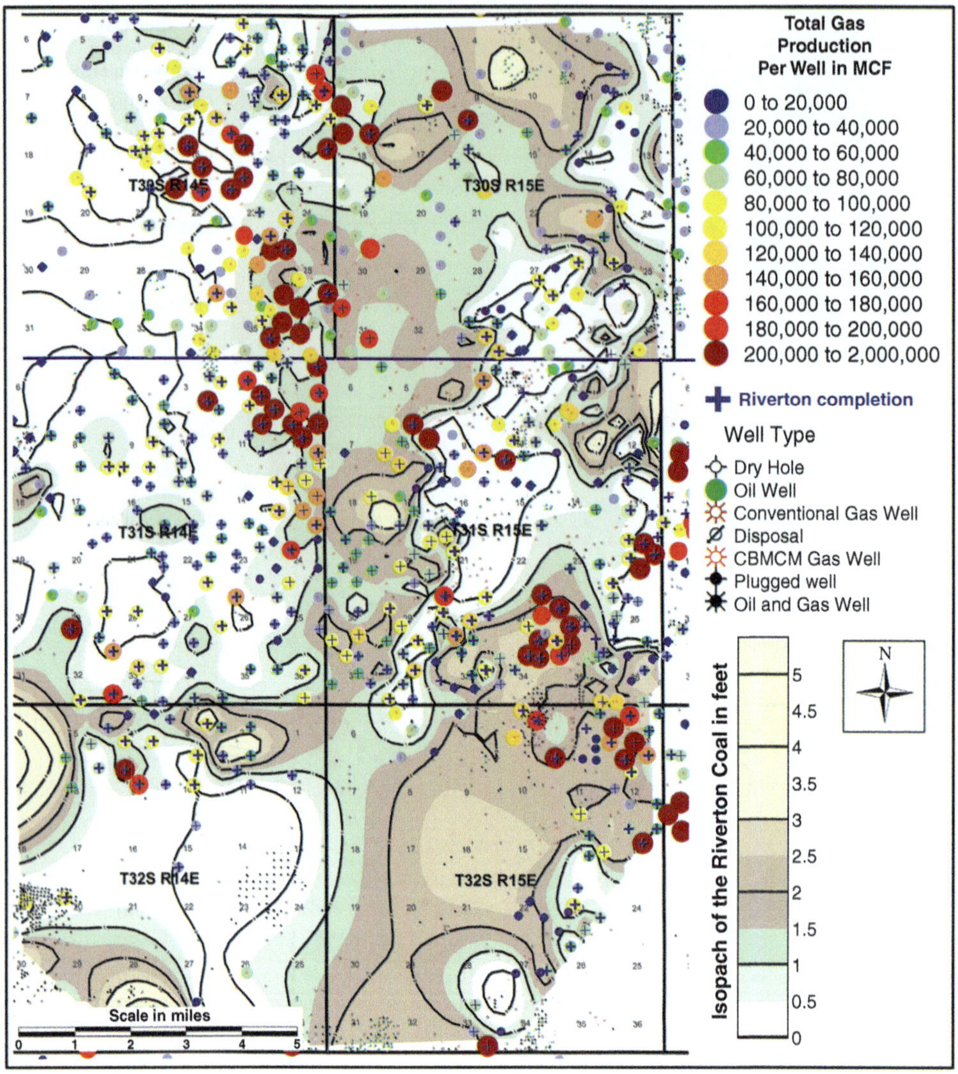

Figure 4.14 Isopach of the Riverton Coal with a classed post of the total CBMCM for the second, third, and fourth years of production in the Jefferson-Sycamore area (modified from Tedesco, 2014).

The cumulative production in Table 4.5 is misleading in that many of the wells produced for different length of time and so analysis is difficult. However, using the second, third, and fourth years' total production can provide a better comparison. The best production was from the Stroud Oil Properties that used cross-link gel to stimulate a single zone. Great Eastern used a combination cross-link gel and slick water. The other operators used primarily slick water and completed multiple zones. The decline rates that were completed in multiple zones resulted in steeper decline rates and potentially lower recovery rates. As stated above, the cross-link gel probably fractured stimulated a good part of the CBMCM reservoirs in a well because the rate of fluid barrels per minute would have fracked multiple zones and placed large volumes of sand in these fractures so they would drain gas from a large area (Tedesco, 2014).

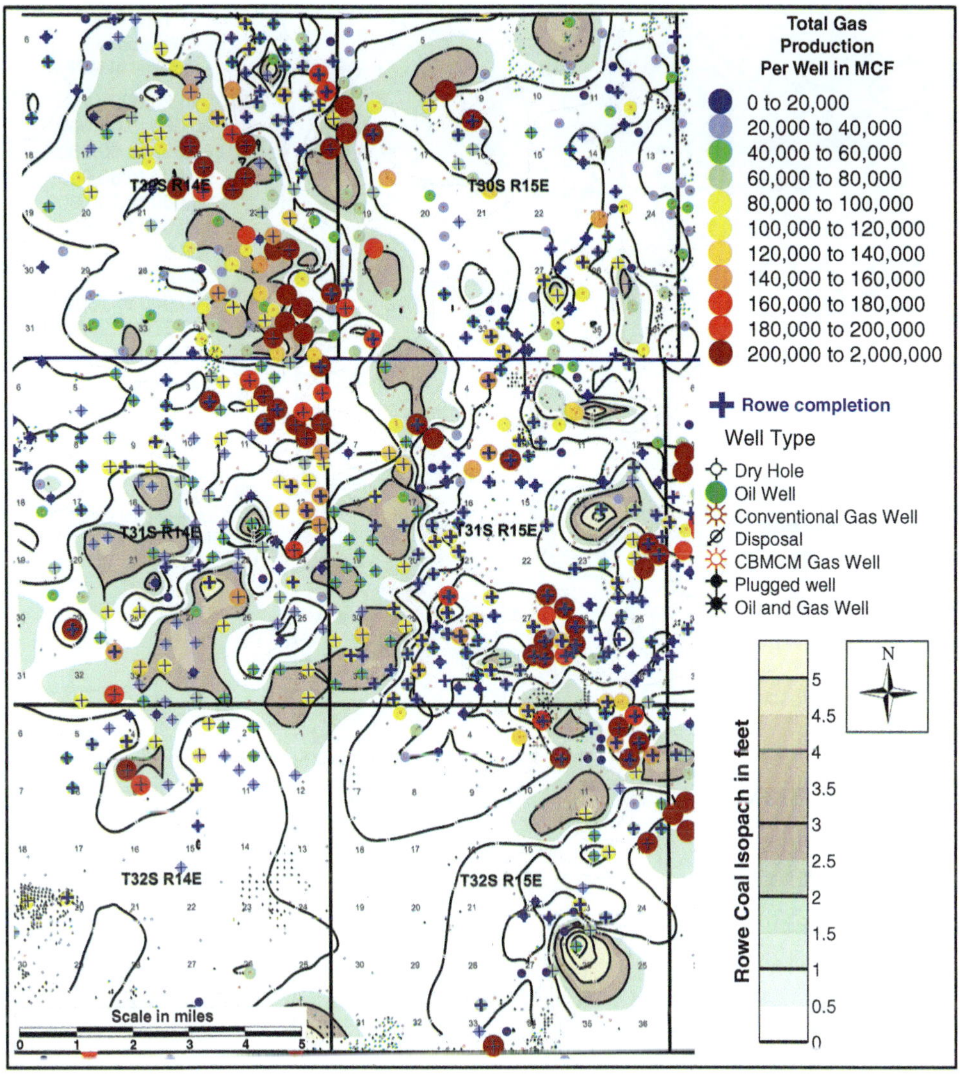

Figure 4.15 Isopach of the Rowe Coal with a classed post of the total CBMCM for the second, third, and fourth years of production in the Jefferson-Sycamore area (modified from Tedesco, 2014).

4.6 Discussion

The productive CBMCM reservoirs in the Cherokee Basin overlie the ancient Chautauqua Arch that formed in Devonian time. Low temperature hydrothermal fluids (LTHF) migrated from the Arkoma and Anadarko basins to the west and south as well as from the Precambrian basement underlying the Cherokee Basin that thermally matured the CBMCM to generate gas. The presence of sphalerite and galena in the coal cleats and presence of several MVT deposits including the world class Tri-State Zinc district support the presence of the LTHF. The CBMCM production is relatively erratic, which is consistent with the results of many other CBMCM plays in other geologic basins. The carbonaceous mudstones, the Excello, V, and Tebo are generally laterally extensive, but only

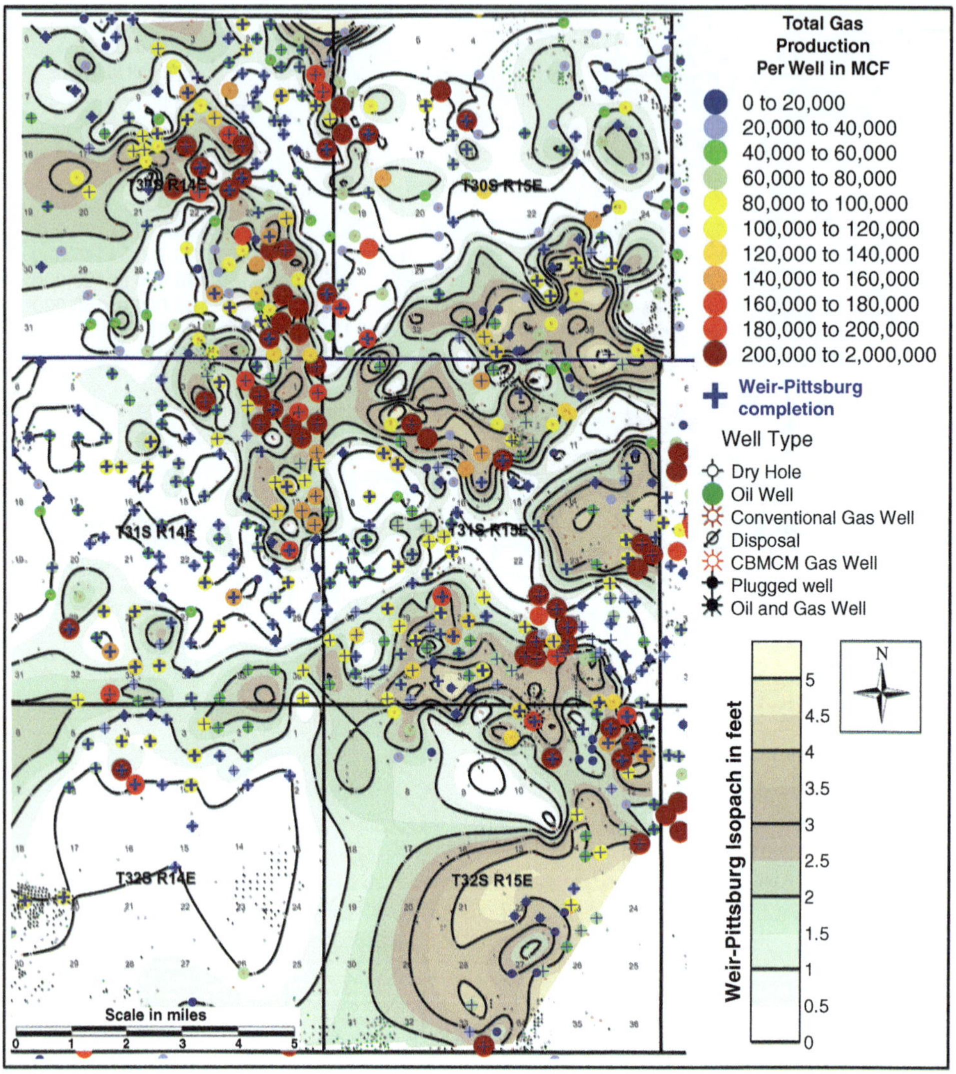

Figure 4.16 Isopach of the Weir-Pittsburg Coal with a classed post of the total CBMCM for the second, third, and fourth years of production in the Jefferson-Sycamore area (modified from Tedesco, 2014).

the Excello produces in some select areas. The CBMCM reservoirs production profile is more similar to a conventional gas well than a typical CBMCM reservoir in other basins. As a result, the desorption data for the CBMCM reservoirs in the Cherokee Basin contain a significant amount of gas in the fracture system. Dewatering was never an issue as the coals produced very little compare to other CBMCM plays.

Only certain reservoirs seem to be commercial. Specifically, the Riverton, Rowe, Drywood, Weir-Pittsburg, and Mulky coals and the Excello shale. The other remaining CBMCM reservoirs completed during multiple zone completion or later in a well's production life did not seem contributed to significant amount of gas. Fracking multiple zones in a well as shown in the Jefferson-Sycamore Field requires individual stimulation to be effective. Five or more fracture stimulations seem to be as effective as using cross-link gel on one zone. The cross-link gel fracture

Table 4.2 The three different types of fracture stimulation and their components with the resulting production used in the Jefferson-Sycamore area

Second, third, and fourth years		Production	Total sand	Total fluid	Daily production	No. of fracks	Total pay	No. of zones	No. of wells
Slick Water									542
	Average	45,991	16,778	49,679	42	4	13	5	
	Median	30,962	15,300	51,366	28	4	12	5	
Cross-link									66
	Average	50,629	23,957	45,360	46	2	7	2	
	Median	31,178	20,750	37,842	28	1	5	1	
Nitrogen									19
	Average	71,401	29,107	18,999	65	4	10	4	
	Median	66,924	28,755	9450	61	5	11	5	

Note: The production column is the total for the second, third, and fourth years of production.
Source: Tedesco (2014).

Table 4.3 Comparison of the resulting total and second, third, and fourth years' production versus the number of CBMCM reservoirs or groups of reservoirs completed.

Component	1	2	3	4	5	6	7
Cumulative production	122,646	75,875	64,839	70,595	106,695	102,877	128,816
Cumulative production second, third, and fourth years	48,191	36,501	36,929	41,432	65,357	54,380	64,456
Sand	12,207	16,952	17,719	16,905	18,525	19,054	17,777
Sand per foot	3577	1798	1250	1493	1576	1465	1275
Fluid	306	842	1175	1236	56,433	1650	2168
Fluid per foot	104	65	75	112	118	135	158
HCL	380	1623	1693	1524	1553	1930	3129
Average total pay	4	14	16	13	12	13	14
No. of fracks	1	2	3	4	5	6	7
No. of zones open	1	4	5	5	5	6	6
Average daily production second, third, and fourth years	44	33	34	38	60	50	59
No. of wells	80	94	117	128	128	50	10

Source: Tedesco (2014).

Table 4.4 Comparison of the production from various zones and zone combinations, the fracture stimulation components and zones either present or not present in completion.

Zone	Single Riverton completion	With Riverton	With Rowe – Drywood	Single Weir – Pittsburg	Excello – Little Osage	With Marmaton – Excello – Little Osage	Without Riverton	Without Rowe – Drywood	Without Weir – Pittsburg	Without Excello – Little Osage	Without Marmaton
COMPONENT											
Sand	7955	15,070	16,402	12,400	8000	15,810	14,740	16,978	15,300	12,400	15,500
Sand per foot	1960	1349	1421	3100	2040	1422	1450	1442	1422	2339	1472
Fluid	9972	1250	1325	307	413	1263	1,018	39,130	1215	389	991
Fluid per foot	2761	95	95	102	138	98	95	4311	95	91	89
HCL	339	1675	1608	250	300	1500	1200	1144	1500	350	1200
Average total pay	3.6	11.0	12.5	3.5	3.5	12.0	9.5	10.2	11.0	11	10.8
No. of fracks	1	4	4	1	1	4	3	3	4	4	3
No. of zones open	1	5	5	1	1	5	4	4	5	5	4
Cumulative production	140,471	62,597	61,138	108,003	37,164	55,339	60,005	50,326	49,802	77,798	61,046
Cumulative production second, third, and fourth years	30,692	31,306	34,209	41,475	14,462	30,962	32,023	26,160	28,368	31,107	29,936
Averages daily production second, third, and fourth years	28	29	31	38	13	28	29	24	26	28	27
No. of wells	12	134	134	49	8	459	553	304	333	116	361

Source: Tedesco (2014).

Table 4.5 Comparison of the different fracture stimulation components for each operator and resulting production in the Jefferson-Sycamore area.

Operator		Bluejay	Dart	Great Eastern	Jones Gas	Layne	Stroud	All
No. of wells		20	302	42	33	166	20	610
COMPONENT								
Sand in pounds								
	Average	26,834	15,795	19,628	19,386	16,857	28,753	17,312
	Median	27,500	14,640	15,000	19,600	16,800	28,000	15,405
	Standard Deviation	9386	7394	12,761	7774	6068	16,451	8531
Sand in pounds per foot								
	Average	1931	1570	2663	2321	962	7288	1703
	Median	2000	1503	1916	2000	885	6750	1430
	Standard deviation	622	558	2101	1242	470	6451	1779
Fluid in barrels								
	Average		1253	812	534	1443	526	1134
	Median		1205	639	548.418	1394	473	1175
	Standard deviation		1141.32	548.418	242.171	584.8	268.28577	617.853
Fluid in barrels per foot								
	Average		125	102	66	83	158	101
	Median		123	85	58	74	158	94
	Standard Deviation		101	64	37	41	86	54
HCL in gallons								
	Average		1267	728	502	2620	386	1595
	Median		1300	413	500	2190	350	1500
	Standard deviation		730	852	221	1755	279	1353
Average total pay (in feet)								
	Average	14.8	10.2	7.4	9.5	19.5	4.9	12.5
	Median	14.0	9.5	6.0	9.0	18.5	4.0	11.0
	Standard deviation	5.0	3.7	4.2	4.6	7.5	4.0	7.5

(Continued)

Table 4.5 (Continued)

Operator		Bluejay	Dart	Great Eastern	Jones Gas	Layne	Stroud	All
No. of fracture stimulations								
	Average	3	4	7	2	3	2	4
	Median	3	5	6	2	3	1	4
	Standard deviation	1	1	4	1	1	1	2
No. of zones open								
	Average	4	5	2	3	7	2	5
	Median	5	5	2	2	7	1	5
	Standard deviation	2	1	1	2	3	1	3
Cumulative production in MCF								
	Average	29,225	94,882	140,495	81,399	62,719	159,027	87,924
	Median	25,482	61,229	96,068	70,205	52,915	192,122	58,848
	Standard deviation	17,167	98,314	143,471	52,212	45,999	105,015	94,774
Cumulative production second, third, and fourth years in MCF								
	Average	16,981	55,387	53,905	43,090	37,648	48,565	46,891
	Median	15,800	33,322	30,246	32,774	31,817	51,512	30,999
	Standard deviation	9671	64,661	73,470	32,591	29,400	37,375	54,042
Average daily production second, third, and fourth years in MCF								
	Average	16	31	49	39	34	44	43
	Median	14	28	28	30	29	47	28
	Standard deviation	9	21	67	30	27	34	49

Source: (Tedesco, 2014).

stimulation would essentially leave the zone and open up several overlying zones for production. So, in a sense this a multi-zone fracture stimulation.

The Cherokee Basin exploration model can be applied to other Pennsylvanian age basins that have gas saturation caused by local thermal anomalies. The opportunity to find CBMCM production in Pennsylvanian strata that do not require significant dewatering is a more economically attractive than waiting a year or more for the gas to desorb from the CBMCM reservoirs.

References

Bartow, E. (1906). Examination of coal dust after explosions in mines of southeastern Kansas. Kansas Academic of Science Transactions 20 (1), pp. 228–229.

Bostic, J., Brady, L., Howes, M., et al. (1993). Investigation of the coal properties and the potential for coal-bed methane in the Forest City Basin: U.S. Geological Survey, Open File Report 93–576, p. 16.

Brady, L.L., Newell. D., & Recoy, K. (2011). 34th forum of the coal geologists of the western Interior Coal Region, Pittsburg, Kansas, Kansas Geological Survey, Open File Reprot 2011-6, 61p.

Cardott, B. J. (1999). Coalbed methane activity in Oklahoma. In: Oklahoma Coalbed-Methane Workshop, Oklahoma Geological Survey Open-File Report, Open File 6-99, p. 20.

Cardott, B. J. (2000). Coalbed methane activity in Oklahoma. In: Oklahoma Coalbed-Methane Workshop, Oklahoma Geological Survey Open-File Report, Open File 2-2000, p. 20.

Charles, H. H. & Page, J. H. (1929). Shale-gas industry of eastern Kansas. American Association of Petroleum Geologists Bulletin 13 (4), pp. 367–381.

Coveney, R. M. Jr. (1992). Evidence for expulsion of hydrothermal fluids and hydrocarbons in the Midcontinent during the Pennsylvanian. In: Source rocks in the Southern Midcontinent, 1992 Symposium (ed. K. S. Johnson & B. J. Cardott), pp. 133–143. Norman: Oklahoma Geologic Survey, Circular 93.

Coveney, R. M. Jr. & Sangster, D. F. (1995). Metals in Pennsylvanian black shales of the Midwestern U.S.A.: product of regional Mississippi-Valley-Type hydrothermal systems. In: Extended Abstracts: International Field Conference on Carbonate Hosted Lead-Zinc Deposits (ed. D. L. Leach and M. B. Goldhaber), pp. 56–57. St. Louis Meeting.

Deason, K. L. (1984). Studies of the Parkville. Prairie Point and Riverside Oil and Gas Fields, Platte County, Missouri, Missouri Department of Natural Resources Open File Report, 84-27-OG, p. 20.

Deason, K. L. & Bohm, R. (1988). Studies of the Oil and Gas Fields of Caldwell, Clay, Clinton and Ray Counties, Missouri. Missouri Department of Natural Resources, Open File Report, 88-69-OG, p. 43.

Green, F. C. (1933). Oil and Gas Pools of Western Missouri. Missouri Bureau of Geology and Mines, Appendix II, 57th Biennial Report, p. 68.

Johnson, T. A. (2004). Stratigraphy, depositional environments and coalbed gas potential of Middle Pennsylvanian (Desmoinesian Stage) coals – Bourbon Arch Region, eastern Kansas, Kansas Geological Survey, Open File Report 2004-38, 309p.

Lange, J. P. (2003). Stratigraphy, depositional environments and coalbed gas resources of the Cherokee Group coals (Middle Pennsylvanian) – southeastern Kansas, Kansas Geological Survey, Open File Report 2003-82, 158p.

Merriam, D. F. (1963). The geologic history of Kansas. Kansas Geological Survey Bulletin 162, p. 317.

Newell, K. D. (1997). Comparison of maturation data and fluid-inclusion homogenization temperatures to simple thermal models: Implications for thermal history and fluid flow in the Midcontinent. Kansas Geological Survey Bulletin 240, part 2, p. 15.

Newell, K. D. & Birdie, T. R. (2021). Geothermal anomalies on the eastern flank of the Cherokee Basin, southeastern Kansas, USA. Midcontinent Geoscience 2, pp. 54–85.

Newell, K. D., Johnson, T. A., Brown, W. M., et al. (2004). Geological and geochemical factors influencing the emerging coalbed gas play in the Cherokee and Forest City basins in eastern Kansas. Kansas Geological Survey Open-File Report, 2004–17, p. 12.

Rascoe, B. Jr. & Adler, F. J. (1983). Permo-Carboniferous hydrocarbon accumulation, Mid-continent, USA. American Association of Petroleum Geologists Bulletin 67 (6), pp. 979–1001.

Smith, D. C. (2001). Coalbed methane Potential of Harrison County, Missouri. In: Missouri Department of Natural Resources Open-File Report, OFR-01-99-GS, p. 28.

Stanton, M. D. (1984). Stratigraphy and depositional environments of the Cherokee Group (Middle Pennsylvanian), Central Cherokee Basin, Southeastern Kansas. Master of Science Thesis. Lawrence, Kansas: University of Kansas.

Stoeckinger, W. T. (1990). Kansas coalbed methane comes on stream, Oil and Gas Journal. Houston, Texas: PennWell Publishing, p. 88–90.

Tedesco, S. A. (1992). Coalbed methane potential assessed in Forest City Basin, Oil and Gas Journal. PennWell Publications, February 10, p. 4.

Tedesco, S. A. (2014). Reservoir characterization and geology of the coals and carbonaceous shale of the Cherokee Group in the Cherokee Basin, Kansas, Missouri and Oklahoma, U.S.A. PhD. Golden: Colorado School of Mines, p. 2201.

United States Geological Survey (USGS) (2007). U.S. Coalbed Methane: Past, Present and Future. U.S. Energy Information Administration, http://www.eia.gov/oil_gas/rpd/cbmusa2.pdf.

Walton, A. W. (1995). Sequences in the Cherokee Group (Desmoinesian, Middle Pennsylvanian) of southeastern Kansas, unpublished manuscript, University of Kansas, p. 13.

Wanless, H. R. (1964). Local and regional variations in Pennsylvanian cyclic sedimentation. In: Symposium on cyclic sedimentation, Kansas Geological Survey Bulletin 169 (ed. D. F. Merriam), pp. 593–606.

Wojcik, K. M., McKibben, M. E., Goldstein, R. H., et al. (1995). Diagenesis, thermal history, and fluid migration, middle and upper Pennsylvanian rocks, Southeastern Kansas. In: Structural Styles in the Southern Styles in the Southern Midcontinent, 1995 Symposium (ed. K. S. Johnson). Norman: Oklahoma Geologic Survey Circular, 97, pp. 144–159.

5

Geophysical Research and Monitoring Within the Framework of a Block-Layered Model with Inclusions of a Hierarchical Structure

Olga Hachay[1], *Andrey Khachay*[2]

[1] *Institute of Geophysics, UB RAS, Ekaterinburg, Russia*
[2] *Ural Federal University, Ekaterinburg, Russia*

5.1 Review

Currently, there have been enough observations that indicate the two most important features of the modern evolution of the geological environment: mechanical movements of the Earth's substances occur on any spatial and temporal scale; the substance of the earth's crust, accessible to the study, forms a block-hierarchical structure (Goldin, 2002). An important role in understanding the formation and development of the hierarchy of structural levels of deformation in solid bodies is played by theoretical and experimental results obtained on samples (Panin et al., 1985). With their help, an approach is developed that uses ideas about dissipative structures in non-equilibrium systems (Nikolis and Prigogin, 1979), for which the processes of self-organization at each of the hierarchical levels take place. As shown in Nicolis (1989), self-organization occurs in the presence of a hierarchical structure. This approach can be applied to the study of such natural technological systems as arrays of rocks that are in the process of working out. To describe them, the model of an open dynamic system is applicable (Nikolis and Prigogin, 1990). An analysis of manifestations of self-organizational processes can give an idea of the stability of the system and contribute to the development of criteria for the stability of the state of the array as a whole regarding the dynamic phenomena of a given energy class. This echoes the production expressed in the work (Goldin, 2002), which consists in the hypothesis of the divisibility of the scale of the environment. Whereas the destruction of the smaller scale fits into the concept of a non-stationary random process, for which the prediction of individual events is not possible.

In the works (Hachay, 2004, 2007; Hachay et al., 2003) using 3D electromagnetic induction Khachay, 2006a, 2006b; Hachay and Khachay, 2015, 2016a, 2016b, 2017a, 2017b, 2018a, 2018b; Hari et al., 2001), it was possible to show that the model of a hierarchical discrete environment is applicable to describe the structure of the rocks of various material compositions. As part of a specific modification of the method, two hierarchical levels managed to trace. The zones of disintegration (Shemyakin et al., 1986, 1992) in the near-working space are located asymmetrically in the

Geophysical Exploration for Hydrocarbon Reservoirs, Geothermal Energy, and Carbon Storage: New Technologies and AI-based Approaches, First Edition. Edited by Said Gaci.

soil and roof, which may be evidence of the non-equilibrium state of the system. These zones are located discretely, that is, there are intervals of their complete absence in the near-working space. The maximum changes in the array, which are under technogenic influence, are manifested precisely in the change in the time of the morphology of the spatial position of these zones. In recent years, new physical and mathematical models of material environments have been intensively developed, which can be considered far-reaching generalizations of classical theories of elasticity (Shemyakin et al., 1986; Panin et al., 1985; Prigogin, 2001; Kunin, 1975). The science of the plasticity and strength of solid bodies is experiencing the stage of change of paradigm. For a long time, the description of the plastic deformation and destruction of solid bodies developed within the framework of linear approaches of the mechanics of the continuous environment (macro-cutting level) and physics of deformation defects in a hard body loaded (micro-cutting level). However, in recent decades, it has become obvious that the deformable solid body is a multi-level hierarchically organized system, which should be described within the framework of nonlinear mechanics and nonequilibrium thermodynamics (Panin et al., 2012). The fundamental problems that arise when applying the second law of thermodynamics to the analysis of systems at the macroscopic and microscopic levels are considered. It is shown that the inequality of the state of the system can cause order in it and that irreversible processes can lead to the emergence of a new type of dynamic states of matter, called "dissipative structures" (Prigogin, 2001). Kunin's work (1975) is devoted to a relatively narrow issue: the study of models of elastic media with microstructure. Historically, one of the first models of the elastic environment, which cannot be described in the framework of the classical theory of elasticity, is the continuum of the Cossera (1909). However, for a long time, the memoir E. and F. Cossera remained unnoticed, and only since about 1958-1960. The generalized models of the Cossera continuum began to develop intensively: the theory of oriented media, asymmetric, moment, multipolar, micro morphic, etc. Theory of elasticity (for brevity, we will call them the moment theories) to date, several hundred work on this topic have been published, and their number continues to increase rapidly. We will try to very schematically classify the various non-cabin theories of elastic environments. The characteristic feature of all such theories is their obvious or implicit nonlinearity. The latter, in turn, is manifested in the fact that theories contain parameters with the dimension of the length. These large-scale parameters can have a different physical meaning: for example, the distance between particles in discrete structures, grain or cell size, a characteristic radius of correlation or range forces, etc. But we will always assume that large-scale parameters are small in comparison with the characteristic size of the body. Cases of strong and weak non-powerfulness should be distinguished. If the "resolution" of the model has the order of a large-scale parameter, i.e., within the framework of the corresponding theory, it is physically acceptable to consider the lengths of waves, commensurate with a large-scale parameter, then we will call the theory of non-local, or very non-ink (if desired, this is emphasized), in such models we can consider elements of the environment of a large-scale parameter, but, as a rule, the distance, much smaller large-scale parameter do not have physical meaning. In non-local models, the velocity of waves depends on their length, so the term "environment with spatial dispersion" is often used. We emphasize that non-locality or spatial dispersion can be of different origin. They can be caused by the microstructure of the environment (in particular, the discreteness of the micro model) or close accounting of parameters such as the thickness of the rod or plate. Accordingly, we can talk about the physical or geometric nature of non-locality. Only effective elastic modules "know" about the structure of the original micro model, but it is impossible to extract this information from them. It follows that the explicit accounting of the effects of the microstructure and, in particular, the internal degrees of freedom is possible only while taking into account non-locality, i.e., the consistent theory of elastic media with microstructure must necessarily be non-local (Panin et al., 2012).

In recent years, much attention has been paid to the study of the spatial stress-deformed state of the rock massif. These studies are carried out as the purpose of the practical development of large depths with the help of deep mines and with the aim of studying stresses in the earth's crust. The main attention is paid to experimental methods for observing deformations and moving in the rock range in the vicinity of production and voltage estimates. The development of instrumental methods requires thorough penetration of physical ideas into mining sciences. The detected effect of zonal disintegration of rocks in the vicinity of underground workings at large depths is one of the natural states of the rock massif, which is under the influence of gravitational and tectonic forces (Shemyakin et al., 1986). In the work (Tazhibayev and Tashmamatov, 2014, see bibliography), the problem of determining residual stresses with experimental methods, the nature of the occurrence and preservation of residual stresses in rocks, metal structures, and multicomponent materials is described in detail. When measuring stresses in the rock mass, unloading methods, the total result of deformations from removing the residual stresses and stresses of external forces is recorded. However, to predict the possible destruction of the array, there is a need to divide the stresses of the array into voltage of external forces and residual internal stresses.

In recent decades, a new science has been born – the physics of non-equilibrium processes associated with concepts such as irreversibility, self-organization, and dissipative structures (Prigogin and Stengers, 2009). It is known that irreversibility leads to many new phenomena, such as the formation of vortices, oscillatory chemical reactions or laser radiation. Irreversibility plays a significant constructive role. It is impossible to imagine life in a world devoid of relationships created by irreversible processes. The prototype of the universal law of nature is the law of Newton, which can briefly formulate this: acceleration is proportional to the force. This law has two fundamental features. It is determined: since the initial conditions are known, we can predict movement. And he will turn in time: there is no difference between the prediction of the future and the restoration of the past; Movement to the future condition and the reverse movement from the current state to the initial are equivalent. Newton's law underlies classical mechanics, the science of the movement of matter, on trajectories. Since the beginning of the 20th century, the boundaries of physics have expanded significantly. Now we have quantum mechanics and the theory of relativity. But, as we will see from the further, the main characteristics of Newton's law – determinism and reversibility in time – have been preserved. Is it possible to modify the very concept of physical laws in such a way as to include in our fundamental description of nature irreversibility, event and arrow of time? The adoption of such a program entails a thorough review of our formulation of the laws of nature, and it became possible thanks to remarkable successes related to the ideas of instability and chaos (Prigogin and Stengers, 2009; Hawking, 1990). Returning to the results obtained for unstable mountainous earthly massifs, we can note that monitoring studies should be conducted in active mode, i.e., There should be a source of excitement (seismic or other nature), and the response is recorded from it for a not very long time, then the effect should be repeated and for this process, the phase diagrams of the state of the mountain massif can be built. Experimental and theoretical studies of vibrational stimulation of oil recovery of overwhelming deposits using vibration management lead to the conclusion about the development of resonant vibrations in fracture-block layers (Dryagin, 2018). These fluctuations, caused by weak, but long and stable and stable influences, create conditions for generating ultrasound waves in the layers of ultrasound, which can destroy thickened oil films in the cracks of collectors (Lavrov and Skuratnik, 2005). The measurement of high-frequency seismic-acoustic fields in wells near the regions of the concentration of stresses in the rocks of the rocks, as well as in crack-porous fluid-saturated layers with low-frequency vibrations, indicates the occurrence of ultrasound vibrations of hundreds and thousands of hertz. In crystalline massife, the existence of these fields is associated with the process of formation of new cracks, in fluid-saturated media, along with elastic, hydrodynamic causes can occur. For fractured porous collectors, which are during operation by the method of high-weight oil suppression with water, the possibility of intensification of ultrasound vibrations can be of important technological value. Even very weak ultrasound is able to destroy viscous oil films that form in cracks between the blocks for a long time, which can cause a decrease in the permeability of the layers and reducing oil recovery (Alekseev et al., 2001).

The hierarchical structure is characteristic of many systems, especially for the lithosphere of the Earth, where more than 30 hierarchical levels are distinguished from tectonic plates of thousands of kilometers to mineral grains (Prangishvilli et al., 2001). Thus, the earth's crust is a discrete system of blocks and, like any synergistic discrete ensemble, has properties of hierarchy and self-likeness (Kocharian and Spivak, 2003). The processes of the development of oil and gas fields are associated with the movement of multi-phase multicomponent media, which are characterized by nonequilibrium and nonlinear rheological properties. The real behavior of the formation systems is determined by the complexity of the rheology of moving liquids and the morphological structure of the porous environment, as well as a variety of interaction processes between the liquid and the porous environment (Khasanov and Bulgakova, 2003). Accounting for these factors is necessary for a substantial description of the processes of filtering fluids in the presence of nonlinearity, disequilibrium and heterogeneity inherent in real systems. At the same time, new synergistic effects are revealed (loss of resistance with the occurrence of fluctuations, the formation of ordered structures). This requires the development of new methods of controlling and managing complex natural systems that are configured to account for the marked phenomena. Thus, the reservoir system from which it is necessary to extract oil is a complex dynamic hierarchical system. When constructing a mathematical model of a real object, it is necessary to use the data of active and passive monitoring, obtained during the current operation of the object, as a priori information.

5.2 Conclusions

As you can see from the written overview, the new block models that contain hierarchical inclusions are more informative, forecasting the nonlinear complicated and synergetic effects in the environments that are rock shocks and earthquakes. These models are also used in new technologies by outworking oil and gas boreholes.

References

Alekseev, A. S., Tsetsokho, V. A., Belonosov, A. V., et al. (2001). Forced oscillations of fractured block fluid-saturated layers in vibrating and technical impacts. Geomechanics Physico-Technical Problems of Mineral Development 6, pp. 3–12.

Dryagin, V.V. (2018). Use of induced acoustic emission of reservoirs for the detection and recovery of hydrocarbons. Georesursy = Georesources 20(3), Part 2, pp. 246–260. https://doi.org/10.18599/grs.2018.3.246-260.

Goldin, S. V. (2002). Lithosphere destruction and physical mesomechanics. Physical Mesomechanics 5 (5), pp. 5–22.

Hachay, O. A. (2004). On the issue of studying the structure and state of the geological heterogeneous nonstationary medium within the framework of the discrete hierarchical model. Russian Geophysical Magazine 33 (34), pp. 32–37.

Hachay, O. A. (2007). The study of the development of instability in the massif of rocks using the active electromagnetic monitoring method. Physics of the Earth 4, pp. 65–70.

Hachay, O. A. & Khachay, A. Y. (2015). Determination of the surface of the fluid-saturated porous inclusion in the hierarchical layered block medium according to electromagnetic monitoring. Mining Information and Analytical Bulletin 4, pp. 150–154.

Hachay, O. A. & Khachay, A. Y. (2016a). Modeling the propagation of the seismic field in a layered block elastic medium with hierarchical plastic inclusions. Mining Information and Analytical Bulletin 12, pp. 318–326.

Hachay, O. A. & Khachay, A. Y. (2016b). Determination of the surface of anomal intense inclusion in a hierarchical layered block medium according to acoustic monitoring. Mining Information and Analytical Bulletin 4, pp. 354–356.

Hachay, O. A. & Khachay, A. Y. (2017). Acoustic wave monitoring of fluid dynamic. In: The Rock Massif with Anomaly Density, Stressed and Plastic Hierarchic Inclusion. Computational and Experimental Studies of Acoustic Waves, Intechopen. 4, pp. 63–80. http://dx.doi/org/10/5772/intechopen 70590.

Hachay, O. A., Khachay, A. Y., & Khachay, O. Y. (2018a). On the question of modeling acoustic monitoring of a hierarchical two-phase geological environment with various physical and mechanical properties. Monitoring. Science and Technology 1, pp. 45–49.

Hachay, O. A., Khachay, A. Y., & Khachay, O. Y. (2018b). Modeling algorithm of acoustic waves penetrating through a medium with composite hierarchical inclusions. AIP Conference Proceedings 2053, pp. 030023. https://doi.org/10.1063/1.5084384.

Hachay, O. A., Khachay, O. Y., & Khachay, A. Y. (2015). New methods of geoinformatics of monitoring of wave fields in hierarchical environments. Geoinformatics 3, pp. 45–51.

Hachay, O. A., Khachay, O. Y., & Khachay, A. Y. (2016). New methods of geoiformatics for the complexation of seismic and gravitational fields in hierarchical environments. Geoinformatics 3, pp. 25–29.

Hachay, O. A., Khachay, O. Y., & Khachay, A. Y. (2017a). Complex Research of acoustic, gravitational, and geomechanical fields in hierarchical environments. Mining Information and Analytical Bulletin 4, pp. 328–336.

Hachay, O. A., Khachay, O. Y., & Khachay, A. Y. (2017b). On the question of the inverse problem of active electromagnetic and acoustic monitoring of the hierarchical geological environment. Geophysical Studies B 18 (4), pp. 71–84. https://doi.org/10.21455/gr2017.4-6.

Hachay, O. A., Novgorodova, E. N., & Khachay, O. Y. (2003). A new method of detecting disintegration zones in the voyage-minded space of rocks of rocks. Mining Information and Analytical Bulletin 11, pp. 26–29.

Hari, O. A., Vlokh, N. P., Novgorodova, E. N., et al. (2001). Three-dimensional electromagnetic monitoring of the state of the rocks massif. Physics of the Earth 2, pp. 85–92.

Hawking, C. (1990). From a large explosion to black holes. Brief History of Time. Peace. Part.8.

Khachay, A. Y. (2006a). The algorithm for solving the direct dynamic problem of seismic when excited by horizontal point force located in an arbitrary layer of N-layer elastic isotropic medium. Informatics and Mathematical Modeling. Ekaterinburg, pp. 170–278.

Khachay, A. Y. (2006b). The algorithm for solving the direct dynamic problem of seismic when excited by a point source of vertical force located in an arbitrary layer of N-layer elastic isotropic medium. Informatics and Mathematical Modeling. Ekaterinburg, pp. 279–310.

Khasanov, M. M. & Bulgakova, G. T. (2003). Nonlinear and non-equilibrium effects in rheological environments. Moscow. Izhevsk: Institute of Computer Research, p. 288.

Kocharian, G. G. & Spivak, A. A. (2003). Dynamics of Deformation of Block Arrays. ICC "Academkniga," p. 424.

Kunin, I. A. (1975). Theory of elastic media with a microstructure, a nonlocal theory of elasticity. The main editorial office of the physics-mathematical literature of the Publishing House "Science," M., p. 416.

Lavrov, A. V. & Skuratnik, V. L. (2005). Akustic emission during the deformation and destruction of rocks (review). Acoustic Magazine V.51. Execution, pp. 6–18.

Nicolis, J. (1989). Dynamics of Hierarchical Systems. Peace, p. 486.

Nikolis, G. & Prigogin, I. (1979). Self-Organization in Nonequilibrium Systems. MIR, p. 300.

Nikolis, G. & Prigogin, I. (1990). Knowledge of Complex. MIR, p. 344.

Panin, V. E., Likhachev, V. A., & Grinjaev, Y. V. (1985). Structural Levels of Deformation of Solids. Novosibirsk: From the USSR Academy of Science, pp. 226, 254.

Panin, V. E., Egorushkin, V. E., & Panin, A. V. (2012). Nonlinear wave processes in a deformable solid body as a multi-level hierarchically organized system. UFN 182 (12), pp. 1351–1357. https://doi.org/10.336//UFNr.0182/201212I.1351.

Prangishvilli, I. V., Pashchenko, F. F., & Busygin, B. P. (2001). System laws and patterns in electrodynamics, nature and society. Science, p. 525.

Prigogin, I. & Stengers, I. (2009). Time, Chaos, Kvant: to the decision of the time paradox. Book House Librok, p. 232.

Prigogin, I. (2001). Introduction to the thermodynamics of irreversible processes. In: Time, Structure and Fluctuations. Nobel Lecture in Chemistry 1977. Scientific and Publishing Center "Regular and Chaotic Dynamics", pp. 123–155.

Shemyakin, E. I., Fisenko, G. L., Kurlenya, M. V., et al. (1986). The effect of zonal disintegration of rocks around underground holes. Dan USSR 289 (5), pp. 1088–1094.

Shemyakin, E. I., Kurlenya, M. V., Oparin, V. N., et al. (1992). Opening No. 400. The phenomenon of zonal disintegration of rocks around underground holes. Bulletin of Inventions 1, pp. 7–15.

Tazhibayev, K. T. & Tashmamatov, A. S. (2014). Residual stresses in rocks and method for their definition. I.TS "Teknik" Bishkek, p. 126.

Section II

Reservoir Characterization Concepts and Workflows

6

A Review on Shear Wave Velocity Estimation Methods

Said Gaci[1], *Mohammed Farfour*[2]

[1] *Sonatrach - Central Directorate of Research & Development (DC R&D), Boumerdes, Algeria*
[2] *Department of Earth Sciences, College of Science, Sultan Qaboos University, Muscat, Oman*

6.1 Introduction

Body waves are of great interest in oil and gas exploration studies and are of two types: primary (or compressional) waves and secondary (or shear) waves. In contrast with S-wave that is sensitive only to variations in rock's rigidity, and density, P-wave is also affected to compressibility.

Shear wave cannot travel through the fluids and is slower than compressional waves, an important property that is used to better characterize reservoirs. Besides, with compressional wave, particles move in the same direction of the wave propagation, whereas shear wave propagates perpendicularly to this direction. Another key difference is that the P-wave compresses volumes, whereas S-wave only changes their shapes. This feature is of a great interest for geoscientists in distinguishing fluid-saturated formation expressions from expressions relating to lithological changes, and to depict gas cloud areas where gas traces disguise subsurface formations (Farfour and Yoon, 2016).

The velocities V_p and V_s can be determined from samples of rock in a laboratory, from well logs, or from other independent measurements, such as vertical seismic profile (VSP), or P-wave and S-wave seismic sections.

Over the past decades, shear wave (S-wave) velocity (V_s) logs have gained a great interest in petroleum geoscience and engineering workflows. Specifically, V_s log data is essential in pre-stack AVO modeling, and analysis, and inversion (Downton, 2005), Lithology indication (Pickett, 1963), fluid discrimination (Smith and Gidlow, 2000; Hamada, 2004), formation evaluation (Aranibar et al., 2013), geomechanical properties (Ameen et al., 2009; Chang et al., 2006), pore type identification (Anselmetti and Eberli, 1993, 1999; Eberli et al., 2003), as well as in the exploration of unconventional reservoirs (Vernik et al., 2017). With the advances in amplitude variation with offset (AVO) studies and the latest tendency toward unconventional resources that provided geomechanics much interest in the industry, shear velocity has become progressively an important parameter that plays a key role in numerous applications.

Shear velocity data are not readily available as P-wave logs, especially in old wells (Tan et al., 2015). In addition, it is worth noting that in contrast with the large available laboratory data related to V_p, only limited information dealing with V_s measurements are available from laboratory

Geophysical Exploration for Hydrocarbon Reservoirs, Geothermal Energy, and Carbon Storage: New Technologies and AI-based Approaches, First Edition. Edited by Said Gaci.

experiments (Ji et al., 2002). This statement is principally explained by the difficulty to measure V_s at low pressures, since the transducers and the end surfaces of the specimen should be firmly connected in order to ensure a good transmission of shear wave through the sample. Besides, the laboratory conditions cannot perfectly mimic downhole field environments (such as state of in situ stresses and fluid content) and other physical parameters (rock type, mechanical properties and loading conditions) that could impact shear wave velocity. In addition, due to the limited core samples undergone mechanical lab experiments, the obtained information is less trustworthy statistically. Consequently, it is required to predict the S-wave velocity from the available data: well log measurements, such as the P-wave velocity (V_p) and density, as well as core data, etc.

Generally, the estimation of S-wave velocity is performed using three categories of methods: empirical relationships, intelligent systems, and rock physical models. The first category consists on predicting V_s with empirical relations linking the P-wave velocity, porosity, shale content, and density. Some examples of this category were published by Castagna et al. (1985), Han et al. (1986), Krief (1990), Greenberg and Castagna (1992), and Kwak et al. (2015). Nevertheless, it is worth mentioning that these inferred empirical relationships are not applicable everywhere without calibration check with the available data and studies (Nourafkan and Kadkhodaie-Ilkhchi, 2015). In addition, these relations between S-wave velocity and the other physical properties of sedimentary rocks could not be interpreted from a physical point of view, and the different physical mechanisms of rocks have not been taken into consideration while formulating these relationships (Wang, 2000). Another drawback is that their application conditions may not be well defined.

Recently, the tremendously increase in data volumes and the advances in sophisticated computing hardware and software have led to the development of a second category of S-wave generation methods using artificial intelligence (Farfour and Mesbah, 2020). Note that artificial intelligent systems and machine learning applications became very widely accepted in geosciences, specifically in petrophysical analysis as well as in log generation and prediction (Eskandari et al., 2003; Rajabi et al., 2010; Tabari et al., 2011; Asoodeh and Bagheripour, 2012; Asoodeh and Bagheripour, 2014; Akhundi et al., 2014; Ranjbar-Karami et al., 2014; Spichak and Goidina, 2016).

The last category of S-wave estimation is using rock physics. The rock physical models present a more realistic frame to estimate V_s data. Some of these models have been constructed under some specific assumptions (Mindlin, 1949; Dvorkin and Nur, 1996; Jørstad et al., 1999; Xu and White, 2010). The model suggested by Xu and White (2010) to estimate S-wave velocity combines the differential effective-medium theories and Kuster-Toksöz model (Kuster and Toksoz, 1974a, 1974b). Then, using this approach, several approaches have been suggested to predict S-wave velocity (Keys and Xu, 2002; Yan et al., 2008; Bai et al., 2013).

The above three categories will be discussed in more details in the following sections.

6.2 Empirical Relationships for Estimating S-Wave Velocity

The work of Pickett (1963) consists of establishing a relationship between the V_p/V_s ratio and lithology. The V_p/V_s values are 1.8 for dolomites, 1.9 for limestones, and varying from 1.6 to 1.75 clean to calcareous sandstone.

In the same view, based on comparison between in-situ and laboratory data with theoretical model data, Castagna et al. (1985) determined a global V_p-to- V_s ratio (V_p/V_s) for clastic silicate rocks. In their research, velocity information was derived from data related to water-saturated mud rocks and sandstones, and general tendencies of the variation of V_p-V_s with depth were determined for Gulf Coast clastic rocks. It was demonstrated that V_p/V_s are mainly governed by mineralogy. Besides, in-situ sonic and field measurements highlight that Vs linearly evolves with V_p according to the following relation: $V_s = 0.862\ V_p - 1.172$.

This equation illustrates the dependence between V_p/V_s and the two parameters: porosity and volume of clay, and an increase of one parameter implies an increase of V_p/V_s. In addition, it is shown that porosity changes are the key factor affecting the velocity ratio. Another noteworthy finding is that V_s is approximately linearly correlated to V_p for both water saturated and dry clastic silicate sedimentary rocks. For a specified V_p value, mud rocks have a tendency to reach V_p/V_s values somewhat greater than those associated with clean porous sandstones, whereas V_p/V_s ratio is approximately steady for dry sandstones, and is decreasing with increasing V_p for wet sandstones and mudstones.

By using Picket experimental data, Domenico (1984) asserts that porosity effect on wave velocity, and porosity effect on carbonates is less than on sandstones. Besides, porosity effect on V_s is more than on V_p in carbonate rocks. He suggested a relation for computing V_s based on porosity ϕ at different effective pressures: $1/V_s = A + B\ \phi$, with porosity in %, V_s in μs/ft, A and B are Domenico equation coefficients based on different pressures: A is linked to rock matrix and B is associated to pore geometry and formation effective pressure.

By conducting laboratory ultrasonic experiments on brine saturated and shaly sandstone samples extracted from cores, Han et al. (1986) introduced a model comparable to that of Castagna': $V_s = 0.794\ V_p - 0.787$.

Later, a linear method was proposed by Greenberg and Castagna (1992) to estimate and calibrate shear wave velocity data for numerous lithology for the Gulf of Mexico: $V_s = 0.80416\ V_p - 0.85588$ for sands, $V_s = 0.76969\ V_p - 0.86735$ for shales, $V_s = -0.05508\ V_p^2 + 1.01677\ V_p - 1.03049$ for limestones, and $V_s = 0.58321\ V_p - 0.07775$ for dolomites.

Wang (2000) presented an empirical formula for predicting V_s from the bulk density of the saturated rock, the pore-fluid modulus as well as V_p value. Besides, Eskandari et al. (2004) suggest the following quadratic relation: $V_s = -0.1236\ V_p^2 + 1.6126\ V_p - 2.3057$. They showed also the effect of petrophysical parameter on the predicted V_s in carbonate rocks and stated that V_s diminishes with increasing porosity and deep resistivity, and increases with increasing bulk density.

Brocher (2005) combined very large datasets for an extensive variety of common lithology including well logs, laboratory, vertical seismic profiling (VSP) and field tomography to define a nonlinear polynomial relationship between compressional wave velocity and shear wave velocity: $V_s = 0.0064\ V_p^4 - 0.1238\ V_p^3 + 0.7949\ V_p^2 - 1.2344\ V_p + 0.7858$.

Al-Kattan (2008) specified that in carbonate rocks, the porosity measured by V_p or V_s should be identical for the same sample. Based on Wyllie equation and a statistical analysis, she suggested for estimating V_s the following relationship: $V_s = 0.5734\ V_p - 90$.

Some researchers have incorporated a density term into the equations. Anselmetti and Eberli (1993) also propose an empirical correlation between density $\rho(\mathrm{g/cm^3})$ and V_s (m/s) for carbonate rocks: $V_s = 199\ \rho^{2.84}$.

By combining the results of laboratory studies on brine saturated porous rocks illustrating that V_s is strongly correlated with V_p, and the fact that V_s is inversely proportional to $\sqrt{\rho}$ as revealed by elastic theory (Hamada, 2004), V_s can be given as a function of $V_p/\sqrt{\rho}$. Taking into account the fact that most experimental relationships between rock strength and rock petrophysical properties are governed by a power or an exponential laws, Chang et al. (2006) suggested a power law function of V_s with respect to $V_p/\sqrt{\rho}$: $V_s = A\left[V_p/\sqrt{\rho}\right]^m$.

Yusuf (2019) modified the power law relation to take into consideration the situations where V_p and ρ equal zero, and are not non-zero positive values when $V_s = 0$ (in case of fluids): $V_s = A\left[V_p/\sqrt{\rho}\right]^m - B$.

The relation is used to predict V_s for wells located at the Grand Banks and improved V_s estimations are reached compared to those derived using the former empirical formulas. However, the relationship becomes less valid as ρ approaches zero. After calibration of the relation to experimental data and available Studies, the values of A, B, and m are determined using the least square regression, the final equation is reduced to: $V_s = 2.41\left[V_p/\sqrt{\rho}\right]^{0.98} - 2.35$.

The new resulting V_s prediction model is revealed more reliable for the Grand Banks area when compared to the previous models introduced by Han et al. (1986), Brocher (2005), and Castagna et al. (1985).

To mention also that some researches introduced empirical relations underlining porosity-depth trends that can be established for several lithology to estimate the velocity functions $\left(V_p \text{ and } V_s\right)$ with respect to depth (Avseth et al., 2001). Later, Avseth et al. (2009) developed a relation between the spatial and rock property changes with depth. A summary of the empirical formula used to predict V_s is given in Table 6.1.

Table 6.1 Main empirical relationships for estimating S-wave velocity established by previous researches.

Authors	Empirical relationships
Pickett (1963)	$V_p/V_s = 1.8$ for dolomites, $= 1.9$ for limestones, $= 1.6$ to 1.75 clean to calcareous sandstone.
Carroll (1969)	$V_s = 0.937562\ V_p^{0.81846}$ (V_p and V_s are in 10^3 ft/s) (for granitic and volcanic sediments)
Castagna et al. (1985)	$V_s = 0.862\ V_p - 1.172$ (V_p and V_s are in km/s) (for clastic silicate rocks)
Dominico (1984)	$1/V_s = A + B\,\phi$ (porosity in %, V_s in μs/ft) A is linked to rock matrix and B is associated to pore geometry and formation effective pressure.
Han et al. (1986)	$V_s = 0.794\ V_p - 0.787$ (V_p and V_s are in km/s) (for brine saturated and shaly sandstone)
Krishna et al. (1989)	$V_s = 2.924\ V_p - 4170.9$ (V_p and V_s are in km/s) (for superficial seawater rocks)
Williams (1990)	$V_p/V_s = C + D\Delta ts$ (V_p and V_s are in ft/μs, Δt_s in μs/ft) (for water-borne shales and sandstones. The constants C and D are linked to lithology).
Krief et al. (1990)	$V_s^2 = A\ V_p^2 - B$ (V_p and V_s are in km/s) (For consolidated clean sediments, The constants A and B are related to lithology).
Miller and Stewart (1991)	$V_s = 0.8\ V_p - 861$ (for water-saturated shaly sandstones samples analyzed by Han et al. (1986)) $V_s = 0.448\ V_p + 496$ (for water-saturated limestone samples analyzed by Kuiper et al. (1959)) (V_p and V_s are in m/s)

Authors	Empirical relationships
Greenberg and Castagna (1992)	$V_s = 0.80416\ V_p - 0.85588$ for sands, $V_s = 0.76969\ V_p - 0.86735$ for shales, $V_s = -0.05508\ V_p^2 + 1.01677\ V_p - 1.03049$ for limestones, $V_s = 0.58321\ V_p - 0.07775$ for dolomites. (V_p and V_s are in km/s)
Castagna et al. (1993)	$V_s = -0.232\ V_p^2 + 1.542\ V_p - 1.214$ for coals. (V_p and V_s are in km/s)
Vernik et al. (2002)	$V_s^2 = \mathrm{A}\ V_p^4 + \mathrm{B}\ V_p^2 + \mathrm{C}$ for water-saturated sandstones and shales. A, B, C constants depending on lithology. (V_p and V_s are in km/s)
Eskandari et al. (2004)	$V_s = -0.1236\ V_p^2 + 1.6126\ V_p - 2.3057$ for carbonate samples (V_p and V_s are in km/s)
Brocher (2005)	$V_s = 0.0064\ V_p^4 - 0.1238\ V_p^3 + 0.7949\ V_p^2 - 1.2344\ V_p + 0.7858$ for all lithological types (excepting calcium-rich and mafic rocks, gabbros, and serpentinites) $V_s = (V_p - 1.36)/1.16$ for or calcium-rich rocks (defined for $5.25 < V_p < 7.25$ km/s) (V_p and V_s are in km/s)
Al-Kattan (2008)	$V_s = 0.5734\ V_p - 90.337$ for carbonate rocks (V_p and V_s are in m/s)
Anselmetti and Eberli (1993)	$V_s = 199\ \rho^{2.84}$ for carbonates (V_s in m/s and ρ in g/cm^3)
Chang et al. (2006)	$V_s = \mathrm{A}\left[V_p/\sqrt{\rho}\right]^m$ (ρ in g/cm^3, V_p and V_s in km/s)
Hossain et al. (2012)	$V_s = 0.95\ V_p - 1.27$ for glauconitic greensand based on is-frame model $V_s = 0.76\ V_p - 0.76$ for water-bearing glauconitic greensand based on core data $V_s = 0.86\ V_p - 0.96$ for glauconitic greensand based on log data (V_p and V_s are in km/s)
Bailey and Dutton (2012)	$V_s = 0.75\ V_p - 562.5$ for consolidated Kimmeridge clay (V_p and V_s are in km/s)
Lee (2013)	$V_s = 0.59\ V_p - 0.6$ for water-filled clay (V_p and V_s are in km/s)
Al-Kattan (2015)	$V_s = 0.699\ V_p^{0.969}$ for carbonates (V_p and V_s are in m/s)
Yusuf (2019)	$V_s = 2.41\left[V_p/\sqrt{\rho}\right]^{0.98} - 2.35.$ for shaly marine sediments (ρ in g/cm^3, V_p and V_s in km/s)

6.3 Intelligent Systems for Estimating S-Wave Velocity

The advent of the intelligent systems and the development of artificial intelligence in geophysics has significantly contributed to enhancing the prediction of shear wave velocity (Eskandari et al., 2003; Rajabi et al., 2010; Tabari et al., 2011; Asoodeh and Bagheripour, 2012;

Asoodeh and Bagheripour, 2014; Akhundi et al., 2014; Ranjbar-Karami et al., 2014; Spichak and Goidina, 2016; Gaci, 2017; Bukar et al., 2019) (Table 6.2).

Eskandari et al. (2003) suggested a multiple variable regression to predict V_s from petrophysical well logs: V_p, neutron porosity (NPHI), bulk density (RHOB), gamma ray (GR), and deep laterolog (LLD). It is demonstrated that the most significant statistical variable to this regression are the NPHI, RHOB, and V_p. The suggested regression equation allows to estimate V_s for carbonate reservoir with a correlation coefficient R^2 of about 0.94.

Rajabi et al. (2010) predicted compressional, shear, and Stonely wave velocities from well log data from an Iranian carbonate reservoir, using genetic algorithms, fuzzy logic, and neuro-fuzzy techniques.

Table 6.2 Main previous studies using intelligent systems for estimating S-wave velocity.

Authors	Artificial intelligence method	Used inputs
Eskandari et al. (2003)	A multiple variable regression	V_p, neutron, bulk density, gamma ray and deep laterolog.
Rajabi et al. (2010)	Genetic algorithms, fuzzy logic and neuro-fuzzy techniques	V_p, neutron, bulk density, gamma ray and deep laterolog.
Tabari et al. (2011)	Back propagation neural networks	V_p, density, neutron and Gamma ray logs
Rajabi and Tingay (2013)	genetic algorithms, fuzzy logic and neuro fuzzy systems	Gamma ray, neutron and density
Akhundi et al. (2014)	Empirical relationships, multivariate regression methods and neural networks	V_p, neutron, bulk density, gamma ray, deep and shallow laterologs.
Gholami et al. (2014)	Independent component analysis (ICA)	Seismic attributes
Gaci (2017)	A multiple layer perceptron artificial neural network (MLP ANN).	V_p, its Hölderian regularity log, and the three components (high frequency, low frequency and trend) obtained from V_p by the fine-to-coarse reconstruction algorithm
Bukar et al. (2019)	Different supervised learning algorithms: Linear Regression Models, Regression Trees, Support Vector Machines, Gaussian Process Regression Models, Ensembles of Trees.	V_p, gamma ray, bulk density, neutron, resistivity, porosity and water saturation logs
Anemangely et al. (2019)	Least Square Support Vector Machine (LSSVM) combined with Particle Swarm Optimization (PSO), Genetic Algorithm (GA), and Cuckoo Optimization Algorithm (COA)	Data from two oilfields located in the southwest of Iran (Ahvaz field): depth, neutron porosity (NPHI), density (RHOB), photoelectric (PEF), gamma ray (GR), caliper, true resistivity (RT), and V_p
Kim et al. (2023)	Gradient Boosting (GB), Random Forest (RF), and Artificial Neural Network (ANN)	Borehole databases from two seismograph networks of Japan: Kyoshin Network (K-NET) and Kiban Kyoshin Network (KiK-net). Various factors have been considered: depth, N-value, density, slope angle, elevation, geology, soil/rock type, and site coordinates

The authors highlight that although prediction in carbonate rocks is a hard task due to diagenetic processes, the used intelligent techniques, that are based on different concepts, lead to trustworthy results, particularly those obtained from fuzzy logic.

Tabari et al. (2011) predicted V_S from V_P, density, neutron, and Gamma-ray logs using back propagation neural networks. The obtained results are excellent compared to those derived using the models proposed by Greenberg and Castagna (1992), Domenico (1984), and Pickett (1963).

Rajabi and Tingay (2013) estimated geomechanical parameters from well logs for diverse lithologies using genetic algorithms, fuzzy logic and neuro fuzzy systems. It is illustrated that the quantity of input points determines the reliability of the predict model, that is, more available input data lead to better results, and to enhance the input-output relation.

For V_S prediction purpose, Akhundi et al. (2014) used empirical relationships, multivariate regression methods and neural networks for an Iranian carbonate reservoir. The multiple regression method is utilized to define parameters affecting estimation V_S. Then, these identified parameters are used as inputs for neural networks. The application was successful with a correlation coefficient of V_S prediction is approximately 92% and 95% for multiple regression and neural network method, respectively.

Gholami et al. (2014) used seismic attributes for evaluating V_S in a located in southern Iranian field. In order to identify the interdependency of all the available attributes and to select the most relevant ones, independent component analysis (ICA) was carried out. Afterward, the selected seismic attributes were introduced as inputs into a multi-layer feed forward neural network (MLFN) for predicting V_S (output). The suggested approach is proved to be efficient for handling seismic attributes and delivering a 2D V_S model, which can be improved using a tough algorithm like support vector machine. The authors conclude that suggested model can bring meaningful information related to the variation of V_S in subsurface.

In his research, Gaci (2017) introduced a prediction model for estimating vs from V_P using a multiple layer perceptron artificial neural network (MLP ANN). This model considers, as inputs, in addition to V_P, its Hölderian regularity log, and the three components (high frequency, low frequency, and trend) obtained from V_P by the fine-to-coarse reconstruction algorithm based on the empirical mode decomposition (EMD). The predictions of V_S are more accurate than those given by the empirical models: Castagna et al. (1985), Han et al. (1986), and Mavko et al. (2009).

Bukar et al. (2019) applied machine learning to V_S prediction using different supervised learning algorithms. The inputs used for these neural networks models are the following logs: V_P as the main predictor, gamma ray, bulk density, neutron, resistivity, porosity, and water saturation logs. It is proved that Gaussian process regression models yield the best V_S estimations compared to the other applied machine learning algorithms: linear regression models, regression trees, support vector machines, and ensembles of trees. In addition, the obtained V_S predictions are more accurate than those yield by the linear regression model based on the Greenberg and Castagna (1992) method. Contrary to Greenberg-Castagna method which is only implemented for brine-saturated rocks, the suggested approach is applicable in hydrocarbon-bearing intervals.

6.4 Rock Physics Models for Estimating S-Wave Velocity

In addition to the empirical models and intelligent systems, a third category based on physical and semi-physical models is suggested to predict share wave velocity using formation parameters (Table 6.3). For this purpose, a semi-physical model was suggested by Greenberg and

Table 6.3 Main previous studies based on rock physics models for estimating S-wave velocity.

Authors	Rock physics model
Xu and White (1995)	Gassmann's equations used to determine the fluid saturated bulk and shear modulus.
Keys and Xu (2002)	Linear ordinary differential equations of a dry rock approximation introduced to Xu –White model.
Lee (2006)	Values of Physical parameters determined empirically with uncertainty (e.g. pore-aspect ratio and differential pressure) introduced to Xu–White model.
Shao et al. (2009)	Inversion method from well logs combining rock physics and multi-mineral analysis
Dvorkin and Nur (1996); Mavko et al. (2009)	Granular effective medium (GEM) found on a mixture of geologic insights with laborious theories and experimental methods
Tan et al. (2015)	A petrophysical model built on core analysis results, considering the models of Krief (1990), Nur (1992), and Pride et al. (2004) to deduce the elastic moduli of dry rock.
Wang et al. (2019)	A granular media model for predicting V_s combining soft-, intermediate and stiff-sand models with an effective coordination parameter
Azadpour et al. (2020)	Combination of rock physics and machine learning techniques to estimate S-wave velocity.
Avseth et al. (2021)	Data-driven rock physics analysis, consisting of: facies classification using elastic well log data from several wells, followed by facies-constrained regression analysis to establish local rock physics relations for prediction of V_p and V_s.

Castagna (1992) to estimate V_s in porous rocks. In this model, the mixed grain incompressibility is computed using Wood's equation and the Voigt-Reuss-Hill equation, and the elastic properties of fluid-saturated rocks were estimated using the Biot-Gassmann theory (Gassmann, 1951). In addition, the suggested model is based on an empirical relation between V_p and V_s. Good results were obtained from data taken from various locations.

Later, a new physical model was proposed by Xu and White (1995) for clay-sand mixture.

For specific dry rock elastic moduli, this model applies Gassmann's equations to determine the fluid saturated bulk and shear modulus. The latter computed physical parameters are then used to estimate P- and S-wave velocities (V_p and V_s) as follows:

$$K_{sat} = K_{dry} + \frac{\left(1 - \frac{K_{dry}}{K_0}\right)^2}{\frac{\emptyset}{K_0} + \frac{1-\emptyset}{K_0} - \frac{K_{dry}}{K_0^2}}$$

$$\mu_{sat} = \mu_{dry}$$

$$V_p = \sqrt{\frac{K_{sat} + \frac{4}{3}\mu_{sat}}{\rho_{sat}}}$$

$$V_s = \sqrt{\frac{\mu_{sat}}{\rho_{sat}}}$$

with ϕ the porosity, K_{dry} the bulk modulus of the dry frame, K_0 the solid rock bulk modulus, K_f the bulk modulus of the pore fluid, and K_{sat} the bulk modulus of fluid-saturated rock, while μ_{sat}, ρ_{sat}, V_p, and V_s denote the shear modulus, the density, P-, and S-wave velocity of the fluid-saturated rock, respectively.

The methodology proposed by Xu and White (1995) was successfully implemented to evaluate V_p and V_s of shaley sands, and was implemented in several studies (Dvorkin and Nur, 1996; Jørstad et al., 1999; Mavko et al., 2009; Liu and Sun, 2015; Yang et al., 2019). This model takes into account the effects of matrice nature, fluids, shale content, porosity dimension, and pore geometry (Xu and White, 1996). The authors consider that sandstone velocities are affected mainly by two physical rock parameters: porosity and clay content. The effect of the latter is due to pore geometry or pore aspect ratio dissimilarities of shale and sandstone (0.1–0.12 for standstone, and 0.03–0.04 for shale). Therefore, according to this model, the pores are split in a sandstone pore associated to a higher aspect ratio, and a shale pore with a lower aspect ratio, where the typical aspect ratio values are 0.12 and 0.035 for shale and sandstone pores, respectively. However, the limitation of this approach is that very low porosity values are requested to fulfill Kuster–Toksoz conditions, and pore space should be partitioned; that implies a large time computational consuming (Xu and White 1995). To speed up calculations, Keys and Xu (2002) introduced linear ordinary differential equations of a dry rock approximation.

Lee (2006) underlined the implementation of Xu and White (1995) model needs the introduction of physical parameters values which are experimentally determined with uncertainty, such as: pore-aspect ratio, and differential pressure, to calculate V_s from measured V_p and porosity. Mostly, the pore-aspect ratio is so variable that it cannot well describe the discrepancy of pore sizes (Brown and Korringa, 1975).

Another approach, initiated by Shao et al. (2009), is built on inversion technique from well logs using rock physics and multi-mineral analysis for V_s estimation in shaly-sand, and leads to predictions more precise than those obtained by Xu-White model.

Besides the inclusion scheme on which Xu and White (1995) model is founded, granular effective medium (GEM) rock physics models are extensively applied in modeling. GEM models are generally a mixture combining numerous geologic perceptions with laborious theories and experimental methods (Dvorkin and Nur, 1996; Mavko et al., 2009). These GEM models have been fruitfully implemented to apprehend rock microstructures using the existing relations between elastic modulus and porosity. The analysis using this model is executed by adjusting an effective-medium theoretical model to a trend of real data by assuming that the rock microstructure is comparable to that used in the model (Avseth et al., 2000, 2009; Hossain et al., 2011). Therefore, the advantage of these GEM rock physics models is that they provide physical meanings explanation of the burial and diagenesis process of sandstones.

Recently, in the same view, Wang et al. (2019) used a granular media model to predict V_s from V_p. They suggested that the soft-, intermediate-, and stiff-sand models can be combined using an effective coordination parameter which allows to link the shear modulus of dry rocks to bulk modulus, and to predict V_s from V_p via the unified model. The results show the estimated V_s values match those deduced from laboratory experiments and well logs. The obtained

Table 6.4 Reservoir properties computed from V_p and V_s.

Parameter	Expression
Bulk modulus (incompressibility)	$\kappa = \rho V_p^2 - \frac{4}{3}\rho V_s^2$
First Lame parameter	$\lambda = \rho V_p^2 - 2\rho V_s^2$
Second Lame parameter (rigidity)	$\mu = \rho V_s^2$
Poisson ratio	$\sigma = \dfrac{V_p^2 - 2V_s^2}{2V_p^2 - 2V_s^2}$
Young modulus (stiffness modulus)	$E = \rho V_s^2 \dfrac{3V_p^2 - 4V_s^2}{V_p^2 - V_s^2}$

numerical results show that the predicted V_s match well the measured velocities for the laboratory data and well logging data. Besides, the precision of predicted velocities increases as differential pressures of rocks reach high values. The authors conclude that this model is very appropriate for estimating V_s for shaly sandstone, mainly for turbidite deposits in deep-water environment.

An integrated approach of rock physics simulation was initiated by Tan et al. (2015) to evaluate V_s for unconventional reservoirs, specifically gas-bearing shale. The latter study is based on a petrophysical model constructed from core analysis results, and considers the models of Krief (1990), Nur (1992), and Pride et al. (2004) to determine the elastic moduli of dry rock, and the best critical physical parameters (porosity and consolidation coefficient). For the purpose of *implementation* of predicting V_s from well logs, a corresponding log interpretation method for shaly gas-bearing reservoirs is analyzed. It was shown that the computed volumetric concentrations of minerals match the results obtained from the X-ray diffraction (XRD) analysis of cores. In addition, the final V_s values are in agreement with dipole sonic imaging (DSI) log, and the suggested workflow is proved to be appropriate for gas-bearing shale formations.

It is worth mentioning that a good estimation of S-wave using any of the methods discussed above will be first success in determining reservoir properties. This is because all quantitative reservoir characterizations require P- and S-waves. When data related to both waves are available, a number of reservoir attributes and properties can be extracted to help discriminate fluid and lithology types of the reservoir (Yoon and Farfour, 2012). Table 6.4 lists some reservoir properties which can be extracted if the V_p and V_s are provided.

6.5 Example

Seismic data with logging data from gas field in Alberta, Canada, are used herein to show the importance of S-wave. The logging data does not contain S-wave. Thus, it was important to derive it using one of the approaches discussed above. We used fluid replacement modeling to generate the S-wave from P-wave. First the Biot-Gassmann equations were used to transfer the elastic properties from the in-situ saturation to 100% brine saturation where Greenberg-Castagna equation can be applied. At this saturation, the S-wave was derived deterministically. Then the S-wave was computed at the reservoir saturation using Biot-Gassmann equations (Figure 6.1). The generated log was used along with P-wave for AVO analysis and inversion. The latter analysis and inversion helped extracting a number of features that could not be derived without the S-wave. Figure 6.2 shows the AVO attributes, which were derived from the AVO analysis. The attributes could show the reservoir and distinguish it from its surrounding. Then the S-wave along with the P-wave were using to build an initial model to help invert the seismic data for impedances. S-impedance and P-impedance sections were successfully derived and displayed in Figure 6.3. From the two quantities other attributes were calculated. This includes but not limited to V_p to V_s ratio, Poisson ratio, Lame parameters which have helped taking the data interpretation steps further toward quantitative interpretation (Figure 6.4).

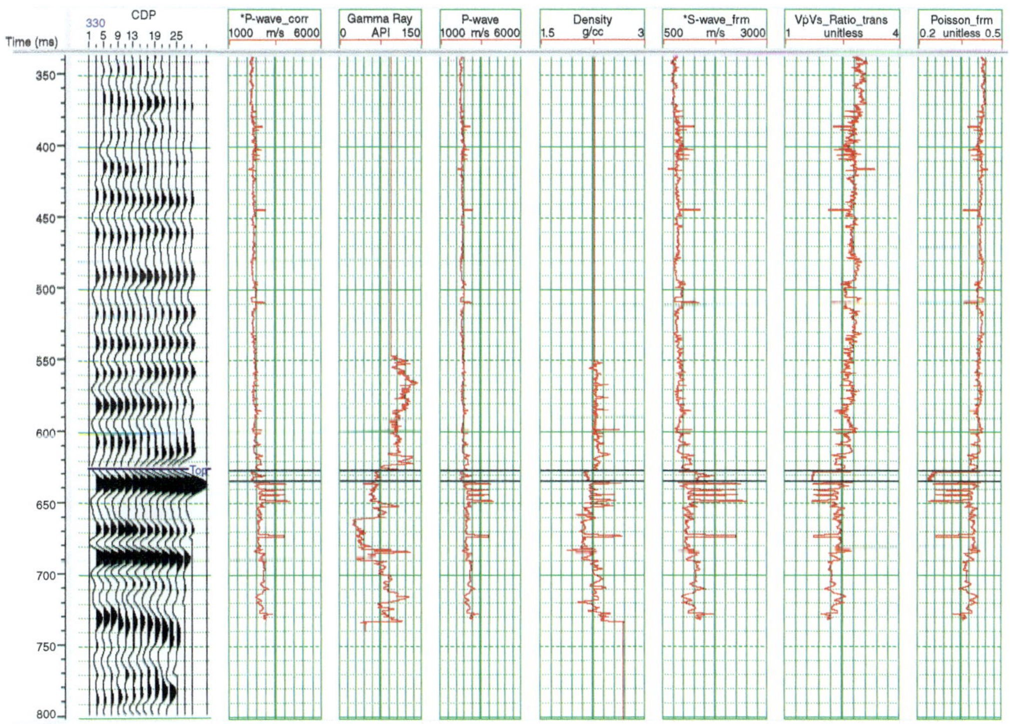

Figure 6.1 P-wave and S-wave log derived from it. Both logs are used to generate synthetic gathers and compute V_pV_s and Poisson ratio at the well location. The computed parameters along with AVO modeling have helped predict the reservoir seismic response and discriminate it against surrounding rocks.

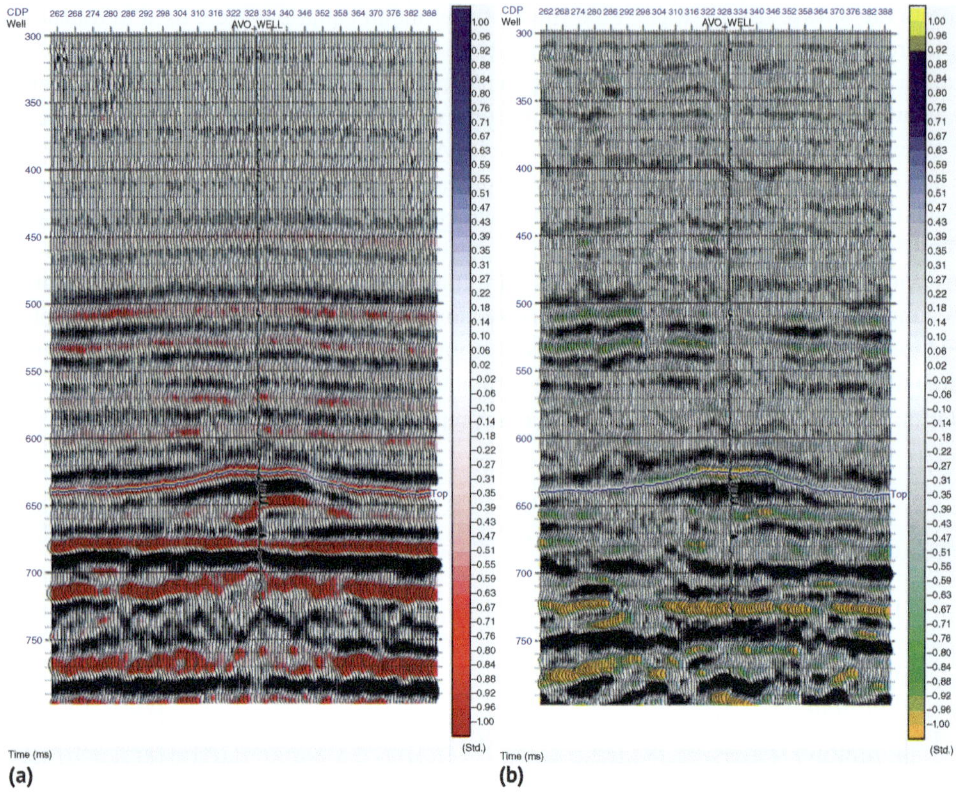

Figure 6.2 Stack section (a) and fluid factor (b) derived from AVO analysis.

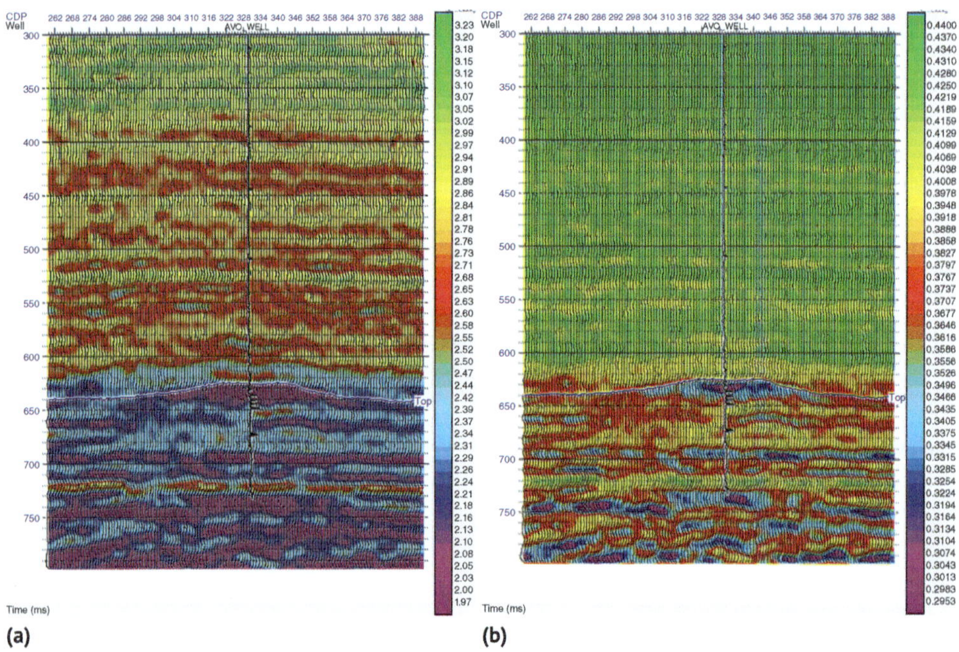

Figure 6.3 V_pV_s ratio derived from AVO inversion (a) and Poisson ratio determined from it (b).

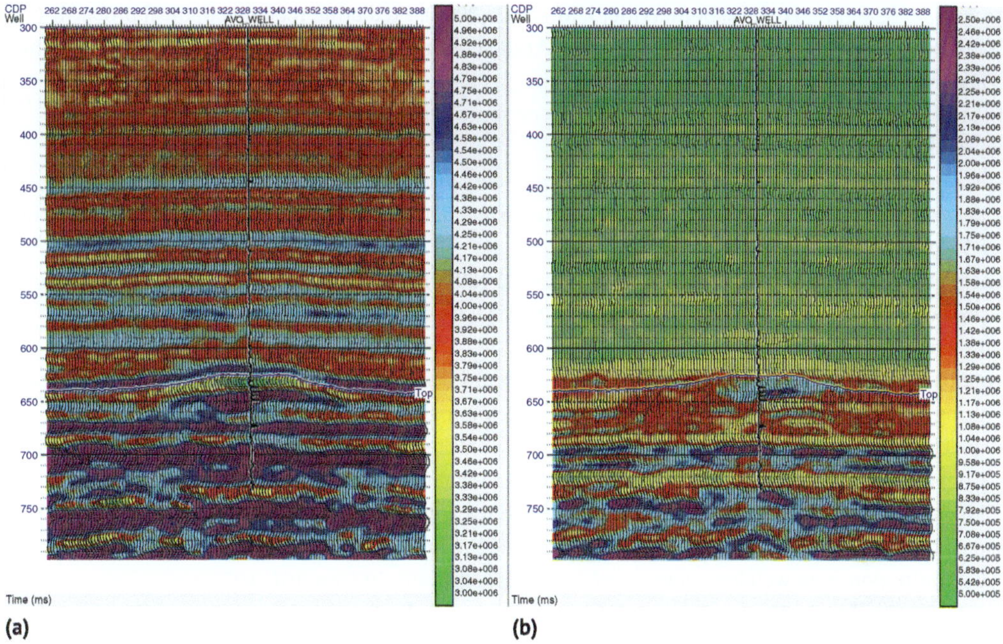

Figure 6.4 Lambda-Rho (a) and Mu-Rho (b) calculated from the impedances. The first shows the fluid incompressibility while the second shows the rigidity of matrix.

6.6 Conclusions

Shear wave velocity (V_S) is of a high interest and importance in hydrocarbon reservoir evaluation, fracture evaluation, geophysical and geomechanical studies. The selection of a model to estimate it depends on the available data (laboratory core analysis results, well logs, seismic data, numerical models, etc.). The best approach consists on combining all the data through intelligent systems, in order to improve the quality of V_S predictions. Once V_S data becomes available, numerous reservoir properties can be derived to help accomplish a comprehensive quantitative reservoir characterization.

References

Al-Kattan, W. M. (2008). A comprehensive approach to the interlock between the carbonate rock mechanical properties, and cementation factor. College of Engineering, Petroleum Dept., University of Baghdad.

Al-Kattan, W. M. (2015). Prediction of shear wave velocity for carbonate rocks. Iraqi Journal of Chemical and Petroleum Engineering 6 (4), pp. 45–49.

Akhundi, H., Ghafoori, M., & Lashkaripour, G.-R. (2014). Prediction of shear wave velocity using artificial neural network technique, multiple regression and petrophysical data: A case study in Asmari reservoir (SW Iran). Open Journal of Geology 4, pp. 303–313. http://dx.doi.org/10.4236/ojg.2014.47023.

Ameen, M. S., Smart, B. G. D., Somerville, J. M., et al. (2009). Predicting rock mechanical properties of carbonates from wireline logs (A case study: Arab-D reservoir, Ghawar field, Saudi Arabia). Marine and Petroleum Geology 26, pp. 430–444.

Anselmetti, F. S. & Eberli, G. P. (1993). Controls on sonic velocity in carbonates. Pure and Applied Geophysics 141, pp. 287–323.

Anselmetti, F. S. & Eberli, G. P. (1999). The velocity-deviation log: A tool to predict pore type and permeability trends in carbonate drill holes from sonic and porosity or density logs. AAPG Bulletin 83, pp. 450–466.

Anemangely, M., Ramezanzadeh, A., Amiri, H., et al. (2019). Machine learning technique for the prediction of shear wave velocity using petrophysical logs. Journal of Petroleum Science and Engineering 174, pp. 306–327.

Aranibar, A., Saneifar, M., & Heidari, Z. (2013). Petrophysical rock typing in organic-rich source rocks using well logs. In: Unconventional Resources Technology Conference pp. 1154–1162.

Asoodeh, M. & Bagheripour, P. (2012). Prediction of compressional, shear, and Stoneley wave velocities from conventional well log data using a committee machine with intelligent systems. Rock Mechanics and Rock Engineering 45, pp. 45–63. https://doi.org/10.1007/s00603-011-0181-2.

Asoodeh, M. & Bagheripour, P. (2014). ACE stimulated neural network for shear wave velocity determination from well logs. Journal of Applied Geophysics 107, pp. 102–107.

Avseth, P., Dvorkin, J., Mavko, G., et al. (2000). Rock physics diagnostic of North Sea sands: Link between microstructure and seismic properties. Geophysical Research Letters 27, pp. 2761–2764.

Avseth, P., Jorstad, A., van Wijngaarden, A. J., et al. (2009). Rock physics estimation of cementation, sorting and net-to-gross in North Sea sandstones. The Leading Edge, 28, pp. 98–109.

Avseth, P., Lehocki, I., Kjøsnes, Ø., et al. (2021). Data-driven rock physics analysis of North Sea tertiary reservoir sands. Geophysical Prospecting 69, pp. 608–621. https://doi.org/10.1111/1365-2478.12986.

Avseth, P., Mukerji, T., Mavko, G., et al. (2001). Rock physics and AVO analysis for lithofacies and pore fluid prediction in a North Sea oil field. The Leading Edge 20 (4), pp. 429–434.

Azadpour, M., Saberi, M. R., Javaherian, A., et al. (2020). Rock physics model-based prediction of shear wave velocity utilizing machine learning technique for a carbonate reservoir, southwest Iran. Journal of Petroleum Science and Engineering 195, pp. 107864. https://doi.org/10.1016/j.petrol.2020.107864.

Bai, J. Y., Yue, C. Q., Liang, Y. Q., et al. (2013). Variable aspect ratio method in the Xu-White model for shear-wave velocity estimation. Journal of Geophysics and Engineering 10 (3), p. 035008.

Bailey, T. & Dutton, D. (2012). An empirical *Vp*/Vs shale trend for the Kimmeridge clay of the Central North Sea. In: 74th EAGE Conference and Exhibition Incorporating EUROPEC 2012.

Brocher, T. M. (2005). Empirical relations between elastic wavespeeds and density in the earth's crust. Bulletin of the Seismological Society of America 95 (6), pp. 2081–2092.

Brown, R. J. S. & Korringa, J. (1975). Dependence of elastic properties of a porous rock on compressibility of pore fluid. Geophysics 40, pp. 608–616.

Bukar, D., Adamu, M. B., & Hassan, U. (2019). A machine learning approach to shear sonic log prediction. Society of Petroleum Engineers SPE-198764-MS.

Carroll, R. D. (1969). The determination of the acoustic parameters of volcanic rocks from compressional velocity measurements. International Journal of Rock Mechanics and Mining Sciences & Geomechanics Abstracts. Elsevier, pp. 557–579.

Castagna, J. P., Batzle, M. L., & Eastwood, R. L. (1985). Relationship between compressional and shear-wave velocities in classic silicate rocks. Geophysics 50 (4), pp. 571–581.

Castagna, J. P., Batzle, M., & Kan, T. (1993). Rock physics – The link between rock properties and AVO response. In: Offset-Dependent Reflectivity – Theory and Practice of AVO Analysis (eds. J. P. Castagna & M. Backus). Investigations in Geophysics No. 8. Tulsa, Oklahoma: Society of Exploration Geophysicists, pp. 135–171.

Chang, C., Zoback, M. D., & Khaksar, A. (2006). Empirical relations between rock strength and physical properties in sedimentary rocks. Journal of Petroleum Science and Engineering 51, pp. 223–237.

Domenico, S. N. (1984). Rock lithology and porosity determination from shear and compressional wave velocity. Geophysics 49, pp. 1188–1195.

Downton, J. E. (2005). *Seismic Parameter Estimation from AVO Inversion*. University of Calgary, Department of Geology and Geophysics.

Dvorkin, J. P. & Nur, A. M. (1996). Elasticity of high-porous sandstones: Theory for two North Sea data sets. Geophysics 61 (5), pp. 1363–1370.

Eberli, G. P., Baechle, G. T., Anselmetti, F. S., et al. (2003). Factors controlling elastic properties in carbonate sediments and rocks. The Leading Edge 22 (7), pp. 654–660.

Eskandari, H., Rezaee, M. R., Javaherian, A., et al. (2003). Shear wave velocity estimation utilizing wireline logs for a carbonate reservoir, South-West Iran. Iranian International Journal of Science 4(2), pp. 209–221.

Eskandari, H., Rezaee, M. R., & Mohammadnia, M. (2004). Application of multiple regression and artificial neural network techniques to predict shear wave velocity from wireline log data for a carbonate reservoir, South Iran. Canadian Society of Exploration Geophysicists (CSEG) Recorder 29, pp. 42–48.

Farfour, M. & Mesbah, M. (2020). Machine intelligence vs. human intelligence in geological interpretation of seismic data. 2020 International Conference on Decision Aid Sciences and Application (DASA), Sakheer, Bahrain, pp. 996–999. https://doi.org/10.1109/DASA51403.2020.9317032.

Farfour, M. & Yoon, W. J. (2016). A review on multicomponent seismology: A potential seismic application for reservoir characterization. Journal of Advanced Research 7 (3), pp. 515–524. https://doi.org/10.1016/j.jare.2015.11.004.

Gaci, S. (2017). A novel model to estimate S -wave velocity integrating Hölderian regularity, empirical mode decomposition, and multilayer perceptron neural networks. In: Oil and Gas Exploration: Methods and Application (ed. S. Gaci and O. Hachay), ch. 12, pp. 181–200. Wiley Editions- USA.

Gassmann, F. (1951). Über die elastizität poröser medien: Vier. der Natur. Gesellschaft Zürich 96, pp. 1–23.

Gholami, R., Moradzadeh, A., Rasouli V., et al. (2014). Shear wave velocity prediction using seismic attributes and well log data. Acta Geophysica 62 (4), pp. 818–848. https://doi.org/10.2478/s11600-013-0200-7.

Greenberg, M. & Castagna, J. (1992). Shear wave velocity estimation in porous rocks: Theoretical formulation, prelimining verification and applications. Geophysical Prospecting 40 (2), pp. 195–209.

Ji, S. C., Wang, Q., & Xia, B. (2002). Handbook of Seismic Properties of Minerals, Rocks and Ores, Polytechnic International Press, Montreal, 630 pp.

Jørstad, A., Mukerji, T., & Mavko, G. (1999). Model-based shear-wave velocity estimation versus empirical regressions. Geophysical Prospecting 47 (5), pp. 785–797.

Hamada, G. M. (2004). Reservoir fluids identification using *Vp*/*Vs* ratio. Oil and Gas Science and Technology – Rev IFP. 59, pp. 649–654.

Han, D. H., Nur, A., & Morgan, D. (1986). Effect of porosity and clay content on wave velocity in sandstones. Geophysics 51 (11), pp. 2093–2107. https://doi.org/10.1190/1.1442062.

Hossain, Z., Mukerji, T., Dvorkin, J., et al. (2011). Rock physics model of glauconitic greensand from the North Sea. Geophysics 76, p. 199.

Hossain, Z., Mukerji, T., & Fabricius, I. L. (2012). V p-V s relationship and amplitude variation with offset modelling of glauconitic greensand. Geophysical Prospecting 60, pp. 117–137. https://doi.org/10.1111/j.1365-2478.2011.00968.x.

Keys, R. G. & Xu, S. Y. (2002). An approximation for the Xu-White velocity model. Geophysics 67 (5), pp. 1406–1414.

Kim, J., Kang, J.-D., & Kim, B. (2023). Machine-learning models to predict P and S-wave velocity profiles for Japan as an example. Frontiers in Earth Science 11, pp. 1267386. https://doi.org/10.3389/feart.2023.1267386.

Krief, M. (1990). A petrophysical interpretation using the velocities of P and S waves (fullwavefoem sonic). The Log Analyst 31, pp. 355–369.

Krief, M., Garat, J., Stellingwerf, J., et al. (1990). A petrophysical interpretation using the velocities of P and S waves (full-waveform Sonic). The Log Analyst 31, pp. 355–369.

Krishna, K. S., Rao, D. G., Murty, G. P. S., et al. (1989). Sound velocity, density, and related properties along a transect across the Bay of Bengal. Geo-Marine Letters 9, pp. 95–102.

Kuiper, J., Van Ryan, W.M.L., & Kocfoed, O. (1959). Laboratory determinations of elastic properties of some limestones. Geophysical Prospecting 7, pp. 38–14.

Kuster, G. T. & Toksoz, M. N. (1974a). Velocity and attenuation of seismic waves in two-phase media, part I. Theoretical formulations. Geophysics 39 (5), pp. 587–606.

Kuster, G. T. & Toksoz, M. N. (1974b). Velocity and attenuation of seismic waves in two-phase media: Part II. Experimental results. Geophysics 39 (5), pp. 607–618.

Kwak, D. Y., Brandenberg, S. J., Mikami, A., et al. (2015). Prediction equations for estimating shear-wave velocity from combined geotechnical and geomorphic indexes based on Japanese data set. Bulletin of the Seismological Society of America 105 (4), pp. 179–188.

Lee, M. W. (2006). A simple method of predicting S-wave velocity. Geophysics 71, pp. F161–F164.

Lee, M. W. (2013). Comparison of methods for predicting shear-wave velocities of unconsolidated shallow sediments in the Gulf of Mexico. United States Department of the Interior and the United States Geological Survey.

Liu, Z. & Sun, S. Z. (2015). The differential Kuster–Toksöz rock physics model for predicting S-wave velocity. Journal of Geophysics & Engineering 12, pp. 839–848.

Mavko, G., Mukerji, T., & Dvorkin, J. (2009). The Rock Physics Handbook (Tools for Seismic Analysis of Porous Media). Cambridge University Press.

Miller, S. L. M. & Stewart, R. R. (1991). The relationship between elastic-wave velocities and density in sedimentary rocks: A proposal. CREWES Research Report pp. 260–273.

Mindlin, R. D. (1949). Compliance of elastic bodies in contact. Journal of Applied Mechanics 16 (3), pp. 259–268.

Nourafkan, A. & Kadkhodaie-Ilkhchi, A. (2015). Shearwave velocity estimation from conventional well log data by using a hybrid ant colony–fuzzy inference system: A case study from Cheshmeh–Khosh oilfield. Journal of Petroleum Science and Engineering 127, pp. 459–468.

Nur, A. (1992). Critical Porosity and the seismic velocities in rocks. EOS Transactions, American Geophysical Union 73, pp. 43–66.

Pickett, G. R. (1963). Acoustic character logs and their application information evaluation. Journal of Petroleum Technology 15, pp. 650–667.

Pride, S. R., Berryman, J. G., & Harris, J. M. (2004). Seismic attenuation due to wave-induced flow. Journal of Geophysical Research Solid Earth 109, (BI), pp. B012OI.

Rajabi, M. & Tingay, M. (2013). Applications of intelligent systems in petroleum geomechanics prediction of geomechanical properties in different types of sedimentary rocks. EAGE International Workshop on Geomechanics and Energy, Lausanne, Switzerland Tu P01 02.

Rajabi, M., Bohloli, B., & Ahangar, E. G. (2010). Intelligent approaches for prediction of compressional, shear and Stoneley wave velocities from conventional well log data: A case study from the Sarvak carbonate reservoir in the Abadan Plain (Southwestern Iran). Computers & Geosciences 36 (5), pp. 647–664.

Ranjbar-Karami, R., Kadkhodaie-Ilkhchi, A., & Shiri, M. (2014). A modified fuzzy inference system for estimation of the static rock elastic properties: A case study from the Kangan and Dalan gas reservoirs, South Pars gas field, the Persian Gulf. Journal of Natural Gas Science and Engineering 21 (2), pp. 962–976.

Shao, C. R., Yin, X. Y., Zhang, F. M., et al. (2009). Shear wave velocity inversion with rountine well logs based on rock physics and multi-mineral analysis. Earth Science – Journal of China University of Geosciences 34, pp. 696–699.

Smith, G. C. & Gidlow, M. (2000). A comparison of the fluid factor with λ and μ in AVO analysis, SEG Technical Program Expanded Abstracts 2000. Society of Exploration Geophysicists, pp. 122–125.

Spichak, V. V. & Goidina, A. G. (2016). Neural network estimate of seismic velocities and resistivity of rocks from electromagnetic and seismic sounding data. Izvestiya, Physics of the Solid Earth 52 (3), pp. 371–381.

Tabari K., Tabari, O., & Tabari M. (2011). A fast method for estimating shear wave velocity by using neural network. Australian Journal of Basic and Applied Sciences 5 (11), pp. 1429–1434, ISSN 1991-8178.

Tan, M., Peng, X., Cao, H., et al. (2015). Estimation of shear wave velocity from wireline logs in gas-bearing shale. Journal of Petroleum Science and Engineering 133, pp. 352–366.

Vernik, L., Fisher, D., & Bahret, S. (2002). Estimation of net-to-gross from P and S impedance in deepwater turbidites. The Leading Edge 21, p. 380. https://doi.org/10.1190/1.1471602.

Vernik, L., Castagna, J., & Omovie, S. J. (2017). Shear-wave velocity prediction in unconventional shale reservoirs. Geophysics 83 (1), pp. 1–43.

Wang, Z. (2000). Velocity relationships in granular rocks. In: Seismic and Acoustic Velocities in Reservoir Rocks (ed. Z. Wang & A. Nur), pp. 145–158.

Wang, J., Wu, S., Zhao, L., et al. (2019). An effective method for shear-wave velocity prediction in sandstones. Marine Geophysical Research 40, pp. 655–664. https://doi.org/10.1007/s11001-019-09396-4.

Williams, D. M. (1990). The acoustic log hydrocarbon indicator. In: SPWLA 31st Annual Logging Symposium. Society of Petrophysicists and Well-Log Analysts.

Xu, S. & White, R. E. (1995). A new velocity model for clay-sand mixture. Geophysical Prospecting 43 (473), pp. 91–118.

Xu, S. & White, R. E. (1996). A physical model for shear-wave velocity prediction. Geophysical Prospecting 475 (44), pp. 687–717.

Xu, S. & White, R. E. (2010). A physical model for shear-wave velocity prediction. Geophysical Prospecting 44 (4), pp. 687–717.

Yan, J., Lubbe, R., & Pillar, N. (2008). Modified velocity model for seismic study via AVO. Journal of Seismic Exploration 17 (4), pp. 371–390.

Yang, Y., Yin, X., Gao, G., et al. (2019). Shear-wave velocity estimation for calciferous sandy shale formation. Journal of Geophysics and Engineering 16, pp. 105–115.

Yoon, W. J. & Farfour, M. (2012). Spectral decomposition aids AVO analysis in reservoir characterization: A case study of Blackfoot field, Alberta, Canada. Computers & Geosciences 46, pp. 60–65.

Yusuf, B. O. (2019). Elastic properties prediction for drilling and subsurface evaluation for the Grand Banks and Niger Delta. PhD Thesis, Memorial University of Newfoundland, Canada.

7

Geomechanics in Petroleum Exploration, Development, and Energy Transition

Ghoulem Ifrene[1], Kuldeep Singh[2]

[1] *University of North Dakota, Grand Forks, North Dakota, USA*
[2] *Earth Science Department, Kent State University, Kent, Ohio, USA*

7.1 Introduction

The realm of subsurface exploration and exploitation has long been pivotal in meeting global energy demands, especially in a world where economic expansion, technological advancements, and population growth surge relentlessly. These pursuits are increasingly aligned with the imperative to decarbonize energy systems to achieve carbon neutrality. The subsurface offers vast potential for energy storage, production, and waste disposal, serving as a critical component in the energy transition landscape. However, the intrinsic complexity of subsurface characterization, coupled with the interplay of multi-physical processes, presents significant challenges. Addressing these requires a nuanced understanding of geomechanics, an evolving field that now integrates principles from geotechnics, physics, geochemistry, and bio-geology.

Furthermore, the increasing importance of unconventional resources like shale gas highlights the evolving role of geomechanics throughout the reservoir life cycle (Figure 7.1). Developments in drilling technologies and hydraulic fracturing have revolutionized shale gas production, underscoring the critical role of geomechanical evaluations in optimizing fracture stimulations and improving recovery rates. Geomechanical properties, such as rock strength and elastic properties, are essential for constructing accurate subsurface models that predict behaviors under different stress conditions and contribute to safer, more efficient operations.

This chapter aims to synthesize the cutting-edge research and trends within geomechanics as they pertain to both conventional and unconventional energy systems. It will explore how geomechanical studies enhance our understanding of the subsurface, facilitate the design and execution of energy extraction operations, and ensure the environmental integrity of these processes. By examining the latest advancements and the ongoing challenges, this discourse seeks to forge pathways toward more sustainable and efficient utilization of subsurface resources, thus playing a crucial role in the broader context of energy transition.

Geophysical Exploration for Hydrocarbon Reservoirs, Geothermal Energy, and Carbon Storage: New Technologies and AI-based Approaches, First Edition. Edited by Said Gaci.

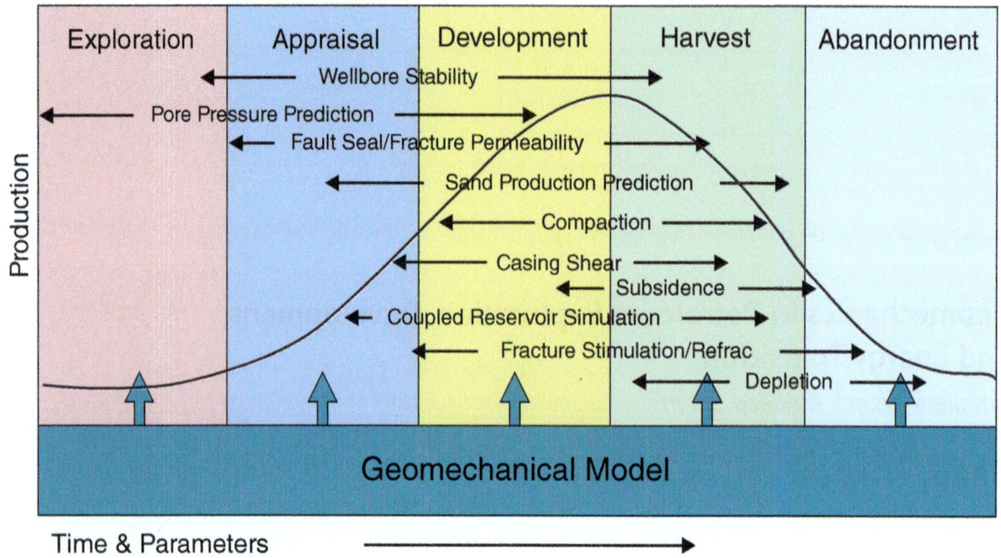

Figure 7.1 Geomechanics life cycle model. *Source*: ZOBACK, 2016 / Stanford University.

7.2 Role of Geomechanics in Exploration and Development

The exploration and development of subsurface resources, such as oil and gas, rely heavily on the integration of geomechanics into the planning and execution phases. Geomechanics plays a crucial role in understanding and managing the subsurface conditions that affect drilling operations and reservoir performance. Key aspects such as formation pressure prediction, wellbore stability, and fracture initiation are vital for safe and efficient drilling practices. Accurate predictions and analyses in these areas prevent drilling hazards, optimize well placement, and enhance hydrocarbon recovery.

Formation pressure prediction is essential for the successful planning and execution of drilling operations. It helps in preventing drilling hazards like wellbore instability, blowouts, and lost circulation. Various methods have been developed to predict pore pressure using geological and geophysical data, each offering unique insights depending on the geological context and data quality. Integrating these methods allows for a comprehensive understanding of subsurface pressure conditions, aiding in safer and more efficient drilling operations.

Wellbore stability is another critical component of drilling operations. Instabilities can arise from uneven stress distributions around the borehole due to hydraulic, mechanical, chemical, and thermal influences. Ensuring wellbore stability is crucial for both the initial drilling phase and subsequent re-entry into mature oil fields. Addressing both mechanical and chemical instabilities through appropriate drilling practices and fluid selection is vital for maintaining borehole integrity.

Tensile fracture initiation mechanisms are also significant in various engineering applications, including geothermal energy extraction, oil and gas recovery, and underground storage. These mechanisms are influenced by high in situ stress, temperature variations, and chemical interactions. Understanding these factors helps in managing and optimizing fracture initiation and propagation, which is essential for the effective exploitation of subsurface resources.

7.2.1 Formation Pressure Prediction

Understanding and predicting pore pressure in subsurface formations is crucial for the successful planning and execution of drilling operations in the oil and gas industry. Accurate pore pressure predictions help in preventing drilling hazards such as wellbore instability, blowouts, and lost circulation. Moreover, reliable pore pressure estimates are essential for optimizing well placement and enhancing hydrocarbon recovery. Over the years, various methods have been developed to predict pore pressure using geological and geophysical data.

Pore pressure prediction methods can be broadly categorized based on the type of data they utilize, including resistivity measurements, sonic velocity, and seismic data. Each method offers unique insights and has its own set of assumptions and applicability depending on the geological context and the quality of the available data. The integration of these methods allows for a more comprehensive understanding of subsurface pressure conditions, aiding in safer and more efficient drilling operations.

There are several widely used pore pressure prediction methods, including resistivity-based methods, interval velocity and transit time-based methods, and empirical models. Some of the methods are covered in Table 7.1 including the Eaton method, Bowers method, Miller method, Tau model, and the adapted Eaton's method with depth-dependent normal compaction trendline.

Table 7.1 Pore pressure prediction methods.

Method	Formula	Explanation
Hottmann and Johnson (1965)	$p_f = \sigma_V - \frac{(\alpha_V - \beta)(A_1 - B_1 \ln \Delta t)^3}{Z^2}$	p_f is the formation fluid pressure (psi); σ_V (psi); α_V is the overburden stress gradient (psi/ft); β is the fluid pressure gradient (psi/ft); Z is the depth (ft); Δt is the slowness (μs/ft) ; A and B are constants, A1 = 82 776 and B1 = 15 695.
Eaton (1972)	$P_{pg} = OBG - (OBG - P_{ng})\left(\frac{R}{R_n}\right)^n$	P_{pg} is the formation pore pressure gradient; OBG is the overburden stress gradient; P_{ng} is the hydrostatic pore pressure gradient; R is the shale resistivity; R_n is the shale resistivity at the normal (hydrostatic) pressure; n is the exponent varied from 0.6 to 1.5.
Eaton (1975)	$P_{pg} = OBG - (OBG - P_{ng})\left(\frac{\Delta t_n}{\Delta t}\right)^3$	Δt_n is the slowness in shales at the normal pressure; Δt is the sonic transit time in shales.
Bowers (1995)	$p = \sigma_V - \left(\frac{10^6\left(\frac{1}{\Delta t} - \frac{1}{\Delta t_{ml}}\right)}{A}\right)^{\frac{1}{B}}$	Δtml is the compressional transit time in the mudline; A and B are constants where A = 10–20 and B = 0.7–0.75; Δt is the slowness; σ_V is the overburden stress.
Zhang et al. (2008)	$p = \sigma_V - \frac{1}{\lambda} \ln\left(\frac{v_m - v_{ml}}{v_m - v_p}\right)$	v_m is the sonic interval velocity in the matrix of the shale; v_p is the compressional velocity; λ is the rate of increase in velocity with effective stress (0.00025).
Lopez et al. (2004)	$p = \sigma_V - A_s\left(\frac{C - \Delta t}{\Delta t - D}\right)^{B_s}$	A_s and B_s are the fitting constants; Δt is the compressional transit time; C is the constant related to the mudline transit time; and D is the constant related to the matrix transit time.

7.2.2 Wellbore Stability Analysis

Wellbore instability lacks a precise definition but refers to a condition encountered during drilling operations where a borehole cannot preserve its original condition. Specifically, it happens when the borehole's diameter enlarges beyond that of the drill bit (Ibrahim, 2021). Such instability in the borehole can result from uneven stress distributions surrounding the borehole, which may stem from various factors, including hydraulic, mechanical, chemical, and thermal influences (Aslannezhad et al., 2020).

Securing the stability of boreholes emerged as a critical component of drilling operations during the early 1980s, especially with the development of extended, highly inclined wells aimed at efficient exploration and exploitation in both offshore and onshore settings of sizable reservoirs. Furthermore, the practice of re-entering both active and abandoned wells through side-tracking in mature oil fields highlights the increased importance. This is due to prior hydrocarbon extraction and interactions with drilling fluids (Figure 7.2), which have altered the mechanical properties of the rocks, thereby emphasizing the need for comprehensive wellbore stability research.

Wellbore instabilities can generally be categorized into two primary types: mechanical and chemical instabilities. Mechanical instabilities are managed by adjusting the stress state around the wellbore. This is achieved by selecting suitable mud weights, planning optimal drilling trajectories, and adhering to rigorous drilling and tripping practices. Chemical instabilities, on the other hand, are mitigated through the use of drilling fluids that are chemically compatible with the formation being drilled, supplemented with the necessary additives in appropriate amounts. While mechanical instability issues can arise in various geological formations, chemical instabilities are predominantly observed when drilling through reactive shale formations.

Chemically induced instability typically results from the interaction of aqueous drilling fluids with shale formations (Chen et al., 2003). This interaction creates a chemical potential that drives the movement of water and solutes into and out of the formation through chemical osmosis and diffusion processes Hale et al. (1993) highlighted that the flow of fluids between shales and drilling fluids is primarily influenced by chemical and hydraulic pressure differences. They noted that highly compacted shale acts like a semi-permeable membrane, facilitating osmotic water transport driven by differences in chemical potential, often described in terms of water activity. At equilibrium, it is usually assumed that the chemical potential of the drilling fluid equals that of the pore fluid. The gradient in chemical potential leads to a chemically induced stress known as osmotic pressure.

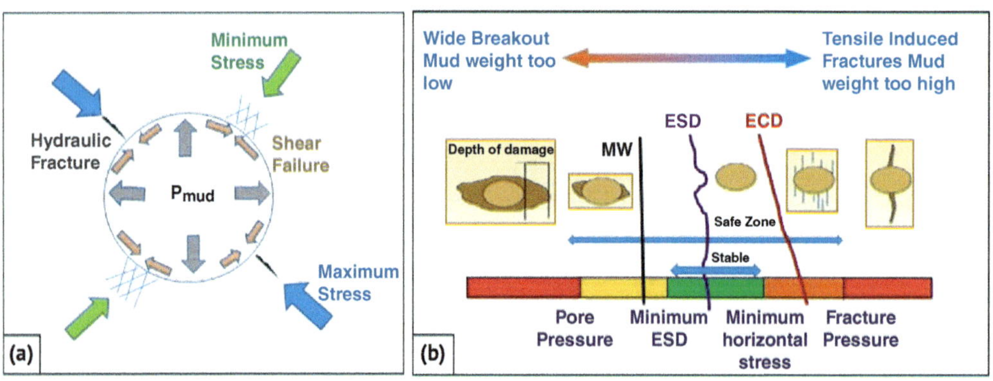

Figure 7.2 This diagram illustrates stress and wellbore damage. It includes (a) directions of maximum and minimum horizontal stress around the borehole, (b) the mud weight, along with different bottom-hole pressures. Here, ESD and ECD represent the respective static and dynamic densities within a configuration, and MW stands for the mud window. *Source*: Radwan, 2022 / MDPI / CC BY 4.0.

The osmotic pressure theoretically expected is determined by the chemical potential difference as described by the equation (7.1) where water activities of the drilling fluid and shale, as well as the wellbore and pore fluid pressures, are considered. When the water activity is higher in the formation, water molecules migrate toward it, and selecting a drilling fluid with lower activity can help reduce instability. Alternatively, it can trigger an osmotic backflow from the formation into the wellbore, countering the hydraulic filtrate flow caused by overbalance pressure during drilling.

$$\frac{RT}{V}\ln\frac{a_{wdf}}{a_{wsh}} = \pm\Delta P = \left(P_{nw} - P_{pf}\right) \tag{7.1}$$

where R is gas law constant; T is thermodynamic temperature; V is the molar volume of water; a_{wdf} is water drill fluid activity; a_{wsh} is shale drill fluid activity; P_{pf} is formation pore pressure; and P_{nw} is pore pressure under different chemical potentials.

However, the osmotic pressures observed experimentally often differ from those predicted theoretically, as shown by studies like those conducted by (Mody and Hale, 1993), and others. The ratio of observed to theoretical osmotic pressure is termed "membrane efficiency" or "reflection coefficient." This efficiency is never perfect in real-world scenarios, where both solutes and water may pass through the membrane. Shale formations, for instance, are considered non-ideal membranes. Their membrane efficiency, which reflects their ability to block solute migration, is influenced by factors such as the hydrated size of solutes and the surface area of the shale. Research by Oleas et al. (2008) revealed that increases in shale porosity and permeability could adversely affect membrane efficiency.

7.2.3 Tensile Fracture Initiation Mechanisms

Tensile fracture initiation in rocks, particularly under deep extreme in situ conditions, is a critical aspect of rock mechanics, influencing various engineering applications such as geothermal energy extraction, oil and gas recovery, and underground storage. The mechanisms driving tensile fracture initiation are influenced by multiple factors including high in situ stress, temperature variations, and chemical interactions. Here is an overview of the key mechanisms:

Stress Concentration:

- High *in situ* stress levels, typical in deep geological formations, significantly influence the initiation of tensile fractures. Stress concentration around pre-existing flaws or heterogeneities within the rock can create conditions favorable for fracture initiation.
- Microcrack Closure and Opening: Under high stress, microcracks within the rock matrix can close, leading to increased rock density and strength. However, when tensile stress exceeds a critical threshold, these microcracks can reopen and propagate, initiating tensile fractures.

Temperature Effects:

- Thermal Expansion: Temperature variations cause differential thermal expansion of mineral grains within the rock. This expansion can induce thermal stresses, particularly at grain boundaries, leading to crack initiation (Zhao et al., 2024).
- Thermal Shock: Rapid temperature changes, such as those caused by liquid nitrogen cooling and super critical CO_2, can create high thermal gradients. This induces thermal shock, increasing the likelihood of tensile fracture initiation (Liu et al., 2022; Mardoukhi et al., 2017).

- Brittle to Ductile Transition: At certain temperature ranges, rocks may exhibit a transition from brittle to ductile behavior (Yin et al., 2015). This transition influences the mode and ease of fracture initiation, with higher temperatures potentially increasing ductility and delaying fracture initiation.

Chemical Interactions:

- Chemical Corrosion: The interaction of rock with chemically active fluids (e.g. acidic or alkaline solutions) can weaken the rock's structural integrity. This chemical weakening reduces the tensile strength, making it easier for fractures to initiate and propagate (Lin et al., 2020; Mei et al., 2022).
- Fluid-Rock Interaction: *In situ* fluids can infiltrate rock fractures, altering the mechanical properties through processes like dissolution, precipitation, and chemical alteration of minerals (Gratier and Gueydan, 2007; Sausse et al., 2001). These interactions can either enhance or impede fracture initiation depending on the chemical nature of the fluids.

7.3 Enhancing Reservoir Performance Through Geomechanics

Geomechanics is pivotal in various sectors of the oil industry, particularly in areas like borehole stability, hydraulic fracturing, and production-induced compaction and subsidence. Key parameters in these domains include in-situ stresses, rock deformations, and fluid flow behavior. The interaction between geomechanics and multiphase fluids is well-recognized in hydraulic fracturing. Research by (Advani et al., 1986)) and (Settari et al., 1992)) emphasized the significance of fracture-induced in-situ stress variations and deformations on reservoir behavior, demonstrating the benefits of integrating hydraulic fracturing with reservoir simulations. However, in other applications, geomechanics is often treated separately from multiphase fluid flow, leading to oversimplified assumptions in each field. This separation is due to the challenges of addressing geomechanics and multiphase fluid flow as intertwined processes.

Interest in incorporating geomechanics into reservoir simulations has recently grown, especially for heavy oil and bituminous sand reservoirs (Zandi et al., 2010), water injection in fractured fields (Liu et al., 2021), and compacting/subsiding areas (Giani et al., 2018). Various approaches have emerged, differing in how geomechanical aspects are integrated with multiphase fluid flow and the degree of coupling between them.

7.3.1 Performance Evaluation of Fractured Reservoirs

The evaluation of fractured reservoirs is crucial for understanding the performance and optimizing the development of unconventional gas resources. The complexity and efficiency of fracture networks, as well as the geomechanical properties of the reservoir, significantly influence the economic viability of these projects.

Effective development of unconventional gas reservoirs, such as shale gas, hinges on maximizing the reservoir area connected to the wellbore through a complex fracture network. Traditional hydraulic fracturing methods, which create planar fractures, are often inadequate for these low-permeability formations. Instead, large-volume, high-rate hydraulic-fracture treatments using water and small-mesh proppant are employed to stimulate existing natural fractures and the rock matrix, creating a large and complex fracture network (Cipolla, 2009; Yu and Sepehrnoori, 2013).

The conductivity of these fracture networks is a key determinant of gas recovery. Conductivity is influenced by the presence and type of proppant used in the fractures. Laboratory studies show that unpropped fractures have very low conductivity, especially under high closure stress typical of shale reservoirs. However, the introduction of high-strength proppant, such as sintered bauxite, can significantly improve fracture conductivity (Fredd et al., 2001).

The geomechanical properties of the reservoir rock, including its Young's modulus and the stress-dependence of fracture conductivity, play a critical role in the performance of fractured wells. Stiffer rocks, which have higher Young's modulus, tend to maintain fracture conductivity better under stress compared to softer rocks. For instance, in the Barnett Shale, the stress-dependent conductivity data indicate that stiff shale samples exhibit a much lower reduction in conductivity under high stress compared to soft shale samples (Alramahi and Sundberg, 2012).

Numerical reservoir simulations that incorporate geomechanical effects provide valuable insights into the performance of shale gas wells. These simulations help in understanding the impact of different fracture conductivities on ultimate gas recovery. Studies have shown that the decline rate of cumulative gas production is significantly influenced by the initial fracture conductivity and the reservoir's geomechanical properties. For instance, higher initial fracture conductivity leads to lower contributions from geomechanical effects over the well's lifetime (Yu and Sepehrnoori, 2013).

Optimization of hydraulic fracturing treatments involves balancing the fracture network complexity with the conductivity of the created fractures. The goal is to maximize gas recovery by creating an extensive fracture network while ensuring sufficient conductivity to maintain gas flow. Reservoir simulations and field data have shown that complex fracture networks with higher conductivity result in improved well performance and gas recovery (Cipolla, 2009).

The performance evaluation of fractured reservoirs requires a detailed understanding of fracture network complexity, geomechanical properties, and the optimization of fracture treatments. Incorporating these factors into reservoir simulations enables the development of more effective strategies for unconventional gas reservoir exploitation, ultimately leading to better economic outcomes.

7.4 Predictive Analyses and Production Optimization

In the realm of oil and gas extraction, optimizing production while mitigating risks is paramount. One of the critical aspects of this optimization involves predictive analyses, which allow operators to foresee potential issues and enhance operational efficiency. Here we delve into the methodologies and mechanisms employed in predictive analyses and production optimization, focusing on two significant challenges: sand production and casing failure in well operations.

7.4.1 Sand Production Mechanism Analysis

Sand production in oil and gas wells occurs when the fluid flow exceeds a certain threshold, influenced by the consistency of the reservoir rock, the stress state, and the type of well completion. This phenomenon can result in problems ranging from minor issues, such as small amounts of solids in the fluid, to major issues like wellbore blockage and erosion (Vaziri et al., 2002).

Various methods have been developed to predict sand production, including physical model testing, analytical and empirical relationships, and numerical models. These methods aim to understand and mitigate the problems associated with sand production (Vaziri et al., 2002).

- Physical Model Testing: These tests are capable of predicting the onset and, in some cases, the volume of sand production. However, they are expensive and time-consuming (Papamichos et al., 2001).
- Analytical Models: These are fast and easy to use but typically only predict the onset of sand production under simplified conditions (Gholami et al., 2016; Papamichos and Furui, 2019).
- Numerical Models: These models are the most powerful tools for predicting sand production, capable of capturing complex behaviors by combining analytical correlations and experimental results for calibration and validation (Fjaer et al., 2004).
- Continuum Models: These treat materials as continuous entities, using differential equations to describe their behavior. These models can be divided into two categories: those that calculate the onset of sand production and those that also predict the sand production rate (Vardoulakis et al., 1996).
- Constitutive Models: Various constitutive models have been used, from simple elastic brittle failure models to more complex elastoplastic models. The choice of model depends on the desired accuracy and the available computational resources.
- Sanding Criteria: Several criteria are used to predict sand production, including tensile failure, shear failure, critical pressure gradient, and erosion criteria. These criteria help determine the conditions under which sand production will occur (Papamichos et al., 2001).
- Coupled Models: Sand production involves both mechanical instability around the wellbore and hydromechanical instability due to flow-induced pressure gradients. Models that couple these mechanisms are more accurate but also more complex.
- Phases Involved: Some models consider only the fluid and solid phases, while others include a fluidized solid phase, representing particles in suspension.
- Permeability Alteration: The permeability of the sanded zone changes over time, affecting the flow and sand production. Different models handle these changes in various ways, often based on empirical relationships (Wang and Xue, 2002).
- Element Removal: In some models, elements that satisfy sanding criteria are removed, simulating cavity growth. Others keep the elements but alter their properties to reflect degraded material (Vardoulakis et al., 1996).

The Discrete Element Method (DEM) is used to model the disintegration of granular media. This approach treats sand production as a dynamic, microscopic process, capturing the detachment of individual particles (Cundall, 1971).

- DEM Models: DEM models simulate particles as individual entities, each with its own physical properties and interactions. This method is particularly useful for understanding the fundamental mechanics of sand production (Cundall, 1971).
- Coupling with Fluid Flow: Some DEM models incorporate fluid flow calculations, combining the continuity equation and Darcy's law to simulate the effects of fluid on particle detachment and transport (Jensen and Preece, 2000).
- Simulation of Hollow Cylinder Tests: DEM has been used to simulate hollow cylinder tests, providing insights into the behavior of sand production under various stress conditions (Li et al., 2006).

While significant advancements have been made in sand production prediction models, challenges remain in accurately predicting sand mass and rate for all field conditions. Both continuum and discontinuum approaches offer valuable insights, and future developments will likely involve hybrid models that combine the strengths of both methods.

7.4.2 Early Warning of Casing Failure

Casing failures in well operations is a significant challenge, leading to various economic and environmental consequences. Early warning systems for casing failure are essential to mitigate these risks and ensure well integrity. Casing, as the primary structural barrier in a well, is subjected to a range of stressors including material degradation, local loads, induced stresses during stimulation, natural fractures, and shear forces (Figure 7.3). These factors contribute to different failure modes, necessitating robust early warning systems to identify potential issues before they escalate.

Casing failures can be attributed to several factors, including high hoop stress, thermal stresses, and cyclic loading. For instance, high hoop stress was identified as a significant factor in casing failures during or after hydraulic fracturing operations. In the Western Anadarko Basin, a study found that 85% of failed casings occurred during these operations due to high hoop stress (Noshi et al., 2018). Additionally, thermal stresses in thermal recovery wells, such as those in China, have been shown to cause casing expansion and contraction, leading to material yielding beyond its elastic limit and subsequent failures (Han et al., 2018).

The primary mechanisms of casing failure include shear failure, buckling, collapse, burst, and fatigue. Shear failure is often due to formation slippage and stress variations. Buckling is influenced by local stresses and axial compression loads, which can lead to sinusoidal or helical buckling depending on the extent of the axial load (Chen et al., 1989; Dusseault et al., 2001). Collapse and burst failures are typically due to unequal external loads and internal pressures that exceed the casing's material yield strength (Abdideh and Hedayati Khah, 2018).

Figure 7.3 (a and b) Common case of a casing joint failure. (c and d) Reveals thread deformation on the pin and box ends. *Source*: Mohammed et al., 2019 / with permission of Elsevier.

To preempt casing failures, various assessment techniques are employed. These include logging tools to detect early signs of material degradation and stress distribution anomalies. For example, multi-arm caliper logs can identify casing deformation, which is a precursor to more severe failures (Roberts, 2014; Tanveer et al., 2023). Additionally, advances in artificial intelligence have enabled the development of automated systems that provide consistent and accurate monitoring of casing integrity. These systems minimize human error and improve the reliability of early warnings (Mohamadian et al., 2021; Noshi et al., 2019; Wang et al., 2023).

Recent developments in casing integrity assessment include the use of automated power tongs for precise make-up connections, which significantly reduce connection-related leaks and failures (Thiemann, 2018). However, limitations remain in the predictive capabilities of existing models, particularly in accounting for complex stress interactions and the varying material properties of casing under different operational conditions (Davies et al., 2014; King and King, 2013).

Implementing effective early warning systems for casing failure involves understanding the diverse factors contributing to failures and employing advanced monitoring and assessment techniques. While significant progress has been made, ongoing research and development are essential to enhance the reliability and accuracy of these systems, ensuring the long-term integrity and safety of well operations.

7.5 Unconventional Hydrocarbon Reservoirs and Geomechanics

Unconventional hydrocarbon reservoirs, particularly shale gas, shale oil, heavy oil sands, and natural gas hydrates, have emerged as critical components of the global energy landscape. The exploration and exploitation of these resources require a comprehensive understanding of their unique geomechanical properties. These properties influence the efficiency and safety of extraction methods such as hydraulic fracturing, thermal recovery processes, and gas hydrate dissociation techniques.

Geomechanics plays a pivotal role in optimizing the extraction of hydrocarbons from unconventional reservoirs. The mechanical behavior of the rock formations, characterized by parameters such as Young's modulus, Poisson's ratio, and compressive strength, dictates the success of techniques like hydraulic fracturing in shale formations and thermal stimulation in heavy oil sands. Moreover, the stability of natural gas hydrate-bearing sediments is heavily dependent on their geomechanical properties, which must be carefully managed to prevent geohazards during gas extraction.

This section explores the geomechanical characteristics of these unconventional reservoirs, exploring the intricacies of shale gas and shale oil formations, the behavior of heavy oil sands under various recovery processes, and the stability considerations for gas hydrate-bearing sediments. Through a detailed examination of experimental methods, numerical modeling, and field applications, the chapter provides a comprehensive overview of the critical role geomechanics plays in the efficient and safe exploitation of unconventional hydrocarbon resources.

7.5.1 Shale Gas and Shale Oil

The exploration and production of shale gas and oil are pivotal to the global energy landscape due to the increasing demand for energy sources. Shale formations, characterized by their fine-grained sedimentary nature, serve as both reservoirs and seals owing to their unique geomechanical properties. Understanding these properties is essential for optimizing hydraulic fracturing techniques, crucial for efficient hydrocarbon extraction from these unconventional resources.

The mineral composition of shale significantly influences its mechanical behavior. Typically, shale formations consist of varying proportions of quartz, carbonates, clay minerals, and organic matter, each contributing differently to the rock's overall geomechanical characteristics. High concentrations of brittle minerals such as quartz and carbonates enhance the rock's brittleness, making it more susceptible to fracturing. This brittleness is advantageous for hydraulic fracturing, as it facilitates the creation of extensive fracture networks essential for hydrocarbon flow (Loucks et al., 2009).

Conversely, an increase in clay minerals within the shale matrix imparts plasticity, reducing the rock's brittleness and making it less amenable to fracturing. Clay minerals, particularly phyllosilicates, weaken the rock structure by decreasing the uniaxial compressive strength. As the content of clay minerals rises, Poisson's ratio increases, indicating higher ductility, while Young's modulus decreases, reflecting reduced stiffness and greater deformability under stress (Sone and Zoback, 2013; Vernik and Milovac, 2011).

A comprehensive understanding of the geomechanical properties of shale is achieved through various experimental methods, each offering unique insights into the rock's behavior under different conditions. These methods include uniaxial and triaxial compression tests, ultrasonic testing, and nano-indentation techniques.

Uniaxial and Triaxial Compression Testing:

- Uniaxial Compression Tests: This fundamental mechanical test involves applying axial stress to a cylindrical rock sample until failure. It measures properties such as Young's modulus and uniaxial compressive strength, providing a basic understanding of the rock's response to stress (Chuanliang et al., 2015; Zhao et al., 2018).
- Triaxial Compression Tests: In these tests, radial stress is added to the axial stress, simulating subsurface conditions more accurately. Triaxial tests help in understanding the rock's behavior under the combined stresses it would experience in situ, providing crucial data for geomechanical modeling (Amann et al., 2012; Chong et al., 1984)

Ultrasonic Testing:

- This method measures the velocities of compressional (P-wave) and shear (S-wave) waves traveling through rock samples. The wave velocities are directly related to the rock's elastic properties, making ultrasonic testing a precise and repeatable method for assessing both static and dynamic geomechanical properties (Josh et al., 2012; Wang et al., 2021)

Nano-indentation:

- Nano-indentation involves applying a controlled force to a small indenter to measure the mechanical properties of the rock at the nanoscale. This technique provides high-resolution data on hardness and elastic modulus, which are critical for understanding the behavior of shale at the microscopic level (Kumar et al., 2012; Liu et al., 2016).

Hydraulic fracturing is a key technique for extracting hydrocarbons from shale formations. It involves injecting high-pressure fluid into the rock to create fractures, which then provide pathways for oil and gas to flow to the wellbore. The efficiency of hydraulic fracturing depends heavily on the geomechanical properties of the shale, particularly its elastic properties.

Elastic properties such as Young's modulus and Poisson's ratio are critical for designing effective hydraulic fracturing operations. Young's modulus indicates the stiffness of the rock, while Poisson's ratio reflects its ductility. Together, these parameters help in predicting how the rock will behave under stress and how fractures will propagate and close during and after the fracturing process (Fjaer et al., 2008; McCormack et al., 2021).

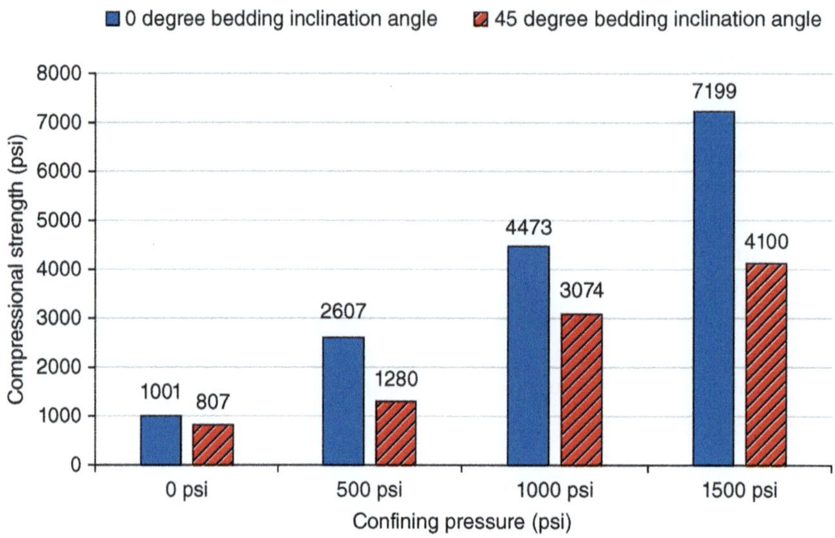

Figure 7.4 UCS of various bedding inclination angles of shale.

Shale formations often exhibit anisotropy due to their layered structure (Figure 7.4), with different properties in different directions (Heng et al., 2015). This anisotropy must be accounted for in geomechanical models to accurately predict the behavior of the rock during hydraulic fracturing. Anisotropic rock-physics models incorporate the variations in elastic properties due to anisotropy, providing more accurate predictions of fracture responses and aiding in the optimization of fracturing techniques (Sayers, 1994; Sondergeld and Rai, 2011; Thomsen, 1986)

The practical application of geomechanical models to real-world scenarios involves calibrating these models using field data. This calibration process includes adjusting model parameters to match observed data from wells, such as compressional and shear wave velocities, and other elastic properties. Proper calibration ensures the accuracy of the models, which is critical for predicting the geomechanical behavior of shale formations under various conditions (Vorobiev et al., 2024).

7.5.2 Heavy Oil Sands

Heavy oil sands, particularly prevalent in Canada, are a vital source of hydrocarbons, necessitating effective extraction methods. The geomechanics of heavy oil sands significantly influences the efficiency of various recovery techniques such as Cold Heavy Oil Production with Sand (CHOPS), Cyclic Steam Stimulation (CSS), and Steam-Assisted Gravity Drainage (SAGD). Understanding the mechanical behavior of these sands under different stress conditions is crucial for optimizing recovery processes and ensuring the stability of reservoirs.

Heavy oil sands are primarily composed of unconsolidated sand grains bound together by bitumen. These sands exhibit unique geomechanical properties, including high porosity and permeability, influenced by the presence of bitumen. The geomechanical behavior of heavy oil sands is characterized by strain softening (Li and Chalaturnyk, 2005), dilatancy (Lin et al., 2015), and significant changes in permeability under stress (Maini and Okazawa, 1987).

Strain softening is a critical aspect of the geomechanical behavior of heavy oil sands. As the sand deforms, the material's strength decreases, leading to further deformation under constant stress (Morales-Monsalve et al., 2019). This behavior is particularly evident during CSS and SAGD operations, where thermal and mechanical stresses induce significant strain in the reservoir. Dilation, the increase in volume due to shear deformation, is another important characteristic. During fluid injection, such as in SAGD, the sands exhibit substantial dilation, which can enhance permeability. Yale et al. (2010) found that oil sands show significant dilation under extensional stress paths, with permeability to water increasing dramatically as a result of volumetric strain.

Numerical models are essential tools for simulating the behavior of heavy oil sands under various recovery processes. These models incorporate geomechanical properties derived from laboratory tests to predict reservoir behavior. For example, the FLAC numerical simulator was used to match laboratory tests and develop a representative strain softening model for oil sand (Li and Chalaturnyk, 2005).

Geomechanical models help design effective recovery strategies and ensure the stability of the reservoir. They account for factors such as modulus of elasticity, peak and post-peak friction angles, and dilation angles, crucial for simulating the coupled geomechanical and thermal processes in heavy oil recovery.

The unique geomechanical properties of heavy oil sands significantly impact the effectiveness of recovery processes. In CHOPS, the production of sand along with oil creates wormholes that enhance permeability and oil recovery. However, this also leads to complex deformation patterns and stress redistributions in the reservoir (Mohamad-Hussein et al., 2022).

In CSS and SAGD, thermal expansion and contraction, along with mechanical loading, induce significant stress changes in the reservoir. These processes require careful management of injection pressures and temperatures to avoid excessive deformation and ensure efficient oil recovery. The stress paths during fluid injection reflect a reduction in mean effective stress and an increase in differential stress, as vertical total stress increases lead to vertical effective stress decreasing more slowly than horizontal effective stress.

7.5.3 Gas Hydrates

The escalating global demand for energy has led to the excessive exploitation of traditional fossil fuels such as coal, petroleum, and natural gas, rendering them unsustainable and economically unfeasible. Natural gas hydrates (NGH) present a promising alternative as they are considered a cleaner energy source compared to conventional fossil fuels (Collett and Kuuskraa, 1998; Kumar and Sonawane, 2004; Kvenvolden, 1999; Kvenvolden and Lorenson, 2001; Moridis et al., 2011; Seol and Lee, 2013; Whiticar et al., 1986). NGH typically form in sediments beneath permafrost regions and ocean beds (Figure 7.5), where gases like methane, ethane, and propane become trapped in the crystal lattice of water molecules under conditions of high pressure and low temperature within the hydrate stability zone (HSZ) (Collett et al., 2014; Joseph et al., 2016; Long et al., 2009; Ning et al., 2013; Priest et al., 2005; Soga et al., 2006)

GHBS are stable due to their enhanced geomechanical properties such as shear strength, stiffness, compressibility, and Poisson's ratio. However, during gas extraction, the dissociation of hydrates can cause instability, potentially leading to submarine slope failures, seabed subsidence, and failures of seafloor installations (Collett et al., 2015; Collett, 2002; Ning et al., 2012; Nixon and Grozic, 2007). It is crucial to assess the stability of GHBS through field and laboratory investigations, although these methods are often expensive and time-consuming.

The geomechanical properties of GHBS are influenced by factors (Figure 7.6) such as hydrate saturation (SH), hydrate morphology, and confining pressure-temperature conditions. Notably, shear strength and stiffness are pivotal for GHBS stability and exhibit a power-law relationship with SH. Cohesion is the primary component of shear strength and shows significant improvement for SH values between 15-50% (Ghiassian and Grozic, 2013; Miyazaki et al., 2011; Yoneda et al., 2015).

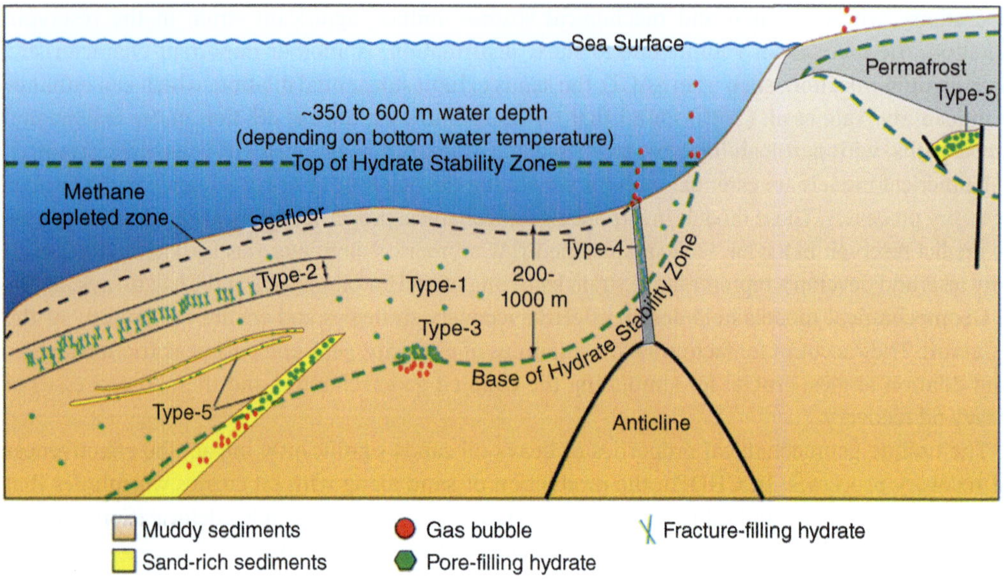

Figure 7.5 Types of natural gas hydrate deposits. *Source*: You et al., 2019 / John Wiley & Sons.

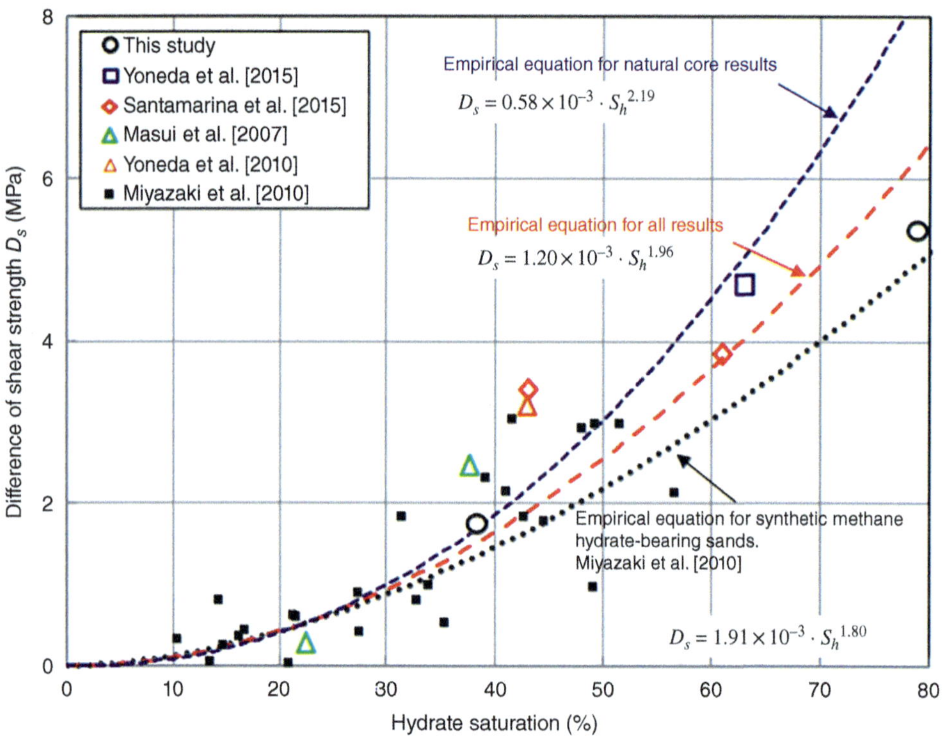

Figure 7.6 Shear strength as a function of hydrate saturation. *Source*: Yoneda et al., 2015 / with permission of Elsevier.

Hydrate morphology significantly affects the geomechanical behavior of GHBS. In coarse-grained sediments, hydrate morphologies include pore-filling, load bearing, and cementing (Figure 7.7), while in fine-grained sediments, they form nodules, lenses, and veins (Jang and Santamarina, 2016; Lei and Santamarina, 2018; Masui et al., 2008). Permeability is another critical parameter, affected by hydrate saturation and morphology. The reduction in permeability with increasing SH is well-documented, and various empirical models have been proposed to predict permeability based on SH and hydrate morphology (Delli and Grozic, 2014; Kumar et al., 2010; Liang et al., 2011).

Triaxial tests are commonly used to determine the geomechanical properties of GHBS, using both natural and synthetic samples. These tests help in understanding the stress-strain behavior under different conditions. Studies have shown that the shear strength, stiffness, and dilation angle of GHBS increase with SH, while the friction angle remains relatively unaffected (Figure 7.8) (Chen et al., 2022; Choi et al., 2018).

Geomechanical modeling of GHBS is essential for simulating stress-strain characteristics and understanding fluid flow during gas production. Models based on Mohr-Coulomb, Duncan-Chang hyperbolic, and critical-state theories have been developed to simulate the geomechanical behavior of GHBS (Dong et al., 2022; Kim et al., 2012; Klar et al., 2010; Pinkert et al., 2015). These models incorporate parameters such as cohesion, friction angle, stiffness, and dilation angle to predict the response of GHBS to mechanical loading.

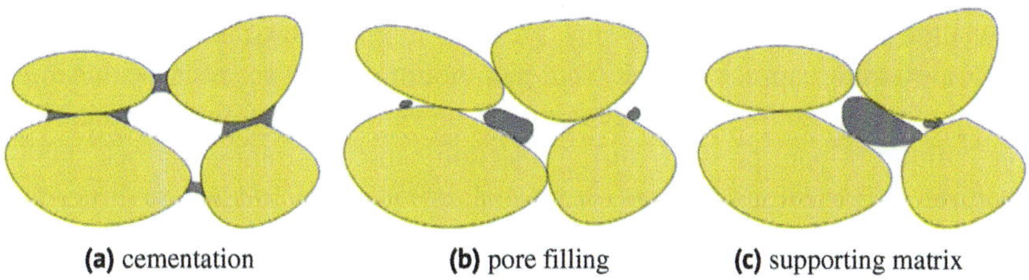

Figure 7.7 Morphological classifications of hydrates (a) cementation (b) pore-filling and (c) load bearing. *Source*: Sánchez et al., 2017 / with permission of Elsevier.

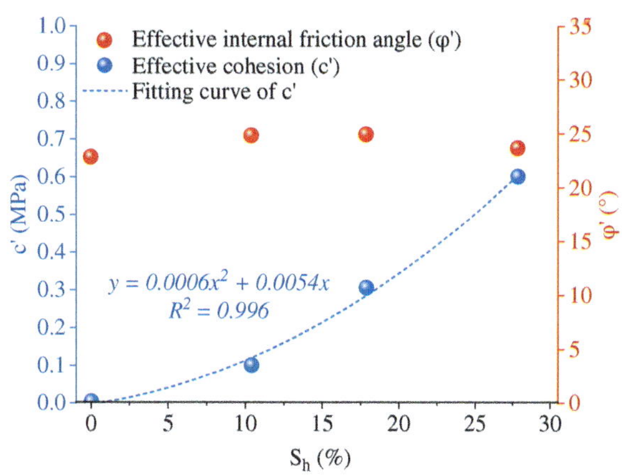

Figure 7.8 Variation of cohesion and friction angle with hydrate saturation degree. *Source*: Chen et al., 2022 / with permission of Elsevier.

Understanding the geomechanical properties of GHBS is crucial for mitigating geohazards associated with gas hydrate dissociation during extraction. Further experimental and modeling efforts are necessary to develop reliable methods for predicting the behavior of GHBS under varying conditions, ensuring safe and efficient gas extraction processes.

7.6 Geomechanics in Geological Carbon Storage

Under the Paris Agreement, nations have committed to a path of "low greenhouse gas emissions development," aiming to limit the increase in global average temperature to 1.5 degrees Celsius above pre-industrial levels (Hulme, 2016). To help achieve this, anthropogenic CO_2 emissions are set to decrease through the application of carbon capture, utilization, and storage (CCUS) technologies. In the approach known as carbon capture and storage (CCS), or geological carbon storage (GCS), CO_2 is captured and then deposited into specific geological formations like exhausted hydrocarbon fields, saline aquifers, and subterranean caverns. CCS represents a practical solution for the large-scale reduction and storage of CO_2 emissions.

7.6.1 Importance and Challenges

Geomechanics plays a crucial role in storage CCS projects (Rutqvist, 2012), as it addresses significant challenges arising from pressure changes in the subsurface during CO_2 injection (Figure 7.9). Maintaining caprock integrity is of paramount importance because this impermeable layer forms a natural seal above the storage reservoir, preventing CO_2 from escaping. However, the unavoidable buildup of pore pressure during injection can compromise caprock stability. If the caprock fractures, or if existing faults are reactivated, stored CO_2 could leak to the surface or contaminate underground aquifers, resulting in environmental hazards and public health concerns. Additionally, such failures may threaten public acceptance of CCS projects due to the perceived risks.

Fault reactivation is another critical challenge linked to geomechanics. When existing faults are exposed to increased pore pressure from CO_2 injection, they might slip, resulting in induced seismicity (Rutqvist, 2012). This can cause structural damage, disrupt operations, and further challenge public acceptance. Therefore, accurately understanding and managing the potential for fault reactivation is essential to prevent catastrophic outcomes.

Well integrity is a third significant issue in CO_2 storage. Well casings and cement can degrade due to prolonged exposure to corrosive CO_2 and brine mixtures, increasing the risk of leakage along the wellbore (Carroll et al., 2016). If well integrity is compromised, stored CO_2 may migrate through compromised wells, possibly leading to groundwater contamination or surface leakage. Continuous monitoring and enhancement of well integrity using tools like permanent downhole gauges (PDGs) can help prevent these risks.

Surface uplift is another important factor in CCS operations. Increased pressure from CO_2 injection can lead to subsurface deformation that extends to the surface, causing ground uplift (Fokker et al., 2011). This can damage infrastructure and negatively affect public perception. Therefore, careful management of injection pressure and accurate modeling of reservoir deformation are crucial to minimizing this risk.

Effective pore pressure management, achieved through controlled injection rates, advanced monitoring systems, and optimized injection well design, is critical to address these geomechanical challenges. Strategies such as using relief wells, conducting comprehensive modeling and

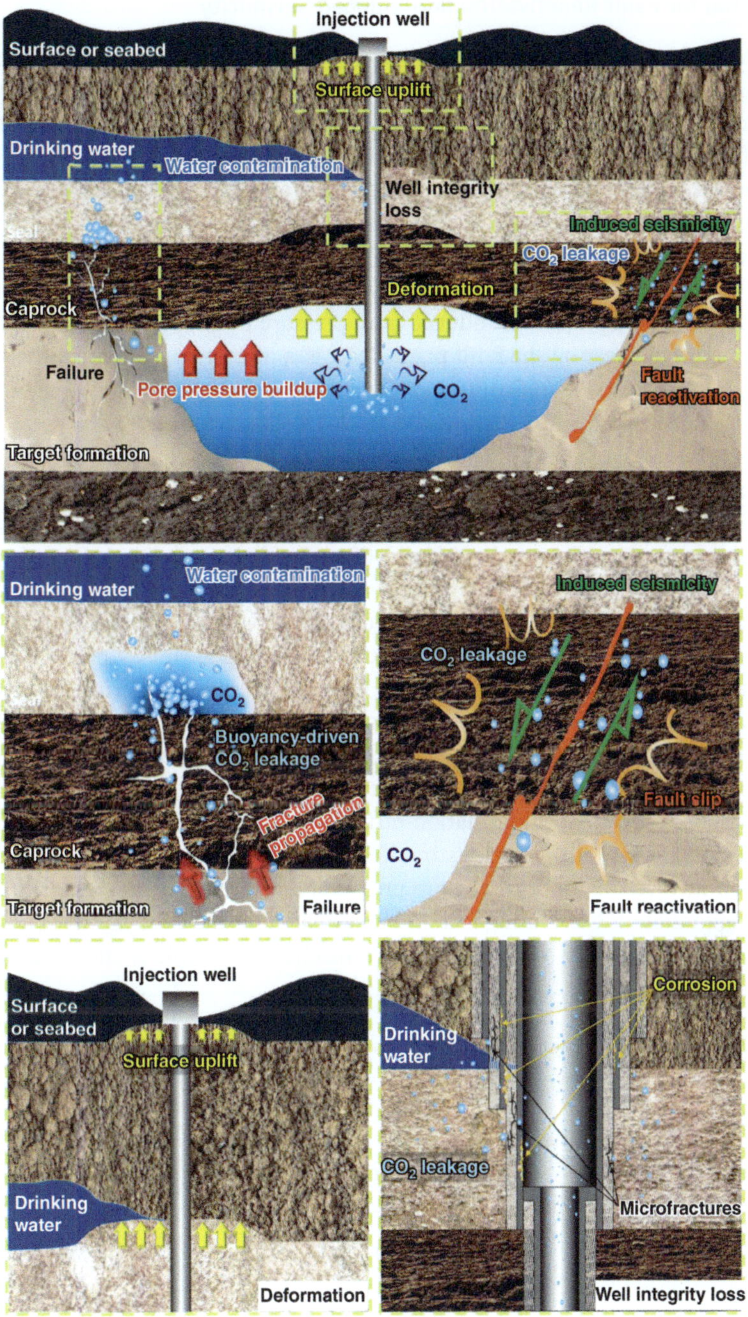

Figure 7.9 Geomechanical risks associated with CO_2 storage. *Source*: Song et al., 2023a / with permission of Elsevier.

simulation studies, and ensuring effective well stimulation can help alleviate pressure buildup and minimize risks (Song et al., 2023b). Thus, incorporating rigorous geomechanical analysis into the design and implementation of CCS projects is essential to ensure safe and efficient CO_2 storage.

7.6.2 Potential for Fault Reactivation and Induced Seismicity

A fault can act either as a barrier or a pathway, a crucial consideration in the design of CCS processes. Geomechanically, faults may reactivate, potentially causing unintended CO_2 leakage and unwanted seismic activity (White and Foxall, 2016). This reactivation is a type of shear failure occurring at faults, utilizing the Mohr-Coulomb failure criterion, similar to the analysis used for caprock failure (Rinaldi et al., 2014). Fault reactivation analysis necessitates additional data on fault characteristics such as inclination angle, shear strength of the fault plane, and fault permeability. Therefore, faults demand a more detailed Mohr-Coulomb failure analysis compared to intact rocks (Cappa and Rutqvist, 2011). Initially, natural faults remain dormant under existing reservoir pressure, but CO_2 injection-induced pore pressure buildup may destabilize these faults as the effective normal stress on the fault plane decreases inversely with pore pressure increases. If the pore pressure change is significant, shear stress on the fault plane may exceed the threshold, leading to fault slip (Lee et al., 2017; Rutqvist et al., 2016; Vasco et al., 2018).

Numerical modeling has been used widely to forecast fault reactivation during CO_2 Storage. GEM, IPARS, COMSOL, and ECLIPSE are often coupled with simulators such as TOUGH and FLAC3D, JAS3D, CMG, and PTREL to integrate these flow dynamics with the geomechanical responses of the formation (Gao et al., 2022; Li, 2016; Minkoff et al., 2003; Rutqvist et al., 2007; Zheng et al., 2021; Zheng et al., 2022)

Fault activation and induced seismicity are critical and unavoidable concerns for Carbon Capture and Storage (CCS) from both safety and public acceptance viewpoints. The process of injecting substances can lead to pressure build-up within the reservoir, altering the existing natural stress conditions. When faults are triggered or rock integrity is compromised, the energy that has accumulated over time due to crustal stress is discharged as seismic waves. Many studies showed that underground CO_2 injection can cause fault reactivation (Table 7.2).

These studies also show that the initiation of faults due to CO_2 injection is closely associated with a complex interaction of thermal, hydraulic, mechanical, and chemical (THMC) processes. This interaction involves disruptions in pore pressure, temperature variations, and stress corrosion. Uneven migration of CO_2 through the reservoir can alter the existing stress field as a result of pore pressure changes, thus initiating a hydro-mechanical coupling effect. Additionally, the

Table 7.2 CO_2 storage projects with fault reactivation potential.

Project name and nature	Country	References
Ohio River Valley, CGS	West Virginia, USA	Lucier et al. (2006)
In Salah, GCS	In salah, Algeria	Cao et al. (2021b, 2021a)
Teapot Dome, CO_2-EOR	Wyoming, USA	Chiaramonte et al. (2011)
Saint-Martin de Bossenay, CO_2-EOR	Paris, France	Vidal-Gilbert et al. (2009)
Brazil, CO_2-EOR	Brazil	Caro et al. (2023)
South-west of Iran, CO2-EOR	Iran	Amiri et al. (2019)
Po River, CO_2-EOR	Italy	Ferronato et al. (2010)
Otway, GCS	Australia	Rogers et al. (2008)
Ketzin, GCS	Germany	Röhmann et al. (2013)
Tundra, CGS	North Dakota, USA	(Peck et al., 2019)

injection of supercritical CO_2, which is considerably much cooler than the reservoir temperature, can lower the temperature of surrounding rocks. This sudden cooling can induce thermal stresses and fractures, modifying the stress field and potentially affecting fault stability through a process known as thermo-mechanical coupling. Over time, the continuous injection of CO_2 may change the chemical composition of the reservoir. The reaction between CO_2 and rock minerals can weaken the rocks, alter the friction coefficient on faults, and change stress conditions, leading to increased fracture formation and fault activation, a sign of chemical-mechanical coupling. These coupling processes typically interact rather than occur independently. Therefore, analyzing THMC coupling requires a comprehensive evaluation of how these combined factors affect fault stability, the development of normal and shear stresses, and the potential for fault activation and associated seismic events. This holistic approach is crucial for a deeper understanding of the dynamics behind fault activation (Figure 7.10).

Induced seismicity in CO_2 storage projects is a complex phenomenon directly influenced by the dynamics of fluid injection and the geomechanical response of the subsurface environment. Seismic activities can be triggered by either creating new failure planes or reactivating existing faults due to excessive pore pressure buildup during the carbon capture and storage (CCS) processes (Mortezaei and Vahedifard, 2015; Nicol et al., 2011; Verdon and Stork, 2016). Notable incidents, such as the ones in Oklahoma in 2011 and 2012, have highlighted the potential significant impacts of such activities, which included damage to structures and injuries to individuals (Ellsworth, 2013). These events underscore the necessity of managing injection processes to mitigate the risks associated with induced seismicity, which is crucial for maintaining public acceptance of CCS projects.

The scientific community continues to explore various predictive models to better understand and anticipate the occurrence of induced seismic events (Furre et al., 2015; Hill et al., 2024; Mazzoldi et al., 2012; Romeo, 2000; Streit and Siggins, 2005; Templeton et al., 2023; Verdon, 2016; Yoon et al., 2023). While no method has proven fully reliable yet, advancements have been made in statistical (McGarr, 2014; Rothert and Shapiro, 2007), physical parameter-based (Verdon and Stork, 2016), and hybrid

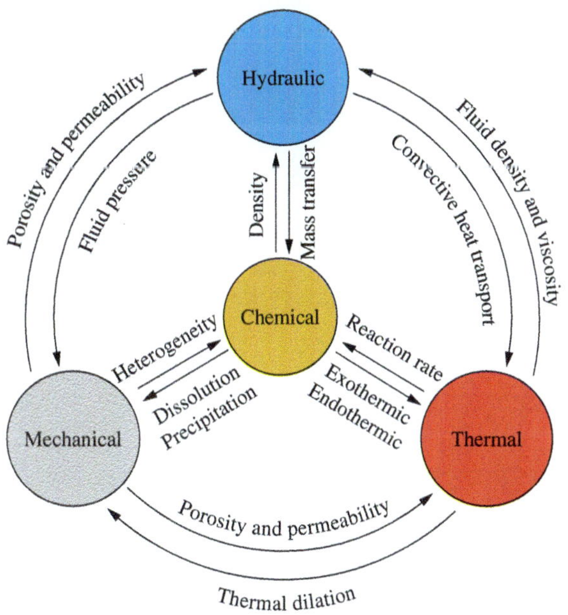

Figure 7.10 Cross-couplings between THMC processes. *Source:* Tao et al., 2019 / John Wiley & Sons / CC BY 4.0.

prediction methods (Shapiro et al., 2011). For instance, statistical models often utilize seismic records to establish relationships between the timing, location, and magnitude of seismic events during fluid injection. Physical models, on the other hand, require detailed geological and geophysical data to simulate the subsurface conditions that lead to seismic activity. These models consider factors such as the regional stress field changes and fault activation due to fluid perturbations. Hybrid models attempt to combine these approaches to provide a more comprehensive understanding of the seismic risks.

One widely recognized statistical approach proposed by (McGarr, 2014) uses the injection volume to predict the maximum seismic magnitude, suggesting a linear relationship governed by the Gutenberg-Richter earthquake distribution law (Figure 7.11). However, incidents like the Pohang earthquake demonstrated limitations of this model, especially when unusual hydrogeological conditions are present (Yeo et al., 2020). On the other hand, physical models emphasize the importance of accurately modeling the geological setting and the stress perturbations induced by injection to predict seismic events more reliably.

Furthermore, operational strategies such as the traffic-light system have been implemented in various locations to provide a regulatory framework that adjusts the injection operations based on detected seismic activities (Baisch et al., 2019; Kao et al., 2018; Wei et al., 2020). This approach helps in minimizing pore pressure buildup, thereby reducing the likelihood of triggering seismic events.

The challenge of predicting and managing induced seismicity in CO_2 storage involves an intricate understanding of geological, hydrogeological, and operational factors. Continued research and development of more accurate prediction models and effective management protocols are essential to ensure the safety and success of CCS initiatives (Khan et al., 2024). This involves not only improving the existing models but also integrating diverse data sources and refining operational strategies to respond dynamically to the evolving subsurface conditions during CO_2 injection.

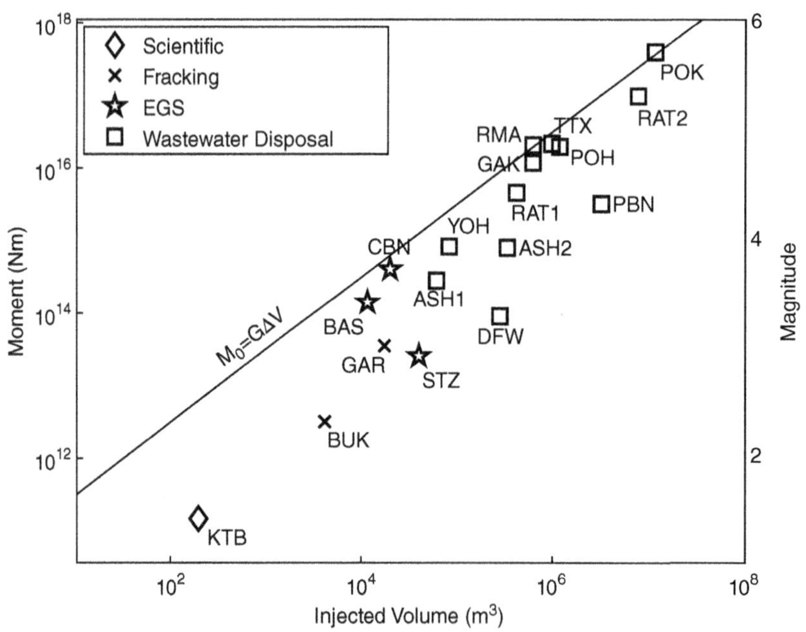

Figure 7.11 The equation represented by the solid line establishes an upper bound for seismic moment, correlating it directly to the product of the modulus of rigidity and the cumulative volume of fluid injected from the start of injection up to the occurrence of the largest induced earthquake. This represents the maximum seismic moment and magnitude as functions of the total injected fluid volume. *Source*: McGarr, 2014 / John Wiley & Sons.

7.6.3 Numerical Modeling for Site Suitability and Injection Parameters

Carbon capture and storage (CCS) is a promising technology for mitigating greenhouse gas emissions. However, injecting carbon dioxide (CO_2) into deep geological formations, such as oil and gas reservoirs and saline aquifers, brings significant geomechanical risks. These risks include fault activation, fracture propagation, and potential contamination of shallow drinking water due to leakage. The complexity of geological formations and injection-induced seismicity further complicates storage operations, making numerical modeling crucial for assessing the geomechanical suitability of potential sites and determining safe injection parameters.

Numerical modeling is invaluable for understanding and predicting how geological formations respond to CO_2 injection. By simulating interactions between various physical processes, such as thermal, hydraulic, mechanical, and chemical changes, models provide insights into pressure buildup, stress distribution, and ground deformation. Several numerical methods are commonly employed in geomechanical modeling, including finite element, finite difference, and discrete element methods (Jiang, 2011). These models are often coupled with reservoir simulators to capture the complex interactions between fluid flow and mechanical behavior.

Fault zones present significant challenges in modeling due to their structural complexity and impact on CO_2 storage. Accurate representation of faults requires specialized modeling techniques that account for anisotropy and stress distribution changes. Moreover, the Mohre-Coulomb failure criterion is used to assess the risk of fault activation, while slip-weakening models simulate the potential for induced seismicity (Cappa and Rutqvist, 2011). However, predicting fault activation and associated seismic events remains difficult due to the variability of fault properties and stress conditions.

Fracture propagation modeling is also essential, as new fractures can provide potential pathways for CO_2 leakage. Although numerical methods like the extended finite element method (XFEM) offer powerful tools for simulating fracture networks, challenges persist in accurately capturing 3D fracture growth. The choice of failure criteria, including the Mohre-Coulomb criterion or toughness-based fracturing criteria, depends on site-specific conditions and injection scenarios.

In summary, numerical modeling plays a critical role in assessing the geomechanical feasibility of CCS sites and optimizing injection parameters. Future advancements should focus on improving fault representation, refining criteria for fracture propagation, and addressing uncertainties in stress and geological properties. Ultimately, this will lead to safer and more effective CO_2 storage strategies that minimize environmental risks.

7.7 Geomechanics of Hydrogen Storage and Production

The global pursuit of sustainable energy solutions has intensified interest in hydrogen as a potential energy carrier. As renewable energy sources like solar and wind become more prevalent, the need for efficient energy storage solutions continues to grow. Hydrogen, produced through renewable sources or hydrocarbon reformation with carbon capture and storage, offers a promising avenue for addressing this challenge (Hassanpouryouzband et al., 2021). Underground Hydrogen Storage (UHS) in geological formations has emerged as a viable solution for large-scale storage (Figure 7.12), crucial for balancing renewable energy intermittency. Subsurface geological storage options, including salt caverns, depleted hydrocarbon reservoirs, and deep saline aquifers, provide significantly larger storage capacities than surface-level storage (Andersson and Grönkvist, 2019).

However, hydrogen's unique properties—low density, high diffusivity, and reactivity—present specific geomechanical challenges that must be addressed to ensure safe and efficient storage operations. These properties require larger storage volumes compared to natural gas and increase leakage

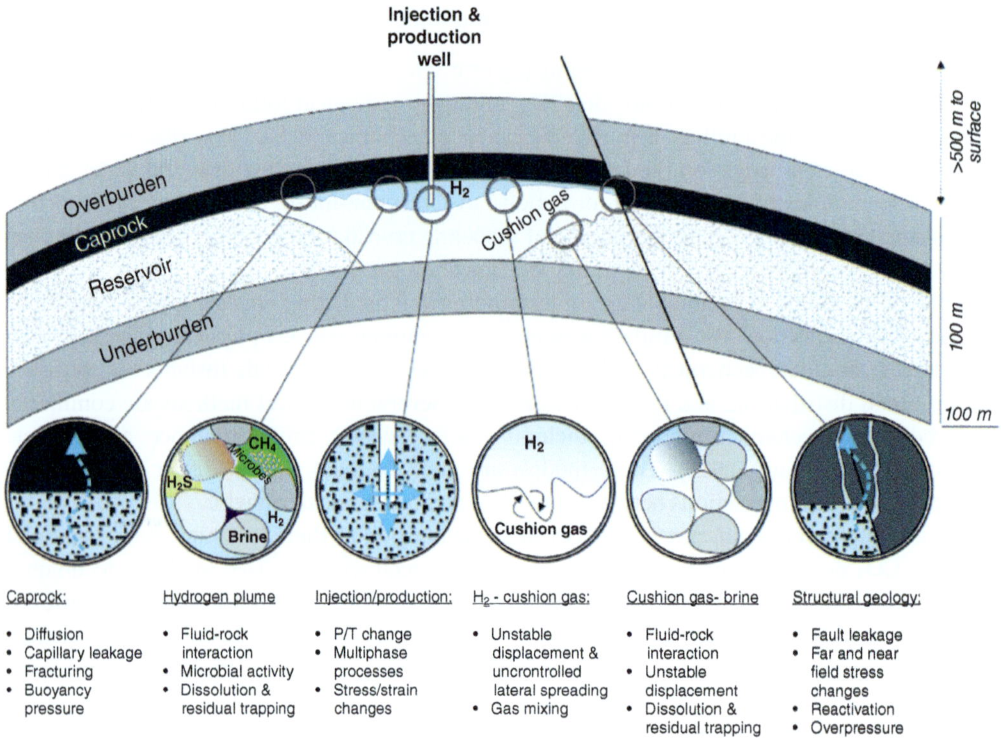

Figure 7.12 Reservoir mechanisms during UHS. *Source*: Heinemann et al., 2021 / The Royal Society of Chemistry / CC BY NC 4.0.

risk, necessitating rigorous assessment of caprock sealing efficiency and consideration of geochemical reactions between hydrogen, formation fluids, and minerals (Muhammed et al., 2022). Geomechanical considerations are critical in UHS operations, as pressure and temperature changes during hydrogen injection and extraction can impact the mechanical behavior of the storage formation, potentially leading to caprock integrity failure or fault reactivation (Heinemann et al., 2021).

This section provides an overview of the current state of knowledge and challenges in UHS, focusing on the geomechanical aspects of hydrogen storage and production. By examining recent research and case studies, we explore the geomechanical responses of storage formations, potential risks, and ongoing efforts to ensure the long-term feasibility and safety of UHS as a key component of a sustainable energy future (Tarkowski and Uliasz-Misiak, 2022).

7.7.1 Geomechanical Properties of Hydrogen Storage Reservoirs

Depleted oil and gas reservoirs offer promising solutions for UHS due to their large storage volumes, proven sealing capabilities, and extensive data from previous hydrocarbon extraction operations. However, the geomechanical properties of these reservoirs are crucial for ensuring UHS success (Figure 7.13).

Key geomechanical considerations include pressure and stress changes, reservoir compaction and subsidence, caprock integrity, fault reactivation, and hydrogen-mineral interactions. Pressure fluctuations during hydrogen injection and withdrawal alter the stress state within the storage complex, potentially leading to deformation beyond the immediate vicinity of the injection well (Muhammed et al., 2022). These changes can impact caprock integrity and cause reservoir compaction and subsidence

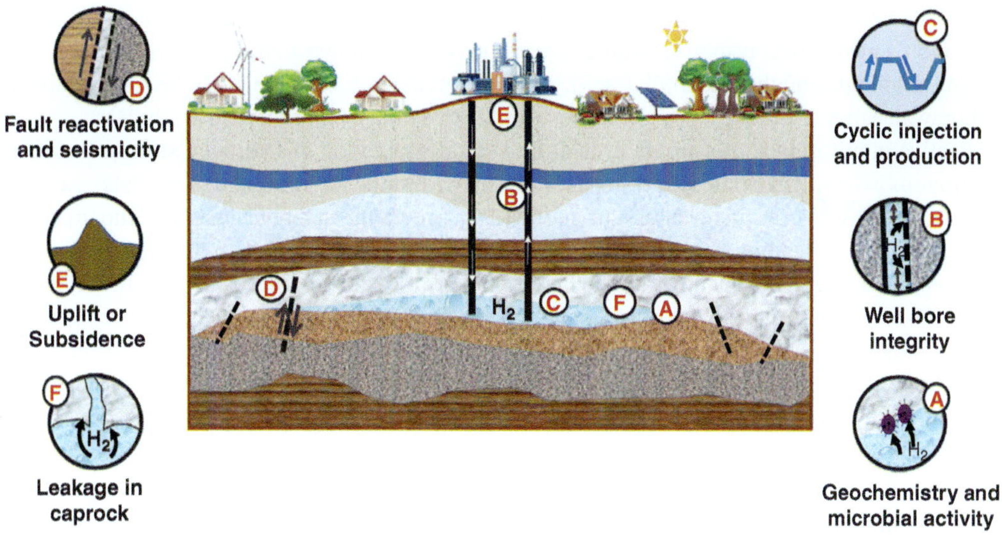

Figure 7.13 A schematic of UHS in depleted gas fields illustrating geomechanical processes at play. *Source*: Kumar et al., 2023 / with permission of Elsevier.

(Andersson and Grönkvist, 2019). Cyclical stress fluctuations can compact the reservoir, reducing porosity and fluid flow, and potentially causing surface subsidence (Saeed and Jadhawar, 2023).

Maintaining caprock integrity is crucial to prevent hydrogen leakage. Reservoir compaction and stress changes can flex the caprock, potentially creating fractures and compromising its sealing ability (Tenthorey et al., 2024). The cyclic nature of hydrogen storage operations exacerbates this issue, as repeated stress changes can progressively weaken the caprock structure. Additionally, stress changes from hydrogen injection can reactivate pre-existing faults, increasing the risk of fault instability and potential leakage pathways (Sekar et al., 2023). Fault reactivation can also lead to microseismic events, impacting the long-term stability of the storage system (Muhammed et al., 2022).

Interactions between hydrogen and reservoir minerals can significantly affect the mechanical response of the storage system. Hydrogen can cause embrittlement of steel-based barrier elements, potentially compromising wellbore integrity, and interact with minerals, altering the geomechanical properties of the reservoir (Heinemann et al., 2021). These interactions may influence fluid-assisted deformation processes and have long-term impacts on clay swelling and sorption processes, critical for maintaining the sealing capacity of the caprock (Heinemann et al., 2021).

7.7.2 Stress and Strain Considerations During Hydrogen Storage

The cyclic nature of hydrogen injection and withdrawal causes stress fluctuations in reservoirs and surrounding rock formations. This process can lead to potential issues if not properly managed, affecting reservoir integrity and caprock sealing capacity (Ciancimino et al., 2024; Jadhawar and Saeed, 2024).

Caprock integrity is a key area of concern as cyclic stress variations can form pathways in the caprock, potentially compromising its ability to trap hydrogen effectively. Understanding the fatigue behavior of caprock materials under such conditions is crucial (Saeed and Jadhawar, 2023). Cyclic loading, especially in clayey caprocks, can lead to a gradual degradation of the material's shear strength, often manifesting as the formation of shear bands within the material, ultimately reducing its ability to withstand stress (Ciancimino et al., 2024).

The strain rate, influenced by the rate of hydrogen injection and withdrawal, significantly affects a reservoir's mechanical response. Studies on stiff clays have shown a strong correlation between loading strain rate and the material's fragility. Higher strain rates, often linked to faster injection or withdrawal, can make the reservoir more susceptible to damage (Tenthorey et al., 2024).

Effective monitoring strategies are crucial for the safe operation of hydrogen storage facilities. Continuous assessment of stress and strain evolution within the reservoir and caprock is essential, with techniques like acoustic emission monitoring providing valuable insights into material responses to cyclic loading (Kumar et al., 2023). Further research specific to hydrogen storage is necessary, building upon existing knowledge from oil and gas operations. This includes in-depth studies on the long-term behavior of reservoir and caprock materials under cyclic stress conditions unique to hydrogen storage, investigating hydrogen's effects on rock failure criteria, friction angle, and cohesion (Saeed and Jadhawar, 2024). Developing advanced monitoring systems for real-time tracking of stress and strain evolution is crucial for early detection of potential issues (Kumar et al., 2023).

7.7.3 Geomechanical Modeling of Hydrogen Injection and Storage

Geomechanical modeling is crucial for ensuring the safe and efficient operation of UHS in depleted reservoirs and salt caverns (Kumar et al., 2023). It helps evaluate the mechanical response of rocks subjected to cyclic hydrogen injection and production, predict fault reactivation potential, and assess the impact of hydrogen on rock properties (Tenthorey et al., 2024).

Coupled hydro-mechanical-chemical (HMC) simulations are essential for UHS modeling, considering subsurface heterogeneity and lithology to provide valuable insights into the governing physics across various timescales and reservoir regions (Ciancimino et al., 2024). Modeling approaches can vary depending on project needs, including one-way coupling, two-way coupling, and fully coupled simulations (Heinemann et al., 2021).

Successful geomechanical modeling for UHS requires consideration of several challenging aspects, such as stress and deformation analysis, fault reactivation assessment, wellbore integrity analysis, and caprock integrity evaluation (Ciancimino et al., 2024; Kumar et al., 2023; Sekar et al., 2023; Tenthorey et al., 2024).

Factors influencing the accuracy of UHS modeling include hydrogen density, energy density ratio (H_2/CH_4), cushion gas, and the impact of reservoir heterogeneity (Kumar et al., 2023). Understanding the long-term effects of hydrogen storage, particularly cyclical stress changes and hydrogen-mineral interactions, is vital for ensuring storage integrity. Further research is needed to quantify these impacts and develop strategies to mitigate potential risks, such as viscous fingering and gravity override (Sekar et al., 2023).

Several software packages are commonly used for geomechanical modeling in UHS, including FLAC3D-TOUGH2MP, OpenGeoSys, CODE-BRIGHT, VISAGE, and STARS. The choice of software depends on factors such as the code's advantages and limitations, the type of geomechanical problem, and the available data.

7.7.4 Current Challenges and Future Directions

Addressing UHS challenges requires interdisciplinary collaboration across reservoir engineering, chemistry, geology, and microbiology, aiming to ensure the safe, efficient, and economically viable implementation of large-scale UHS solutions (Figure 7.14). Key areas of focus include the impact of injected hydrogen pressure and temperature on reservoir stability, biofilm growth and chemical

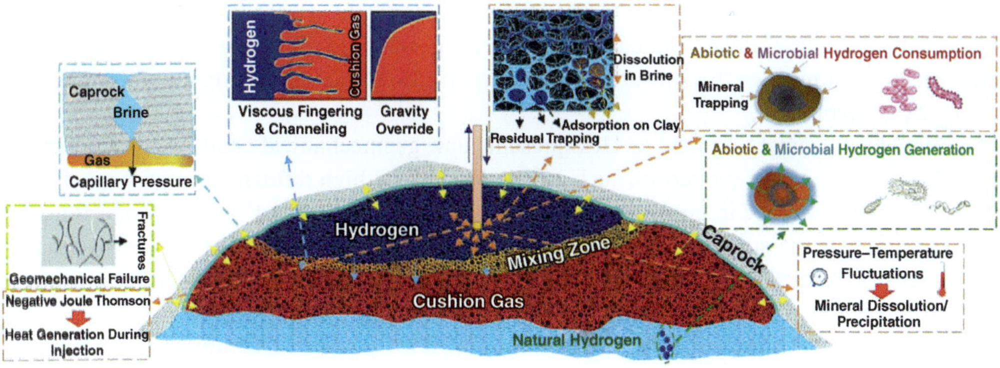

Figure 7.14 Graphical illustration of various physical, geochemical, and microbial processes associated with UHS. *Source*: Hassanpouryouzband et al., 2021 / American Chemical Society.

interactions, the influence of cyclic loading on salt cavern integrity, and the behavior of stiff clays under cyclic loading conditions (Ciancimino et al., 2024; Heinemann et al., 2021; Kumar et al., 2023; Tenthorey et al., 2024).

Developing accurate predictive models for UHS remains challenging due to the complex interactions among geochemical, geomechanical, and hydrodynamic processes (Saeed and Jadhawar, 2024). Limited field data on specific loading conditions associated with green hydrogen storage and intermittent production complicates the validation of geomechanical models (Sekar et al., 2023). Future research should focus on developing accurate multi-phase fluid flow models, quantifying the impact of hydrogen on geochemical reactions, characterizing microbial communities in storage formations, and investigating the long-term mechanical behavior of reservoir rocks under cyclic hydrogen injection and production (Andersson and Grönkvist, 2019; Heinemann et al., 2021; Muhammed et al., 2022).

7.8 Conclusions

In the evolving landscape of energy exploration and development, geomechanics stands as a cornerstone, enabling the efficient and safe extraction of subsurface resources. This chapter has elucidated the multifaceted role of geomechanics in both conventional and unconventional hydrocarbon reservoirs, highlighting its critical importance in ensuring operational integrity and optimizing resource recovery.

Geomechanics is pivotal in predicting and managing formation pressure, wellbore stability, and fracture initiation, which are essential for safe drilling and maximizing hydrocarbon recovery. Techniques such as hydraulic fracturing have revolutionized the extraction of unconventional resources like shale gas and shale oil, underscoring the need for accurate geomechanical evaluations to enhance fracture stimulation and improve recovery rates.

The intricate geomechanical properties of heavy oil sands and gas hydrate-bearing sediments present unique challenges and opportunities for extraction. Understanding the mechanical behavior of these formations is crucial for designing effective recovery processes and mitigating geohazards associated with their extraction.

Moreover, the application of geomechanics in GCS is paramount for addressing the environmental imperatives of our time. The integrity of caprocks, the reactivation of faults, and the induced seismicity associated with CO_2 injection are significant concerns that require rigorous geomechanical analysis and modeling. By incorporating advanced numerical simulations, we can better predict and manage the subsurface responses to CO_2 storage, ensuring the safety and efficacy of these operations.

Additionally, the emerging field of UHS presents new geomechanical challenges and opportunities. Hydrogen's unique properties, such as low density and high diffusivity, require specialized considerations for storage in geological formations. The cyclic nature of hydrogen injection and withdrawal affects the stress state and mechanical behavior of storage reservoirs, necessitating comprehensive geomechanical evaluations to ensure caprock integrity, prevent fault reactivation, and address hydrogen-mineral interactions.

The integration of geomechanics with multiphase fluid flow and reservoir simulations offers a more comprehensive understanding of subsurface dynamics, facilitating more informed decision-making and improved operational outcomes. As the energy sector continues to transition towards more sustainable practices, the insights and methodologies discussed in this chapter will be instrumental in optimizing the exploitation of subsurface resources while minimizing environmental impacts.

The advancements in geomechanics are not just enhancing the efficiency and safety of petroleum exploration and development but are also pivotal in the broader context of energy transition. By leveraging the principles and techniques of geomechanics, we can navigate the complexities of subsurface exploration, ensuring a more sustainable and resilient energy future.

References

Abdideh, M. & Hedayati Khah, S. (2018). Analytical and numerical study of casing collapse in Iranian oil field. Geotechnical and Geological Engineering 36 (3), pp. 1723–1734.

Advani, S. H., Lee, J. K., Khattab, H. A., et al. (1986). Fluid flow and structural response modeling associated with the mechanics of hydraulic fracturing. SPE Formation Evaluation 1 (3), pp. 309–318.

Alramahi, B. & Sundberg, M. I. (2012). Proppant embedment and conductivity of hydraulic fractures in shales. ARMA US Rock Mechanics/Geomechanics Symposium, ARMA-2012.

Amann, F., Kaiser, P., & Button, E. A. (2012). Experimental study of brittle behavior of clay shale in rapid triaxial compression. Rock Mechanics and Rock Engineering 45, pp. 21–33.

Amiri, M., Lashkaripour, G. R., Ghabezloo, S., et al. (2019). Mechanical earth modeling and fault reactivation analysis for CO_2-enhanced oil recovery in Gachsaran oil field, south-west of Iran. Environmental Earth Sciences 78, pp. 1–22.

Andersson, J. & Grönkvist, S. (2019). Large-scale storage of hydrogen. International Journal of Hydrogen Energy 44 (23), pp. 11901–11919.

Aslannezhad, M., Keshavarz, A., & Kalantariasl, A. (2020). Evaluation of mechanical, chemical, and thermal effects on wellbore stability using different rock failure criteria. Journal of Natural Gas Science and Engineering 78 p. 103276.

Baisch, S., Koch, C., & Muntendam-Bos, A. (2019). Traffic light systems: To what extent can induced seismicity be controlled? Seismological Research Letters 90 (3), pp. 1145–1154.

Bowers, G. L. (1995). Pore pressure estimation from velocity data: Accounting for overpressure mechanisms besides undercompaction. SPE Drilling & Completion 10 (2), pp. 89–95.

Cao, W., Shi, J. -Q., Durucan, S., et al. (2021a). Evaluation of shear slip stress transfer mechanism for induced microseismicity at in Salah CO_2 storage site. International Journal of Greenhouse Gas Control 107, p. 103302.

Cao, W., Shi, J., Durucan, S., & Korre, A. (2021b). Investigation into the Hydro-Geomechanical Mechanisms Related to Induced Microseismicity at in Salah Storage Site. SSRN Electronic Journal. https://papers.ssrn.com/sol3/papers.cfm?abstract_id=3814029. Cappa, F., & Rutqvist, J. (2011). Modeling of coupled deformation and permeability evolution during fault reactivation induced by deep underground injection of CO_2. International Journal of Greenhouse Gas Control 5 (2), pp. 336–346.

Caro, D. M. R., Batezelli, A., & Leite, E. P. (2023). Fault reactivation potential in a carbonate field in Brazil based on geomechanical analysis. Marine and Petroleum Geology 150, p. 106131.

Carroll, S., Carey, J. W., Dzombak, D., et al. (2016). Role of chemistry, mechanics, and transport on well integrity in CO_2 storage environments. International Journal of Greenhouse Gas Control 49, pp. 149–160.

Chen, G., Chenevert, M. E., Sharma, M. M., et al. (2003). A study of wellbore stability in shales including poroelastic, chemical, and thermal effects. Journal of Petroleum Science and Engineering 38 (3–4), pp. 167–176.

Chen, H., Du, H., Shi, B., et al. (2022). Mechanical properties and strength criterion of clayey sand reservoirs during natural gas hydrate extraction. Energy 242, p. 122526.

Chen, V. C., Lin, V. H., & Cheatham, J. B. (1989). An analysis of tubing and casing buckling in horizontal wells. Offshore Technology Conference, OTC-6037.

Chiaramonte, L., Zoback, M., Friedmann, J., et al. (2011). Fracture characterization and fluid flow simulation with geomechanical constraints for a CO_2–EOR and sequestration project Teapot Dome Oil Field, Wyoming, USA. Energy Procedia 4, pp. 3973–3980.

Choi, J., Dai, S., Lin, J., et al. (2018). Multistage triaxial tests on laboratory-formed methane hydrate-bearing sediments. Journal of Geophysical Research: Solid Earth 123 (5), pp. 3347–3357.

Chong, K. P., Chen, J. -L., Dana, G., et al. (1984). Triaxial testing of Devonian oil shale. Journal of Geotechnical Engineering 110 (10), pp. 1491–1497.

Chuanliang, Y., Jingen, D., Lianbo, H., et al. (2015). Brittle failure of shale under uniaxial compression. Arabian Journal of Geosciences 8, pp. 2467–2475.

Ciancimino, A., Cosentini, R. M., Foti, S., et al. (2024). A preliminary investigation on the mechanical behaviour of a stiff Italian clay in the context of hydrogen storage. Geomechanics for Energy and the Environment 38, p. 100562.

Cipolla, C. L. (2009). Modeling production and evaluating fracture performance in unconventional gas reservoirs. Journal of Petroleum Technology 61 (9), pp. 84–90.

Collett, T., Bahk, J. -J., Baker, R., et al. (2015). Methane Hydrates in Nature Current Knowledge and Challenges. Journal of Chemical & Engineering Data 60 (2), pp. 319–329.

Collett, T. S. (2002). Energy resource potential of natural gas hydrates. AAPG Bulletin, 86 (11), pp. 1971–1992.

Collett, T. S., Boswell, R., Cochran, J. R., et al. (2014). Geologic implications of gas hydrates in the offshore of India: Results of the National Gas Hydrate Program Expedition 01. Marine and Petroleum Geology 58, p. 3–28.

Collett, T. S. & Kuuskraa, V. A. (1998). Hydrates contain vast store of world gas resources. Oil and Gas Journal 96 (19), pp. 90–95.

Cundall, P. A. (1971). A computer model for simulating progressive, large-scale movement in blocky rock system. Proceedings of the International Symposium on Rock Mechanics 8, pp. 129–136.

Davies, R. J., Almond, S., Ward, R. S., et al. (2014). Oil and gas wells and their integrity: Implications for shale and unconventional resource exploitation. Marine and Petroleum Geology 56, pp. 239–254.

Delli, M. L. & Grozic, J. L. H. (2014). Experimental determination of permeability of porous media in the presence of gas hydrates. Journal of Petroleum Science and Engineering 120, pp. 1–9.

Dong, B. -C., Xiao, P., Sun, Y. -F., et al. (2022). Coupled flow and geomechanical analysis for gas production from marine heterogeneous hydrate-bearing sediments. Energy 255, p. 124501.

Dusseault, M. B., Bruno, M. S., & Barrera, J. (2001). Casing shear: causes, cases, cures. SPE Drilling & Completion 16 (2), pp. 98–107.

Eaton, B. A. (1972). The effect of overburden stress on geopressure prediction from well logs. Journal of Petroleum Technology 24 (8), pp. 929–934.

Eaton, B. A. (1975). The equation for geopressure prediction from well logs. SPE Annual Technical Conference and Exhibition? SPE-5544.

Ellsworth, W. L. (2013). Injection-induced earthquakes. Science 341 (6142), p. 1225942.

Ferronato, M., Gambolati, G., Janna, C., et al. (2010). Geomechanical issues of anthropogenic CO_2 sequestration in exploited gas fields. Energy Conversion and Management 51 (10), pp. 1918–1928.

Fjaer, E., Cerasi, P., Li, L., et al. (2004). Modeling the rate of sand production. ARMA North America Rock Mechanics Symposium, ARMA-04.

Fjaer, E., Holt, R. M., Horsrud, P., et al. (2008). Petroleum related rock rechanics 2nd edition.. In *Developments in petroleum science*. https://doi.org/10.1016/s0376-7361(07)x5300-7.

Fokker, P. A., Orlic, B., Van der Meer, L. G. H., et al. (2011). Geomechanical modeling of surface uplift around well KB-502 at the In Salah CO_2 storage site. 73rd EAGE Conference and Exhibition Incorporating SPE EUROPEC 2011 cp-238.

Fredd, C. N., McConnell, S. B., Boney, C. L., et al. (2001). Experimental study of fracture conductivity for water-fracturing and conventional fracturing applications. SPE Journal 6 (3), pp. 288–298.

Furre, A. -K., Kiær, A., & Eiken, O. (2015). CO_2-induced seismic time shifts at Sleipner. Interpretation 3 (3), pp. SS23–SS35.

Gao, Q., Han, S., Cheng, Y., et al. (2022). Flow-coupled-geomechanical modelling of CO_2 transport in depleted shale from a microscopic perspective. Energy 257, p. 124727.

Ghiassian, H. & Grozic, J. L. H. (2013). Strength behavior of methane hydrate bearing sand in undrained triaxial testing. Marine and Petroleum Geology 43, pp. 310–319.

Gholami, R., Aadnoy, B., Rasouli, V., et al. (2016). An analytical model to predict the volume of sand during drilling and production. Journal of Rock Mechanics and Geotechnical Engineering 8 (4), pp. 521–532.

Giani, G., Orsatti, S., Peter, C., et al. (2018). A coupled fluid flow: Geomechanical approach for subsidence numerical simulation. Energies 11 (7), p. 1804.

Gratier, J. -P. & Gueydan, F. (2007). Deformation in the presence of fluids and mineral reactions: Effect of fracturing and Fluid–Rock interaction on seismic cycles. In The MIT Press eBooks (pp. 319–356). https://doi.org/10.7551/mitpress/6703.003.0014.

Hale, A. H., Mody, F. K., & Salisbury, D. P. (1993). The influence of chemical potential on wellbore stability. SPE Drilling & Completion 8 (3), pp. 207–216.

Han, L., Wang, H., Wang, J., et al. (2018). Strain-based casing design for cyclic-steam-stimulation wells. SPE Production & Operations 33 (2), pp. 409–418.

Hassanpouryouzband, A., Joonaki, E., Edlmann, K., et al. (2021). Offshore geological storage of hydrogen: is this our best option to achieve net-zero? ACS Energy Letters 6 (6), pp. 2181–2186.

Heinemann, N., Alcalde, J., Miocic, J. M., et al. (2021). Enabling large-scale hydrogen storage in porous media–the scientific challenges. Energy & Environmental Science 14 (2), pp. 853–864.

Heng, S., Guo, Y., Yang, C., et al. (2015). Experimental and theoretical study of the anisotropic properties of shale. International Journal of Rock Mechanics and Mining Sciences 74, pp. 58–68.

Hill, R. G., Weingarten, M., Langenbruch, C., et al. (2024). Mitigation and optimization of induced seismicity using physics-based forecasting. ArXiv Preprint ArXiv:2403.10675.

Hottmann, C. E. & Johnson, R. K. (1965). Estimation of formation pressures from log-derived shale properties. Journal of Petroleum Technology 17 (6), pp. 717–722.

Hulme, M. (2016). 1.5 C and climate research after the Paris Agreement. Nature Climate Change 6 (3), pp. 222–224.

Ibrahim, A. (2021). A review of mathematical modelling approaches to tackling wellbore instability in shale formations. Journal of Natural Gas Science and Engineering 89, p. 103870.

Jadhawar, P. & Saeed, M. (2024). Mechanistic evaluation of the reservoir engineering performance for the underground hydrogen storage in a deep North Sea aquifer. International Journal of Hydrogen Energy 50, pp. 558–574.

Jang, J. & Santamarina, J. C. (2016). Hydrate bearing clayey sediments: Formation and gas production concepts. Marine and Petroleum Geology 77, pp. 235–246.

Jensen, R. P. & Preece, D. S. (2000). Modeling sand production with Darcy-flow coupled with discrete elements. Sandia National Lab.(SNL-NM), Albuquerque, NM (United States); Sandia.

Jiang, X. (2011). A review of physical modelling and numerical simulation of long-term geological storage of CO_2. Applied Energy 88 (11), pp. 3557–3566.

Joseph, J., Singh, D. N., Kumar, P., et al. (2016). State-of-the-art of gas hydrates and relative permeability of hydrate bearing sediments. Marine Georesources & Geotechnology 34 (5), pp. 450–464.

Josh, M., Esteban, L., Delle Piane, C., et al. (2012). Laboratory characterisation of shale properties. Journal of Petroleum Science and Engineering 88, pp. 107–124.

Kao, H., Visser, R., Smith, B., et al. (2018). Performance assessment of the induced seismicity traffic light protocol for northeastern British Columbia and western Alberta. The Leading Edge 37 (2), pp. 117–126.

Khan, S., Khulief, Y., Juanes, R., et al. (2024). Geomechanical Modeling of CO_2 Sequestration: A Review Focused on CO_2 Injection and Monitoring. Journal of Environmental Chemical Engineering, p. 112847.

Kim, J., Moridis, G. J., Yang, D., et al. (2012). Numerical studies on two-way coupled fluid flow and geomechanics in hydrate deposits. SPE Journal 17 (2), pp. 485–501.

King, G. E. & King, D. E. (2013). Environmental risk arising from well-construction failure—differences between barrier and well failure, and estimates of failure frequency across common well types, locations, and well age. SPE Production & Operations 28 (4), pp. 323–344.

Klar, A., Soga, K., & Ng, M. Y. A. (2010). Coupled deformation–flow analysis for methane hydrate extraction. Geotechnique 60 (10), pp. 765–776.

Kumar, A., Maini, B., Bishnoi, P. R., et al. (2010). Experimental determination of permeability in the presence of hydrates and its effect on the dissociation characteristics of gas hydrates in porous media. Journal of Petroleum Science and Engineering 70 (1–2), pp. 114–122.

Kumar, K. R., Honorio, H., Chandra, D., et al. (2023). Comprehensive review of geomechanics of underground hydrogen storage in depleted reservoirs and salt caverns. Journal of Energy Storage 73, p. 108912.

Kumar, P. & Sonawane, A. K. (2004). Engineering significance and consequences of gas hydrates in deep and ultra deep water development. ISOPE International Ocean and Polar Engineering Conference, ISOPE-I.

Kumar, V., Sondergeld, C. H., & Rai, C. S. (2012). Nano to macro mechanical characterization of shale. SPE Annual Technical Conference and Exhibition?, SPE-159804.

Kvenvolden, K. A. (1999). Potential effects of gas hydrate on human welfare. Proceedings of the National Academy of Sciences 96 (7), pp. 3420–3426.

Kvenvolden, K. A. & Lorenson, T. D. (2001). The global occurrence of natural gas hydrate. Geophysical Monograph Series 124, pp. 3–18.

Lee, H., Shinn, Y. J., Ong, S. H., et al. (2017). Fault reactivation potential of an offshore CO_2 storage site, Pohang Basin, South Korea. Journal of Petroleum Science and Engineering 152, pp. 427–442.

Lei, L. & Santamarina, J. C. (2018). Laboratory strategies for hydrate formation in fine-grained sediments. Journal of Geophysical Research: Solid Earth 123 (4), pp. 2583–2596.

Li, G. (2016). Numerical investigation of CO_2 storage in hydrocarbon field using a geomechanical-fluid coupling model. Petroleum 2 (3), pp. 252–257.

Li, L., Papamichos, E., & Cerasi, P. (2006). Investigation of sand production mechanisms using DEM with fluid flow. Proceedings of the International Symposium of the International Society for Rock Mechanics: Eurock 241–247.

Li, P. & Chalaturnyk, R. J. (2005). Geomechanical model of oil sands. SPE International Thermal Operations and Heavy Oil Symposium, SPE-97949.

Liang, H., Song, Y., Chen, Y., et al. (2011). The measurement of permeability of porous media with methane hydrate. Petroleum Science and Technology 29 (1), pp. 79–87.

Lin, B., Chen, S., You, H., et al. (2015). Experimental investigation on dilation mechanism of ultra-heavy oilsands from Xinjiang oilfield. ISRM Congress, ISRM-13CONGRESS.

Lin, Y., Zhou, K., Li, J., et al. (2020). Weakening laws of mechanical properties of sandstone under the effect of chemical corrosion. Rock Mechanics and Rock Engineering 53, pp. 1857–1877.

Liu, H., Zhang, K., Liu, T., et al. (2022). Experimental and numerical investigations on tensile mechanical properties and fracture mechanism of granite after cyclic thermal shock. Geomechanics and Geophysics for Geo-Energy and Geo-Resources 8 (1), p. 18.

Liu, K., Ostadhassan, M., & Bubach, B. (2016). Applications of nano-indentation methods to estimate nanoscale mechanical properties of shale reservoir rocks. Journal of Natural Gas Science and Engineering 35, pp. 1310–1319.

Liu, Y., Liu, L., Leung, J. Y., et al. (2021). Coupled flow/geomechanics modeling of interfracture water injection to enhance oil recovery in tight reservoirs. SPE Journal 26 (1), pp. 1–21.

Long, D., Lovell, M. A., Rees, J. G., et al. (2009). Sediment-hosted gas hydrates: new insights on natural and synthetic systems. Geological Society, London, Special Publications 319 (1), pp. 1–9.

Lopez, J. L., Rappold, P. M., Ugueto, G. A., et al. (2004). Integrated shared earth model: 3D pore-pressure prediction and uncertainty analysis. The Leading Edge 23 (1), pp. 52–59.

Loucks, R. G., Reed, R. M., Ruppel, S. C., et al. (2009). Morphology, genesis, and distribution of nanometer-scale pores in siliceous mudstones of the Mississippian Barnett Shale. Journal of Sedimentary Research 79 (12), pp. 848–861.

Lucier, A., Zoback, M., Gupta, N., et al. (2006). Geomechanical aspects of CO_2 sequestration in a deep saline reservoir in the Ohio River Valley region. Environmental Geosciences 13 (2), pp. 85–103.

Maini, B. B. & Okazawa, T. (1987). Effects of temperature on heavy oil-water relative permeability of sand. Journal of Canadian Petroleum Technology 26 (3). https://doi.org/10.2118/87-03-03.

Mardoukhi, A., Mardoukhi, Y., Hokka, M., et al. (2017). Effects of heat shock on the dynamic tensile behavior of granitic rocks. Rock Mechanics and Rock Engineering 50, pp. 1171–1182.

Masui, A., Miyazaki, K., Haneda, H., et al. (2008). Mechanical properties of natural gas hydrate bearing sediments retrieved from eastern Nankai trough. Offshore Technology Conference, OTC-19277.

Mazzoldi, A., Rinaldi, A. P., Borgia, A., et al. (2012). Induced seismicity within geological carbon sequestration projects: Maximum earthquake magnitude and leakage potential from undetected faults. International Journal of Greenhouse Gas Control 10, pp. 434–442.

McCormack, K. L., Zoback, M. D., & Kuang, W. (2021). A case study of vertical hydraulic fracture growth, stress variations with depth and shear stimulation in the Niobrara Shale and Codell Sand, Denver-Julesburg Basin, Colorado. Interpretation 9 (4), pp. SG59–SG69.

McGarr, A. (2014). Maximum magnitude earthquakes induced by fluid injection. Journal of Geophysical Research: Solid Earth 119 (2), pp. 1008–1019.

Mei, C., Fang, Z., & Wu, W. (2022). Slip transition of rock fractures due to chemical corrosion. Engineering Geology 308, pp. 106801.

Minkoff, S. E., Stone, C. M., Bryant, S., et al. (2003). Coupled fluid flow and geomechanical deformation modeling. Journal of Petroleum Science and Engineering 38 (1–2), pp. 37–56.

Miyazaki, K., Masui, A., Sakamoto, Y., et al. (2011). Triaxial compressive properties of artificial methane-hydrate-bearing sediment. Journal of Geophysical Research: Solid Earth *116*, B06102. https://doi.org/10.1029/2010JB008049.

Mody, F. K. & Hale, A. H. (1993). Borehole-stability model to couple the mechanics and chemistry of drilling-fluid/shale interactions. Journal of Petroleum Technology 45 (11), pp. 1093–1101.

Mohamad-Hussein, A., Mendoza, P. E. V., Delbosco, P. F., et al. (2022). Geomechanical modelling of cold heavy oil production with sand. Petroleum 8 (1), pp. 66–83.

Mohamadian, N., Ghorbani, H., Wood, D. A., et al. (2021). A geomechanical approach to casing collapse prediction in oil and gas wells aided by machine learning. Journal of Petroleum Science and Engineering 196, p. 107811.

Mohammed, A. I., Oyeneyin, B., Atchison, B., et al. (2019). Casing structural integrity and failure modes in a range of well types-a review. Journal of Natural Gas Science and Engineering 68, p. 102898.

Morales-Monsalve, C. B., Alzate-Espinosa, G. A., & Arbelaez-Londoño, A. (2019). Stress–Strain Behavior in Heavy Oil Reservoirs Using Mohr–Coulomb Constitutive Model. Geotechnical and Geological Engineering 37 (4), pp. 3343–3353.

Moridis, G. J. J., Collett, T. S. S., Pooladi-Darvish, M., et al. (2011). Challenges, uncertainties, and issues facing gas production from gas-hydrate deposits. SPE Reservoir Evaluation & Engineering 14 (1), pp. 76–112.

Mortezaei, K. & Vahedifard, F. (2015). Numerical simulation of induced seismicity in carbon capture and storage projects. Geotechnical and Geological Engineering 33, pp. 411–424.

Muhammed, N. S., Haq, B., Al Shehri, D., et al. (2022). A review on underground hydrogen storage: Insight into geological sites, influencing factors and future outlook. Energy Reports 8, pp. 461–499.

Nicol, A., Carne, R., Gerstenberger, M., et al. (2011). Induced seismicity and its implications for CO_2 storage risk. Energy Procedia 4, pp. 3699–3706.

Ning, F., Yu, Y., Kjelstrup, S., et al. (2012). Mechanical properties of clathrate hydrates: status and perspectives. Energy & Environmental Science 5 (5), pp. 6779–6795.

Ning, F., Zhang, K., Wu, N., et al. (2013). Invasion of drilling mud into gas-hydrate-bearing sediments. Part I: effect of drilling mud properties. Geophysical Journal International 193 (3), pp. 1370–1384.

Nixon, M. F. & Grozic, J. L. H. (2007). Submarine slope failure due to gas hydrate dissociation: a preliminary quantification. Canadian Geotechnical Journal 44 (3), pp. 314–325.

Noshi, C. I., Noynaert, S. F., & Schubert, J. J. (2018). Casing failure data analytics: A novel data mining approach in predicting casing failures for improved drilling performance and production optimization. SPE Annual Technical Conference and Exhibition?, D011S001R003.

Noshi, C., Noynaert, S., & Schubert, J. (2019). Data mining approaches for casing failure prediction and prevention. International Petroleum Technology Conference, D021S018R001.

Oleas, A., Osuji, C. E., Chenevert, M. E., et al. (2008). Entrance pressure of oil based mud into shale: effect of shale, water Activity, and Mud Properties. SPE Annual Technical Conference and Exhibition?, SPE-116364.

Papamichos, E. & Furui, K. (2019). Analytical models for sand onset under field conditions. Journal of Petroleum Science and Engineering 172, pp. 171–189.

Papamichos, E., Vardoulakis, I., Tronvoll, J., et al. (2001). Volumetric sand production model and experiment. International Journal for Numerical and Analytical Methods in Geomechanics 25 (8), 789–808.

Peck, W. D., Ayash, S. C., Klapperich, R. J., et al. (2019). The North Dakota integrated carbon storage complex feasibility study. International Journal of Greenhouse Gas Control 84, pp. 47–53.

Pinkert, S., Grozic, J. L. H., & Priest, J. A. (2015). Strain-softening model for hydrate-bearing sands. International Journal of Geomechanics 15 (6), pp. 04015007.

Priest, J. A., Best, A. I., & Clayton, C. R. I. (2005). A laboratory investigation into the seismic velocities of methane gas hydrate-bearing sand. Journal of Geophysical Research: Solid Earth 110, B04102. https://doi.org/10.1029/2004JB003259.

Radwan, A. E. (2022). Drilling in complex pore pressure regimes: analysis of wellbore stability applying the depth of failure approach. Energies 15 (21), pp. 7872.

Rinaldi, A. P., Rutqvist, J., & Cappa, F. (2014). Geomechanical effects on CO_2 leakage through fault zones during large-scale underground injection. International Journal of Greenhouse Gas Control 20, pp. 117–131.

Roberts, B. L. (2014). Casing analysis for well integrity investigation. Sandia National Lab.(SNL-NM), Albuquerque, NM (United States).

Rogers, C., Van Ruth, P. J., & Hillis, R. R. (2008). Fault reactivation in the Port Campbell Embayment with respect to carbon dioxide sequestration, Otway Basin, Australia. Geological Society, London, Special Publications 306 (1), pp. 201–214.

Röhmann, L., Tillner, E., Magri, F., et al. (2013). Fault reactivation and ground surface uplift assessment at a prospective German CO_2 storage site. Energy Procedia 40, pp. 437–446.

Romeo, R. (2000). Seismically induced landslide displacements: a predictive model. Engineering Geology 58 (3–4), pp. 337–351.

Rothert, E. & Shapiro, S. A. (2007). Statistics of fracture strength and fluid-induced microseismicity. Journal of Geophysical Research: Solid Earth 112 (B04309), pp. 1–16, https://doi.org/10.1029/2005JB003959 Rutqvist, J. (2012). The geomechanics of CO_2 storage in deep sedimentary formations. Geotechnical and Geological Engineering 30, pp. 525–551.

Rutqvist, J., Birkholzer, J., Cappa, F., et al. (2007). Estimating maximum sustainable injection pressure during geological sequestration of CO_2 using coupled fluid flow and geomechanical fault-slip analysis. Energy Conversion and Management 48 (6), pp. 1798–1807.

Rutqvist, J., Rinaldi, A. P., Cappa, F., et al. (2016). Fault activation and induced seismicity in geological carbon storage–Lessons learned from recent modeling studies. Journal of Rock Mechanics and Geotechnical Engineering 8 (6), pp. 789–804.

Saeed, M. & Jadhawar, P. (2023). Modelling underground hydrogen storage: A state-of-the-art review of fundamental approaches and findings. Gas Science and Engineering 121, p. 205196.

Saeed, M. & Jadhawar, P. (2024). Optimizing underground hydrogen storage in aquifers: The impact of cushion gas type. International Journal of Hydrogen Energy 52, pp. 1537–1549.

Sánchez, M., Gai, X., & Santamarina, J. C. (2017). A constitutive mechanical model for gas hydrate bearing sediments incorporating inelastic mechanisms. Computers and Geotechnics 84, pp. 28–46.

Sausse, J., Jacquot, E., Fritz, B., et al. (2001). Evolution of crack permeability during fluid–rock interaction. Example of the Brezouard granite (Vosges, France). Tectonophysics, 336 (1–4), 199–214.

Sayers, C. M. (1994). The elastic anisotrophy of shales. Journal of Geophysical Research: Solid Earth 99 (B1), pp. 767–774.

Sekar, L. K., Kiran, R., Okoroafor, E. R., et al. (2023). Review of reservoir challenges associated with subsurface hydrogen storage and recovery in depleted oil and gas reservoirs. Journal of Energy Storage 72, p. 108605.

Seol, J. & Lee, H. (2013). Natural gas hydrate as a potential energy resource: From occurrence to production. Korean Journal of Chemical Engineering 30, pp. 771–786.

Settari, A., Ito, Y., & Jha, K. N. N. (1992). Coupling of a fracture mechanics model and a thermal reservoir simulator for tar sands. Journal of Canadian Petroleum Technology 31 (9). https://doi.org/10.2118/92-09-02.

Shapiro, S. A., Krüger, O. S., Dinske, C., et al. (2011). Magnitudes of induced earthquakes and geometric scales of fluid-stimulated rock volumes. Geophysics 76 (6), pp. WC55–WC63.

Soga, K., Lee, S. L., Ng, M. Y. A., et al. (2006). Characterisation and engineering properties of methane hydrate soils. Characterisation and Engineering Properties of Natural Soils 4, pp. 1642–2591.

Sondergeld, C. H. & Rai, C. S. (2011). Elastic anisotropy of shales. The Leading Edge 30 (3), pp. 324–331.

Sone, H. & Zoback, M. D. (2013). Mechanical properties of shale-gas reservoir rocks—Part 2: Ductile creep, brittle strength, and their relation to the elastic modulus. Geophysics 78 (5), pp. D393–D402.

Song, Y., Jun, S., Na, Y., et al. (2023a). Geomechanical challenges during geological CO_2 storage: A review. Chemical Engineering Journal 456, pp. 140968.

Song, Y., Jun, S., Na, Y., et al. (2023b). Geomechanical challenges during geological CO_2 storage: A review. Chemical Engineering Journal 456, p. 140968.

Streit, J. E. & Siggins, A. F. (2005). Predicting, monitoring and controlling geomechanical effects of CO_2 injection. In: Greenhouse Gas Control Technologies 7, Elsevier Science Ltd, Oxford, pp. 643–651. 643–651. https://doi.org/10.1016/b978-008044704-9/50065-3.

Tanveer, M. S., Bari, A., Zubair, T., Khan, M. R., et al. (2023). Comprehensive Liner Inspection Using Smart Integration of Acoustic and Caliper Logs. SPE/PAPG Pakistan Section Annual Technical Conference, SPE-219506.

Tao, J., Wu, Y., Elsworth, D., et al. (2019). Coupled thermo-hydro-mechanical-chemical modeling of permeability evolution in a CO_2-circulated geothermal reservoir. Geofluids, 5210730, p. 15. https://doi.org/10.1155/2019/5210730.

Tarkowski, R. & Uliasz-Misiak, B. (2022). Towards underground hydrogen storage: A review of barriers. Renewable and Sustainable Energy Reviews 162, pp. 112451.

Templeton, D. C., Schoenball, M., Layland-Bachmann, C. E., et al. (2023). A project lifetime approach to the management of induced seismicity risk at geologic carbon storage sites. Seismological Society of America 94 (1), pp. 113–122.

Tenthorey, E., Hsiao, W. M., Puspitasari, R., et al. (2024). Geomechanics of hydrogen storage in a depleted gas field. International Journal of Hydrogen Energy 60, pp. 636–649.

Thiemann, B. (2018). Automation of tubular running services. SPE Norway Subsurface Conference?, D012S013R004.

Thomsen, L. (1986). Weak elastic anisotropy. Geophysics 51 (10), pp. 1954–1966.

Vardoulakis, I., Stavropoulou, M., & Papanastasiou, P. (1996). Hydro-mechanical aspects of the sand production problem. Transport in Porous Media 22, pp. 225–244.

Vasco, D. W., Bissell, R. C., Bohloli, B., et al. (2018). Monitoring and modeling caprock integrity at the In Salah carbon dioxide storage site, Algeria. Geological Carbon Storage: Subsurface Seals and Caprock Integrity 5, pp. 243–269.

Vaziri, H., Barree, B., Xiao, Y., et al. (2002). What is the magic of water in producing sand? SPE Annual Technical Conference and Exhibition?, SPE-77683.

Verdon, J. P. (2016). Using microseismic data recorded at the Weyburn CCS-EOR site to assess the likelihood of induced seismic activity. International Journal of Greenhouse Gas Control 54, pp. 421–428.

Verdon, J. P. & Stork, A. L. (2016). Carbon capture and storage, geomechanics and induced seismic activity. Journal of Rock Mechanics and Geotechnical Engineering 8 (6), pp. 928–935.

Vernik, L. & Milovac, J. (2011). Rock physics of organic shales. The Leading Edge 30 (3), pp. 318–323.

Vidal-Gilbert, S., Nauroy, J. -F., & Brosse, E. (2009). 3D geomechanical modelling for CO_2 geologic storage in the Dogger carbonates of the Paris Basin. International Journal of Greenhouse Gas Control 3 (3), pp. 288–299.

Vorobiev, S., Zharnikov, T., Vorobiev, V., et al. (2024). Anisotropic Rock-physics Model And Its Applications To Laminated Reservoirs Analysis. International Petroleum Technology Conference, D021S064R001.

Wang, P., Zhong, C., Fan, S., et al. (2023). Prediction of Collapsing Strength of High-Strength Collapse-Resistant Casing Based on Machine Learning. Processes 11 (10), pp. 3007.

Wang, Y. & Xue, S. (2002). Coupled reservoir-geomechanics model with sand erosion for sand rate and enhanced production prediction. SPE International Conference and Exhibition on Formation Damage Control, SPE-73738.

Wang, Y., Zhao, L., Han, D., et al. (2021). Anisotropic dynamic and static mechanical properties of organic-rich shale: The influence of stress. Geophysics 86 (2), pp. C51–C63.

Wei, X., Li, Q., Li, X., et al. (2020). Traffic Light Detection Method for Underground CO 2 Injection–Induced Seismicity. International Journal of Geomechanics 20 (2), pp. 4019162.

White, J. A. & Foxall, W. (2016). Assessing induced seismicity risk at CO_2 storage projects: Recent progress and remaining challenges. International Journal of Greenhouse Gas Control 49, pp. 413–424.

Whiticar, M. J., Faber, E., & Schoell, M. (1986). Biogenic methane formation in marine and freshwater environments: CO_2 reduction vs. acetate fermentation—isotope evidence. Geochimica et Cosmochimica Acta 50 (5), pp. 693–709.

Yale, D. P., Mayer, T., & Wang, J. (2010). Geomechanics of oil sands under injection. ARMA US Rock Mechanics/Geomechanics Symposium, ARMA-10.

Yeo, I. W., Brown, M. R. M., Ge, S., et al. (2020). Causal mechanism of injection-induced earthquakes through the Mw 5.5 Pohang earthquake case study. Nature Communications 11 (1), pp. 2614.

Yin, T., Li, X., Cao, W., et al. (2015). Effects of thermal treatment on tensile strength of Laurentian granite using Brazilian test. Rock Mechanics and Rock Engineering 48, pp. 2213–2223.

Yoneda, J., Masui, A., Konno, Y., et al. (2015). Mechanical properties of hydrate-bearing turbidite reservoir in the first gas production test site of the Eastern Nankai Trough. Marine and Petroleum Geology 66, pp. 471–486.

Yoon, S., Lee, H., & Kim, J. (2023). The modeling of fault activation, slip, and induced seismicity for geological CO_2 storage at a pilot-scale site in the Janggi Basin, South Korea. International Journal of Rock Mechanics and Mining Sciences 170, pp. 105441.

You, K., Flemings, P. B., Malinverno, A., Collett, T. S., et al. (2019). Mechanisms of methane hydrate formation in geological systems. Reviews of Geophysics 57 (4), 1146–1196.

Yu, W. & Sepehrnoori, K. (2013). Numerical evaluation of the impact of geomechanics on well performance in shale gas reservoirs. ARMA US Rock Mechanics/Geomechanics Symposium, ARMA-2013.

Zandi, S., Renard, G., Nauroy, J. -F., et al. (2010). Numerical coupling of geomechanics and fluid flow during steam injection in SAGD. SPE Improved Oil Recovery Conference?, SPE-129739.

Zhang, J., Standifird, W., & Lenamond, C. (2008). Casing ultradeep, ultralong salt sections in deep water: a case study for failure diagnosis and risk mitigation in record-depth well. SPE Annual Technical Conference and Exhibition?, SPE-114273.

Zhao, B., Liu, D., Li, Z., et al. (2018). Mechanical behavior of shale rock under uniaxial cyclic loading and unloading condition. Advances in Civil Engineering, 9750480, p. 8. https://doi.org/10.1155/2018/9750480.

Zhao, Z., Ma, J., Zheng, S., et al. (2024). Modeling temperature dependence of tensile fracture strength for rocks considering phase transition and the direct effect of thermal damage. International Journal of Damage Mechanics 33 (1), pp. 57–82.

Zheng, F., Jahandideh, A., Jha, B., et al. (2021). Geologic CO_2 storage optimization under geomechanical risk using coupled-physics models. International Journal of Greenhouse Gas Control 110, pp. 103385.

Zheng, H., Cao, S., Yuan, W., et al. (2022). A time-dependent hydro-mechanical coupling model of reservoir sandstone during CO_2 geological storage. Rock Mechanics and Rock Engineering 55 (10), pp. 5845–5861.

Zoback, M. D. (2016): Reservoir Geomechanics.—unpublished lecture document: pp. 1–75.

8

Size Scaling and Spatial Clustering of Natural Fracture Networks Using Fractal Analysis

Sofiane Djezzar, Aldjia Boualam

University of North Dakota, Grand Forks, USA

8.1 Introduction

The application of fractal geometry serves to analyze and quantify the recurring geometric patterns of these fractures across different scales (Mandelbrot, 1982). This approach has become instrumental in understanding the spatial distribution of fractures, adopting various geometric factors like orientation, length distribution, density, aperture, and shear displacement into the analysis (Barton et al., 1995; Balberg et al., 1983, 1991; Cacas et al., 2001; Bonnet et al., 2001; Odling et al., 1999). Fractal analysis provides insights into the scaling and clustering of complex fracture networks, which is crucial in fields like hydrocarbon exploration. Fractal analysis aids in predicting the presence and characteristics of subseismic fractures, which can have profound effects on the quality of reservoirs and cap rocks, based on the attributes of seismically resolved fractures. The focus of this study is the Mouydir basin's south edge in the Saharan platform, where the Cambro-Ordovician analog becomes critical alternative data for characterizing and understanding fracture distribution, connectivity, and kinematics in a region known as less explored with limited data available from a few drilling attempts in the 1950s and sparse 2D seismic surveys (Djezzar, 2019; Djezzar et al., 2019).

This understanding is pivotal for predicting fracture occurrence and frequency in the subsurface. The Cambro-Ordovician formations, marked by a stiff tectonic style, display a dense network of fractures and folds impacting both the Cambro-Ordovician and Paleozoic series. Utilizing digital elevation models (DEMs), various attributes like curvature, slopes, and illumination helps in detecting both major and minor fractures (Djezzar et al., 2020). These fractures are digitized and analyzed for their orientation, length distribution, and density. Fractal dimension analysis, applied to these 2D fracture networks, employs both center distance and box-counting algorithms to estimate the fractal dimensions of major and minor fracture networks, alongside individual fracture sets. This approach acknowledges that shorter fractures are less connected than longer ones, and various field studies confirm a power-law distribution in fracture lengths, with the exponent "a" typically ranging between 1 and 3 (Djezzar and Boualam, 2021). This exponent value indicates the relative abundance of short versus long fractures in a given population.

Geophysical Exploration for Hydrocarbon Reservoirs, Geothermal Energy, and Carbon Storage: New Technologies and AI-based Approaches, First Edition. Edited by Said Gaci.

By integrating curvature attributes, illumination, geological maps, satellite imagery, and DEMs, this study generates a comprehensive fracture map that aids in identifying major and minor fracture sets and understanding the fractography of each formation, ultimately providing insights into the Cambro-Ordovician reservoir analog's subsurface behavior. This enriched understanding is crucial for guiding future exploration efforts in the underexplored Mouydir basin, aiming to unveil its potential hydrocarbon resources. The concept of fractal dimension has been utilized to examine the 2D fracture networks impacting both the basement and Cambro-Ordovician reservoir analogs in the northern Hoggar shield. This area encompasses the Ajjers, In-Tahouite, and Tamadjert units, which characterize the Cambro-Ordovician reservoir analogs (Bensalah et al., 1972; Ait Kaci, 1990). These units are notable for their stiff tectonic style, characterized by a dense network of fractures that penetrates the entire Paleozoic series. Statistical analyses conducted on the fracture length distribution within the Cambro-Ordovician reservoir analog reveal a variety of fault set orientations, with fracture lengths exhibiting a power-law distribution (Arab and Djezzar, 2011). The estimation of the fractal dimension, denoted as Dm, employed both the center distribution method and the box-counting algorithms (Djezzar, 2019). Fractures are categorized into major and minor types based on their geometric and structural characteristics, facilitating a fractal analysis conducted across multiple scales and employing various methodologies. This comprehensive analysis confirms the fractal nature of the fracture networks affecting the basement and Cambro-Ordovician units in the studied region, with the Dm ranging between 1 and 2, indicating a complex pattern of geological features.

8.2 Geological Settings

The study area is geographically delineated by significant geological features: to the north by the Mouydir Basin, to the west by the Idjerane Spur and the Ahnet Basin, to the east by the Amguid-El-Biod Spur and the Illizi Basin, and to the south by the uplift of the Hoggar Shield (Figure 8.1). This extensive area includes regions where Paleozoic strata are exposed at the surface, specifically in Ain-Tadjoubar, Arak, Iftessene, and Khanget-El-Hadid, showcasing the geological diversity of the region (Bensalah et al., 1972). The Cambro-Ordovician period is represented by three primary units: the Ajjers, In-Tahouite, and Tamadjert formations. These units not only underlie much of the Saharan platform basins but also crop out in the northern reaches of the Hoggar Shield. The Ordovician formations span an extensive geological timeframe from the Tremadocian to the Ashgill periods. Key formations from this era include the El-Atchane Sandstones (Tremadocian); the Hamra Quartzites (Arenigian); the Ouargla Sandstone (Llanvirnian); the Oued-Saret Sandstone (Lower Caradocian); and the M'kratta Sandstone (Upper Caradocian-Ashgillian) (Bennacef et al., 1971; Zieliński, 2011). The region's geological evolution has been influenced by multiple tectonic phases, including potential strike-slip displacements of Precambrian lineaments during the Cambro-Ordovician period. A notable regional geological event was the basin inversion in the late Silurian to early Devonian, attributed to the Caledonian Orogeny, which significantly altered the landscape. The Middle Devonian period marked the reactivation of Precambrian faults bordering the basin, introducing a basin-and-ridge topography that further complicated the geological structure. The Variscan Orogeny later introduced a series of north-south trending reverse faults into the Paleozoic rocks, marking another phase of tectonic activity. In the Mesozoic era, the uplift of the Hoggar Shield, preceding the Cretaceous period and associated with the Alpine Orogeny, had profound impacts on the region (WEC, 2007; Boualam et al., 2019; Follot, 1952; Fekirine and Abdallah, 1998; Haddoum et al., 2001; Djezzar, 2012; Djezzar and Boualam, 2021; Coward and Ries, 2003). This event led to the partial inversion of sedimentary basins along the Hoggar Shield's

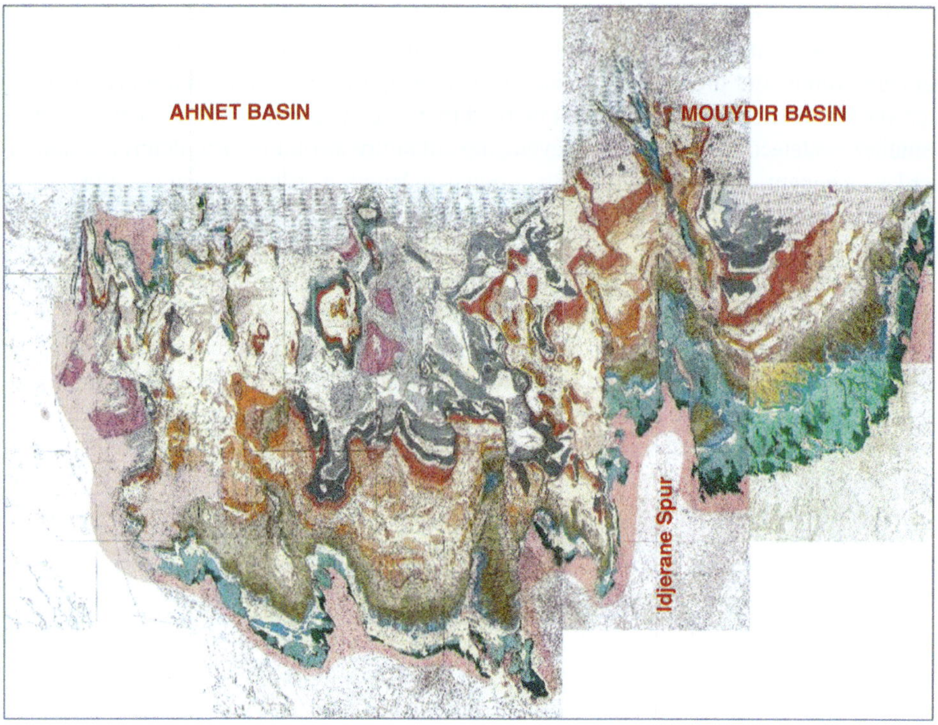

Figure 8.1 Composed Geological Maps in Ahnet and Mouydir basins.

northern margin and resulted in the erosion of a significant amount of the Paleozoic sedimentary cover (Wendt et al., 2006; Zieliński, 2011). The tectonic narrative continued into the Cretaceous and early Cenozoic periods, with the basins experiencing further inversion due to the collision between Europe and Africa, which induced the reactivation of ancient faults (Follot, 1952; Beuf et al., 1971; Fekirine and Abdallah, 1998; Liégeois et al., 2005). This complex tectonic history has significantly influenced the geological formation and evolution of the study area, shaping its current geological characteristics of the Saharan platform and potential hydrocarbon for resources in the area (Djezzar and Boualam, 2020b).

8.3 Methods and Approaches

In geological studies, fracture detection and digitalization have evolved with the integration of sophisticated techniques and diverse data inputs, including curvature and illumination attributes, geological maps, slope attributes, DEMs, and satellite imagery. To accurately characterize the orientation, length distribution, and fractal dimensions of fracture networks, a dual-scale approach is implemented. At the macro-scale, a 1:100 000 ratio facilitates the digitization and analysis of lineaments indicative of major fractures impacting the Paleozoic series. Conversely, a more detailed 1:25 000 scale is applied to identify and assess minor fractures affecting the basement and the Cambro-Ordovician units, offering a detailed view of the fracture landscape. The methodology encompasses four distinct approaches to ascertain the characteristics of fracture sets, including their length and fractal dimensions within 2D fracture networks. Fractal dimensions are calculated using both the center distance and box-counting algorithms, methods that provide a quantitative measure of the fracture networks' complexity (Davy et al., 1990, 1992, 1993; Segall and Pollard, 1983). This analysis is multi-faceted: initially, the orientation of fractures unveils the network's structural composition and preferential directions. Following this, length distributions are modeled via power law distributions, a statistical method that extends the observable fracture data to include smaller, undetectable fractures by leveraging curvature attributes. The determination of the power law exponent, achieved by plotting cumulative length distributions on a log-log scale, yields insights into the scale invariance within the fracture distribution (Darcel et al., 2003; Bour and Davy, 1997, 1998). Furthermore, the fractal analysis technique is specifically designed to correlate smaller fractures with those that are readily visible, enhancing the comprehensiveness of the study. To support these analyses, two pivotal software tools are employed: Global Mapper for map manipulation, providing a robust platform for the visualization and interpretation of geological data, and FracaFlow for in-depth fracture characterization, allowing for the detailed examination of fracture properties and behaviors. This integrated approach, combining advanced technological tools with sophisticated analytical methods, not only enhances the accuracy of fracture detection and digitalization but also significantly contributes to the understanding of geological structures and their implications (Beicip, 2018).

8.4 Fractal Analysis

Fractal theory, a pivotal tool in characterizing natural formations, underlines the inherent self-similarity of structures like fracture networks across various scales (Verscheure et al., 2012; Berkowitz et al., 2000). This theory highlights how numerous natural phenomena manifest fractal characteristics, distinguishable by their non-integer fractal dimensions, which, for a 2D network, oscillates between 1 and 2. A fractal dimension nearing 1 typically indicates a network composed mainly of discrete, linear entities. Conversely, a dimension approaching 2 suggests that the fractures densely populate the 2D space, offering near-uniform coverage. Despite the initial broad definition of fractals as structures devoid of a characteristic length scale, Mandelbrot's 1982 clarification introduces the necessity for natural fractals to possess both upper and lower bounds, thereby tailoring the definition to accommodate natural occurrences. In this research, the FracaFlow software was used in evaluating fracture networks, employing both the center distance and box-counting algorithms to ascertain the Dm. The methodology integrates four distinct approaches, contingent upon the fractures' length and the lithological context of the basement and the Cambro-Ordovician units, to evaluate the fracture networks impacting the study region comprehensively (Djezzar and Boualam, 2020a).

8.4.1 First Approach

This approach primarily focuses on the major fractures impacting the study area, including the basement and the entire Paleozoic series extending into the Mouydir basin's subsurface. These fractures are characterized by Mode II fracture type, these significant fractures exhibit long lengths, strike-slip kinematics, and span across the Paleozoic series. According to Haddoum (2001), the primary fracture sets affecting the Paleozoic series are predominantly strike-slip and reverse faults, with major structures including N-S trending dextral strike-slip faults and NNW-SSE to N-S trending reverse faults. Minor fractures, featuring NE-SW trending dextral or NW-SE trending sinistral strike-slip faults, intersect these major faults. These structural patterns are a legacy of the Pan-African orogeny, later reactivated during the Hercynian and Alpine orogeneses (Beuf et al., 1971).

Table 8.1 Fractal dimensions related to the major faults affecting the area of study.

Data	Points	Fract dim. (center distance)	Fract dim. (box counting)	Rmin
All faults	394	1.7114	1.57	425.956
Major Faults-N000	100	1.5939	1.3798	1409.9569
Major Faults-N040	38	1.0573	1.0091	851.6644
Major Faults-N060	44	1.0637	no value	843.3506
Major Faults-N090	59	1.2378	no value	524.139
Major Faults-N110	37	1.2903	no value	1493.9429
Major Faults-N130	59	1.5116	1.0802	1670.2399
Major Faults-N150	57	1.4246	1.1566	2543.0053

Pole diagrams and stereo-diagrams are employed to efficiently analyze and visualize the orientation of each fracture set. The analysis of 394 major faults within the study area has delineated seven distinct fracture sets, with orientations ranging from N000 to N150. Notably, the sets at N000, N090, and N130 emerge as the most significant in terms of fracture number and length. Fracture lengths span from 1.5 km to 165 km, with log-log length distribution analyses indicating that these major fractures follow a power law function, evidenced by high correlation coefficients between 0.91 to 0.97 and power law coefficients from 2.37 to 2.81. The fractal dimensions calculated for these major faults, via center distance and box-counting algorithms, yield values of 1.7114 and 1.57, respectively (Table 8.1, Figure 8.2). Further analysis extends to the individual fracture sets within the major fracture network, revealing fractal dimensions ranging from 1.0536 to 1.5939 (center distance algorithm) and 1.0091 to 1.3798 (box-counting algorithm). It's noteworthy that three fracture sets, specifically N060, N090, and N110, do not exhibit a fractal dimension when analyzed through the box-counting algorithm, underscoring the complexity and variability inherent in geological fractal analyses (Figure 8.3).

8.4.2 Second Approach

This approach is meticulously applied across various geological formations, encompassing both the basement and the Cambro-Ordovician units. Each unit undergoes an individualized study aimed at comprehensively understanding its unique fractography. This detailed examination, grounded in the specific lithological characteristics of each formation within the Cambro-Ordovician reservoir and the basement, is pivotal for elucidating the intricate relationships between lithofacies and the various fracture networks.

For each geological unit, the fracture sets are identified, and their length distributions are quantified. This process involves a thorough analysis of the entire fracture network as well as the individual fracture sets that constitute the network. The estimation of fractal dimensions is a key component of this analysis, utilizing both the center dimension and box-counting algorithms. This dual-method approach allows for a nuanced quantification of the network's complexity, offering insights into the spatial distribution and scaling properties of fractures within each geological formation.

By dissecting the fracture networks of these formations with such granularity, researchers can glean valuable insights into the structural dynamics and potential reservoir characteristics of the underlying geology. This comprehensive analysis not only aids in understanding the geological

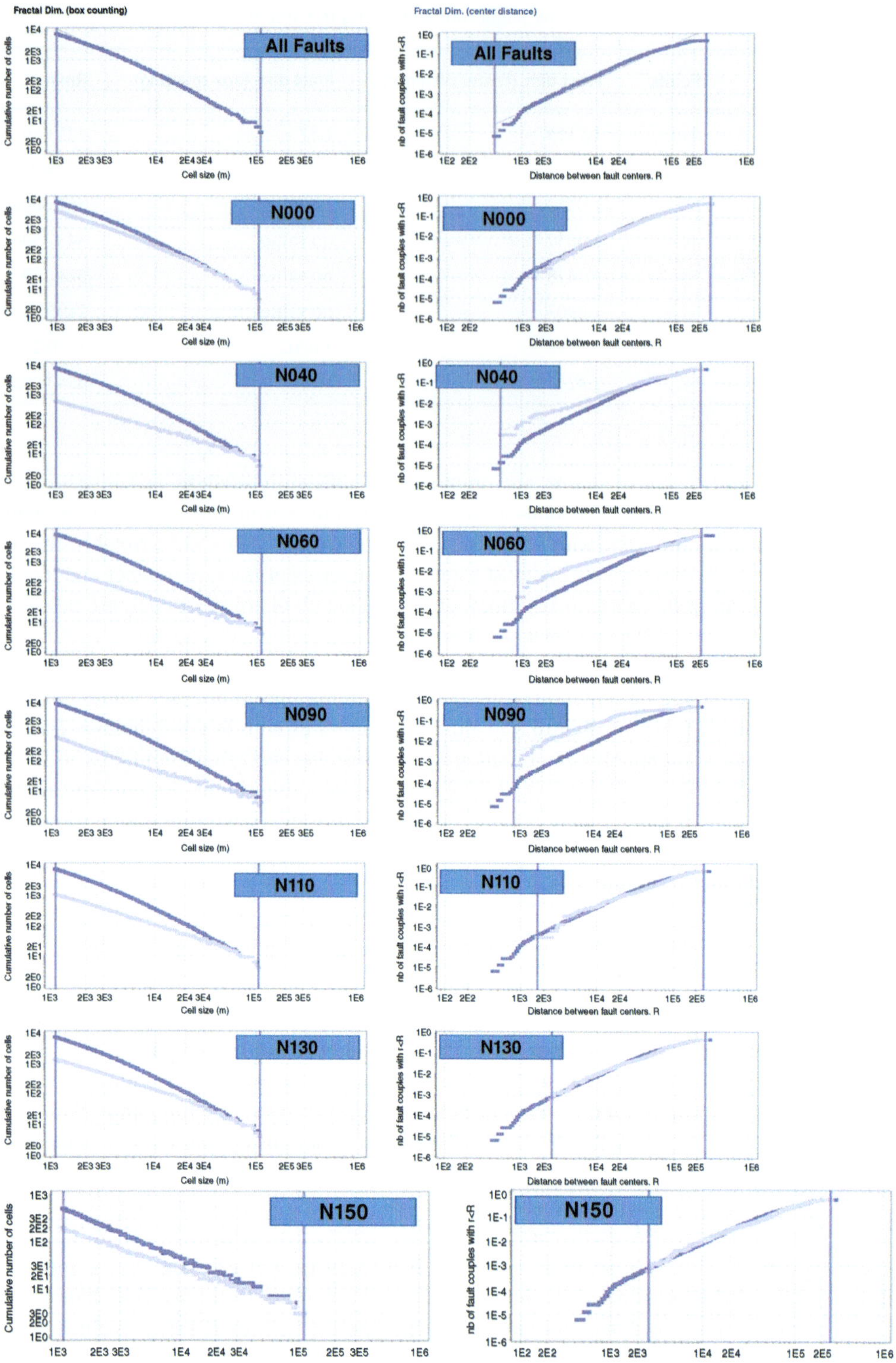

Figure 8.2 Fractal dimension distributions for the major faults by fault set using the center distance and the box-counting algorithms.

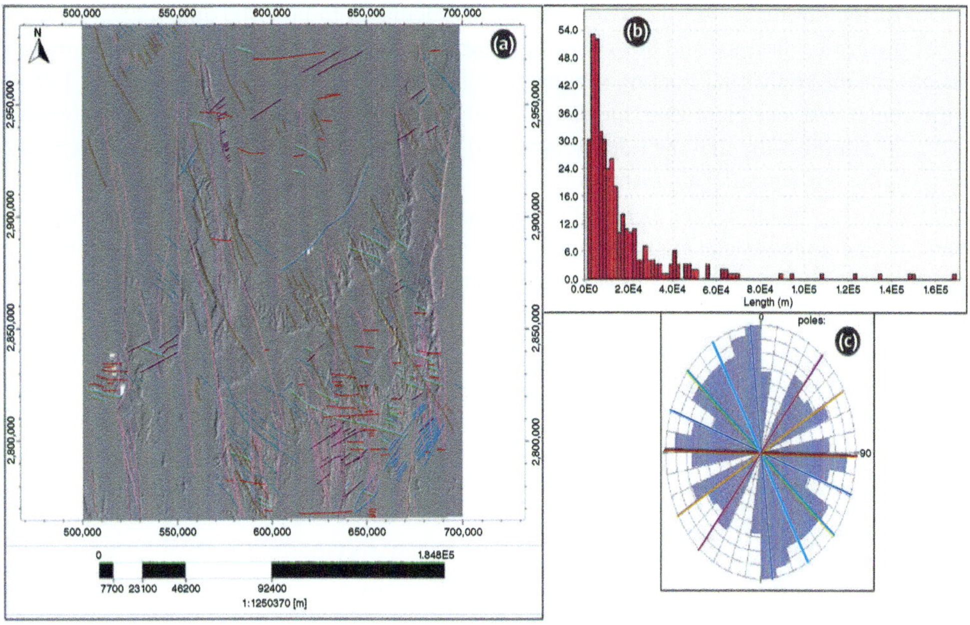

Figure 8.3 Major faults characteristics. (a) Major faults map in the area of study; (b) major faults' length distribution; (c) major fault sets orientation. *Source*: Sofiane Djezzar (2019).

past and present of these formations but also in predicting their behavior and utility in future exploration and exploitation research. Through this methodical examination, the intricate interplay between geological facies and fracture networks is unveiled, enriching our understanding of subsurface characteristics, and enhancing the efficacy of geological and reservoir modeling.

8.4.2.1 The Basement Formation

The Hoggar Shield, a prominent geological feature, is predominantly composed of metamorphic and igneous rocks, illustrating the complex geological history and composition of this region (Black et al., 1994). This Cenozoic swell covers an expansive area of approximately 550 000 km^2 and is primarily made up of Precambrian rocks, which are encircled by Paleozoic sediments laid down after the end of the Pan-African orogeny (Liégeois, 2019). A notable feature of the Hoggar Shield is the identification of twenty-three distinct terranes, each demarcated by either subvertical mega-shear zones or thrust fronts (Liégeois, 2019). These terranes exhibit unique characteristics in terms of lithology, metamorphism, magmatism, or tectonics, highlighting the region's diverse geological makeup (Black et al., 1994).

The basement is particularly noteworthy for its extensive fracturing, with 761 fractures recorded, marking it as one of the most fractured formations within the study area. This fracturing is organized into five primary sets: N000, N050, N090, N120, and N160, with the N-S and N160 sets standing out due to their low number and the lengths of the fractures. These fractures vary significantly in length, ranging from 500 m to 22 km. Analysis via log-log diagrams of length distribution reveals that the fracture lengths within the basement adhere to a power law function, evidenced by high correlation coefficients between 0.91 to 0.97 and power law coefficients ranging from 2.37 to 2.81.

The Dm of the entire fracture network have been calculated to be 1.705 and 1.3993 through the use of the center distance and box-counting algorithms, respectively. Additionally, the fractal dimensions for the individual fracture sets have been determined to range from 1.376 to 1.4693 using the center distance algorithm and from 1.0091 to 1.3798 with the box-counting algorithm (Table 8.2). It's noteworthy that the fracture set N090 was found to lack a fractal dimension when analyzed via the box-counting algorithm, underscoring the unique characteristics and challenges in quantifying the fractal nature of specific fracture sets within the Hoggar Shield's basement formation. This detailed fractal analysis contributes significantly to our understanding of the geological structure and the potential implications for geological and reservoir studies in this region (Figures 8.4 and 8.5).

Table 8.2 Fractal dimensions related to the minor faults affecting basement formation.

Data	Points	Fract dim. (center distance)	Fract dim. (box counting)	Rmin
All faults	761	1.705	1.3993	107.415
Basement-N000	256	1.4693	1.2163	385.8096
Basement-N050	123	1.3783	1.0512	453.2459
Basement-N090	121	1.376	no value	661.2585
Basement-N120	116	1.4425	1.0149	632.9037
Basement-N160	145	1.4009	1.086	314.4352

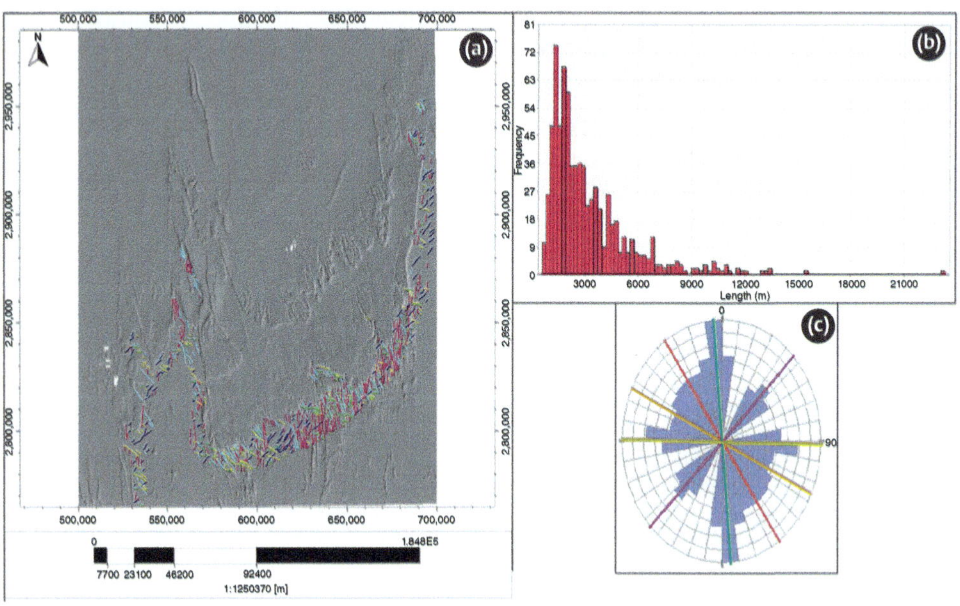

Figure 8.4 Basement faults characteristics. (a) Minor faults map in the basement formation; (b) minor faults' length distribution; (c) minor fault sets orientation. *Source*: Sofiane Djezzar (2019).

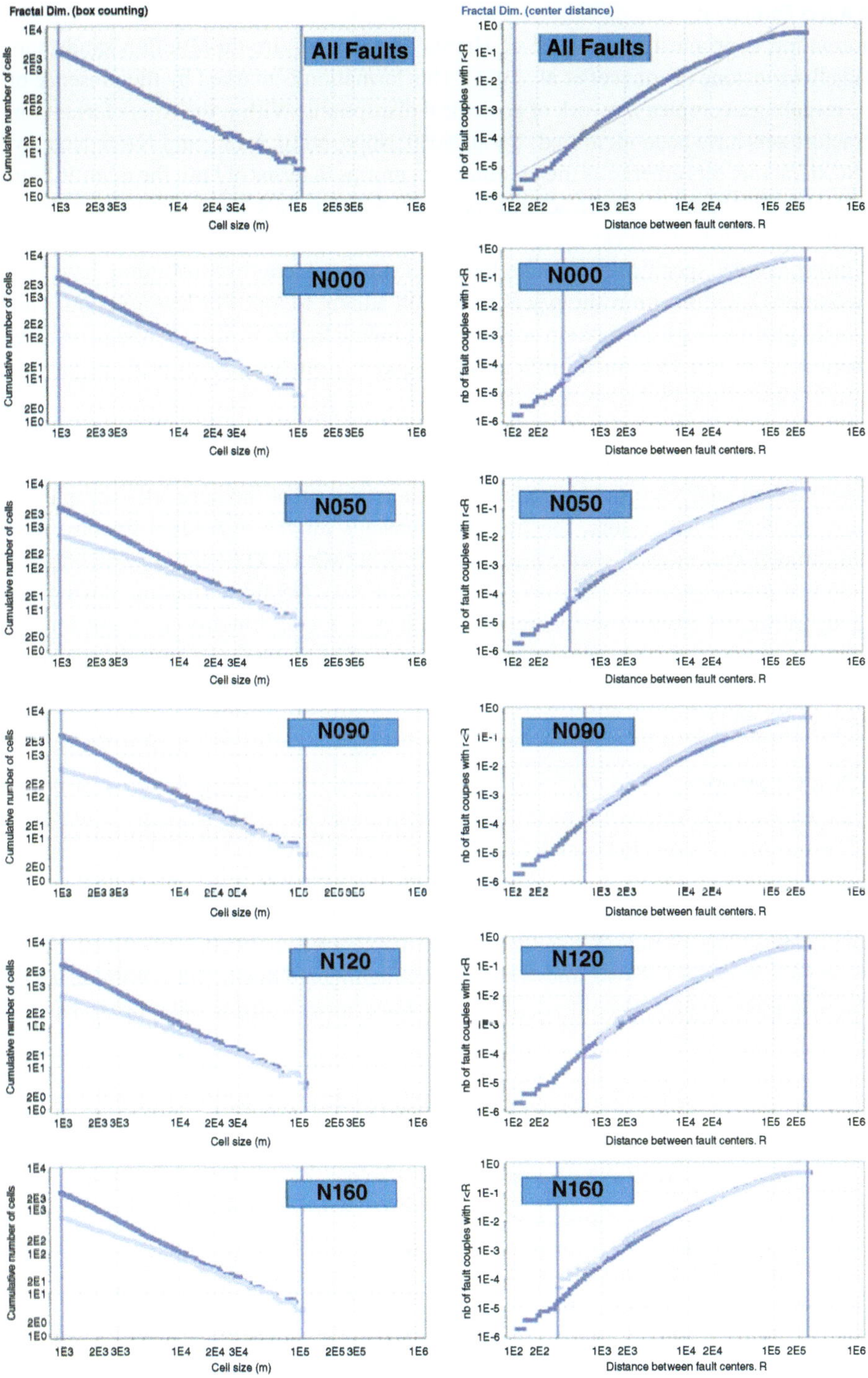

Figure 8.5 Fractal dimension distributions for the minor faults in the basement formation by fault set using the center distance and the box-counting algorithms.

8.4.2.2 The Ajjers Formation

The Ajjers formation is primarily composed of Hamra Quartzites, Vire-du-Mouflon sandstone, and Tin-Taradjelli sandstone (Bennacef et al., 1971). This formation is marked by the presence of 492 fractures, revealing a complex network of geological disruptions. Within the Ajjers formation, six distinct fracture sets have been identified: N000, N050, N090, N110, N130, and N150. Notably, the N-S and N090 fracture sets emerge as the predominant groups in terms of both the quantity and length of fractures. The lengths of these fractures span from 300 m to 16 km, suggesting significant variability within the fracture network.

Analytical investigations, specifically through log-log diagrams of length distribution, have verified that the fracture lengths within the Ajjers formation adhere to a power law function. This adherence is highlighted by high correlation coefficients, ranging from 0.91 to 0.97, and power law coefficients between 2.37 and 2.81. Such findings underscore the predictable yet intricate pattern of fracture distribution within the Ajjers formation.

The fractal dimensions of the entire fracture network within the Ajjers formation have been quantified as 1.5945 and 1.2125, utilizing the center distance and box-counting algorithms, respectively. This dual-method analysis provides a nuanced understanding of the network's complexity and spatial arrangement. Furthermore, the fractal dimensions for the individual fracture sets, as determined through the center distance algorithm, vary between 1.1119 and 1.3777, offering insight into the heterogeneity of the fracture systems (Table 8.3). Notably, when employing the box-counting algorithm, all fracture sets were found to lack a fractal dimension, except for the N000 set, which recorded a fractal dimension of 1. This exception highlights the unique geometrical properties of the N000 fracture set within the broader context of the Ajjers formation's fractal landscape (Figures 8.6 and 8.7).

8.4.2.3 In-Tahouite Formation

The In-Tahouite formation is primarily constituted of bioturbated siltstones along with very-fine to fine-grained sandstone. This geological unit also encompasses the Azzel shale, Tiferouine shale, and Ouargla sandstone (Bennacef et al., 1971). With 172 fractures recorded, the In-Tahouite unit is noted as being less fractured in comparison to other formations, a characteristic attributed mainly to its shale components. This formation is delineated by five fracture sets: N000, N050, N060, N090, and N130, among which the N-S, N090, and N130 sets are highlighted due to their prominence in terms of both the number and lengths of fractures. These fractures exhibit a broad range in length, stretching from 500 m to 10 km.

Table 8.3 Fractal dimensions related to the minor faults affecting Ajjers formation.

Data	Points	Fract dim. (center distance)	Fract dim. (box counting)	Rmin
All faults	492	1.5945	1.2125	113.9189
Ajjer-N000	120	1.3777	1	293.8084
Ajjer-N50	36	1.1119	no value	716.8985
Ajjer-N90	123	1.2062	no value	285.2209
Ajjer-N110	52	1.2444	no value	439.7315
Ajjer-N130	92	1.1953	no value	518.2868
Ajjer-N150	69	1.2012	no value	592.9249

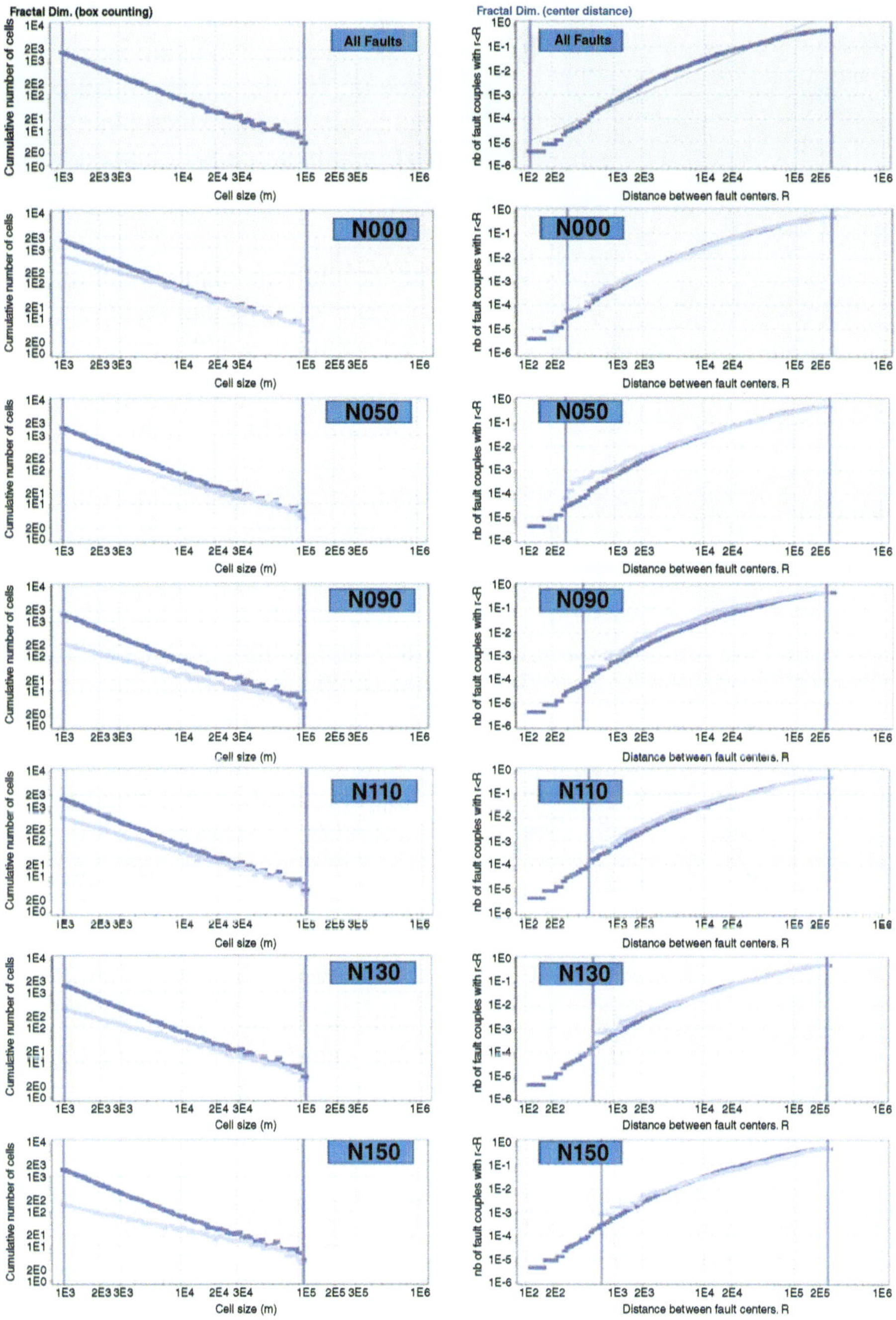

Figure 8.6 Fractal dimension distributions for the minor faults in the Ajjers formation by fault set using the center distance and the box-counting algorithms.

Analysis via log-log diagrams of fracture length distribution has validated that the fracture lengths within the In-Tahouite formation adhere to a power law function. This adherence is indicated by high correlation coefficients, ranging between 0.93 to 0.97, and power law coefficients

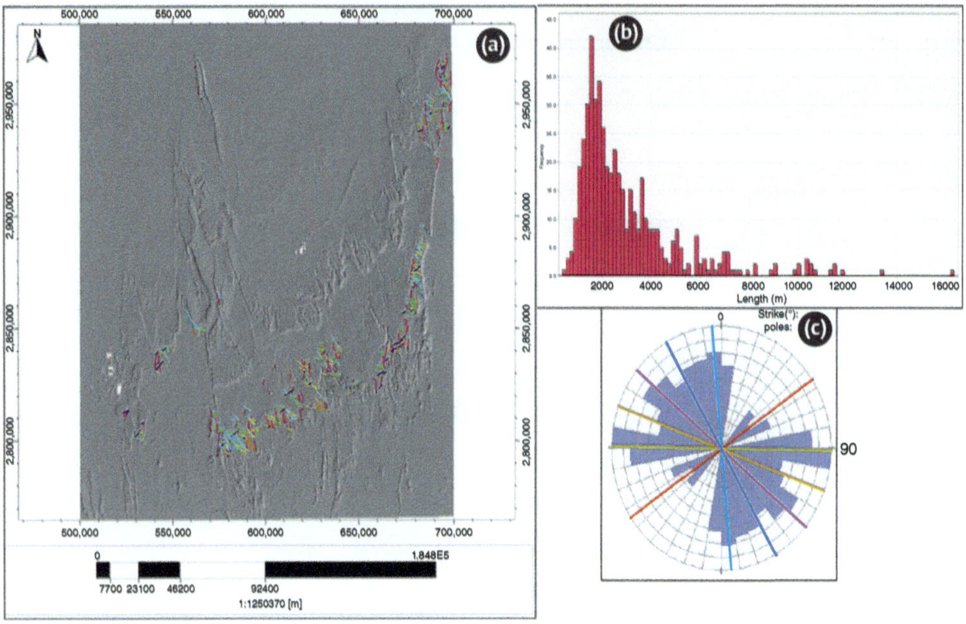

Figure 8.7 Ajjers faults characteristics. (a) Minor faults map in the Ajjers formation; (b) minor faults' length distribution; (c) minor fault sets orientation. *Source*: Sofiane Djezzar (2019).

Table 8.4 Fractal dimensions related to the minor faults affecting In-Tahouite formation.

Data	Points	Fract dim. (center distance)	Fract dim. (box counting)	Rmin
All faults	172	1.3722	1.04	220.1225
In-Tahouite-N000	40	no value	no value	266.3499
In-Tahouite-N050	15	no value	no value	997.2087
In-Tahouite-N060	17	no value	no value	549.5823
In-Tahouite-N090	49	1.0946	no value	615.7739
In-Tahouitew-N130	51	1.1392	no value	600.7082

spanning from 2.27 to 3. Such findings illustrate a significant complex pattern of fracture distribution reflective of the formation's geological characteristics.

The Dm of the entire fracture network within the In-Tahouite formation have been calculated as 1.3722 and 1.04, utilizing the center distance and box-counting algorithms, respectively (Table 8.4). These values provide a quantified measure of the fracture network's complexity and spatial distribution. However, a notable aspect of this analysis is that the individual fracture sets, except for the N090 and N130 sets, did not exhibit calculable fractal dimensions when assessed through both the center distance and box-counting algorithms. This outcome suggests a distinctive geometric or distributional pattern for the N090 and N130 fracture sets within the broader context of the In-Tahouite formation's fracture network, pointing towards the heterogeneity and the nuanced geological dynamics shaping this less fractured unit (Figures 8.8 and 8.9).

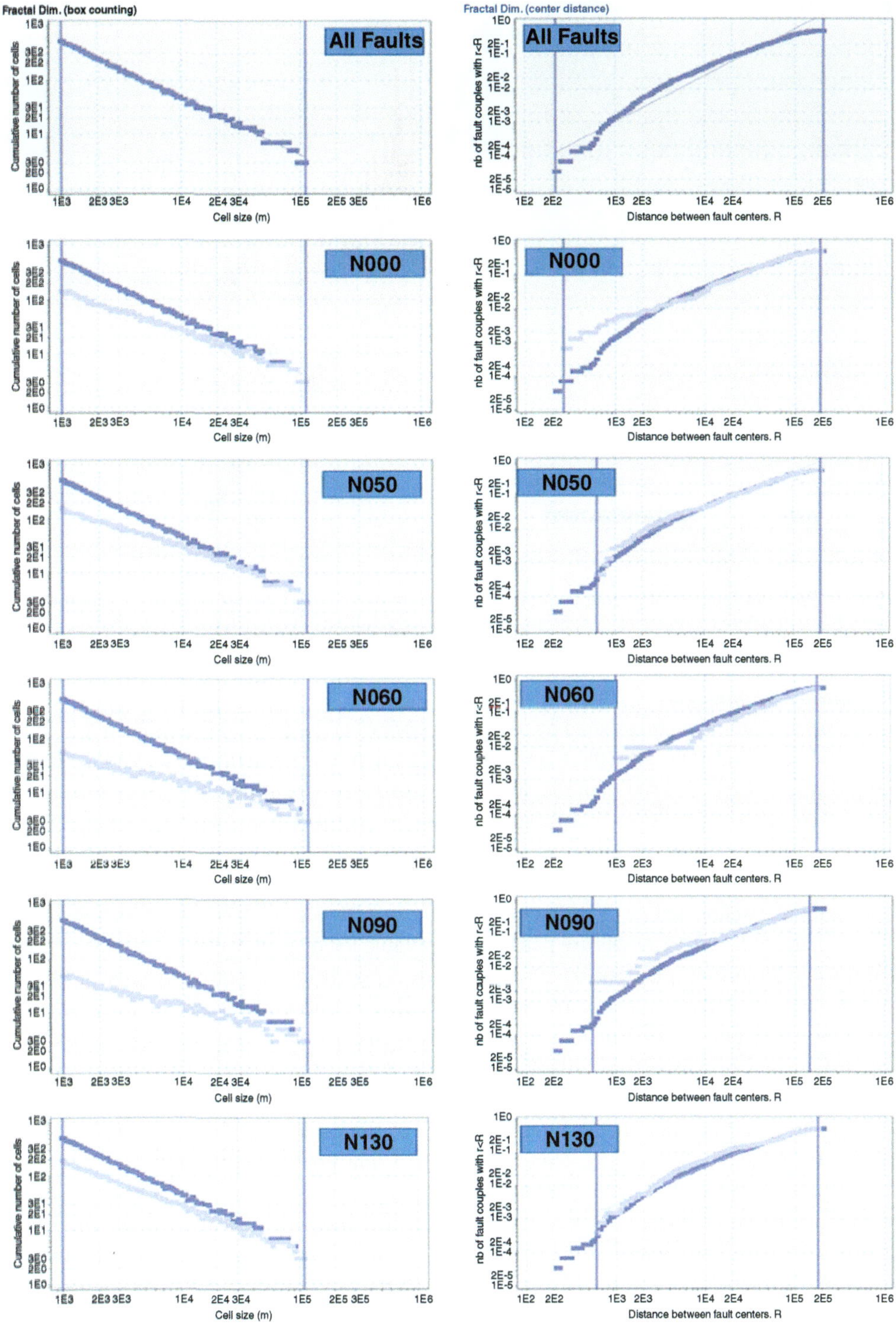

Figure 8.8 Fractal dimension distributions for the minor faults in the In-Tahouite formation by fault set using the center distance and the box-counting algorithms.

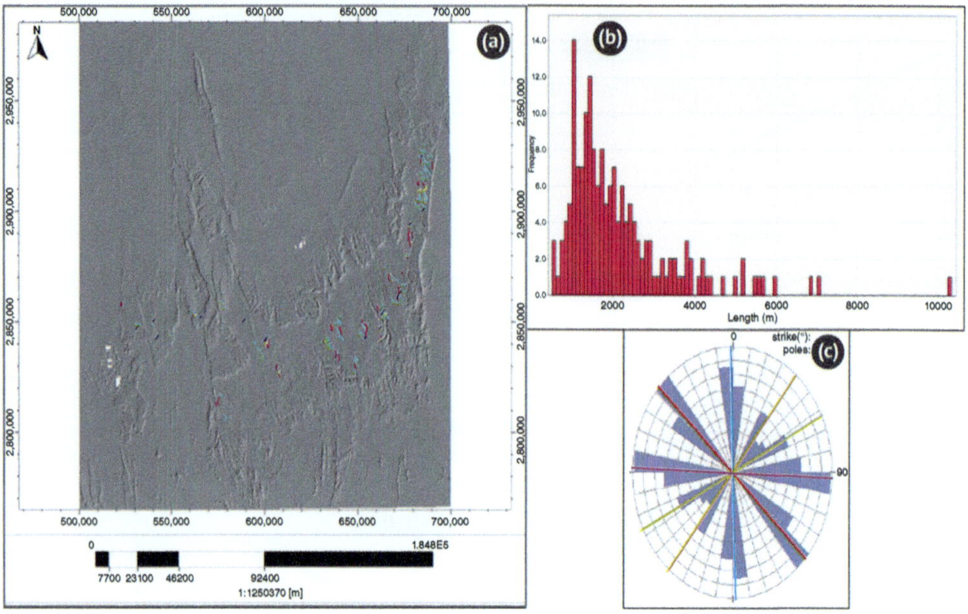

Figure 8.9 In-Tahouite faults characteristics. (a) Minor faults map in the In-Tahouite formation; (b) minor faults' length distribution; (c) minor fault sets orientation. *Source*: Sofiane Djezzar (2019).

8.4.2.4 Tamadjert Formation

The Tamadjert formation, distinguished by its heterogeneity, serves as a significant reservoir composed of M'Kratta sandstone, El-Goléa sandstone, and micro-conglomerate shale (WEC, 2007). The unique stratigraphic and sedimentological characteristics of the Tamadjert glacial formation set it apart from other primary detrital Paleozoic units across the Sahara (Bennacef et al., 1971). With 843 fractures identified, the Tamadjert formation is marked as the most fractured among the studied units, a trait attributed to the inherent stiffness of its constituent materials.

This formation is delineated by five principal fracture sets: N000, N050, N090, N120, and N160. Among these, the N-S and N160 sets are particularly notable for their prevalence in terms of both the quantity and lengths of fractures, which span a wide range, from 500 m to an extensive 79 km. Analysis through log-log diagrams of fracture length distribution underscores that Tamadjert's fractures adhere to a power law function, evidenced by high correlation coefficients between 0.91 and 0.97 and power law coefficients ranging from 2.37 to 2.81.

The fractal dimensions of the entire fracture network within the Tamadjert formation have been quantified as 1.705 and 1.3993, through the use of the center distance and box-counting algorithms, respectively (Table 8.5). This dual-method analysis yields insights into the network's complexity and spatial arrangement. Additionally, the individual fracture sets exhibit fractal dimensions ranging from 1.376 to 1.4693 using the center distance algorithm, and from 1.0091 to 1.3798 with the box-counting algorithm. It is noteworthy that the N090 fracture set did not display a fractal dimension when analyzed with the box-counting algorithm, indicating a unique geometric characteristic or distribution pattern within the Tamadjert formation. This detailed fractal analysis enhances the understanding of the formation's fracture network, contributing to the broader knowledge of geological structures and their implications for reservoir studies in the Sahara (Figures 8.10 and 8.11).

Table 8.5 Fractal dimensions related to the minor faults affecting Tamadjert formation.

Data	Points	Fract dim. (center distance)	Fract dim. (box counting)	Rmin
All faults	843	1.6962	1.4099	82.0379
Tamadjert-N000	212	1.618	1.1874	427.2758
Tamadjert-N040	84	1.2823	no value	969.4864
Tamadjert-N070	57	1.0858	no value	787.2807
Tamadjert-N090	178	1.4913	1.0213	142.9118
Tamadjert-N120	171	1.4854	1.0853	466.2224
Tamadjert-N150	141	1.4881	1.1093	634.168

8.4.3 The Third Approach

This methodology focuses exclusively on the gathered minor fracture networks within both the basement and the Cambro-Ordovician units. The aim is to delve into the behavior and fractography of each geological unit, shedding light on how the basement fractures influence the various components of the Cambro-Ordovician reservoir analog. By examining and comparing the orientation, length, kinematics, and fractal dimensions of the different fracture sets, insights into the relationships and impacts among these geological features are gained. The fracture networks of the basement, Ajjers, In-Tahouite, and Tamadjert formations are amalgamated to form a comprehensive global fracture network. This aggregated network, consisting of 1891 fractures, is segmented into five principal fracture sets: N000, N050, N090, N120, and N150, with the N-S and N150 sets, being particularly significant in terms of the number and length of fractures, which vary from 360 m to 79 km in length.

The analysis of the global fracture network's length distribution through log-log diagrams reveals adherence to a power law function, with a high correlation coefficient ranging from 0.91 to 0.97 and power law coefficients between 2.55 and 3.12 (Table 8.6). This pattern highlights the predictable yet complex distribution of fracture lengths across the combined network. The calculated fractal dimensions of the entire Global fracture network stand at 1.7946 and 1.5754, as determined by the center distance and box-counting algorithms, respectively. These values indicate a significant level of complexity and spatial heterogeneity within the network.

Furthermore, the fractal dimensions for the individual fracture sets within this Global network vary, with values ranging from 1.4796 to 1.6716 using the center distance algorithm, and from 1.1149 to 1.3984 via the box-counting algorithm. These fractures are further classified into Mode I and Mode II types, providing a deeper understanding of their mechanical characteristics and implications for the geological structure and reservoir properties of the region. Through this approach, a more nuanced comprehension of the interplay between minor and major fracture networks across different geological formations is achieved, enhancing our overall understanding of subsurface potential reservoir behaviors (Figures 8.12 and 8.13).

8.4.4 The Fourth Approach

This approach integrates a combined analysis of both Major and Minor fracture networks within the study area, aiming to elucidate the dynamics and interrelations between these fracture types across both spatial and temporal dimensions. This approach provides valuable insights into how

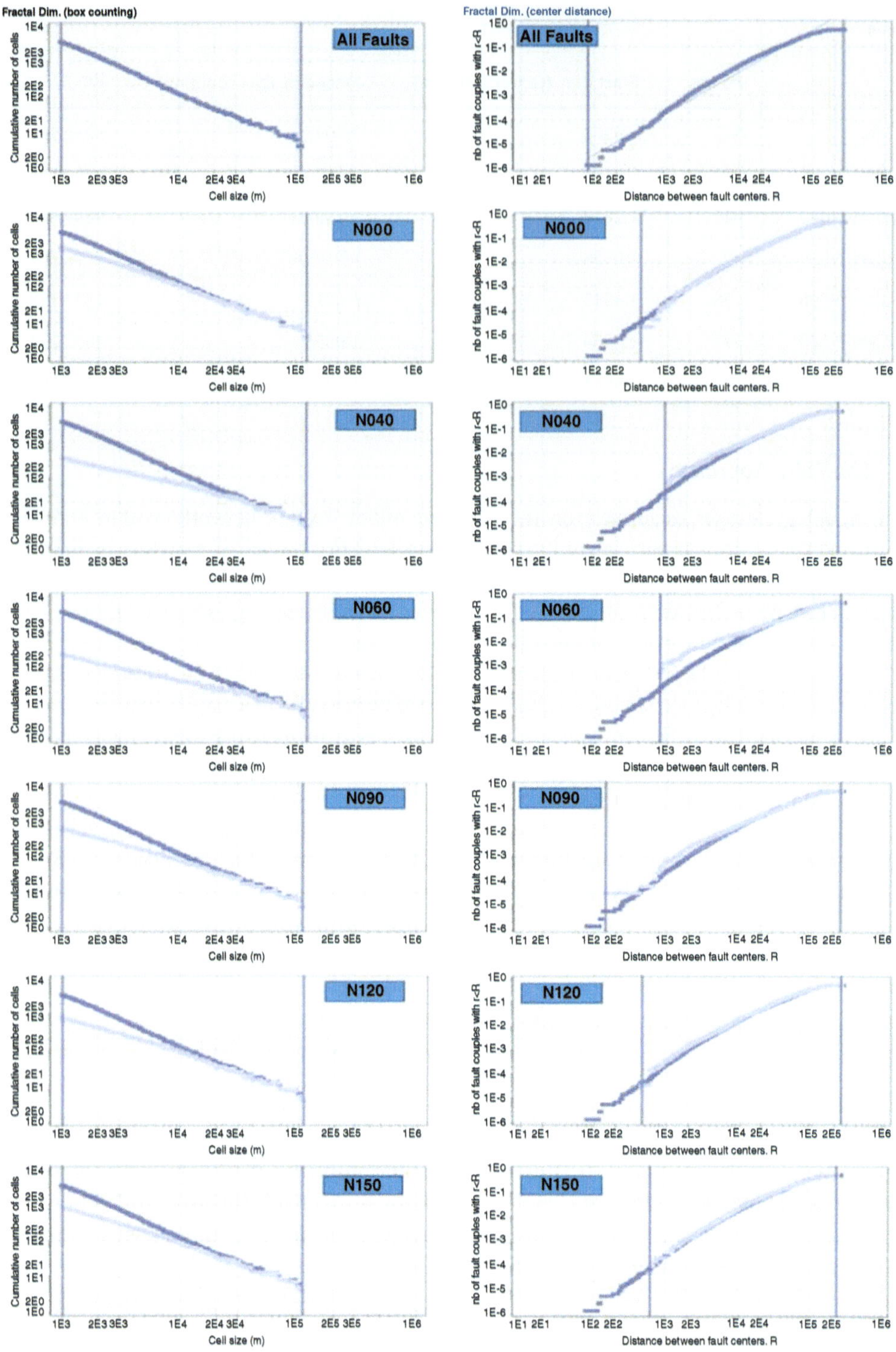

Figure 8.10 Fractal dimension distributions for the minor faults in the Tamadjert formation by fault set using the center distance and the box-counting algorithms.

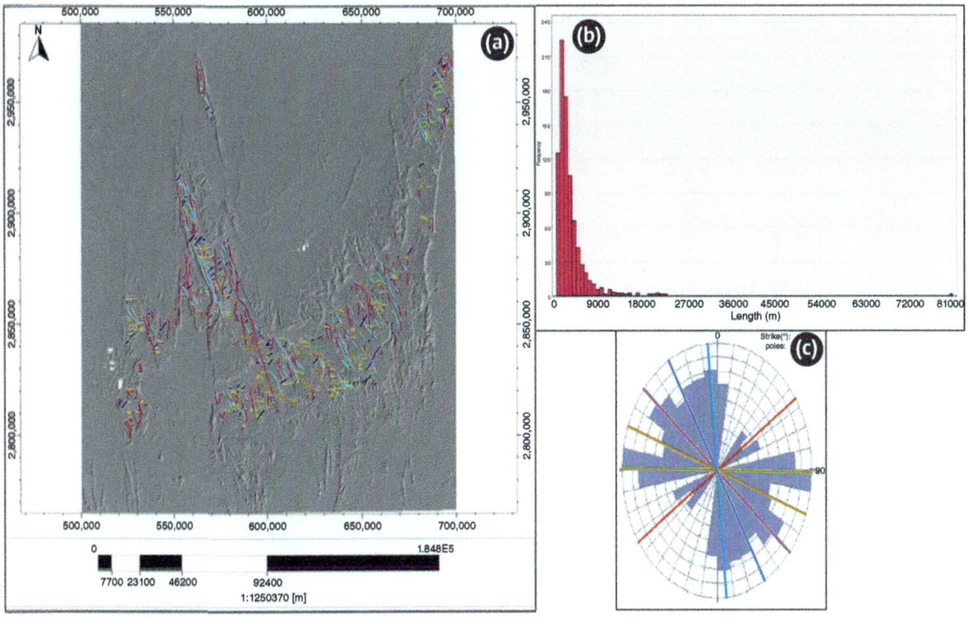

Figure 8.11 Tamadjert faults characteristics. (a) Minor faults map in the Tamadjert formation; (b) minor faults' length distribution; (c) minor fault sets orientation. *Source*: Sofiane Djezzar (2019).

Table 8.6 Fractal dimensions related to the global faults.

Data	Points	Fract dim. (center distance)	Fract dim. (box counting)	Rmin
All faults	1891	1.7946	1.5754	88.0436
Global-Faults-N000	628	1.6716	1.3984	91.1372
Global-Faults-N050	222	1.4796	1.1149	453.5019
Global-Faults-N090	348	1.5941	1.1774	143.0563
Global-Faults-N120	338	1.6271	1.2284	466.4582
Global-Faults-N150	355	1.6259	1.2821	88.0078

basement fractures influence the overlying sedimentary cover, the geological age of various fracture sets, and the role of major fractures in the actual architecture of the Mouydir basin. By integrating the Minor and Major fracture networks into a singular total fracture network, a comprehensive perspective on the fractured landscape is achieved.

The total fracture network, comprising 2544 fractures, is segmented into five primary fracture sets: N000, N050, N090, N120, and N150. Notably, the N-S, N090, and N120 sets emerge as the most significant, both in terms of the quantity and lengths of fractures, which exhibit a wide range, extending from 360 m to an impressive 160 km. The analysis through log-log diagrams of fracture length distribution underscores that the total fracture network adheres to a power law function, demonstrated by high correlation coefficients between 0.97 to 0.99 and power law coefficients from 2.36 to 2.57 (Table 8.7). This pattern indicates a consistent yet intricate fractal distribution across the network.

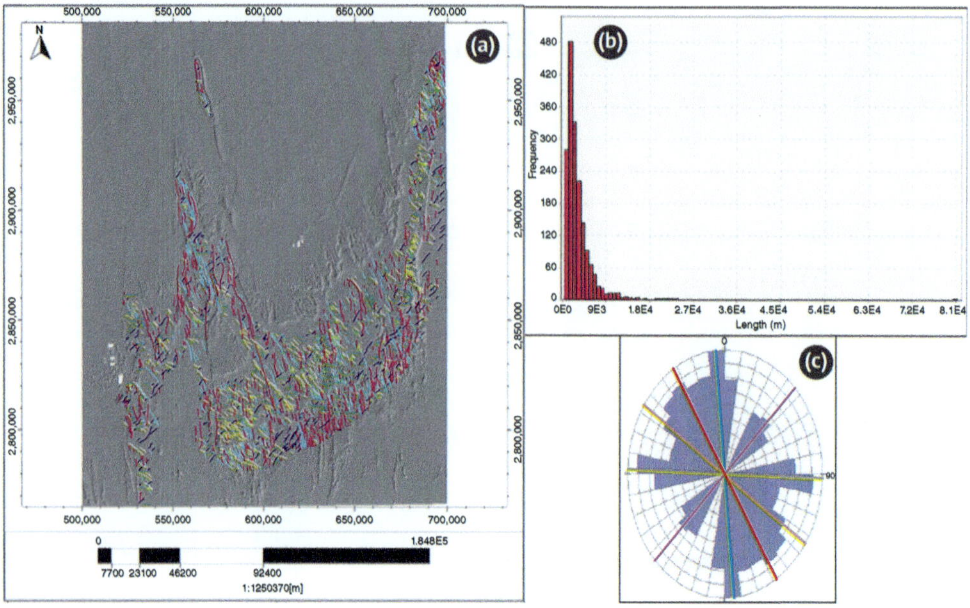

Figure 8.12 Global faults characteristics. (a) Minor global faults map (basement and Cambro-Ordovician formations); (b) minor faults' length distribution; (c) minor fault sets orientation. *Source*: Sofiane Djezzar (2019).

The calculated fractal dimensions of the entire total fracture network stand at 1.7798 and 1.6742, as determined by the center distance and box-counting algorithms, respectively. These values reflect the network's complexity and spatial heterogeneity. Moreover, the individual fracture sets within this comprehensive network exhibit fractal dimensions ranging from 1.4115 to 1.6825 using the center distance algorithm, and between 1.2158 and 1.5061 with the box-counting algorithm. These findings highlight the varied geometric characteristics and distribution patterns among the fracture sets, enhancing our understanding of the geological processes at play and their implications for the architecture and evolution of the Mouydir basin. Through this integrated approach, a deeper comprehension of the interplay between minor and major fracture networks is achieved, contributing to a broader knowledge of subsurface formations and potential implications for geological and reservoir studies (Figures 8.14 and 8.15).

8.5 Conclusions

The northern Hoggar shield's Paleozoic series and Cambro-Ordovician units are defined by a complex and dense fracture network. Through detailed fracture analysis, the study elucidates various fracture sets, their densities, spatial interconnections, and temporal relationships. This analytical framework aids in deciphering the chronology and origins of these fractures, including their discontinuities or continuities beneath the surface. It's revealed that the Cambro-Ordovician formations comprise distinct mechanical stratigraphy units, differentiated by their lithology and thickness, deviating from a non-stratabound fracture model.

This deterministic fracture model delineates the distribution of fractures, enabling a clearer understanding of the mechanical properties of Cambro-Ordovician units and their interrelations.

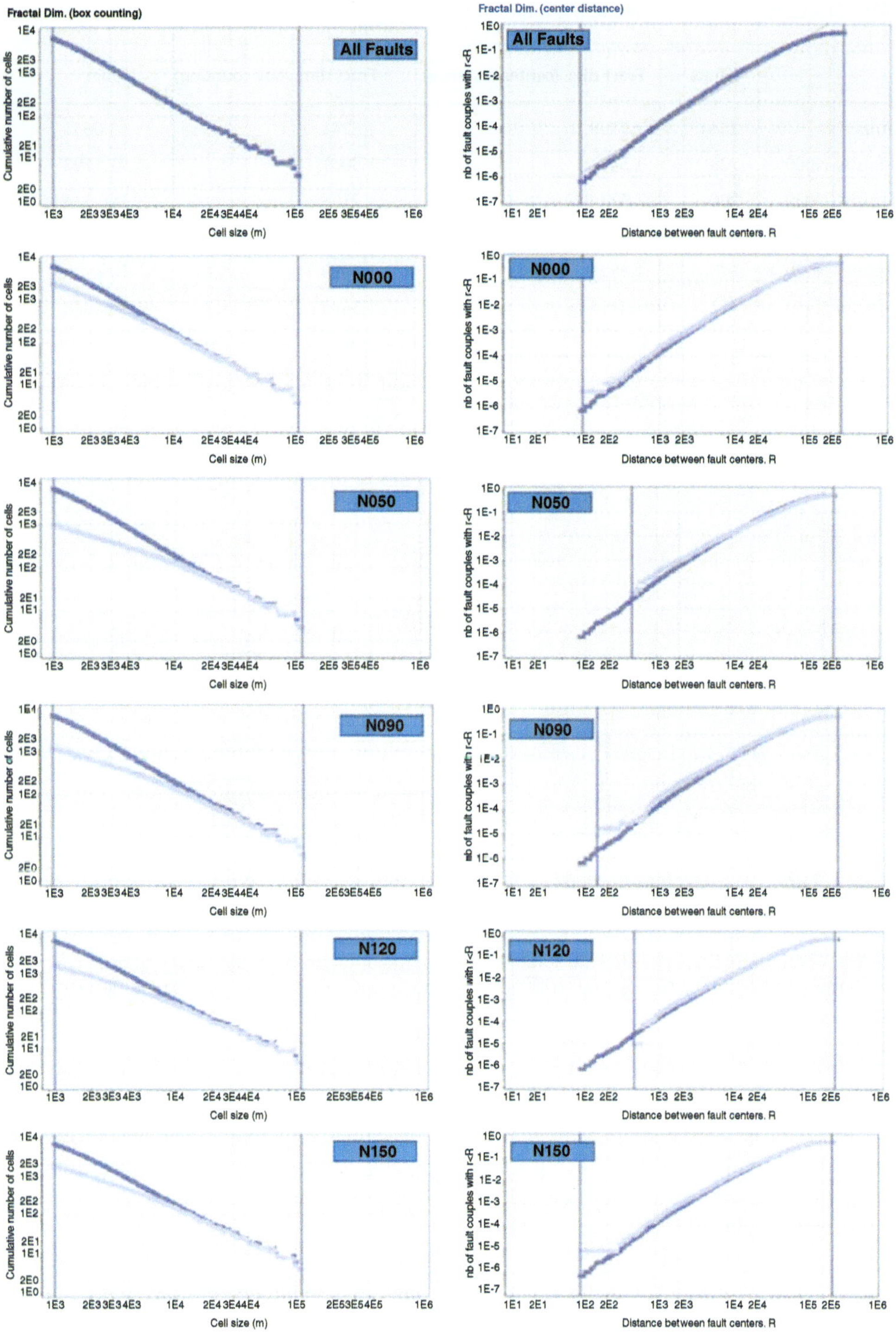

Figure 8.13 Fractal dimension distributions for the minor global faults in the area of study by fault set using the center distance and the box-counting algorithms.

Table 8.7 Fractal dimensions related to the total faults.

Data	Points	Fract dim. (center distance)	Fract dim. (box counting)	Rmin
All faults	2544	1.7798	1.6742	65.0631
Total-faults-N000	728	1.5451	1.5061	91.1792
Total-faults-N050	296	1.5234	1.2158	250.3348
Total-faults-N090	530	1.4725	1.2691	143.0568
Total-faults-N120	578	1.6825	1.3583	209.6989
Total-faults-N150	412	1.4115	1.3484	88.0078

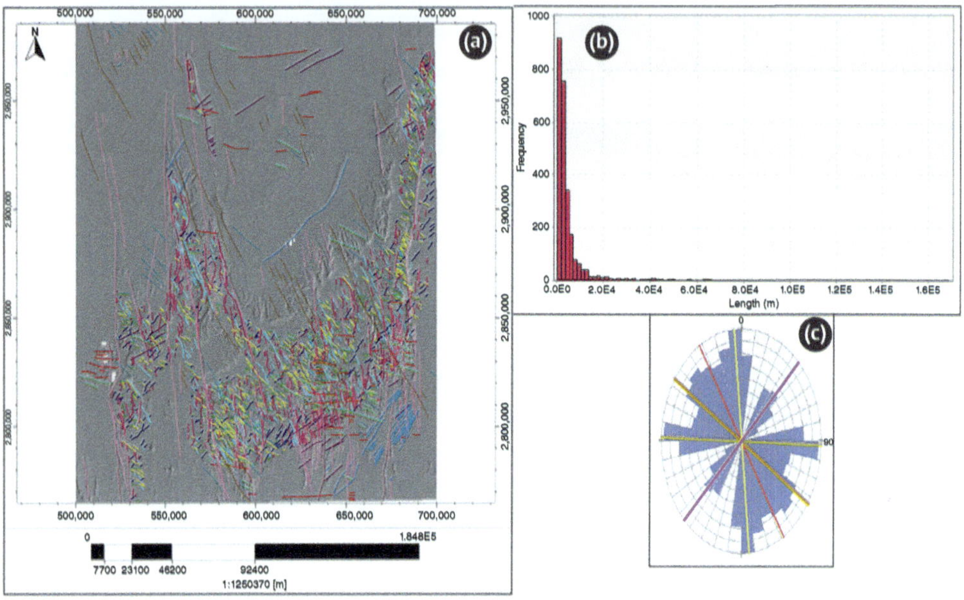

Figure 8.14 Total faults characteristics. (a) Total faults map (major faults, basement, and Cambro-Ordovician formations); (b) faults' length distribution; (c) fault sets orientation. *Source*: Sofiane Djezzar (2019).

The Tamadjert formation, attributed to its lithological attributes, emerges as the most fractured unit, whereas the In-Tahouite formation's shaly composition makes it the least fractured. The analysis suggests that basement fractures significantly influence the sedimentary cover, structuring the Cambro-Ordovician units and introducing a predominantly brittle tectonic style. This brittleness is linked to major basement fractures inherited from the Pan-African orogeny, fostering a network of major vertical fractures.

The geological mapping further reveals that the Paleozoic era was profoundly affected by dense fracture networks, with major structures manifesting as N-S trending dextral strike-slip fractures and NNW-SSE trending sinistral strike-slip fractures. Two major fracture corridors are identified, delineating the western and eastern boundaries of the Mouydir basin. These corridors, originating from the Hoggar shield's basement and diverging northwards across the Saharan platform, are defined by NNE to NS and NNW to NS orientations, representing the Amguid and Idjerane spurs, respectively.

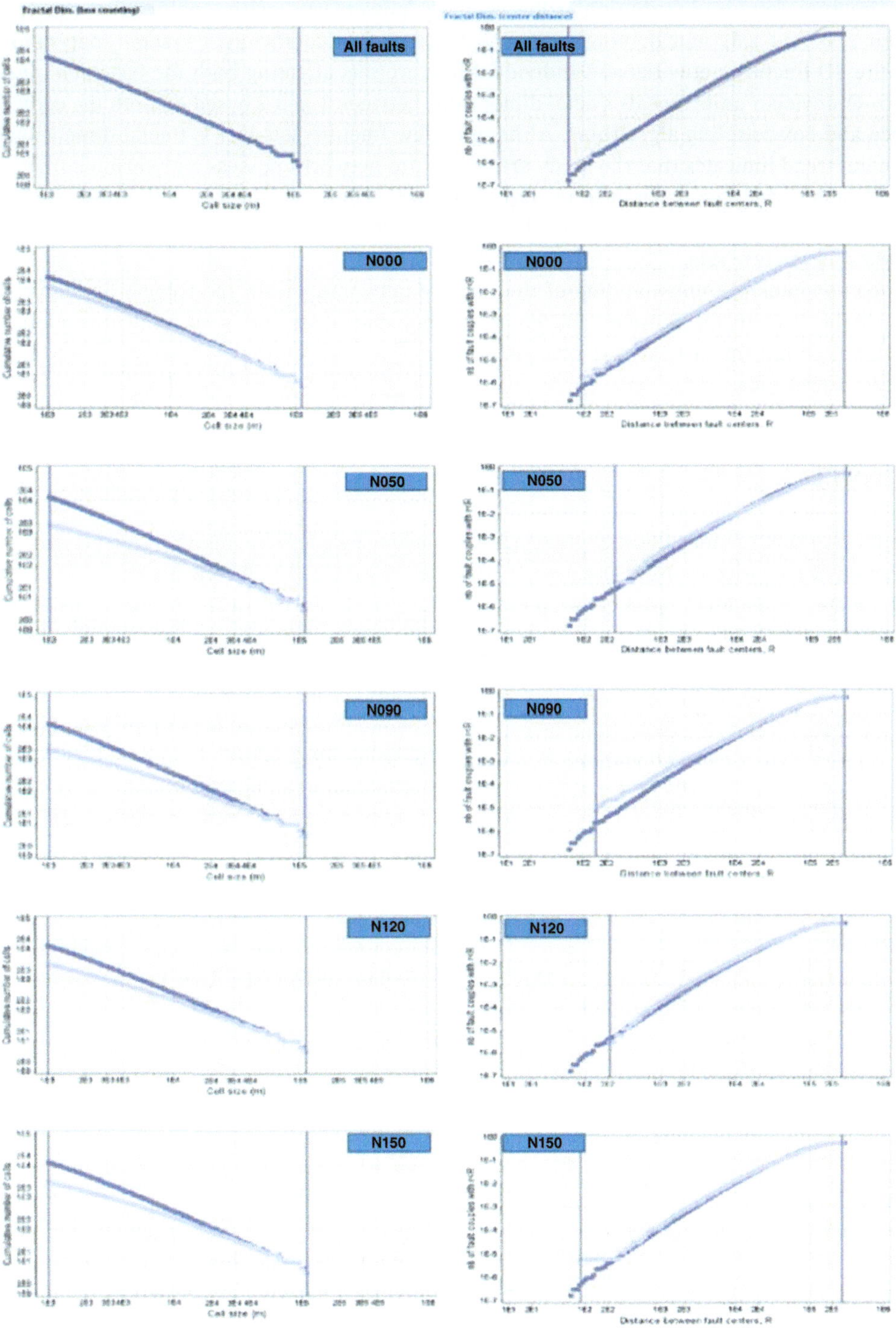

Figure 8.15 Fractal dimension distributions for the major and minor faults in the area of study by fault set using the center distance and the box-counting algorithms.

The fracture length distribution adheres to a power law, with coefficients ranging between 2.31 and 2.69 and demonstrating high correlation coefficients. Fractal analysis of the entire 2D fracture networks and individual fracture sets affecting both the basement and Cambro-Ordovician units reveals fractal dimensions between 1 and 2 based on both the center distance and box-counting algorithms. Although a few fracture sets lack a fractal dimension, the general trend indicates that the study area's fracture networks possess a discernible fractal nature. Notably, the fractal dimensions derived from the box-counting algorithm are consistently 0.2 to 0.3 lower than those obtained from the center distance algorithm across different networks and fracture sets.

This comprehensive understanding of the fracture networks within the Hoggar Shield provides valuable insights for future oil and gas exploration endeavors in this relatively unexplored basin, highlighting the critical role of geological fractures in reservoir characterization and exploration strategies.

References

Ait Kaci, A. (1990). Evolution lithostratigraphique et sédimentologiquedes monts d'Ougarta pendant le Cambrien (Sahara Algérien Nord Occidental). Thèse de 3ème cycle, IST-USTHB, p. 193.

Arab, M. & Djezzar, S. (2011). Hydrocarbon potential in Reggane Basin. Oil Mediterranean Conference (OMC) Italy, OMC-2011-046.

Balberg, I., Berkowitz, B., & Drachsler, G. E. (1991). Application of a percolation model to flow in fractured hard rocks. Journal of Geophysical Research 96 (B), pp. 10015–10021.

Balberg, I. & Binenbaum, N. (1983). Computer study of the percolation threshold in a two-dimension anisotropic system of conducting sticks. Physical Review 28, pp. 3799–3812.

Barton, C. & La Pointe, P. (1995). Fractals in Petroleum Geology and Earth Processes. New York: Plenum Press.

Beicip-Franlab. (2018). FracaFlow software tutorial.

Bennacef, A., Beuf, S., Biju-Duval, B., et al. (1971). Example of cratonic sedimentation: Lower Palaezoic of Algerian Sahara. The American Association of Petroleum Geologists Bulletin 55 (12), pp. 2225–2245.

Bensalah, A., Beuf, S., Gabriel, O., et al. (1972a). Carte Géologique de l'Algérie: Arak, feuille NG-31-X et NG-31-IV. Ministère de l'Industrie et de l'Energie.

Bensalah, A., Beuf, S., Gabriel, O., et al. (1972b). Carte Géologique de l'Algérie: Ain Tadjoubar, feuille NG-31-XVI. Ministère de l'Industrie et de l'Energie.

Bensalah, A., Beuf, S., Gabriel, O., et al. (1972c). Carte Géologique de l'Algérie: Khanget El Hadid & Ifetessene, feuille NG-31-XVII. Ministère de l'Industrie et de l'Energie.

Berkowitz, B., Bour, O., Davy, P., et al. (2000). Scaling of fracture connectivity in geological formations. Geophysical Research Letters 27 (14), pp. 2061–2064.

Beuf, S., B., D., Biju-Duval, De Charpal, R., Rognon, A., Bennacef, (1971). Les grès du Paléozoïque inférieur au Sahara. Sédimentation et discontinuité: évolution structurale d'un craton. In: Technip-Institut Français du pétrole, vol. 18. Collection Sciences et Techniques du Pétrole, Paris, p. 464.

Black, R., Latouche, L., Liégeois, J. P., et al. (1994). Pan-African displaced terranes in the Hoggar shield (Central Sahara). Geology 22, pp. 641–644.

Bonnet, E., Bour, O., Odling, N. E., et al. (2001). Scaling of fracture systems in geological media. Review of Geophysics 39 (3), pp. 347–383.

Boualam, A., Djezzar, S., Rasouli, V., et al. (2019). 3D modeling and natural fractures characterization in Hassi Guettar Field, Algeria. 53rd American Rock Mechanics Association Symposium, NY, USA. 19-A-373-ARMA

Bour, O. & Davy, P. (1997). On the connectivity of random fault networks following a power law length distribution. Water Resources Research 33, pp. 1567–1583.

Bour, O. & Davy, P. (1998). On the connectivity of three-dimensional fault networks. Water Resources Research 34, pp. 2611–2622.

Cacas, M. C., Daniel, J. M., & Letouzy, J. (2001). Nested geological modeling of naturally fractured reservoirs. Petroleum Geoscience 7, pp. 43–52.

Coward, M. P. & Ries, A. C. (2003). Tectonic development of North Africa basins. In: New Themes, and Developing Technologies (eds. Arthur, T. J., Macgregor, D. S., & Cameron, N. R.), pp. 61–83. Geological Society of London, Special Publication 207.

Darcel, C., Bour, O., & Davy, P. (2003). Stereological analysis of fractal fracture networks. Journal of Geophysical Research 108 (B9), p. 2451.

Davy, P. (1993). On the fault-Length frequency distribution of the San Andreas Fault system Journal of Geophysical Research 98 (12), pp. 141–151

Davy, P., Sornette, A., & Sornette, D. (1990). Some consequences of a proposed fractal nature of continental faulting. Nature 348, pp. 56–58.

Davy, P., Sornette, A., & Sornette, D. (1992). Experimental discovery of scaling laws relating fractal dimensions and the length distribution exponent of fault systems. Geophysical Research Letters 19, pp. 361–363.

Djezzar, S. (2012). Impact de la fracturation naturelle sur la productivité des réservoirs du Dévonien inférieur dans le bassin de Reggane, Sahara Occidental, Algérie. Thèse Magister. Ed. USTHB. -Alger: USTHB, 2012.-93 p.: ill.; 30 cm. + CD-Rom.

Djezzar, S. (2019). Fracture detection and prediction in unconventional reservoirs for finding sweet spot. Theses and Dissertations. 2551. https://commons.und.edu/theses/2551/

Djezzar, S. & Boualam. A. (2020a). Analysis and modeling of tight oil fractured reservoir. SEG 2020 Annual Meeting, TX, USA. https://doi.org/10.1190/segam2020-3425011.1

Djezzar, S. & Boualam. A. (2020b). Fractures Characterization and Their Impact on the Productivity of Hamra Quartzite Reservoir (Haoud-Berkaoui field, Oued-Mya basin, Algeria). 54th American Rock Mechanics Association Symposium, Co, USA. 20-A-1438-ARMA

Djezzar, S. & Boualam, A. (2021). Impact of Natural Fractures on the Productivity of Lower Devonian Reservoirs in Reggane Basin, Algeria. 55th US Rock Mechanics/Geomechanics Symposium. 21-A-1295-ARMA

Djezzar, S., Rasouli, V., Boualam, A., et al. (2019). Size scaling and spatial clustering of natural fracture networks using fractal analysis. 53rd American Rock Mechanics Association Symposium, NY, USA. 19-A-375-ARMA

Djezzar, S., Rasouli, V., Boualam, A., et al. (2020). An integrated workflow for multiscale fracture analysis in reservoir analog. Arabian Journal of Geosciences 13, p. 161, https://doi.org/10.1007/s12517-020-5085-6

Follot, J. (1952). Ahnet et Mouydir. 19eme Congres international de Geologie, Alger. Monographie Régionale, Série 1, 1, pp. 1–6.

Fekirine, B. & Abdallah, H. (1998). Palaeozoic lithofacies correlatives and sequence stratigraphy of the Saharan Platform, Algeria. Geological Society, London, Special Publications, 132, pp. 97–108.

Haddoum, H., Guiraud, R., & Moussine-Pouchkine, M. A. (2001). Hercynian compressional deformations of the Ahnet-Mouydir Basin, Algerian Saharan platform: far-field stress effects of the late Palaezoic orogeny. Terra Nova 13, pp. 220–226.

Liégeois, J. P. (2019). A new synthetic geological map of the hoggar shield: an overview of its global structure and geological evolution. In: *The Geology of the Arab World-An Overview*, (ed. A. Bendaoud et al.), pp. 83–107. Springer Geology.

Liégeois, J.P, Benhallou, A., Azzouni-Sekkal, A. et al. (2005). The Hoggar swell and volcanism: Reactivation of the Precambrian Tuareg shield during Alpine convergence and West African Cenozoic volcanism. Plates, plumes, and paradigms (Eds. Foulger, G.R., Natland, J.H., Presnall, D. C., & Anderson, D.L.). The Geological Society of America. ISBN print: 9780813723884 Publication date: January 01, 2005.

Mandelbrot, B. (ed.) (1982). The Fractal Geometry of Nature Freeman. Echo Point Books & Media, LLC, San Francisco, p. 461.

Odling, N. E., Gillespie, P., Bourgine, B., et al. (1999). Variations in fracture system geometry and their implications for fluid flow in fractured hydrocarbon reservoirs. Petroleum Geoscience 5, pp. 373–384.

Segall, P. & Pollard, D. D. (1983). Joint formation in granitic rock of the Sierra Nevada. Geological Society of America Bulletin 94, pp. 563–575.

Verscheure, M., Fourno, A., & Chilès, J. -P. (2012). Joint inversion of fracture model properties for CO2 storage monitoring or oil recovery history matching. Oil & Gas Science and Technology – Rev. IFP Energies Nouvelles 67(2), pp. 221–235. https://doi.org/10.2516/ogst/2011176

Well Evaluation Conference (WEC) (2007). Well Evaluation Conference, Internal report, "Joint documents of Sonatrach and Schlumberger." Algeria. p. 489.

Wendt, J., Kaufmann, B., Belka, Z., et al. (2006). Sedimentary evolution of a Palaeozoic basin and ridge system: The Middle and Upper Devonian of the Ahnet and Mouydir (Algerian Sahara). Geological Magazine 143 (3), 269–299.

Zieliński, M. (2011). Reconstruction of thermal history of the Ahnet and Mouydir basins (southern Algeria). PhD Thesis. Poznan: Faculty of Geographical and Geological Sciences, Adam Mickiewicz University. Institute of Geology.

9

Application of Seismic Attributes on Digital Elevation Model: Fractures Detection and Reservoir Implication

Sofiane Djezzar, Aldjia Boualam

University of North Dakota, Grand Forks, USA

9.1 Introduction

In recent years, the exploration and development of unconventional reservoirs have witnessed a remarkable evolution, underscored by a significant uptick in hydrocarbon discoveries where natural fractures play a pivotal role in hydrocarbon production. This advancement has deepened the understanding of how fracturing influences fluid flow mechanisms, highlighting the importance of natural fractures in enhancing reservoir productivity (Boualam et al., 2019; Arab and Djezzar, 2011; Djezzar, 2012; Hamidou, 1999; Djezzar and Boualam, 2020b; Issad, 1996 and 2000). As field development progresses, the placement of wells within the implementation strategy is informed by the spatial distribution of fractures throughout the reservoir. Geoscientists aim to target zones with a dense network of interconnected fractures, facilitating more efficient drainage from the surrounding rock matrix. Consequently, the orientation of wells is strategically aligned with both geometrical and geomechanical parameters, including the orientation of fracture planes and the prevailing direction of maximum horizontal stress, to optimize hydrocarbon production (Djezzar, 2019; Djezzar et al., 2019; Conrad and Lemosquet, 2007; Djezzar and Boualam, 2020a). Modeling and predicting the complex networks of natural fractures at various scales, from surface outcrops to deep reservoir levels, present substantial challenges. To accurately characterize and model these fractures, advanced 3D seismic data and well logs are indispensable. This process demands an enormous budget and extensive effort in data collection and analysis. As an alternative, studying reservoir analogs offers a more accessible approach to understanding natural fractures, requiring fewer specific data. This methodology is exemplified by the study of the Ordovician formation in the Mouydir Basin, known for its modest reservoir properties where natural fractures significantly influence production (Djezzar and Boualam, 2021). The Ordovician analog study in the Mouydir Basin underscores the critical role of natural fractures in guiding future exploration and drilling endeavors. Unlike subsurface studies, which necessitate specialized data, outcrop analog studies provide a more straightforward means to examine fracture networks (Figure 9.1). These analog studies are invaluable for modeling fractures across different scales, from macroscopic to microscopic levels, facilitating insights into fracture distribution, connectivity, and kinematics

Geophysical Exploration for Hydrocarbon Reservoirs, Geothermal Energy, and Carbon Storage: New Technologies and AI-based Approaches, First Edition. Edited by Said Gaci.

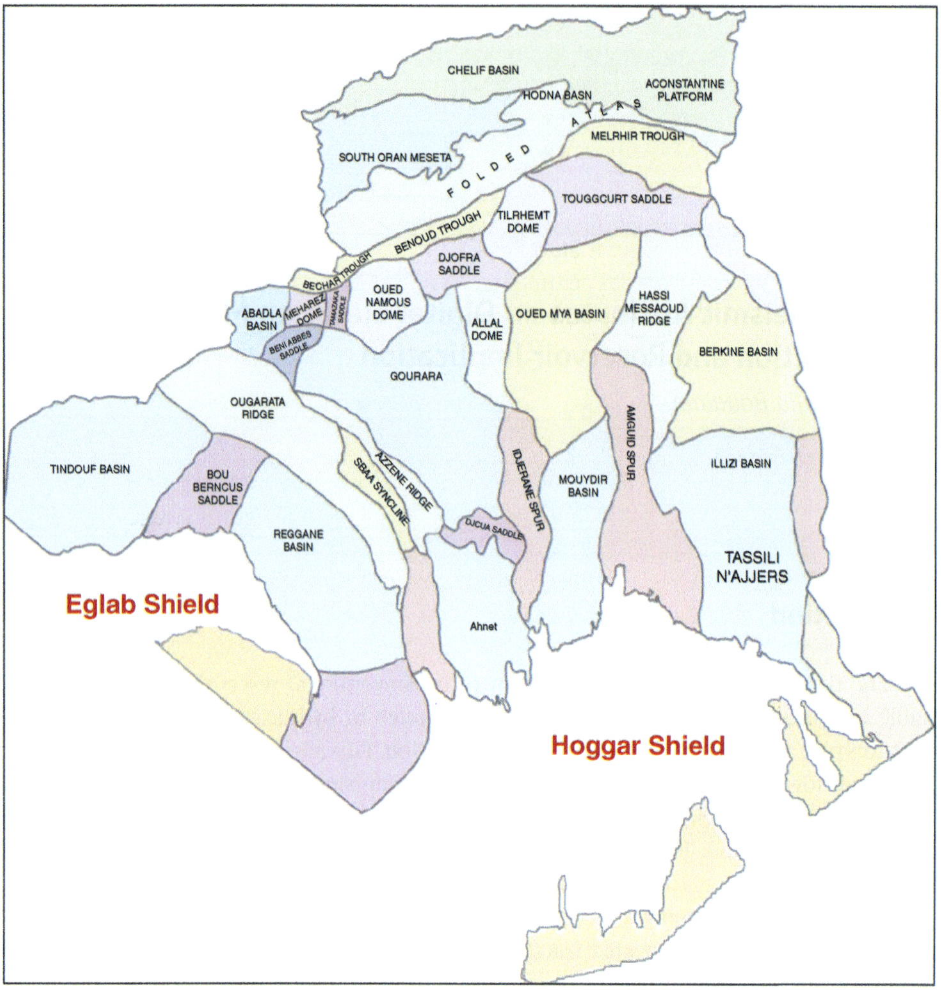

Figure 9.1 Algerian basins (Sonatrach document).

(Haddoum, 1997; Haddoum et al., 2001). This research introduces a comprehensive workflow for characterizing and modeling natural fractures, employing specific methods such as seismic curvature and illumination attributes on a digital elevation model (DEM). The focus on the Ordovician reservoir analog at the southern edge of the Mouydir Basin exemplifies this methodology, offering a promising approach to enhancing our understanding and exploitation of naturally fractured reservoirs.

9.2 Problematic

The Mouydir Basin, situated in Algeria, remains one of the least explored basins due to the absence of 3D seismic data and the limited availability of low-quality 2D seismic surveys (Djezzar, 2019; Djezzar et al., 2019). Historical drilling efforts in the basin have been unsuccessful, with all wells drilled to date yielding no positive outcomes. The Cambrian-Ordovician reservoirs in this region are typically tight, making their productivity heavily reliant on the presence of natural fractures. To address the challenge presented by the scarce data, an innovative and tailored workflow has been developed. This workflow is designed to thoroughly analyze and characterize the natural fractures within the Cambro-Ordovician reservoir analog located at the southern edge of the Mouydir Basin. The objective is to provide fresh insights and directions for future petroleum exploration efforts in the basin. By employing a deterministic fracture model, the workflow not only delineates major and minor fractures but also offers a deeper understanding of the basin's geological development. Furthermore, it elucidates the influence of basement fractures on the structural formation of the basin and its sedimentary layers, potentially unlocking new avenues for exploration and development in this underexplored basin.

9.3 Workflow and Methodology

In this workflow, a synergistic use of satellite imagery, geological maps, and DEMs forms the backbone for identifying and delineating fault structures. This methodology leverages attributes like curvature, slopes, and illumination using the DEM to uncover both significant fault lines and nuanced fracture patterns. Subsequently, these features are digitized and scrutinized for their orientation, length distribution, and density employing advanced software. The analysis bifurcates into two primary segments: Initially, it concentrates on the major fault lines impacting the targeted zone. This step integrates various data sources, including geological maps, satellite imagery, and DEMs, for a comprehensive digital representation of these faults. The latter segment looks into the intricate fracture patterns within the basement and Cambro-Ordovician strata, utilizing identical data sets but scrutinized at varying scales to ensure a detailed parameter and output alignment. The culmination of this process is the creation of diverse visual data representations, such as stereonet, cross plots, and histograms. These visual tools are instrumental in decoding the orientation of fault sets, along with their length attributes, distribution metrics, correlation, and power-law coefficients, offering profound insights into the fracture morphology and typology pertinent to the reservoir analog (Figure 9.2). Consequently, the development of 3D deterministic fault models at both the formation scale and specific area of interest provides a dynamic illustration of fracture distribution across both temporal and spatial dimensions, enhancing the understanding of the geological framework (Djezzar and Boualam, 2020a; Djezzar et al., 2020). In this advanced workflow, geological maps at a scale of 1:100,000, satellite imagery with a 30 m resolution, and DEMs with a 65 m resolution serve as the primary data sources. These diverse datasets are matched and integrated using Global Mapper software, ensuring a coherent geospatial framework for subsequent analyses. The process of detecting and digitizing faults leverages satellite imagery alongside curvature and illumination attributes derived from the DEM, utilizing FracaFlow software for attribute mapping. This integration facilitates the creation of histograms, maps, rose diagrams, and graphs that elucidate fault orientations, length distributions, and key statistical parameters (Figures 9.3–9.7).

The generation of 3D deterministic fault models marks a significant step forward in the analysis, offering a tangible representation of fracture distribution across both the Cambro-Ordovician and basement units. For each stratigraphic layer within the Cambro-Ordovician sequence, a dedicated 3D model is constructed, shedding light on the interplay between stratigraphic composition and fracture emergence. Moreover, an in-depth characterization of basement faults provides insights into their influence on the overlying Cambro-Ordovician sedimentary cover, with kinematics analysis primarily guided by geological mapping.

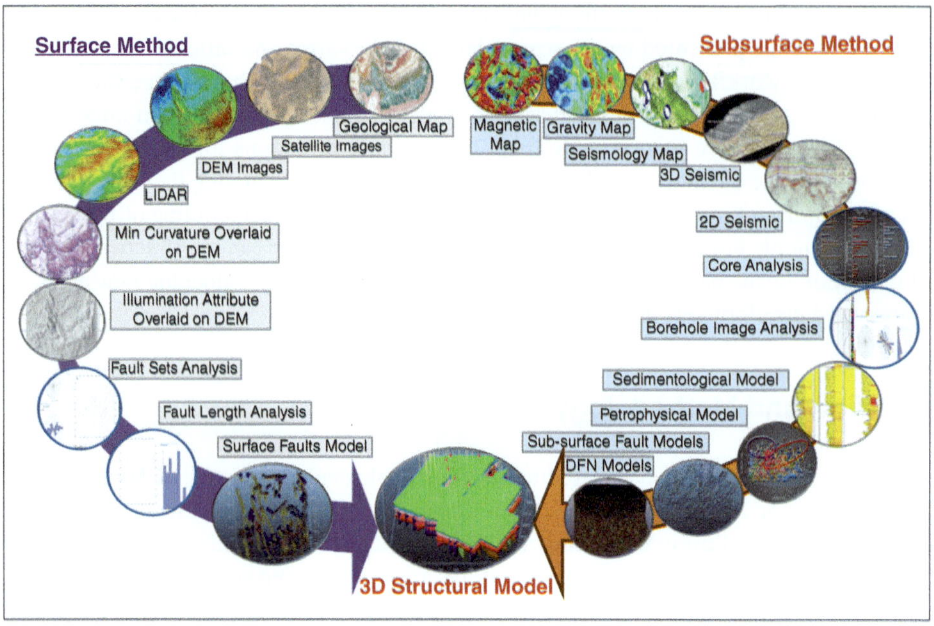

Figure 9.2 The proposed fracture analysis workflow. *Source*: Sofiane Djezzar (2019).

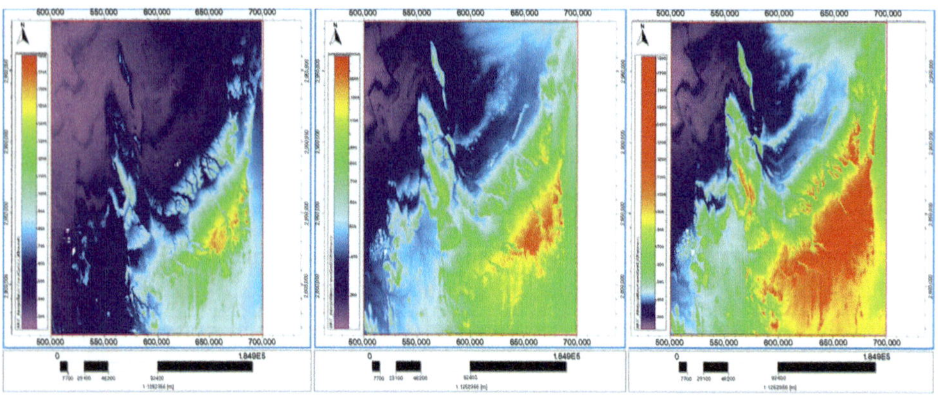

Figure 9.3 Digital elevation models.

The study adopts a dual-scale observational approach to ascertain fracture typology and distribution laws: a macroscale (1:100,000) for digitalizing and examining major faults impacting the study area, which spans from the northern reaches of the Hoggar shield to the southern fringes of the Mouydir Basin, and a mesoscale (1:25,000) for detailing minor lineaments at the formation level. This methodology enables a comprehensive definition of various network configurations and the calculation of fractal dimensions, considering lithological characteristics, fault orientation, and lengths, thus enhancing our understanding of the region's geological complexities.

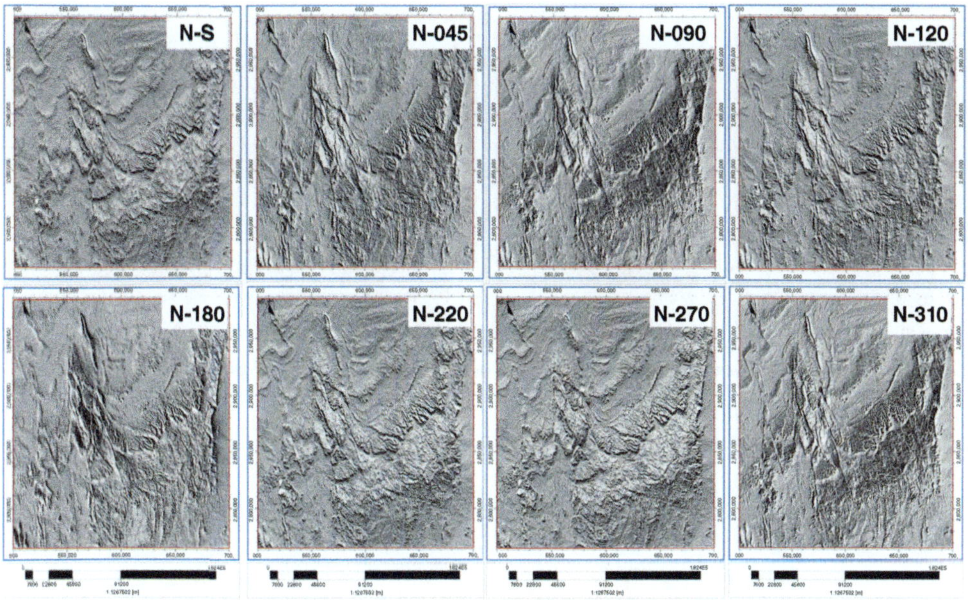

Figure 9.4 Illuminations maps with different light direction. *Source*: Sofiane Djezzar (2019).

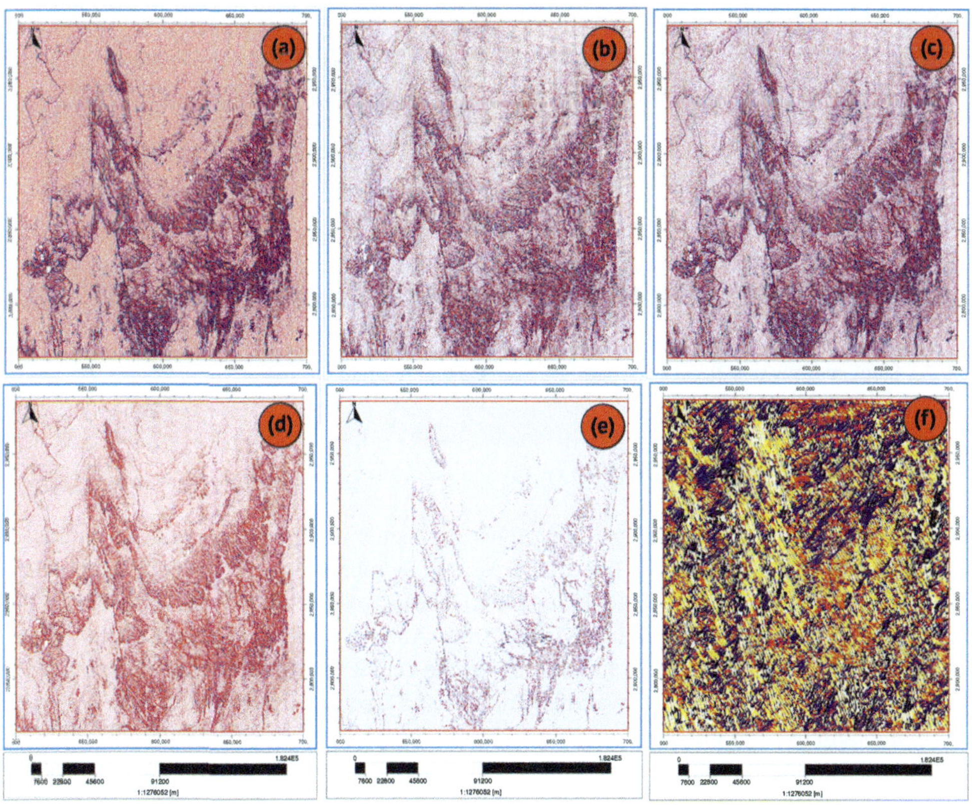

Figure 9.5 Curvature attributes. (a) Maximum; (b) minimum; (c) mean; (d) first principal; (e) Gaussian; (f) Azimuthal. *Source*: Sofiane Djezzar (2019).

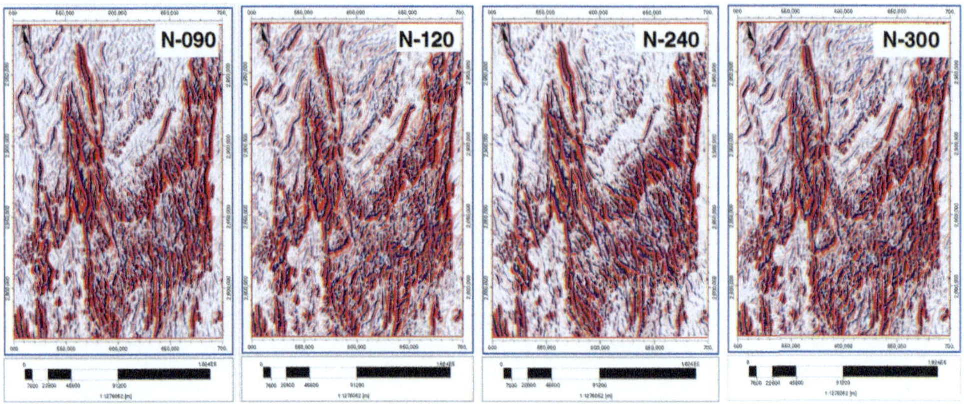

Figure 9.6 Oriented curvature attributes. *Source*: Sofiane Djezzar (2019).

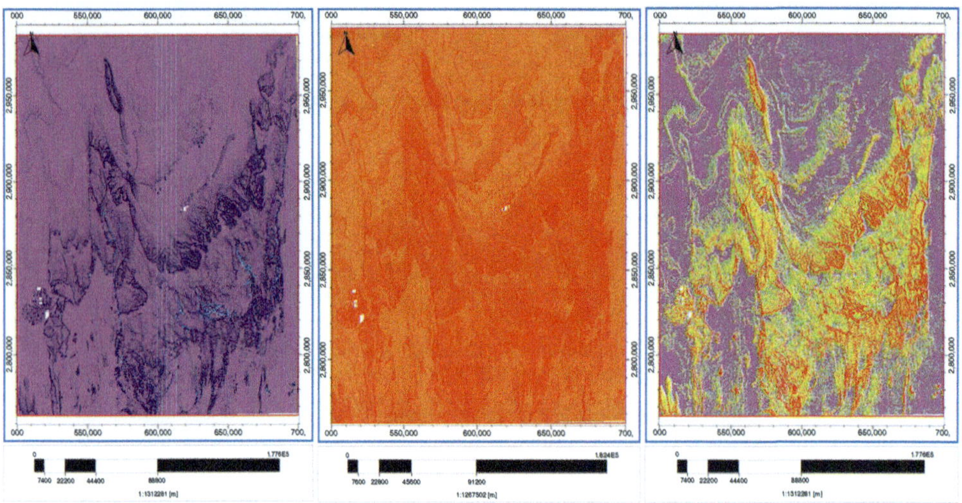

Figure 9.7 Slope maps with different distribution. *Source*: Sofiane Djezzar (2019).

9.4 Fault Detection Techniques

Seismic attributes have become indispensable tools in seismic interpretation, offering nuanced visualizations of subsurface features like fractures and geomorphological structures including channels (Chopra and Marfurt, 2007; Richard, 1994; Roberts, 2001; Collomb and Donzeau, 1974; Segall and Pollard, 1983; Sanderson and Nixon, 2015). Among these, geometric seismic attributes like curvature play a pivotal role, in quantifying the degree of bending at specific points on surfaces. This correlation between curvature attributes and fracture identification has been validated through both outcrop measurements and seismic data analysis, underlining their utility in revealing small-scale geological deformations such as flexures, folds, and differential compaction areas not readily visible in standard coherence imagery.

Advancements in volumetric curvature attributes now allow for a more comprehensive delineation of these subtle features across continuous horizons, enhancing the interpreter's ability to detect otherwise obscured geological formations. Similarly, the illumination attribute, simulating low-angle light casting to accentuate topographical relief, further aids in fracture detection by providing a visual contrast that highlights surface irregularities. This method calculates an artificial lighting effect over the terrain, assuming parallel and horizontal light rays from a single source, with the light's azimuth direction informing the computation. The interaction between the incident light and the surface normal at each triangle on the map is used to define specular illumination, offering a detailed depiction of surface gradients and orientations.

The integration of slope gradients as an attribute—representing the maximum change in elevation from a given point—complements this approach by elucidating the directional inclination of terrain features, which can indicate potential fracture zones.

For enhanced fracture detection, geological maps, satellite imagery, and DEMs serve as the primary data inputs. These are processed using specialized software to apply various attributes, including illumination and curvatures, across multiple orientations and scales. Illumination attributes, applied in specific directions, generate distinct visualizations that emphasize linear features potentially indicative of fractures. Concurrently, the application of curvature attributes—spanning maximum, minimum, mean, and specialized forms of curvature—alongside slope analysis based on DEMs, further refines the identification of lineaments.

Digitalization of detected fractures leverages these attribute-enhanced maps, with traces meticulously drawn using line tools in FracaFlow software. This multilayered approach, combining directional lighting, curvature analysis, and slope mapping, significantly improves the accuracy and efficiency of fracture detection within the area of study.

9.5 Fault Analysis

A fracture represents a discontinuity in a rock mass, prompted by stress, and can significantly influence fluid dynamics and other critical rock properties. Fractures manifest across a diverse scale spectrum, from microscopic fissures to extensive fault lines spanning vast distances. These discontinuities are broadly categorized into two main types: Mode I (tensile or opening mode), characterized by displacement perpendicular to the fracture walls, and Mode II (in-plane shear mode), where displacement occurs along the plane of the discontinuity (Balberg and Binenbaum, 1983; Balberg et al., 1991; Bazalgette, 2004; Berkowitz et al., 2000; Long et al., 1982; Manzocchi, 2002; Mauldon et al., 2001; Suzuki et al., 1998).

Recent advancements and comprehensive studies have further refined our understanding of fracture mechanics and their implications. The orientation of fractures is now more precisely analyzed using enhanced pole diagrams and stereonet, facilitating quicker and more accurate interpretations of fault orientations. Moreover, the recognition of the importance of fracture length distribution has deepened, acknowledging that the connectivity of fracture networks is dependent not solely on their density and orientation but also on the length of the individual fractures (Beicip, 2008; Brown and Bruhn, 1998).

Empirical research highlights that fracture sets typically follow a power-law distribution in terms of their lengths, with the power-law exponent "a" most commonly found between 1 and 3. This exponent value provides insights into the prevalence of short versus long fractures within a given set, influencing the overall connectivity and fluid flow characteristics of the rock mass (Beicip, 2018; Davy, 1993; Moore and Lockner, 1995; Oda, 1983; Odling et al., 1999).

In the specific context of the Cambro-Ordovician strata under investigation, both faults and joints constitute the primary types of discontinuities. These are analyzed by plotting their length populations against a normalized cumulative frequency distribution, which, when graphed on logarithmic axes, reveals the power-law distribution through the slope of the line. This quantitative approach not only elucidates the distribution and dimensions of fracture sets but also assists in differentiating between major and minor fault sets within the study area, enhancing the understanding of each formation's fractography.

To compile a comprehensive fault map, an integrated methodology leveraging curvature attributes, illumination effects, geological mapping, satellite imagery, and DEMs is employed. This multifaceted approach allows for the delineation of fracture sets, the measurement of fracture lengths, and the estimation of fractal dimensions across different fracture networks. Consequently, detailed 3D deterministic fault models for each geological formation have been constructed. These models provide a spatial representation of fracture distribution, shedding light on their origins, interrelations, and potential continuity into the subsurface, thus offering valuable predictive insights for exploration and development strategies.

9.5.1 Major Faults Analysis

This refined approach focuses on the major fractures impacting the study area, spanning across the basement and the entirety of the Paleozoic strata. These significant geological features, primarily Mode II fractures, are characterized by their extensive lengths and strike-slip kinematics, with their influence extending into the Mouydir Basin's subsurface layers. Following the assessments by Haddoum (2001), the primary fracture sets disrupting the Paleozoic layers are identified as strike-slip and reverse faults. Specifically, the dominant structural patterns include north-south (N-S) trending dextral strike-slip faults and a range from north-northwest-southeast (NNW-SSE) to N-S trending reverse faults. Additionally, minor northeast-southwest (NE-SW) trending dextral or northwest-southeast (NW-SE) trending sinistral strike-slip faults intersect these major N-S trending faults, tracing back to tectonic reactivations during the Pan-African, Hercynian, and Alpine orogeneses. Stereonet serves as an analytical tool in this updated methodology, enabling a more efficient and precise determination of fracture orientations. By representing each fracture set on a stereo-diagram, the orientation analysis becomes streamlined, facilitating the identification of global lineaments at a 1:100,000 scale. This process has led to the digitization and analysis of approximately 400 faults through the integration of geological data, curvature, and illumination attributes, unveiling a complex network of fracture orientations classified into seven distinct sets: N000, N040, N060, N090, N110, N130, and N150 (Figure 9.8).

The length distribution of these fractures is visualized through bar charts, which compare the fracture lengths to their frequency, showcasing the dominant, minimal, and average lengths observed. The power-law distribution is evident, with sets N000, N090, and N130 emerging as the most significant, covering lengths ranging from 4 km to 166 km, 2 km to 37 km, and 1.6 km to 47 km, respectively. The extensive nature of these fractures suggests potential intersections at various depths, particularly for the N000 set, which is highlighted as the most critical fault network impacting the region. This network spans the basement and traverses all Paleozoic series, including Cambro-Ordovician, Silurian, and Devonian formations, confirmed as strike-slip faults through geological map analyses.

Further statistical evaluation of global fracture lengths and individual fracture sets has been conducted, focusing on parameters such as frequency distributions, correlation coefficients, and the

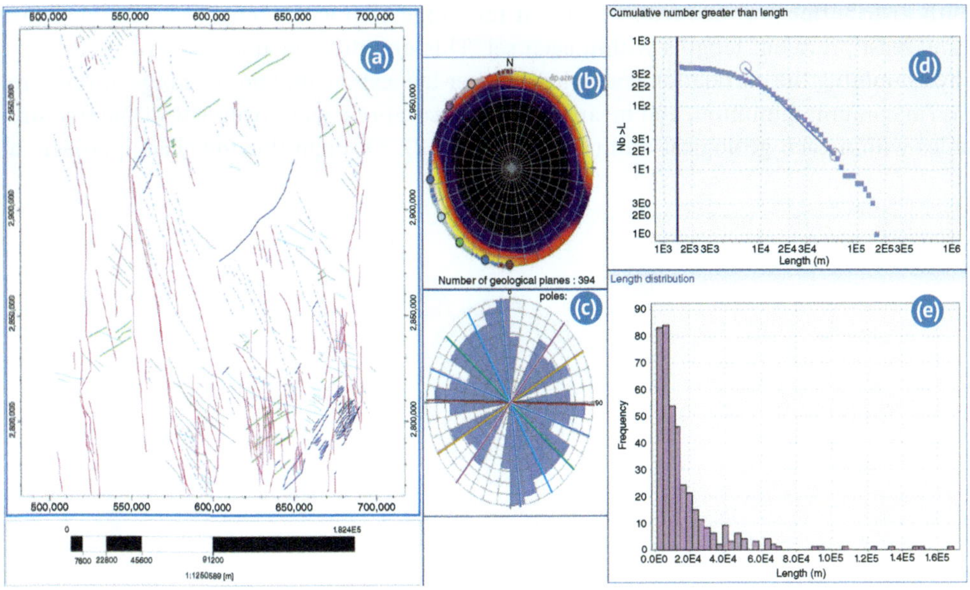

Figure 9.8 Major faults characteristics. (a) Major faults map in the area of study; (b) major faults' pole orientation; (c) major faults sets orientation; (d) length power law distribution; (e) length distribution.

power-law exponent. Correlation coefficients have been observed to range between 0.93 and 0.97, with the power-law coefficient fluctuating between 1.97 and 2.42, underscoring the complexity and the critical nature of these fault systems within the study area.

9.5.2 Minor Faults Analysis

This enhanced approach is meticulously applied across various geological layers, encompassing the basement and the three Cambro-Ordovician units, with each unit undergoing a distinct examination to unravel the intricacies of its fractography. This granular analysis aids in delineating the interplay between facies variations and the intricate fracture networks within each geological formation, thereby shedding light on the structural and compositional complexities of the Cambro-Ordovician reservoir and the basement formation.

For every geological layer under scrutiny, specific fracture sets are identified, and their length distributions are meticulously computed. This step is pivotal for understanding the spatial arrangement and connectivity of fractures within each formation. Moreover, the fractal dimensions of the entire fracture network, as well as the individual fracture sets constituting this network, are quantified employing both the center dimension and box-counting methods. These dimensions offer insights into the complexity and the hierarchical organization of the fracture networks, reflecting the geological processes that have shaped these formations.

The comprehensive analysis extends to fractures intersecting the basement and the Cambro-Ordovician formations, emphasizing the critical nature of understanding these interactions for a holistic fracture analysis workflow. Utilizing curvature and illumination maps alongside other pertinent data, fractures are digitized to precisely define their orientations and calculate fracture density for each stratigraphic layer. This data is then synthesized through the creation of rose diagrams and the application of statistical analyses to convey fracture orientation and density across the stratigraphic units.

Length distributions are plotted to pinpoint the range of fracture lengths, identifying the most and least prevalent lengths within each set. This quantitative analysis is further deepened by examining the correlation coefficient, power-law exponent, and the range of fracture lengths (maximum, minimum, and mean), offering a comprehensive overview of the fracture dynamics within each geological unit (Figures 9.9–9.12). Through this nuanced approach, a

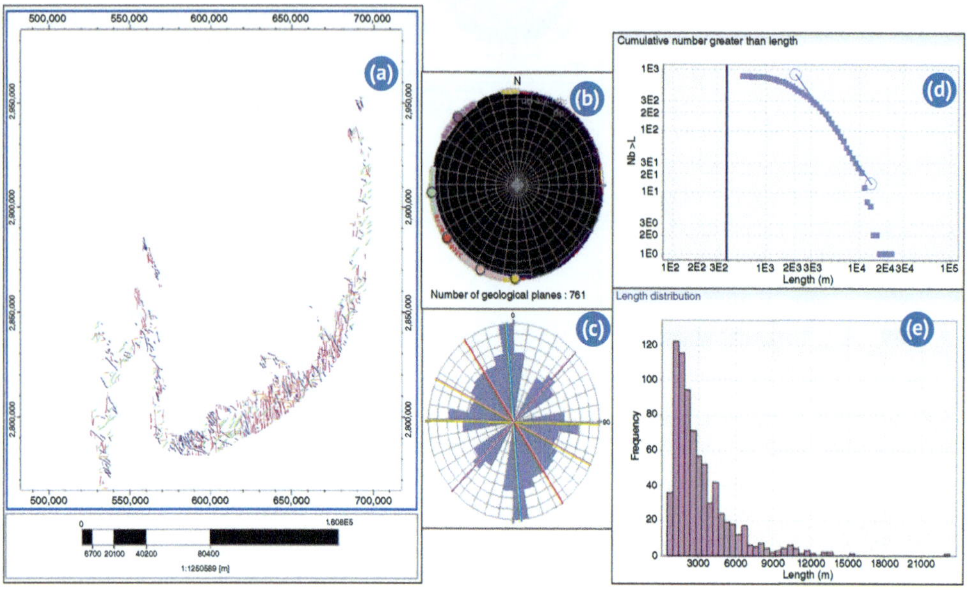

Figure 9.9 Basement Minor faults characteristics. (a) Minor faults map in the basement formation; (b) minor faults' pole orientation; (c) minor faults sets orientation; (d) length power law distribution; (e) length distribution.

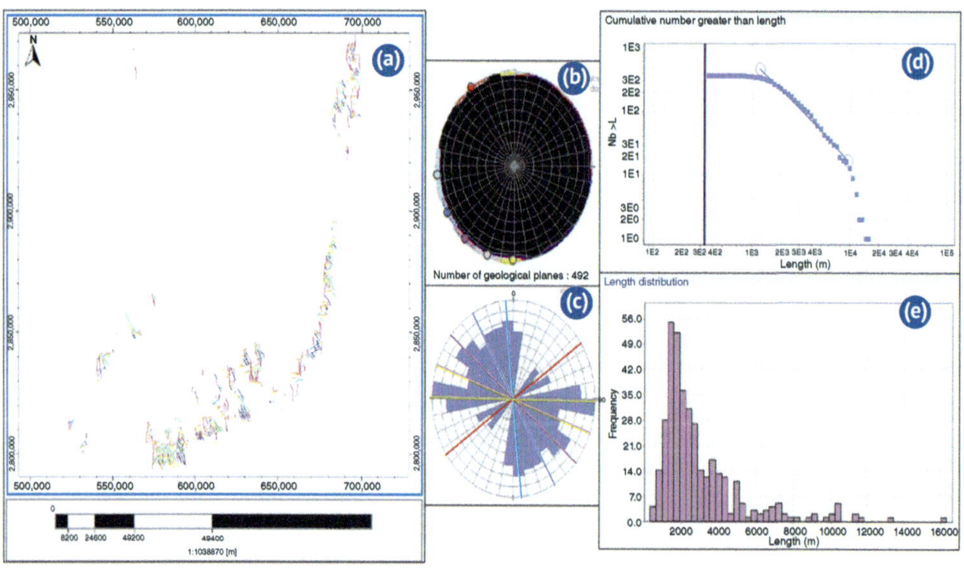

Figure 9.10 Ajjers Minor faults characteristics. (a) Minor faults map in the Ajjers formation; (b) minor faults' pole orientation; (c) minor faults sets orientation; (d) length power law distribution; (e) length distribution.

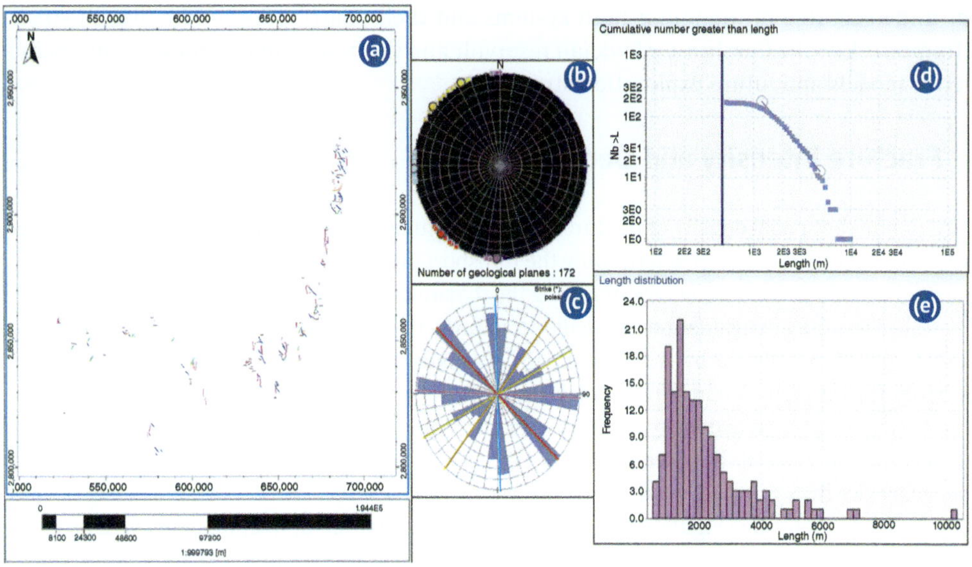

Figure 9.11 In-Tahouite Minor faults characteristics. (a) Minor faults map in the In-Tahouite formation; (b) minor faults' pole orientation; (c) minor faults sets orientation; (d) length power law distribution; (e) length distribution.

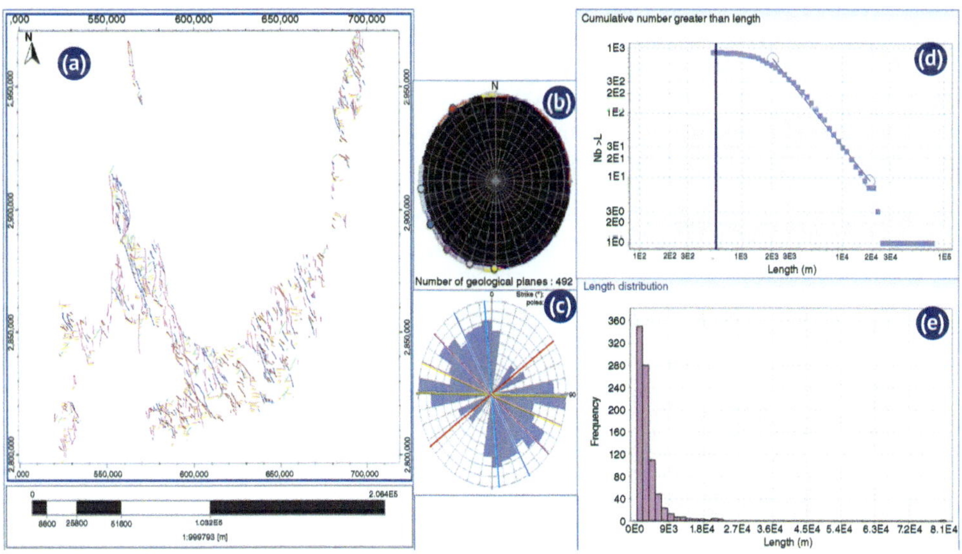

Figure 9.12 Tamadjert Minor Faults characteristics. (a) Minor faults map in the Tamadjert formation; (b) minor faults' pole orientation; (c) minor faults sets orientation; (d) length power law distribution; (e) length distribution.

deeper understanding of the minor fault systems and their contribution to the overall structural framework of the Cambro-Ordovician reservoir and the basement is achieved, enhancing predictive models and future exploration strategies.

9.6 Fracture Intensity and Density Analysis

The variation in the spatial density of fractures, which fluctuates based on proximity to larger geological structures, plays a key role in determining the transport properties of a rock mass. Understanding these spatial variations is essential for unraveling the dynamics of shear fracture development, which often results from the amalgamation of smaller fractures. In this context, the FracPaQ™ toolbox, a Matlab™ application designed by Healy et al. (2017), stands out by offering robust quantification of fracture patterns through comprehensive analysis of 2D fracture data. This toolbox facilitates the calculation of two critical spatial density measures: fracture intensity (P21) and fracture density (P20), as originally categorized by Dershowitz and Herda (1992). Fracture intensity, expressed in units of meters inverse, quantifies the aggregate length of fractures within a given area, thereby offering insights into the prevalence of fracturing (Figures 9.13 and 9.14). On the other hand, fracture density, denoted in units of meters squared inverse, measures the frequency of fractures per unit area, providing a direct gauge of fracture concentration (Rizzo et al., 2017a, 2017b).

Utilizing the circular scan window methodology introduced by Mauldon et al. (2001), the study generates refined estimates of both fracture density and intensity for the specified area, leading to the production of detailed "Estimated Density, P20" and "Estimated Intensity, P21" maps. These maps illuminate the distribution and intensity of fractures within the zone of interest, particularly highlighting areas of high fracture intensity around the periphery of the Mouydir Basin (Figure 9.15). This distribution pattern aligns with the predominant strike-slip fractures oriented north-south (N-S) and northwest-southeast (NW-SE).

By digitizing the entire fracture network using Neuralog software and importing the data as an ASCII file into the FracPaQ toolbox, the analysis identifies two primary fracture trends corresponding to the major faults labeled N000 and N150. The generated maps not only showcase the zones of heightened fracture intensity but also delineate the directional trends of these fractures, correlating them with the major structural fault orientations in the area. This enhanced

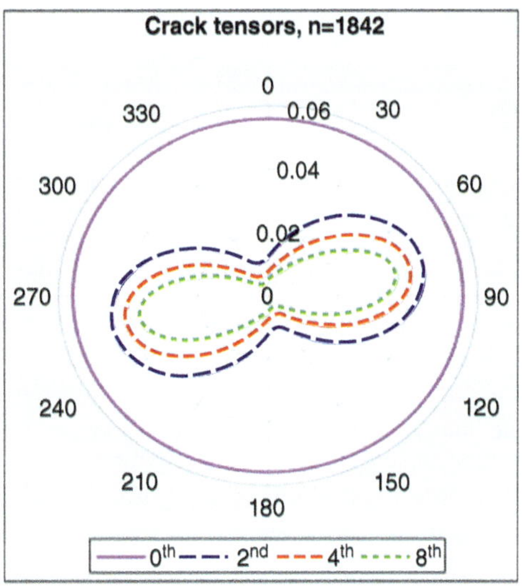

Figure 9.13 Crack tensors for the fracture pattern. Zero, second, fourth, and eighth rank crack tensors are shown in different colors.

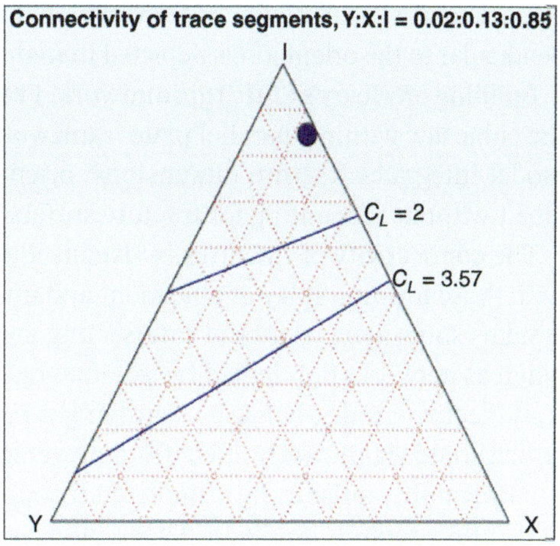

Figure 9.14 A ternary plot of fracture segment connectivity for the fracture pattern in the area of study. Two contour lines for Connections per Line (CL) are also shown, using indicative values described by Sanderson & Nixon, 2015 / with permission of Elsevier.

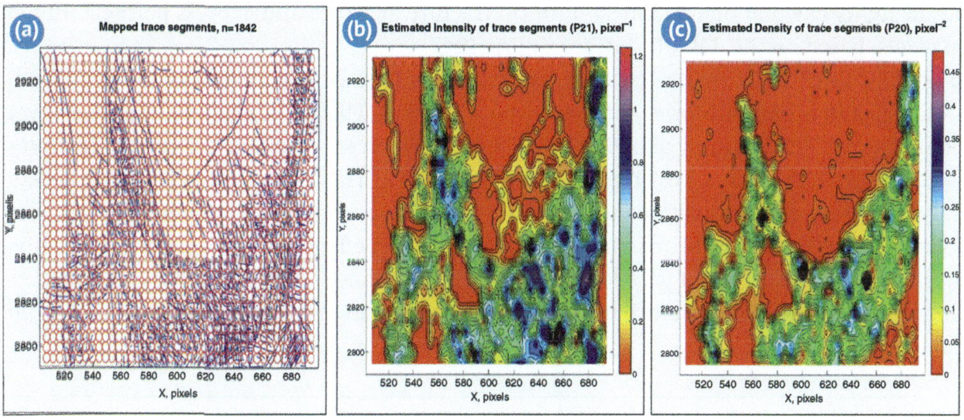

Figure 9.15 (a) Scan circle using the method of Mauldon et al.,2001 / with permission of Elsevier. (b) Estimated fracture intensity (P21) and (c) Estimated fracture density (P20) maps in the area of study.

methodology provides a nuanced understanding of fracture distribution, aiding in the assessment of geological formations for their potential impact on fluid transport and reservoir characteristics.

9.7 Fracture Connectivity, Permeability, and Wavelet Analysis

The FracPaQ toolbox simplifies the analysis of fracture traces, presuming they lie on a uniform 2D surface, thereby sidestepping the need for topographical adjustments. This assumption streamlines the quantification of fracture lengths and orientations into straightforward geometric operations. Understanding these fracture characteristics is essential for deciphering the tectonic history of rock formations and assessing their behavior in terms of permeability and mechanical strength.

To synthesize fracture data, crack tensor analysis transforms fracture traces into 2-dimensional tensors, integrating information on orientation, length, and spatial density. This method yields a multidimensional measure of fracture patterns, correlating with physical properties like bulk permeability, elasticity, and acoustic velocities, including anisotropy in P- and S-waves. The orientation data utilized in crack tensor calculations are derived from the poles to fracture segments, offering a perspective perpendicular to the orientations depicted in traditional rose diagrams.

Building on Healy's (2017) groundwork, FracPaQ also approximates 2D permeability by applying the cubic law within a parallel plate framework, complemented by the crack tensor approach. This model integrates fracture dimensions, orientations, and distributions to calculate permeability, albeit without accounting for fracture surface roughness.

The connectivity of fractures is visualized through a ternary plot, contrasting isolated fractures with those forming splay or abutment and intersection nodes within the network. Well-connected systems show a prevalence of intersecting and abutting fractures, crucial for enhanced fluid flow, whereas networks dominated by isolated fractures indicate poor connectivity. The analysis reveals a connectivity index below 1, signifying a predominance of isolated fractures in the study area, which could significantly affect fluid movement through the rock mass (Figures 9.16 and 9.17).

Permeability anisotropy is depicted through ellipse plots, oriented based on the flow direction and fluid pressure gradient. These ellipses are scaled to represent maximum and minimum permeability directions, highlighting the influence of fracture sets on permeability orientation. The permeability calculations assume a uniform fracture aperture and express permeability in square pixels, based on the default settings in FracPaQ.

Additionally, FracPaQ incorporates 2D wavelet analysis for a detailed examination of fracture maps, utilizing filters like the Morlet and Mexican hat wavelets. This method offers advantages over traditional Fourier Transform by allowing adjustable wavelet sizes to match the scale of fractures and selecting geometries that best represent crack patterns. The Morlet wavelet, in particular, excels in identifying scale transitions within the fracture maps, providing valuable insights into the structural complexities of the rock mass through wavelet coefficient maps, optimal orientation rose diagrams, and histograms of wavelet coefficients. This advanced analysis enhances our understanding of fracture networks, contributing to more accurate assessments of rock mass behavior and fluid flow properties.

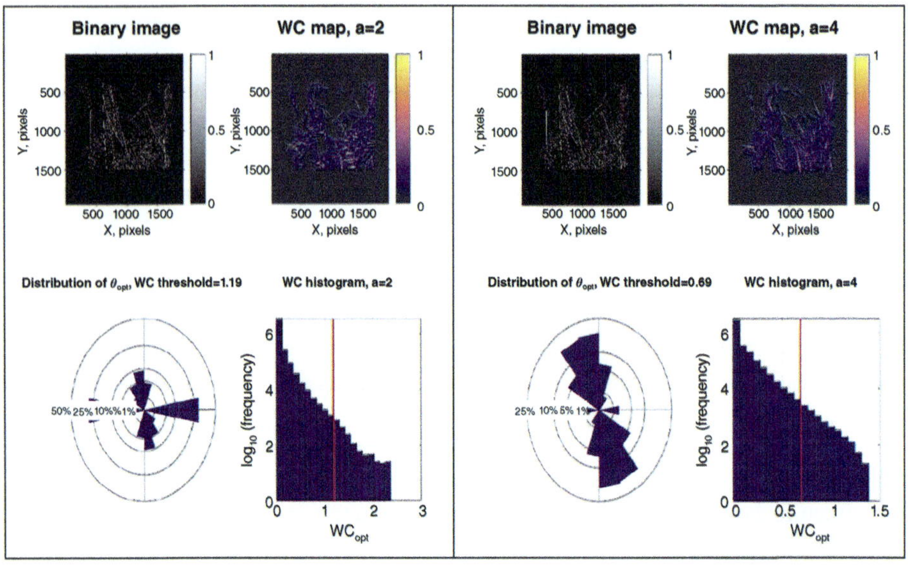

Figure 9.16 Wavelet analysis in FracPaQ. The left side shows results from Morlet wavelet analysis at a length scale of a = 2, and right side shows the same analysis for a length scale a = 4. Note how the orientation distribution (rose diagrams) changes at bigger length scales. *Source*: Sofiane Djezzar (2019).

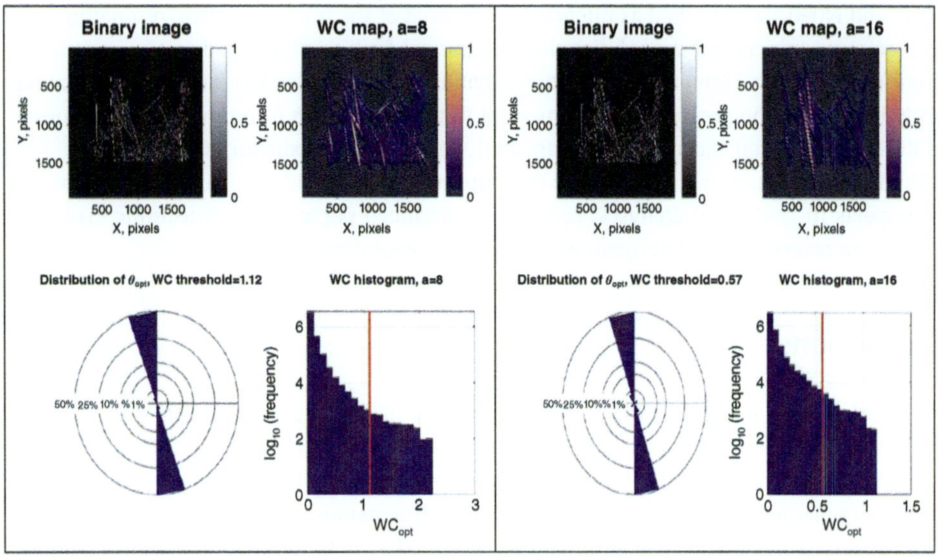

Figure 9.17 Examples of output from Wavelet analysis in FracPaQ. The left side shows results from Morlet wavelet analysis at a length scale of a = 8, and the right side shows the same analysis for a length scale a = 16. Note how the orientation distribution (rose diagrams) change. *Source*: Sofiane Djezzar (2019).

9.8 Discussion

The Cambro-Ordovician formations within the study area exhibit distinctive mechanical stratigraphy, defined by their lithological composition and thickness variations (Beuf et al., 1968). The integration of fault models across these formations has led to the development of a unified deterministic fault model, offering comprehensive insights into the spatial distribution of faults, and elucidating the interactions between the foundational basement faults and the overlying sedimentary layers. This integrated analysis reveals that the fracture networks permeating both the basement and the Cambro-Ordovician units generally exhibit fractal characteristics, albeit with some deviations attributed to the limited number of fractures observed.

These formations predominantly showcase a brittle tectonic regime, closely associated with significant basement faults that are remnants of the Pan-African orogeny, characterized by prominent vertical faults. Notably, the primary structural features align with north-south (N-S) trending dextral strike-slip faults and from north-northwest-southeast (NNW-SSE) to N-S trending reverse faults and associated folds, indicative of the complex tectonic fabric of the region (Fabre, 2005; Ait-Kaci, 1990; Conrad and Lemosquet, 1984).

Two principal faults emerge as pivotal in delineating the Mouydir Basin's margins, extending from the Hoggar shield and diverging across the Saharan platform in directions from north-northeast to north (NNE-NS) and from north-northwest to north (NNW-NS), known respectively as the Amguid and Idjerane spurs. These faults are not only significant due to their length but also due to their implication in the basin's geomorphology, serving as major conduits for fracturing and deformation, as evidenced by the presence of secondary faults and folds discernible on geological maps (Barouline et al., 1971).

Haddoum (2001) proposed that the basin experienced a compressional phase aligned from north-northeast-south-southwest (NNE-SSW) to east-northeast-west-southwest (ENE-WSW) around the Carboniferous/Permian transition, likely extending into the Early Permian. The fault orientations

N040, N060, and N090 are identified as integral components of the fracture system, resonating with the Hercynian orogeny's compressional direction. Additional fault sets identified may correspond to later tectonic activities post-Hercynian, suggesting a dynamic and complex tectonic evolution that has significantly influenced the geological and structural framework of the Cambro-Ordovician formations and the wider Mouydir Basin area (Zazoun, 2001, WEC, 2007, Khodjaoui, 2008).

This enhanced understanding underscores the importance of considering both lithological and structural aspects in geological analyses, providing a nuanced perspective on tectonic evolution and its impact on reservoir characterization and potential exploitation strategies.

9.9 Conclusions

This study meticulously delineates the intricate networks of fractures permeating the Paleozoic series and Cambro-Ordovician units within the northern region of the Hoggar Shield. Through comprehensive fracture analysis, this research has identified a variety of fracture sets, densities, and interconnections across both spatial and temporal dimensions. The investigation into stratabound and non-stratabound fracture models has revealed that the Cambro-Ordovician formations are composed of distinct mechanical stratigraphy units, differentiated by their lithological properties and thickness.

A unified deterministic fracture model developed through this study has elucidated the distribution patterns of fractures, enabling a deeper understanding of the mechanical units within the Cambro-Ordovician formations and their interrelations. Notably, the Tamadjert formation has emerged as the most fractured unit, attributed to its lithological makeup, while the In-Tahouite formation exhibits fewer fractures, likely due to its shale composition.

The impact of basement fractures on the sedimentary cover is profound, structuring the Cambro-Ordovician units and introducing a complex fracture network primarily characterized by brittle tectonic behaviors. These behaviors are closely linked to major basement fractures stemming from the Pan-African orogeny, which have spawned a vast network of predominantly vertical fractures. The geological mapping further underscores the presence of a dense fracture network affecting the Paleozoic era, with significant structures aligning with north-south (N-S) trending dextral strike-slip and north-northwest-south-southeast (NNW-SSE) trending sinistral strike-slip fractures.

Distinguishing two major fracture corridors that define the Mouydir Basin's western and eastern boundaries, these corridors originate from the Hoggar shield basement and extend northward in divergent directions across the Saharan platform. The primary fault sets identified within the Cambro-Ordovician reservoirs, namely N000, N050, N090, N120, and N160, highlight the dominance of N-S trending dextral strike-slip faults and reverse faults. This intricate fault system indicates the profound influence of basement fractures on the sedimentary architecture of the Cambro-Ordovician rocks.

A notable variation in the basement fractures' impact on the Cambro-Ordovician sedimentary layers is evident, with the same fracture sets impacting both the basement and Cambro-Ordovician units, suggesting a significant role of basement faults in basin structuring. The fractures are markedly prevalent in denser formations such as Ajjers and Tamadjert but are less common in the In-Tahouite formation due to their shaly composition. The fracture length distribution within the area of interest adheres to a power-law, with coefficients ranging between 2.31 and 2.69 and high correlation coefficients, indicating a complex, yet systematic distribution of fracture lengths.

The connectivity analysis within the fracture network reveals that only a small fraction of fractures are splay, with a majority being isolated, signifying a connectivity index below 1. These intricate fractures, their distribution, and the mechanical stratigraphy of the Cambro-Ordovician formations offer invaluable insights into the geological settings and structural architecture of the region, underscoring the critical role of tectonic activities in shaping the subsurface reservoir characteristics.

References

Ait-Kaci, A. (1990). Evolution lithostratigraphique et sédimentologiquedes monts d'Ougarta pendant le Cambrien (Sahara Algérien Nord Occidental). Thèse de 3ème cycle, IST-USTHB, p. 193.

Arab, M. & Djezzar, S. (2011). Hydrocarbon potential in Reggane Basin. Oil Mediterranean Conference (OMC), OMC-2011-046. Italy.

Balberg, I., Berkowitz, B., & Drachsler, G. E. (1991). Application of a percolation model to flow in fractured hard rocks. Journal of Geophysical Research 96B, pp. 10015–10021.

Balberg, I. & Binenbaum, N. (1983). Computer study of the percolation threshold in a two-dimension anisotropic system of conducting sticks. Physical Review 28 (7), pp. 3799–3812.

Barouline, G., Iakovlev, B., Korj, M., et al. (1971). In: Structures geologiques et perspectives en petrole et en gaz du Sahara Algerien (ed. Altamira-Rotopress, S. A.), Sonatrach Éditions.

Bazalgette, L. (2004). Relations plissement/fracturation multi échelle dans les multicouches sédimentaires du domaine élastique/fragile Thèse doct. Université Montpellier II, p. 252.

Beicip. (2008). Fracture characterization of the Reggane field, final report, prepared for Repsol.

Beicip. (2018). FracaFlow software tutorial.

Beuf, S., Biju-Duval, B., De Charpal, O., et al. (1968). Une conséquence directe de la structure du bouclier africain: l'ébauche des bassins de l'Ahnet et du Mouydir au Paléozoïque inférieur: Publications de Service Géologique de l'Algérie. Bulletin 38, pp. 103–131.

Berkowitz, B., Bour, O., Davy, P., et al. (2000). Scaling of fracture connectivity in geological formations. Geophysical Research Letters 27 (14), pp. 2061–2064.

Boualam, A., Djezzar, S., Rasouli, V., et al. (2019). 3D modeling and natural fractures characterization in Hassi Guettar Field, Algeria. In: 53rd American Rock Mechanics Association Symposium, 19-A-373-ARMA. NY, USA. OnePetro.

Brown, S. R. & Bruhn, R. L. (1998). Fluid permeability of deformable fracture networks. Journal of Geophysical Research: Solid Earth 103 (B2), pp. 2489–2500.

Chopra, S. & Marfurt, K. J. (2007). Seismic attributes for prospect identification and reservoir characterization. Publisher Society of Exploration Geophysicist, copyright year 2007, p. 481.

Collomb, P. & Donzeau, M. (1974). Relation entre kink-bands décamétriques et fractures du socle dans Hercynien des monts d'Ougarta (Sahara Occidental, Algérie). Tectonophysique 24 (3), pp. 213–242.

Conrad, J. & Lemosquet, Y. (1984). Evolution sédimentaire et structurale du bassin Ahnet-Timimoun-Bechar (sahara Algérien) au cours du Carbonifère; données paleoclimatiques. Bulletin de la Société Géologique de France 7 T XXVI, (6), pp. 987–994.

Davy, P. (1993). On the fault-Length frequency distribution of the San Andreas fault system. Journal of Geophysical Research 98 (12), pp. 141–151.

Dershowitz, W., S. & Herda., H., H. (1992). Interpretation of fracture spacing and intensity. Paper presented at the 33rd U.S. Symposium on Rock Mechanics (USRMS), Santa Fe, New Mexico, June 1992.

Djezzar, S. (2012). Impact de la fracturation naturelle sur la productivité des réservoirs du Dévonien inférieur dans le bassin de Reggane, Sahara Occidental, Algérie. Thèse Magister. Ed. USTHB. Alger: USTHB. p. 93: ill.; 30 cm. + CD-Rom.

Djezzar, S. (2019). Fracture detection and prediction. In: Unconventional Reservoirs for Finding Sweet Spot. Theses and Dissertations, 2551. https://commons.und.edu/theses/2551/.

Djezzar, S. & Boualam, A. (2020a). Analysis and modeling of tight oil fractured reservoir. In: SEG 2020 Annual Meeting, TX, USA. https://doi.org/10.1190/segam2020-3425011.1.

Djezzar, S. & Boualam, A. (2020b). Fractures characterization and their impact on the productivity of Hamra Quartzite reservoir (Haoud-Berkaoui field, Oued-Mya basin, Algeria). In: 54th American Rock Mechanics Association Symposium, Co, 20-A-1438-ARMA. USA.

Djezzar, S. & Boualam, A. (2021). Impact of natural fractures on the productivity of lower Devonian reservoirs in Reggane basin, Algeria. In: 55th US Rock Mechanics/Geomechanics Symposium 21-A-1295-ARMA.

Djezzar, S., Rasouli, V., Boualam, A., et al. (2019). Size scaling and spatial clustering of natural fracture networks using fractal analysis. In: 53rd American Rock Mechanics Association Symposium, 19-A-375-ARMA. NY, USA.

Djezzar, S., Rasouli, V., Boualam, A., et al. (2020). An integrated workflow for multiscale fracture analysis in reservoir analog. Arabian Journal of Geosciences 13, 161. https://doi.org/10.1007/s12517- 020-5085-6.

Fabre, J. (2005). Géologie du Sahara occidentale et centrale. Musée royal de l'Afrique centrale (Ed), Tervuren, Belgique. p. 572.

Haddoum, H. (1997). Le Sahara central dans les régions de I'Ahnet et de Bled El Mass: Une zone mobile hercynienne. Résumé, 3ème Séminaire de Géologie Pétrolière, Algérie, pp. 82–83.

Haddoum, H., Guiraud, R., & Moussine-Pouchkine, A. (2001). Hercynian compressional deformations of the Ahnet-Mouydir Basin, Algerian Saharan platform: far-field stress effects of the late palaezoic orogeny. Terra Nova 13, pp. 220–226.

Hamidou, M. (1999). évaluation des prospects des périmètres de Tasfalout, Reggane. (Rapport interne SH/EXP).

Healy, D., Rizzo, R. E., Cornwell, D. C., et al. (2017). FracPaQ: A MATLAB™ toolbox for the quantification of fracture patterns. Journal of Structural Geology 95, pp. 1–16.

Issad, M. (1996). Géodynamiques du Bassin de Reggane. Evolution structurale, typologie des structures et étude de la subsidence. Rapport interne SONATRACH.

Issad, M. (2000). Potentiel petrolier du bassin de Reggane. Rapport Sonatrach interne.

Khodjaoui A. (2008). Le Dévonien inférieur du bassin de Reggane (Sahara occidental algériens): sédimentologie, biostratigraphie et stratigraphie séquentielle. Mémoire de Magister, Université M'hamed Bouguerra de Boumerdes, p. 126.

Long, J. C. S., Remer, J. S., Wilson, C. R., et al. (1982). Porous media equivalents for networks of discontinuous fractures. Water Resources Research 18 (3), pp. 645–658.

Manzocchi, T. (2002). The connectivity of two-dimensional networks of spatially correlated fractures. Water Resources Research 38 (9), pp. 1–1.

Mauldon, M., Dunne, W. M., & Rohrbaugh, M. B. (2001). Circular scanlines and circular windows: new tools for characterizing the geometry of fracture traces. Journal of Structural Geology 23 (2), pp. 247–258.

Moore, D. E. & Lockner, D. A. (1995). The role of microcracking in shear-fracture propagation in granite. Journal of Structural Geology 17 (1), pp. 95–114.

Oda, M. (1983). A method for evaluating the effect of crack geometry on the mechanical behavior of cracked rock masses. Mechanics Materials 2 (2), pp. 163–171.

Odling, N. E. Gillespie, P. Bourgine, B, et al. (1999). Variations in fracture system geometry and their implications for fluid flow in fractured hydrocarbon reservoirs. Petroleum Geoscience 5, pp. 373–384.

Repsol. (2009). Feasibility study of Reggane Nord project, Reggane field (internal document, unpublishable).

Richard, J. L. (1994). Detection of zones of abnormal strains in structures using Gaussian curvature analysis: AAPG Bulletin 78 (12), pp. 1811–1819.

Roberts, A. (2001). Curvature attributes and their application to 3-D interpreted horizons. First Break 19 (2), pp. 85–100.

Rizzo, R. E., Healy, D., & De Siena, L. (2017a). Benefits of a Maximum Likelihood Estimator for fracture attribute analysis. Journal of Structural Geology 95, pp. 17–31.

Rizzo, R. E., Healy, D., Farrell, N. J., et al. (2017b). Riding the right wavelet: Quantifying scale transitions in fractured rocks. Geophysical Research Letters 44 (23), pp. 11–808.

Sanderson, D. J. & Nixon, C. W. (2015). The use of topology in fracture network characterization. Journal of Structural Geology 72, pp. 55–66.

Segall, P. & Pollard, D. (1983). Joint formation in granitic rock of the Sierra Nevada. Geological Society of America Bulletin 94, pp. 563–575.

Suzuki, K., Oda, M., Yamazaki, M., et al. (1998). Permeability changes in granite with crack growth during immersion in hot water. International Journal of Rock Mechanics and Mining Sciences 35 (7), pp. 907–921.

Well Evaluation Conference (WEC, 2007). Well Evaluation Conference, Internal report. "Joint documents of Sonatrach and Schlumberger." Algeria, pp. 489

Zazoun, R. (2001). La tectogenèse hercynienne dans la partie occidentale du bassin de I'Ahnet et la région de Bled El-Mass, Sahara Algérien: un continuum de déformation. Journal of African Earth Sciences 32 (4), pp. 869–887.

10

Structural Analysis and Fracture Kinematics Using Seismic 2D and Geological Maps

Sofiane Djezzar, Aldjia Boualam

University of North Dakota, Grand Forks, USA

10.1 Introduction

The Paleozoic analog, situated in the northern part of the Hoggar shield and along the southern edges of the Ahnet, Mouydir, and Illizi basins, is primarily composed of Cambro-Ordovician, Silurian, and Devonian formations (Figure 10.1). These formations exhibit a brittle tectonic style, significantly influenced by major basement fractures originating from the Pan-African orogeny, as identified in studies by Arab et al. (2011) and Djezzar et al. (2020). These ancient fractures were reactivated during the Hercynian orogeny, leading to the present-day horst and graben architecture prevalent across a substantial portion of the Saharan platform. Specifically, the Mouydir Basin, characterized by this horst and graben structure, has its periphery defined by two main horsts, with additional horsts within the basin center acting as positive structures where numerous wells have been drilled. Further structural analysis highlights the Saharan platform's complex geological history, marked by multiple tectonic phases, with the Hercynian orogeny at the end of the Carboniferous period being particularly pivotal in shaping the platform's structure (Boualam et al., 2019). By employing a comparative approach that integrates seismic 2D profiles with surface geological maps, the subsurface architecture of the Mouydir Basin is examined, elucidating the kinematics, dip, and geometry of major fault sets impacting both the Ahnet and Mouydir basins.

Surface geological mapping reveals a dense fracture network within the Paleozoic series, comprising three main fracture sets oriented N-S, NNE-SSW, and NW-SE. The N-S-oriented fault network forms extensive fracture corridors that delineate the two basins' boundaries, specifically the Idjerane and Amguid spurs, exhibiting both dextral and sinistral strike-slip movements. Similarly, the NNE-SSW and NW-SE fault sets are characterized by strike-slip faults with dextral and sinistral trends.

Additionally, the 2D seismic profile analysis uncovers that the deep strike-slip faults affecting the Ahnet and Mouydir basins possess vertical components, manifesting as normal and reverse faults, a result of transpressional and transtensional strain regimes during the Hercynian orogeny (Djezzar and Boualam, 2020b). These findings suggest that the study area underwent at least two tectonic events, oriented N040 and N120, leading to the formation of compressive and distensive

Geophysical Exploration for Hydrocarbon Reservoirs, Geothermal Energy, and Carbon Storage: New Technologies and AI-based Approaches, First Edition. Edited by Said Gaci.

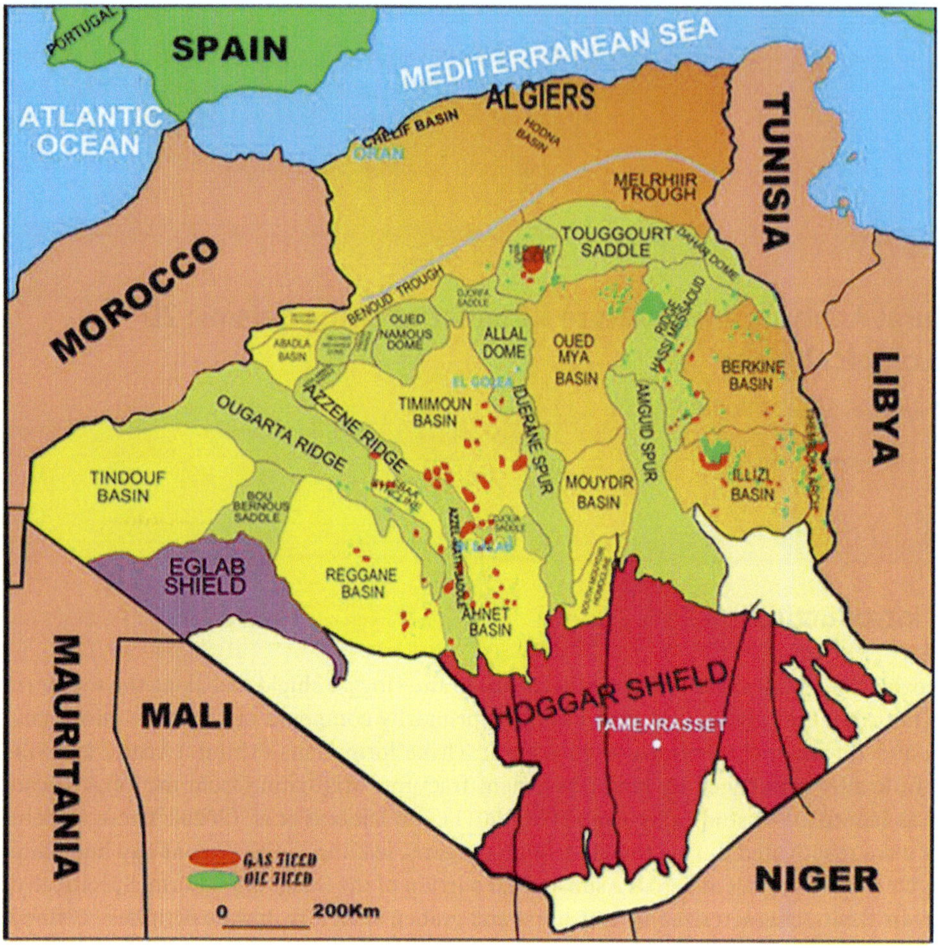

Figure 10.1 Main geological domains in Algeria.

structures, including anticlines/reverse faults and synclines/normal faults. This evidence supports the notion that the Hercynian orogeny represents a phase of continuous deformation rather than distinct, separate tectonic events.

This perspective aligns with the observed evidence of overlapping compressional and extensional phases within the structural evolution of the Saharan platform, highlighting the intricate interplay of tectonic forces that have shaped the region over geological time.

10.2 Material and Methods

The methodology encompasses the utilization of gravity data, geological maps, and 2D seismic profiles to conduct a comprehensive analysis aimed at elucidating the kinematics of various fracture sets within the study area. This analysis involves both outcrop observations and subsurface investigations to achieve precise determinations of fracture dynamics (Djezzar et al., 2019). A critical step involves comparing gravity data with geological maps to ascertain any correlations or analogies between these datasets, thereby enhancing the understanding of subsurface structures. Subsequently, the focus shifts to the fault network within the Mouydir Basin. Here, the analysis entails identifying different fault sets and assessing their lengths, and orientation. This model is instrumental in tracing the origin and extension of the fault network from surface observations to the subsurface realm, offering insights into the structural complexity of the two basins. Further analysis targets the major fault sets impacting the study area. Geological maps (Basseto et al., 1974a, 1974b; Bensalah et al., 1972a, 1972b, 1972c, 1972d) serve as a foundational resource for determining the age and kinematics of these faults, while 2D seismic profiles crossing the faults are analyzed to interpret their subsurface kinematics.

10.3 Geological Settings

The Cambro-Ordovician is well-known in the Saharan Platform where the Ordovician reservoirs are considered the second most important oil and gas-bearing formations that produce naturally due to the existence of natural fractures (WEC, 2005; Legrand and Nabos, 1962; Djezzar and Boualam, 2020a). Three main formations, Ajjers, In-Tahouite, and Tamadjert compose the Cambro-Ordovician. The Ordovician formations have an age span varying from the Tremadocian to Ashgill. The Ordovician reservoirs are composed of El Atchane sandstones (Tremadocian); Hamra Quartzites (Arenigian); Ouargla sandstone (Llanvirnian); Oued Saret sandstone (Lower Caradocian); Ramade or M'kratta sandstone (Upper Caradocian-Ashgillian). (WEC, 2005). The deposition of the Ajjers formation is followed by the Ordovician transgression and deposit of the In-Tahouite formation (Bennacef et al., 1971; Boudjema, 1987). During the latest Ordovician, due to polycyclic glaciations, the shallow marine/fluvial system changed into the glaciomarine system (Craig et al., 2008). Its record consists of diamictites, and glacial-fluvial sandstones, with a wide spectrum of sedimentary structures described as the Tamadjert Formation (Beuf et al., 1971). The Cambro-Ordovician falls under two types of classification. The first classification is mainly used to describe the Cambro-Ordovician in the subsurface and divides it into six units: Unit I, including El Moungar conglomerate; Unit II, including Hassi Leila series; Unit III-1; age Tremadocian, including Meribel sandstone; El-Gassi shale; and El Atchane sandstone. Unit III-2, age Lower to Middle Arenig, is represented by Hamra Quartzites. Unit III-3, age Llanvirnian-Llandeilian-Caradoc, is composed of Ouargla sandstone, Azzel shale, and the Oued-Saret sandstone. Unit IV, age Ashgill, is represented by a silico-clastic complex corresponding to the Micro-conglomeratic shale and the M'Kratta sandstone (WEC, 2005). The second classification is used to describe the Cambro-Ordovician at outcrops. It splits the Cambro-Ordovician into three different geological formations, including Ajjers, In-Tahouite, and Tamadjert. The Ajjers formation is composed of El-Moungar conglomerate, Tin-Tin-Taradjelli sandstone, Vire-du-Mouflon sandstone, and Hamra-Quartzites (WEC, 2005).

They correspond to the units I & II (Cambrian), III-1, and III-2, respectively. The second formation is In-Tahouite, which corresponds to unit III-3. The third formation is Tamadjert, which is equivalent to unit IV (Figure 10.2). The area of study is prone to several tectonic phases where some of the Precambrian lineaments may have experienced strike-slip displacements during the Cambro-Ordovician (Coward and Ries, 2003). In the Late Silurian-Early Devonian regionally, extensive inversion of the basins occurred due to the Caledonian orogeny (Follot, 1952; Beuf et al., 1971; Fekirine and Abdallah, 1998; Issad et al., 2011; Donzeau et al., 1981, 1983; Haddoum et al., 2001). Middle Devonian tectonic movements reactivated the Precambrian faults that bordered the basin creating basin-and-ridge topography (Wendt et al., 2006). The Variscan compression led also to the origin of numerous, mainly N-S trending, reverse faults into Paleozoic rocks (Zieliński, 2011). During the Mesozoic, the Hoggar Shield was uplifted, and this took place

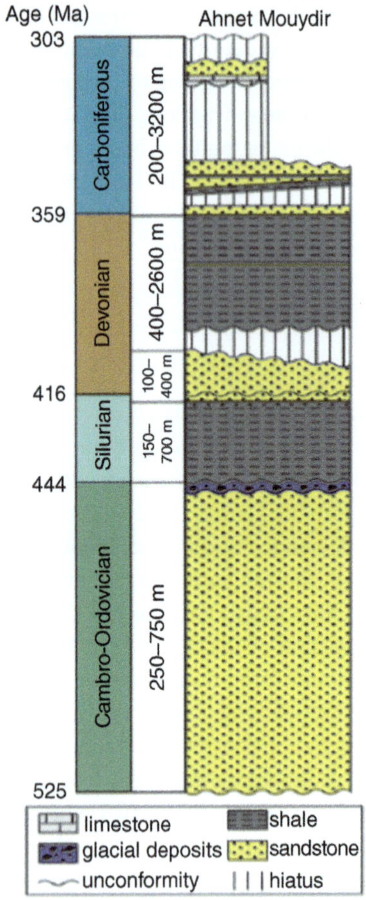

Figure 10.2 Schematic lithostratigraphic column of the Ahnet and Mouydir basins (compiled after Follot, 1952; Beuf et al., 1971; Wendt et al., 2006, 2009; Vecoli et al., 2008 in Zieliński, 2011; Zegrir, 2014).

before the Cretaceous time and relates to the Alpine tectonics (Liégeois et al., 2005; Liégeois, 2019; Haddoum et al., 2001; Zazoun, 2001). This movement resulted in a partial inversion of the sedimentary basins along the northern margin of the Hoggar Shield and eroded a significant portion of the Paleozoic sedimentary cover. Between the Cretaceous and the early Cenozoic, the sedimentary basins underwent tectonic inversion caused by a collision between Europe and Africa inducing the reactivation of the Precambrian faults (Zieliński, 2011).

10.4 Gravity Data

Gravity surveys offer a valuable, though indirect, method for assessing the density characteristics of subsurface materials. The principle underlying this approach is straightforward: higher gravity readings typically indicate the presence of denser subsurface rocks. A detailed analysis of the gravity data from the Mouydir Basin reveals two notable north-south trending positive anomalies. These are not arbitrary features but correspond closely to the basin's margins, delineated by the Idjerane spur to the west and the Amguid-Biod spur to the east. Within the central sector of the basin, another north-south trending positive anomaly was identified, near which the HL well was strategically drilled (Figure 10.3). These areas of increased gravity, suggesting denser geological formations, are interspersed with zones of negative anomalies, locations that, unfortunately, have seen the drilling of numerous unsuccessful wells (Djezzar, 2019).

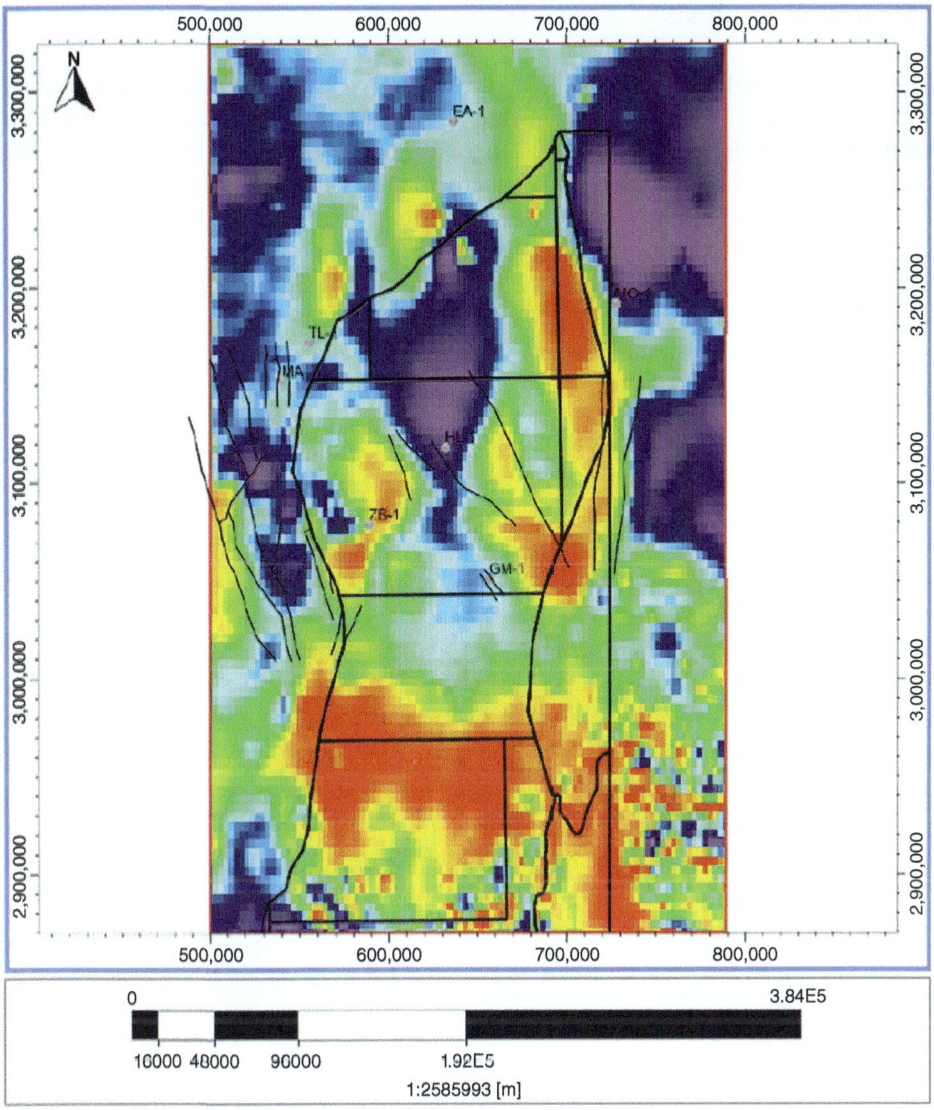

Figure 10.3 Gravity map and seismic faults in the Mouydir Basin.

Further integrating this gravity data with seismic information has elucidated the basin's structural complexity. The correlation is striking: the orientation of major faults within the basin, as well as their depicted representation on geological maps, aligns remarkably with the positive gravity anomalies. These faults predominantly follow a north-south and northwest-southeast trajectory, thereby accentuating the Mouydir Basin's edges. This integration of gravity and seismic data not only reinforces our understanding of the basin's geological framework but also provides a coherent structural context that is invaluable for hydrocarbon exploration and other geoscientific investigations. The convergence of these datasets enhances our comprehension of subsurface structures, facilitating more informed decisions in both exploration and development endeavors within the basin.

10.5 Structural Analysis

The Paleozoic series overlaying the northern part of the Hoggar Shield, primarily located along the southern edges of the Ahnet, Mouydir, and Illizi basins, significant insights have been gained into the geological complexities of the region. These series, encompassing the Cambro-Ordovician, Silurian, and Devonian formations, are profoundly influenced by Mode II fractures. These fractures, characterized by their considerable length and strike-slip kinematics, pervade through the entire Paleozoic strata, and extend into the subsurface within the Mouydir and Ahnet basins (Figure 10.4). This fracturing, particularly within the Cambro-Ordovician formations, exhibits a brittle tectonic style directly linked to major basement fractures stemming from the Pan-African orogeny. These ancient fractures underwent reactivation during the Hercynian orogeny, thereby establishing the distinctive horst and graben architecture observed in both the Ahnet and Mouydir basins. The Hercynian orogeny, occurring at the Carboniferous period's conclusion, emerges as a pivotal event that meticulously structured the Saharan platform into these horsts and grabens, as outlined by Boudjema (1987) and further elucidated by Le Heron, (2008).

The impact of the Hercynian orogeny extends to the deformation and structuring of the Paleozoic cover across the study area, forming distinct geological features such as the horsts of Idjerane M'Zab in the west and Amguid-Biod in the east, and the Mouydir basin graben, a configuration supported by findings from Zazoun (2001), and the comprehensive studies by Djezzar and Boualam (2021). The prevalent fracture network affecting the Paleozoic series comprises sets oriented N-S, NNE-SSW, and NW-SE, with the N-S fault network delineating the basins' boundaries, particularly the Idjerane and Amguid spurs, as these faults emanate from the Hoggar shield's basement and diverge northwards across the Saharan platform, displaying both dextral and sinistral strike-slip movements (Figure 10.5).

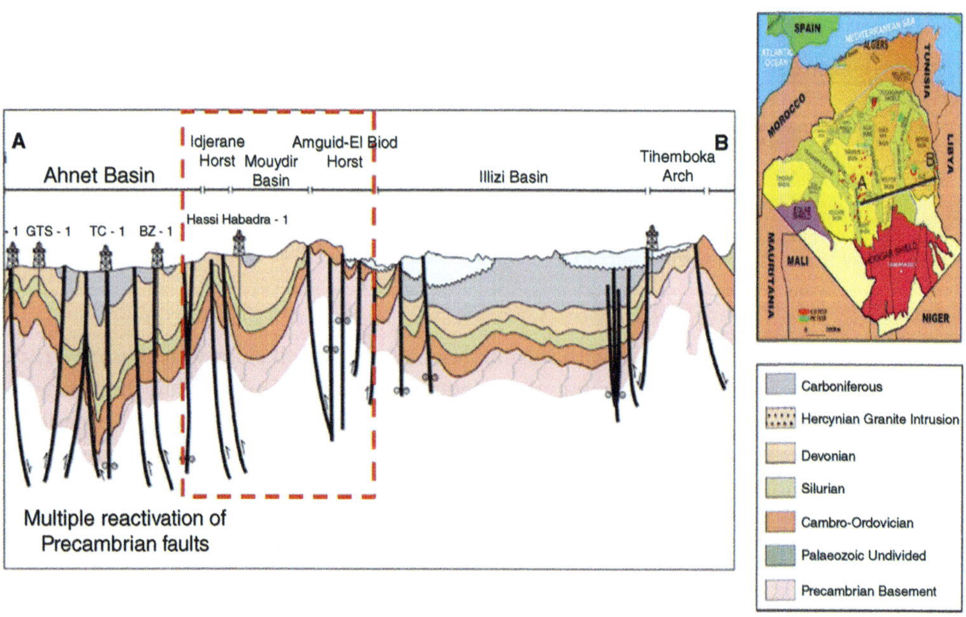

Figure 10.4 Geological cross section illustrating the main structural units in the Saharan platform (Craig et al., 2006).

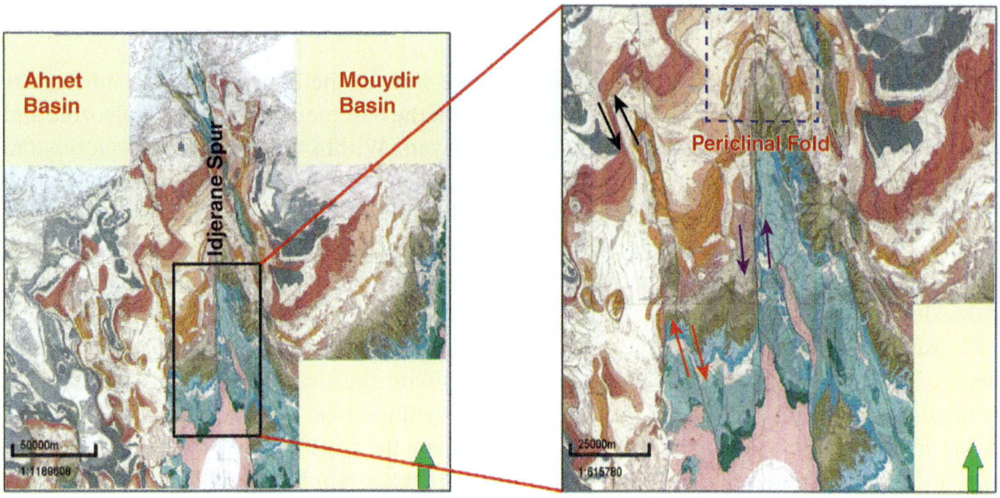

Figure 10.5 N-S and NW-SE faults in the area of study.

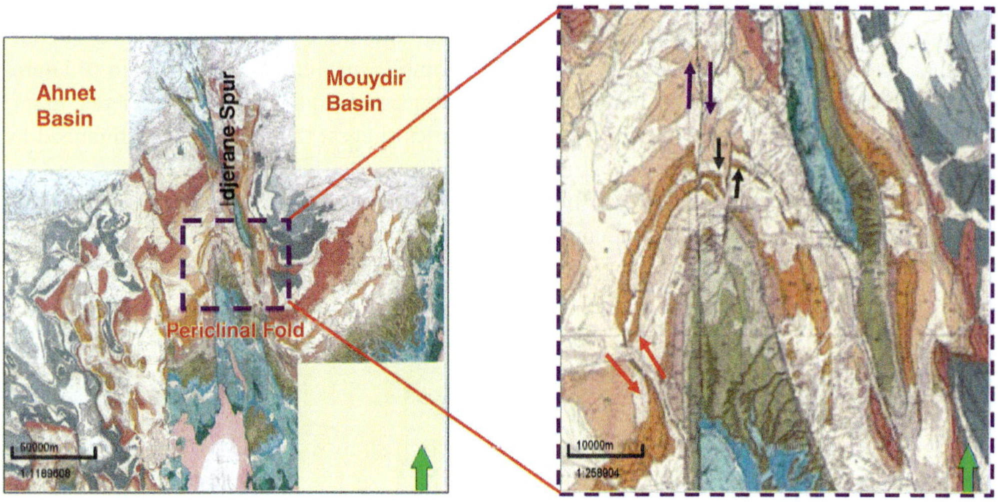

Figure 10.6 Kinematics of the N-S and NW-SE faults in the area of study.

The Hercynian orogeny's significance on the Saharan platform is further detailed through two main tectonic events oriented N040 and N120, with the former occurring during the Visean period and the latter spanning from post-Namurian to post-Permian (Figure 10.6). These events have been identified across various locations by numerous authors, including the recognition of the N040 event by Blès (1969) in Algeria's Ougarta Mountains and by Allary et al (1976), Pique and Michard (1989) in Morocco, highlighting a broad consensus on the Hercynian orogeny's role across the region. Moreover, the continuous deformation associated with the Hercynian orogeny, as proposed by Zazoun (2001) and explored through the extensive research of Donzeau et al. (1981), Donzeau (1983), and Djezzar (2019), Djezzar et al (2020a, 2021), underscores the orogeny's enduring influence on the structural framework of the Saharan platform, offering profound insights into the complex geological history of this vast region.

10.6 Seismic Data Analysis

The analysis of 2D seismic data provides an intricate view of the Mouydir Basin's subsurface, revealing a complex sedimentary cover that spans from the basement through the Paleozoic and Mesozoic formations to the relatively thin Cenozoic layers. Within this stratigraphic succession, several key unconformities demarcate the major geological epochs: the Hercynian, separating the Paleozoic from the Mesozoic; the Austrian, delineating the Mesozoic from the Cenozoic; and the Mio-Pliocene, marking the transition within the Cenozoic era. These unconformities highlight significant periods of non-deposition or erosion that punctuate the geological history of the Mouydir Basin.

The subsurface architecture of the Mouydir Basin, as revealed through 2D seismic analysis, supports a horst and graben configuration that aligns with the structural patterns identified in gravity data studies. The 2D seismic profiles particularly illuminate two major horsts that define the basin's margins, along with additional horsts that form the positive axis within the basin. This horst and graben structure is capped by the Hercynian unconformity, suggesting that the basin's present-day structure was largely established by the end of the Paleozoic era.

The integration of 2D seismic profiles with geological mapping has been instrumental in delineating the subsurface geometry, kinematics, and orientation of major fault systems impacting both the Ahnet and Mouydir basins. By aligning the seismic profiles perpendicular to these fault systems, researchers can accurately assess their real dip and kinematic behavior. This analysis indicates that the region's deep strike-slip faults exhibit complex movements, not limited to lateral (sinistral or dextral) displacements but also incorporating significant vertical components, manifesting as normal and reverse faults. These characteristics suggest the faults were influenced by both transpressional and transtensional stress regimes during the Hercynian orogeny.

Furthermore, the seismic data reveals that the area has undergone at least two major tectonic events, as evidenced by the creation of compressive structures (anticlines and reverse faults) and

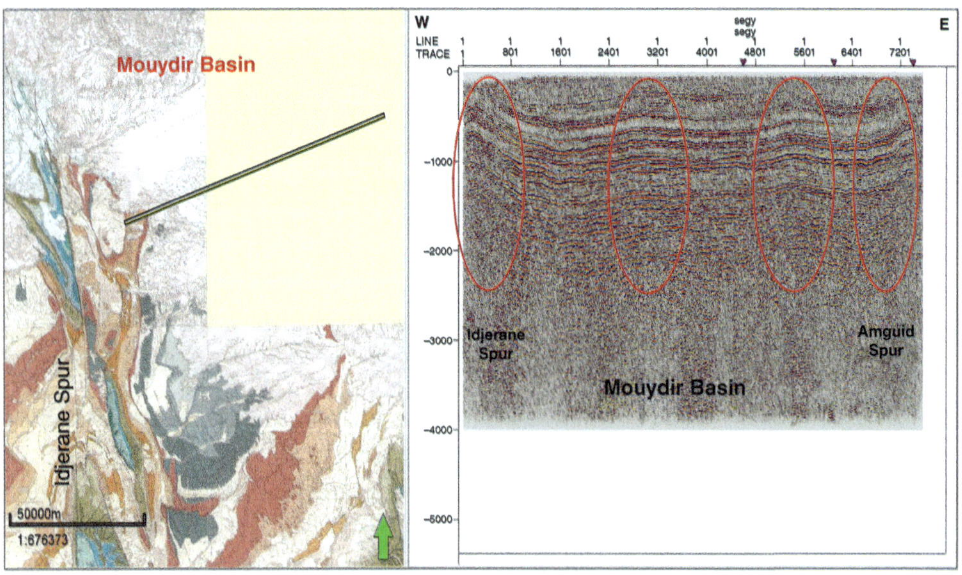

Figure 10.7 2D seismic profile showing the different positive and negative structures in the Mouydir Basin.

distensive features (synclines and normal faults). This dual nature of tectonic activity highlights the dynamic and multifaceted geological history of the Mouydir Basin, underlining the importance of 2D seismic analysis in unraveling the complex interplay of structural forces that have shaped the region over geological time (Figures 10.7–10.13).

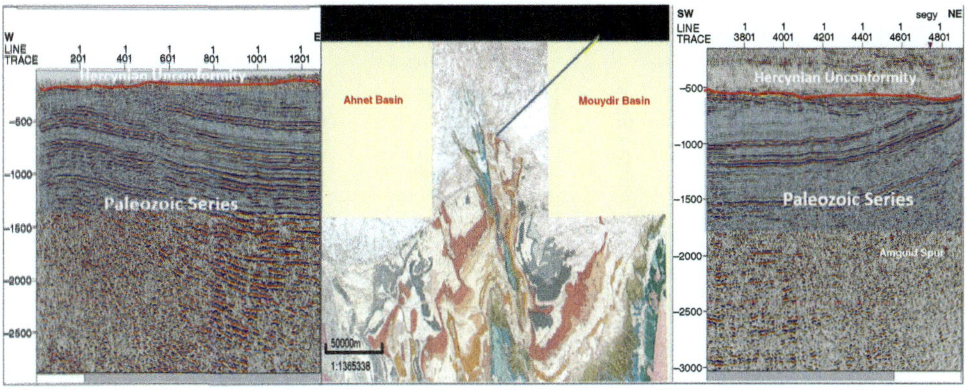

Figure 10.8 Composed 2D seismic profile showing the horsts architecture sealed by the Hercynian Unconformity in Mouydir Basin.

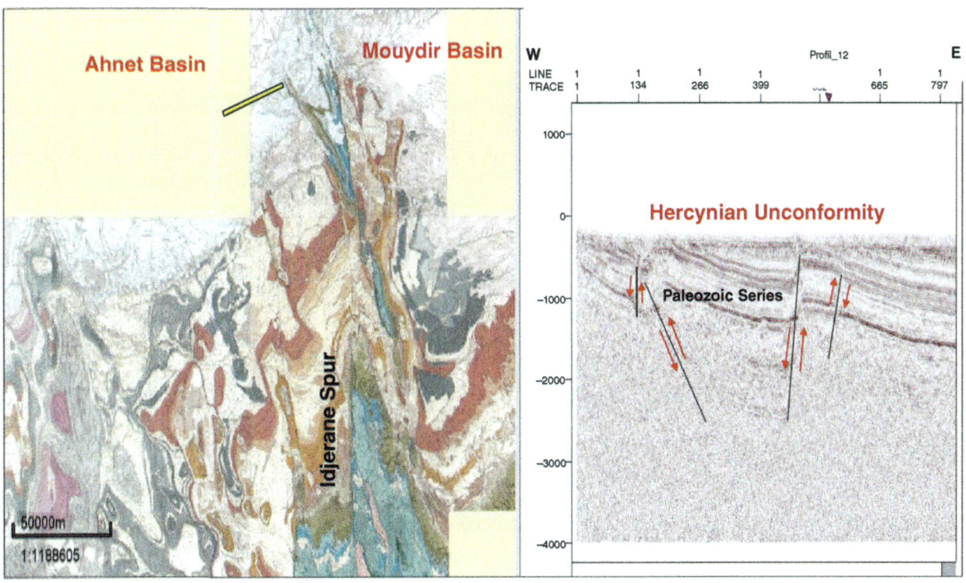

Figure 10.9 2D seismic profile showing a compressive structure in a syncline sealed by the Hercynian unconformity.

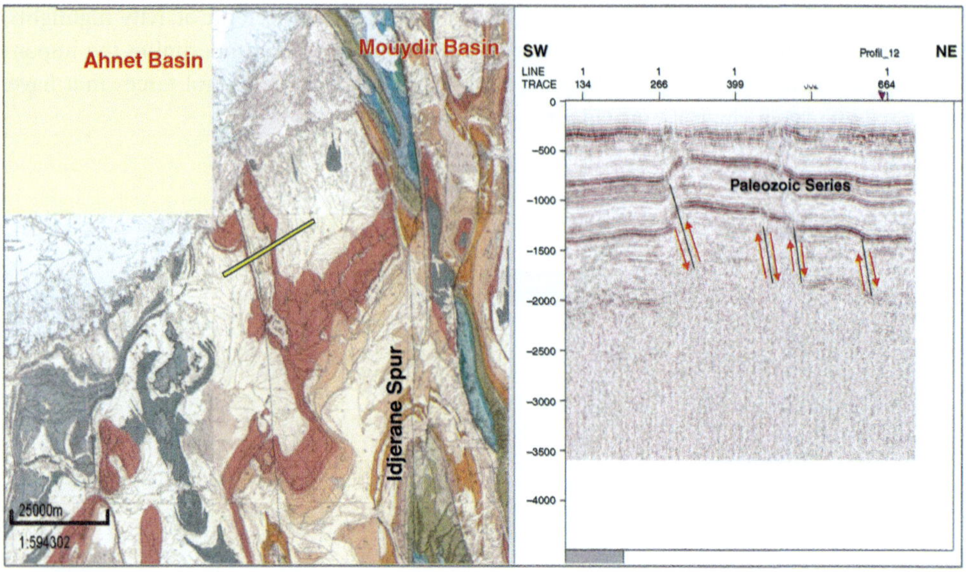

Figure 10.10 2D seismic profile showing compressive and distensive structures.

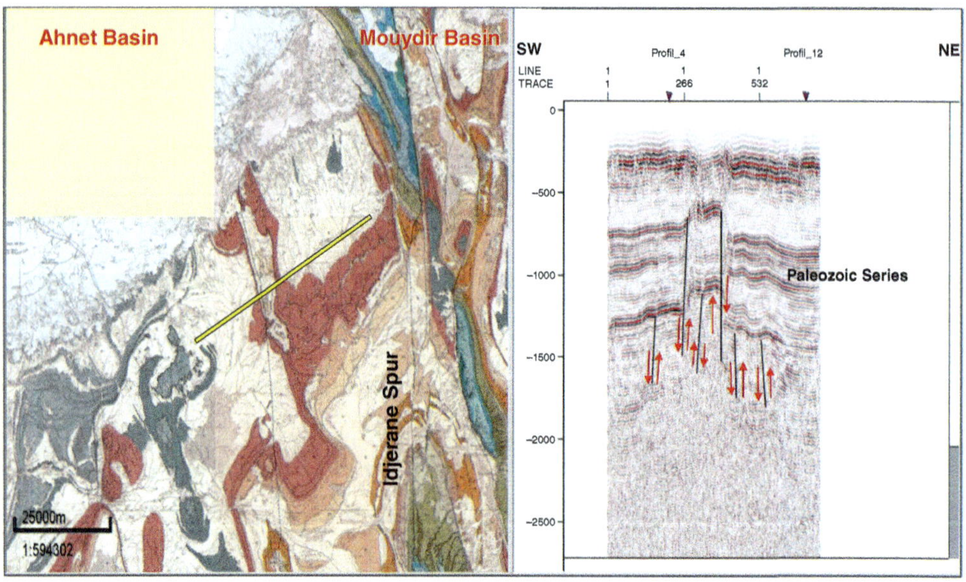

Figure 10.11 2D seismic profile showing a pop-up structure.

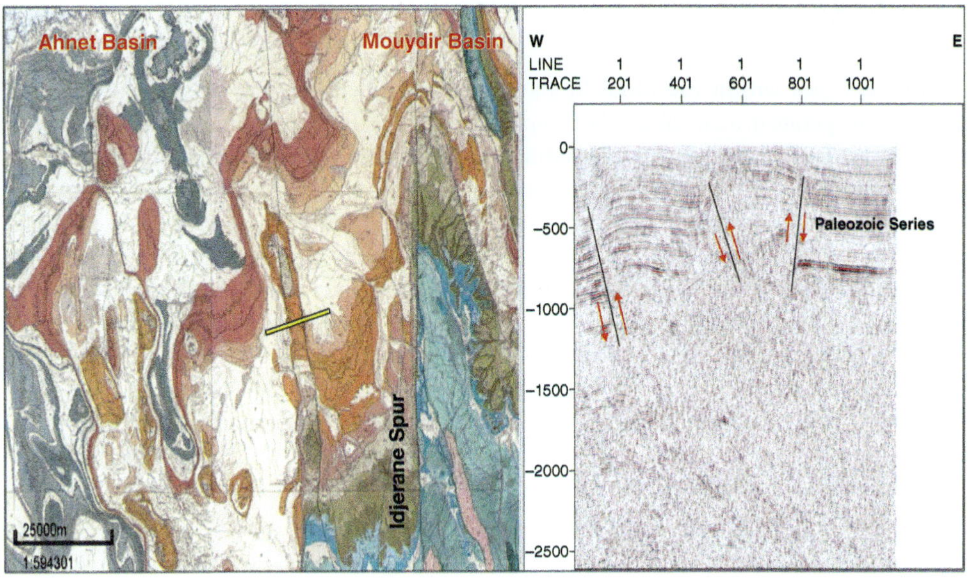

Figure 10.12 2D seismic profile showing a succession of a compressive structures.

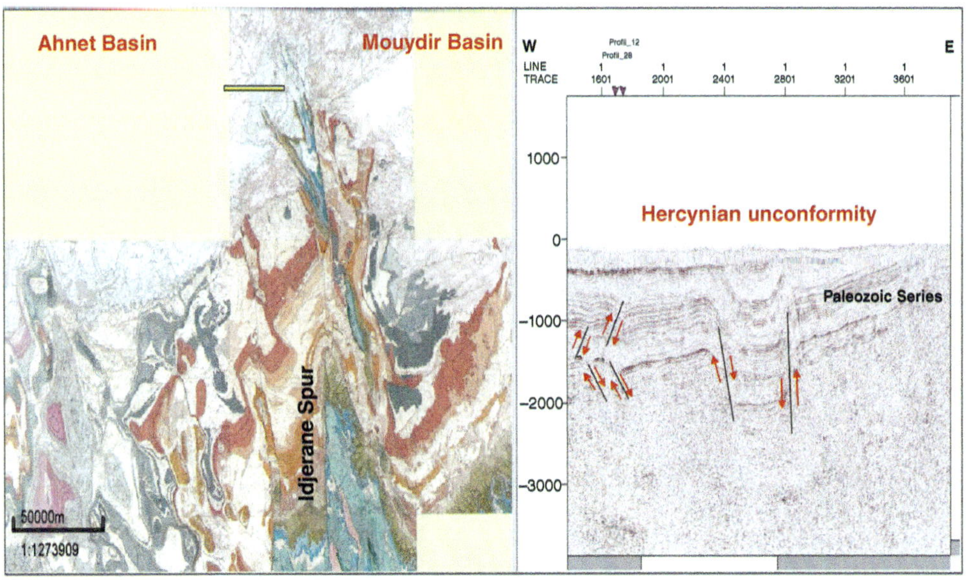

Figure 10.13 2D seismic profile showing distensive and compressive structures sealed by the Hercynian Unconformity.

10.7 Fault Analysis

The fracture analysis within the Mouydir Basin primarily identifies the presence of mode II fractures, which operate in an in-plane shear mode. This means the displacements occur within the plane of the fractures themselves, encompassing a range of fault types, including normal/reverse faults, strike-slip faults, or a combination thereof, that lead to the development of transpression and transtension faults. To streamline and enhance the fault orientation analysis, pole diagrams are employed. These diagrams are visualized on stereo-diagrams representing clusters of poles, either constructed manually or generated automatically, to depict fault sets distinctly. Despite the challenges posed by the relatively lower resolution of 2D seismic data in the Mouydir Basin, notable work by Brahimi (2015) identified top Ordovician and cataloged 27 distinct faults. These were organized into four main fault sets with orientations of N000, N020, N140, and N160.

The Cambro-Ordovician formations, in particular, exhibit a predominantly brittle tectonic style, closely linked to significant basement fractures derived from the Pan-African orogeny. This orogenetic activity has given rise to a comprehensive network of major vertical fractures, which were

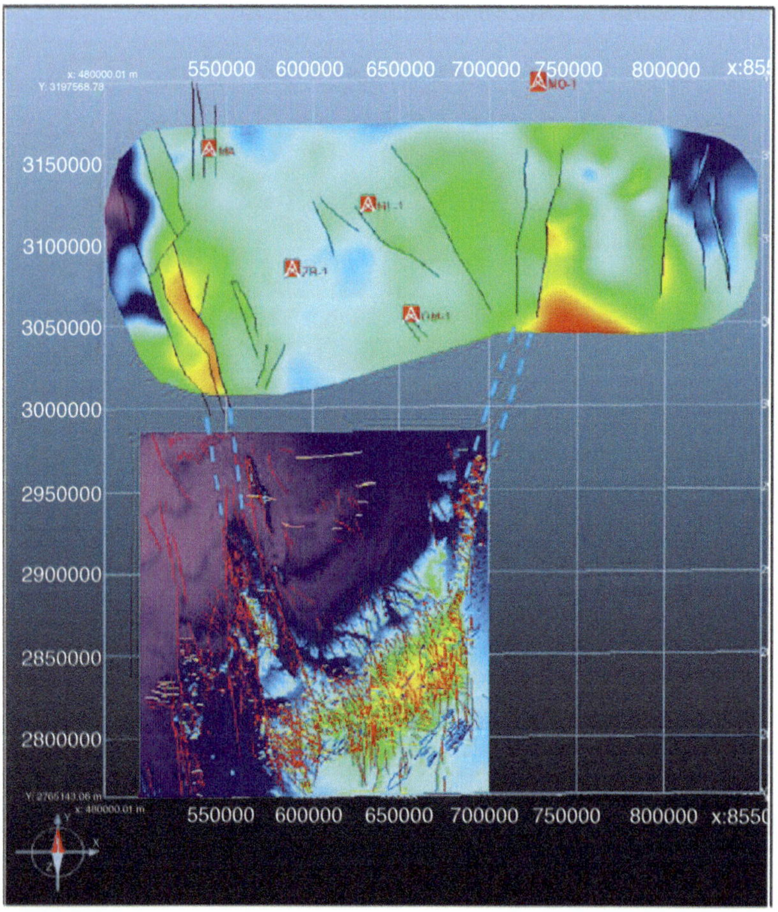

Figure 10.14 3D merged fault model using outcrops and subsurface faults.

subsequently reactivated during the Hercynian orogeny, shaping the current horst and graben configuration of the Ahnet and Mouydir basins.

Geological mapping efforts, as detailed by Djezzar (2012) and Djezzar et al. (2019), reveal a dense network of fractures impacting the Paleozoic strata. Notably, two major fracture corridors delineate the western and eastern boundaries of the Mouydir Basin. Originating from the Hoggar Shield's basement, these corridors extend northward in diverging directions across the Saharan platform. Oriented from NNE to NS and NNW to NS, they form the structural backbones of the Amguid and Idjerane spurs, respectively, highlighting the intricate fault dynamics and their profound influence on the basin's geological architecture (Figure 10.14).

10.8 Conclusions

The structural intricacies observed within the Ahnet and Mouydir basins can be largely attributed to two significant Hercynian tectonic episodes, both of which are demarcated by the widespread Hercynian unconformity, an angular discordance marking a pivotal geological transition. This orogenic activity has sculpted the Paleozoic strata into a distinct landscape of horsts and grabens, a configuration underscored by detailed geological mapping that reveals an extensive network of fractures impacting the Paleozoic series. These fractures, characterized by orientations of N-S, NNE-SSW, and NW-SE, form robust fault networks that define the peripheries of both basins, featuring strike-slip faults with both dextral and sinistral movements.

Further scrutiny into these fault systems through the lens of 2D seismic profiles illuminates the multifaceted nature of the deep strike-slip faults affecting the Ahnet and Mouydir basins. Notably, these faults are not purely strike-slip in nature but also exhibit significant vertical movements, behaving as both normal and reverse faults. These complex fault dynamics can be attributed to transpressional and transtensional stress regimes that prevailed during the Hercynian orogeny, contributing to the region's geological complexity.

The 2D seismic data elucidate that the structural evolution of the Mouydir and Ahnet basins was influenced by at least two major tectonic events. These events, manifesting as sequences of distension and compression, led to the formation of synclines and normal faults, as well as anticlines and reverse faults, respectively. The Hercynian orogeny, a defining geological phenomenon in the Saharan platform, encompasses two principal tectonic phases aligned N040 and N120, dating from the Visean period through to the post-Namurian to post-Permian era. This orogeny is better understood as a period of continuous deformation rather than isolated tectonic occurrences.

Such an analytical perspective sheds light on the geological mechanisms shaping the study area, revealing the origins of complex fracture patterns. This understanding is pivotal for predicting fracture occurrences in regions lacking direct data, thereby informing future exploration strategies within the Mouydir Basin. Specifically, this knowledge underscores the significance of natural fractures in enhancing the productivity of the Ordovician reservoirs, guiding exploratory efforts towards more effective and strategic endeavors in hydrocarbon exploration and development.

References

Allary, A., Lavenu, A., & Ribeyrolles, M. (1976). Etude tectonique et microtectonique d'un segment de chaine hercycienne dans la partie sud-orientale du Maroc central. (n.p.): Ed. du Service géol. du Maroc.

Arab, M. & Djezzar, S. (2011). Hydrocarbon potential in Reggane Basin. Oil Mediterranean Conference (OMC), Italy. OMC-2011-046.

Basseto, D., Ben Salah, A., Beuf, S., et al. (1974a). Carte geologique de Aoulef El- Arab. Ahnet et Mouydir. Service de la Carte Geologique, Algerie, jeu de 13 feuilles, Bchelle 1: 200 000, NG 31, No. 941, p. 14.

Basseto, D., Ben Salah, A., Beuf, S., et al. (1974b). Carte geologique de Reggane. Ahnet et Mouydir. Service de la Carte Geologique, Algerie, jeu de 13 feuilles, Bchelle 1: 200 000, NG 31, No. 941, p. 13.

Bennacef, A., Beuf, S., Biju-Duval, B., et al. (1971). Example of cratonic sedimentation: lower Palaezoic of Algerian Sahara. American Association of Petroleum Geologists Bulletin 55 (12), pp. 2225–2245.

Bensalah, A., Beuf, S., Gabriel, O., et al. (1972a). Carte Géologique de l'Algérie: Arak, feuille NG-31-X et NG-31-IV. Ministère de l'Industrie et de l'Energie.

Bensalah, A., Beuf, S., Gabriel, O., et al. (1972b). Carte Géologique de l'Algérie: Ain Tadjoubar, feuille NG-31-XVI. Ministère de l'Industrie et de l'Energie.

Bensalah, A., Beuf, S., Gabriel, O., et al. (1972c). Carte Géologique de l'Algérie: Khanget El Hadid, feuille NG-31-XVII. Ministère de l'Industrie et de l'Energie.

Bensalah, A., Beuf, S., Gabriel, O., et al. (1972d). Carte Géologique de l'Algérie: Ifetessene, feuille NG-31-XI. Ministère de l'Industrie et de l'Energie.

Beuf, S., Biju-Duval, B., De Charpal, D., et al. (1971). Les grès du Paléozoïque inférieur au Sahara. Sédimentation et discontinuité: évolution structurale d'un craton. In: Technip-Institut Français du pétrole, vol. 18. Collection Sciences et Techniques du Pétrole, Paris, p. 464.

Blès, J. L. (1969). Les relations des microfractures avec le plissement dans la région du Djebel Ben Tadjine et au'km 30'(Chaînes d'Ougarta-Sahara occidental, Algérie). *Publication Service Géologique Algérie*, *39*, 193-204.

Boualam, A., Djezzar, S., Rasouli, V., et al. (2019). 3D Modeling and natural fractures characterization in Hassi Guettar Field, Algeria. 53rd American Rock Mechanics Association Symposium, NY, USA. 19-A-373-ARMA.

Boudjema, A. (1987). Evolution structurale du bassin petrolier 'triasique' du Sahara Nord Oriental (Algerie). These Doctorat Etat, Paris XI Orsay, France, 290 p.

Brahimi, F. (2015). Potential Petrolier du Bassin du Mouydir (Rapport interne). Sonatrach (unpublished report).

Coward, M. P. & Ries, A. C. (2003). Tectonic development of North Africa basins. In: New Themes and Developing Technologies (ed. T. J. Arthur, D. S. Macgregor & N. R. Cameron), pp. 61–83. Geological Society of London. Special Publication 207.

Craig, J., Rizzi, C., Said, F., et al. (2008). Structural styles and prospectivity in the Precambrian and Palaeozoic hydrocarbon systems of North Africa. In: The geology of East Libya, vol. IV (ed. M. J. Salem), pp. 51–122. Malta: Gutenberg Press.

Djezzar, S. (2012). Impact de la fracturation naturelle sur la productivité des réservoirs du Dévonien inférieur dans le bassin de Reggane, Sahara Occidental, Algérie. Thèse Magister. Ed. USTHB. -Alger: USTHB, 2012.-93 p.: ill.; 30 cm. + CD-Rom.

Djezzar, S. (2019). Fracture detection and prediction in unconventional reservoirs for finding sweet spot. Theses and Dissertations. 2551. https://commons.und.edu/theses/2551/.

Djezzar, S. & Boualam, A. (2020a). Analysis and modeling of tight oil fractured reservoir. SEG 2020 Annual Meeting, TX, USA. https://doi.org/10.1190/segam2020-3425011.1.

Djezzar, S. & Boualam, A. (2020b). Fractures characterization and their impact on the productivity of Hamra Quartzite reservoir (Haoud-Berkaoui field, Oued-Mya basin, Algeria). 54th American Rock Mechanics Association Symposium, Co, USA. 20-A-1438-ARMA.

Djezzar, S. & Boualam, A. (2021). Impact of natural fractures on the productivity of Lower Devonian reservoirs in Reggane Basin, Algeria. 55th US Rock Mechanics/Geomechanics Symposium. 21-A-1295-ARMA.

Djezzar, S., Rasouli, V., Boualam, A., et al. (2019). Size scaling and spatial clustering of natural fracture networks using fractal analysis. 53rd American Rock Mechanics Association Symposium, NY, USA. 19-A-375-ARMA.

Djezzar, S., Rasouli, V., Boualam, A., et al. (2020). An integrated workflow for multiscale fracture analysis in reservoir analog. Arabian Journal of Geosciences 13, p. 161. https://doi.org/10.1007/s12517-020-5085-6.

Donzeau, M. (1983). Tectonique des monts de'Ougerta. In: *Afrique de l'Ouest. Introduction Geologique et Termes Startigraphiques, Lexique Stratigraphique International*, Nouvelle Serie 1 (ed. J. Fabre), pp. 118–120. Oxford, UK: Pergamon Press.

Donzeau, M., Fabre, J., & Moussine Pouchkine, A. (1981). Comportement de la dalle saharienne et orogenese varisque. Essai d'interpretation. Bulletin de la Société d'Histoire Naturelle de l'Afrique du Nord 69 (3–4), pp. 171–184.

DONZEAU, M. (1983). Tectonique des monts d'Ougarta. In Afrique de l'Ouest; Introduction géologique et termes stratigraphiques, pp. 118–120.

Fekirine, B. & Abdellah, H. (1998). Palaeozoic lithofacies correlatives and sequence stratigraphy of the Saharan Platform, Algeria. Geological Society, London, Special Publications 132 (1), pp. 97–108.

Follot, J. (1952). Ahnet et Mouydir. 19' Congres international de Geologie, Alger. Monographie Regionale, Serie 1, 1, pp. 1–6.

Haddoum, H., Guiraud, R., & Moussine-Pouchkine, A. (2001). Hercynian compressional deformations of the Ahnet-Mouydir Basin, Algerian Saharan Platform: far-field stress effects of the late palaezoic orogeny. Terra Nova 13, pp. 220–226.

Issad, M., Amia, Z., Djezzar, S., et al. (2011). Reevaluation du potential petrolier du basin du Mouydir. Rapport Interne Sonatrach.

Legrand, P. & Nabos, G. (1962). Contribution à la stratigraphie du Cambro-Ordovicien dans le bassin saharien occidental. Bulletin de la Société Géologique de France 7, pp. 123–131.

Le Heron, D.P., J. Craig, J. (2008). First-order reconstructions of a Late Ordovician Saharan ice sheet. *Journal of the Geological Society* 2008; 165 (1): 19–29. https://doi.org/10.1144/0016-76492007-002.

Liégeois, J. P. (2019). A new synthetic geological map of the Hoggar shield: an overview of its global structure and geological evolution. In: The Geology of the Arab World—An Overview (ed. A. Bendaoud et al.). Springer Geology, pp. 83–107.

Liégeois, J. P., Benhallou, A., Azzouni-Sekkal, A., et al. (2005). The Hoggar swell and volcanism: reactivation of the Precambrian Hoggar shield during Alpine convergence and West African Cenozoic volcanism. In: Plates, Plumes and Paradigms (ed. G. R. Foulger, J. H. Natland, D. C. Presnall, et al.), vol. 388, pp. 379–400, Geological Society of America Special Papers.

Pique, L. & Michard, A. (1989). Morroccan Hercynides: A synopsis. The Paleozoic sedimentary and tectonic evolution at the Northern margin of west Africa. American Journal of Science, Vol. 289, March 1989, pp. 286–330.

Well Evaluation Conference (WEC) (2005). Well evaluation conference, internal report. Joint documents of Sonatrach and Schlumberger, Algeria, p. 489.

Vecoli, M., Videt, B., & Paris, F. (2008). First biostratigraphic (palynological) dating of Middle and Late Cambrian strata in the subsurface of northwestern Algeria, North Africa: Implications for regional stratigraphy. Review of Palaeobotany and Palynology, 149 (1–2), 57–62.

Wendt, J., Kaufmann, B., Belka, Z., et al. (2006). Sedimentary evolution of a Palaeozoic basin and ridge system: the Middle and Upper Devonian of the Ahnet and Mouydir (Algerian Sahara). Geological Magazine 143 (3), pp. 269–299.

Zazoun, R. S. (2001). La tectogenèse hercynienne dans la partie occidentale du basin de l'Ahnet et la région de Bled El-Mass, Sahara Algérien: un continuum de déformation. Journal of African Earth Sciences32 (4), pp. 869–887.

Zegrir, H. (2014). Distribution de la fracturation dans les reservoirs compacts du Cambro-Ordovicien: Méthodologie d'intégration des données de surface au modèle de sub-surface (Bassin de l'Ahnet plate forme saharienne-Algérie). Memoire de Magister, Faculté des Hydrocarbures et de la Chimie. Université M'hamed Bougara, Boumerdès, Algérie.

Zieliński, M. (2011). Reconstruction of thermal history of the Ahnet and Mouydir basins (southern Algeria). PhD thesis, Institute of Geology, Faculty of Geographical and Geological Sciences, Adam Mickiewicz University, Poznan.

11

A New Method for Reservoir Fracture Characterization and Modeling Using Surface Analog

Sofiane Djezzar, Aldjia Boualam

University of North Dakota, Grand Forks, USA

11.1 Introduction

The Cambro-Ordovician formations in the Algerian Saharan platform are notable for their compact sandstone formations with minimal petrophysical properties, making natural fractures critical for their production (Arab and Djezzar, 2011; Djezzar, 2012). The Mouydir Basin is the least-explored basin in Algeria with very few low-quality 2D seismic surveys and a limited number of dry wells drilled in the 1950s. Despite this lack of data, the Mouydir Basin is surrounded by significant gas provinces such as the Ahnet and Timimoun basins to the west, and the prospective oil provinces of the Oued Mya and Illizi basins to the north and east, respectively, suggests a promising yet challenging exploration frontier. These adjacent basins share the same petroleum system with the Mouydir basin, where the Ordovician reservoirs, known for their considerable oil and gas production, depend significantly on natural fractures due to their inherent compaction (Boualam et al., 2019; Djezzar and Boualam, 2020a, 2020b; Boudjema, 1987).

To address the challenges posed by the lack of data and the critical role of natural fractures in reservoir productivity, a novel and comprehensive workflow is proposed for analyzing and characterizing natural fractures in the Cambro-Ordovician reservoir analogs at the Mouydir Basin's southern edge. This workflow amalgamates various datasets to construct a deterministic fracture model, elucidating both major and minor fractures to shed light on the basin's geological evolution, the influence of basement fractures on basin structuration, and the sedimentary cover's impact (Djezzar, 2019; Djezzar and Boualam, 2021).

The development of a 3D fracture model under this workflow aims to deepen understanding of fracture distribution, connectivity, and kinematics, providing valuable insights for predicting subsurface fracture extension and occurrence. This model could also offer explanations for the historically non-productive wells in the area. In the realm of 3D seismic interpretation, attributes such as coherence, dip, and edge are instrumental in identifying fracture planes, though they fall short in detecting smaller-scale fractures (Djezzar et al., 2019).

Applying various attributes to digital elevation models (DEMs) enables the detection of both significant and minor fractures affecting the Cambro-Ordovician analog in the Mouydir Basin.

Geophysical Exploration for Hydrocarbon Reservoirs, Geothermal Energy, and Carbon Storage: New Technologies and AI-based Approaches, First Edition. Edited by Said Gaci.

These fractures are meticulously digitized and analyzed based on orientation, length distribution, fractal dimension, and density, using a combination of geological maps, satellite imagery, and DEMs. This analysis encompasses a detailed study of major fractures impacting the zone of interest and a focused examination of fractures affecting the basement and Cambro-Ordovician units across different scales.

The resulting data, presented through diagrams, cross plots, and histograms, offers a comprehensive understanding of the fractures' fractography and typology within the reservoir analog. By building 3D deterministic fracture models for both the formation and the zone of interest, this innovative approach highlights the fractures' spatiotemporal distribution and the basement fractures' significant impact on the sedimentary cover, paving the way for informed and strategic future petroleum exploration efforts in the Mouydir Basin.

11.2 Methodology

This study introduces a novel methodology and a detailed workflow for the characterization and modeling of natural fractures, with a focus on the Cambro-Ordovician reservoir analog within the Mouydir basin in Algeria (Figure 11.1). The fracture analysis employs a dual-pronged approach. Initially, a comprehensive evaluation of the broader fracture networks impacting the study area is

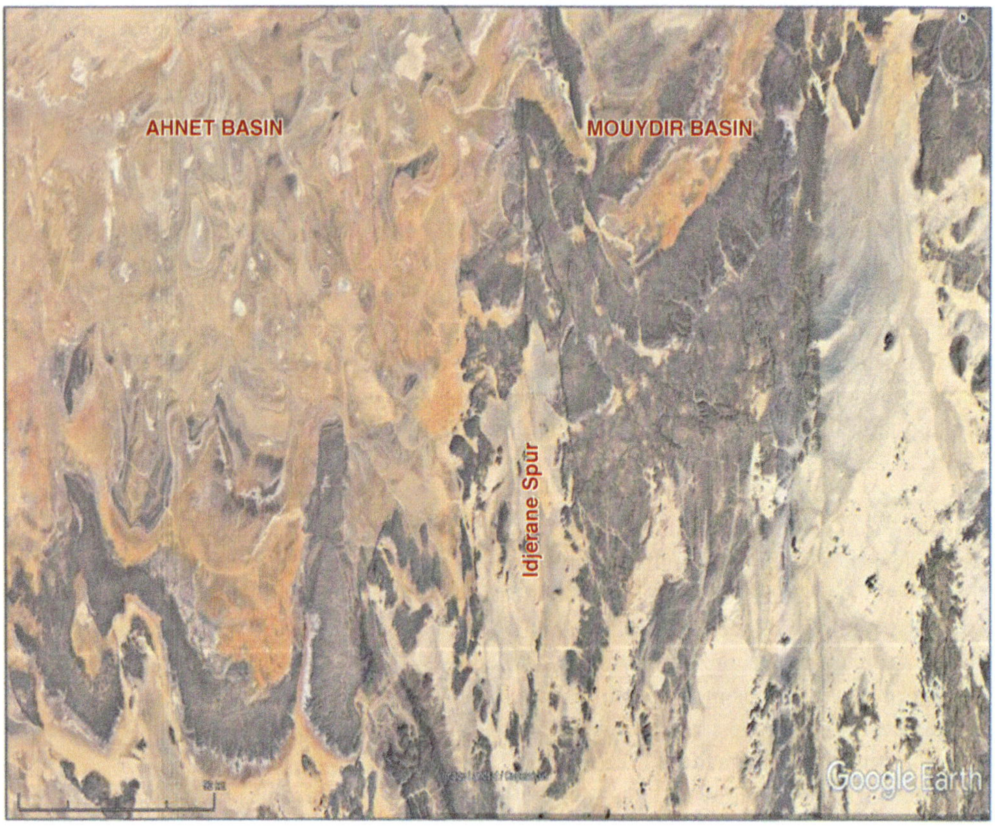

Figure 11.1 Satellite image from Google Earth.

conducted (Beicip, 2018). This involves generating a fracture map through the integration of various data sources, including curvature and illumination attributes from seismic data, geological maps, satellite imagery, and DEMs. The analysis aims to ascertain key fracture characteristics, such as the configuration of fracture sets, the distribution of fracture lengths, correlation coefficients, power law coefficients, and fractal dimensions.

The second approach zooms in on the fractures affecting the basement rocks and the diverse Cambro-Ordovician stratigraphic units. This finer-scale analysis utilizes the same array of input data to deduce analogous parameters and outcomes, ensuring a coherent and comprehensive fracture characterization across different geological contexts.

Subsequently, 3D deterministic fracture models are constructed for each geological unit under investigation. These models are meticulously developed to depict the spatial distribution of fractures and, through their layered superposition, facilitate the elucidation of fractures' origins, their interconnected relationships, and their kinematic behaviors. A critical aspect of this analysis is the illustration of how fractures within the basement influence the overlying sedimentary cover, thereby providing insights into the structural evolution of the basin and guiding future exploration efforts.

This innovative approach stands out by offering a multidimensional perspective on fracture analysis, combining broad-scale mapping with detailed, unit-specific modeling. The integration of diverse data sources and analytical techniques underpins the development of robust, deterministic fracture models. These models not only enhance our understanding of the fracture networks within the Mouydir Basin but also contribute significantly to the strategic planning and optimization of future petroleum exploration endeavors in this and similar geological settings.

11.3 Geological Background

The geological setting of the study area is situated in the northern portion of the Hoggar Shield in Algeria, delineated by latitudes 24°30′N to 28°00′N and longitudes 3°00′E to 6°00′E. This region is bordered by the Mouydir Basin to the north, the Idjerane Spur and Ahnet Basin to the west, the Amguid Spur and Illizi Basin to the east, and the uplift of the Hoggar Shield to the south (Bensalah et al., 1972a, 1972b; Black et al., 1994). The study area encompasses a vast terrain where Paleozoic formations are prominently exposed in regions such as Ain-Tadjoubar, Arak, Iftessene, and Khanget-El-Hadid, segmented by significant sub-meridian fractures that played a crucial role in the structural compartmentalization of the Mouydir Basin across various orogenic events (Figure 11.2).

The Cambro-Ordovician sequence, a subject of considerable interest within Algerian geological studies, is recognized for hosting the second most significant tight oil and gas reservoirs, primarily driven by natural fracturing processes. This stratigraphic interval comprises three principal formations: Ajjers, In-Tahouite, and Tamadjert, which are consistently present below the surface across the Saharan Platform basins and are visible in outcrops in the northern reaches of the Hoggar Shield. The temporal range of the Ordovician formations extends from the Tremadocian to the Ashgillian periods, encompassing various lithological units such as the El Atchane sandstone of Tremadocian age, Hamra Quartzite of Arenigian age, Ouargla sandstone of Llanvirnian age, Oued Saret sandstone of Lower Caradocian age, and the Ramade or M'kratta sandstone spanning the Upper Caradocian to Ashgillian periods (WEC, 2007).

Depositional environments of the Cambro-Ordovician formations were predominantly fluvial and fluvio-glacial, positioning these strata as secondary objectives in the quest for hydrocarbons within Algerian basins. Notably, the M'Kratta sandstone reservoir in the Oued Mya Basin, along with the Hamra Quartzite reservoir, have been identified as prolific sources of oil and gas

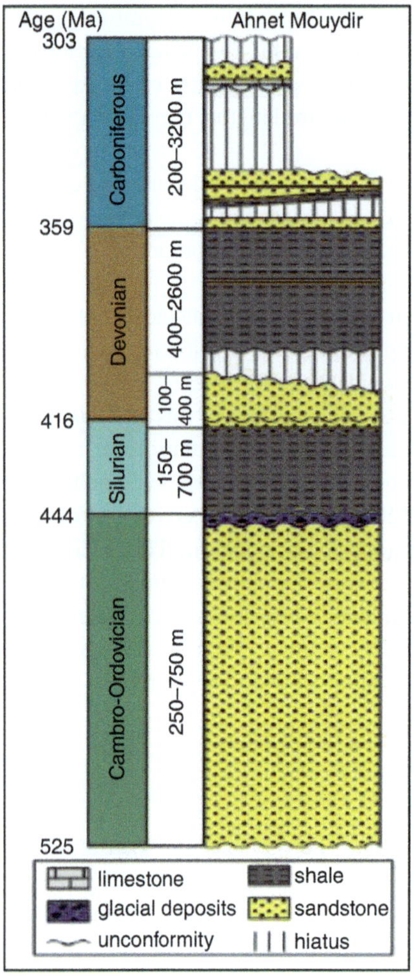

Figure 11.2 Schematic lithostratigraphic column of the Ahnet and Mouydir basins (compiled after Follot, 1952; Beuf et al., 1971; Wendt et al., 2006; Vecoli et al., 2008 in Zieliński, 2011; Zegrir, 2014).

accumulations, extending their significance to the Ahnet Basin and the Hassi Messaoud field. Despite the compact nature of the Ordovician reservoirs, characterized by an average porosity of 7% and permeabilities below 0.1 millidarcies, their hydrocarbon productivity is significantly enhanced by the presence of open fractures, underscoring the critical role of fracture systems in the hydrocarbon extraction process from these tight reservoirs (WEC, 2007).

11.4 Material and Methods

Fractures are categorized into two primary modes: Mode I and Mode II. Mode I fracture, also known as tensile or opening mode fractures, feature displacement that is perpendicular to the discontinuity walls, such as joints (Bonnet et al., 2001; Bour et al., 1997, 1998; Cacas et al., 2001; Davy, 1993; Davy et al., 1990; Davy and Sornette, 1992). Conversely, Mode II fractures operate under an in-plane shear mode, with displacement occurring along the plane of the discontinuity, typically observed in faults. These fractures can span a vast range of sizes, from mere microns to extensive lengths of hundreds of kilometers, significantly influencing fluid flow and various other rock properties (Roberts, 2001; Lisle, 1994; Gao, 2013). In the context

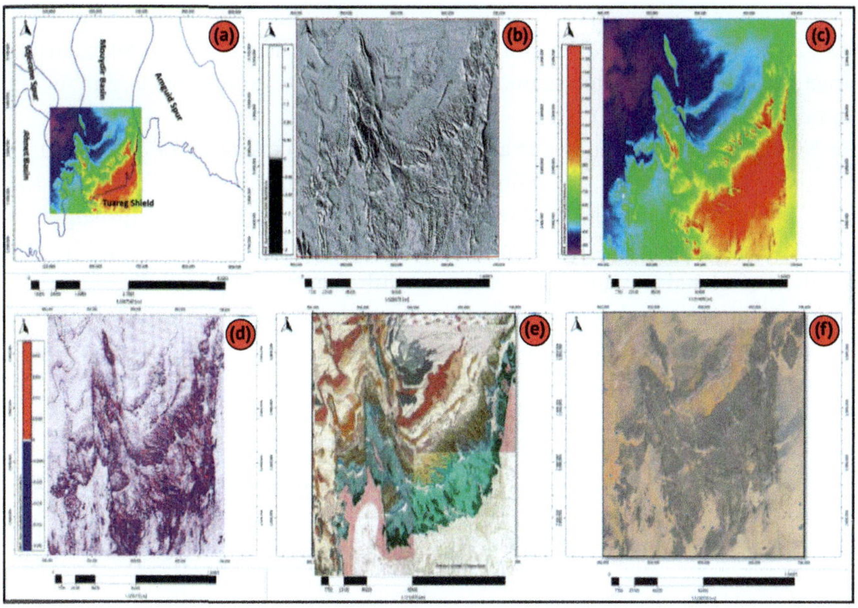

Figure 11.3 Workflow application on the major faults. (a) Location of the zone of interest (ZOI) with the basin's location; (b) illumination of the ZOI. *Source*: Sofiane Djezzar (2019) / https://commons.und.edu/theses/2551; (c) digital elevation model of the ZOI; (d) maximum curvature map of the ZOI; (e) geological maps of the ZOI, (f) satellite image of the ZOI. *Source*: Sofiane Djezzar (2019).

of the Cambro-Ordovician formations, quantitative analyses of fractures are meticulously performed, focusing both on major fractures impacting the area of interest and on minor fractures affecting the basement and Cambro-Ordovician strata. To achieve this, fracture maps are constructed by leveraging the curvature and illumination attributes, facilitating the identification of distinct fracture sets, measurement of fracture lengths, and calculation of fractal dimensions across the entire network and within specific fracture sets (Beicip, 2018). This approach enables a nuanced differentiation between major and minor fracture sets, providing a deeper insight into the fractography characteristic of each geological formation. Such analyses are essential for understanding the structural complexity of the Cambro-Ordovician reservoirs, offering valuable perspectives on their potential for hydrocarbon recovery and guiding exploration and development strategies in these tight reservoir systems (Figure 11.3).

11.5 Data Analysis

The process of identifying major fractures within the study area and the wider Mouydir Basin has been enhanced through the creation of a comprehensive lineaments map at a scale of 1:100,000. This map vividly delineates the primary fracture sets impacting the region, with close to 400 fractures meticulously digitized and analyzed. The fracture network in this locale exhibits a diverse array of orientations, systematically categorized into seven main sets: N000, N040, N060, N090, N110, N130, and N150.

The distribution of fracture lengths is effectively visualized using a bar chart, which plots the length of fractures in meters against their frequency. This visualization method elucidates the most, least, and average lengths observed in the study field. A notable characteristic of the length distribution is its adherence to the power law, with the N000, N090, and N130 sets emerging as the predominant fracture sets. The lengths within these sets vary significantly, ranging from 4 km to 166 km for N000, 2 km to 37 km for N090, and 1.6 km to 47 km for N130. Given the extensive lengths of these fractures, it is plausible to assume their intersection at certain depths, enhancing their geological significance (Figure 11.4).

Particularly, the N000 fracture set stands out as the most crucial, consisting of 100 fractures that penetrate deep into the basement, traversing through the Paleozoic strata, including the Cambro-Ordovician, Silurian, and Devonian formations. Analysis of geological maps supports the interpretation of these fractures as strike-slip fractures, highlighting their structural importance. The examination of global fracture lengths, along with the specific lengths within each fracture set, was underpinned by statistical analyses that considered various parameters. These parameters include the most, least, and average frequent lengths, alongside the correlation coefficient and power law exponent. The findings reveal that the correlation coefficient spans from 0.93 to 0.97, indicating a strong linear relationship within the data. Furthermore, the power law coefficient fluctuates between 1.97 and 2.42, underscoring the fractal nature of fracture distribution within the study area. This detailed analysis of major fractures not only sheds light on the

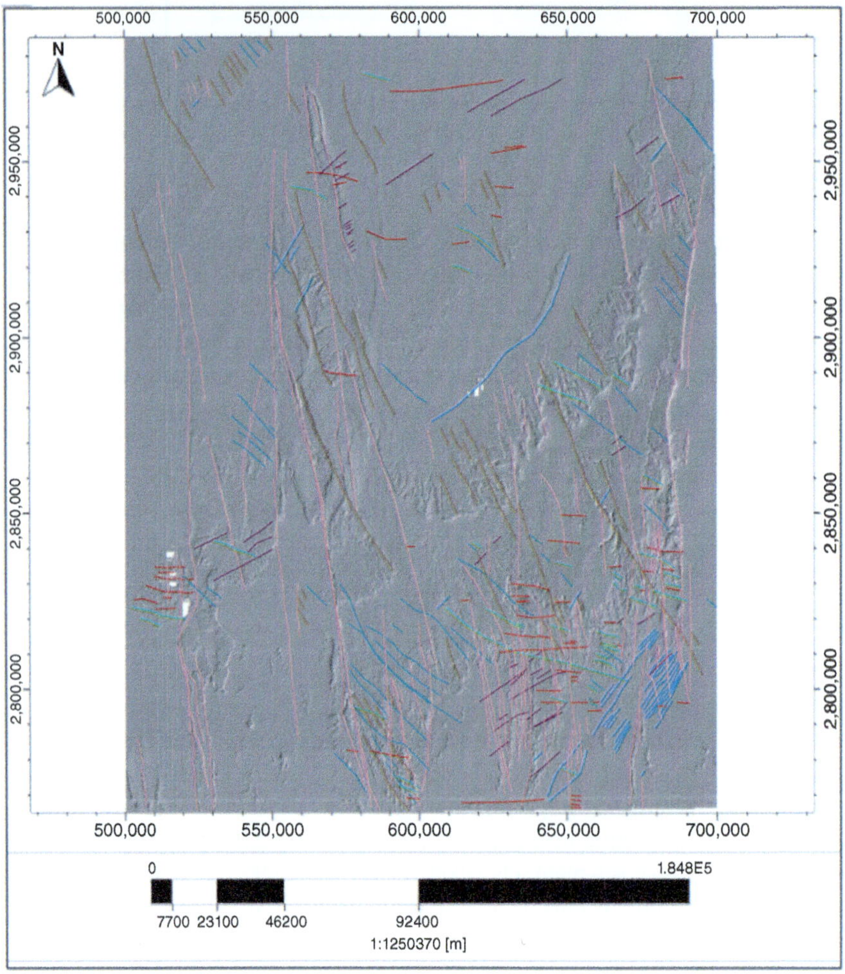

Figure 11.4 Major faults characteristics. *Source:* Sofiane Djezzar (2019).

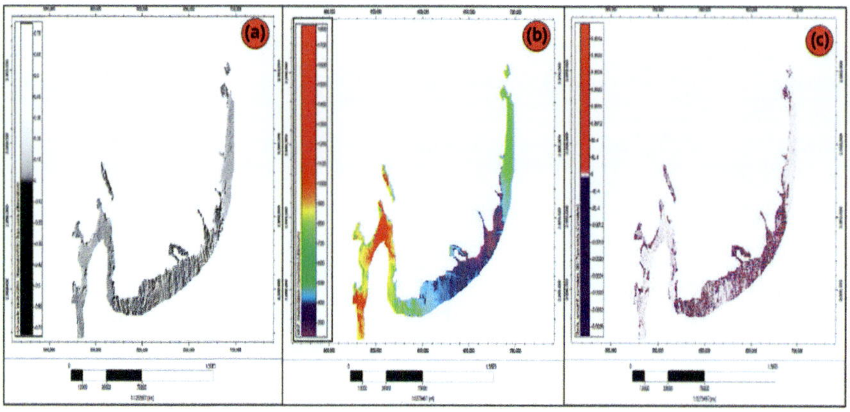

Figure 11.5 Basement minor faults. (a) Illumination map of the basement formation; (b) digital elevation model of the basement formation; (c) curvature map of the basement formation.

structural complexity of the region but also enhances the understanding of geological processes influencing the Mouydir Basin and its surrounding areas.

The detection and analysis of minor fractures within the study area have been carried out through detailed digitalization based on curvature and illumination maps (Figure 11.5). This meticulous process facilitated the determination of fracture orientation and density across various stratigraphic units, as inferred from the analysis of rose diagrams and other statistical metrics. Length distribution analyses were conducted to pinpoint both the longest and shortest frequently occurring fractures. Further examination of each fracture set included assessments based on the correlation coefficient, power law exponent, and the range of lengths (maximum, minimum, and mean). Focusing on the basement formation, which is a part of the Hoggar Shield, this foundational geological layer is predominantly composed of metamorphic and igneous rock types. The analysis identified 640 fractures impacting the basement, organized into five principal orientation sets: N000, N050, N090, N120, and N160. Notably, the N000 and N160 sets emerge as the most significant, containing 256 and 145 fracture planes, respectively. The fractures within these sets display a considerable range in length, from as short as 500 m to as long as 13 km for the N000 set, and from 1 km to 10 km for the N160 set (Figure 11.6). A log-log plot of the fracture lengths within the basement supports the conclusion that their distribution follows a power law, with correlation coefficients ranging from 0.89 to 0.97. This indicates a highly structured, scale-invariant pattern of fracture distribution. The power law coefficients observed range from 2.36 to 2.89, reinforcing the notion that the fracture network within the basement formation exhibits a complex, fractal nature. This detailed analysis of minor fractures, alongside the earlier discussion on major fractures, provides a comprehensive view of the fracture dynamics within the study area, offering crucial insights into the geological and structural intricacies of the Hoggar Shield and its associated formations.

The Ajjers Formation is primarily composed of the Hamra Quartzite, Vire-du-Mouflon sandstone, and Tin-Taradjelli sandstone, presenting a complex geological structure (Figure 11.7). This formation is characterized by the presence of 492 fractures, organized into six orientation sets: N000, N050, N090, N110, N130, and N150. Among these, the N000 and N090 sets emerge as the most prominent, comprising 120 and 123 fracture planes, respectively. The length distribution of fractures within the Ajjers Formation indicates significant variability, with the N-S (N000) and E-W (N090) orientations displaying lengths between 750 m to 10 km and 800 m to 5.5 km, respectively, following a power law trend in their distribution (Figure 11.8).

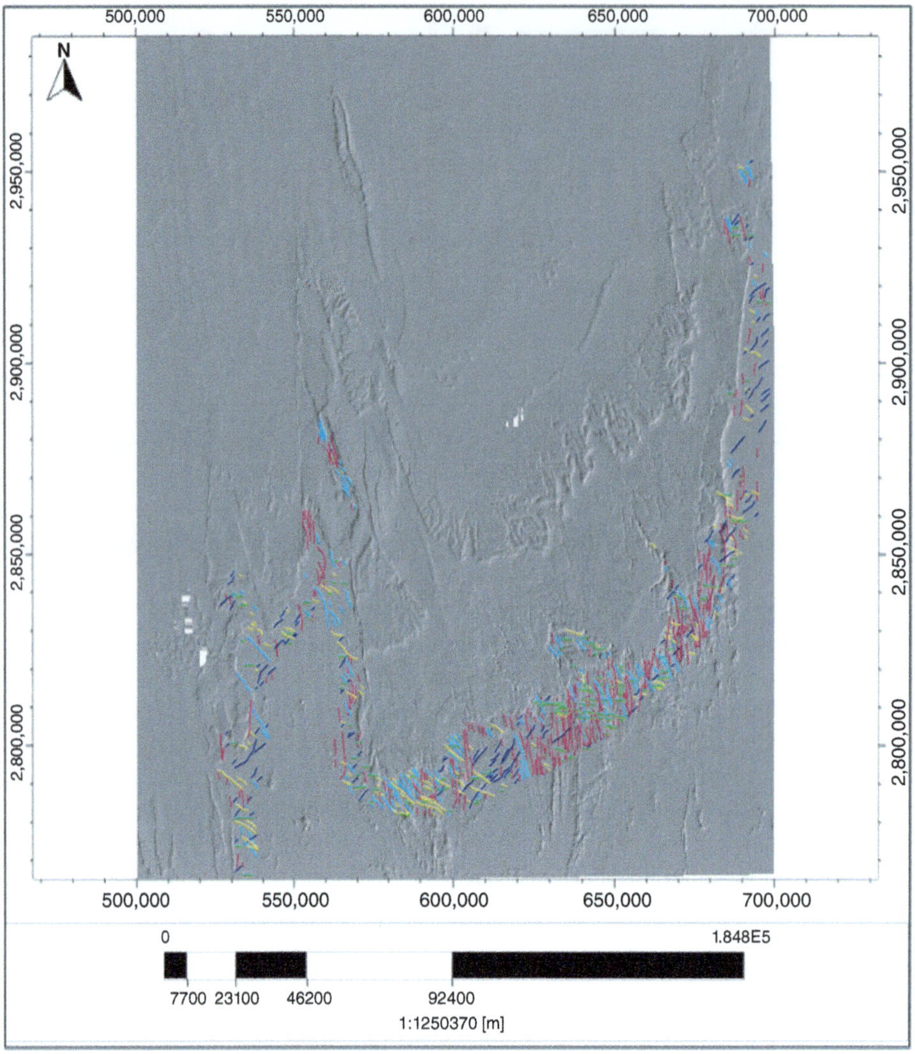

Figure 11.6 Basement fault characteristics. *Source*: Sofiane Djezzar (2019).

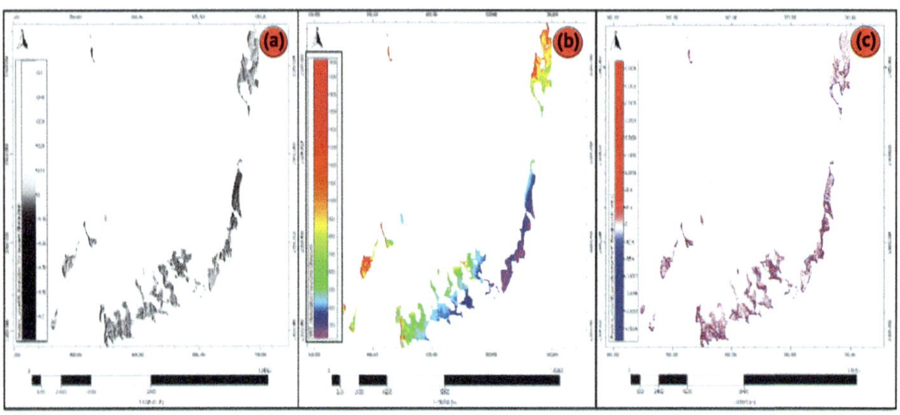

Figure 11.7 Ajjers minor faults. (a) Illumination map of the Ajjers formation; (b) digital elevation model of the Ajjers formation; (c) curvature map of the Ajjers formation.

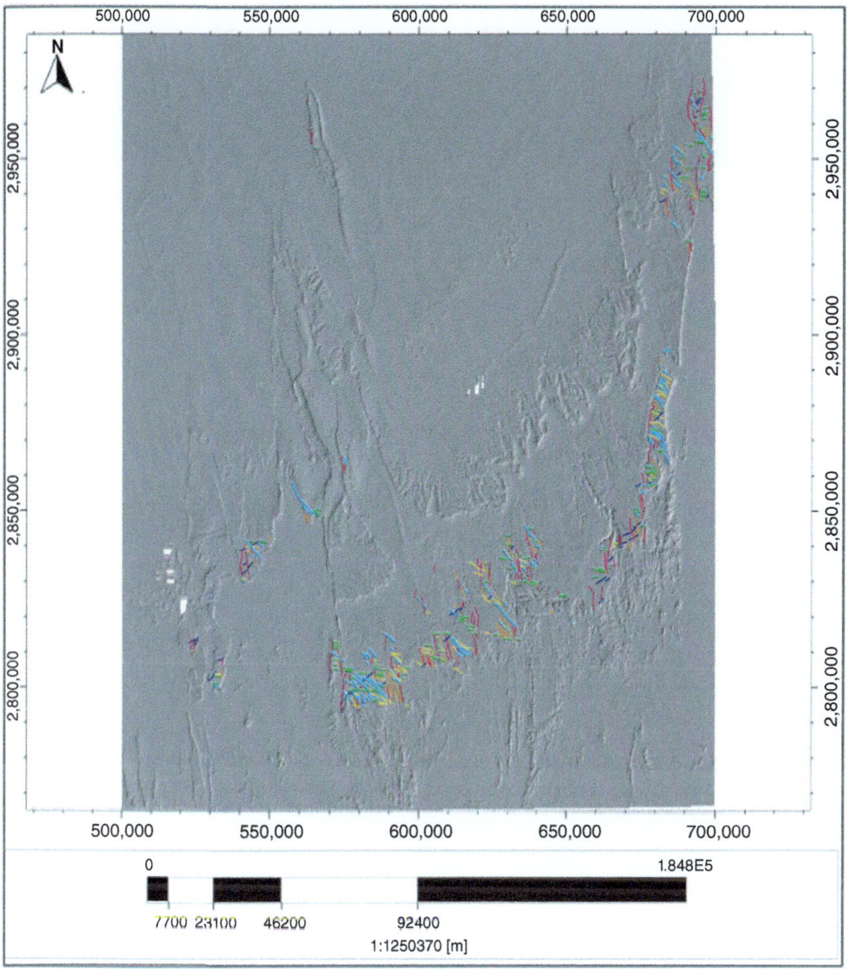

Figure 11.8 Ajjers faults characteristics. *Source*: Sofiane Djezzar (2019).

The In-Tahouite Formation, distinguished by 172 fractures, exhibits a similar fracture set organization to the Ajjers Formation, with orientations classified into six sets: N000, N040, N060, N090, N120, and N130 (Figure 11.9). The dominant fracture sets within this formation are N000, N090, and N130, featuring 40, 49, and 51 fracture planes, respectively. The length analysis for the In-Tahouite Formation reveals that the largest fractures are found in the N-S (N000), E-W (N090), and N130 orientations, ranging from 350 m to 16 km, 650 m to 10 km, and 660 m to 13 km, respectively (Figure 11.10). The distribution of fracture lengths in this formation also adheres to a power law pattern, with correlation coefficients and power law coefficients ranging from 0.89 to 0.95 and 2.19 to 3.12, respectively.

The Tamadjert Formation, recognized for its heterogeneous reservoir qualities, includes the M'Kratta sandstone, El-Goléa sandstone, and micro-conglomerate shale. Affected by 843 fractures, this formation's fracture network is divided into five sets: N000, N040, N070, N090, and N130 (Figure 11.11). The dominant fracture sets are N000, N090, and N130, with 212, 178, and 171 fractures, respectively. The N070 and N090 sets exhibit the longest fractures, ranging from 500 m to 9 km and 530 m to 8.7 km, respectively (Figure 11.12). The fracture length distribution in the

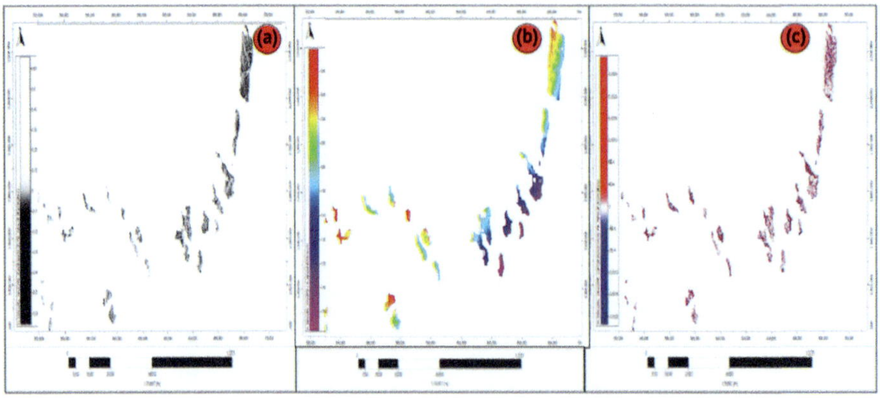

Figure 11.9 In-Tahouite minor faults. (a) Illumination map of the In-Tahouite formation; (b) digital elevation model of the In-Tahouite formation; (c) curvature map of the In-Tahouite formation. *Source*: Sofiane Djezzar (2019).

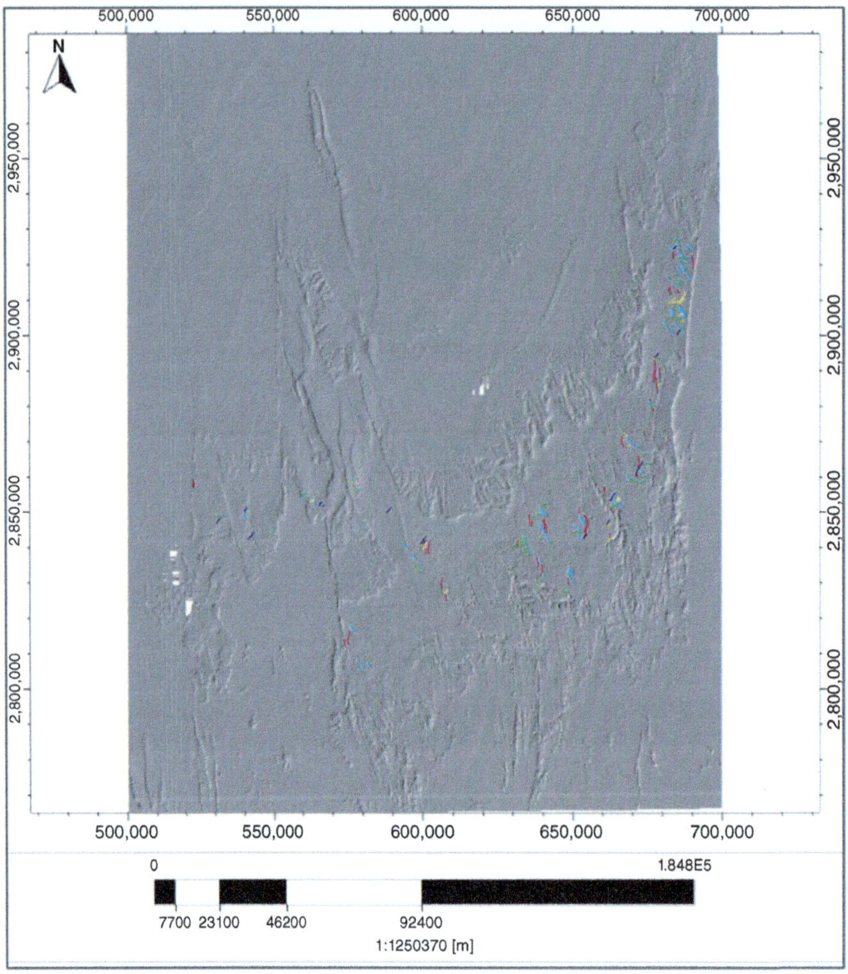

Figure 11.10 In-Tahouite faults characteristics. *Source*: Sofiane Djezzar (2019).

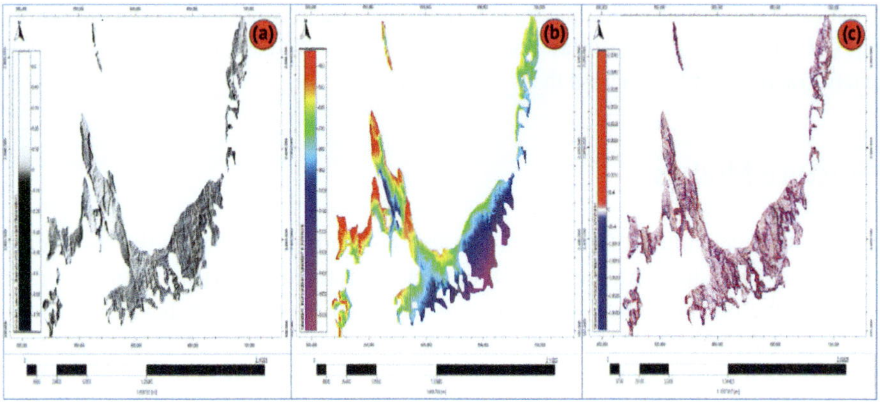

Figure 11.11 Tamadjert minor faults. (a) Illumination map of the Tamadjert formation; (b) digital elevation model of the Tamadjert formation; (c) curvature map of the Tamadjert formation.

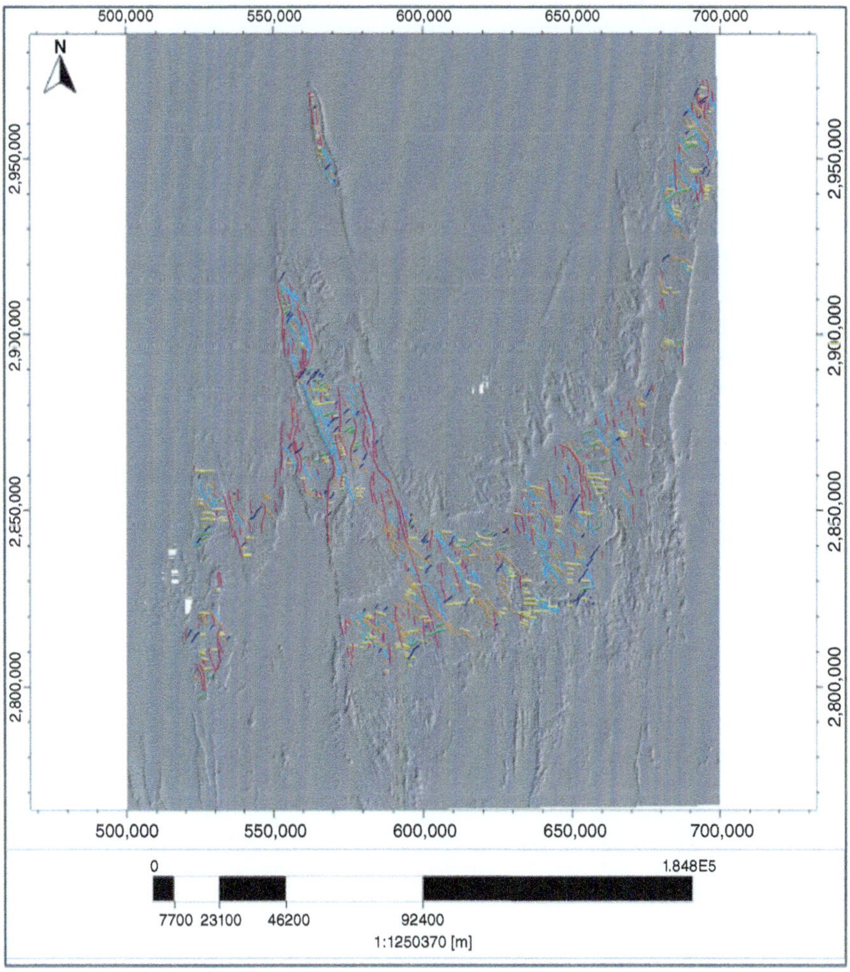

Figure 11.12 Tamadjert faults characteristics. *Source*: Sofiane Djezzar (2019).

Tamadjert Formation is consistent with a power law function, underscored by a high correlation coefficient range of 0.91–0.97 and a power law coefficient range of 2.37–2.81, indicating a complex and scale-invariant fracture network across these formations.

11.6 3D Fracture Models

The development of 3D deterministic fracture models for this study encompasses the basement, Ajjers, In-Tahouite, and Tamadjert stratigraphic units, each reflecting the intricate nature of fracture networks within the northern part of the Hoggar Shield, Algeria. By amalgamating these individual models into a comprehensive 3D framework, the resulting unified model serves as a powerful tool for visualizing the diverse fracture sets, their spatial distribution, and density (Figures 11.13 and 11.14). This holistic view facilitates a deeper understanding of the fractures'

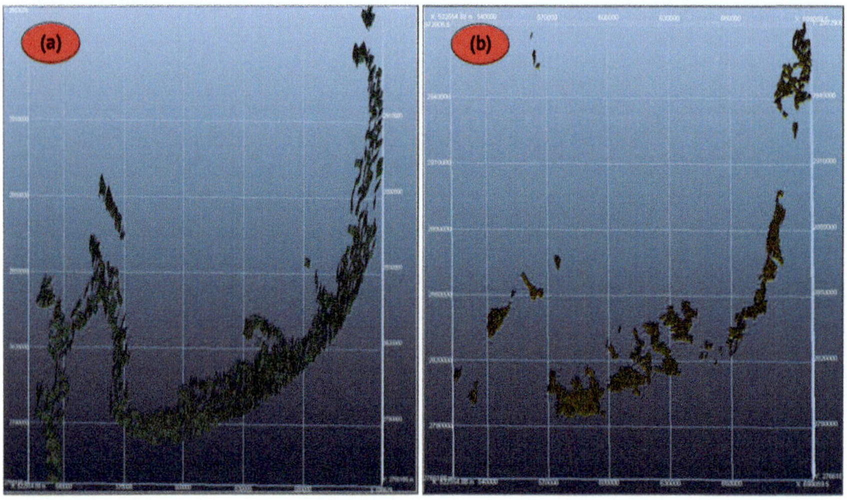

Figure 11.13 (a) The 3D basement faults model; (b) the 3D Ajjers' faults model.

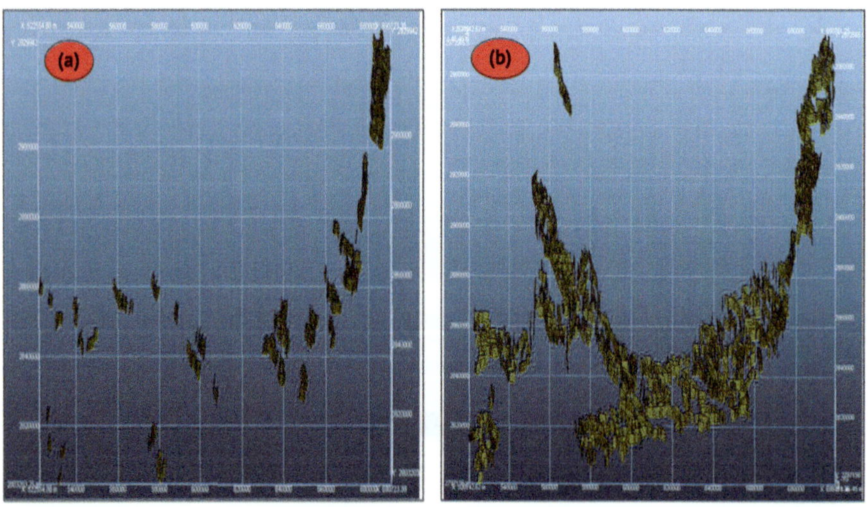

Figure 11.14 (a) 3D In-Tahouite faults model; (b) 3D Tamadjert faults model.

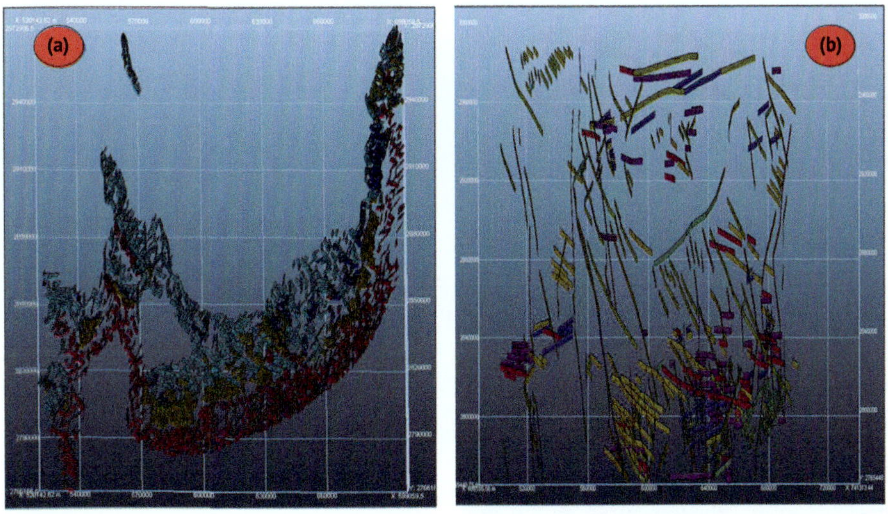

Figure 11.15 (a) The 3D global faults model; (b) the 3D Major faults model.

relationships, chronology, origins, and their continuity or discontinuity beneath the surface. The comparative analysis of these models reveals that the Cambro-Ordovician formations exhibit varied mechanical stratigraphy, influenced significantly by their thickness and lithological composition. Notably, the Tamadjert Formation is identified as the most fractured unit, a characteristic attributed to its specific lithological properties (Figure 11.15). Conversely, the In-Tahouite Formation is marked as the least fractured, likely due to its shaly composition, which may inhibit fracture development. The fractures within the basement exert a pronounced impact on the overlying Cambro-Ordovician sedimentary cover, suggesting a fundamental role in structuring these units. The comprehensive 3D deterministic fracture model not only delineates fracture distribution but also elucidates the mechanical distinctions between the Cambro-Ordovician units and their interrelations. Major structural features identified include N-S-trending dextral strike-slip fractures and fractures extending from NNW-SSE to N-S. Two principal fracture corridors emerge as defining features of the Mouydir Basin's edges, originating from the Hoggar Shield and diverging northward across the Saharan platform, aligned from NNE-SSW to NNW-SSE, corresponding to the Amguid and Idjerane spurs, respectively. The significant N-S and N160 fractures impacting both the basement and the Paleozoic series are highlighted as remnants of the Pan-African orogeny, subsequently reactivated during various tectonic events, including the Hercynian and Alpine phases (Figure 11.16). This intricate network of fractures, depicted through the 3D deterministic fracture model, offers invaluable insights into the geological evolution of the area and its structural complexity, providing a solid foundation for understanding regional geodynamics and informing exploration strategies.

11.7 Discussion and Conclusions

The fracture analysis of the Cambro-Ordovician reservoir analog at the southern edge of the Mouydir Basin has yielded insights crucial for understanding the subsurface characteristics and behavior of this reservoir, thereby informing the placement of prospective wells in the area. The

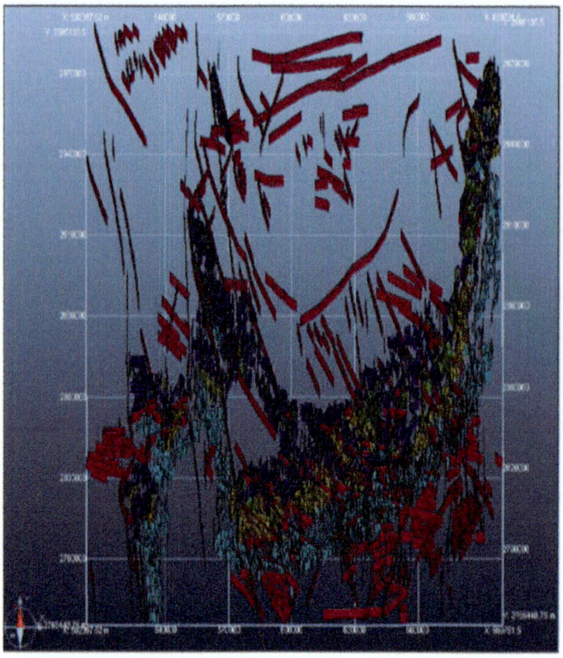

Figure 11.16 3D total faults model.

findings indicate that the principal fracture sets align along N000, N050, N090, N120, and N160, with major structures trending N-S and NNW-SSE, identified as strike-slip fractures. This variation in fracture orientation and distribution underscores the significant influence of basement fractures on the structuring of the Cambro-Ordovician sedimentary layers, suggesting a non-stratabound fracture system across both the basement and the Cambro-Ordovician units.

Among the formations analyzed, the Tamadjert Formation emerged as the most fractured, primarily due to its composition of tight sandstone units, whereas the In-Tahouite Formation displayed the least fracturing, attributed to its shaly constitution. The fractal distribution analysis of the fracture networks further reveals that both the basement and Cambro-Ordovician units exhibit fractal dimensions, with values ranging between 1 and 2 as determined by the center distance and box-counting algorithms. This fractal nature indicates a complex, self-similar pattern of fractures across different scales.

Comparative analysis with the Cambro-Ordovician reservoirs in the subsurface of the Ahnet Basin shows similar fracture orientations and distributions, suggesting regional consistency in fracture patterns across the Saharan platform. Moreover, borehole imagery analysis from three wells in the Hamra Quartzite reservoir of the Ahnet Basin identified three main fracture sets oriented N030, N140, and N170, with these fractures being conductive to semi-conductive. The maximum horizontal stress (SHmax) in this region, inferred from wellbore breakout widths, aligns NW-SE, consistent with the broader stress patterns observed across the Saharan platform.

The implication of fractures aligned with the maximum horizontal stress being generally open, and thus enhancing petrophysical properties, versus those perpendicular to SHmax being resistive, indicates strategic drilling directions. Specifically, a horizontal pilot well targeted at the Cambro-Ordovician reservoir and oriented perpendicular to the SHmax in the Mouydir Basin could represent a strategic approach to unlocking the potential of this yet-to-be-fully explored basin. Such an approach may signal the beginning of a new exploration era, leveraging the detailed fracture analysis to optimize reservoir exploitation strategies.

References

Arab, M. & Djezzar, S. (2011). Hydrocarbon potential in Reggane Basin, Oil Mediterranean Conference (OMC), Italy. OMC-2011-046.

Beicip. (2018). FracaFlow tutorial.

Bensalah, A., Beuf, S., Gabriel, O., et al. (1972a). Carte Géologique de l'Algérie: Khanget El Hadid et Arak, feuille NG-31-XVII, NG-31-IV et NG-31-XVI. Ministère de l'Industrie et de l'Energie.

Bensalah, A., Beuf, S., Gabriel, O., et al. (1972b). Carte Géologique de l'Algérie: Ifetessene, feuille NG-31-XI. Ministère de l'Industrie et de l'Energie.

Beuf, S., Biju-Duval, B., De Charpal, D., et al. (1971). Les grès du Paléozoïque inférieur au Sahara. Sédimentation et discontinuité: évolution structurale d'un craton. In: Technip-Institut Français du pétrole, vol. 18. Collection Sciences et Techniques du Pétrole, Paris, p. 464.

Black, R., Latouche, L., Liégeois, J. P., et al. (1994). Pan-African displaced terranes in the Hoggar shield (central Sahara). Geology 22, pp. 641–644.

Bonnet, E., Bour, O., Odling, N. E., et al. (2001). Scaling of fracture systems in geological media. Review of Geophysics 39 (3), pp. 347–383.

Boualam, A., Djezzar, S., Rasouli, V. et al. (2019). 3D modeling and natural fractures characterization in Hassi Guettar field, Algeria. 53rd American Rock Mechanics Association Symposium, NY, USA. 19-A-373-ARMA.

Boudjema, A. (1987). Evolution structurale du bassin petrolier 'triasique' du Sahara Nord Oriental (Algerie). These Doctorat Etat, Paris XI-Orsay, France, p. 290.

Bour, O. & Davy, P. (1997). On the connectivity of random fracture networks following a power law length distribution. Water Resources Research 33, pp. 1567–1583.

Bour, O. & Davy, P. (1998). On the connectivity of three-dimensional fracture networks. Water Resources Research 34, pp. 2611–2622.

Cacas, M. C., Daniel, J. M., & Letouzy, J. (2001). Nested geological modeling of naturally fractured reservoirs. Petroleum Geoscience 7, pp. 43–52.

Davy, P. (1993). On the fracture-length frequency distribution of the San Andreas Fracture system. Journal of Geophysical Research 98 (12), pp. 141–151.

Davy, P. & Sornette, D. (1992). Experimental discovery of scaling laws relating fractal dimensions and the length distribution exponent of fracture systems. Geophysical Research Letters 19, pp. 361–363.

Davy, P. A., Sornette, D., & Sornette, D. (1990). Some consequences of a proposed fractal nature of continental fractureing. Nature 348, pp. 56–58.

Djezzar, S. (2012). Impact de la fracturation naturelle sur la productivité des réservoirs du Dévonien inférieur dans le bassin de Reggane, Sahara Occidental, Algérie. Thèse Magister. Ed. USTHB. -Alger: USTHB, 2012.-93 p.: ill.; 30 cm. + CD-Rom.

Djezzar, S. (2019). Fracture detection and prediction in unconventional reservoirs for finding sweet spot. Theses and Dissertations. 2551. https://commons.und.edu/theses/2551/.

Djezzar, S. & Boualam, A. (2020a). Analysis and modeling of tight oil fractured reservoir. SEG 2020 Annual Meeting, TX, USA. https://doi.org/10.1190/segam2020-3425011.1.

Djezzar, S. & Boualam. A. (2020b). Fractures characterization and their impact on the productivity of Hamra Quartzite Reservoir (Haoud-Berkaoui field, Oued-Mya basin, Algeria). 54th American Rock Mechanics Association Symposium, Co, USA. 20-A-1438-ARMA.

Djezzar, S. & Boualam, A. (2021). Impact of natural fractures on the productivity of lower Devonian reservoirs in Reggane Basin, Algeria. 55th US Rock Mechanics/Geomechanics Symposium. 21-A-1295-ARMA.

Djezzar, S., Rasouli, V., Boualam, A., et al. (2019). Size scaling and spatial clustering of natural fracture networks using fractal analysis. 53rd American Rock Mechanics Association Symposium, NY, USA. 19-A-375-ARMA.

Djezzar, S., Rasouli, V., Boualam, A., et al. (2020). An integrated workflow for multiscale fracture analysis in reservoir analog. Arabian Journal of Geosciences 13, p. 161. https://doi.org/10.1007/s12517-020-5085-6.

Follot, J. (1952). Ahnet et Mouydir. 19' Congres international de Geologie, Alger. Monographie Regionale, Serie 1, 1, pp. 1–6.

Gao, D. (2013). Integrating 3D seismic curvature and curvature gradient attributes for fracture detection: Methodologies and interpretational implications. Geophysics 78 (2), pp. O21–O31.

Lisle, R. J. (1994). Detection of zones of abnormal strains in structures using Gaussian curvature analysis. AAPG Bulletin 78 (12), pp. 1811–1819.

Roberts, A. (2001). Curvature attributes and their application to 3-D interpreted horizons. First Break 19 (2), pp. 85–100.

Vecoli, M., Videt, B., & Paris, F. (2008). First biostratigraphic (palynological) dating of Middle and Late Cambrian strata in the subsurface of northwestern Algeria, North Africa: Implications for regional stratigraphy. Review of Palaeobotany and Palynology, 149 (1–2), 57–62.

Well Evaluation Conference (WEC). (2007). Well Evaluation Conference, Internal report, "Joint documents of Sonatrach and Schlumberger," Algeria, p. 489.

Wendt, J., Kaufmann, B., & Belka, Z., et al. (2006). Sedimentary evolution of a Palaeozoic basin and ridge system: the Middle and Upper Devonian of the Ahnet and Mouydir (Algerian Sahara). Geological Magazine 143 (3), pp. 269–299.

Zegrir, H. (2014). Distribution de la fracturation dans les reservoirs compacts du Cambro-Ordovicien: Méthodologie d'intégration des données de surface au modèle de sub-surface (Bassin de l'Ahnet-plate forme saharienne-Algérie). Memoire de Magister, Faculté des Hydrocarbures et de la Chimie. Université M'hamed Bougara, Boumerdès, Algérie.

Zieliński, M. (2011). Reconstruction of thermal history of the Ahnet and Mouydir basins (southern Algeria). PhD thesis, Institute of Geology, Faculty of Geographical and Geological Sciences, Adam Mickiewicz University, Poznan.

12

An Integrated Workflow for Multiscale Fracture Analysis in Reservoir Analog

Sofiane Djezzar, Aldjia Boualam

University of North Dakota, Grand Forks, USA

12.1 Introduction

Studying naturally fractured reservoirs presents significant challenges, primarily in modeling and predicting fracture networks that span various scales, from outcrops and shallow depths to the depths of reservoirs and basins (Cacas et al., 2001 Rives et al., 1992). The need for advanced 3D seismic and well data to characterize and model these fractures introduces a time-consuming process that demands extensive data collection and analysis (Djezzar, 2019). An alternative approach might involve studying reservoir analogs, which can offer more readily accessible data. This method facilitates the characterization and modeling of fractures across different scales through several available workflows, enabling the understanding of fracture distribution, connectivity, and kinematics. In this context, an integrated workflow is introduced, aimed at characterizing and modeling natural fractures within the Cambro-Ordovician reservoir analogs located at the southern edge of the Mouydir basin on the Saharan platform in Algeria (Djezzar et al., 2020). This workflow leverages both temporal and spatial fracture distributions observed at outcrops to anticipate their extension into the subsurface. The analytical process unfolds in two primary phases: initially, a comprehensive fault map for the study area is compiled, integrating curvature and illumination attributes with geological maps, satellite imagery, and digital elevation models. This step yields insights into fault sets, length distributions, correlation coefficients, power law coefficients, and fractal dimensions. Subsequently, employing the same datasets and analytical parameters, the study evaluates the impact of the basement fractures on the Cambro-Ordovician units. The methodologies proposed herein facilitate the identification of fault sets across different scales, enhancing our understanding of the fractography and typology of fractures impacting the Cambro-Ordovician units. The development of 3D deterministic fault models for each formation provides a visual representation of fracture distribution over time and space. Through the superposition of these models, it can help to elucidate the origins of fractures, their interrelationships, and kinematics, and determine the influence of basement faults on the sedimentary cover and overall basin architecture (Djezzar and Boualam, 2020a, Boualam et al., 2019). The current workflow integrates satellite imagery, geological maps, and digital elevation models (DEMs) as

Geophysical Exploration for Hydrocarbon Reservoirs, Geothermal Energy, and Carbon Storage: New Technologies and AI-based Approaches, First Edition. Edited by Said Gaci.

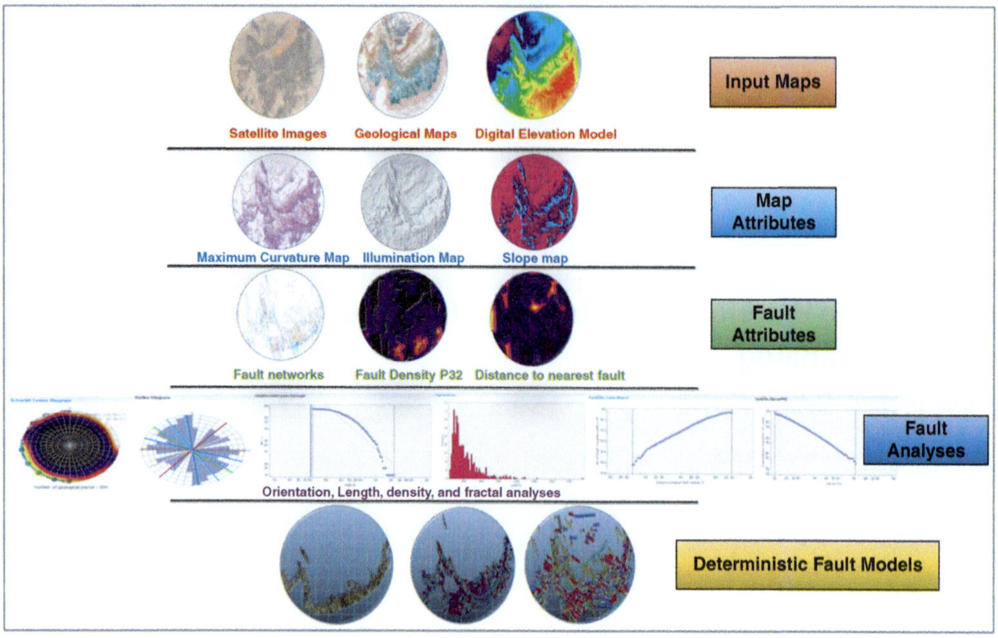

Figure 12.1 Workflow for reservoir fracture characterization and modeling. *Source*: Sofiane Djezzar (2019).

main data to accurately identify and map fault planes. This process utilizes a suite of seismic attributes—curvatures, slopes, and illumination—applied to DEMs, enabling the detection of both major faults and minor fractures (Figure 12.1). These detected features are then meticulously digitized and analyzed through their orientation, length distribution, fractal dimension, and density with the assistance of FracaFlow software, facilitating a comprehensive fracture analysis (Beicip, 2018, Djezzar et al., 2019, Djezzar and Boualam, 2020b). This analysis is bifurcated into two primary facets: the investigation of significant faults influencing the target zone and the detailed examination of fractures impacting the basement and the diverse Cambro-Ordovician stratigraphic units. This dual approach employs consistent datasets across varying scales to ascertain similar parameters and results, culminating in the generation of various graphical representations such as diagrams, cross plots, and histograms. These graphical outputs are instrumental in decoding the fractography and typology of fractures, thus enhancing the understanding of the reservoir analog. Subsequently, 3D deterministic fault models are constructed at different scales, showcasing the spatial and temporal distribution of fractures, their origins, interrelationships, kinematics, and the influence of basement faults on the sedimentary overlay.

12.2 Geological Background

The Cambro-Ordovician strata within the Algerian Saharan platform, characterized by tight sandstone formations with low petrophysical characteristics, underscore the pivotal role of natural fractures in reservoir production (Djezzar, 2012; Donzeau, 1983; Donzeau et al., 1981; Djezzar and Boualam, 2021). The Mouydir basin, noted for its underexplored status due to the absence of 3D seismic data and limited existing wells and low-quality 2D seismic surveys, necessitates a focused study on natural fractures to guide future exploration and drilling activities. The lithology and structural heritage of the

Cambro-Ordovician layers in the northern part of the Hoggar Shield illustrate those in the Saharan platform's basins, offering significant insights for subsurface reservoir characterization (Craig et al., 2008; Coward and Ries, 2003; Issad et al., 2011; Arab and Djezzar 2011). This analog study of the Cambro-Ordovician aims to provide a detailed fracture characterization and modeling, concentrating on fracture orientation, length distribution, and the correlation between fracture density and lithology across the Paleozoic series. In the Mouydir basin, this series encompasses the Cambro-Ordovician, Silurian, Devonian, and Carboniferous formations, each contributing to the comprehensive analysis of natural fractures and their implications on reservoir dynamics (Legrand, 1962; Follot, 1952; Fabre, 2005; Legrand, 1985; Haddoum et al., 2001; Fekirine and Abdellah, 1998). The study area is situated between the southern fringes of the Mouydir basin and the northern reaches of the Hoggar shield, spanning latitudes from 24°30′N to 28°00′N and longitudes from 3°00′E to 6°00′E (Figure 12.2). This geographical setting is framed by the Mouydir basin to the north, the Idjerane spur and Ahnet basin to the west, the Amguid spur and Illizi basin to the east, and the uplift of the Hoggar shield to the south. Characterized by expansive regions where Paleozoic formations are exposed, such as Ain-Tadjoubar,

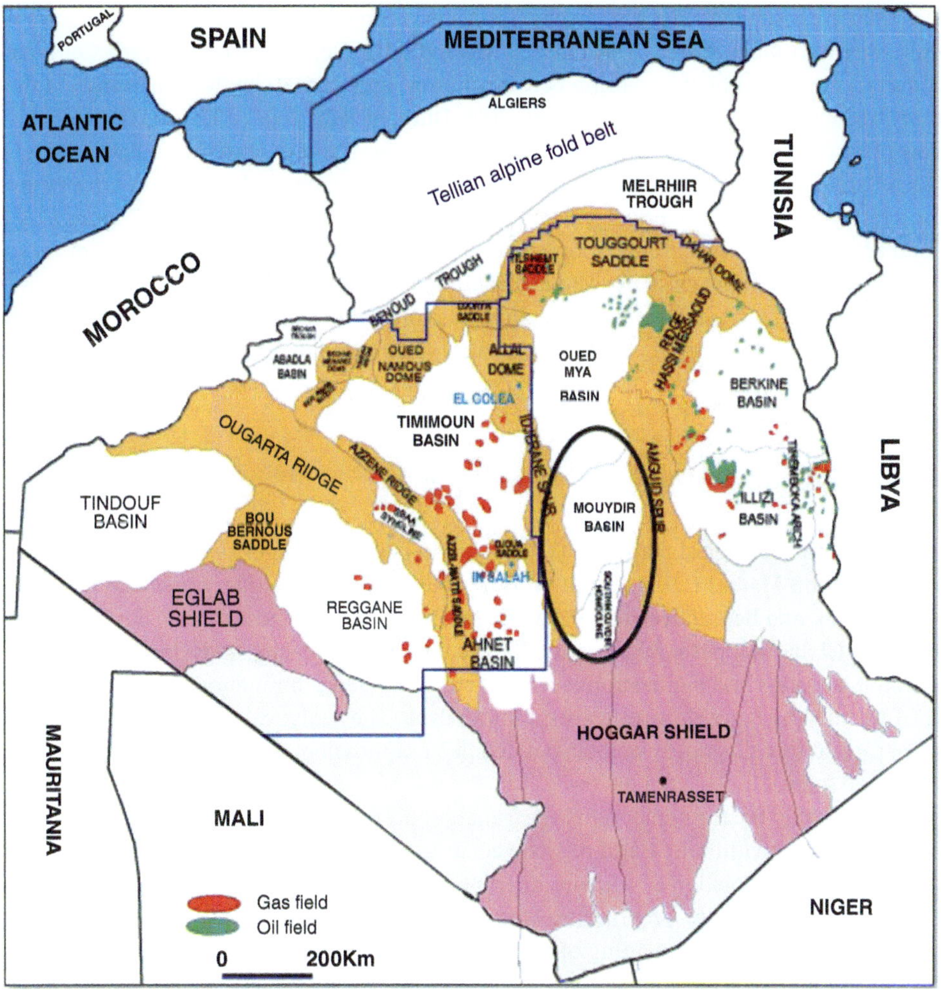

Figure 12.2 Location of the area of study.

Arak, Iftessene, and Khanget-El-Hadid, the study zone is segmented by substantial sub-meridian faults that have played a critical role in the tectonic shaping of the Mouydir basin across various orogenic events (Basseto et al., 1974a, 1974b; Bensalah et al., 1972a, 1972b, 1972c, 1972d; Bennacef et al., 1971).

The Cambro-Ordovician period is marked by Ordovician reservoirs that rank as the second most crucial source of oil and gas, primarily due to the presence of natural fractures (Boudjema, 1987). This stratigraphic interval is composed of three principal formations: Ajjers, In-Tahouite, and Tamadjert, which manifest both at the subsurface across the Saharan Platform basins and at surface exposures in parts of the Hoggar Shield. These formations span the Tremadocian to Ashgill periods, with recent palynological studies suggesting the Ajjers Formation largely fell into the Middle Cambrian age. The stratigraphy further includes various sandstones and quartzites, notable for their contributions to the Ordovician reservoirs and their geological significance through different depositional phases, from fluviatile to glacio-marine systems as a response to polycyclic glaciations during the late Ordovician (Vecoli et al., 2008).

The tectonic history of the area reveals multiple phases of activity, including the reactivation of Precambrian faults during the Cambro-Ordovician period, regional basin inversion during the Caledonian orogeny, and Middle Devonian tectonic movements that reactivated ancient faults, contributing to a diverse basin-and-ridge topography. The Variscan orogeny introduced numerous reverse faults, while Mesozoic uplifts related to Alpine tectonics and subsequent Cretaceous to early Cenozoic tectonic inversions, spurred by the collision between Europe and Africa, significantly impacted the sedimentary basins (Liégeois, 2019; Liégeois et al., 2005, Durham et al., 2004; WEC, 2005).

In terms of reservoir dynamics, the El Atchane sandstone, Hamra Quartzites, Ouargla sandstone, Oued Saret sandstone, and M'Kratta sandstone form the core of the Cambro-Ordovician formation's reservoirs, predominantly deposited in fluvial and fluvio-glacial settings (Figure 12.3). While compact in nature, with average porosity and low permeability, the productivity of these reservoirs heavily relies on the existence and distribution of open fractures, underscoring the critical role of natural fracture systems in enhancing reservoir performance in the Algerian basins (Biju-Duval et al., 1974; Black et al., 1994; Beuf et al., 1971; Haddoum et al., 2001; Zazoun, 2001, 2008; Aliev et al., 1971; Wendt et al., 2006, 2009; Zegrir, 2014; Zieliński, 2001).

12.3 Material and Method

The dataset comprises geological maps at a 1:100,000 scale, satellite imagery with 30 m resolution, and Digital Elevation Models (DEMs) with 65 m resolution, encompassing works from Basseto et al. (1974a, 1974b) and Bensalah et al. (1972a, 1972b, 1972c, 1972d). These diverse data sets were integrated using Global Mapper software, ensuring accurate georeferencing and consolidation. The process of fault detection and digitization leveraged curvature and illumination attributes derived from the DEM, utilizing FracaFlow software for the analysis. The analytical outputs, including histograms, maps, and rose diagrams, facilitated a detailed examination of fault orientations, length distribution, and so on.

The outcome of this fracture analysis is the generation of 3D deterministic fault models for each stratigraphic layer within the Cambro-Ordovician and the underlying basement units. These models provide a vivid depiction of fracture distribution, offering visions into the interplay between stratigraphy and fracture occurrence. Moreover, the characterization of basement faults presents a comprehensive understanding of their influence on the overlying Cambro-Ordovician sedimentary cover and the basin architecture.

The study employed two distinct observational scales to capture the complexity of fracture typology and distribution laws. The mega-scale, at 1:100,000, was utilized to digitalize and

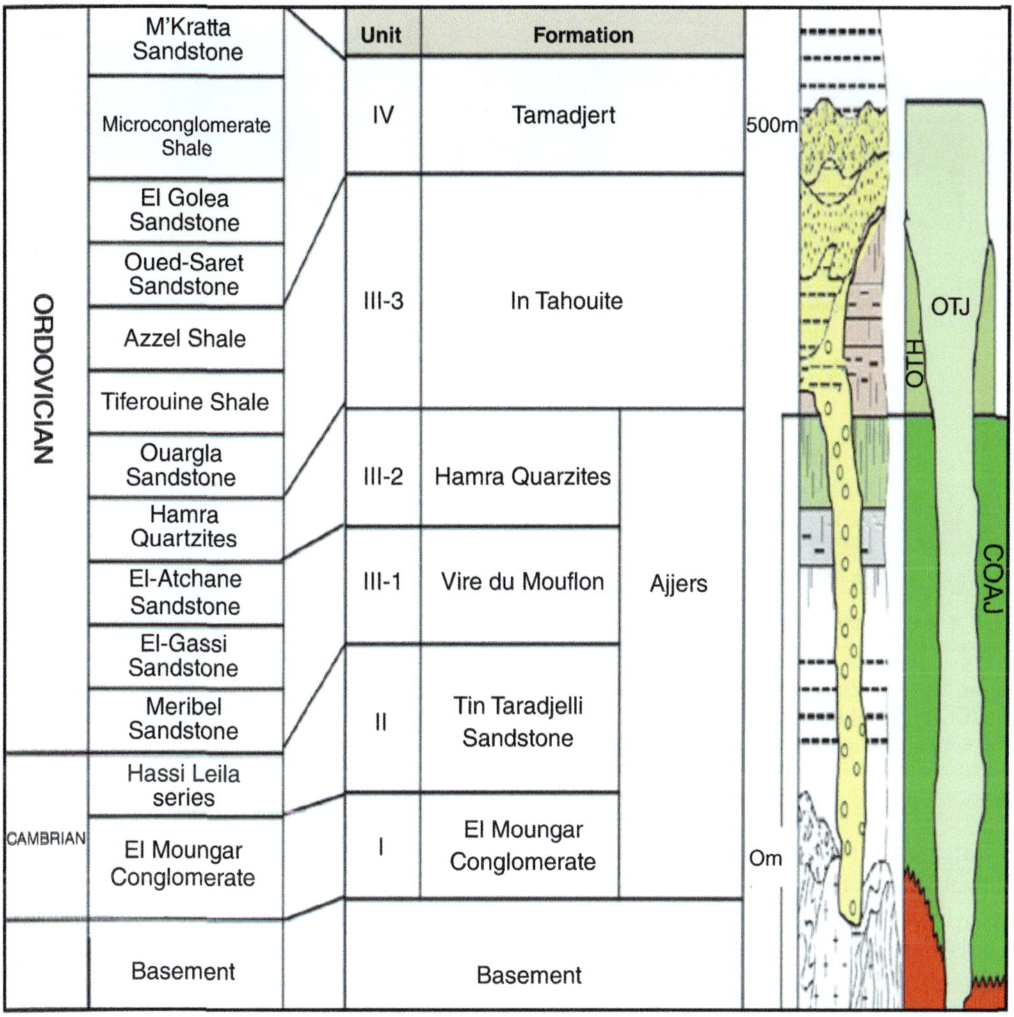

Figure 12.3 Cambro-Ordovician units in the area of the Mouydir basin (Adapted from Zeqrir, 2014).

scrutinize the major faults intersecting the study area, bridging the northern part of the Hoggar shield with the southern periphery of the Mouydir basin. On a finer scale, a mesoscale of 1:25,000 facilitated the detailed mapping of minor lineaments at the formation level. The integration of Global Mapper™ for map manipulation and georeferencing, alongside FracaFlow™ for the detailed fracture analysis and modeling, underscores the comprehensive approach taken in this study to elucidate the structural and stratigraphic intricacies of the region (Figure 12.4).

12.4 Fracture Characterization

Fractures are defined as any discontinuity developed in response to stress and can be classified into faults (mode II fractures) with in-plane shear displacement, and joints (mode I fractures) displaying displacement normal to the discontinuity walls. This distinction, elaborated by Bour et al. (2002), underlines the importance of understanding fracture mechanics. The analysis of fault

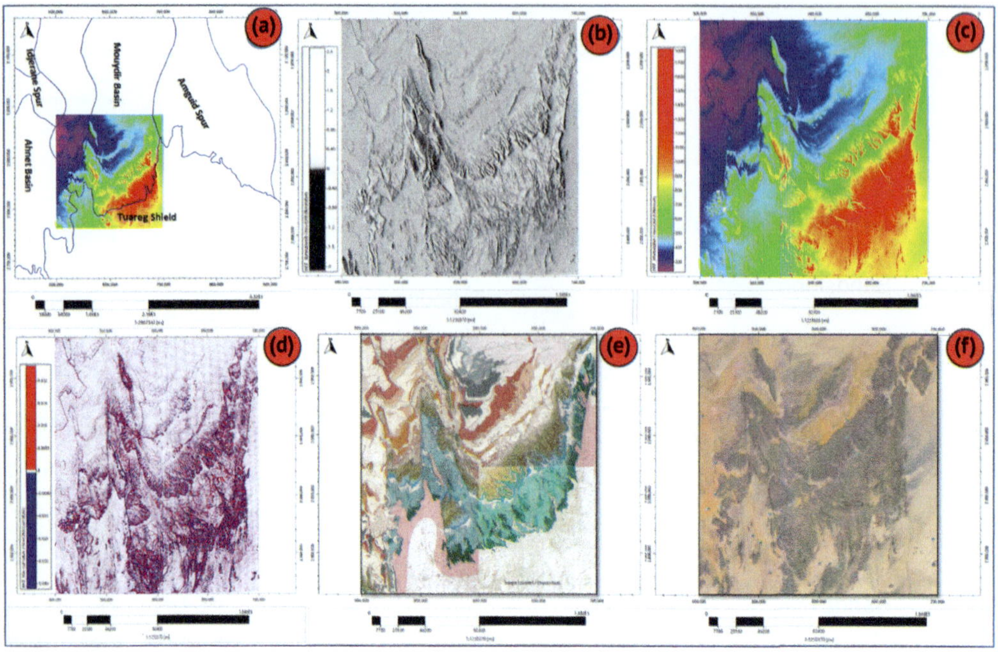

Figure 12.4 (a) Location of the zone of interest (ZOI) concerning the basin's location; (b) illumination of the ZOI; (c) digital elevation model of the ZOI; (d) maximum curvature map of the ZOI; (e) geological maps of the ZOI; (f) satellite image of the ZOI. *Source*: Sofiane Djezzar (2019).

orientations is facilitated by pole diagrams, which visually aggregate fault sets on stereonet for both manual and automated interpretation. The significance of length distribution in fracture connectivity, as highlighted by Balberg and Binenbaum (1983) and further expanded by Balberg et al. (1991), indicates that collections of long fractures offer greater connectivity than those composed of shorter fractures. This concept is supported by numerous field studies (Davy et al., 1990, 1992; Davy, 1993; Pickering et al., 1995; Bour and Davy, 1997, 1998; Odling et al., 1999; Bonnet et al., 2001; Darcel et al., 2003) demonstrating that fracture sets often follow a power law length distribution, with the power law exponent "a" typically ranging between 1 and 3 (Segall and Pollard, 1983; Berkowitz et al., 2000).

In the study area, both faults and joints constitute the observed discontinuities, with their length populations analyzed via normalized cumulative frequency distribution. This analysis, rooted in the methodologies of Odling et al. (1999) and others, reveals the power law nature of fracture length distribution and its implications for subsurface fracture analysis. The fracture analysis in the Cambro-Ordovician, informed by a combination of geological maps, satellite imagery, and digital elevation models, leverages curvature and illumination attributes to delineate fracture sets and understand their fractography across different formations. This comprehensive approach yields 3D deterministic fault models that elucidate fracture distribution and their geological implications (Figure 12.5).

Subsurface data analyses, including Bouguer gravity anomaly mapping and petrophysical modeling, further delineate structural zones and reservoir characteristics, highlighting the influence of fractures on permeability and fluid flow within the Mouydir basin. Borehole imagery data (e.g. UBI-OBMI-FMI) facilitate the characterization of fracture attributes and their impact on reservoir productivity, emphasizing the role of interconnected fracture networks in low permeability

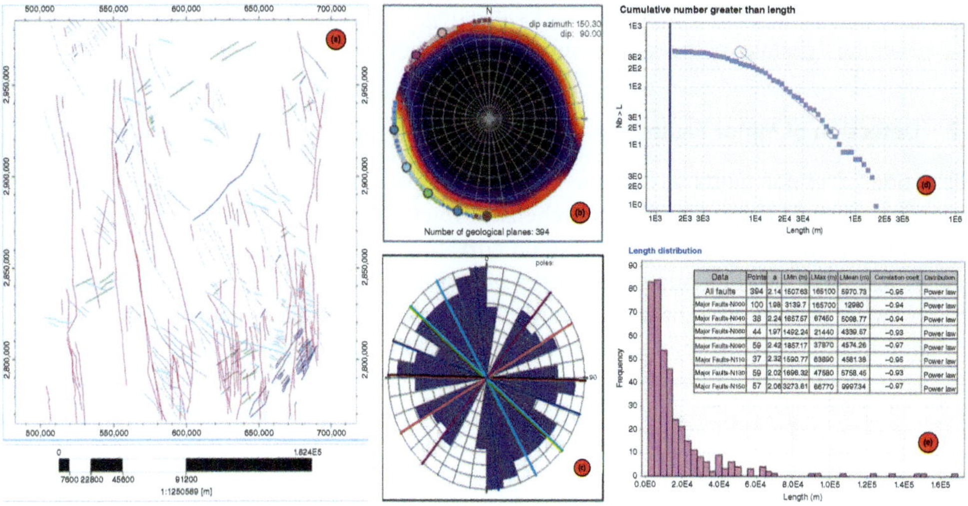

Data	Points	a	LMin (m)	LMax (m)	LMean (m)	Correlation coeff.	Distribution
All faults	394	2.14	1507.63	165100	5970.73	–0.95	Power law
Major Faults-N000	100	1.98	3139.7	165700	12980	–0.94	Power law
Major Faults-N040	38	2.24	1657.57	67450	5098.77	–0.94	Power law
Major Faults-N060	44	1.97	1492.24	21440	4339.57	–0.93	Power law
Major Faults-N090	59	2.42	1857.17	37870	4574.26	–0.97	Power law
Major Faults-N110	37	2.32	1590.77	63890	4581.38	–0.95	Power law
Major Faults-N130	59	2.02	1696.32	47580	5758.45	–0.93	Power law
Major Faults-N150	57	2.06	3273.61	66770	9997.34	–0.97	Power law

Figure 12.5 (a) Fault map in the zone of interest (ZOI); (b) faults pole plan projection; (c) fault set orientations; (d) faults power law distribution; (e) faults' length distribution.

reservoirs. Core analysis and borehole imagery complement seismic data interpretation, allowing for a nuanced understanding of the type of fracture, fracture kinematics, and the relationship between fractures and faults.

This analysis poised to inform oil and gas exploration in the Mouydir basin, underscores the complex interplay between geological structures and reservoir characteristics. The integration of surface and subsurface fault models offers critical insights into fault continuity, density, and the fractures-faults relationship at various scales.

12.4.1 Detection of Major Faults

In the detection of major faults within the study area, a global lineaments map at a 1:100,000 scale was developed, providing a detailed view of major fault sets and their length distributions. Through the digitization and analysis of approximately 400 faults using geological maps, curvature, and illumination maps, a comprehensive fracture system with diverse orientations emerged (Roberts, 2001; Gao, 2013). This system was organized into seven distinct sets: N000, N040, N060, N090, N110, N130, and N150, showcasing the complex geological fabric of the region.

The analysis of length distribution, presented as a bar chart, emphasizes the range of fracture lengths and their frequency, identifying the most, least, and average lengths observed. Notably, the N000, N090, and N130 fracture sets stood out as the most dominant, with fault lengths spanning 4–166 km for N000, 2–37 km for N090, and 1.6–47 km for N130. The extensive lengths of these fractures suggest potential intersections at various depths, particularly for the N000 set, which is highlighted as a crucial fault network impacting the entire Paleozoic series, including Cambro-Ordovician, Silurian, and Devonian formations. Geological map analysis further indicates that these fractures predominantly function as strike-slip faults.

The global and individual fracture lengths were subjected to statistical analysis, evaluating the most, least, and average frequent lengths, alongside correlation coefficients and power law exponents. The correlation coefficients for these analyses ranged between 0.93 and 0.97, while the power law coefficients varied from 1.97 to 2.42. This quantification of fracture characteristics underscores the complexity and significance of the fault networks affecting the study area, providing critical insights into the structural dynamics of the region's subsurface geology.

12.4.2 Detection of Minor Faults

The detection of minor faults within the Cambro-Ordovician formations and the underlying basement is a pivotal component of comprehensive fracture analysis. This process involves the meticulous digitization of fractures, drawing on curvature and illumination maps among other datasets, to ascertain the orientation and density of fractures across each geological stratum. This analysis is augmented by rose diagrams and various statistical measures to establish a detailed understanding of fracture distribution patterns. Specifically, the length distribution of fracture sets is scrutinized to pinpoint prevalent fracture lengths, employing correlation coefficients, power-law exponents, and statistical measures of length such as maximum, minimum, and average.

12.4.2.1 Basement Formation

The basement formation, integral to the study area, is part of the Hoggar shield, characterized predominantly by metamorphic and igneous rock types. This geological feature, a Cenozoic uplift covering an expanse of approximately 550,000 km^2, is primarily constituted of Precambrian rocks framed by Paleozoic sediments deposited post-Pan-African orogeny. Research by Liégeois (2019) and others highlights the complex structural makeup of the Hoggar Shield, identifying 23 distinct terranes demarcated by mega-shear zones or thrust fronts, each showcasing unique lithological, metamorphic, magmatic, or tectonic properties.

Within this geological setting, the basement's fracture network is organized into five principal orientation sets: N000, N050, N090, N120, and N160. The N000 and N160 orientations emerge as the predominant fault sets, encompassing 256 and 145 fault planes respectively, with their lengths spanning from 500 m to 13 km for the N000 set, and 1–10 kilometers for the N160 set (Table 12.1). This distribution is depicted in log-log diagrams, evidencing that the basement fractures adhere to a power law distribution. Correlation coefficients ranging from 0.89 to 0.97, alongside power law coefficients between 2.36 and 2.89, further substantiate the power law distribution of fracture lengths within the basement, reflecting the complex and varied fracture dynamics that influence the geological and hydrogeological properties of the region.

Table 12.1 Length distribution parameters per fault set in the basement formation.

Data	Points	a	LMin (m)	LMax (m)	LMean (m)	Correlation coeff.	Distribution
All faults	761	2.89	551.54	22,950	1132.38	−0.93	Power law
Basement-N000	256	2.36	609.52	15,290	1601.27	−0.89	Power law
Basement-N050	123	2.6	549.43	10,710	1226.41	−0.91	Power law
Basement-N090	121	2.83	718.86	22,980	1497.33	−0.97	Power law
Basement-N120	116	2.43	646.6	9570.19	1508.62	−0.94	Power law
Basement-N160	145	2.62	885.08	11,870	1877.66	−0.95	Power law

12.4.2.2 Ajjers Formation

The Ajjers Formation is characterized by its rich sandstone composition, including notable units such as the Hamra Quartzites, Vire-du-Mouflon sandstone, and Tin-Taradjelli sandstone. This formation is marked by a complex network of 492 faults, organized into six distinct orientation sets: N000, N050, N090, N110, N130, and N150. The N000 and N090 fault sets emerge as particularly significant, comprising 120 and 123 fault planes, respectively, with their lengths spanning between 750 m and 10 km for the N000 set, and between 800 m and 5.5 km for the N090 set. Analysis reveals that the fracture length distribution within the Ajjers Formation adheres to a power law, supported by correlation coefficients ranging from 0.93 to 0.96 and a power law coefficient between 2.27 and 3 (Table 12.2).

12.4.2.3 In-Tahouite Formation

The In-Tahouite Formation, primarily composed of bioturbated siltstones and very fine to fine-grained sandstones, along with Azzel shale, Tiferouine shale, and Ouargla sandstone, features a fracture network intersected by 172 faults. This network is categorized into six sets with orientations of N000, N040, N060, N090, N120, and N130. The most pronounced fault networks within this formation are N000, N090, and N130, consisting of 40, 49, and 51 fault planes, respectively. Fractures within these sets exhibit significant lengths, ranging from 350 m to 16 km for the N000 set, from 650 m to 10 km for the N090 set, and from 660 m to 13 km for the N130 set. The length distribution of these fractures also follows a power law pattern, as indicated by correlation coefficients between 0.89 and 0.95, and power law coefficients from 2.19 to 3.12 (Table 12.3).

Table 12.2 Length distribution parameters per fault set in the Ajjers formation.

Data	Points	a	LMin (m)	LMax (m)	LMean (m)	Correlation coeff.	Distribution
All faults	492	2.47	357.83	16,030	936.88	−0.9	Power law
Ajjers-N000	120	2.19	357.83	16,030	1168.08	−0.89	Power law
Ajjers-N050	36	2.52	843.7	6966.39	1713.95	−0.95	Power law
Ajjers-N090	123	3.12	645.12	9842.41	1166.97	−0.96	Power law
Ajjers-N110	52	3.05	1110.37	8108.58	1933.39	−0.95	Power law
Ajjers-N130	92	2.33	663.03	13,160	1703.08	−0.93	Power law
Ajjers-N150	69	2.74	1003.68	11,270	1997.4	−0.95	Power law

Table 12.3 Length distribution parameters per fault set in the In-Tahouite formation.

Data	Points	a	LMin (m)	LMax (m)	LMean (m)	Correlation coeff.	Distribution
All faults	172	2.9	518.09	10,180	1021.16	−0.94	Power law
In-Tahouite-N000	40	2.67	752.79	10,200	1569.66	−0.96	Power law
In-Tahouite-N050	15	2.54	579.87	4593.54	1162.1	−0.93	Power law
In-Tahouite-N060	17	2.55	532.86	6983.19	1157.13	−0.93	Power law
In-Tahouite-N090	49	3	807.65	5494.03	1407.55	−0.99	Power law
In-Tahouite-N130	51	2.27	516.52	5563.09	1211.6	−0.94	Power law

Table 12.4 Length distribution parameters per fault set in the Tamadjert formation.

Data	Points	*a*	LMin (m)	LMax (m)	LMean (m)	Correlation coeff.	Distribution
All Faults	843	2.73	516.16	79,370	1195.95	−0.97	Power law
Tamajert-N00	212	2.37	516.16	79,370	1614.87	−0.97	Power law
Tamajert-N40	84	2.58	648.57	6233.58	1328.5	−0.9 1	Power law
Tamajert - N70	57	2.56	514.66	9283.62	1161.36	−0.94	Power law
Tamajert-N90	178	2.75	530.2	8750.73	1093.91	−0.91	Power law
Tamajert-N120	171	2.81	684.66	17,110	1422.51	−0.95	Power law
Tamajert-N150	141	2.53	890.15	21,920	2118.6	−0.97	Power law

12.4.2.4 Tamadjert Formation

The Tamadjert Formation presents a heterogeneous reservoir environment, distinguished by its unique stratigraphic and sedimentological characteristics linked to glacial phenomena. It features a base marked by either an angular unconformity or a disconformity, leading to abrupt thickening and lithologic variations. This formation includes M'Kratta sandstone, El-Goléa sandstone, and microconglomerate shale, and is traversed by 843 faults divided into five orientation sets: N000, N040, N070, N090, and N130. The dominant fault sets, N000, N090, and N130, are represented by 212, 178, and 171 faults, respectively. Fracture length analysis highlights the N070 and N090 fault sets, with lengths ranging from 500 m to 9 km and from 530 m to 8.7 km, respectively. The fracture length distribution within the Tamadjert Formation confirms a power law function, underscored by correlation coefficients of 0.91 to 0.97 and power law coefficients between 2.37 and 2.81 (Table 12.4).

12.5 Fracture Analysis

The quantitative fracture analysis within the Cambro-Ordovician formations encompasses a detailed examination of both major and minor fault systems impacting the basement and Cambro-Ordovician strata. Fault maps are synthesized from curvature and illumination attribute data, enabling the classification of fracture sets, lengths, and fractal dimensions across the entire network and within individual sets. This comprehensive approach facilitates the differentiation between major and minor fault sets, offering insights into the unique fractography of each geological formation.

At a 1:100,000 scale, the mapping of major lineaments unveils the predominant fault sets and their respective length distributions, while the identification of minor faults leverages curvature and illumination maps to articulate the fracture orientation and density across various stratigraphic units. Employing rose diagrams and statistical analyses, the study delineates the length distribution of fractures, assessing each set's correlation coefficient, power law exponent, and length metrics.

This analysis elucidates the diversity of fracture sets, their densities, and interconnections, contributing to an understanding of fracture chronology, origins, and subsurface continuity. The Cambro-Ordovician region is marked by both stratabound and non-stratabound fracture models, with mechanical stratigraphy units identified based on thickness and lithology. A deterministic fracture model illustrates the spatial distribution of fractures, highlighting the mechanical distinctions between Cambro-Ordovician units and their interrelations. Notably, the Tamadjert formation emerges as the most fractured unit, attributed to its lithological properties, whereas the In-Tahouite formation exhibits lesser fracture density, likely due to its shaly composition (Figure 12.6).

The study reveals that the basement fractures significantly influence the sedimentary cover, structuring the Cambro-Ordovician units with a predominant brittle tectonic style. This structural framework is rooted in major basement fractures derived from the Pan-African orogeny, leading to a comprehensive fracture network characterized by major vertical fractures. These networks are further defined by N-S trending dextral strike-slip and NNW-SSE trending sinistral strike-slip fractures, with two primary fracture corridors delineating the Mouydir basin's western and eastern edges. Fracture length distribution within the area follows a power law, with coefficients ranging between 2.31 and 2.69 and high correlation coefficients, supporting the fractal nature of these networks. The 3D surface fault models, developed from digitized fault lineaments, facilitate a deeper understanding of fault length, connectivity, orientation, and kinematics, assisting in the construction of a unique deterministic fault model that illustrates the spatial distribution of faults and the interaction between basement fractures and the sedimentary cover. The Cambro-Ordovician formations' fracture networks, predominantly styled by brittle tectonics linked to major basement faults from the Pan-African orogeny, comprise an extensive network of major vertical faults. These structures, alongside N-S trending dextral strike-slip faults, NNW-SSE to N-S trending reverse faults and folds, shape the geological and hydrogeological dynamics of the region. This intricate network of fractures, characterized by varying degrees of connectivity and implications for fluid flow, underscores the complex interplay between geological structures and the movement of fluids within the rock mass.

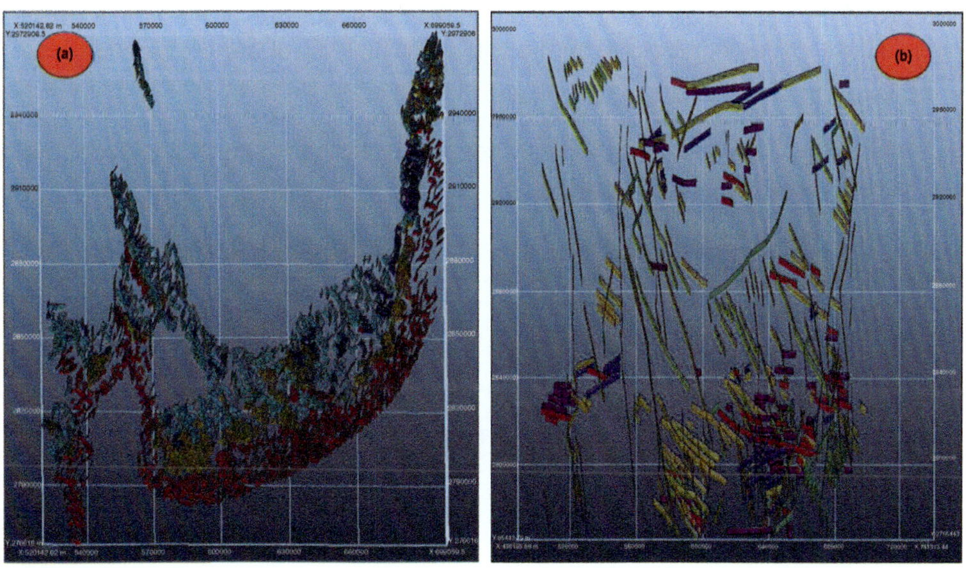

Figure 12.6 (a) 3D minor faults that affect the basement and the Cambro-Ordovician units; (b) 3D major faults 3D minor faults that affect the basement and the Cambro-Ordovician units.

12.6 Fractal Analysis

The concept of fractals, introduced by Benoit Mandelbrot in 1975, has significantly influenced the study of fracture networks, providing insights into their complex, two-dimensional geometry. This fractal approach has been employed by numerous researchers (e.g. Mandelbrot, 1975, 1982; Allègre et al., 1982; Davy et al., 1990, 1992; Davy, 1993; Cowie et al., 1995, 1996; Bour and Davy, 1997, 1998; Bour, 1999; Bour et al., 2002; Bonnet et al., 2001; Darcel et al., 2003) to characterize the intricate patterns of 2D fracture networks across different scales, noting a broad spectrum of fractal dimensions ranging from 1 to 2. The fractal dimension is typically derived through two methodologies: the center distance algorithm, which assesses the fractal nature based on the distribution of fault centers and utilizes the correlation dimension from the two-point correlation function, and the box-counting algorithm, which evaluates the fractal dimension across a set of points or fault traces, considering the entire network as fractal. The box-counting method, in particular, involves overlaying various grid sizes on a 2D fault trace map, counting cells intersected by at least one fracture trace, and plotting these counts against grid cell size on a log-log scale to ascertain the fractal dimension. In this study, FracaFlow software is employed to calculate the fractal dimensions of major faults and the overall basement and Cambro-Ordovician units, along with the dimensions for specific fracture sets. Initial findings reveal a global fractal dimension of Dm = 1.7114 (center distance) and Dm = 1.57 (box-counting) for the fracture network, with variations across different fracture sets indicating a range of 1.0573 < Dm < 1.5939 (center distance) and 1.0091 < Dm < 1.3798 (box-counting). The basement fractures exhibit a fractal dimension of Dm = 1.5945 (center distance) and Dm = 1.2125 (box-counting), with individual fracture sets showing a range of 1.0573 < Dm < 1.5939 (center distance) and 1.0149 < Dm < 1.2163 (box-counting). The Ajjers fracture network demonstrates fractal behavior with dimensions of Dm = 1.7114 (center distance) and Dm = 1.57 (box-counting), although certain fracture sets deviate from these findings due to a limited number of fractures. The In-Tahouite fracture network, characterized by a fractal dimension of Dm = 1.3722 (center distance) and Dm = 1.04 (box-counting), and the Tamadjert fracture network, with Dm = 1.6962 (center distance) and Dm = 1.4099 (box-counting), also display fractal properties. However, variations in the fractal dimensions for specific fracture sets within these formations underscore the diversity and complexity of fracture distribution patterns across different geological units. These fractal analyses contribute to a deeper understanding of the spatial distribution and interconnectivity of fractures within the Cambro-Ordovician and basement formations, highlighting the inherent complexity and varied characteristics of these geological features (Figure 12.7).

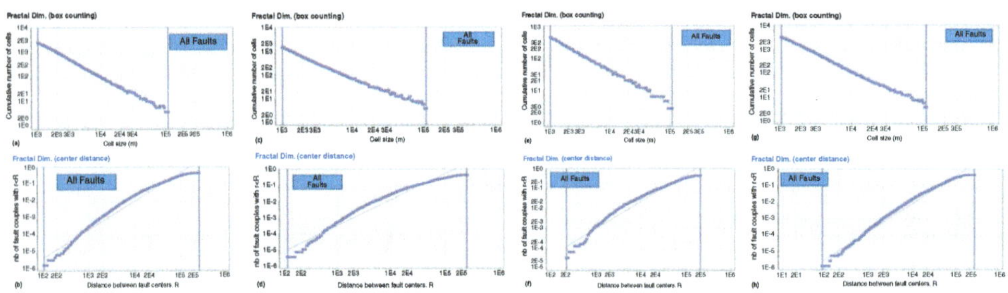

Figure 12.7 (a, b) Fractal dimension for basement faults; (c, d) fractal dimension for Ajjers faults; (e, f) fractal dimensions for In-Tahouite faults; (g, h) fractal dimension for Tamadjert faults.

12.7 3D Fault Models

The construction of 3D fault models for the area of study integrates data across various stratigraphic units, including the basement, Ajjers, In-Tahouite, and Tamadjert formations. These models serve as a foundational tool to visualize and analyze the complex interplay of different fracture sets, their density, and spatial interconnections. Through this comprehensive visualization, the models facilitate a deeper understanding of fracture relationships, chronology, origins, and their continuity or discontinuity within the subsurface.

A holistic 3D deterministic fault model, synthesized from the amalgamation of individual unit models, provides a detailed representation of fracture distributions. This integrated model is instrumental in elucidating the mechanical stratigraphy of the Cambro-Ordovician formations, which is discerned through variations in thickness and lithological properties. It highlights the Tamadjert formation as notably fractured, attributable to its specific lithological features, whereas the In-Tahouite formation exhibits a lower fracture density, likely due to its shaly composition.

The impact of basement fractures on the Cambro-Ordovician sedimentary cover is varied, indicating a structural influence on the organization of Cambro-Ordovician units. These formations predominantly exhibit a brittle tectonic style, shaped by significant basement faults originating from the Pan-African orogeny. This has led to the development of a widespread network of major vertical faults. Geological mapping further reveals a dense network of faults and folds impacting the Paleozoic era, with prominent structures including N-S trending dextral strike-slip faults and NNW-SSE to N-S faults.

Two principal fault corridors are identified, delineating the Mouydir basin edges, extending from the Hoggar shield and diverging northward into the Saharan platform. These corridors, oriented from NNE to NS and from NNW to NS, are known as the Amguid and Idjerane spurs, respectively. This spatial arrangement and the delineation of fault corridors underscore the complex geological dynamics at play, offering insights into the structural forces shaping the basin and the broader region's geology (Figure 12.8).

12.8 Discussion

This study spans approximately 67 255 km^2, incorporating various maps that encompass the basement and all Paleozoic units within the region. Utilizing curvature and illumination attributes derived from digital elevation models has proven to be an effective technique for identifying fractures visible at outcrops, thereby enabling the digitization of fault networks. These networks are essential for constructing temporal and spatial fracture models. 3D surface fault models, generated from digitized fault lineaments, reveal fault length, connectivity, and orientation, enriching our understanding of their continuity, kinematics, and chronological development.

Each Cambro-Ordovician formation displays distinct mechanical stratigraphy, differentiated by thickness and lithology. The integration of fault models from these formations facilitates the creation of a unified deterministic fault model. This comprehensive model maps out spatial fault distributions and elucidates the interactions between basement faults and the sedimentary cover. It reveals the Cambro-Ordovician formations are mainly characterized by a brittle tectonic style, stemming from significant basement faults inherited from the Pan-African orogeny. This geological heritage has given rise to a vast network of major vertical faults, as evidenced by the analysis of geological maps.

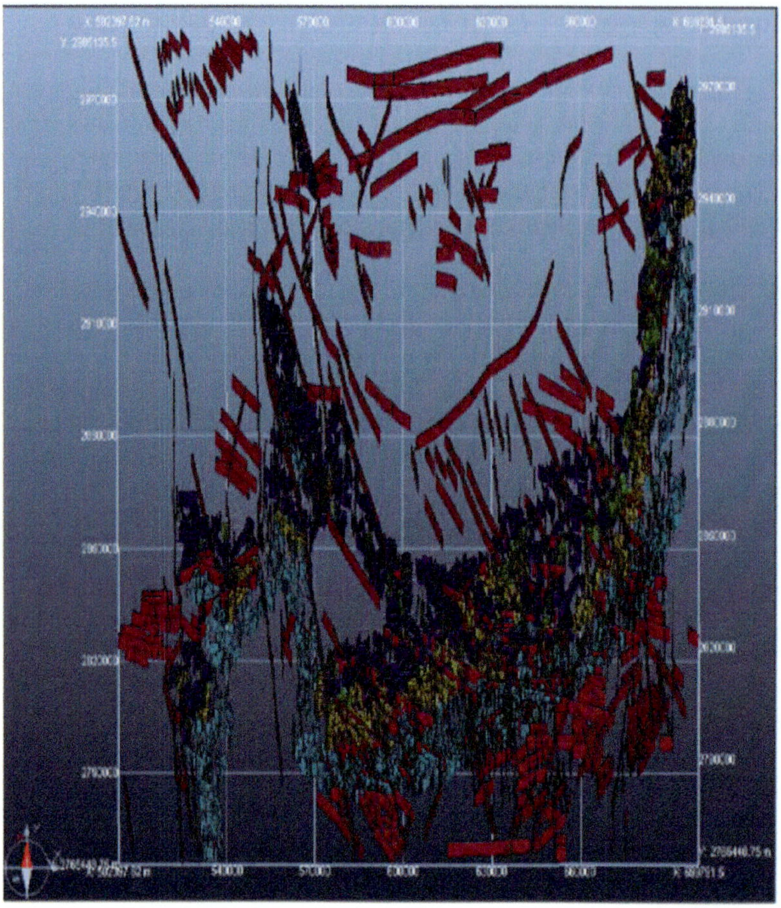

Figure 12.8 3D faults model in the area of study.

The fracture networks identified indicate that the Paleozoic series was significantly impacted by a dense array of faults and folds. Major structures are predominantly N-S trending dextral strike-slip fractures and NNW-SSE trending sinistral strike-slip fractures. Two principal fracture corridors demarcating the Mouydir basin's boundaries emerge from this analysis, originating from the Hoggar shield and diverging northward into the Saharan platform, identified as the Amguid and Idjerane spurs.

Furthermore, gravity mapping highlights two significant north-south-oriented positive anomalies at the basin's edges, corresponding to the Idjerane and Amguid-Biod spurs, with another smaller positive structure where well HL was drilled. The geological mapping reveals a complex network of strike-slip faults, indicating multiple reactivation phases throughout the Paleozoic era.

The Hercynian orogeny, identified through geological mapping and analysis, played a pivotal role in shaping the Paleozoic formations into horsts and grabens. This orogenic phase is recognized for its dual tectonic events, directed N040 and N120, spanning from the Visean to the post-Permian period. The Hercynian deformation is considered more a continuous process than distinct tectonic events, contributing to the formation of N-S trending dextral strike-slip faults and NNW-SSE to N-S trending reverse faults and folds.

These major faults create significant fracturing corridors that induce intense deformation, leading to the formation of secondary faults and folds observable in geological maps. This deformation pattern suggests an NNE-SSW to ENE-WSW shortening around the Carboniferous/Permian transition, or more likely during the Early Permian. Fault sets N040, N060, and N090 align with the Hercynian shortening direction, while other fault sets may be associated with post-Hercynian tectonic activities. This intricate geological history underscores the dynamic and complex tectonic evolution of the Mouydir basin and its surrounding regions.

12.9 Conclusions

The comprehensive fracture analysis conducted in this study highlights several key findings regarding the Cambro-Ordovician reservoirs. It reveals that the predominant fault sets are oriented N000, N050, N090, N120, and N160, with major structures consisting of N-S-trending dextral strike-slip faults and NNW-SSE to N-S trending reverse faults. This structural and stratigraphic investigation underscores a significant variation in the influence of basement fractures on the Cambro-Ordovician sedimentary layers.

The Cambro-Ordovician formations are primarily characterized by non-stratabound fracture systems, indicating a widespread impact of basement faults on both basin structure and the overlying sedimentary cover. This influence is particularly pronounced in the more compact formations such as Ajjers and Tamadjert, which exhibit a higher fracture density. Conversely, the In-Tahouite formation displays a lower fracture prevalence, likely due to its composition of shaly units. Notably, the Tamadjert formation, with its dense sandstone composition, emerges as the most fractured, whereas the In-Tahouite formation is less affected by fracturing due to the presence of shale units like Azzel and Tiferoulne.

The N-S and N160 fault orientations play a pivotal role across the basement and the entire Paleozoic series, tracing their origins back to the Pan-African orogeny but undergoing subsequent reactivation through various tectonic phases. The fault sets N040, N060, and N090 align with the Riedel shear system, reflecting the Hercynian orogeny's compressive stress direction oriented N040.

Fracture network analyses across the basement and Cambro-Ordovician units reveal a fractal distribution pattern, as determined through both center distance and box-counting algorithms. This distribution is characterized by power law coefficients ranging from 2.31 to 2.69, accompanied by high correlation coefficients, suggesting a complex and interconnected fracture network. However, certain fracture sets do not exhibit a fractal dimension, attributed to a limited number of fractures or their dispersed distribution within the study area. This observation points to the nuanced and varied nature of fracture systems across different geological formations, underscoring the complexity of subsurface structures in the region.

References

Aliev, M., Ait Laoussine, N., Avrov, V., et al. (1971). Structures geologiques et perspectives en petrole et en gaz du Sahara Algerien (ed. S. A. Altamira-Rotopress). Sonatrach éditions.

Allègre, J. C., Lemouel, J. L., & Provost, A. (1982). Scaling rules in rock fracture and possible implications for earthquake prediction. Nature 297, pp. 47–49.

Arab, M. & Djezzar, S. (2011). Hydrocarbon potential in Reggane Basin. Oil Mediterranean Conference (OMC), Italy. OMC-2011-046.

Balberg, I., Berkowitz, B., & Drachsler, G. E. (1991). Application of a percolation model to flow in fractured hard rocks. Journal of Geophysical Research 96B, pp. 10015–10021.

Balberg, I. & Binenbaum, N. (1983). Computer study of the percolation threshold in a two-dimension anisotropic system of conducting sticks. Physical Review 28 (7), pp. 3799–3812.

Basseto, D., Ben Salah, A., Beuf, S., et al. (1974a). Carte geologique de Aoulef El-Arab. Ahnet et Mouydir. Service de la Carte Geologique, Algerie, jeu de 13 feuilles. Bchelle 1: 200 000, NG 31, (941), p. 14.

Basseto, D., Ben Salah, A., Beuf, S., et al. (1974b). Carte geologique de Reggane. Ahnet et Mouydir. Service de la Carte Geologique, Algerie, jeu de 13 feuilles. Bchelle 1: 200 000, NG 31, (941), p. 13.

Beicip. (2018). FracaFlow tutorial.

Bennacef, A., Beuf, S., Biju-Duval, B., et al. (1971). Example of cratonic sedimentation: Lower Palaezoic of Algerian Sahara. American Association of Petroleum Geologists 55 (12), pp. 2225–2245.

Bensalah, A., Beuf, S., Gabriel, O., et al. (1972a). Carte Géologique de l'Algérie: Arak, feuille NG-31-X et NG-31-IV. Ministère de l'Industrie et de l'Energie.

Bensalah, A., Beuf, S., Gabriel, O., et al. (1972b). Carte Géologique de l'Algérie: Ain Tadjoubar, feuille NG-31-XVI. Ministère de l'Industrie et de l'Energie.

Bensalah, A., Beuf, S., Gabriel, O., et al. (1972c). Carte Géologique de l'Algérie: Khanget El Hadid, feuille NG-31-XVII. Ministère de l'Industrie et de l'Energie.

Bensalah, A., Beuf, S., Gabriel, O., et al. (1972d). Carte Géologique de l'Algérie: Ifetessene, feuille NG-31-XI. Ministère de l'Industrie et de l'Energie.

Berkowitz, B., Bour, O., Davy, P., et al. (2000). Scaling of fracture connectivity in geological formations. Geophysical Research Letters 27 (14), pp. 2061–2064.

Beuf, S., Biju-Duval, B., De Charpal, D., et al. (1971). Les grès du Paléozoïque inférieur au Sahara. Sédimentation et discontinuité: évolution structurale d'un craton. In: Technip-Institut Français du pétrole, vol. 18, p. 464. Collection Sciences et Techniques du Pétrole, Paris.

Biju-Duval, B. M., Deynoux, M., & Rognon, P. (1974). Essai d'interprétation des "fractures en gradins" observées dans les formations glaciaires Précambriennes et ordoviciennes du Sahara. Rev Géogra Phys Géol Dyn Paris 16 (1974), pp. 503–512.

Black, R., Latouche, L., Liégeois, J. P., et al. (1994). Pan-African displaced terranes in the Hoggar shield (central Sahara). Geology 22, pp. 641–644.

Bonnet, E., Bour, O., Odling, N. E., et al. (2001). Scaling of fracture systems in geological media. Reviews of Geophysics 39 (3), pp. 347–383.

Boualam, A., Djezzar, S., Rasouli, V., et al. (2019). 3D Modeling and Natural Fractures Characterization in Hassi Guettar Field, Algeria. In: 53rd American Rock Mechanics Association Symposium, NY, USA. 19-A-373-ARMA.

Boudjema, A. (1987). Evolution structurale du bassin petrolier 'triasique' du Sahara Nord Oriental (Algerie). These Doctorat Etat, Paris XI Orsay, France, p. 290.

Bour, D. (1999). Clustering and size distributions of fault patterns: Theory and measurements. Geophysical Research Letters 26 (13), pp. 2001–2004. https://doi.org/10.1029/1999GL900419.

Bour, O. & Davy, P. (1997). Connectivity of random fault networks following a power law fault length distribution. Water Resources Research 33, pp. 1567–1583.

Bour, O. & Davy, P. (1998). On the connectivity of three-dimensional fault networks. Water Resources Research 34 (10), pp. 2611–2622.

Bour, O., Davy, P., Darcel, C., et al. (2002). A statistical scaling model for fracture network geometry, with validation on a multiscale mapping of a joint network (Hornelen Basin, Norway). Journal of Geophysical Research 107 (B6), pp. ETG-4. https://doi.org/10.1029/2001JB000176.

Cacas, M. C., Daniel, J. M., & Letouzy, J. (2001). Nested geological modeling of naturally fractured reservoirs. Petroleum Geoscience 7 (S), pp. 43–52.

Coward, M. P. & Ries, A. C. (2003). Tectonic development of North Africa basins. In: New themes, and Developing Technologies (eds. T. J. Arthur, D. S. Macgregor, & N. R. Cameron), pp 61–83. Geological Society of London: Special Publication 207.

Cowie, P. A., Knipe, R. K., & Main, I. G. (1996). Introduction to the special issue: scaling laws for fault and fracture population-analysis and applications. Journal of Structural Geology 18, pp. 1–5.

Cowie, P. A., Sornette, D., & Vanneste, C. (1995). Multifractal scaling properties of growing fault population. Geophysical Journal International 122, pp. 457–469.

Craig, J., Rizzi, C., Said, F., et al. (2008). Structural styles and prospectivity in the Precambrian and Palaeozoic hydrocarbon systems of North Africa. In: The geology of East Libya (ed. M. J. Salem), vol. IV, pp. 51–122. Malta: Gutenberg Press.

Darcel, C., Bour, O., & Davy, P. (2003). Stereological analysis of fractal fracture networks. Journal of Geophysical Research 108 (B9), p. 2451.

Davy, P. (1993). On the fault—length frequency distribution of the San Andreas fault system. Journal of Geophysical Research 98 (12), pp. 141–151.

Davy, P., Sornette, A., & Sornette, D. (1990). Some consequences of a proposed fractal nature of continental faulting. Nature 348 (6296), pp. 56–58.

Davy, P., Sornette, A., & Sornette, D. (1992). Experimental discovery of scaling laws relating fractal dimensions and the length distribution exponent of fault systems. Geophysical Research Letters 19 (4), pp. 361–363.

Djezzar, S. (2012). Impact de la fracturation naturelle sur la productivité des réservoirs du Dévonien inférieur dans le bassin de Reggane, Sahara Occidental, Algérie. Thèse Magister. Ed. USTHB. -Alger: USTHB. p. -93: ill.; 30 cm. + CD-Rom.

Djezzar, S. (2019). Fracture detection and prediction. In: Unconventional Reservoirs For Finding Sweet Spot. Theses and Dissertations. 2551. https://commons.und.edu/theses/2551/.

Djezzar, S. & Boualam, A. (2020a). Analysis and modeling of tight oil fractured reservoir. SEG 2020 Annual Meeting, Tx, USA. https://doi.org/10.1190/segam2020-3425011.1.

Djezzar, S. & Boualam, A. (2020b). Fractures Characterization and Their Impact on the Productivity of Hamra Quartzite Reservoir (Haoud-Berkaoui field, Oued-Mya basin, Algeria). 54th American Rock Mechanics Association Symposium, Co, USA. 20-A-1438-ARMA.

Djezzar, S. & Boualam, A. (2021). Impact of Natural Fractures on the Productivity of Lower Devonian Reservoirs in Reggane Basin, Algeria. In: 55th US Rock Mechanics/Geomechanics Symposium. 21-A-1295-ARMA.

Djezzar, S., Rasouli, V., Boualam, A., et al. (2019). Size scaling and spatial clustering of natural fracture networks using fractal analysis. In: 53rd American Rock Mechanics Association Symposium. NY, USA. 19-A-375-ARMA.

Djezzar, S., Rasouli, V., Boualam, A., et al. (2020). An integrated workflow for multiscale fracture analysis in reservoir analog. Arabian Journal of Geosciences 13, p. 161. https://doi.org/10.1007/s12517-020-5085-6.

Donzeau, M. (1983). Tectonique des monts d'Ougerta. In: Afrique de l'Ouest. Introduction Geologique et Termes Stratigraphiques, Lexique Stratigraphique International, Nouvelle Serie 1 (ed. J. Fabre), pp. 118–120. Oxford, UK: Pergamon Press.

Donzeau, M., Fabre, J., & Moussine Pouchkine, A. (1981). Comportement de la dalle saharienne et orogenese Varisque. Essai d'interpretation. Bulletin de la Société d'histoire naturelle de l'Afrique du Nord 69 (3–4), pp. 171–184.

Durham, M. J., Eales, M. H., Beswetherick, S., et al. (2004). Structural styles and prospectivity in the Precambrian and Palaeozoic hydrocarbon systems of North Africa. In: Conference Proceedings, Geology of East Libya Symposium, p. 110, Benghazi, Libya: Maghreb Petroleum Research Group.

Fabre, J. (2005). Géologie du Sahara occidental et central. Musée Royale de l'Afrique Centrale (Ed.), p. 572. Tervuren, Belgique.

Fekirine, B. & Abdellah, H. (1998). Palaeozoic lithofacies correlatives and sequence stratigraphy of the Saharan Platform, Algeria (ed. D. S. Macgregor & R. T. J. Moody). Geological Society, London, Special Publications 132 (1), pp. 97–108.

Follot, J. (1952). Ahnet et Mouydir. 19' Congres international de Geologie, Alger. Monographie Regionale, Serie 1 (1), pp. 1–6.

Gao, D. (2013). Integrating 3D seismic curvature and curvature gradient attributes for fracture detection: methodologies and interpretational implications. Geophysics 78 (2), pp. O21–O31

Haddoum, H., Guiraud, R., & Moussine-Pouchkine, A. (2001). Hercynian compressional deformations of the Ahnet-Mouydir Basin, Algerian Saharan Platform: far-field stress effects of the late palaeozoic orogeny. Terra Nova 13, pp. 220–226.

Issad, M., Amia, Z., Djezzar, S., et al. (2011). Reevaluation du potential petrolier du basin du Mouydir. Rapport Interne Sonatrach.

Legrand, P. (1962). Comparaison des séries Cambro-Ordoviciennes reconnues en affleureument dans la région d'Amguid et en forage au centre du basin saharien occidental. Bulletin de la Société Géologique de France 7 (IV), pp. 132–135.

Legrand, P. (1985). Lower Palaeozoic rocks of Algeria. In: Lower Palaeozoic of North-Western and West-Central Africa (ed. C. H. Holland), pp. 6–84. New York: John Wiley and Sons.

Liégeois, J. P. (2019). A new synthetic geological map of the Hoggar shield: an overview of its global structure and geological evolution. In: The geology of the Arab world—an overview (eds. A. Bendaoud et al.), pp. 83–107. Springer Geology.

Liégeois, J. P., Benhallou, A., Azzouni-Sekkal, A., et al. (2005). The Hoggar swell and volcanism: Reactivation of the Precambrian Hoggar shield during Alpine convergence and West African Cenozoic volcanism. In: Plates, plumes and paradigms (eds. G. R. Foulger, J. H. Natland, D. C. Presnall et al.), vol. 388, pp. 379–400. Geol Soc America Spec, Paper.

Lisle, R. J. (1994). Detection of zones of abnormal strains in structures using Gaussian curvature analysis. AAPG Bull 78 (12), pp. 1811–1819.

Mandelbrot, B. (1975). Les objets fractals. Flammarion (Ed.). Paris, p. 212.

Mandelbrot, B. (1982). The fractal geometry of nature. Freeman (Ed.). San Francisco, p. 461.

Odling, N. E., Gillespie, P., Bourgine, B., et al. (1999). Variations in fracture system geometry and their implications for fluid flow in fractured hydrocarbon reservoirs. Petroleum Geoscience 5 (4), pp. 373–384.

Pickering, G., Bull, J. M., & Sanderson, D. J. (1995). Sampling power-law distributions. Tectonophysics 248, pp. 1–20.

Rives, T., Razack, M., Petit, J. P., et al. (1992). Joint spacing: Analogue and numerical simulations. Journal of Structural Geology 14 (8–9), pp. 925–937.

Roberts, A. (2001). Curvature attributes and their application to 3-D interpreted horizons. First Break 19 (2), pp. 85–100.

Segall, P. & Pollard, D. (1983). Joint formation in granitic rock of the Sierra Nevada. Geological Society of America Bulletin 94 (5), pp. 563–575.

Vecoli, M., Videt, B., & Paris, F. (2008). First biostratigraphic (palynological) dating of Middle and Late Cambrian strata in the subsurface of northwestern Algeria, North Africa: Implications for regional stratigraphy. Review of Palaeobotany and Palynology 149 (1–2), pp. 57–62.

Well Evaluation Conference (WEC) (2005). Well evaluation conference, internal report. Joint documents of Sonatrach and Schlumberger, Algeria, p. 489.

Wendt, J., Kaufmann, B., Belka, Z., et al. (2006, 2009). Sedimentary evolution of a Palaeozoic basin and ridge system: the Middle and Upper Devonian of the Ahnet and Mouydir (Algerian Sahara). Geological Magazine 143 (3), pp. 269–299.

Zazoun, R. S. (2001). La tectogenèse hercynienne dans la partie occidentale du basin de l'Ahnet et la région de Bled El-Mass, Sahara Algérien: un continuum de déformation. Journal of African Earth Sciences 32 (4), pp. 869–887.

Zazoun, R. S. (2008). The Fadnoun area, Tassili-n-Azdjer, Algeria: Fracture network geometry analysis. Journal of African Earth Sciences 50 (5), pp. 273–285.

Zegrir, H. (2014). Distribution de la fracturation dans les reservoirs compacts du Cambro-Ordovicien: Méthodologie d'intégration des données de surface au modèle de sub-surface (Bassin de l'Ahnet plate forme saharienne-Algérie). Memoire de Magister, Faculté des Hydrocarbures et de la Chimie. Université M'hamed Bougara, Boumerdès, Algérie.

Zieliński, M. (2011). Reconstruction of thermal history of the Ahnet and Mouydir basins (southern Algeria). PhD thesis, Institute of Geology, Faculty of Geographical and Geological Sciences, Adam Mickiewicz University, Poznan.

Section III

Artificial Intelligence Applied to Reservoir Characterization

13

Exploring the Depths: Satellite Image Processing and Artificial Intelligence in the Oil and Gas Industry

Hasna Yazid[1], *Said Gaci*[2]

[1] *Department of Computer Science, University of M'Hamed Bougerra (UMBB), Boumerdes, Algeria*
[2] *Director of Scientific and Technical Support to Research at Sonatrach's Central Research and Development Directorate (DC R&D)*

13.1 Introduction

The use of satellite imagery and artificial intelligence (AI) in the oil and gas industry has revolutionized the way companies explore, monitor, and extract resources. Satellite technology, coupled with advanced image processing techniques and AI algorithms, has provided valuable insights and improved operational efficiency in the industry (Huang and Wang, 2014; Guo et al., 2017; Ball et al., 2017; Nieto-Hidalgo et al., 2018; Gallego et al., 2018; Zeng and Wang, 2020; McGrath et al., 2021; Yang et al., 2022). In this chapter, we delve into the various techniques and algorithms used to process satellite images and extract valuable insights for the oil and gas industry.

We begin with an overview of satellite technology, discussing its evolution over the years and its significance in the oil and gas industry. By understanding the types of satellites, such as geostationary, polar orbiting, and low Earth orbit (LEO) satellites, we can explore their specific applications and benefits in the industry. Additionally, we delve into the different sensors and data products available, including optical sensors, synthetic aperture radar (SAR) sensors, thermal infrared (IR) sensors, and hyperspectral sensors, and examine their contributions to the oil and gas industry.

Next, we focus on satellite image processing techniques, which are essential for extracting meaningful information from the vast amounts of data collected by satellites. Preprocessing and image enhancement techniques, such as radiometric and atmospheric corrections, as well as geometric corrections, play a crucial role in optimizing the quality and accuracy of satellite imagery. Furthermore, we explore feature extraction and analysis methods, such as texture analysis and object-based image analysis (OBIA), which enable the identification and characterization of relevant features in the oil and gas industry. Classification algorithms, including supervised and unsupervised classification, are also discussed, along with image segmentation techniques like region growing and watershed segmentation.

The integration of AI and machine learning (ML) algorithms in satellite imagery processing has further advanced the capabilities of analysis and interpretation. We introduce the fundamentals of AI and ML and highlight the significance of deep learning, particularly convolutional neural networks (CNNs), for object detection and feature identification in satellite imagery. We also address

Geophysical Exploration for Hydrocarbon Reservoirs, Geothermal Energy, and Carbon Storage: New Technologies and AI-based Approaches, First Edition. Edited by Said Gaci.

the challenges associated with AI-based processing, such as data availability and algorithm robustness, and provide strategies for mitigating these challenges.

Furthermore, we explore the practical applications of satellite imagery and AI in the oil and gas industry. These applications encompass various domains, including exploration and prospect identification, drilling and well planning, pipeline monitoring and leak detection, environmental monitoring and compliance, as well as operational optimization and cost reduction. By leveraging satellite imagery and AI, companies can enhance decision-making, improve safety measures, and optimize their operations, ultimately leading to increased efficiency and reduced costs.

13.2 Overview of Satellite Technology

13.2.1 Definition and Basic Principles

Satellite technology involves the use of artificial satellites placed in orbit around the Earth to perform various functions such as communication, weather monitoring, and remote sensing. Satellites are equipped with sensors and instruments that collect data about the Earth's surface, atmosphere, and other parameters. The fundamental principles of satellite technology include orbital mechanics, which govern the movement and positioning of satellites in space, and communication links, which enable the transmission of data between the satellite and the ground station. By understanding these basic principles, we can appreciate the significance of satellite technology in the oil and gas industry (Guo et al., 2020).

13.2.2 Types of Satellites

Satellites used in the oil and gas industry can be categorized into three main types: geostationary satellites, polar orbiting satellites, and LEO satellites (Guo et al., 2020).

13.2.2.1 Geostationary Satellites

Geostationary satellites are positioned in a fixed spot above the Earth's equator, maintaining a synchronized orbit with the Earth's rotation. This allows them to remain stationary relative to a specific geographic location, providing continuous coverage of that region. Geostationary satellites are commonly used for communication services and weather monitoring, as they enable uninterrupted transmission and reception of signals. In the oil and gas industry, geostationary satellites play a role in real-time communication between offshore platforms and onshore facilities, facilitating efficient operations and emergency response.

13.2.2.2 Polar Orbiting Satellites

Polar orbiting satellites circle the Earth from pole to pole in a sun-synchronous orbit, covering the entire globe during each orbit. As the Earth rotates underneath them, these satellites provide comprehensive coverage of the planet's surface. Polar orbiting satellites are equipped with sensors that capture high-resolution images and collect data across different spectral bands. This makes them suitable for applications such as environmental monitoring, land cover mapping, and disaster response. In the oil and gas industry, polar orbiting satellites contribute to various activities, including monitoring pipeline routes, assessing land subsidence, and detecting oil spills.

13.2.2.3 Low Earth Orbit Satellites

LEO satellites orbit at relatively low altitudes, typically a few hundred kilometers above the Earth's surface. They move at high speeds, completing several orbits per day. LEO satellites offer advantages such as high spatial resolution imagery and improved revisit rates. These satellites are commonly used for imaging purposes, capturing detailed information about the Earth's surface. In the oil and gas industry, LEO satellites assist in tasks such as geological exploration, site selection for drilling operations, and monitoring offshore platforms.

13.2.3 Sensors and Data Products

Satellites are equipped with various sensors and instruments that capture data about the Earth's surface and atmosphere. These sensors play a crucial role in acquiring remote sensing data for the oil and gas industry.

13.2.3.1 Optical Sensors

Optical sensors operate in the visible, near-IR, and shortwave IR regions of the electromagnetic spectrum. They capture images by detecting and recording the reflected or emitted light from the Earth's surface. Optical sensors provide valuable information about land cover, vegetation health, and geological features. In the oil and gas industry, optical sensors are used to identify potential hydrocarbon reservoirs, map surface features, and monitor environmental changes.

13.2.3.2 Synthetic Aperture Radar Sensors

SAR sensors use microwave signals to capture images of the Earth's surface. Unlike optical sensors, SAR can penetrate clouds, haze, and darkness, allowing for all-weather imaging capabilities. SAR sensors measure the time taken for radar signals to travel to the target and return, providing information on surface roughness, terrain topography, and detecting changes over time. In the oil and gas industry, SAR sensors contribute to activities such as pipeline monitoring, offshore asset detection, and oil spill identification.

13.2.3.3 Thermal Infrared Sensors

Thermal IR sensors detect and measure the thermal energy emitted by objects on the Earth's surface. They operate in the long-wave IR region, which captures heat signatures. Thermal IR sensors provide information about temperature variations, heat loss, and thermal anomalies. In the oil and gas industry, thermal IR sensors are used to monitor industrial facilities, detect pipeline leaks, and assess the thermal properties of reservoirs.

13.2.3.4 Hyperspectral Sensors

Hyperspectral sensors capture data across numerous narrow and contiguous spectral bands, offering detailed spectral information about the Earth's surface. Hyperspectral sensors enable the identification and analysis of materials based on their unique spectral signatures. In the oil and gas industry, hyperspectral sensors assist in tasks such as mineral exploration, detecting hydrocarbon seepages, and monitoring vegetation health in pipeline corridors.

13.3 Evolution of Satellite Technology in the Oil and Gas Industry

Satellite technology has a rich history of early applications in the oil and gas industry. In the past, satellite imagery was primarily used for regional-scale mapping, geological surveys, and exploration activities. Satellites provided a bird's-eye view of the Earth's surface, allowing geologists and exploration teams to identify potential hydrocarbon-rich areas. Satellite imagery aided in mapping geological formations, identifying structural features, and locating potential drilling sites. Early applications also included monitoring activities such as pipeline routes, offshore platforms, and the environmental impact of oil and gas operations.

Advancements in satellite technology have led to significant improvements in spatial resolution, allowing for more detailed and precise imaging. Early satellites provided imagery with lower spatial resolution, limiting their use in specific applications. However, with the advent of high-resolution satellites and advancements in sensor technology, satellites can now capture imagery with exceptional detail. High-resolution imagery enables the identification of smaller features, such as individual infrastructure components, geological structures, and surface anomalies. The improved spatial resolution has greatly enhanced exploration and monitoring activities in the oil and gas industry, providing geologists and engineers with valuable insights and enabling more accurate decision-making (Guo et al., 2020).

Satellite sensors have also witnessed improvements in spectral resolution and accuracy over the years. Earlier sensors captured data in limited spectral bands, restricting the ability to differentiate and analyze different materials and features accurately. However, modern satellites employ sensors with increased spectral resolution, capturing data across a wider range of spectral bands. This enhancement enables more precise identification and characterization of materials and surface properties. By analyzing the unique spectral signatures of various substances, including hydrocarbons, minerals, and vegetation, satellite imagery can aid in identifying potential oil and gas reserves, assessing environmental impacts, and monitoring changes in land cover and vegetation health (Tysiąc et al., 2022).

Temporal resolution refers to the frequency at which satellite images are acquired over a specific location. Early satellites provided infrequent coverage, limiting the ability to monitor dynamic changes and capture time-sensitive events. However, the evolution of satellite technology has led to improvements in temporal resolution, allowing for more frequent revisits and monitoring capabilities. Satellites equipped with rapid revisit rates enable the monitoring of environmental changes, natural disasters, and the progression of oil and gas operations over time. This temporal information is invaluable for assessing trends, detecting anomalies, and facilitating effective decision-making in the industry (Akhtar et al., 2023).

13.4 Satellite Image Processing Techniques

Satellite imagery processing plays a crucial role in extracting valuable information from the vast amounts of data collected by satellites. This section focuses on the different image processing techniques employed in the analysis of satellite images for the oil and gas industry. It covers preprocessing and image enhancement techniques, including radiometric and atmospheric corrections, as well as geometric corrections to improve the quality and accuracy of the imagery. It also discusses feature extraction and analysis techniques, such as texture analysis and OBIA, which aid in identifying relevant features. Additionally, classification algorithms, both supervised and unsupervised, used to categorize the extracted features, are explored. Moreover, image segmentation techniques like region growing and watershed segmentation, which assist in partitioning the images into meaningful segments, are discussed (Asokan et al., 2020).

13.4.1 Preprocessing and Image Enhancement

Preprocessing and image enhancement techniques are crucial steps in satellite image processing to improve the quality and usability of the data. This stage involves correcting and enhancing the acquired satellite images to minimize distortions and optimize the information content.

13.4.1.1 Radiometric and Atmospheric Corrections

Radiometric corrections aim to normalize the brightness and contrast of satellite imagery by removing sensor-specific effects and variations in illumination conditions. This correction process ensures that the images accurately represent the true reflectance or radiance values of the Earth's surface. Atmospheric corrections, on the other hand, compensate for the scattering and absorption of light caused by the Earth's atmosphere. By accounting for atmospheric effects, the corrected imagery provides more accurate information about surface features and properties (Fan et al., 2015).

13.4.1.2 Geometric Corrections

Geometric corrections are applied to satellite imagery to rectify distortions caused by satellite motion, Earth's curvature, and sensor orientation. Georeferencing techniques align the images to a specific coordinate system, enabling accurate spatial analysis and integration with other geospatial data. Geometric corrections ensure that the satellite imagery aligns with the Earth's surface, enabling precise measurements and analysis of features and objects (Fan et al., 2015).

13.4.2 Feature Extraction and Analysis

Feature extraction and analysis techniques focus on identifying and characterizing specific objects or patterns within the satellite imagery. These techniques enable the extraction of relevant information from the imagery for further analysis and interpretation.

13.4.2.1 Texture Analysis

Texture analysis is used to quantify the spatial patterns and variations within an image. It involves analyzing the arrangement and distribution of pixels to extract textural features such as roughness, smoothness, and homogeneity. Texture analysis aids in identifying distinctive patterns related to geological structures, land cover types, or surface properties within the satellite imagery (Cheng and Han, 2016).

13.4.2.2 Object-Based Image Analysis

OBIA takes a holistic approach to image analysis by grouping pixels into meaningful objects or regions based on their spatial and spectral characteristics. Instead of analyzing individual pixels, OBIA considers contextual information and the relationships between neighboring pixels. This technique allows for the extraction of objects such as buildings, vegetation patches, or infrastructure networks, facilitating more accurate classification and interpretation of the satellite imagery (Cheng and Han, 2016).

13.4.3 Classification Algorithms

Classification algorithms are employed to categorize pixels or objects within the satellite imagery into distinct classes or categories. Classification is a fundamental task in satellite image processing as it enables the identification and mapping of specific features or land cover types.

13.4.3.1 Supervised Classification

Supervised classification involves training a classification algorithm using a set of representative training samples, where each sample is labeled with the corresponding class. The algorithm learns the spectral signatures associated with each class and applies this knowledge to classify the remaining pixels or objects in the image. Supervised classification is particularly useful when specific features or land cover types need to be identified in the imagery (Cheng and Han, 2016).

13.4.3.2 Unsupervised Classification

Unsupervised classification does not rely on pre-labeled training samples. Instead, it automatically groups pixels or objects based on their spectral similarity. The algorithm identifies clusters or classes within the image data, and the analyst assigns appropriate labels to each cluster during the post-classification interpretation. Unsupervised classification is valuable when the number or nature of classes in the imagery is unknown or when exploring patterns and relationships within the data (Cheng and Han, 2016).

13.4.4 Image Segmentation Techniques

Image segmentation techniques partition the satellite imagery into distinct regions or objects based on certain criteria. Image segmentation aids in delineating boundaries, separating objects of interest, and extracting meaningful information at different scales.

13.4.4.1 Region Growing

Region growing is a bottom-up approach to image segmentation where neighboring pixels with similar properties are iteratively grouped together. This technique allows for the identification of homogeneous regions within the imagery based on specified criteria, such as spectral similarity or texture. Region growing is useful for segmenting objects with well-defined boundaries or regions with uniform characteristics (Cheng and Han, 2016).

13.4.4.2 Watershed Segmentation

Watershed segmentation is a technique inspired by hydrology, where the image intensity is treated as a topographic relief. It identifies segment boundaries by simulating the filling of basins with water, resulting in the delineation of catchment basins. Watershed segmentation is effective in segmenting objects with irregular shapes and strong intensity gradients. It is commonly used in satellite image processing for delineating features like rivers, coastlines, or geological formations (Cheng and Han, 2016).

By applying preprocessing and image enhancement techniques, performing feature extraction and analysis, and utilizing classification algorithms and image segmentation techniques, satellite imagery can be processed and analyzed effectively in the oil and gas industry. These techniques allow for the extraction of valuable information and the identification of relevant features and patterns, facilitating informed decision-making and improved operational efficiency.

13.5 Artificial Intelligence in Satellite Imagery Processing

AI has emerged as a powerful tool in satellite imagery processing, enabling more efficient and accurate analysis. This section introduces the fundamentals of AI and ML and their application in satellite imagery processing for the oil and gas industry. It explores deep learning techniques, particularly CNNs, which excel in identifying features in satellite imagery. It also highlights the role of AI in object detection and feature identification tasks. However, it acknowledges the challenges associated with AI-based processing, such as data availability, model training, and algorithm interpretation. Strategies for mitigating these challenges are discussed to facilitate effective AI integration in satellite imagery processing (Jude Hemanth, 2019).

13.5.1 Introduction to AI and ML

AI refers to the development of intelligent systems that can perform tasks without explicit programming by learning from data and experiences. ML is a subset of AI that focuses on algorithms and models that enable computers to learn patterns and make predictions or decisions based on data. ML techniques have become increasingly important in satellite imagery processing, as they allow for automated analysis, feature extraction, and classification of large volumes of satellite data (Ma et al., 2019).

13.5.2 Deep Learning and Convolutional Neural Networks

Deep learning is a subfield of ML that utilizes artificial neural networks (ANNs) with multiple layers to extract high-level features and patterns from complex data. CNNs are a type of deep learning architecture specifically designed for image processing tasks. CNNs excel in identifying spatial patterns and features in satellite images, enabling tasks such as object detection, land cover classification, and change detection. By leveraging the hierarchical structure of CNNs, complex relationships and spatial dependencies within the satellite imagery can be effectively captured and utilized for analysis and decision-making (Zhang et al., 2016).

13.5.3 Object Detection and Feature Identification

Object detection and feature identification are essential tasks in satellite imagery processing, enabling the identification and extraction of specific objects or features of interest. By training CNN models on labeled datasets, they can learn to identify and localize objects with high accuracy and efficiency. Object detection techniques are applied to various applications in the oil and gas industry, such as detecting oil spills, identifying infrastructure components (e.g. wells, pipelines), and monitoring vegetation encroachment in pipeline corridors. These AI-based approaches enhance the speed and accuracy of object detection, enabling more efficient and informed decision-making in the industry (Li et al., 2020).

13.5.4 Challenges and Mitigation Strategies in AI-Based Processing

The integration of AI-based processing in satellite imagery analysis presents unique challenges that need to be addressed to ensure reliable and accurate results. There are benefits and drawbacks to integrating AI with satellite processing. The potential of AI to improve satellite processing and communication, including resource management, network control, and energy consumption, is highlighted by Ghiglione and Serra (2022), and Fourati and Alouini (2021). However, the processing power of radiation-hardened hardware (Ghiglione and Serra, 2022) and the requirement for more processing power for onboard AI/ML processing (Ortiz et al., 2023) limit the deployment of AI. Wijata et al. (2023) highlight how crucial it is to choose AI applications for onboard deployment while taking into account resource frugality, trust-building, hardware and energy limitations. For small satellite earth observation missions, Chintalapati et al. (2024) demonstrated that the use of AI system on board the satellite encountered a number of issues, including small satellites' limited computational capacity, data transmission restrictions, and inconsistent algorithm performance. Taken together, these studies highlight how, in order to fully exploit AI's promise in satellite processing, advances in hardware and computational capacities, together with the creation of standardized procedures for AI application selection, are required.

The application of AI techniques, including deep learning and CNNs, in satellite imagery processing has revolutionized the capabilities and efficiency of analysis in the oil and gas industry (Huang and Wang, 2014; Guo et al., 2017; Ball et al., 2017; Nieto-Hidalgo et al., 2018; Gallego et al., 2018; Zeng and Wang, 2020; McGrath et al., 2021; Yang et al., 2022). By harnessing the power of AI, complex patterns and features within satellite images can be automatically detected, classified, and analyzed. Object detection techniques enable the identification of specific objects or features of interest, aiding in tasks such as oil spill detection, infrastructure monitoring, and environmental assessment. However, the adoption of AI-based processing in the oil and gas industry comes with challenges, including availability of labeled training data, ensuring model generalizability across different regions and conditions, interpretability of AI models, addressing biases and uncertainties in the training process. Through careful consideration and mitigation strategies, these challenges can be overcome, leading to more accurate, efficient, and reliable satellite imagery analysis for the oil and gas industry.

13.6 Practical Applications and AI in the Oil and Gas Industry

This section explores the practical applications of satellite imagery and AI in the oil and gas industry. It showcases how satellite technology has revolutionized various aspects of the industry, including exploration, drilling, pipeline monitoring, and environmental monitoring. For instance, satellite imagery aids in identifying potential oil and gas reserves, planning drilling operations, detecting leaks in pipelines, and assessing environmental impacts. Moreover, it highlights the potential of satellite imagery and AI to optimize operations and reduce costs, leading to enhanced efficiency and sustainability in the industry.

13.6.1 Exploration and Mapping

Satellite imagery and AI play a vital role in the exploration phase of the oil and gas industry. By leveraging satellite data, exploration teams can identify potential hydrocarbon prospects and assess their viability before costly ground operations commence. Satellite imagery provides a broad overview of geological structures, including sedimentary basins, fault lines, and potential reservoirs. AI techniques, such as ML algorithms and deep learning models, can be trained to analyze the imagery and identify patterns associated with hydrocarbon deposits. This enables exploration teams to prioritize and target areas with a higher likelihood of containing valuable resources, optimizing exploration efforts and reducing costs (Cracknell et al., 2014; Francis et al., 2014; Scafutto et al., 2017; Kwok et al., 2018; Golubov and Lubskyi, 2020; Sales et al., 2022; Usery, 2020; Sang et al., 2020; Wu et al., 2022; Newaz et al., 2023).

13.6.1.1 Geological and Topographic Mapping

For geological mapping and lithology classification, Cracknell et al. (2014) use widely available and geographically constrained remotely sensed geophysical data via five machine leaning algorithms: Naive Bayes, k-Nearest Neighbors, random forests, support vector machines (SVMs), and ANN. They showed that ML methods are valuable tools for generating reliable first-pass predictions for practical geological mapping applications that combine widely available geophysical data with explicit spatial information. In this view, more recent instances of this potential are given by Usery (2020) and Sang et al. (2020), who explore the use of deep learning algorithms and ML to topographic mapping and, in Sang's case, provide a method for high-resolution geological mapping utilizing unmanned aerial vehicles (UAVs). Together, these findings show how AI is becoming more and more important in improving the precision and effectiveness of geology mapping.

13.6.1.2 Mapping and Identification of Rock Outcrops

Since outcrops have characteristics similar to oil and gas reservoirs in subsurface, they are intensively investigated and analyzed to describe deep reservoirs. With ML, these rocky exposures can be readily identified in satellite photos. The potential of ML to recognize and map rock outcrops from satellite data has been shown in recent studies. Francis et al. (2014) employed color photography and multiclass linear discriminant analysis to examine rock outcrops. In order to locate rock outcrops in Hong Kong, Kwok et al. (2018) created a deep learning algorithm that combines remote sensing methods with CNNs. Sales et al. (2022) used the multilayer perceptron layer (MLP) technique for carbonate outcrop image classification in satellite data related to the Jandaíra formation, in northeast Brazil. These experiments demonstrate how ML may be used to automate the mapping and identification of rock outcrops, which could be especially helpful in isolated or difficult-to-reach locations.

13.6.1.3 Forecasting Hydrocarbon Deposit Locations

Golubov and Lubskyi (2020) successively analyzed Sentinel-2 satellite data during the project to statistically quantify the likelihood of finding vegetation-related spectrum abnormalities. Hydrocarbon deposits may be the source of these anomalies. The method used to calculate the values of several spectral indices of vegetation for 20 pairs of polygons built on a homogeneous area covered in vegetation and within known hydrocarbon deposits was this one. The challenge of detecting spectral and temperature abnormalities associated with hydrocarbon deposits was found to be effectively solved by utilizing the spatial data, sensitivity, and frequency of Sentinel-2 and Landsat-8 satellite imagery data.

The process of discovering new petroleum deposits beneath the surface of the earth is usually challenging because it is very costly and has poor precision. In this regard, Newaz et al. (2023) suggested a new approach to forecast petroleum deposit locations, and applied it on Middle Eastern region—Iraq—to locate where petroleum deposits are expected to be found there using the suggested methodology. This prediction technique is based on data that is freely accessible and was collected by an open satellite known as the Gravity Recovery and Climate Experiment (GRACE). It consists on computing the earth's gravity gradient tensor over the investigated region and its environs using GRACE data, then combining ML, graph-based analysis, and an OR-nAND approach to make the predictions of potential petroleum deposits in the studied region.

Due of its accuracy and speed, hyperspectral remote sensing has been used to estimate the component content of soil in recent years for the hydrocarbon exploration purposes. Indeed, remote identification and mapping of petroleum hydrocarbons (PHCs) in situ in continental areas is remains an issue, due to small extension of the occurrences and the mixture of spectral signatures of PHCs and mineral substrates in imaging pixels. Hyperspectral images' high spectral resolution makes it possible to directly identify PHCs on the surface and offers useful data on the location and spread of oil spills, which can help with containment and cleanup efforts. Through the qualitative and quantitative assessment of oil occurrences, combining the spectrum information with ML approaches also has the potential to improve exploration programs aimed at the finding of new exploratory fields (Scafutto et al., 2017; Wu et al., 2022).

The application of ML and hyperspectral remote sensing to the petroleum exploration has been the subject of numerous studies. The estimation models between hyperspectral characteristics and the component content of soil based on ML algorithms are extensively researched due to the swift development of AI algorithms and technology (Gholizadeh et al., 2018; Bi et al., 2019; Hamidisepehr et al., 2019; Paoletti et al., 2019; Peng et al., 2019; Gruszczynski and Gruszczynski, 2022). For the estimation of soil component composition, ANNs, classification and regression trees (CART), SVMs, and extreme learning machines (ELMs) are currently commonly used (Song et al., 2018; Wei et al., 2019; Xue et al., 2020; Wang et al., 2022). Recently, ensemble learning is increasing interest in hyperspectral remote sensing (Ang and Seng, 2021; Nguyen et al., 2021; Taghizadeh-Mehrjardi et al., 2020; Wu et al., 2022).

13.6.1.4 Managing Offshore Exploration and Development Operations

During oil and gas exploration, offshore structures may be impacted by hazardous maritime environments (strong storms, cyclones, etc.). The Meteorological Department's estimations of offshore waves and expected wind speed must be used to quickly decide whether to continue with oil and gas development or to halt it in the interest of public safety. In this view, ML algorithms (ANN, SVM, and random forest) applied by Azad and Uddin (2022) to metocean data to estimate the offshore wave height and period from the wind speed.

13.6.2 Drilling and Well Planning

Satellite imagery and AI contribute to the efficient planning and execution of drilling operations in the oil and gas industry. Prior to drilling, satellite data can assist in identifying optimal well locations by considering factors such as surface features, geological formations, and proximity to existing infrastructure. AI-based analysis of satellite imagery aids in identifying potential drilling hazards, such as underground cavities or areas prone to landslides. By integrating satellite imagery with other geospatial data, AI algorithms can generate detailed terrain models and subsurface maps, allowing for more accurate well planning and reducing the risk of costly drilling complications (Song et al., 2020; Temizel et al., 2021; Wang et al., 2021, 2023).

Song et al. (2020) employed a two-stage approach for a Faster regional convolutional neural network (R-CNN) based model to identify the oil wells from remote sensing photos. Wang et al. (2021) used deep learning models based on techniques for object detection from optical remote sensing images. Experimental findings indicate that the most advanced deep learning models obtain excellent precision for remote sensing oil well detection. Wang et al. (2023) suggested an enhanced Faster R-CNN-based model with significant changes to increase the detection accuracy of clustered oil wells.

13.6.3 Pipeline Monitoring and Leak Detection

Satellite imagery, combined with AI techniques, offers powerful capabilities for pipeline monitoring and leak detection. Satellite data can be used to track the routes of pipeline networks, ensuring they remain free from encroachments and identifying any potential vulnerabilities. AI algorithms can analyze satellite imagery to detect changes in vegetation patterns near pipelines, indicating possible leaks or unauthorized access. Additionally, thermal IR sensors onboard satellites can identify temperature anomalies along pipelines, which could indicate leaks or structural issues. Rapid detection of leaks using satellite imagery and AI enables prompt response, minimizing environmental damage and ensuring the integrity of the pipeline infrastructure (Keramitsoglou et al., 2006; Nirchio et al., 2005, 2009; Zhong et al., 2019; Zeng and Wang, 2020; McGrath et al., 2021).

For SAR oil spill detection based on traditional ML, several classification algorithms, including the linear model (Nirchio et al., 2005), fuzzy logic (Keramitsoglou et al., 2006), Decision Trees (Solberg and Solberg, 1996; Topouzelis and Psyllos, 2012; Singha et al., 2013), ANNs (Topouzelis et al., 2007; Del et al., 2000; Stathakis et al., 2006; Singha et al., 2013; Ravulakollu et al., 2024), SVMs (Brekke and Solberg, 2008; Xu et al., 2014; Singha et al., 2016), Bayesian Classifiers (Solberg and Volden, 1997; Solberg et al., 1999), Ensemble Learning (Ravulakollu et al., 2024), Deep Learning (DL) methods (Huang and Wang, 2014; Guo et al., 2017; Ball et al., 2017; Nieto-Hidalgo et al., 2018; Gallego et al., 2018; Zeng and Wang, 2020; McGrath et al., 2021; Yang et al., 2022).

In addition, ML techniques have been combined with satellite imagery for oil tank detection. A conventional ML strategy using the SVM classifier and Speeded up Robust Features (SURF) technique was employed by Jivane and Soundrapandiyan (2017). Due to the high computational complexity of remote sensing photos, deep learning-based techniques have been used more recently to detect oil tanks (Zalpour et al., 2020. In order to distinguish oil tanks from complex backdrops, Zhang et al. (2015) developed a recognition algorithm based on deep environment features by combining the CNN model with SVM classifier. Using the Faster R-CNN deep learning network, some researchers (Zhang et al., 2018) have successfully identified oil facilities using high-resolution satellite photos. Zeng and Wang (2020) proposed a Deep Convolutional Neural Network (DCNN) using SAR satellite images for oil spill detection, called the Oil Spill Convolutional Network (OSCNet), which can perform the last two steps of the three-step processing framework, and McGrath et al. (2021) suggested a way to ascertain the destiny and route of the oil by fusing autonomous satellite detections with numerical modeling.

Bayramov et al. (2020) present an application of AI on satellite images for mapping vegetation stress in pipeline right-of-ways (Figure 13.1). Indeed, Pipeline right-of-ways require regular monitoring to detect vegetation encroachment, which can pose a risk to pipeline integrity. This study showcases the use of satellite imagery and AI techniques to map and monitor vegetation stress along pipeline corridors. Satellite data, including optical and multispectral imagery, is analyzed using ML algorithms to identify areas of stressed vegetation. By quantifying the vegetation health index and detecting changes over time, companies can prioritize and implement targeted vegetation management strategies, reducing the need for manual inspections and ensuring compliance with safety and regulatory requirements (Bayramov et al., 2020).

13.6.4 Environmental Monitoring and Compliance

Satellite imagery and AI are valuable tools for environmental monitoring and ensuring compliance with environmental regulations in the oil and gas industry. Satellite data allows for the assessment of environmental impacts associated with oil and gas operations, such as deforestation, land subsidence, and changes in water quality. AI algorithms can analyze satellite imagery to detect and monitor changes in vegetation health, identify areas of pollution or contamination, and assess the extent of environmental damage. This information helps oil and gas companies meet compliance requirements, mitigate environmental risks, and develop sustainable practices (Adler et al., 2018; Power et al., 2018; DeLancey et al., 2019; Dileep et al., 2020; Hungund et al., 2023; Akhtar et al., 2023; Lu, 2023; Hernandez et al., 2023; Yu et al., 2023).

Adler et al. (2018) analyze satellite imagery using ML Analysis for Detection of Permian Basin Water Impoundments, and found a significant positive correlation between the total frac pond surface area in the satellite imagery and IHS data on aggregate water use for hydraulic fracturing in the Midland Basin area studied. In addition, a dataset of 5000 ship and iceberg targets was taken from Sentinel-1 multi-channel data by Power et al. (2018) in order to test if new ML techniques

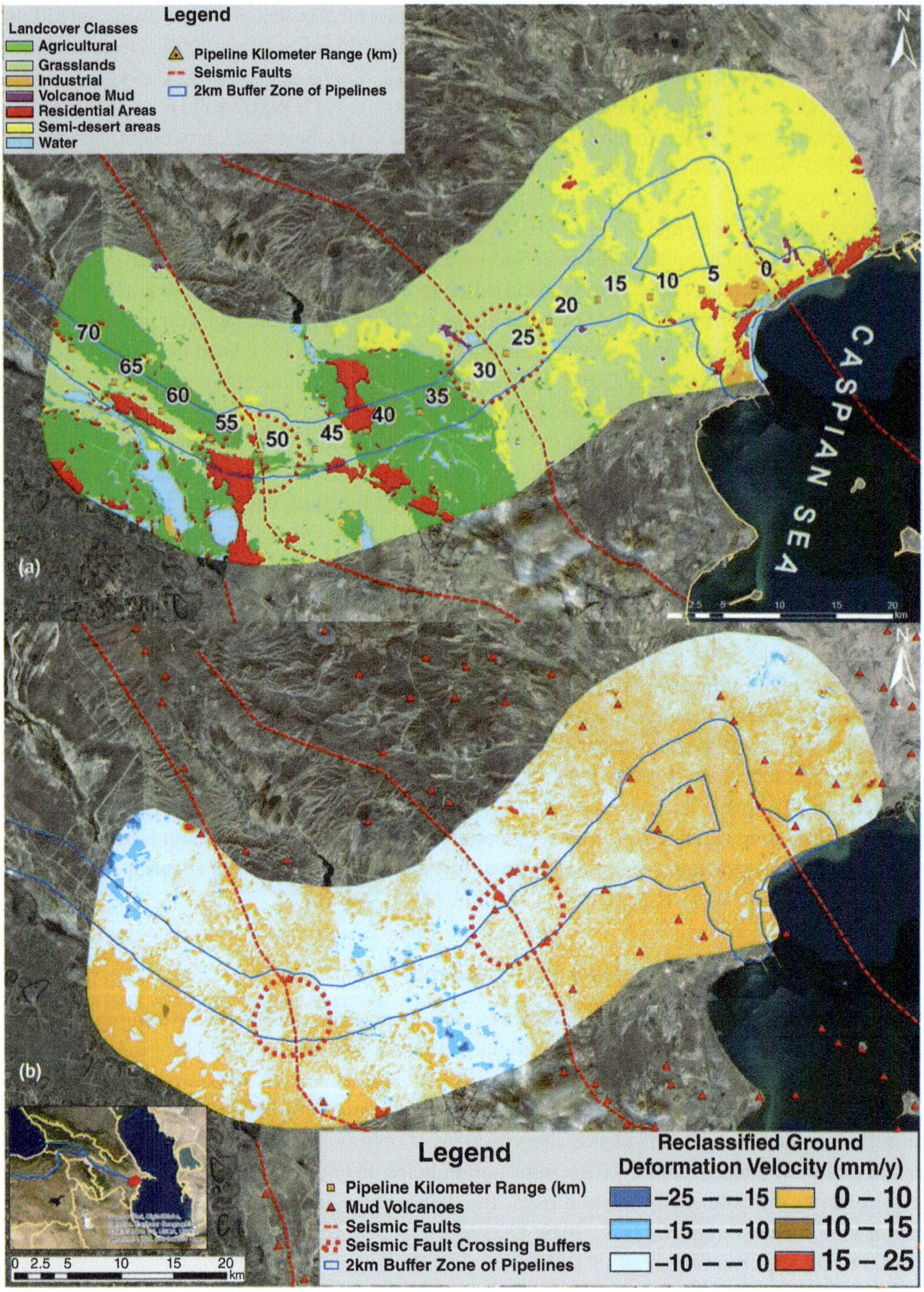

Figure 13.1 (a) Maps of the landcover; (b) categorized ground velocity along pipes carrying gas and petroleum (Bayramov et al., 2020 / Taylor & Francis / CC BY 4.0).

could be applied to boost classifier accuracy. The authors have successfully trained target classifiers using computer vision techniques. Dileep et al. (2020) demonstrated the possibility of automated detection using CNN model from satellite data, leading to more precise and comprehensive estimates of production emissions. The results showed an average well pad detection accuracy of 95% on the Denver-Julesburg basin in northeastern Colorado.

Large-scale, high-resolution landcover mapping on a regular basis is now feasible because to freely accessible satellite data streams and the capacity to handle them on cloud computing platforms like Google Earth Engine. DeLancey et al. (2019) use these technologies in conjunction with ML to map peatlands in the Boreal Forest Natural Region of Alberta, Canada. Peatlands are a type of landcover that is essential for maintaining biodiversity, mitigating the effects of climate change, and providing ecosystem services like carbon storage. This data-driven method can be used for wetland and landcover inventories that support long-term, ethical resource management at large geopolitical scales (e.g. provincial, national).

Hungund et al. (2023) anticipate and quantify ground-level methane (CH_4) and Nitrogen dioxide (NO_2) concentrations using high-resolution satellite data, climate data, and production data, via ML techniques; random forest, and Gradient Boosting (Light and Extreme) models in the Permian Basin in the state of Texas between January 2022 and December 2022. In order to identify oil-gas leaks in multi-source remote sensing data, Zhao et al. (2023) suggested combining spectral shape information (SSI) with multi-scale encoding (MSE) technique.

A satellite retrieval technique for seawater pCO_2 (partial pressure of carbon dioxide) was developed by Yu et al. (2023) who integrated the semi-mechanistic algorithm and machine learning approach (MeSAA-ML-ECS), based on in situ seawater pCO_2 data gathered on 51 cruises/legs over the previous 20 years (2003–2019) in the Bohai Sea, Yellow Sea, and East China Sea. MeSAA-ML-ECS provided semi-analytical parameters to characterize the combined influence of atmospheric CO_2 forcing, thermodynamic effects, and different mixing mechanisms on seawater pCO_2.

Modern remote sensing technologies have certainly transformed the management of the palm oil sector by combining environmental and commercial advantages on one platform. The continuous trend shows that high-resolution image processing allows remote sensing using satellite and aerial data to deliver accurate and timely information for large palm oil plantation areas. Using the most recent geo-information tools and technologies with AI algorithms and image processing could significantly improve the palm oil industry by assisting in the identification of illegal deforestation, tree count, tree height, and the early detection of diseased leaves. Akhtar et al. (2023) presents some of the most recent technologies for managing palm oil plantations that are outfitted with AI, image processing, and remote sensing capabilities.

Nowadays, producers of generated gas in the petroleum business frequently flare it rather than commodifying it. In certain states, there are significant flaring magnitudes, which lead to issues with energy waste and CO_2 emissions. Lu (2023) addressed three issues with Bayesian learning through Markov chain Monte Carlo (MCMC) methods: distribution estimation, flaring time series analytics, and county level model construction. Additionally, he showed a nearest-neighbor-based strategy for operator level analytics and monitoring.

In light of the growing interest in soybean composition among farmers, agronomists, and commodity traders, Hernandez et al. (2023) aims to use remote-sensing satellite data to spatially predict soybean seed protein and oil concentrations at the field level before harvest time. This will give farmers and agronomists a competitive advantage to segregate quality and create an economic advantage to position the production at both domestic and global markets.

The intricate nature of air pollution prediction is very challenging, due to varying land uses and sources of pollution. In order to forecast and create a particulate matter pollution susceptibility map, Abu El-Magd et al. (2023) used the ML technique (random forest) using time series of particulate matter pollution records related to Ras Ghareb City, Egypt, in order to implement stringent regulations and improve particulate matter pollution management. Five air quality stations were used to gather data on air pollution for the years 2018 to 2021. The obtained prediction of pollutant concentration maps could be useful for monitoring and enhancing air quality in the future, and can enhance decision-making.

For illustrating the satellite-based oil spill detection combined with AI technique, the study conducted by Tysiąc et al. (2022) is presented. It demonstrates the effectiveness of satellite technology in detecting and monitoring oil spills in marine environments. Satellite imagery, particularly SAR sensors, can penetrate cloud cover and capture images regardless of weather conditions or time of day. By analyzing SAR images, AI algorithms can identify the distinctive signatures of oil spills on the water's surface, enabling rapid detection and response (Figure 13.2). This satellite-based oil spill detection system enhances environmental protection efforts by enabling timely containment and cleanup operations, minimizing the spread of oil and reducing the ecological impact on marine ecosystems.

13.6.5 Operational Optimization and Cost Reduction

Satellite imagery and AI techniques have the potential to optimize operations and reduce costs in the oil and gas industry. By continuously monitoring assets such as offshore platforms, refineries, and storage facilities using satellite imagery, companies can identify maintenance needs, detect anomalies, and prevent costly breakdowns. AI-based predictive analytics can leverage historical data and satellite imagery to forecast equipment failures or performance issues, allowing for proactive maintenance

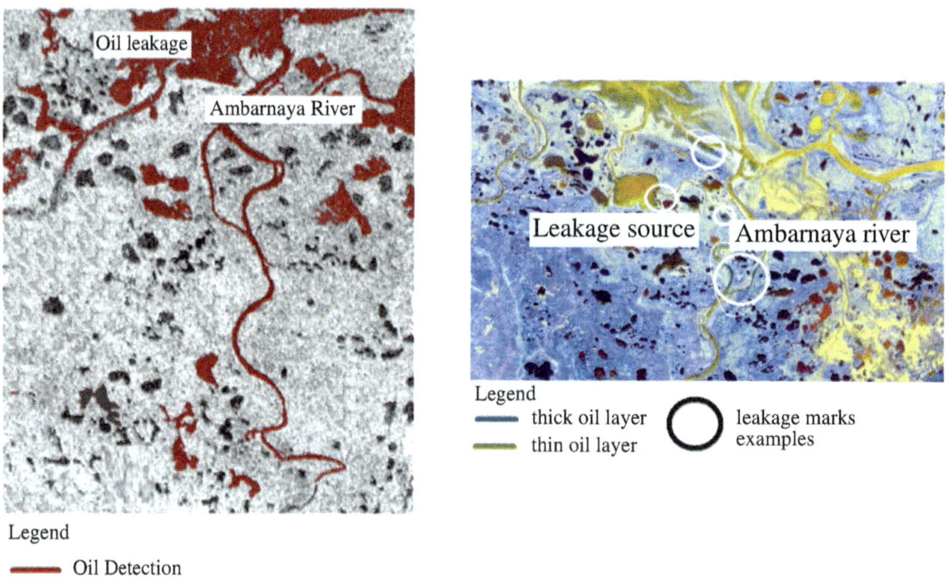

Figure 13.2 Results for oil detection. (a) Sentinel 1 imagery (oil is indicated in red); (b) Sentinel 2 imagery (oil is indicated in blue and green-blue) (Tysiąc et al., 2022 / MDPI / CC BY 4.0).

and reducing downtime. Additionally, satellite data can provide valuable information for supply chain optimization, such as tracking vessel movements, monitoring inventories, and optimizing transportation routes. These insights improve operational efficiency, minimize disruptions, and result in cost savings (De Kerf et al., 2020; Magana-Mora et al., 2021; Pan et al., 2020; Steel et al., 2024).

The practical applications of satellite imagery and AI in the oil and gas industry span multiple domains. From exploration and prospect identification to drilling planning, pipeline monitoring, environmental compliance, and operational optimization, satellite data and AI techniques offer valuable insights that enhance decision-making, reduce costs, and promote sustainable practices in the industry. By leveraging the power of satellite imagery and AI, oil and gas companies can improve efficiency, safety, and environmental stewardship across various operational aspects.

For oil spill detection, De Kerf et al. (2020) describe a novel framework that uses a thermal IR camera in conjunction with UAVs to forecast the location and amount of the oil spill inside port environments using CNN. After its training, this network will allow for low-cost and frequent inspections.

Large-scale infrastructure systems must be well-evaluated in order to guarantee the safety of their operations. Pan et al. (2020) propose to use deep learning techniques for image-based conditional evaluation of large-scale systems, to extract sensitive information from satellite photos for evaluation. The results demonstrate that, in comparison to conventional methods, ML techniques show considerable potential for assessing infrastructure, including high bridges and oil and gas pipelines, at both geographical and temporary scales.

By combining the advantages of extended coverage (both temporally and spatially) from numerical modeling and satellite remote sensing data with the asset-level accuracy of point-based observations, ML techniques have the potential to completely transform the provision of metocean forecasts that are essential to the safe and successful operation of offshore infrastructure. Steele et al. (2024) modified and applied a promising framework to the waters of the Gulf of Mexico that was first put forth by the present authors for the prediction of wave conditions on the European North West Shelf.

The crew's experience is crucial to maintaining safe and effective operations in the demanding environment of drilling operations. Even though the drilling rigs are currently equipped with sensors, the majority of the data analysis is still done manually. Magana-Mora et al. (2021) presented the Well Control Space Out technology, an internet-of-things (IoT) environment that combines cameras with an edge server to enable the real-time analysis of drillstring video pictures using cutting-edge deep-learning models. The implemented IoT technology analyzes camera data and offers an advising system for a specific drilling operation. The computer models automatically identify objects in order to monitor important drilling rig parts. Without requiring any laborious, manual computations, the video analysis results are shown on a dashboard that explains the situation and what to do in a well control issue. The potential benefits of using IoT for such industrial applications include improved asset utilization, enhanced safety and security, increased productivity, improved process efficiency, and cost reduction.

13.7 Conclusions

13.7.1 Summary of Key Points

Throughout the chapter, the significance of satellite imagery and AI in the oil and gas industry became evident. By leveraging satellite data and AI techniques, companies can enhance exploration activities, optimize drilling operations, monitor pipelines, ensure environmental compliance, and optimize overall operational efficiency. The integration of satellite technology and AI enables more informed decision-making, reduces costs, improves safety measures, and promotes sustainable practices within the industry.

13.7.2 Future Directions and Emerging Technologies

Looking ahead, several future directions and emerging technologies hold promise for the further advancement of satellite imagery processing and AI integration in the oil and gas industry. Firstly, the utilization of high-resolution hyperspectral sensors will enable more detailed spectral analysis, enhancing the identification and characterization of surface materials and improving accuracy in classification tasks. The integration of multi-sensor data fusion, combining optical, SAR, thermal, and hyperspectral data, will provide a comprehensive and holistic understanding of the Earth's surface (Zhang et al., 2016).

Advancements in data analytics and AI algorithms will continue to enhance the capabilities of satellite image processing. For instance, the integration of advanced ML techniques, such as reinforcement learning and generative adversarial networks (GANs), can further improve object detection, anomaly detection, and predictive analytics. Explainable AI techniques will gain importance, enabling better transparency, interpretability, and trust in AI models. Moreover, the deployment of small satellites and constellations will enhance temporal resolution and revisit rates, facilitating near-real-time monitoring and analysis. The integration of satellite data with other data sources, such as ground-based sensors, Internet of Things (IoT) devices, and aerial imagery, will lead to comprehensive and integrated monitoring systems for improved decision-making.

As the industry progresses, challenges related to data accessibility, scalability, and algorithm robustness will need to be addressed. Collaboration among stakeholders, including satellite operators, AI researchers, and industry professionals, will be essential for developing standardized frameworks, sharing data, and promoting best practices.

In the oil and gas industry, emerging technologies are being adopted by in order to overcome obstacles and enhance operations. Upstream activities are using AI to expedite and de-risk procedures (Koroteev and Tekic, 2021). A number of upstream industries are investigating and implementing Industry 4.0 technologies, such as edge computing, AI, and drones (Elijah et al., 2021). These technologies have the potential to upend the market; Wishnow et al. (2019) highlighted digital twins, AI, and aerial data collection as the top three areas for disruption. The usage of UAVs for inspection jobs is growing, providing answers to problems that traditional inspection methods encounter in dangerous areas (Sudevan et al., 2018). In the future, the digitization of the oil and gas sector is expected to be greatly assisted by the integration of robotics, sensors, and process control technology. This will result in lower operating costs, more productivity from the workforce, and enhanced efficiency (Sudevan et al., 2018).

In conclusion, satellite imagery processing and AI have revolutionized the oil and gas industry by providing valuable insights, optimizing operations, and supporting sustainable practices. The integration of satellite technology and AI enables accurate and efficient analysis of vast amounts of data, facilitating informed decision-making across various domains. With ongoing advancements in technology, the future of satellite imagery processing and AI integration in the oil and gas industry holds great potential, paving the way for improved efficiency, safety, and environmental stewardship in the sector.

References

Abu El-Magd, S., Soliman, G., Morsy, M. et al. (2023). Environmental hazard assessment and monitoring for air pollution using machine learning and remote sensing. International Journal of Environmental Science and Technology 20, pp. 6103–6116. https://doi.org/10.1007/s13762-022-04367-6.

Adler, J., Scanlon, B., & Reedy, R. (2018). Machine learning analysis of satellite imagery for detection of permian basin water impoundments. Paper presented at the SPE Annual Technical Conference and Exhibition, Dallas, Texas, USA, September 2018. Paper Number: SPE-191743-MS

Akhtar, M. N., Ansari, E., Alhady, S. S. N., & Abu Bakar, E. (2023). Leveraging on advanced remote sensing- and artificial intelligence-based technologies to manage palm oil plantation for current global scenario: a review. Agriculture 13 (2), p. 504. https://doi.org/10.3390/agriculture13020504.

Ang, K. L. M. & Seng, J. K. P. (2021). Big data and machine learning with hyperspectral information in agriculture. IEEE Access 9, pp. 36699–36718.

Asokan, A., Anitha, J., Ciobanu, M., et al. (2020). Image processing techniques for analysis of satellite images for historical maps classification—an overview. Applied Sciences 10 (12), pp. 4207. https://doi.org/10.3390/app10124207.

Azad, M. & Uddin, M. A. (2022). Prediction of offshore wave at east coast of Malaysia—a comparative study. Electronics 11 (16), p. 2527. https://doi.org/10.3390/electronics11162527.

Ball, J., Anderson, D., & Chan, C. (2017). Comprehensive survey of deep learning in remote sensing: Theories, tools, and challenges for the community. Journal of Applied Remote Sensing 11, p. 042609.

Bayramov, E., Buchroithner, M., & Kada, M. (2020). Quantitative assessment of ground deformations for the risk management of petroleum and gas pipelines using radar interferometry. Geomatics, Natural Hazards and Risk 11 (1), pp. 2540–2568. https://doi.org/10.1080/19475705.2020.1853611.

Bi, Q. F., Goodman, K. E., Kaminsky, J., et al. (2019). What is machine learning? A primer for the epidemiologist. American Journal of Epidemiology 188, pp. 2222–2239.

Brekke, C. & Solberg, H. (2008). Classifiers and confidence estimation for oil spill detection in ENVISAT ASAR images. IEEE Geoscience and Remote Sensing Letters 5 (1), pp. 65–69.

Cheng, G. & Han, J. (2016). A survey on object detection in optical remote sensing images. ISPRS Journal of Photogrammetry and Remote Sensing 117, pp. 11–28, ISSN 0924-2716, https://doi.org/10.1016/j.isprsjprs.2016.03.014.

Chintalapati, B., Precht, A., Hanra, S., et al. (2024). Opportunities and challenges of on-board AI-based image recognition for small satellite earth observation missions. Advances in Space Research ISSN 0273-1177, https://doi.org/10.1016/j.asr.2024.03.053.

Cracknell, M. J. & Reading, A. M. (2014). Geological mapping using remote sensing data: A comparison of five machine learning algorithms, their response to variations in the spatial distribution of training data and the use of explicit spatial information. Computers & Geosciences 63, pp. 22–33, ISSN 0098-3004, https://doi.org/10.1016/j.cageo.2013.10.008.

De Kerf, T., Gladines, J., Sels, S., et al. (2020). Oil spill detection using machine learning and infrared images. Remote Sensing 12 (24), p. 4090. https://doi.org/10.3390/rs12244090.

Del, F., Petrocchi, A., & Lichtenegger, J. (2000). Neural networks for oil spill detection using ERS-SAR data. IEEE Transactions on Geoscience and Remote Sensing 38, pp. 2282–2287.

DeLancey, E. R., Kariyeva, J., Bried, J. T., et al. (2019). Large-scale probabilistic identification of boreal peatlands using Google Earth Engine, openaccess satellite data, and machine learning. PLoS ONE 14 (6), p. e0218165. https://doi.org/10.1371/journal.pone.0218165.

Dileep, S., Zimmerle, D., Beveridge, J. R., et al. (2020). Automated identification of oil field features using CNNs. NeurIPS 2020 Workshop on Tackling Climate Change with Machine Learning. https://s3.us-east-1.amazonaws.com/climate-change-ai/papers/neurips2020/32/paper.pdf.

Elijah, O., Ling, P., Rahim, S. K., et al. (2021). A survey on industry 4.0 for the oil and gas industry: upstream sector. IEEE Access 9, pp. 144438–144468, https://doi.org/10.1109/ACCESS.2021.3121302.

Fan, J., Zhang, F., Zhao, D., et al. (2015). Oil spill monitoring based on SAR remote sensing imagery. Aquatic Procedia 3, pp. 112–118, ISSN 2214-241X, https://doi.org/10.1016/j.aqpro.2015.02.234.

Fourati, F. & Alouini, M. S. (2021). Artificial intelligence for satellite communication: A review. In: Intelligent and Converged Networks 2 (3), pp. 213–243, https://doi.org/10.23919/ICN.2021.0015.

Francis, R., McIsaac, K. A., Thompson, D. R., et al. (2014). Autonomous mapping of outcrops using multiclass linear discriminant analysis. *Proceedings of the 12th International Symposium on Artificial Intelligence, Robotics and Automation in Space* (i-SAIRAS 2014), Montreal, Canada, June 2014.

Gallego, A. J., Gil, P., & Pertusa, A. (2018). Segmentation of oil spills on side-looking airborne radar imagery with autoencoders. Sensors 18, p. 797.

Ghiglione, M. & Serra, V. (2022). Opportunities and challenges of AI on satellite processing units. *Proceedings of the 19th ACM International Conference on Computing Frontiers*, Turin,Italy. May 17-22, 2022, pp. 221–224, https://doi.org/10.1145/3528416.3530985.

Gholizadeh, A., Saberioon, M., Ben-Dor, E., et al. (2018). Monitoring of selected soil contaminants using proximal and remote sensing techniques: background, state-of-the-art and future perspectives. Critical Reviews in Environmental Science and Technology 48, pp. 243–278.

Golubov, S. I. & Lubskyi, M. S. (2020). Application of visible and long-wave infrared satellite data for hydrocarbon deposits prospecting. Geoinformatics: Theoretical and Applied Aspects 2020 2020, pp. 1–5, https://doi.org/10.3997/2214-4609.2020geo102.

Gruszczynski, S. & Gruszczynski, W. (2022). Supporting soil and land assessment with machine learning models using the Vis-NIR spectral response. Geoderma 405, pp. 1–17.

Guo, H., Goodchild, M. F., & Annoni, A. (2020). *Manual of Digital Earth*. Singapore: Springer 852, ISBN 978-981-32-9914-6, https:// doi.org/10.1007/978-981-32-9915-3.

Guo, H., Wu, D., & An, J. (2017). Discrimination of oil slicks and lookalikes in polarimetric SAR images using CNN. Sensors 17, p. 1837.

Hamidisepehr, A., Sama, M. P., & Hamidisepehr, A. (2019). Moisture content classification of soil and stalk residue samples from spectral data using machine learning. Transactions of the ASABE 62, pp. 1–8.

Hernandez, C. M., Correndo, A., & Kyveryga, P. (2023). On-farm soybean seed protein and oil prediction using satellite data. Computers and Electronics in Agriculture 212, p. 108096, ISSN 0168-1699, https://doi.org/10.1016/j.compag.2023.108096.

Huang, X. & Wang, X. (2014). The classification of synthetic aperture radar oil spill images based on the texture features and deep belief network. In: Wong, W.E., Zhu, T. (eds) *Computer Engineering and Networking*, vol 277, Springer: Cham, Switzerland, pp. 661–669. https://doi.org/10.1007/978-3-319-01766-2_76.

Hungund, B., Nair, G., & Rathore, P. S. (2023). Quantifying CH_4 and NO_2 emissions using satellite measurements with environmental impact in Texas: A county-wide assessment for oil & gas industry. Paper presented at the ADIPEC, Abu Dhabi, UAE, October 2023. Paper Number: SPE-215973-MS, https://doi.org/10.2118/215973-MS.

Jivane, N. & Soundrapandiyan, R. (2017). Enhancement of an algorithm for oil tank detection in satellite images. International Journal of Intelligent Engineering & Systems 10, pp. 218–225.

Jude Hemanth, D. (2019). Artificial Intelligence Techniques for Satellite Image Analysis, Springer Nature Switzerland AG, 274 pages, ISBN-10: 3030241793; ISBN-13: 978-3030241797

Keramitsoglou, I., Cartalis, C., & Kiranoudis, C. T. (2006). Automatic identification of oil spills on satellite images. Environmental Modelling & Software 21, pp. 640–652.

Koroteev, D. A. & Tekic, Z. (2021). Artificial intelligence in oil and gas upstream: Trends, challenges, and scenarios for the future. Energy and AI 3, p. 100041, https://doi.org/10.1016/j.egyai.2020.100041.

Kwok, C. Y., Wong, M. S., Ho, H. C., et al. (2018). Deep learning approach for rock outcrops identification. *2018 Fifth International Workshop on Earth Observation and Remote Sensing Applications (EORSA)*, Xi'an, China, 2018, pp. 1-6, https://doi.org/10.1109/EORSA.2018.8598577.

Li, K., Wan, G., Cheng, G., et al. (2020). Object detection in optical remote sensing images: A survey and a new benchmark. ISPRS Journal of Photogrammetry and Remote Sensing 159, pp. 296–307.

Lu, R. (2023). Application of machine learning to gas flaring, Doctoral dissertation, Colorado School of Mines, 2020, Submitted on 11 Jan 2023, https://doi.org/10.48550/arXiv.2301.04141.

Ma, L., Liu, Y., Zhang, X., et al. (2019). Deep learning in remote sensing applications: A meta-analysis and review. ISPRS Journal of Photogrammetry and Remote Sensing 152, pp. 166–177, ISSN 0924-2716, https://doi.org/10.1016/j.isprsjprs.2019.04.015.

Magana-Mora, A., Affleck, M., Ibrahim, M., et al. (2021). Well control space out: a deep-learning approach for the optimization of drilling safety operations. IEEE Access 9, pp. 76479–76492, https://doi.org/10.1109/ACCESS.2021.3082661.

McGrath, G. G., Woolridge, T., Dodge, K., et al. (2021). Incorporating automatic satellite detections of oil spills with numerical fate and trajectory modeling. International Oil Spill Conference Proceedings 2021 (1), p. 687930. https://doi.org/10.7901/2169-3358-2021.1.687930.

Newaz, S., Tahmid, M. T., Al-Aboody, N., et al. (2023). A ubiquitous method for predicting underground petroleum deposits based on satellite data. Scientific Reports 13, p. 6638.

Nguyen, K. A., Chen, W., Lin, B., et al. (2021). Comparison of ensemble machine learning methods for soil erosion Pin Measurements. ISPRS International Journal of Geo-Information 10, p. 42.

Nieto-Hidalgo, M., Gallego, A. J., & Gil, P. (2018). Two-stage convolutional neural network for ship and spill detection using SLAR images. IEEE Transactions on Geoscience and Remote Sensing 56, pp. 5217–5230.

Nirchio, F., Pandiscia, G., Ruggieri, G., et al. (2009). COSMO-SkyMed contribution in oil spill monitoring of the Mediterranean Sea. *2009 IEEE International Geoscience and Remote Sensing Symposium (IGARSS)*, Cape Town, South Africa, 2009, *2*, pp. II-781-II-784. https://doi.org/10.1109/IGARSS.2009.5418209.

Nirchio, F., Sorgente, M., & Giancaspro, A. (2005). Automatic detection of oil spills from SAR images. International Journal of Remote Sensing 26, pp. 1157–1174.

Ortiz, F., Monzon Baeza, V., Garces-Socarras, L. M., et al. (2023). Onboard processing in satellite communications using AI accelerators. Aerospace 10, p. 101. https://doi.org/10.3390/aerospace10020101.

Pan, H., Zhang, Z., Cao, Q., et al. (2020). Conditional assessment of large-scale infrastructure systems using deep learning approaches (Conference Presentation).

Paoletti, M. E., Haut, J. M., Plaza, J., et al. (2019). Deep learning classifiers for hyperspectral imaging: A review. ISPRS Journal of Photogrammetry and Remote Sensing 158, pp. 279–317, ISSN 0924-2716, https://doi.org/10.1016/j.isprsjprs.2019.09.006.

Peng, Y. P., Zhao, L., Hu, Y. M., et al. (2019). Prediction of soil nutrient contents using visible and Near-Infrared reflectance spectroscopy. ISPRS International Journal of Geo-Information 8, p. 437.

Power, D. T., Howell, C., Dodge, K., et al. (2018). Towards automation of satellite-based radar imagery for iceberg surveillance—machine learning of ship and iceberg discrimination.

Ravulakollu, K.K., Dewan, R., Verma, K., et al. (2024). Oil spill classification using machine learning, 2024 11th International Conference on Computing for Sustainable Global Development

(INDIACom), New Delhi, India, pp. 553-559, https://doi.org/10.23919/INDIACom61295.2024.10499056.
Sales, V. F., Racolte, G., Marques, A., et al. (2022). Analysis of machine learning techniques for carbonate outcrop image classification in landsat 8 satellite data. IGARSS 2022 - 2022 IEEE International Geoscience and Remote Sensing Symposium 3604–3607.
Sang, X., Xue, L., Ran, X., et al. (2020). Intelligent high-resolution geological mapping based on SLIC-CNN. ISPRS International Journal of Geo-Information 9, p. 99.
Scafutto, R. D. M., de Souza, C. R., & de Oliveira, W. J. (2017). Hyperspectral remote sensing detection of petroleum hydrocarbons in mixtures with mineral substrates: implications for onshore exploration and monitoring. ISPRS Journal of Photogrammetry and Remote Sensing 128, pp. 146–157.
Singha, S., Bellerby, T., & Trieschmann, O. (2013). Satellite oil spill detection using artificial neural networks. IEEE Journal of Selected Topics in Applied Earth Observations and Remote Sensing 6, pp. 2355–2363.
Singha, S., Ressel, R., & Velotto, D. (2016). A combination of traditional and polarimetric features for oil spill detection using TerraSAR-X. IEEE Journal of Selected Topics in Applied Earth Observations and Remote Sensing 9 (11), pp. 4979–4990.
Singha, S., Vespe, M., & Trieschmann, O. (2013). Automatic synthetic aperture radar based oil spill detection and performance estimation via a semi-automatic operational service benchmark. Marine Pollution Bulletin 73, pp. 199–209.
Solberg, H. & Solberg, R. (1996). A large-scale evaluation of features for automatic detection of oil spills in ERS SAR images. In Proceedings of the 1996 IEEE International Geoscience and Remote Sensing Symposium (IGARSS), Lincoln, NE, USA, 31 May 1996; 3, pp. 1484–1486.
Solberg, H. & Volden, E. (1997). Incorporation of prior knowledge in automatic classification of oil spills in ERS SAR images. In Proceedings of the 1997 IEEE International Geoscience and Remote Sensing Symposium (IGARSS), Singapore, 3–8 August 1997; 1, pp. 157–159.
Solberg, H., Storvik, G., & Solberg, R. (1999). Automatic detection of oil spills in ERS SAR images. IEEE Transactions on Geoscience and Remote Sensing 37, pp. 1916–1924.
Song, G., Wang, Z., Bai, L., et al. (2020). Detection of oil wells based on faster R-CNN in optical satellite remote sensing images. Image and Signal Processing for Remote Sensing 11533, pp. 114–121.
Song, Y. Q., Zhao, X., Su, H. Y., et al. (2018). Predicting spatial variations in soil nutrients with hyperspectral remote sensing at regional scale. Sensors 18, pp. 1–18.
Stathakis, D., Topouzelis, K., & Karathanassi, V. (2006). Large-scale feature selection using evolved neural networks. In: *Proceedings of the 2006 Image and Signal Processing for Remote Sensing, International Society for Optics and Phonetics*, Ispra, Italy, 6 October 2006; 6365, pp. 1–9.
Steele, E., Chen, J., Ashton, I., et al. (2024). A Spatiotemporal Machine Learning Framework for the Prediction of Metocean Conditions in the Gulf of Mexico. Paper presented at the Offshore Technology Conference, Houston, Texas, USA, May 2024, https://doi.org/10.4043/35104-MS.
Sudevan, V., Shukla, A., & Karki, H. (2018). Inspection of vertical structures in oil and gas industry: A review of current scenario and future trends. *SEG Global Meeting Abstracts, pp.* 65-68. *RDPETRO 2018: Research and Development Petroleum Conference and Exhibition, Abu Dhabi, UAE, 9-10 May 2018*. https://doi.org/10.1190/RDP2018-41952188.1.
Taghizadeh-Mehrjardi, R., Schmidt, K., Amirian-Chakan, A., et al. (2020). Improving the spatial prediction of soil organic carbon content in two contrasting climatic regions by Stacking machine learning models and rescanning covariate space. Remote Sensing 12, pp. 1–26.

Temizel, C., Canbaz, C. H, Palabiyik, Y., et al. (2021). A thorough review of machine learning applications in oil and gas industry. Paper presented at the SPE/IATMI Asia pacific oil & gas conference and exhibition, virtual, October 2021, https://doi.org/10.2118/205720-MS.

Topouzelis, K., Karathanassi, V., & Pavlakis, P. (2007). Detection and discrimination between oil spills and look-alike phenomena through neural networks. ISPRS Journal of Photogrammetry and Remote Sensing 62, pp. 264–270.

Topouzelis, K. & Psyllos, A. (2012). Oil spill feature selection and classification using decision tree forest on SAR image data. ISPRS Journal of Photogrammetry and Remote Sensing 68, pp. 135–143.

Tysiąc, P., Strelets, T., & Tuszyńska, W. (2022). The application of satellite image analysis in oil spill detection. Applied Sciences 12 (8), p. 4016. https://doi.org/10.3390/app12084016.

Usery, E. L. (2020). GeoAI for topographic mapping feature extraction to knowledge graph. Abstracts of the ICA 2, p. 39, https://doi.org/10.5194/ica-abs-2-39-2020.

Wang, S., Guan, K. Y., Zhang, C. H., et al. (2022). Using soil library hyperspectral reflectance and machine learning to predict soil organic carbon: assessing potential of airborne and spaceborne optical soil sensing. Remote Sensing of Environment 271, pp. 1–15.

Wang, Z., Bai, L., Song, G., et al. (2021). An oil well dataset derived from satellite-based remote sensing. Remote Sensing 13 (6), 1132, https://doi.org/10.3390/rs13061132.

Wang, Z., Bai, L., Song, G., et al. (2023). Optimized faster R-CNN for oil wells detection from high-resolution remote sensing images. International Journal of Remote Sensing 44 (22), pp. 6897–6928, https://doi.org/10.1080/01431161.2023.2275322.

Wei, L. F., Yuan, Z. R., Zhong, Y. F., et al. (2019). An improved gradient boosting regression tree estimation model for soil heavy metal (Arsenic) pollution monitoring using hyperspectral remote sensing. Applied Science 9, pp. 1–16, https://doi.org/10.3390/app9091943.

Wijata, A. M., Foulon, M. F., Bobichon, Y., et al. (2023). Taking artificial intelligence into space through objective selection of hyperspectral earth observation applications: to bring the "brain" close to the "eyes" of satellite missions. *IEEE Geoscience and Remote Sensing Magazine* 11(2), pp. 10–39, https://doi.org/10.1109/MGRS.2023.3269979.

Wishnow, D., Azar, H., & Rad, M. P. (2019). A Deep Dive into Disruptive Technologies in the Oil and Gas Industry. Paper presented at the Offshore Technology Conference Brasil, Rio de Janeiro, Brazil, October 2019, https://doi.org/10.4043/29779-MS.

Wu, M., Lin, N., Li, G., et al. (2022). Hyperspectral estimation of petroleum hydrocarbon content in soil using ensemble learning method and LASSO feature extraction. Environmental Pollutants and Bioavailability 34 (1), pp. 308–320. https://doi.org/10.1080/26395940.2022.2102543.

Xu, L., Li, J., & Brenning, A. (2014). A comparative study of different classification techniques for marine oil spill identification using RADARSAT-1 imagery. Remote Sensing Environment 141, pp. 14–23.

Xue, Y., Zou, B., Wen, Y. M., et al. (2020). Hyperspectral inversion of chromium content in soil using support vector machine combined with lab and field spectra. Sustainability 12, pp. 1–16.

Yang, Y. J., Suman, S., & Mayerle, R. (2022). A deep learning based oil spill detector using Sentinel-1 SAR imagery. International Journal of Remote Sensing 43 (11), pp. 4287–4314, https://doi.org/10.1080/01431161.2022.2109445.

Yu, S., Song, Z., Bai, Y., et al. (2023). Satellite-estimated air-sea CO_2 fluxes in the Bohai Sea, Yellow Sea, and East China Sea: Patterns and variations during 2003–2019. Science of the Total Environment 904, p. 166804, https://doi.org/10.1016/j.scitotenv.2023.166804. Epub 2023 Sep 7. PMID: 37689183.

Zalpour, M., Akbarizadeh, G., & Alaei-Sheini, N. (2020). A new approach for oil tank detection using deep learning features with control false alarm rate in high-resolution satellite imagery. International Journal of Remote Sensing 41, pp. 2239–2262.

Zeng, K. & Wang, Y. (2020). A deep convolutional neural network for oil spill detection from spaceborne SAR images. Remote Sensing 12 (6), p. 1015. https://doi.org/10.3390/rs12061015.

Zhang, L., Shi, Z., & Wu, J. (2015). A hierarchical oil tank detector with deep surrounding features for high-resolution optical satellite imagery. IEEE Journal of Selected Topics in Applied Earth Observations and Remote Sensing 8 (10), pp. 4895–4909.

Zhang, L., Zhang, L., & Du, B. (2016). Deep learning for remote sensing data: a technical tutorial on the state of the art. IEEE Geoscience and Remote Sensing Magazine 4 (2), pp. 22–40, https://doi.org/10.1109/MGRS.2016.2540798.

Zhang, N., Liu, Y., & Zou, L. (2018). Automatic Recognition of Oil Industry Facilities Based on Deep Learning. In Proceedings of the IGARSS 2018—2018 IEEE International Geoscience and Remote Sensing Symposium, Valencia, Spain, 22–27 July 2018; pp. 2519–2522.

Zhao, D. & Tan, B. (2023). Multi-Scale Encoding (MSE) Method with spectral shape information (SSI) for detecting marine oil-gas leakages. Remote Sensing 15, p. 2184.

Zhong, Y., Xu, Y., Wang, X., et al. (2019). Pipeline leakage detection for district heating systems using multisource data in mid- and high-latitude regions. ISPRS Journal of Photogrammetry and Remote Sensing 151, pp. 207–222, ISSN 0924-2716, https://doi.org/10.1016/j.isprsjprs.2019.02.021.

14

Modern AI Usage in the Oil and Gas Industry for Reservoir Characterization and Lithofacies Forecasting (Rock Typing)

Hasna Yazid[1], Said Gaci[2], Mohammed Farfour[3]

[1] *Department of Computer Science, University of M'Hamed Bougerra (UMBB), Boumerdes, Algeria*
[2] *Sonatrach–Central Directorate of Research & Development(DC R&D), Boumerdes, Algeria*
[3] *Department of Earth Sciences, College of Science, Sultan Qaboos University, Muscat, Oman*

14.1 Introduction

The petrophysical evaluation of reservoir rocks is a fundamental part of reservoir characterization and modeling. It aims at identifying different reservoir rocks and facies having both petrophysical and geological significance, and predicting their distribution.

The characterization process is based on an integrated analysis of all available well, core and production data. The technique used to achieve this integrated petrophysical characterization is called "rock typing" (RT). In petroleum reservoirs, "rock typing" refers to a novel petrophysical classification technique that categorizes rock units based on their petrophysical properties, such as porosity, permeability, lithology, pore size distribution, and fluid saturation (Tavakoli, 2018). This technique aims to link geological facies to reservoir characteristics, enhancing the understanding of reservoir behavior on both macroscopic and microscopic levels. Ideal rock type samples exhibit similar geological and reservoir characteristics, including fluid movement and storage (Figure 14.1) (Gunter et al., 2014).

According to this method, reservoir rocks are divided into discrete units that have undergone comparable diagenesis and were deposited under comparable geological circumstances. For the purpose of creating static and dynamic reservoir models for simulation tasks, rock typing can be used on both clastic and carbonate reservoirs.

Rock typing is critical for reservoir characterization and modeling since it allows to understand reservoir heterogeneity and to predict fluid flow behavior by sorting reservoir rocks into different categories or groups based on their petrophysical attributes. It is of high importance to reservoir management and characterization, allowing for more precise reservoir simulations and well-informed decision-making all the way through the field's lifecycle (Chandra et al., 2015; Riazi, 2018).

There are various techniques for rock typing, such as:

1. Core Analysis: Direct measurements of petrophysical characteristics like porosity and permeability are possible through laboratory analysis of core samples that were taken out

Geophysical Exploration for Hydrocarbon Reservoirs, Geothermal Energy, and Carbon Storage: New Technologies and AI-based Approaches, First Edition. Edited by Said Gaci.

of the reservoir. The petrophysical properties of core plugs are determined by tests, from which distinct rock types are classified.

2. Log Analysis: Wireline logs are acquired and utilized to determine the petrophysical parameters of the reservoir rocks. Interpretation methods of log data can be employed to derive parameters for identifying rock types (Figure 14.2).
3. Geological and Petrophysical Modeling: Comprehensive reservoir models can be created by combining petrophysical data with geological information, such as facies, sedimentary formations, and diagenetic characteristics. Subsequently, reservoir development strategies can be optimized and fluid flow behavior can be simulated using these models.
4. Geostatistical Techniques: The application of geostatistical algorithms, such as principal component analysis (PCA) and cluster analysis, to petrophysical data can aid in the identification and characterization of discrete rock types. These techniques help uncover spatial trends and patterns in reservoir attributes.

Rock typing is typically conducted on cores and thin sections derived from them. The identified rock units may show consistent porosity, permeability, capillary pressure, and dynamic behavior. Due to the high financial cost of coring every well in a large field or every zone of interest within a specific well, rock typing is utilized to minimize these substantial expenses.

Several rock typing approaches have been proposed over the years (Kolodzie, 1980; Tiainen et al., 2002; Aguilera, 2004; Gunter et al., 2014; Abedini and Torabi, 2015; Palabiran et al., 2016; Haikel et al., 2018; Tavakoli, 2018; Krivoshchekov et al., 2023). The most commonly used techniques for rock typing include the flow zone indicator (FZI), Winland R35, Pore Geometry Structure (PGS) method, and Lucia approach:

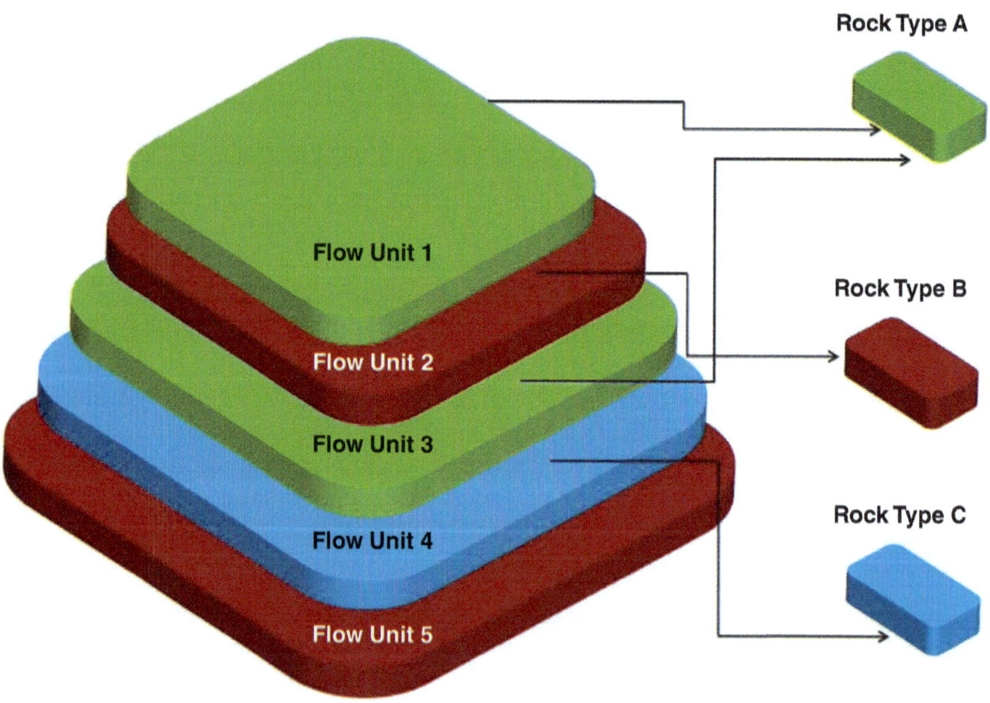

Figure 14.1 Depositional pattern of rock type beneath the surface.

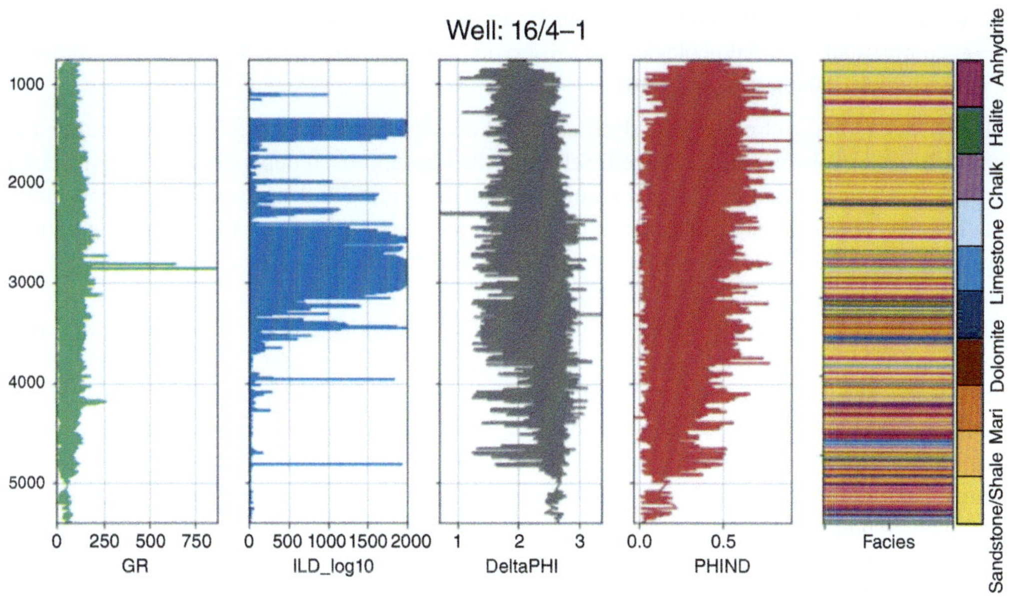

Figure 14.2 Facies identification based on well logs. Panels 1–6 (from left to right) represent: gamma ray (GR, measured in API), resistivity (ILD-log10, logarithmically transformed deep induction log [ILD] values, measured in ohm meters), neutron-density porosity difference, average neutron-density porosity (DeltaPHI and PHIND, both dimensionless quantities), and facies (each lithology is assigned with a different color), respectively.

1. Flow zone indicator (FZI) technique is used to categorize different types of rock according to the flow zones found in reservoir rock (Haikel et al., 2018).
2. Winland R35 examines core data to determine permeability and categorize different types of rocks (Haikel et al., 2018; Krivoshchekov et al., 2023).
3. Structure of Pore Geometry (PGS) evaluates the geometry and pore structure of reservoir rocks to help with classification (Haikel et al., 2018).
4. Lucia approach provides a categorization scheme for reservoir rocks by classifying rocks according to their grain size and rock fabric (Haikel et al., 2018).

Each of them has pros and cons of its own and classifies rocks based on distinct concepts:

1. Drill Cuttings Analysis: A consistent, unbiased, and reasonably priced data source is offered by drill cuttings, particularly when traditional core data is not available. However, the accuracy and representativeness of the analysis of drill cuttings may be lower than that of core data (Tiainen et al., 2002).
2. The flow zone indicator (FZI) method categorizes rock types according to hydraulic flow units (HFU) based on permeability and porosity relationships. However, a drawback of this technique is that the classification is linked with uncertainties (Krivoshchekov et al., 2023).
3. The Winland R35 approach, rooted in core data analysis, aids in permeability estimation and rock type classification. However, its limitation lies in its inability to fully encompass the diverse reservoir characteristics necessary for accurate rock typing (Palabiran et al., 2016).
4. The Pore Geometry Structure (PGS) method assists in classifying rocks by assessing the pore structure and geometry of reservoir rocks. However, it may not comprehensively identify all crucial reservoir features required for comprehensive rock typing (Palabiran et al., 2016).

5. The Lucia method relies on the fabric and grain size of rocks; however, this approach may not encompass all essential geological and reservoir characteristics necessary for comprehensive rock typing (Palabiran et al., 2016).

Every rock typing technique has advantages and disadvantages that affect how accurately and thoroughly reservoirs are characterized. Selecting the best approach depending on the particular geological setting and data availability requires an understanding of these benefits and drawbacks.

In recent years, artificial intelligence (AI) has emerged as a significant tool for reservoir rock typing (Tschannen et al., 2017; Valentin et al., 2019; Fadokun et al., 2020; Mohammadian et al., 2022; Yazid and Gaci, 2022, 2024; Xing et al., 2023).

Tschannen et al. (2017) utilized convolutional neural networks (CNNs) with an inception network architecture to conduct rock typing, yielding promising outcomes. Valentín et al. (2019) introduced a deep residual CNN tailored for the same purpose. Fadokun et al. (2020) investigated supervised and unsupervised machine learning techniques for facies prediction, showcasing their efficacy in reducing uncertainty and bias. Mohammadian et al. (2022) devised a data-driven approach employing machine learning to enhance the precision of petrophysical rock typing and permeability prediction within a heterogeneous carbonate reservoir in Iran, achieving notable accuracy in permeability and pore throat radius prediction. Xing et al. (2023) proposed a machine learning methodology for rock type classification using both core and log data, achieving remarkable accuracy and addressing the challenge of acquiring core samples from wells.

14.2 Workflow of Rock Typing Using Machine Learning

Rock typing prediction using machine learning algorithms can be conducted as follows:

1. Data Collection: It consists of gathering a dataset of rock samples together with related characteristics, such as density, acoustic qualities, porosity, permeability, and mineral composition. Usually, imaging methods, well logging, or laboratory testing are used to get these qualities.
2. Data preprocessing: That incorporates handling missing values, standardizing features, and eliminating anomalies to clean up the data. Ensuring the quality of the dataset and enhancing the machine learning model's performance necessitate this stage.
3. Selecting and extracting features: This step allows choosing pertinent aspects or taking useful features out of the unprocessed data. The most crucial elements for rock typing can be found using feature selection methods like PCA or feature importance analysis.
4. When choosing a machine learning model for rock typing, it's essential to select one that suits the specific requirements of the task. Neural networks, decision trees (DT), random forests (RF), support vector machines (SVM), and KNN are commonly used models. The choice depends on several factors such as the nature of the data and the complexity of the rock typing problem.
5. Model training where the dataset is divided into sets for testing and training. Using techniques that discover patterns and connections between different rock types and characteristics, train the chosen machine learning model on the training set.
6. Evaluation of the performance of the trained model by utilizing metrics such as accuracy, precision, recall, F1-score, and confusion matrix on the testing data. This step helps gauge how effectively the model generalizes to hypothetical rock samples.
7. Model Optimization: To further enhance performance, fine-tune the model's hyperparameters and modify its feature set. For hyperparameter tweaking, methods such as grid search and cross-validation can be applied.

14.3 Application

This section presents the overall architecture of the system designed for implementing rock typing approaches. Various machine learning methods were explored, including SVM, one-class support vector machine (OCSVM), radial basis function (RBF), logistic regression (LR), KNN, multilayer perceptron neural networks (MLP), DT, both unsupervised and supervised self-organizing maps (SOM), RF, gradient boost model (GBM), and adaptive boost (AB).

14.3.1 Used Dataset

The dataset originates from the Council Grove gas reservoir located in southwest Kansas, specifically within the Panoma Council Grove Field, which predominantly functions as a carbon dioxide reservoir spanning an area of 2700 mi^2. This dataset comprises information from 98 wells, containing a collection of four predictive variables alongside rock facies (class) for each example vector. Additionally, validation data shares the same predictive variables in the feature vector. The facies are determined based on core descriptions obtained from these wells, sampled vertically at half-foot intervals, representing continuous variables.

The dataset comprises seven predictor variables derived from wireline logging including gamma ray (GR, measured in API), resistivity record (ILD-log10, logarithmically transformed deep induction log (ILD) values, measured in ohm meters), neutron-density porosity difference, average neutron-density porosity (DeltaPHI and PHIND, both dimensionless quantities – in percent units), photoelectric factor (PEF, unitless), compressional wave travel time (DTC, measured in microseconds per foot), and rate of penetration or drilling progress (ROP, measured in feet per hour). Prior to being utilized by machine learning algorithms, this dataset undergoes a series of pre-processing steps.

14.3.2 Data Preparation

Despite recent advancements in statistics such as LR and SVMs, it remains evident that in most real-world classification scenarios, the success of a model is heavily reliant on the nature and quality of the data rather than the modeling technique employed. Numerous comparisons based on one or more datasets underscore that very minor discrepancies in performance across various strategies are sometimes insignificant when compared to the inherent uncertainties involved (Yazid and Gaci, 2022).

The first uncertainty arises from defining the variable under study, while the second uncertainty pertains to the representativeness of the study sample, from which training and test samples are drawn, compared to a population where not all individuals have been observed (selection bias), or which may have changed since observation (retrospective sampling). The third source of uncertainty relates to the impact of selecting training and test samples, and the resulting variance, particularly when unusual occurrences are observed, on performance. Lastly, the statistician's influence on performance, as they may more effectively optimize a method they are familiar with, represents the fourth source of uncertainty.

In many cases, the most substantial and reliable performance is achieved not only by sophisticated techniques but also by simpler methods with stringent assumptions, such as linear discriminant analysis, which are less prone to overfitting. Due to these uncertainties, the marginal performance gain of a complex model may be misleading, particularly when compared to the potential improvement of simpler models.

14.3.2.1 Data Cleaning

Ambient factors surrounding the sensors (logs) can introduce noise and lead to data loss. Data points affected by such inaccuracies are labeled as anomalies and must be eliminated from the dataset.

14.3.2.2 Data Normalization

Normalization is a pre-processing technique that allows outliers to be removed from the data. Additionally, it is necessary in order to use specific algorithms (Fadokun et al. 2020).

To implement this adjustment, we will normalize the data by dividing it by its standard deviation and subtracting it from its mean. However, it is advisable to avoid this approach when employing alternative methods that rely on the relative significance of each dimension, such as PCA, which will be utilized later on.

14.3.2.3 Principal Component Analysis

Big data sets are getting more and more prevalent, yet they can be challenging to understand. One method for lowering these datasets' dimensionality and improving interpretability while minimizing information loss is PCA (Fadokun et al., 2020).

PCA maximizes variance by creating new, uncorrelated variables. As an adaptive data analysis technique, PCA solves an eigenvalue/eigenvector problem to identify these new variables, which are determined by the available dataset rather than predefined criteria. Additionally, PCA is flexible, with many variations of the method developed and tailored to fit different types and configurations of data (Yazid and Gaci, 2022).

14.3.2.4 Class Distribution in the Dataset

Upon examining the frequency of each class in the dataset, it was observed that the dataset is unbalanced. Some classes contain as few as 100 samples, while others have up to 20 000 samples, leading to biased learning. To balance the classes, we attempted to reduce the size of the larger classes.

14.3.2.5 Choice of Parameters

Several runs were conducted using various combinations of learning parameters, revealing that these parameters significantly affect the results. To address this issue, a learning parameter selection approach was implemented, which notably improved the accuracy of the findings. Figure 14.3 illustrates a graph showing the correlation between the dataset parameters.

14.3.2.6 Recursive Feature Elimination Selection Algorithm

Recursive Feature Elimination (RFE) is an algorithm widely used across various fields, including high-dimensional multi-category data (You et al., 2014), handwritten digit recognition (Zeng et al., 2009), feature selection for SVMs with RBF kernel (Liu et al., 2011), and gene selection in microarray data (Furlanello, 2003). These studies demonstrate that RFE is an effective tool for identifying relevant features and improving classification accuracy.

RFE begins with all features in the training dataset and iteratively eliminates the least important ones until the desired number of features is retained. This process involves fine-tuning the model, removing less significant features, rearranging the remaining features by relevance, and adjusting the machine learning algorithm at the core of the model. This cycle continues until the predetermined number of features remains.

14.3.2.7 Selection of Machine Learning Algorithms

After preparing the dataset for learning, various machine learning methods were chosen: SVM, RBF, OCSVM, LR, KNN, MLP, DT, unsupervised and supervised SOM, RF, GBM, and AB.

14.3.2.7.1 Support Vector Machine

The SVM was selected due to its notable potential from a machine learning perspective. An SVM is a linear or nonlinear classifier, which serves as a mathematical function capable of

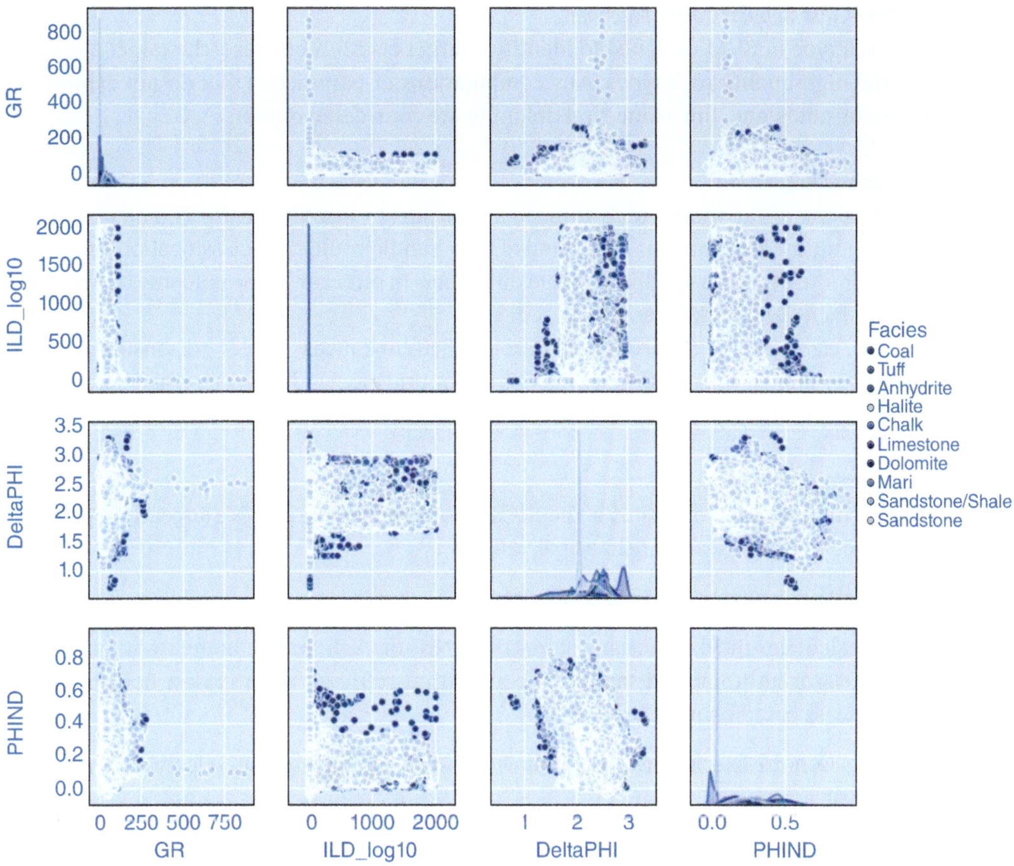

Figure 14.3 Cross plots of different well logs.

distinguishing between two (or more) different classes of objects. In our case study, instead of identifying the best, average, and poor rock types (with the latter two being less useful for petrophysicists and geophysicists), we will focus on the cluster representing the best rock type, characterized by high permeability facies such as limestone, sandstone, and dolomite. The remaining types will be considered poor rock types (Pedersen and Schoeberl, 2006).

14.3.2.7.2 Radial Basis Function Kernel Parameters

The gamma and C parameters of the SVM and RBF kernel significantly impact its performance. The gamma parameter intuitively defines the range of influence for a single training example, with low values indicating a broader influence and high values indicating a narrower influence. Essentially, gamma can be seen as the inverse of the radius of influence for the samples chosen as support vectors by the model.

The C parameter balances the correct classification of training examples against maximizing the margin of the decision function. Higher values of C allow for a smaller margin if it means the decision function can more accurately classify all training points. Conversely, lower values of C prefer a larger margin, resulting in a simpler decision function at the expense of learning precision. In other words, C acts as a regularization parameter in the SVM.

14.3.2.7.3 One-Class Support Vector Machine

The OCSVM is a type of SVM designed to identify a target class. In our case, the target class corresponds to the high-quality rock types. Any combinations of parameters that do not exhibit the characteristics of sandstone, limestone, and dolomite are considered outliers.

14.3.2.7.4 Logistic Regression

LR stands out as one of the most widely utilized algorithms in machine learning, and it was also employed in our study. Multinomial LR extends LR to handle multiclass classification problems, where there are more than two possible discrete outcomes. In our case, these outcomes correspond to the classes of facies and rock types (Albon, 2018).

Classification error rate: proportion of misclassified instances out of all instances. The classification error rate on the training set may be too optimistic. Under no circumstances should test data be used to train the classifier.

14.3.2.7.5 K-Nearest Neighbor

Another algorithm utilized in our study is the KNN algorithm. Primarily, it's a supervised learning method employed for classification (Zhao et al., 2021).

The KNN algorithm operates in two main steps:

1. Initially, it calculates the distance and stores the KNN for each sample in the training set.
2. Subsequently, for an unlabeled sample, the algorithm retrieves the k closest neighbors and predicts its class based on a majority vote.

In our investigation, we executed the program with each distance function, selecting the one that yielded the highest accuracy.

Determination of the K value of the KNN

The algorithm relies on the assigned K value to calculate the distance from the K. Different metrics can be used to calculate similarities under the KNN algorithm. The options vary depending on the type of variables characterizing the observations (quantitative or qualitative).

Distances available for quantitative (metric) data: Euclidean, Minkowski, Manhattan, Chebyshev, and Canberra. The distance used in our work is the Euclidean distance, because it represents the metric with the minimum error rate compared to other metrics. In our work, we tested with all the distances mentioned, and we chose the one with the best precision. Figure 14.4 shows the accuracy as a function of the K value of the KNN (Albon, 2018).

14.3.2.7.6 MLP neural networks

An MLP typically comprises at least three layers of nodes: an input layer, a hidden layer, and an output layer (Hosseini et al., 2020). Each node, except for those in the input layer, functions as a neuron utilizing a nonlinear activation function. In our study, we employed 200 hidden layers. The theoretical underpinning of the MLP pseudo algorithm is detailed in (Albon, 2018).

A model with insufficient capacity may struggle to learn from the training dataset, resulting in underfitting. Conversely, a model with excessive capacity runs the risk of memorizing the training dataset, leading to overfitting or potentially getting stuck or lost. The capacity of a neural network model is determined by adjusting the number of nodes and layers (Albon, 2018).

Alpha serves as a regularization term parameter, also known as a penalty term, aimed at mitigating overfitting by constraining the magnitude of the weights. Elevating alpha can address high variance,

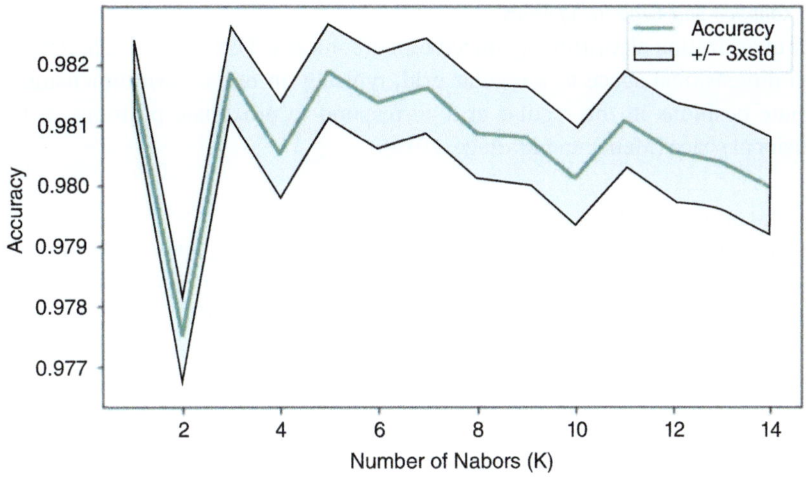

Figure 14.4 Optimization of the *k* value in the KNN model

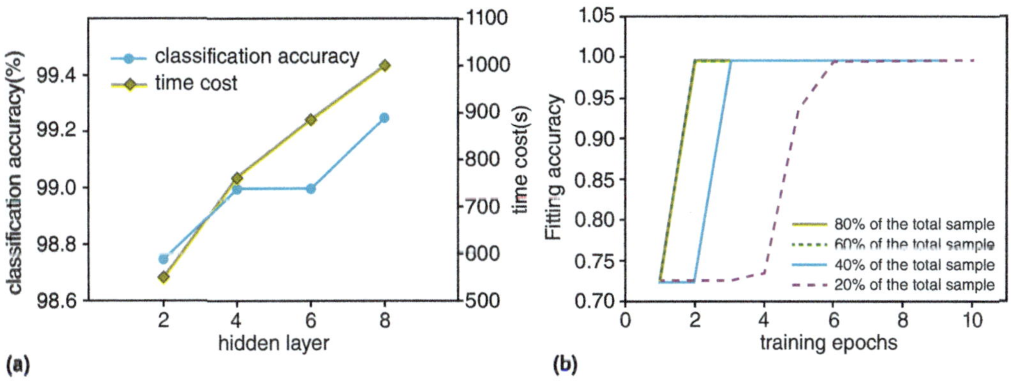

Figure 14.5 Parameter selection for MLP performance optimization (Jiang et al., 2018) (a) Classification accuracy and time cost across varying numbers of hidden layers; (b) Fitting accuracy achieved using different proportions of the training dataset.

indicative of overfitting, by promoting smaller weights. This adjustment leads to decision boundary plots with reduced curvatures. Conversely, reducing alpha can mitigate high bias, indicating underfitting, by encouraging larger weights, potentially resulting in a more intricate decision boundary (Albon, 2018).

An epoch, in the context of machine learning, refers to the number of times the entire training dataset is passed through the algorithm during training. If the batch size encompasses the entire training dataset, the number of epochs equates to the number of iterations. Figure 14.5 depicts the accuracy in relation to both the number of hidden layers and the number of epochs. In our study, we conducted training with 500 epochs (Albon, 2018).

14.3.2.7.7 Decision trees

The DT classifier creates a hierarchical partition of the training data space by using constraints on attribute values to divide the data. This division process is carried out iteratively until the leaf nodes contain a specified number of records, which are then utilized for classification purposes (Saravanan and Sujatha, 2018).

14.3.2.7.8 Kohonen Self-Organizing Maps

It is a vector quantization algorithm organized topographically. This method calculates a mapping from a high-dimensional space to a regular grid, typically in one or two dimensions, ensuring that proximate positions in the regular grid correspond to proximate positions in the original high-dimensional space (Melka and Mariage, 2019).

14.3.2.7.9 SOM Unsupervised

In our study, we employed a 2D grid, known as the map, within a SOM, which can be configured with various topologies. We opted for the simplest topology, consisting of a 2D rectangular grid composed of nrow × ncolumn nodes. This grid is fully connected to the input layer, meaning each node is linked to all n input features via n weights. Additionally, we utilized the Euclidean distance for processing (Asan and Ercan, 2012). Figure 14.6 illustrates an example of unsupervised SOM execution, resulting in three clusters.

14.3.2.7.10 SOM Supervised

To apply the algorithm to supervised learning classification tasks, we incorporate a second SOM alongside the unsupervised SOM. The two SOMs differ in the size and structure of their weights. The weights of the unsupervised SOM match the dimensionality of the input data, meaning that adapting these weights can alter the processing for each input data point. Conversely, the weights of the supervised SOM match the dimensionality of the target variable for the given task. Two scenarios are distinguished: regression and classification. In regression, the weights of the supervised SOM are one-dimensional and contain continuous values. In classification, the weights of the supervised SOM contain class labels. By combining the unsupervised and supervised SOMs, the former selects each data point, while the latter associates the selected points with a specific estimate (Asan and Ercan, 2012).

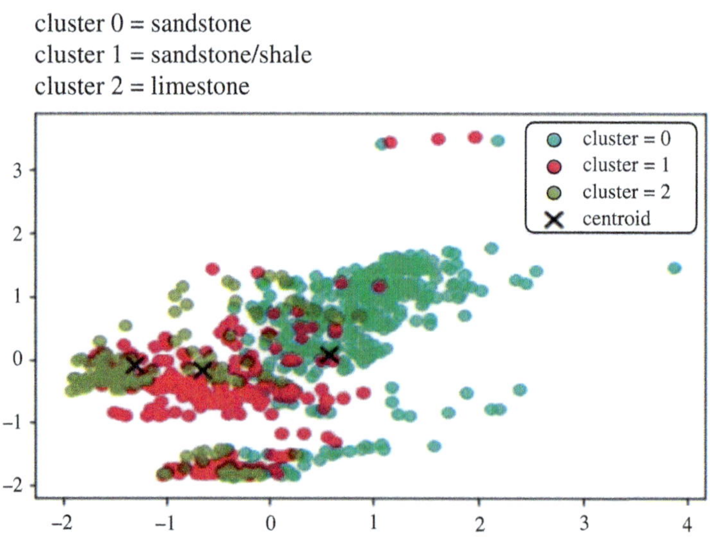

Figure 14.6 Example of unsupervised SOM execution, resulting in three distinct clusters: cluster 0 (sandstone), cluster 1 (sandstone/shale), and cluster 2 (limestone). Cluster centroids are marked with black crosses.

Once the maximum number of iterations, t_{max}, is reached, the unsupervised SOM is fully trained. Subsequently, the unsupervised SOM can compute for each input data point, irrespective of whether the data point was included in the training dataset. Given that the number of nodes on the SOM grid can exceed the number of data points in a given dataset, certain nodes may exist without being associated with a data point in that dataset (Asan and Ercan, 2012).

14.3.2.7.11 Random Forest

Even with limited data, this algorithm yields satisfactory results (Gupta et al., 2022). It comprises multiple DT, and the final result is determined by aggregating the outcomes of all the trees through a voting mechanism. In our study, we employed one hundred trees.

14.3.2.7.12 Gradient Boost Model

Since its introduction in 2014, this model has been considered the best in hackathons and machine learning competitions. From predicting click-through rates to classifying high-energy physical events, it has a proven track record of performance and speed (Aziz et al., 2020). We used it with a hundred DT.

14.3.2.7.13 Adaptive Boost

This technique belongs to the ensemble category, wherein numerous weak learners are amalgamated to form a robust learner (Chengsheng et al., 2017). In this approach, each weak learner is structured as decision stems, resembling trees with single splits and two terminal nodes, which are utilized for classifying observations.

Unlike RF, where each classifier is accorded equal importance, in this technique, classifiers are weighted differently based on their individual performance. More weight is allocated to classifiers exhibiting higher accuracy, and vice versa. Additionally, weights are assigned to observations at the conclusion of each round, with poorly predicted observations receiving increased weight. This augmentation enhances their likelihood of being selected more frequently in subsequent classifier samples. Consequently, the training sets in successive iterations are influenced by their preceding counterparts, creating a correlation between the constructed trees.

This approach heightens predictive accuracy by assigning weights to observations at the conclusion of each tree and by assigning weights (scores) to each classifier. Consequently, each classifier contributes differently to the final prediction, in contrast to RF where all trees hold equal weight.

14.3.3 Data Augmentation

The type of data we possess is numerical, which necessitates transitioning to data supported by data augmentation. In the case of image processing, they are converted into a vector of pixel values to apply learning algorithms later.

In our case, each line of the log dataset represents a vector that can be transformed into an image. Indeed, we have managed to convert our data into an image dataset labeled by the corresponding class. The general workflow is as follows: cleaning and normalizing data, data augmentation, Splitting data (train/test set), model selection, hyper parameter tuning, model training, and data prediction (Albon, 2018).

14.3.3.1 Preparing the Dataset for Data Augmentation

As mentioned before, our dataset is unbalanced, which will influence learning and classification as a result. The transition to image processing allows us to perform data augmentation.

Data Augmentation: It is an integral process of deep learning, as in deep learning, we need large amounts of data, and in some cases, it is not possible to collect thousands or millions of images. Data augmentation thus comes to the rescue.

It is the process of increasing the quantity and diversity of data. We do not collect new data; instead, we transform the existing data. This helps us increase the size of the dataset and introduce variability into the dataset. The most commonly used operations are:

Rotation: This operation, as its name suggests, simply rotates the image by a specified degree.

Shearing: It is also used to transform the orientation of the image.

Zoom: It allows us to zoom in or out.

Cropping: That allows us to crop the image or select a specific area from an image.

Flipping: It allows us to flip the orientation of the image. We can use horizontal or vertical flipping. We must use this feature with caution because there will be scenarios where this operation may not make much sense. For example, suppose we are designing a facial recognition system; it is very unlikely that a person would stand upside down in front of a camera, so we can avoid using the vertical flipping operation.

Brightness Adjustment: This feature helps us cope with lighting changes. We might encounter a scenario where most of your dataset consists of images with a similar brightness level, for example, when collecting images of employees entering the office. By augmenting the images, we ensure that our model is robust and capable of detecting the person even in different environments.

Vectorization of Images: After generating the images from each sample of the log results, separating them into classes, and augmenting the dataset in a balanced way, the images must be transformed back into vectors of numerical values (pixels) to be processed by the deep learning model.

14.3.3.2 Type of Data Used in the Use Case

Log Data for Facies

The dataset comprises 1,000,000 samples, each categorized into one of 12 facies (Figure 14.7), which include shale, sandstone, sandstone/shale, marl, dolomite, limestone, chalk, halite, anhydrite, tuff, and coal.

Log Data for Rock Type

The dataset is made up of 300 samples including six rock types (Figure 14.8), labeled as follows: "1,""2,""3,""4,""5,""6."

14.3.4 Data Analysis

To explore the attributes and patterns of the dataset, a study is conducted on ten wells. The distribution of facies in each well is illustrated in Figure 14.9. It is evident that each well exhibits a distinct facies distribution.

As the selection of parameters for analysis is crucial, we also examined their behavior across the ten wells using box plots. The behavior of each parameter varies from one well to another. For instance, the ROP parameter shows low concentration in the fourth well, unlike in the sixth well. In contrast, the GR parameter exhibits similar behavior in the first four wells, which could lead to classification errors or cases where the model fails to differentiate between classes, while it behaves differently in the remaining wells. Figure 14.10 illustrates the behavior of the seven log parameters across the ten wells.

	wellName	datasetName	DEPTH	GR	ILD	Perm LRA	PetroRockType	PetroRockType_ Training	PHIE-F	PHIT-F	RHOB	ROP
1	Well1	LQC_2021	10,692	107.388	2.152	0.062979805	6	0	0	0.1833613	2.428	123.7386408
2	Well1	LQC_2021	10692.5	104.757	2.089	0.155227236	6	0	0	0.1775234	2.426	33.59177645
3	Well1	LQC_2021	10,693	102.607	1.986	0.315333172	6	0	0	0.1727528	2.429	17.9174007
4	Well1	LQC_2021	10693.5	100.46	1.942	0.639571095	6	0	0.009438666	0.1774275	2.418	11.17835893
5	Well1	LQC_2021	10,694	99.75	1.94	0.855140046	6	0	0.01721056	0.183624	2.405	9.482262249
6	Well1	LQC_2021	10694.5	99.518	1.94	0.968809782	6	0	0.02037032	0.186269	2.392	8.871398178
7	Well1	LQC_2021	10,695	102.171	1.968	0.458036807	6	0	0.02503825	0.1968236	2.384	13.76823835
8	Well1	LQC_2021	10695.5	104.348	2.056	0.231501688	6	0	0.03039691	0.2070128	2.376	22.98840397
9	Well1	LQC_2021	10,696	106.285	2.056	0.112254882	6	0	0	0.1809139	2.388	48.8402981
10	Well1	LQC_2021	10696.5	106.292	2.07	0.113116383	6	0	0	0.1809294	2.403	48.36773248
11	Well1	LQC_2021	10,697	107.985	2.07	0.060009795	6	0	0	0.184686	2.412	137.1145907
12	Well1	LQC_2021	10697.5	108.957	2.084	0.060009795	6	0	0	0.1868427	2.408	137.1145907

Figure 14.7 Dataset facies table showing, from left to right, the well name, dataset name, depth (in feet), GR (in API), ILD (log10 values, in ohm meters), permeability (in millidarcy), petrorock type (discrete value), petrorock type training (training dataset), effective porosity (PHIE-F), total porosity (PHIT-F) (both porosities are in percent units), density (gram per cubic centimeter), and Rop (in feet per hour), respectively.

	Well Name	Depth	Formation	GR	ILD_loglO	DeltaPHI	PHIND	LITHOLOGY
4238	15/9-13	1138.7040	Utsira Fm.	55.892757	1.091499	1.774626	0.765867	30,000
4239	15/9-13	1138.8560	Utsira Fm.	60.929138	1.122706	1.800986	0.800262	65,000
4240	15/9-13	1139.0080	Utsira Fm.	62.117264	1.148141	1.817696	0.765957	65,000
4241	15/9-13	1139.1600	Utsira Fm.	61.010860	1.170984	1.829333	0.702521	65,000
4242	15/9-13	1139.3120	Utsira Fm.	58.501236	1.184080	1.813854	0.639708	65,000
4243	15/9-13	1139.4640	Utsira Fm.	54.480759	1.178056	1.762011	0.588014	65,000
4244	15/9-13	1139.6160	Utsira Fm.	51.365631	1.166245	1.693400	0.563947	65,030
4245	15/9-13	1139.7680	Utsira Fm.	50.678062	1.159258	1.636616	0.582253	65,030
4246	15/9-13	1139.9200	Utsira Fm.	52.008244	1.145730	1.619630	0.603731	65,030
4247	15/9-13	1140.0720	Utsira Fm.	53.246262	1.121395	1.651072	0.632982	65,030
4248	15/9-13	1140.2240	Utsira Fm.	55.392452	1.103845	1.711574	0.694154	30,000
4249	15/9-13	1140.3760	Utsira Fm.	58.438663	1.099062	1.751916	0.733934	65,030
4250	15/9-13	1140.5280	Utsira Fm.	61.225414	1.107686	1.753598	0.690551	65,030
4251	15/9-13	1140.6800	Utsira Fm.	63.147400	1.129260	1.746340	0.576971	65,030
4252	15/9-13	1140.8320	Utsira Fm.	63.658535	1.154226	1.768298	0.503176	65,000
4253	15/9-13	1140.9840	Utsira Fm.	63.739819	1.171425	1.792004	0.521140	65,000
4254	15/9-13	1141.1360	Utsira Fm.	62.337303	1.180301	1.803149	0.563884	65,000
4255	15/9-13	1141.2880	Utsira Fm.	61.367290	1.193747	1.799327	0.572325	65,000

Figure 14.8 Dataset rock type table showing, from left to right, the well name, depth (in feet), dataset name, GR (in API), ILD (log10 values, in ohm meters), neutron-density porosity difference, average neutron-density porosity (DeltaPHI and PHIND, both dimensionless quantities are in percent units), lithology type code, respectively.

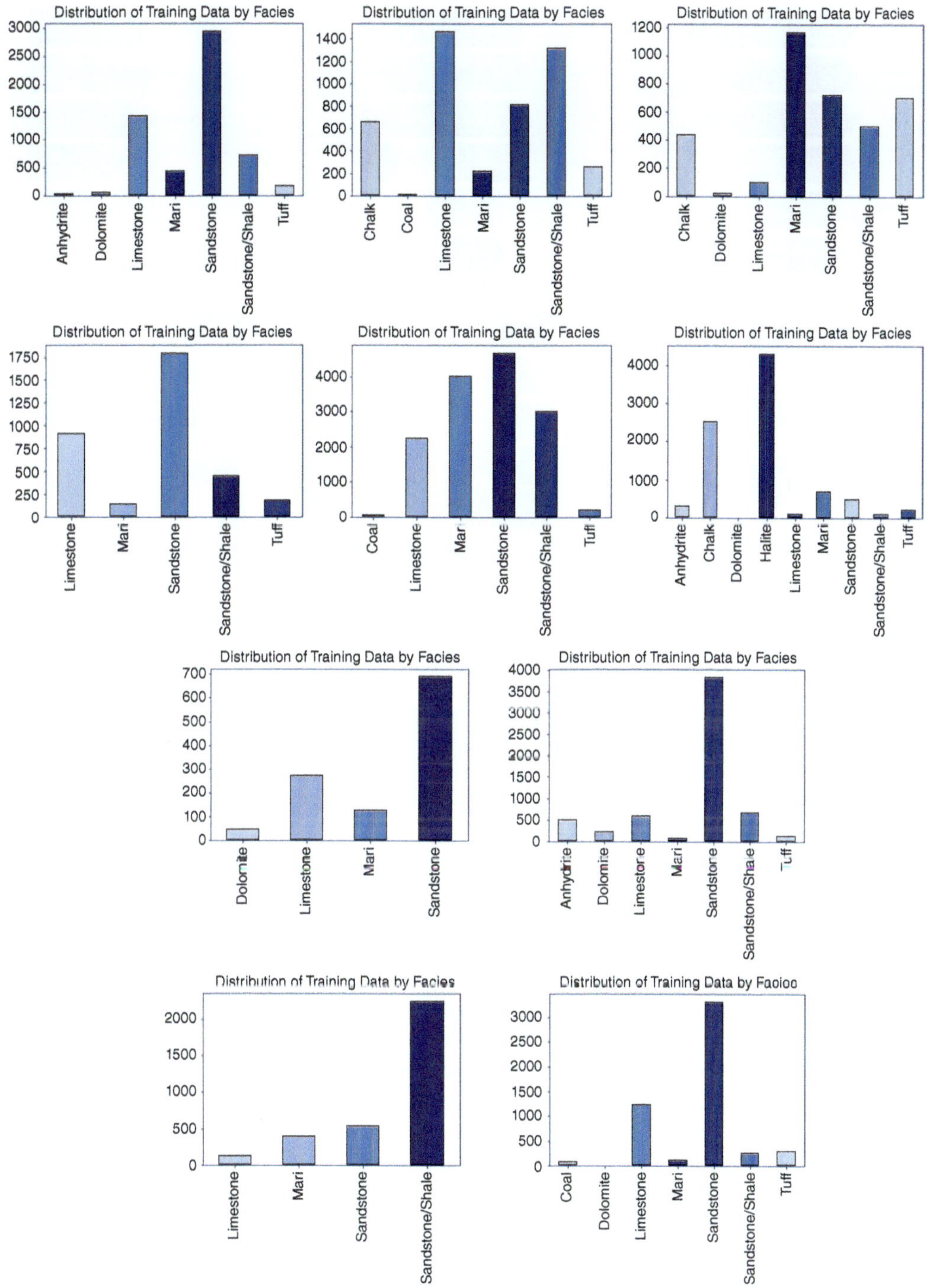

Figure 14.9 Facies distribution in the ten studied wells.

We also analyzed the behavior of these parameters for each facies across all wells and observed similar conclusions. The parameters do not exhibit consistent behavior across the wells, even for the same facies. For instance, the ROP parameter (in blue) is highly variable; at times, its value

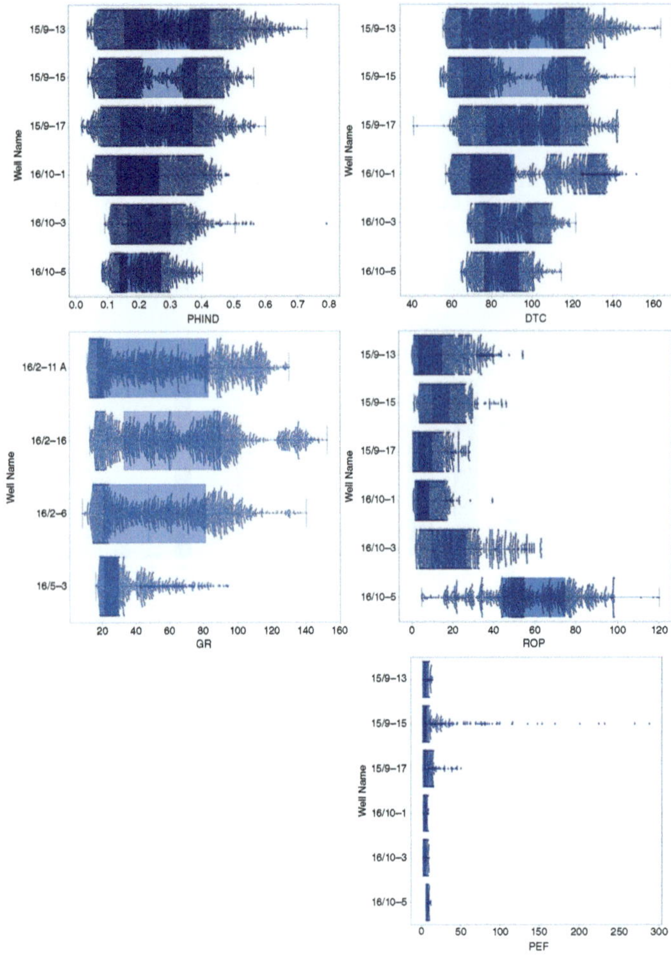

Figure 14.10 Behavior of log parameters across the ten wells.

differs significantly across facies, which is beneficial for ML methods to differentiate between facies during classification. However, in other instances, the values are very similar, potentially causing confusion between different classes. Figure 14.11 shows histogram distributions of the seven well log values across the ten wells.

We continued analyzing the parameters by facies in each well, using box plots to observe their behavior. For example, the DTC parameter for sandstone facies exhibits very different behaviors, while the GR parameter shows similar behavior in the first and third wells but varies significantly between the second and third wells. Ideally, the parameters should display consistent behavior for the same facies to allow the ML model to identify a classification pattern. Unfortunately, these observations complicate the task. Initially, the results were only passable, but after performing pre-processing and recursive parameter selection, the results improved significantly.

14.3.5 Tests and Results

Using cross-validation on data from 98 wells, we tested a range of machine learning algorithms: SVM, RBF, OCSVM, LR, KNN, MLP, DT, unsupervised and supervised SOM, RF, GBM, and AB.

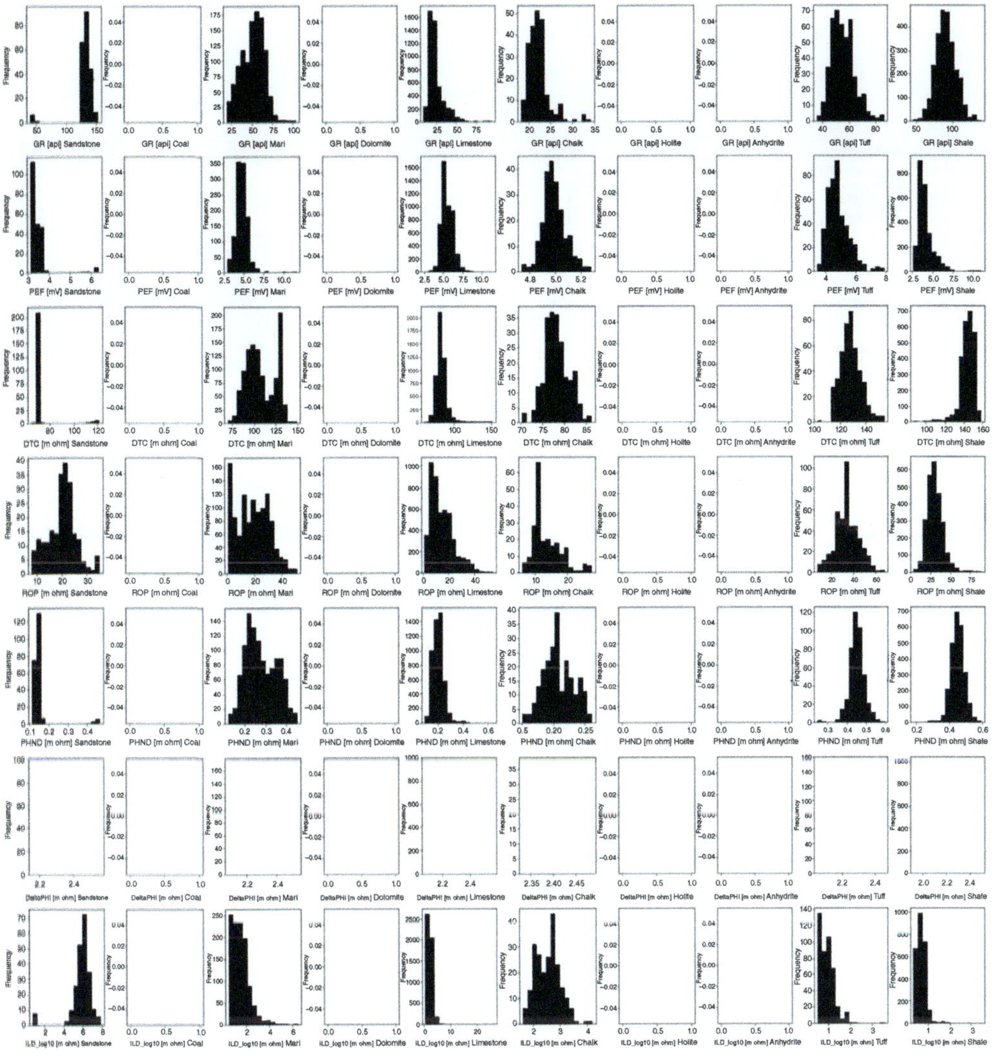

Figure 14.11 Histogram distributions of log parameters across the ten wells.

We studied 11 facies: chalk, dolomite, limestone, sandstone, sandstone/shale, shale, halite, anhydrite, tuff, and coal. For brevity, only the outcomes from the MLP prediction method are shown in Figure 14.12.

The results indicate that all the algorithms achieved a high degree of accuracy in predicting rock types. However, the neural network outperformed the other algorithms. Table 14.1 summarizes the measured performance (accuracy rate) for each of the prediction algorithms considered.

14.4 Conclusions

In oil and gas exploration, identifying wellbore types, completion approaches, and predicting reservoir performance are directly linked to determining rock types and flow units. Accurately understanding the complex variations in pore geometry within different facies is crucial for dynamic modeling of field development and making reliable production forecasts.

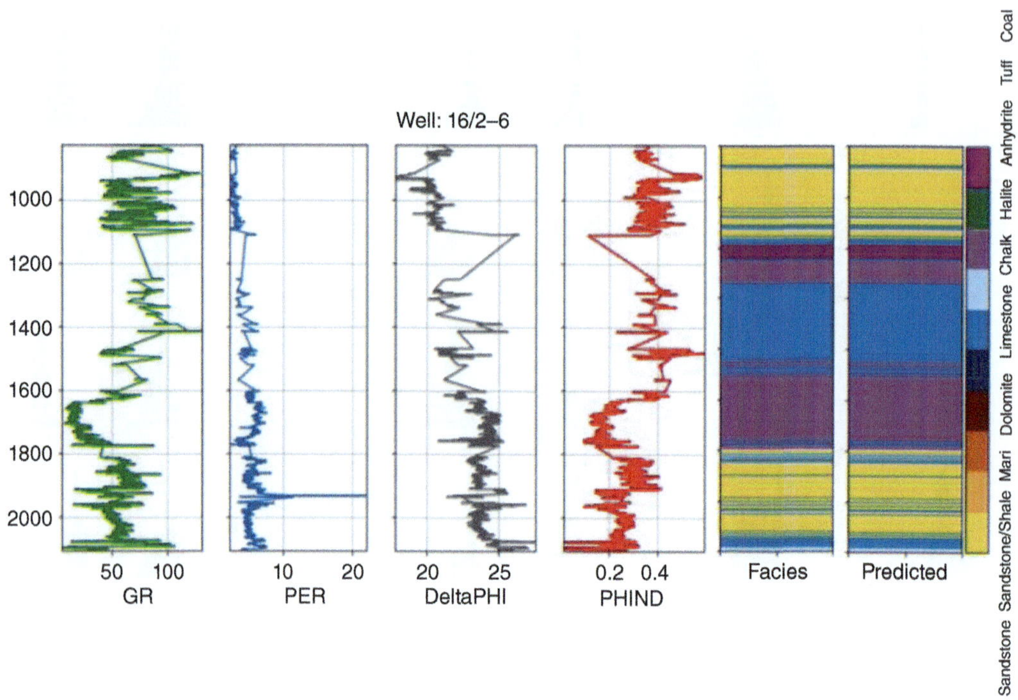

Figure 14.12 Results of lithology prediction at well 1 core using the MLP technique. The panels 1 through 7 (left to right) correspond to GR (in API), PHIND (in percent units), DTC (in us/ft), DeltaPHI (in percent units) logs, the simplified real (facies) and projected lithologies at well 1, respectively.

Table 14.1 Various accuracy metrics were evaluated on the test dataset for the machine learning algorithms used.

ML algorithm	Accuracy (%)
SVM	97
OCSVM	84
LR	97
KNN	97
MLP ANN	98
DT	89
SUP SOM	86
UNSUP SOM	69
RF	89
GBM (Gradient Boost Model)	91
AB (Adaptive boost)	90

Traditionally, rock type classification relies on subjective geological observations and empirical permeability-porosity relationships. However, for a given porosity, permeability can vary by several orders of magnitude, suggesting the presence of multiple flow units.

We applied machine learning methods to well log data to study the characteristics of reservoir wells through rock typing. Our implementation demonstrated that the MLP algorithm was the most effective prediction tool, achieving a prediction accuracy of 98%.

The results demonstrate that ML algorithms can be of high interest to understand the reservoir based on features (rock types), extracted from well log data, which bring additional information to the conventional core analysis.

Acknowledgment

We would like to express our gratitude to the organizers of the "FORCE Machine Learning Competition with Well Logs and Seismic 2020" for providing the data used in this study.

References

Abedini, A. & Torabi, F. (2015). Pore size determination using normalized J-function for different hydraulic flow units. Petroleum 1 (2), pp. 106–111, ISSN 2405-6561. https://doi.org/10.1016/j.petlm.2015.07.004.

Aguilera, R. (2004). Integration of geology, petrophysics, and reservoir engineering for characterization of carbonate reservoirs through Pickett plots. AAPG Bulletin 88 (4), pp. 433–446. https://doi.org/10.1306/12010303071.

Albon, C. (2018). Machine learning cookbook: Practical solutions from preprocessing to deep learning. Published by O'Reilly Media (1st edition), Inc., 1005 Gravenstein Highway North, Sebastopol, CA 95472. USA, 366 pages.

Asan, U. & Ercan, S. (2012). An Introduction to Self-Organizing Maps. In: Computational Intelligence Systems in Industrial Engineering. Atlantis Computational Intelligence Systems (ed. C. Kahraman), vol. 6. Paris: Atlantis Press. https://doi.org/10.2991/978-94-91216-77-0_14.

Aziz, N., Akhir, E. A. P., Aziz, I. A., et al. (2020), A Study on Gradient Boosting Algorithms for Development of AI Monitoring and Prediction Systems. In: 2020 International Conference on Computational Intelligence (ICCI), pp. 11–16. Malaysia: Bandar Seri Iskandar. https://doi.org/10.1109/ICCI51257.2020.9247843.

Chandra, V., Barnett, A., Corbett, P., et al. (2015). Effective integration of reservoir rock-typing and simulation using near-wellbore upscaling. Marine and Petroleum Geology 67, pp. 307–326. https://doi.org/10.1016/j.marpetgeo.2015.05.005.

Chengsheng, T., Huacheng, L., & Bing, X. (2017). AdaBoost typical Algorithm and its application research. MATEC Web of Conferences 139, pp. 00222. https://doi.org/10.1051/matecconf/201713900222.

Fadokun, D. O., Oshilike, I. B., & Onyekonwu, M. O. (2020). Supervised and unsupervised machine learning approach in facies prediction. In: Paper presented at the SPE Nigeria Annual International Conference and Exhibition, Virtual, August 2020. https://doi.org/10.2118/203726-MS.

Furlanello, C., Serafini, M., Merler, S., et al. (2003). Gene selection and classification by entropy-based recursive feature elimination. Proceedings of the International Joint Conference on Neural Networks, 2003, vol. 4, pp. 3077–3082. Portland, OR, USA. https://doi.org/10.1109/IJCNN.2003.1224063.

Gunter, G. W., Spain, D. R., Viro, E. J., et al. (2014). Winland pore throat prediction method—a proper retrospect: new examples from carbonates and complex systems. In: Paper presented at the SPWLA 55th Annual Logging Symposium. Abu Dhabi, United Arab Emirates, May 2014.

Gupta, R., Pierdzioch, C., & Salisu, A. A. (2022). Oil-Price uncertainty and the U.K. unemployment rate: A forecasting experiment with random forests using 150 years of data. Resources Policy 77, pp. 102662.

Haikel, S., Rosid, M. S., & Haidar, M. W. (2018). Study comparative rock typing methods to classify rock type carbonate reservoir Field "s" East Java. Journal of Physics: Conference Series 1120 (1), Article 012047. https://doi.org/10.1088/1742-6596/1120/1/012047S.

Hosseini, S. A., Shirani, A. S., Lotfi, M., et al. (2020). Design and application of supervisory control based on neural network PID controllers for pressurizer system. Progress in Nuclear Energy 130, pp. 103570.

Jiang, W., He, G., Long, T., et al. (2018). Multilayer Perceptron Neural Network for Surface Water Extraction in Landsat 8 OLI Satellite Images. Remote Sensing 10(5), 755. https://doi.org/10.3390/rs10050755.

Krivoshchekov, S., Kochnev, A., Kozyrev, N., et al. (2023). Rock typing approaches for effective complex carbonate reservoir characterization. Energies 16, p. 6559. https://doi.org/10.3390/en16186559.

Kolodzie, S. (1980). Analysis of pore throat size and use of the Waxman-Smits equation to determine OOIP in Spindle field, Colorado. In: Proceedings of the SPE Annual Technical Conference and Exhibition, pp. 21–24. Dallas, TX, USA, September 1980. SPE-9382-MS.

Liu, Q., Chen, C., Zhang, Y., et al. (2011). Feature selection for support vector machines with RBF kernel. Artificial Intelligence Review 36, pp. 99–115. https://doi.org/10.1007/s10462-011-9205-2.

Melka, J. & Mariage, J. J. (2019). Adapting self-organizing map algorithm to sparse data. In: Computational Intelligence. IJCCI 2017. Studies in Computational Intelligence, (eds. C. Sabourin, J.J. Merelo, K. Madani et al.), vol. 829. Springer, Cham. https://doi.org/10.1007/978-3-030-16469-0_8.

Mohammadian, E., Kheirollahi, M., Liu, B., et al. (2022). A case study of petrophysical rock typing and permeability prediction using machine learning in a heterogenous carbonate reservoir in Iran. Scientific Reports 12, p. 4505. https://doi.org/10.1038/s41598-022-08575-5.

Palabiran, M., Sesilia, N., & Akbar, M. N. A. (2016). An analysis of rock typing methods in carbonate rocks for better carbonate reservoir characterization: A case study of Minahaki carbonate formation, Banggai Sula basin, Central Sulawesi. In: 41st Scientific Annual Meeting of Indonesian Associatation of Geophysicists. Lampurg.

Pedersen, R. & Schoeberl, M., (2006). An embedded support vector machine. In: 2006 International Workshop on Intelligent Solutions in Embedded Systems, pp. 1–11. Vienna, Austria. https://doi.org/10.1109/WISES.2006.329117.

Riazi, Z. (2018). Application of integrated rock typing and flow units identification methods for an Iranian carbonate reservoir. Journal of Petroleum Science and Engineering 160, pp. 483–497. https://doi.org/10.1016/j.petrol.2017.10.025.

Saravanan, R. & Sujatha P. (2018). A state of art techniques on machine learning algorithms: A perspective of supervised learning approaches in data classification. In: 2018 Second International Conference on Intelligent Computing and Control Systems (ICICCS), pp. 945–949.

Tavakoli, V. (2018). Rock typing. In: Geological Core Analysis. SpringerBriefs in Petroleum Geoscience & Engineering. Springer, Cham. https://doi.org/10.1007/978-3-319-78027-6_6.

Tiainen, S., King, H., Cubitt, C., et al. (2002). Drill cuttings analysis—A new approach to reservoir description and characterisation; Examples from the Cooper Basin, Australia. The APPEA Journal 42 (1), pp. 495–509.

Tschannen, V., Delescluse, M., Rodriguez, M., et al. (2017). Facies classification from well logs using an inception convolutional network. arXiv preprint arXiv:1706.00613.

Valentín, M. B., Bom, C. R., Coelho, J.M., et al. (2019). A deep residual convolutional neural network for automatic lithological facies identification in Brazilian pre-salt oilfield wellbore image logs. Journal of Petroleum Science and Engineering 179, pp. 474–503.

Xing, Y., Yang, H., & Yu, W. (2023). An Approach for the Classification of Rock Types Using Machine Learning of Core and Log Data. Sustainability 2023, 15, 8868. https://doi.org/10.3390/su15118868.

Yazid, H. & Gaci, S. (2022). Machine learning-based techniques for reservoir characterization using rock-typing from well logs. In: 2022 7th International Conference on Image and Signal Processing and their Applications (ISPA), pp. 1–6. Mostaganem, Algeria, 2022. https://doi.org/10.1109/ISPA54004.2022.9786300.

Yazid, H. & Gaci, S. (2024). Integrating hybrid weak learners for lithofacies classification using well log data. In: 2024 8th International Conference on Image and Signal Processing and their Applications (ISPA), pp. 1–5. Biskra, Algeria. https://doi.org/10.1109/ISPA59904.2024.10536797.

You, W., Yang, Z., & Ji, G. (2014). Feature selection for high-dimensional multi-category data using PLS-based local recursive feature elimination. Expert Systems with Applications 41 (4), pp. 1463–1475, Part 1, 2014, ISSN 0957-4174. https://doi.org/10.1016/j.eswa.2013.08.043.

Zhao, W. L., Wang, H., & Ngo, C. W. (2021). Approximate K-NN graph construction: A generic online approach. IEEE Transactions on Multimedia 24, pp. 1909–1921.

Zeng, X., Chen, Y. W., & Tao, C. (2009). Feature selection using recursive feature elimination for handwritten digit recognition. In: 2009 Fifth International Conference on Intelligent Information Hiding and Multimedia Signal Processing, pp. 1205–1208. Kyoto, Japan, 2009. https://doi.org/10.1109/IIH-MSP.2009.145.

15

Logging-Data-Driven Fluid Prediction in Clastic Reservoir Based on Fractal Attributes and Machine Learning Methods

Abdelbasset Boulassel[1,3], *Soraya Makhlouf*[1], *Fethi Ali Cheddad*[1,3], *Zinelaabidine Boumelit*[1], *Badis Zegagh*[2], *Salah Boufenchouche*[3], *Amar Boudella*[1], *Naima Zaourar*[1], *Said Gaci*[4]

[1] *Geophysics Department–FSTGAT, University of Sciences and Technology Houari Boumediene (USTHB), Algiers, Algeria.*
[2] *Department of Geophysics, Faculty of Hydrocarbon and chemistry, University of Boumerdes M'Hamed BOUGARA (UMBB), Independence Avenue, Boumerdes, Algeria*
[3] *Sonatrach, Exploration Division, Exploration Operations Directorate, Ouargla, Algeria*
[4] *Sonatrach–Central Directorate of Research & Development(DC R&D), Boumerdes, Algeria*

15.1 Introduction

Accurately predicting the fluid nature within clastic reservoirs is crucial for effective reservoir management and optimal resource extraction in the oil and gas industry. Traditional methods sometimes fall short when addressing the inherent complexity and heterogeneity of clastic reservoirs, as they typically fail to capture the intricate geological variations and multiscale properties that significantly influence fluid distribution (Diana and Arnold, 1992; Liu, 2002; Lucia, 2007; Guo et al., 2008; Yu et al., 2017; Fanchi, 2010; Xuqiang et al., 2019; Ma, 2019; Yu et al., 2022; Chengzhou et al., 2023; Ze et al., 2023; Wenbiao et al., 2024). Consequently, advanced techniques are needed to handle these complexities and improve prediction accuracy (Elkatatny et al., 2018; Pyrcz et al., 2017; Ali et al., 2024).

The literature on fluid prediction in clastic reservoirs emphasizes the challenges posed by their complex and heterogeneous nature. Traditional methods, often based on deterministic models and linear relationships, struggle to account for the multiscale variations and intricate geological features. Recent advancements in fractal analysis have shown promise in capturing these complexities, providing a more detailed representation of the reservoir's structure (Shun et al., 2005; Peng et al., 2020; Ke et al., 2020; Dayu et al., 2023).

Fractal and multifractal analyses have been widely used in various geological applications, including reservoir characterization, to describe the spatial distribution and variability of geological properties. These techniques effectively model the irregular and self-similar nature of geological structures (Hardy and Beier, 1994; López and Aldana, 2007; Hernandez-Martinez et al., 2013; Subhakar and Chandrasekhar, 2016a, 2016b; Habib et al., 2017). Integrating machine learning (ML) with fractal analysis represents a significant advancement in fluid prediction techniques (Boulassel et al., 2021). ML algorithms handle large datasets and complex relationships, making

Geophysical Exploration for Hydrocarbon Reservoirs, Geothermal Energy, and Carbon Storage: New Technologies and AI-based Approaches, First Edition. Edited by Said Gaci.

them well-suited for modeling the nonlinear interactions between geological attributes and fluid distributions. The combination of these techniques allows for a more accurate and robust prediction of fluid types within clastic reservoirs.

This study introduces a logging-data-driven approach for predicting fluid distributions in clastic reservoirs by combining fractal attributes of well logs with ML techniques. By leveraging fractal and multifractal analyses, we extract attributes from geophysical well logs to capture multiscale variations in rock properties. These attributes are then utilized in various ML algorithms to model and predict fluid types (oil or gas) within the reservoir. This integration enhances the characterization of hydrocarbon zones and improves fluid differentiation.

The study employs a multi-step methodology to achieve its objectives. Initially, we collect and preprocess geophysical well log data from 93 wells, ensuring thorough data quality control. The dataset comprises 1085 logs from diverse sources, including gamma ray (GR), acoustic logging (AC), density logging (DEN), neutron logging (CNL), electrical resistivity logging (R), photoelectric absorption factor (PF), and natural gamma ray spectroscopy (GRS). Using fractal and multifractal analysis techniques, we extract key attributes from the well logs, such as Box and Regularization dimensions, Width, Right height, Left height, αpeak, αmin, αmax, f(αpeak), f(αmin), f(αmax), and C-value, capturing the geological structure's multiscale variations (Boulassel et al., 2021).

Subsequently, we train and validate multiple ML models using the extracted fractal attributes. The ML techniques evaluated include logistic regression, random forest (RF), K-nearest neighbors (KNN), support vector machines (SVM), and discriminant analysis. Each algorithm's performance is assessed based on metrics such as Accuracy, Precision, Recall, Correct classification, Misclassification, and F-score to identify the most effective technique for fluid prediction. Overall, this research highlights the potential of combining fractal attributes and ML to enhance fluid prediction accuracy in clastic reservoirs. This approach offers a robust and efficient tool for reservoir characterization, aiding in optimal resource extraction and management within the oil and gas industry. It not only improves prediction accuracy but also provides valuable insights into the geological structure of reservoirs, allowing for more precise identification of fluid zones.

The remainder of this chapter is structured as follows: The next section provides a detailed description of the datasets used, including information on the wells, logs, and multi-attribute datasets. This is followed by a brief overview of fractal analysis and the ML techniques employed. Next, we discuss the statistical metrics used to evaluate the performance during both the training and prediction phases. Subsequently, we present a comprehensive analysis of the results obtained from applying these techniques, along with a discussion of the findings. Finally, we summarize the main findings, draw key conclusions, and outline the future perspectives of this work.

15.2 Studied Dataset

The dataset used in this study encompasses geophysical well logs from 93 wells across seven sedimentary basins in Algeria. This dataset, marked by geographic and geological diversity, enhances the generalizability of the findings across various hydrocarbon systems. It captures differences in rock formations, porosity, permeability, fluid compositions, pressure, temperature, and depth, reflecting significant petrophysical and geochemical variability.

The dataset comprises 1085 distinct well logs corresponding to Oil/Gas portions, including GR, AC, DEN, CNL, R, PF, and GRS.

Rigorous data selection and interpretation informed the dataset's compilation. This process involved detailed petrophysical interpretations, pressure measurements, fluid sampling, and Drill Stem Tests (DST). These steps ensured the presence of oil or gas in the zone of interest where the geophysical well logs were recorded, enhancing the reliability and representativeness of the dataset and thereby bolstering the credibility of the statistical analysis.

Fractal and multifractal analysis were employed to describe the variability indicated by the 1085 geophysical well-log fluctuations across multiple scales. The analysis focused on Box and Regularization Dimensions, and Multifractal Singularity Spectrum Attributes such as Width, Right height, Left height, αpeak, αmin, αmax, f(αpeak), f(αmin), f(αmax), and C value. These analyses were used to construct a high-dimensional multi-attribute database.

The obtained multi-attribute dataset was analyzed and evaluated using various ML techniques. Each technique's performance was assessed through comprehensive metrics to identify the most effective method for fluid prediction using fractal attributes. This thorough analysis contributes to optimizing hydrocarbon recovery strategies and supports effective reservoir management by providing accurate and reliable predictions of fluid distributions within the reservoir.

15.3 Overview of Fractal Analysis Steps Employed in Geophysical Well Logs Study

Fractal analysis is a powerful tool for uncovering complex patterns within geophysical well logs, offering deeper insights into subsurface formations. By employing advanced fractal methodologies, we extract meaningful attributes that form the basis of a comprehensive multi-attribute dataset from 1085 well logs. This dataset is subsequently used in ML studies to reveal predictive patterns and enhance our understanding of subsurface properties. The adopted workflow, depicted in Figure 15.1, outlines a systematic process from initial data analysis to the construction of a high-dimensional dataset.

15.3.1 Key Steps in Fractal Analysis of Geophysical Well Logs

A. **Geophysical Well Logs Analysis:** Detailed examination of well logs, focusing on both oil and gas portions, providing the foundational data for subsequent analysis.
B. **Continuous Wavelet Transform (CWT):** Generation of a wavelet coefficient matrix in the time-scale domain to capture detailed signal variations within the logs.
C. **Wavelet Transform Modulus Maxima (WTMM):** Identification of maxima lines along CWT-defined scales to pinpoint local singularities, crucial for understanding data intricacies.
D. **Partition Function (Z (ε, q)):** Quantification of singularities across scales (ε) and moment orders (q), revealing the multifractal nature of the dataset and its variability.
E. **Regression Range Selection:** Identification of linear behavior ranges within the dataset to ensure accurate analysis of scaling behavior under various conditions.
F. **Scaling Exponent (τ(q)):** Computation of scaling exponents for different moment orders (q), providing insights into hierarchical data structure and scaling properties.
G. **Singularity Spectrum (*f*(α)):** Derivation of singularity spectrum through regression, visualizing multifractal behavior and quantifying the distribution and strengths of singularities within the data (see Figure 15.2).
H. **Extraction of Fractal Parameters:** Extraction of various fractal parameters from each well log, including Box dimension, Regularization dimension, Right height, Left height, αmin, αmax, αpeak, $f(\alpha_{min})$, $f(\alpha_{max})$, $f(\alpha_{peak})$, Width, and C value (see Figure 15.2).
I. **Building a Multivariable Dataset:** Creation of a comprehensive multivariable dataset integrating extracted attributes from the 1085 well logs. This dataset serves as the foundation for subsequent ML studies aimed at uncovering deeper insights and predictive patterns.

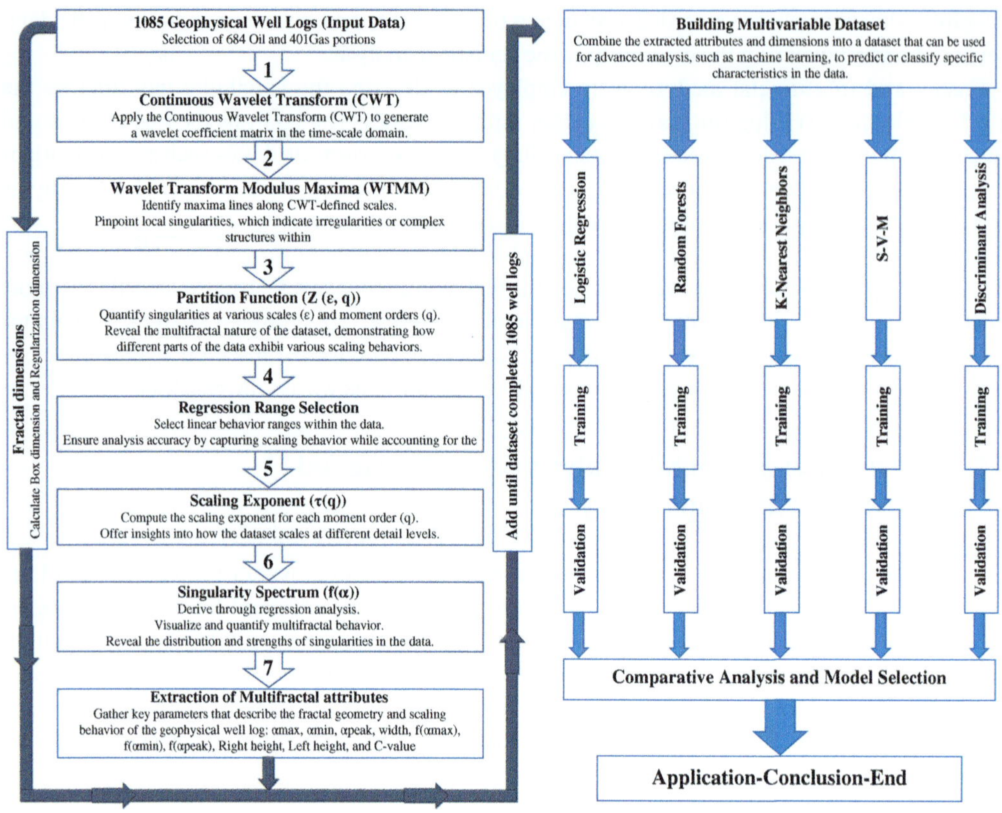

Figure 15.1 Comprehensive workflow for applying multiple ML techniques to a multi-attribute dataset built from well logs using fractal analysis.

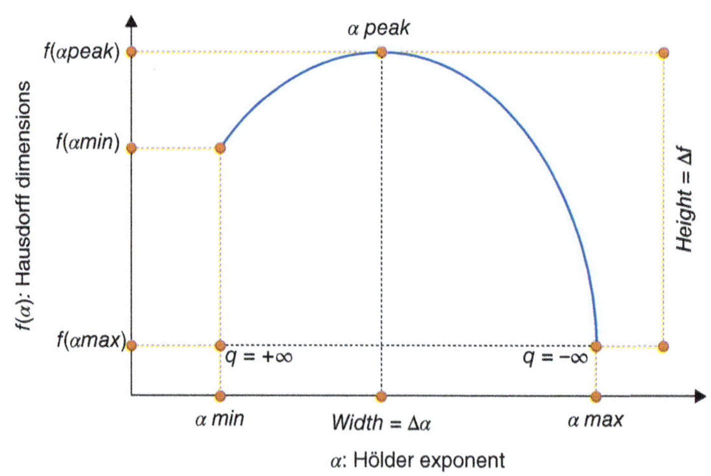

Figure 15.2 Illustration of various attributes extracted from the multifractal singularity spectrum generated from well logs after fractal analysis.

This structured approach ensures rigorous preparation of the dataset, from initial analysis to ML application, establishing a robust foundation for further geophysical studies and practical applications.

15.4 Overview of Employed Machine Learning Methods

In this study, we explore multiple ML techniques to predict changes in fluid characteristics using fractal attributes derived from geophysical well logs. Our goal is to identify the most effective model for precise fluid prediction and thorough reservoir characterization. Evaluated algorithms include logistic regression, RF, KNN, SVM, and discriminant analysis.

Performance of each algorithm is rigorously evaluated using metrics such as Accuracy, Precision, Recall, Correct Classification, Misclassification, and F-score to determine the optimal technique for integrating fractal attributes into fluid nature detection. Our objective is to propose an ML approach that effectively combines fractal attributes to enhance the accuracy and reliability of reservoir characterization studies through advanced data-driven methodologies.

15.4.1 Logistic Regression

Logistic regression is a widely used binary classification technique, essential for various prediction tasks. The core of this method is the logistic function, a sigmoid function characterized by its S-shaped curve. This function maps a weighted linear combination of features into real values between 0 and 1, which can be interpreted as probabilities. Rather than predicting a specific class, logistic regression predicts the probability of belonging to a given class (Choudhary and Gianey, 2017; Saidi et al., 2021). This method has found numerous applications in the medical diagnostic field. For example, Javed et al. (2017) presented an approach for classifying finger movements to control upper limb prostheses using logistic regression, achieving an accuracy of 65%. Additionally, Behadada et al. (2017) proposed a method for detecting cardiac arrhythmia using a multi-class logistic regression classifier, achieving an impressive accuracy of 93.13%.

In the engineering domain, logistic regression has been utilized for predicting failure probabilities. Sami, for instance, presented a model for predicting company failures using partial least squares logistic regression, which attained an accuracy of 94.5% (Ben Jabeur, 2017). Another significant application area of logistic regression is sentiment analysis. (Goyani and Patel, 2017) proposed a method for facial expression recognition using a logistic regression classifier. When tested with the TFEID dataset, their approach achieved a remarkable accuracy of 96.84%.

15.4.2 Random Forest

The RF algorithm, introduced by Breiman (2001), is an ensemble learning method designed to enhance the traditional decision tree approach. Originating from the bagging algorithm (Breiman, 1996), RF consists of numerous decision trees that function independently (Liaw and Wiener, 2002). This algorithm is applicable to both regression and classification problems. RF utilizes bootstrap sampling to randomly generate diverse training set samples from the original dataset, on which decision tree (DT) operations are performed (Hefner et al., 2014). During the construction of these decision trees, no pruning is performed. Instead, trees are created by selecting random column variables combined with random row observations, forming a "random forest." A single decision tree is prone to producing incorrect predictions, leading to a higher error rate. However, when multiple decision trees are combined into a forest, the likelihood of correct predictions increases, and the overall error rate decreases compared to using a single decision tree. Building an RF model involves two main stages: forest generation and decision-making. The final prediction of an RF model is achieved through voting or averaging the results of the individual decision trees. This integration of DT outcomes generally yields superior results compared to single decision trees and has been successfully applied across various fields (Tyralis et al., 2019; Diaz et al., 2021; Xu and Luo, 2021). Recent research by Benali et al. (2019) indicates that RF algorithms perform exceptionally well in high-dimensional, nonlinear problems. While the RF algorithm often produces better results than DT algorithms, its performance is heavily dependent on the amount of labeled data available in the training dataset. A substantial quantity of labeled data is necessary for the RF algorithm to generate accurate predictions (Gislason et al., 2006). Additionally, Kiangala and Wang (2021) have noted that the RF algorithm's robustness and versatility make it a preferred choice in various ML applications.

15.4.3 Support Vector Machine

SVMs are a robust set of supervised learning methods utilized for both classification and regression tasks. Originating from statistical learning theory (Vapnik, 2010), SVM aims to identify a hyperplane in an n-dimensional space that distinctly classifies data points into different categories. The central concept is to find the optimal separating hyperplane that maximizes the margin between different classes (Cortes and Vapnik, 1995; Vapnik, 1998; Urso et al., 2018). In a classification context, SVM works by mapping input vectors into a high-dimensional feature space where it can linearly separate data using decision planes (Scholkopf et al., 1995). The margin is defined by two parallel hyperplanes "pushed up against" the two data sets, with the optimal hyperplane situated midway, ensuring the largest possible distance to the nearest data points from both classes (Cristianini and Shawe-Taylor, 2000). These nearest data points are termed support vectors, and they crucially influence the hyperplane's orientation and position (Platt, 1999). SVMs are particularly advantageous as they can handle both linearly and non-linearly separable data by employing nonlinear mapping techniques. This mapping transforms the original finite-dimensional space into a much higher-dimensional space, facilitating better data separation (Chen et al., 2015). Unlike traditional discriminant analyses, SVM can effectively handle multiple classes and numerical prediction tasks (Wang and Xue, 2014; Wu et al., 2008). Despite being inherently non-probabilistic, meaning they do not provide probability estimates for classifications, SVMs are highly accurate and have been widely applied across various domains including pattern recognition, medical diagnosis, marketing, and text classification (Yingxin and Xiaogang, 2005; Joachims, 1998). The method has shown superior performance over neural networks and decision trees, especially when dealing with smaller training data sets (Hefner and Ousley, 2014; Hefner et al., 2015). SVM's effectiveness in classification tasks, particularly in estimating ancestry between more than two groups, has also been highlighted. The method's reliance on solving large quadratic programming optimization problems ensures that the parameters of the maximum-margin hyperplane are optimally derived (Vapnik, 1998).

15.4.4 K-Nearest Neighbors

The KNN algorithm is a versatile and intuitive method used for both regression and classification tasks. It works by identifying the K closest data points in the training dataset to a new observation, using these points to predict the value or class of the new observation. In KNN regression, the algorithm estimates the value of a new data point based on the values of its KNN. This technique is commonly applied in predicting stock prices, estimating housing prices, and forecasting weather patterns (Naghibi and Dashtpagerdi, 2017). The algorithm calculates the distance between the new observation and all observations in the training dataset, typically using the Euclidean distance metric. After determining these distances, KNN algorithm identifies the K closest data points and predicts the new observation's value by averaging (or taking the median of) their target variable values. For KNN classification, the algorithm assigns a class label to a new observation based on the class labels of its KNN. This method is particularly effective for tasks such as image recognition, handwriting detection, and various other classification problems. The algorithm defers computation until classification, which is why it is considered a lazy learning algorithm. During the training phase, no explicit model is built; instead, the training data populates the search space with instances whose classes are known. When a new instance needs to be classified, the algorithm identifies its KNN and assigns the most common class among these neighbors (Sumayli, 2023).

Both regression and classification with KNN involve two main phases: the training phase and the prediction (or classification) phase. During the training phase, the feature vectors and corresponding labels of the training samples are stored. In the prediction phase, a user-defined constant K is used to determine the number of nearest neighbors considered. For regression, the predicted value is the average of the nearest neighbors' values, while for classification, the predicted class is the most frequent class among the nearest neighbors.

The advantages of KNN include its simplicity, ease of interpretation, and effectiveness in various practical applications (Sumayli, 2023). However, selecting an appropriate value for K is crucial. A value of K that is too small may lead to overfitting, while a value that is too large may result in underfitting. Additionally, KNN can be computationally expensive in terms of both time and memory, particularly in high-dimensional data or when the data has a complex structure (Bishop, 2006).

15.4.5 Discriminant Analysis

Discriminant analysis is a way to build classifiers: that is, the algorithm uses labelled training data to build a predictive model of group membership which can then be applied to new cases. While regression techniques produce a real value as output, discriminant analysis produces class labels. As with regression, discriminant analysis can be linear, attempting to find a straight line that separates the data into categories, or it can fit any of a variety of curves (Hallinan, 2012). It can be two dimensional or multidimensional; in higher dimensions the separating line becomes a plane, or more generally a hyperplane. Discriminant analysis also outputs an equation that can be used to classify new examples.

Discriminant analysis makes the assumptions that the variables are distributed normally, and that the within-group covariance matrices are equal. However, discriminant analysis is surprising robust to violation of these assumptions, and is usually a good first choice for classifier development (Hallinan, 2012).

15.5 Model Evaluation

The developed ML models will be rigorously evaluated using a diverse set of statistical metrics to ensure their accuracy, reliability, and overall performance. These metrics will offer a comprehensive assessment of the models' predictive capabilities and their ability to generalize to new, unseen data. The key evaluation metrics employed in this analysis include:

15.5.1 Confusion Matrix

A confusion matrix is a table used to describe the performance of a classification model. It shows the counts of actual versus predicted classifications, providing a clear visualization of the model's performance, it is a matrix of size 2×2 for binary classification with actual values on one axis and predicted on another. The confusion matrix takes the classification results and groups them into four categories:

- True Positive (TP): when both the actual and predicted values are 1.
- True Negative (TN): when both the actual and predicted values are 0.
- False Positive (FP): when the actual value is 0 but the predicted value is 1.
- False Negative (FN): when the actual value is 1 but the predicted value is 0.

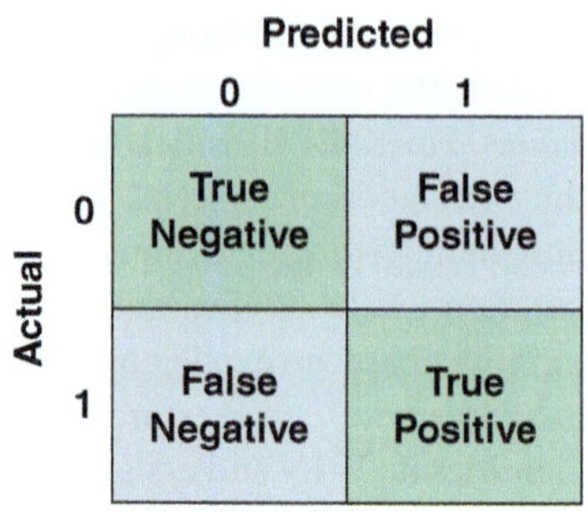

Figure 15.3 Confusion matrix presentation.

Recall that in our case, we refer to the event we want to capture (1 – Spam) as Positive and non-event (0 – Not Spam) as Negative. The confusion matrix for binary classification is a 2-by-2 matrix, where each column represents one class, as in Figure 15.3.

15.5.2 Accuracy

Accuracy measures the proportion of correctly predicted instances (both true positives and true negatives) out of the total instances. It gives an overall measure of the model's performance

$$Accuracy = \frac{Number\ of\ correct\ predictions}{Total\ number\ of\ predictions} = \frac{TP + TN}{TP + TN + FP + FN} \tag{15.1}$$

Accuracy is often used as the measure of classification performance because it is simple to compute and easy to interpret.

15.5.3 Precision

Precision measures the proportion of correctly predicted positive instances out of all instances predicted as positive. It indicates how many of the predicted positives are actually positive.

$$Precision = \frac{True\ Positive\ (TP)}{True\ Positive\ (TP) + False\ Positive\ (FP)} \tag{15.2}$$

The precision value lies between 0 and 1.

15.5.4 Recall (Sensitivity)

Recall, also known as Sensitivity or True Positive Rate, measures the proportion of correctly predicted positive instances out of all actual positive instances. It indicates how well the model can identify positive instances.

$$Recall = \frac{True\ Positive\ (TP)}{True\ Positive\ (TP) + False\ Negative\ (FN)} \tag{15.3}$$

15.5.5 Correct Classification

This metric refers to the total number of instances that are correctly classified, including both true positives and true negatives. It is often used interchangeably with accuracy in discussions about the total count of correct predictions

$$Correct\ Classification = TP + TN \tag{15.4}$$

15.5.6 Misclassification

Misclassification measures the proportion of instances that are incorrectly predicted. It is the complement of accuracy.

$$Misclassification = \frac{FP + FN}{TP + TN + FP + FN} \tag{15.5}$$

15.5.7 F-Score (F1-Score)

The F-score, or F1-score, is the harmonic mean of precision and recall. It provides a balance between precision and recall, especially useful when the class distribution is imbalanced.

$$FScore = \frac{2 * \text{Precision} * \text{Recall}}{\text{Precision} + \text{Recall}}. \tag{15.6}$$

F-Score gives the same weightage to recall and precision.

15.5.8 ROC (Receiver Operating Characteristic) Curve

The ROC curve is a graphical plot that illustrates the diagnostic ability of a binary classifier system as its discrimination threshold is varied. It plots the true positive rate (sensitivity) against the false positive rate (1-specificity).

- A curve closer to the top-left corner indicates a better performance.
- A diagonal line (45 degrees) represents a random classifier.

15.5.9 AUC (Area Under the ROC Curve)

AUC measures the entire two-dimensional area underneath the entire ROC curve. It provides an aggregate measure of performance across all possible classification thresholds.

- AUC ranges from 0 to 1.
- An AUC of 1 represents a perfect model.
- An AUC of 0.5 represents a model with no discriminative ability (equivalent to random guessing).

These metrics collectively provide a comprehensive evaluation of a ML model's performance, highlighting different aspects such as accuracy, precision, recall, and the balance between them.

15.6 Results and Discussion

15.6.1 Training and Validation Results

Based on a comprehensive analysis of the results and visualizations, it is clear that different ML models exhibit varying levels of effectiveness on the training and validation dataset. We employed logistic regression, RF, KNN, SVM, and discriminant analysis to investigate the multi-attribute dataset extracted from fractal analysis of geophysical well logs from various sources (Tables 15.1 and 15.2). Each algorithm's performance was assessed using a comprehensive set of metrics to determine the most effective method for fluid prediction. The dataset was randomly divided into two subsets: 80%

Table 15.1 Summary statistics (quantitative data).

Attribute	Observations	Missing data	Minimum	Maximum	Mean	Std deviation
REG dimension	1085	0	1.015	2.249	1.561	0.198
BOX dimension	1085	0	1.256	1.986	1.776	0.131
αmax	1085	0	−0.880	7.200	2.041	1.341
αmin	1085	0	−3.400	3.800	0.450	0.897
αpeak	1085	0	−1.300	5.100	1.403	1.045
Width	1085	0	0.001	6.600	1.591	1.243
f (αmax)	1085	0	−2.400	1.000	0.420	0.629
f (αmin)	1085	0	−2.600	1.000	0.018	0.719
f (αpeak)	1085	0	0.410	1.000	0.983	0.059
Right height,	1085	0	0.000	3.400	0.563	0.638
Left height	1085	0	0.000	3.600	0.965	0.729
C values	1085	0	0.000	3.200	0.728	0.625

Table 15.2 Summary statistics (qualitative data).

Variable	Categories	Counts	Frequencies	%
Fluid type	GAZ	401	401	36.96
	OIL	685	685	63.04

of the observations were used for training, and the remaining 20% for validation. Additionally, fluid nature predictions were performed for a specific part to test the selected ML technique's predictability. The performances of these models were evaluated using a range of statistical metrics, including confusion matrix, Accuracy, Precision, Recall, Correct Classification, Misclassification, F-Score, ROC, and AUC. The results are summarized in tables and plots below (Tables 15.3–15.8, Figures 15.4 and 15.5).

15.6.2 Results Interpretation

1. Random Forests:
   RF demonstrated the best overall performance among the methods tested. It achieved the highest accuracy (97.2%), perfect recall (1.000), and a high precision (0.936). The misclassification rate is the lowest (6 instances), indicating that it is highly reliable for this dataset. The high F-score (0.967) confirms that it maintains a good balance between precision and recall, making it the most effective model for fluid prediction.
2. Logistic Regression, K Nearest Neighbors, and SVM:
   These methods also performed well, with high accuracy and balanced precision and recall. Logistic Regression and SVM both had an accuracy of 95.4%, while KNN had a slightly higher accuracy at 96.3%. However, they all had a higher misclassification rate compared to RF, indicating slightly lower reliability.

3. Discriminant Analysis:
   Discriminant analysis performed significantly worse than the other methods. It had the lowest accuracy (32.7%), poor precision (0.262), and low recall (0.364). The high misclassification rate (146 instances) indicates that it is not a reliable method for this dataset. The low F-score (0.305) further confirms its poor performance.
4. Conclusion:
   Based on the comprehensive evaluation of these methods, RF emerge as the superior choice for fluid prediction in this multi-attribute dataset, achieving exceptional performance with the highest accuracy, perfect recall, and robust precision. Following RF, KNN, logistic regression, and SVM also demonstrated strong performance, albeit with slightly lower reliability metrics compared to RF. In contrast, Discriminant analysis proved inadequate for this dataset due to its notably inferior accuracy and misclassification rate.

   The detailed analysis will now pivot towards a deeper exploration of RF' results and insights. This focused examination will highlight its strengths in handling complex datasets, its potential applications in real-world scenarios, and further nuances that contribute to its effectiveness in fluid prediction tasks. This emphasis aims to provide actionable insights for leveraging RF optimally in practical applications.

15.6.3 Detailed Interpretation of Random Forest's Results

I. Misclassification Rate
   Table 15.4 provides the misclassification rates for both the Out-Of-Bag (OOB) sample and the validation sample in the context of the RF model.

   The OOB error rate is a unique feature of the RF algorithm. It provides an internal estimate of the generalization error, which is calculated using the "out-of-bag" samples.

   - **OOB Sample**: This consists of all observations not included in the bootstrap sample for a particular tree in the forest. Each tree in the RF is trained on a bootstrap sample, and the OOB sample comprises the data points not selected in the bootstrap sample.
   - **OOB Error Rate**: The OOB error rate (0.050) is the average classification error rate calculated across all trees using their respective OOB samples. This error rate provides a reliable estimate of the model's performance on unseen data without the need for a separate validation set. In this case, the OOB misclassification rate of 0.050 indicates that 5% of the OOB samples are misclassified on average.

Table 15.3 A summary table containing several indices, computed on the validation sample, corresponding to several employed ML methods.

Performance metrics	Logistic regression	Random forests	K nearest neighbors	SVM	Discriminant analysis
Accuracy	0.954	0.972	0.963	0.954	0.327
Precision	0.924	0.936	0.917	0.906	0.262
Recall	0.966	1.000	1.000	0.989	0.364
Correct classification	207.000	211.000	209.000	207.000	71.000
Misclassification	10.000	6.000	8.000	10.000	146.000
F-score	0.944	0.967	0.957	0.946	0.305

Note: The best model has been given by RF method, according to the statistic Misclassification computed on the validation sample, and is given in bold.

Table 15.4 Misclassification rates for both the Out-Of-Bag (OOB) sample and the validation sample.

OOB sample	0.050
Validation sample	0.028

Table 15.5 Confusion matrices for the out-of-bag (OOB) sample derived from the random forest model.

From \ to	GAZ	OIL	Total	% Correct
GAZ	293	23	316	92.722
OIL	20	532	552	96.377
Total	313	555	868	95.046

Table 15.6 Confusion matrices for the validation sample derived from the random forest model.

From \ to	GAZ	OIL	Total	% Correct
GAZ	88	6	94	93.617
OIL	0	123	123	100.000
Total	88	129	217	97.235

Table 15.7 Variable importance scores derived from the random forest model for each attribute.

Attribute	GAZ	OIL	Overall
REG dimension	26.788	15.341	30.848
BOX dimension	36.751	8.631	31.874
αmax	5.400	4.436	6.776
αmin	6.064	2.157	6.850
αpeak	6.059	0.338	6.205
Width	37.646	16.598	40.330
f (αmax)	4.500	−3.122	4.018
f (αmin)	6.933	−2.951	6.848
f (αpeak)	−0.222	2.608	0.929
Right height	2.240	−1.413	1.993
Left height	4.358	−2.700	4.364
C values	3.379	−0.156	3.505

Table 15.8 Prediction fluid nature using RF model.

Observation	Prediction
GAZ	GAZ
OIL	OIL
OIL	OIL
GAZ	GAZ
OIL	OIL
GAZ	GAZ
GAZ	GAZ
OIL	OIL
OIL	OIL
OIL	OIL
OIL	OIL
OIL	OIL
GAZ	GAZ
OIL	OIL
OIL	OIL
OIL	OIL
OIL	OIL
OIL	OIL
OIL	OIL
OIL	OIL

The validation sample misclassification rate provides an external estimate of the model's performance using a separate dataset that was not used during the training process.

- **Validation Sample**: This is a set of data reserved for testing the model's performance after it has been trained. It is completely independent of the training process.
- **Validation Error Rate**: The validation misclassification rate (0.028) is the proportion of validation samples that are incorrectly classified by the model. In this case, the validation misclassification rate of 0.028 indicates that 2.8% of the validation samples are misclassified.

When comparing between them, we find that:

- **Lower Validation Error Rate**: The validation sample misclassification rate (0.028) is lower than the OOB sample misclassification rate (0.050). This suggests that the model performs slightly better on the validation sample compared to the OOB sample.
- **Model Performance**: Both the OOB and validation misclassification rates are quite low, indicating strong generalization capability and high predictive accuracy of the RF model. The low error rates corroborate the model's reliability and robustness in making accurate predictions.

The misclassification rates for both the OOB and validation samples indicate that the RF model is highly effective at predicting the target variable with minimal errors. The slightly lower validation misclassification rate further validates the model's performance on unseen data, confirming its suitability for fluid prediction in the given dataset.

II. Confusion Matrices

The OOB sample confusion matrix (Table 15.5) shows that the model has a high overall accuracy of 95.046%, with a slightly lower accuracy for GAZ (92.722%) compared to OIL (96.377%). The OOB error provides a reliable estimate of model performance during training.

The validation sample confusion matrix (Table 15.6) shows that the model performs even better on the validation set with an overall accuracy of 97.235%. It achieves a perfect classification rate for OIL (100.000%) and a high accuracy for GAZ (93.617%).

As conclusion, both confusion matrices demonstrate the effectiveness of the RF model:

- **OOB Sample**: The overall accuracy is 95.046%, indicating the model generalizes well during the training phase.
- **Validation Sample**: The overall accuracy is 97.235%, showcasing the model's robustness and high performance on unseen data.

The high percentages of correctly classified observations in both matrices reinforce the model's reliability and accuracy in predicting fluid types (GAZ and OIL) within the given dataset.

III. Variable Importance

Table 15.7 and Figure 15.4 provide the variable importance scores derived from the RF model for each attribute, categorized by their impact on predicting GAZ, OIL, and overall fluid behavior.

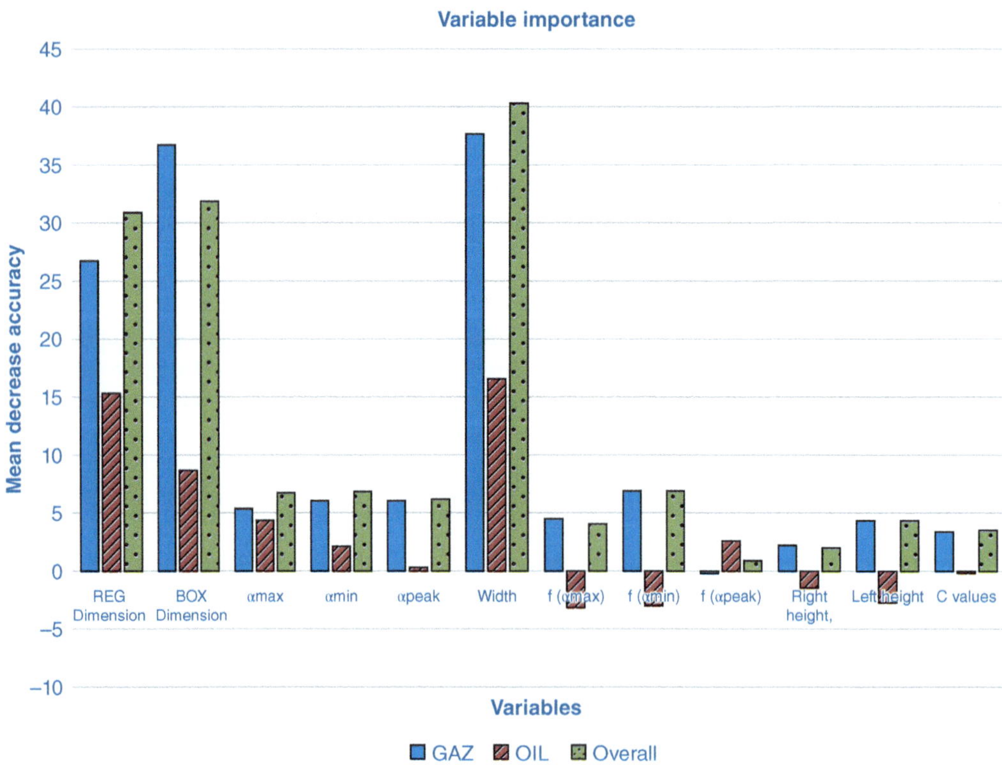

Figure 15.4 Variable importance scores derived from the RF model for each attribute.

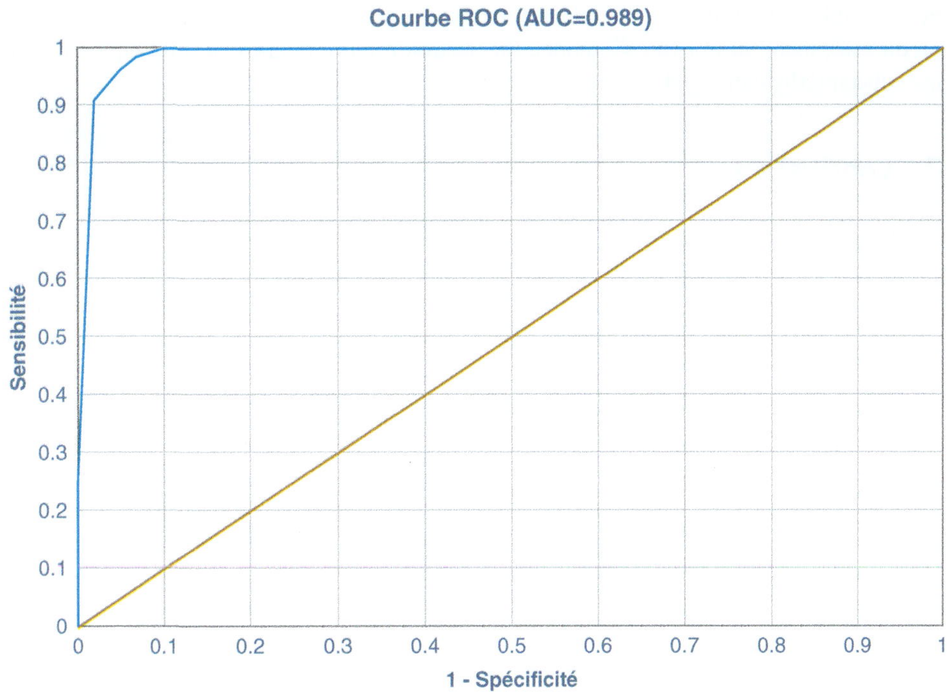

Figure 15.5 Variable importance scores derived from the RF model for each attribute.

- **Attribute**: This column lists the specific attributes or features considered in the analysis.
- **GAZ**: Indicates the importance score of each attribute for predicting GAZ (gas-related fluid behavior).
- **OIL**: Indicates the importance score of each attribute for predicting OIL (oil-related fluid behavior).
- **Overall**: Represents the overall importance score of each attribute across both GAZ and OIL predictions.

IV. ROC (AUC)

The performance of the RF classifier (Figure 15.5) was evaluated using the ROC curve and the Area Under the Curve (AUC). The ROC curve, which illustrates the trade-off between sensitivity and specificity at various threshold levels, showed that our model has a high discriminative ability. The AUC was found to be 98.97%, indicating that the model has a 98.97% chance of correctly distinguishing between positive and negative instances. This high AUC value reflects the model's robustness and its suitability for accurate classification in our study.

V. Predicted Values

Table 15.8 showcases the observed and predicted fluid types derived from the RF model. The results highlight the model's exceptional performance in predicting fluid types in clastic reservoirs using fractal attributes. Specifically, the RF model achieved flawless accuracy by correctly identifying all 20 instances of "GAZ" and "OIL" fluid types without any errors.

This exemplary accuracy underscores the potential of combining fractal analysis with ML for advanced reservoir characterization. Fractal attributes adeptly capture intricate, multiscale variations in geological properties, significantly enhancing the model's ability to differentiate between different fluid types. By integrating these detailed geological features with the robust predictive capabilities of the RF model, the accuracy and reliability of fluid type predictions are substantially enhanced.

15.7 Conclusions

This study combines fractal analysis of geophysical well logs with ML algorithms to predict fluid distributions in clastic reservoirs. By leveraging fractal and multifractal analyses, a diverse set of attributes was extracted from GR, AC, DEN, CNL, R, PF, and GRS logs. These attributes, including Box and Regularization dimensions, and multifractal singularity spectrum attributes, were used in logistic regression, RF, KNN, SVM, and discriminant analysis models.

The RF algorithm emerged as the most effective, achieving an accuracy of 97.2%, precision of 93.6%, recall of 100%, and F-score of 96.7%. This robust performance underscores its capability in handling complex geological data and improving hydrocarbon zone characterization. Importantly, this integrated approach not only enhances prediction accuracy but also provides deeper insights into reservoir structures, facilitating more informed decision-making in reservoir management and resource optimization.

The study's dataset, encompassing 1085 logs from 93 wells across diverse sedimentary basins in Algeria, demonstrates the method's applicability and reliability in varied geological contexts. Rigorous data selection and preprocessing ensure the credibility and representativeness of the findings, supporting its broader adoption in the oil and gas industry.

Acknowledgment

We would like to acknowledge the use of Python and its essential libraries, which were instrumental in our ML model development and data analysis.

References

Ali, M., Zhu, P., Jiang, R., et al. (2024). Data-driven lithofacies prediction in complex tight sandstone reservoirs: A supervised workflow integrating clustering and classification models. Geomechanics and Geophysics for Geo-Energy and Geo-Resources 10, p. 70. https://doi.org/10.1007/s40948-024-00787-5.

Behadada, O., Trovati, M., Kontonatsios, G., et al. (2017). A multinomial logistic regression approach for arrhythmia detection. International Journal of Distributed Systems and Technologies (IJDST) 8, pp. 17–33. https://doi.org/10.4018/ijdst.2017100102.

Ben Jabeur, S. (2017). Bankruptcy prediction using partial least squares logistic regression. Journal of Retailing and Consumer Services 36, pp. 197–202. https://doi.org/10.1016/j.jretconser.2017.02.005.

Benali, L., Notton, G., Fouilloy, A., et al. (2019). Solar radiation forecasting using artificial neural network and random forest methods: Application to normal beam, horizontal diffuse and global components. Renewable Energy 132, pp. 871–884. https://doi.org/10.1016/j.renene.2018.08.044.

Bishop, C. M. (2006). Pattern Recognition and Machine Learning. Springer. 978-0387310732

Boulassel, A., Zaourar, N., Gaci, S., et al. (2021). A new multifractal analysis-based for identifying the reservoir fluid nature. Journal of Applied Geophysics 185, p. 104185. https://doi.org/10.1016/j.jappgeo.2020.104185.

Breiman, L. (1996). Bagging predictors. Machine Learning 24, pp. 123–140. https://doi.org/10.1007/BF00058655.
Breiman, L. (2001). Random Forests. Machine Learning 45, pp. 5–32. https://doi.org/10.1023/A:1010933404324.
Chen, F., Deng, P., Wan, J., et al. (2015). Data mining for the internet of things: Literature review and challenges. International Journal of Distributed Sensor Networks 11(8), pp. 431047. https://doi.org/10.1155/2015/431047.
Chen, K., Chen, H., & Xu, P. (2020). A new relative permeability model of unsaturated porous media based on fractal theory. Fractals 28 (01), p. 2050002. https://doi.org/10.1142/S0218348X20500024.
Choudhary, R. & Gianey, H. K. (2017). Comprehensive review on supervised machine learning algorithms. In 2017 International Conference on Machine Learning and Data Science (MLDS) pp. 37–43. https://doi.org/10.1109/MLDS.2017.11.
Cortes, C. & Vapnik, V. (1995). Support-vector networks. Machine Learning 20, pp. 273–297. https://doi.org/10.1007/BF00994018.
Cristianini, N. & Shawe-Taylor, J. (2000). An Introduction to Support Vector Machines. New York, NY: Cambridge University Press. https://doi.org/10.1017/CBO9780511801389.
Diana, M. T. & Arnold, M. W. (1992). Development Geology Reference Manual. American Association of Petroleum Geologists. ISBN electronic: 9781629811109. ISBN print: 0891816607. https://doi.org/10.1306/Mth10573.
Diaz, P., Salas, J. C., Cipriano, A., et al. (2021). Random forest model predictive control for paste thickening. Minerals Engineering 163, p. 106760. https://doi.org/10.1016/j.mineng.2020.106760.
Elkatatny, S., Mahmoud, M., Tariq, Z., et al. (2018). New insights into the prediction of heterogeneous carbonate reservoir permeability from well logs using artificial intelligence network. Neural Computing and Applications 30, pp. 2673–2683. https://doi.org/10.1007/s00521-017-2850-x.
Fan, X., Wang, G., Dai, Q., et al. (2019). Using image logs to identify fluid types in tight carbonate reservoirs via apparent formation water resistivity spectrum. Journal of Petroleum Science and Engineering 178, pp. 937–947. ISSN 0920-4105. https://doi.org/10.1016/j.petrol.2019.04.006.
Fanchi, J. R. (2010). Integrated reservoir asset management: Principles and best practices. Burlington, MA: Gulf Professional Pub. ISBN 978-0-12-382088-4. https://doi.org/10.1016/C2009-0-62240-6.
Gislason, P. O., Benediktsson, J. A., & Sveinsson, J. R. (2006). Random forests for land cover classification. Pattern Recognition Letters 27 (4), pp. 294–300. https://doi.org/10.1016/j.patrec.2005.08.011.
Goyani, M. M. & Patel, N. M. (2017). Multi-level Haar wavelet based facial expression recognition using logistic regression. The International Journal of Next-Generation Computing 9. https://doi.org/10.17485/IJST/2017/V10I9/108944.
Guo, B., Sun, K., & Ghalambor, A. (2008). Well Productivity Handbook: Vertical, Fractured, Horizontal, Multilateral, and Intelligent Wells. Gulf Publishing Company 2 Greenway Plaza, Suite 1020 Houston, TX 77046, ISBN-13: 978-1-933762-32-6, https://industri.fatek.unpatti.ac.id/wp-content/uploads/2019/03/228-Well-Productivity-Handbook-Boyun-Guo-Kai-Sun-Ali-Ghalambor-Edisi-1-2014.pdf
Habib, M., Guangqing, Y., Xie, C., et al. (2017). Optimizing oil and gas field management through a fractal reservoir study model. Journal of Petroleum Exploration and Production Technology 7, pp. 43–53. https://doi.org/10.1007/s13202-016-0252-8.
Hallinan, J. S. (2012). Data mining for microbiologists. Systems Biology of Bacteria pp. 27–79. https://doi.org/10.1016/b978-0-08-099387-4.00002-8.
Hardy, H. H. & Beier, R. A. (December 1, 1994). Fractals in Reservoir Engineering, World Scientific Publishing Company; Illustrated edition, ISBN-13: 978-9810220693. https://doi.org/10.1142/2574.

He, Z., Tang, J., Xin, Y., et al. (2023). Dolomite reservoir fluid discrimination method based on fluid identification factor. Geoenergy Science and Engineering 226, pp. 211750. ISSN 2949-8910. https://doi.org/10.1016/j.geoen.2023.211750.

Hefner, J. T. & Ousley, S. D. (2014). Statistical classification methods for estimating ancestry using morphoscopic traits. Journal of Forensic Sciences 59 (4), pp. 883–890. https://doi.org/10.1111/1556-4029.12421.

Hefner, J. T., Pilloud, M. A., Black, C. J., et al. (2015). Morphoscopic trait expression in "Hispanic" populations. Journal of Forensic Sciences 60 (5), pp. 1135–1139. https://doi.org/10.1111/1556-4029.12826.

Hefner, J. T., Spradley, M. K., & Anderson, B. (2014). Ancestry assessment using random forest modeling. Journal of Forensic Sciences 59 (3), pp. 583–589. https://doi.org/10.1111/1556-4029.12402.

Hernandez-Martinez, E., Perez-Muñoz, T., Velasco-Hernandez, J. X., et al. (2013). Facies recognition using multifractal hurst analysis: Applications to Well-Log Data. Mathematical Geosciences 45, pp. 471–486. https://doi.org/10.1007/s11004-013-9445-6.

Huang, W., Guo, Y., Qin, J., et al. (2024). Fluid identification and tight oil layer classification for the southwestern Mahu Sag, Junggar Basin using NMR logging-based spectrum decomposition. Energy Geoscience 5 (2), 100253. ISSN 2666-7592. https://doi.org/10.1016/j.engeos.2023.100253.

Javed, A., Tiwana, M. I., Tiwana, M. I., et al. (2017). Recognition of finger movements using EEG signals for control of upper limb prosthesis using logistic regression. Biomedical Research 28 (17), pp. 7361–7369.

Jiang, C., Wang, G., Song, L., et al. (2023). Identification of fluid types and their implications for petroleum exploration in the shale oil reservoir: A case study of the Fengcheng Formation in the Mahu Sag, Junggar Basin, Northwest China. Marine and Petroleum Geology 147, p. 105996. ISSN 0264-8172. https://doi.org/10.1016/j.marpetgeo.2022.105996

Joachims, T. (1998). Text Categorization with Support Vector Machines: Learning with Many Relevant Features. In: Machine Learning: ECML-98, Lecture Notes in Computer Science (Lecture Notes in Artificial Intelligence) (eds. C. Nédellec & C. Rouveirol) vol. 1398. Berlin, Heidelberg: Springer. https://www.cs.cornell.edu/people/tj/publications/joachims_98a.pdf

Kiangala, S. K. & Wang, Z. (2021, June). An effective adaptive customization framework for small manufacturing plants using extreme gradient boosting-XGBoost and random forest ensemble learning algorithms in an Industry 4.0 environment. Machine Learning with Applications 4, p. 100024. ISSN 2666-8270. https://doi.org/10.1016/j.mlwa.2021.100024.

Liaw, A. & Wiener, M. (2002). Classification and regression by randomForest. R news 2 (3), pp. 18–22. ISSN 1609-3631. http://CRAN.R-project.org/doc/Rnews/. https://journal.r-project.org/articles/RN-2002-022/RN-2002-022.pdf

Liu, C. B. (May 2002). New Method for Identifying Fluid Type of Reservoirs in Tough Conditions. Paper presented at the Offshore Technology Conference, Houston. Texas. https://doi.org/10.4043/14078-MS.

López, M. & Aldana, M. (2007). Facies recognition using wavelet-based fractal analysis and waveform classifier at the Oritupano-a Field, Venezuela. Nonlinear Processes in Geophysics 14, pp. 325–335. https://doi.org/10.5194/npg-14-325-2007.

Lucia, F. J. (2007). Carbonate Reservoir Characterization. Heidelberg: Springer Berlin. https://doi.org/10.1007/978-3-540-72742-2.

Ma, Y. Z. (2019). Multiscale Heterogeneities in Reservoir Geology and Petrophysical Properties. In: Quantitative Geosciences: Data Analytics, Geostatistics, Reservoir Characterization and Modeling. Cham: Springer. https://doi.org/10.1007/978-3-030-17860-4_8.

Naghibi, S. A. & Moradi Dashtpagerdi, M. (2017). Evaluation of four supervised learning methods for groundwater spring potential mapping in Khalkhal region (Iran) using GIS-based features. Hydrogeology Journal 25, pp. 169–189. https://doi.org/10.1007/s10040-016-1466-z.

Platt, J. (1999). Advances in kernel methods-support vector learning, chapter fast training of support vector machines using sequential minimal optimization. MIT Press 36, pp. 185–208. https://doi.org/10.7551/mitpress/1130.003.0016.

Pyrcz, M. J., Janele, P., Weaver, D., et al. (2017). Geostatistical methods for unconventional reservoir uncertainty assessments. In: Geostatistics Valencia 2016. Quantitative Geology and Geostatistics (eds. J. Gómez-Hernández, J. Rodrigo-Ilarri, M. Rodrigo-Clavero, et al.) vol 19. Cham: Springer. https://doi.org/10.1007/978-3-319-46819-8_45.

Saidi, A., Ben Othman, S., Dhouibi, M., et al. (2021). FPGA-based implementation of classification techniques: A survey. Integration 81, pp. 280–299. https://doi.org/10.1016/j.vlsi.2021.08.004.

Scholkopf, B., Burges, C., & Vapnik, V. (1995). Extracting support data for a given task. In: Proceedings of the First International Conference on Knowledge discovery and Data Mining (eds. U.M. Fayyad & R. Uthurusamy) pp. 252–257. Menlo Park, CA: AAAI Press. https://cdn.aaai.org/KDD/1995/KDD95-030.pdf

Subhakar, D. & Chandrasekhar, E. (2016a). Reservoir characterization using multifractal detrended fluctuation analysis of geophysical well-log data. Physica A: Statistical Mechanics and its Applications 445, pp. 57–65. ISSN 0378-4371. https://doi.org/10.1016/j.physa.2015.10.103.

Subhakar, D. & Chandrasekhar, E. (2016b). Detrended fluctuation analysis of geophysical well-log data. In: Fractal Solutions for Understanding Complex Systems in Earth Sciences (ed. V. Dimri). Cham: Springer. https://doi.org/10.1007/978-3-319-24675-8_4.

Sumayli, A. (2023). Development of advanced machine learning models for optimization of methyl ester biofuel production from papaya oil: Gaussian process regression (GPR), multilayer perceptron (MLP), and K-nearest neighbor (KNN) regression models. Arabian Journal of Chemistry 16 (7), pp. 104833. ISSN 1878-5352. https://doi.org/10.1016/j.arabjc.2023.104833.

Tyralis, H., Papacharalampous, G., & Langousis, A. (2019). A brief review of random forests for water scientists and practitioners and their recent history in water resources. Water 11 (5), p. 910. https://doi.org/10.3390/w11050910.

Urso, A., Fiannaca, A., La Rosa, M., et al. (2018). Data mining: Classification and prediction. Reference Module in Life Sciences. https://doi.org/10.1016/b978-0-12-809633-8.20461-5.

Vapnik, V. (1998). Statistical Learning Theory. New York, NY: Wiley. ISBN: 0471030031, 9780471030034, 9780585357058

Vapnik, V. (2010). The Nature of Statistical Learning Theory. Springer. ISBN: 1441931600, 9781441931603

Wang, Z. & Xue, X. (2014). Multi-class support vector machine. In: Support Vector Machines Applications, 1e (eds. Y. Ma & G. Guo) pp. 23–48. Cham: Springer International Publishing. https://doi.org/10.1007/978-3-319-02300-7_2.

Wu, X., Kumar, V., Quinlan, R. J., et al. (2008). Top 10 algorithms in data mining. Knowledge and Information Systems 14, pp. 1–37. https://doi.org/10.1007/s10115-007-0114-2.

Wu, Y. Y., Lai, Q., Li, K. X., et al. (2022). Reservoir characterization and fluid typing identification of tight carbonate reservoir by using high-resolution electrical image logs: A case study of Central Sichuan Basin, China. SEG Global Meeting Abstracts pp. 2902–2909. https://doi.org/10.15530/urtec-2022-3722421.

Xu, P., Zhang, L., Rao, B., et al. (2020). A fractal scaling law between tortuosity and porosity in porous media. Fractals 28 (02), p. 2050025. https://doi.org/10.1142/S0218348X23500792.

Xu, R. & Luo, F. (2021). Risk prediction and early warning for air traffic controllers' unsafe acts using association rule mining and random forest. Safety Science 135, p. 105125. https://doi.org/10.1016/j.ssci.2020.105125.

Ye, D., Liu, G., Yu, B., et al. (2023). Evaluation of shale gas exploration by microstructure behavior and shale permeability based on fractal theory and under multi-field effects. Fractals 31 (07), p. 2350079. https://doi.org/10.1142/S0218348X23500792.

Yingxin, L. & Xiaogang, R. (2005). Feature selection for cancer classification based on support vector machine. Jisuanji Yanjiu yu Fazhan (Computer Research and Development) 42 (10), pp. 1796–1801. https://api.semanticscholar.org/CorpusID:13301855

Yu, H., Wang, Z., Rezaee, R., et al. (April 2017). Fluid type identification in carbonate reservoir using advanced statistical analysis. Paper presented at the SPE Oil and Gas India Conference and Exhibition. Mumbai, India. https://doi.org/10.2118/185447-MS.

Zeng, S. P., Chen, J., Yang, X. -W., et al. (2005). Applying the fractal geometry to analyze the unsteady seepage characteristics of gas reservoirs of the double-porosity system. In: Wavelet Analysis and Active Media Technology pp. 1026–1031. https://doi.org/10.1142/9789812701695_0160.

16

Unlocking Deeper Insights: Using Machine Learning to Predict Dynamic Shear Wave Slowness from Well Logs

Abdelbasset Boulassel[1,3]*, Soraya Makhlouf*[1]*, Zinelaabidine Boumelit*[1]*, Badis Zegagh*[2]*, Salah Boufenchouche*[3]*, Fethi Ali Cheddad*[1,3]*, Amar Boudella*[1]*, Naima Zaourar*[1]*, Said Gaci*[4]

[1] *Geophysics Department–FSTGAT, University of Sciences and Technology Houari Boumediene (USTHB), Algiers, Algeria.*
[2] *Department of Geophysics, Faculty of Hydrocarbon and chemistry, University of Boumerdes M'Hamed BOUGARA (UMBB), Boumerdes, Algeria*
[3] *Sonatrach, Exploration Division, Exploration Operations Directorate, Ouargla, Algeria*
[4] *Sonatrach, Central Directorate of Research and Development (DC R&D), Boumerdes, Algeria*

16.1 Introduction

The methodologies employed across the various phases of the petroleum industry, from exploration to exploitation, are largely determined by the physical and mechanical properties of the reservoir and its overlying formations. Geophysical well logs, which provide continuous and cost-effective formation data, are essential tools in acquiring this knowledge. Among these logs, shear wave velocity (Vs) typically derived from Dynamic Shear Wave Slowness (DTS) measurements, is particularly important for a wide range of applications in geophysics (Cregger and Lamb, 1984; Waluyo et al., 1995; Xu and McDonald, 1988; Phadke et al., 2000), petrophysical and reservoir characterization (Pickett, 1963; Eberli et al., 2003), and geomechanics (Eaton, 1976; Kumar, 1976; Chang et al., 2006; Asoodeh and Bagheripour, 2014). This parameter is crucial for calculating geomechanical properties and constructing mechanical earth models (Khatibi and Aghajanpour, 2020), which are essential for understanding wellbore stability, hydraulic fracturing, drilling mud windows, reservoir compaction, and fault reactivation (Grandi et al., 2002; Ramjohn et al., 2018; Suppachoknirun and Tutuncu, 2017; Zhang et al., 2022). Accurate prediction of Vs enhances seismic data interpretation and detailed geological characterization, both vital for efficient resource exploitation.

Direct measurement of shear wave velocity (Vs) using acoustic tools is prohibitively expensive and therefore conducted in only a few wells. This limitation has prompted extensive research into models for estimating Vs from other conventional geophysical well logs, with ongoing studies aiming to improve their accuracy. Existing empirical correlations for Vs estimation (Pickett, 1963; Castagna et al., 1985; Krief et al., 1990; Greenberg and Castagna, 1992; Bastos et al., 1998) require validation against field data and often lack universal applicability. To overcome these limitations, intelligent systems have been increasingly utilized. Researchers have applied machine learning and optimization algorithms to predict Vs from geophysical well log data, demonstrating the superiority of these

Geophysical Exploration for Hydrocarbon Reservoirs, Geothermal Energy, and Carbon Storage: New Technologies and AI-based Approaches, First Edition. Edited by Said Gaci.

methods over traditional empirical correlations (Rezaee et al., 2007; Rajabi et al., 2010; Nourafkan and Kadkhodaie-Ilkhchi, 2015; Anemangely et al., 2019; Olayiwola and Sanuade, 2021).

This chapter explores the application of machine learning techniques to predict shear wave slowness from well logs. Specifically, it evaluates the performance of four machine learning models—extreme gradient boosting (XGBoost), random forest (RF), support vector machine (SVM), and K-nearest neighbors (KNN)—in predicting shear wave slowness. These models are trained on a comprehensive dataset of well logs, incorporating data from multiple sources and origins, to assess their predictive accuracy and reliability.

The remainder of this chapter is structured as follows: Firstly, a brief explanation of the deployed dataset in terms of wells and logs is provided. Next, an overview of the applied machine learning techniques is presented, followed by a discussion of the statistical metrics adopted for evaluating both the training and prediction phases. Subsequently, the analysis of the results obtained from the applications is detailed and discussed. Finally, the main findings, key conclusions, and perspectives of this work are summarized.

16.2 Studied Wells and Dataset

The dataset we used includes a diverse set of geophysical well logs extracted from wide-ranging regions and sources, characterized by a broad spectrum of geological and petrophysical settings. The data were collected from 22 wells spanning seven sedimentary basins across Algeria, each featuring a variety of geological contexts, including diverse lithological settings, depositional environments, structural formations, and tectonic activities. These wells were drilled in varied petrophysical environments, capturing differences in rock formations, porosity levels, permeability, and fluid compositions. They also reflect diverse reservoir conditions such as variations in pressure, temperature, and depth, all of which influence the petrophysical behavior of rocks and fluids within these settings.

Table 16.1 Descriptive statistics of analyzed logs from 22 wells used in machine learning applications.

Log	Observations	Missing data	Minimum	Maximum	Average	Standard deviation
DTS	24,730	0	85.77	203.79	130.12	21.11
GR	24,730	0	12.52	536.63	101.18	47.53
DTC	24,730	0	54.39	105.37	71.73	8.48
NPHI	24,730	0	0.01	58.56	18.48	12.02
RHOB	24,730	0	1.88	4.63	2.59	0.14
PE	24,730	0	1.47	13.77	4.66	2.38
K	24,730	0	0.08	6.57	2.09	1.20
TH	24,730	0	0.86	99.47	13.57	6.66
U	24,730	0	−0.04	42.04	3.68	3.36
M2R1	24,730	0	0.15	3679.65	17.31	62.56
M2R2	24,730	0	0.19	2171.63	16.40	41.54
M2R3	24,730	0	0.17	1615.48	16.88	39.48
M2R6	24,730	0	0.16	1939.80	17.80	45.15
M2R9	24,730	0	0.16	1976.00	18.51	47.64

The dataset comprises various logging natures and sources, including compressional slowness (DTP), shear slowness (DTS), gamma ray (GR), neutron porosity (NPHI), density (RHOB), photoelectric effect (PEF), resistivity logging (R), and natural gamma ray spectroscopy (GRS), (Table 16.1). This broad spectrum of data types provides a multifaceted view of the subsurface conditions, enhancing our analytical depth. Furthermore, rigorous quality control (QC) and quality assurance (QA) measures were applied to ensure that the data used in training and validation are reliable, thereby strengthening the robustness and accuracy of our predictive models.

Additionally, two wells (WP-1 and WP-2) have been selected to test several employed machine learning techniques. This well serves as evidence of the strength and accuracy of the results generated by these techniques, and consequently, the overall validity of this study.

16.3 Overview of Employed Machine Learning Methods

We explore several applied machine learning techniques to predict shear wave slowness from well logs. Specifically, we focus on XGBoost, RF, SVM, and KNN.

16.3.1 Extreme Gradient Boosting

XGBoost is an advanced and efficient machine learning algorithm that extends decision tree methods as its fundamental unit. The resulting XGBoost model is essentially a forest composed of numerous decision trees, enhancing and optimizing the Gradient Boosting Decision Tree (GBDT) technique to deliver more accurate outcomes compared to standalone decision tree algorithms. The effectiveness of XGBoost over various machine learning algorithms has been validated by researchers such as Yang et al. (2017), Wang et al. (2020), Zhao et al. (2018), and Kiangala and Wang (2021).

Initially introduced by Chen and Guestrin (2016), XGBoost aimed to boost the performance and speed of gradient-boosted decision trees. It employs a sequential training process to combine weak predictors (decision trees) into stronger predictors. In this process, each tree relies on the outcomes of its predecessor to develop a more refined predictor (Zhu and Zhu, 2019). To prevent overfitting, XGBoost incorporates a regularization term during the modeling process, significantly accelerating the computation (Mo et al., 2019). As part of the Distributed Machine Learning Community (DMLC), XGBoost is designed for portability, flexibility, and efficiency (Yue et al., 2021). It supports three gradient boosting techniques: Regularized Boosting, Stochastic Boosting, and Gradient Boosting. Additionally, XGBoost optimizes memory usage and handles missing values effectively during training, earning the term "Sparse Aware" (Reinstein, 2017). Although XGBoost is a powerful algorithm for both regression and classification tasks, its performance can decline with imbalanced label data in classification problems, where one or more classes are underrepresented in the training dataset (Wang et al., 2020; Kiangala and Wang, 2021).

Widely adopted in business intelligence modeling, XGBoost excels in efficiently processing scattered data, leading to superior overall model performance. The algorithm supports parallel computing, making it one of the fastest algorithms available. By optimizing CPU memory usage with parallel computing, XGBoost is well-suited for managing extensive data models, classification tasks, and feature recognition. It leverages the boosting ensemble learning principle to enhance predictive performance by increasing the weights of incorrectly classified training samples (high error rate) and decreasing the weights of correctly classified ones. The algorithm focuses on higher-weighted, previously misclassified samples through multiple iterations to reduce the error rate. Consequently, the final XGBoost model achieves greater accuracy compared to single decision tree models (Fan et al., 2019; Qu et al., 2019; Suo et al., 2019; Kiangala and Wang, 2021).

16.3.2 Random Forest

The RF algorithm, introduced by Breiman (2001), is a widely adopted machine learning technique that enhances the traditional decision tree approach by combining the outputs of multiple decision trees to reach a single, more accurate result. It originated from the bagging algorithm (Breiman, 1996), and is applicable to both regression and classification problems. The RF algorithm is renowned for its flexibility and ease of use, making it a preferred choice in various applications (Liaw and Wiener, 2002).

At its core, the RF algorithm generates a multitude of decision trees during the training phase, each constructed using a random subset of the original dataset's rows and columns (Hefner et al., 2014). This randomness in data sampling and feature selection introduces variability among the individual trees, reducing the risk of overfitting—a common issue in traditional decision trees—and improving overall prediction performance.

Unlike a single decision tree, which may suffer from high variance and potentially incorrect predictions, the RF algorithm aggregates the outcomes of all the individual trees to produce a final prediction. For classification tasks, this aggregation is achieved through a majority voting system, while for regression tasks, it is done by averaging the predictions of all the trees. This collaborative decision-making process enhances the stability and accuracy of the model, often yielding superior results compared to using a single decision tree (Diaz et al., 2021; Tyralis et al., 2019; Xu and Luo, 2021).

Furthermore, the RF algorithm has proven to be exceptionally effective in handling high-dimensional, nonlinear problems, as highlighted by recent research (Benali et al., 2019). Despite its robustness and versatility, the RF algorithm's performance is contingent upon the availability of a substantial quantity of labeled data in the training set. The more labeled data the algorithm has, the more accurate its predictions will be (Gislason et al., 2006).

The RF algorithm's ability to mitigate overfitting, its applicability to complex datasets, and its reliable predictive power have solidified its role as a fundamental tool in machine learning, as noted by Kiangala and Wang (2021). Its effectiveness across diverse fields underscores its importance in both academic research and practical applications.

16.3.3 Support Vector Machine

SVMs are a robust and versatile set of supervised learning algorithms used for classification, regression, and outlier detection tasks. Originating from statistical learning theory (Vapnik, 1995), the primary objective of SVM is to identify an optimal hyperplane in an N-dimensional space that effectively separates data points into distinct classes. This optimal hyperplane is determined by maximizing the margin between the closest data points of different classes, known as support vectors, which critically influence the hyperplane's orientation and position (Cortes and Vapnik, 1995; Platt, 1998; Vapnik, 1998; Urso et al., 2019).

In a classification context, SVM operates by mapping input vectors into a high-dimensional feature space, enabling the linear separation of data through decision planes (Scholkopf et al., 1995). The margin is defined by two parallel hyperplanes, with the optimal hyperplane situated midway, ensuring the largest possible distance to the nearest data points from both classes (Cristianini and Shawe-Taylor, 2000). These nearest data points are termed support vectors, and they crucially influence the hyperplane's orientation and position (Platt, 1998). The dimensionality of the hyperplane is contingent on the number of features: for two features, it is a line; for three features, it becomes a 2D plane.

As the number of features increases, visualizing the hyperplane becomes more complex. However, SVM remains highly effective in managing high-dimensional data, making it suitable for a broad spectrum of applications, including text classification, image classification, spam detection, handwriting identification, gene expression analysis, and anomaly detection. SVMs are particularly advantageous because they can handle both linearly and non-linearly separable data by employing nonlinear mapping techniques. This mapping transforms the original finite-dimensional space into a much higher-dimensional space, facilitating better data separation (Chen et al., 2015).

Despite SVM's inherent non-probabilistic nature, which means it does not directly provide probability estimates for classifications, the algorithm has proven to be highly accurate across various domains such as pattern recognition, medical diagnosis, marketing, and text classification (Li and Ruan 2005; Joachims, 1998). The method has also demonstrated superior performance over neural networks and decision trees, especially when working with smaller training data sets (Hefner and Ousley, 2014; Hefner et al., 2015).

The effectiveness of SVM is further highlighted by its ability to handle situations where the number of dimensions exceeds the number of samples. This is achieved by carefully selecting kernel functions and regularization terms to avoid overfitting. The versatility of SVM is enhanced by the availability of various kernel functions for the decision function, which allows for the specification of custom kernels when necessary. However, calculating probability estimates through techniques like five-fold cross-validation can be computationally expensive, which is a notable limitation. SVM's utility is underscored by its application in tasks such as estimating ancestry among multiple groups, where solving large quadratic programming optimization problems ensures that the parameters of the maximum-margin hyperplane are optimally derived (Vapnik, 1998). Overall, SVMs are adaptable and efficient in a wide range of applications due to their capacity to handle high-dimensional data and nonlinear relationships (Wang and Xue, 2014; Wu et al., 2008).

16.3.4 K-Nearest Neighbors

The KNN algorithm is a versatile and intuitive supervised machine learning method employed to tackle both regression and classification tasks. Initially developed by Evelyn Fix and Joseph Hodges in 1951 and later expanded by Thomas Cover, KNN has become a foundational technique in the field of machine learning, particularly in pattern recognition, data mining, and intrusion detection. KNN operates on the principle that similar data points tend to have similar labels or values, making it a non-parametric algorithm that does not assume any specific distribution of the underlying data. This characteristic distinguishes it from other algorithms, such as Gaussian Mixture Models (GMM), which assume a Gaussian distribution. Given its flexibility, KNN finds widespread application in real-life scenarios, including stock price prediction, housing price estimation, and weather forecasting (Naghibi and Dashtpagerdi, 2017).

During the training phase, the KNN algorithm stores the entire training dataset as a reference, populating the search space with instances whose classes or target values are known. When making predictions, KNN calculates the distance between a new observation and all the training examples, typically using a distance metric such as Euclidean distance. The algorithm then identifies the K data points in the training dataset that are closest to the new observation. In the case of regression, it estimates the value of the new data point by averaging (or taking the median of) the target variable values of its KNN. In classification, KNN assigns the most common class label among the K neighbors as the predicted label for the new observation (Sumayli, 2023).

KNN's simplicity and ease of interpretation contribute to its popularity across various domains. However, its performance is highly dependent on the choice of K and the distance metric, necessitating careful parameter tuning for optimal results. A small value of K might lead to overfitting, capturing noise in the data, while a large value of K could result in underfitting, overlooking important details. Additionally, KNN can be computationally expensive in terms of both time and memory, especially when dealing with high-dimensional data or data with complex structures (Bishop and Nasrabadi, 2006).

The algorithm is particularly useful for classification tasks such as image recognition, handwriting detection, and various other classification problems. Its non-parametric nature makes it highly adaptable and widely applicable in the industry, especially for classification problems. However, because KNN is a lazy learning algorithm, no explicit model is built during the training phase; instead, computation is deferred until classification or prediction is needed (Sumayli, 2023). In summary, KNN is a straightforward and powerful machine learning technique that is effective in various practical applications. Its versatility, combined with its simplicity, makes it a popular choice, although it requires careful consideration of parameters such as K and the distance metric to achieve the best performance.

16.4 Model Evaluation

The developed machine learning models will be evaluated using a range of statistical metrics to ensure their accuracy, reliability, and overall performance. These metrics provide a comprehensive understanding of how well the models predict and generalize to new data. The key evaluation metrics used include:

16.4.1 Mean Absolute Error

MAE measures the average magnitude of errors in a set of predictions, without considering their direction. It represents the average of the absolute differences between the predicted values and the actual values, offering a straightforward interpretation of model accuracy. The formula for MAE is:

$$MAE = \frac{1}{n}\sum_{i=1}^{n}|y_i - \hat{y}_i| \tag{16.1}$$

where y_i is the actual value and $\hat{y}_i$ is the predicted value. MAE provides insight into how far the predictions are from the actual outcomes on average.

16.4.2 Mean Squared Error

MSE measures the average of the squares of the errors, emphasizing larger errors due to the squaring of the error terms. It is particularly useful for identifying models that perform poorly with significant deviations. The formula for MSE is:

$$MSE = \frac{1}{n}\sum_{i=1}^{n}(y_i - \hat{y}_i)^2 \tag{16.2}$$

16.4.3 Coefficient of Determination (R^2)

The Coefficient of Determination, or R^2, indicates the proportion of the variance in the dependent variable that is predictable from the independent variables. It is a measure of the goodness of fit of the model:

$$R^2 = 1 - \frac{\sum_{i=1}^{n}(y_i - \hat{y}_i)^2}{\sum_{i=1}^{n}(y_i - \bar{y})^2} \tag{16.3}$$

where $\bar{y}$ is the mean of the actual values. R^2 ranges from 0 to 1, with higher values indicating a better fit of the model to the data.

16.4.4 Adjusted R^2

Adjusted R^2 adjusts the R^2 value for the number of predictors in the model. It accounts for the model complexity and helps in comparing models with different numbers of predictors:

$$Adjusted\ R^2 = 1 - \left(\frac{1 - R^2}{n - k - 1}\right) \tag{16.4}$$

where n is the number of observations and k is the number of predictors. Adjusted R^2 can decrease if unnecessary predictors are added to the model, thus penalizing the inclusion of non-significant variables.

16.4.5 Akaike Information Criterion

Akaike information criterion (AIC) is a measure of the relative quality of a statistical model for a given set of data. It balances the goodness of fit and the simplicity of the model:

$$AIC = 2k - 2\ln(L) \tag{16.5}$$

where k is the number of parameters in the model and L is the likelihood function. Lower AIC values indicate a better model, with a trade-off between accuracy and complexity.

16.4.6 Schwarz Bayesian Criterion or Bayesian Information Criterion

Schwarz Bayesian Criterion (SBC), also known as the Bayesian Information Criterion (BIC), is similar to AIC but includes a stronger penalty for models with more parameters:

$$BIC = k\ \ln(n) - 2\ \ln(L) \tag{16.6}$$

where n is the number of observations, k is the number of parameters in the model, L is the likelihood of the model given the data. Like AIC, lower BIC values indicate a better model, favoring simpler models more strongly compared to AIC.

16.4.7 Interpretation of Metrics

- **MAE**: Measures the average magnitude of the errors in a set of predictions, without considering their direction. Lower values indicate better model performance and good prediction accuracy.
- **MSE**: Measures the average of the squares of the errors, giving more weight to larger errors. Again, lower values indicate better performance and good prediction accuracy.
- **R^2** (R-Squared): Indicates the proportion of the variance in the dependent variable that is predictable from the independent variables. Values closer to 1 indicate a better fit.
- **Adjusted R^2**: Similar to R^2 but adjusted for the number of predictors in the model.

- **AIC**: A measure of the relative quality of a statistical model for a given set of data. Lower values indicate a better fit.
- **SBC**: Similar to AIC but includes a penalty term for the number of parameters in the model. Lower values indicate a better fit.

These evaluation metrics are essential for assessing the performance and robustness of machine learning models, guiding the selection of the most appropriate model for a given dataset and predictive task. They help ensure that the models not only fit the training data well but also generalize effectively to new, unseen data.

16.5 Results and Discussion

Based on a comprehensive analysis of the results and visualizations, it is evident that different machine learning models exhibit varying levels of effectiveness in predicting dynamic shear wave slowness from well logs. This study employed XGBoost, RF, SVM, and KNN techniques to analyze 14 logs from 22 wells. These logs included GR, Compressional slowness (DTC), Shear slowness (DTS), Neutron Porosity (NPHI), Bulk Density (RHOB), Photoelectric Effect (PE), Potassium (K), Thorium (TH), Uranium (U), and multiple resistivity measurements (M2R1, M2R2, M2R3, M2R6, M2R9).

The dataset was randomly divided into two subsets: 80% of the observations were used for training the models, and the remaining 20% were used for validation. Additionally, predictions of dynamic shear wave slowness were performed for wells WP-1 and WP-2 to test the predictability performance of the employed ML techniques. The performances of these models were evaluated using a range of statistical metrics and visual regression analyses, with the results summarized in the tables and plots below (Tables 16.2–16.4, Figures 16.1–16.6).

16.5.1 Training and Validation Results

Based on the detailed findings and visual analyses, it is clear that the RF and XGBoost models are the most effective and reliable for predicting dynamic shear wave slowness from well logs. They provide high predictive accuracy, as evidenced by their low MAE and MSE values, and maintain model efficiency with the lowest Akaike AIC and Schwarz SBC values, indicating a strong fit with the dataset. Additionally, the high R^2 values for both models (0.96) highlight their ability to explain

Table 16.2 Performance indicators of various machine learning models.

Performance indicators	DTS_XGBOOST	DTS_Random forest	DTS_SVM	DTS_KNN
MAE	2.780	2.482	12.278	3.110
MSE	4.196	4.140	17.419	4.924
R^2	0.959	0.961	0.302	0.944
Adjusted R^2	0.959	0.960	0.300	0.944
Akaike AIC	14212.90	14079.70	28292.98	15794.67
Schwarz SBC	14297.50	14164.30	28377.56	15879.25

Note: The performance indicators are presented for various machine learning models computed on the validation sample (20% of the entire dataset). The results demonstrate the effectiveness and robustness of each employed ML technique.

Table 16.3 Performance indicators of various machine learning models.

Performance indicators	DTS_XGBOOST	DTS_Random forest	DTS_SVM	DTS_KNN
MAE	4.92	4.52	16.39	6.53
MSE	51.03	40.60	458.05	110.84
R^2	0.91	0.93	0.17	0.80
Adjusted R^2	0.91	0.93	0.17	0.80
Akaike AIC	16002.23	15461.69	21187.94	17835.10
Schwarz SBC	16013.77	15473.23	21199.48	17846.63

Note: The performance indicators are presented for various machine learning models computed on the prediction sample for WP-1. The results demonstrate the effectiveness and robustness of each employed ML technique.

Table 16.4 Performance indicators of various machine learning models.

Performance indicators	DTS_XGBOOST	DTS_Random forest	DTS_SVM	DTS_KNN
MAE	4.93	4.94	16.59	7.64
MSE	49.47	51.13	407.55	112.95
R^2	0.90	0.89	0.15	0.76
Adjusted R^2	0.90	0.89	0.15	0.76
Akaike AIC	66250.13	66575.84	86980.13	74365.96
Schwarz SBC	66264.52	66590.22	86994.52	74380.34

Note: This table presents the performance indicators for various machine learning models computed on the prediction sample for WP-2. The results demonstrate the effectiveness and robustness of each employed ML technique.

a large proportion of the variance in the DTS values, further underscoring their robustness and reliability. The KNN model, while still performing adequately, falls short in comparison to RF and XGBoost. Its slightly higher error metrics, AIC/SBC values, and R^2 value (0.94) suggest it is less efficient and accurate but can still be considered a viable model for DTS prediction. Conversely, the SVM model is the least effective, with significantly lower predictive accuracy and efficiency. Its high error metrics, AIC/SBC values, and low R^2 value (0.38) highlight its poor performance, making it unsuitable for this specific application (Table 16.2, Figures 16.1 and 16.2).

16.5.2 Analysis and Interpretation of Prediction Results for Well WP-1

The depth range covered in Well WP-1 is from 3665 m to 4025 m. The performance indicators for various machine learning models used to predict dynamic shear wave slowness (DTS) from well logs are summarized in the Table 16.3 and visualized in the plots (Figures 16.3 and 16.4).

- **XGBoost**: This model exhibited strong performance with an MAE of 4.92 and an MSE of 51.03. The coefficient of determination (R^2) was 0.91, indicating a high level of accuracy in predictions. The AIC and SBC values were 16002.23 and 16013.77, respectively. These low values indicate that the XGBoost model strikes a good balance between model complexity and goodness of fit.

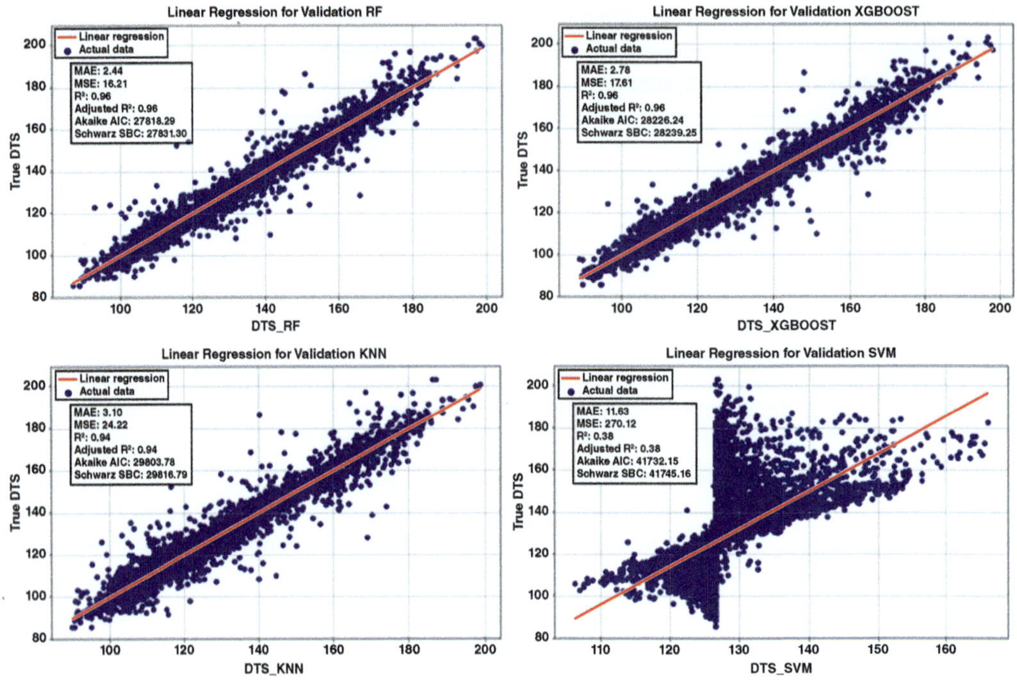

Figure 16.1 Linear regression analyses for DTS (Delta T Shear) predictions using various machine learning models (RF, XGBoost, KNN, and SVM) applied during the validation phase with 20% of the entire dataset. Each plot compares the predicted DTS values against the true DTS values, providing key statistical metrics to evaluate model performance. The figure highlights the remarkable similarities between the actual DTS and the predicted DTS values, demonstrating high performance indicators.

- **Random Forest**: This model outperformed others with an MAE of 4.52 and an MSE of 40.60. It achieved the highest R^2 value of 0.93, showing excellent predictive capability. The AIC and SBC values were 15461.69 and 15473.23, respectively, which are the lowest among all models, indicating a very efficient and balanced model in terms of complexity and fit.
- **K-Nearest Neighbors**: The KNN model showed moderate performance with an MAE of 6.53 and an MSE of 110.84. The R^2 value was 0.80, indicating good predictive ability. The AIC and SBC values were 17835.10 and 17846.63, respectively. These higher values reflect the model's greater complexity and lower efficiency compared to XGBoost and RF.
- **Support Vector Machine**: The SVM model demonstrated the least effective performance with an MAE of 16.39 and a significantly higher MSE of 458.05. The R^2 value was only 0.17, reflecting poor predictive accuracy. The AIC and SBC values were 21187.94 and 21199.48, respectively. These very high values indicate a poor balance between model complexity and goodness of fit, further highlighting the SVM model's inefficiency and inaccuracy for this task.

The comparative analysis of these models reveals that the RF and XGBoost models are the most effective and reliable for predicting DTS in Well WP-1 within the given depth range. The superior performance of these models can be attributed to their advanced algorithms that handle the complex relationships between input features and target variables more efficiently.

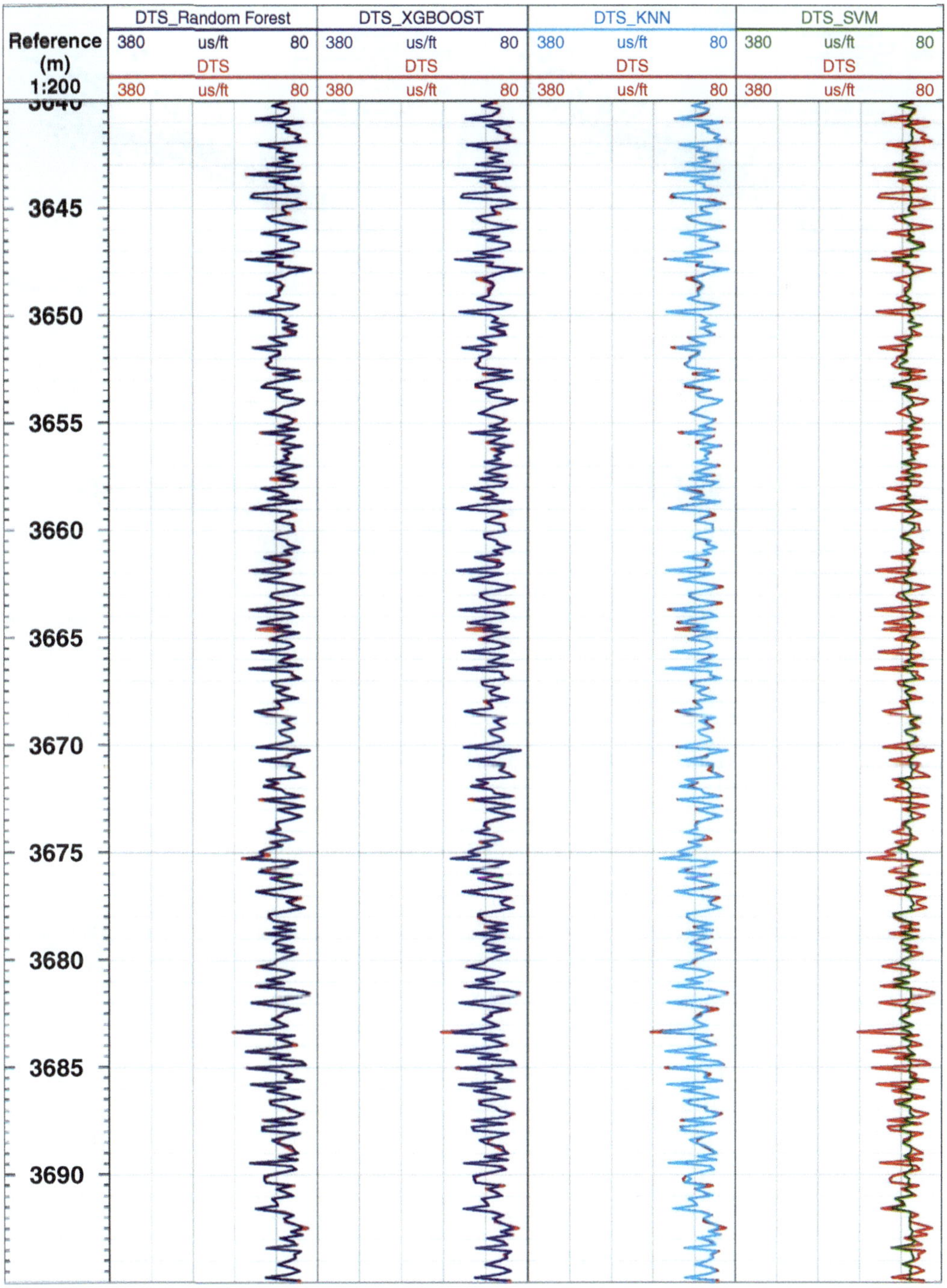

Figure 16.2 A segment of well logs across multiple tracks, including real DTS logs and DTS predictions from various machine learning models (RF, XGBoost, KNN, and SVM) applied during the validation phase with 20% of the entire dataset. The figure highlights the remarkable similarities between the actual DTS and the predicted DTS values.

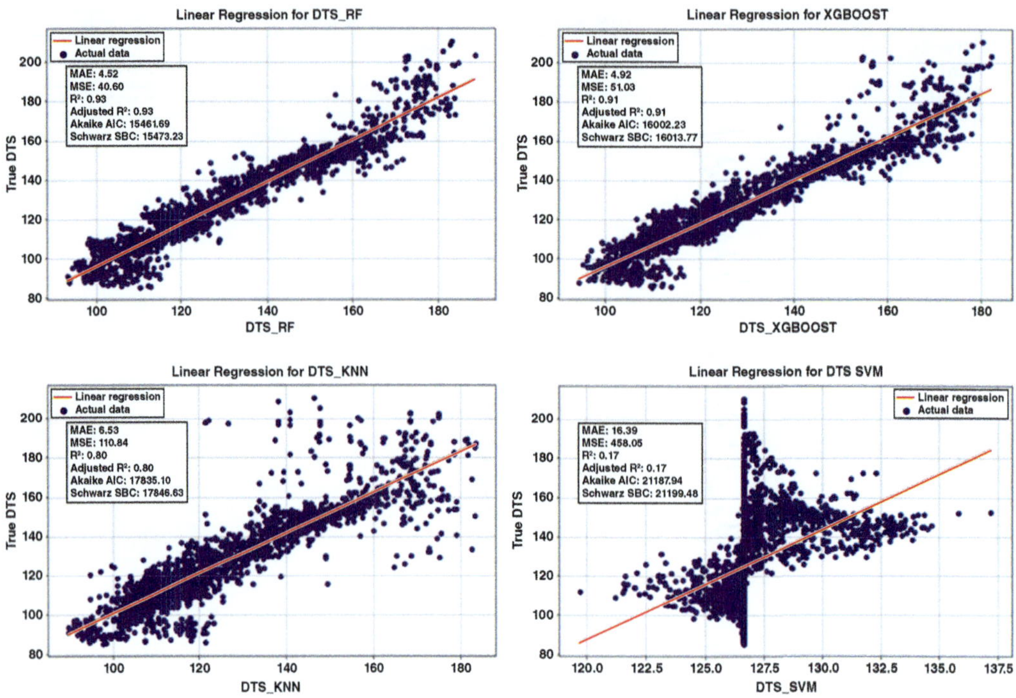

Figure 16.3 Linear regression analysis for DTS predictions: this figure showcases linear regression analyses for DTS (Delta T Shear) predictions using various machine learning models, developed as part of the WP-1 modeling task. Each plot compares the predicted DTS values against the true DTS values, providing key statistical metrics to evaluate model performance.

16.5.3 Analysis and Interpretation of Prediction Results for Well WP-2

The depth range covered in Well WP-2 is from 3200 m to 4600 m. The performance indicators for various machine learning models used to predict dynamic shear wave slowness (DTS) from well logs are summarized in the Table 16.4 and visualized in the regression plots (Figures 16.5 and 16.6).

- **XGBoost**: This model exhibited strong performance with an MAE of 4.93 and an MSE of 49.47. The coefficient of determination (R^2) was 0.90, indicating a high level of accuracy in predictions. The AIC and SBC values were 66250.13 and 66264.52, respectively. These low values indicate that the XGBoost model strikes a good balance between model complexity and goodness of fit.
- **Random Forest**: This model demonstrated good performance with an MAE of 4.94 and an MSE of 51.13. The R^2 value was 0.89, showing strong predictive capability. The AIC and SBC values were 66575.84 and 66590.22, respectively, which are slightly higher than those for XGBoost. This suggests that while the RF model is highly accurate, it may be slightly more complex or less efficient than the XGBoost model.

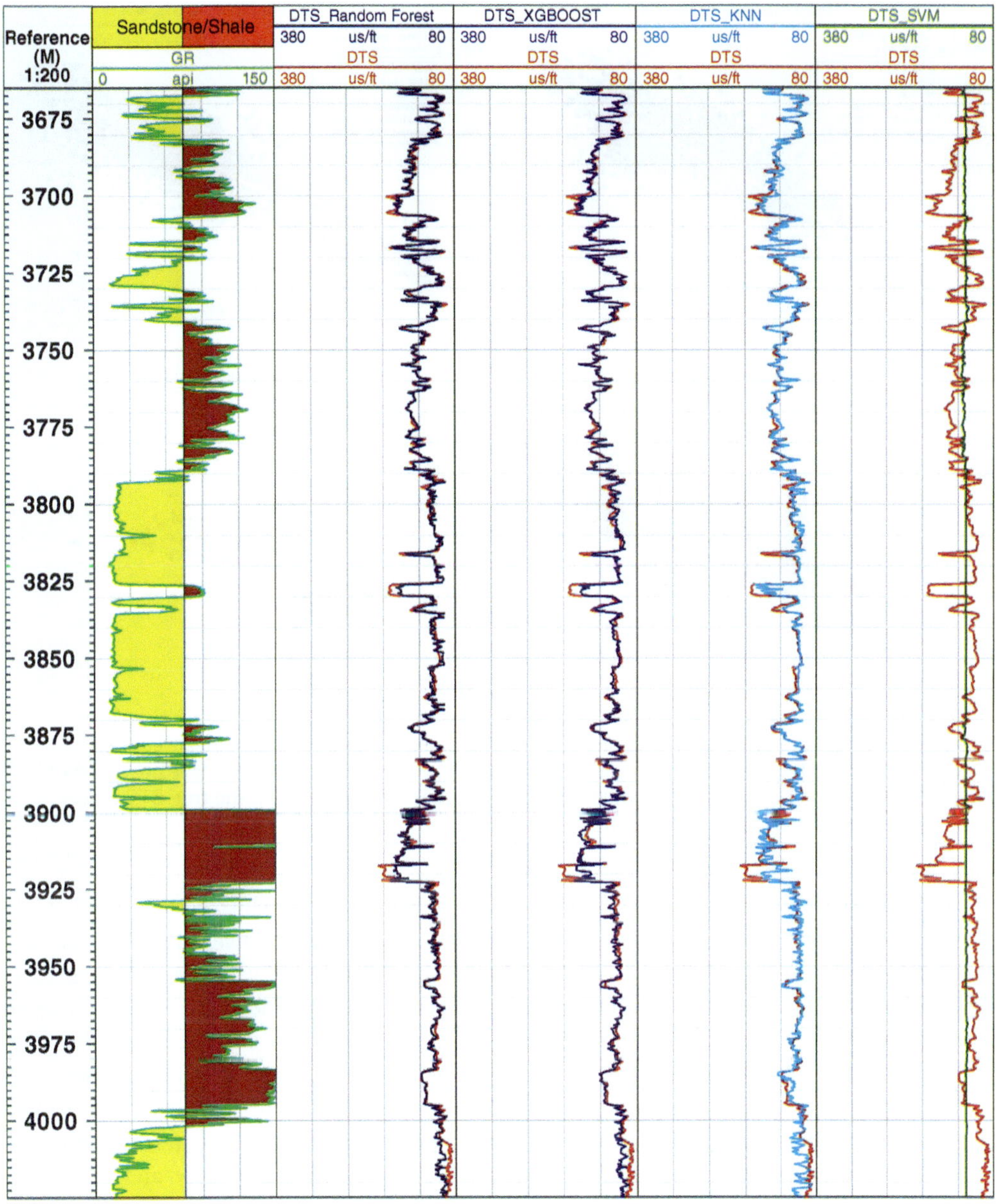

Figure 16.4 Well logs describe subsurface properties along a 360-meter open hole borehole, illustrated through multiple tracks. These tracks include lithology classification using gamma-ray (GR) logs, GR measurements, real DTS data logs, and DTS predictions from various machine learning models (RF, XGBOOST, KNN, and SVM), developed as part of the WP-1 modeling task.

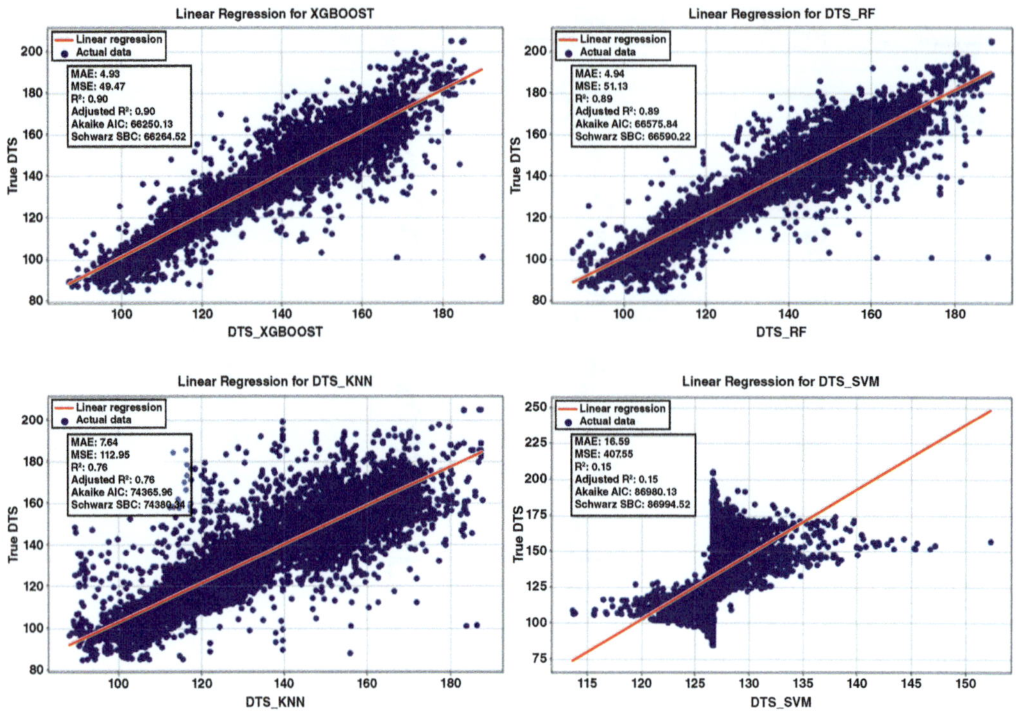

Figure 16.5 Linear regression analysis for DTS predictions: this figure showcases linear regression analyses for DTS (Delta T Shear) predictions using various machine learning models, developed as part of the WP-2 modeling task. Each plot compares the predicted DTS values against the true DTS values, providing key statistical metrics to evaluate model performance.

- **K-Nearest Neighbors**: The KNN model showed moderate performance with an MAE of 7.64 and an MSE of 112.95. The R^2 value was 0.76, indicating good predictive ability. The AIC and SBC values were 74365.96 and 74380.34, respectively. These higher values reflect the model's greater complexity and lower efficiency compared to XGBoost and RF.
- **Support Vector Machine**: The SVM model demonstrated the least effective performance with an MAE of 16.59 and a significantly higher MSE of 407.55. The R^2 value was only 0.15, reflecting poor predictive accuracy. The AIC and SBC values were 86980.13 and 86994.52, respectively. These very high values indicate a poor balance between model complexity and goodness of fit, further highlighting the SVM model's inefficiency and inaccuracy for this task.

The comparative analysis of these models reveals that the XGBoost and RF models are the most effective and reliable for predicting DTS in Well WP-2 within the given depth range. The superior performance of these models can be attributed to their advanced algorithms that handle the complex relationships between input features and target variables more efficiently.

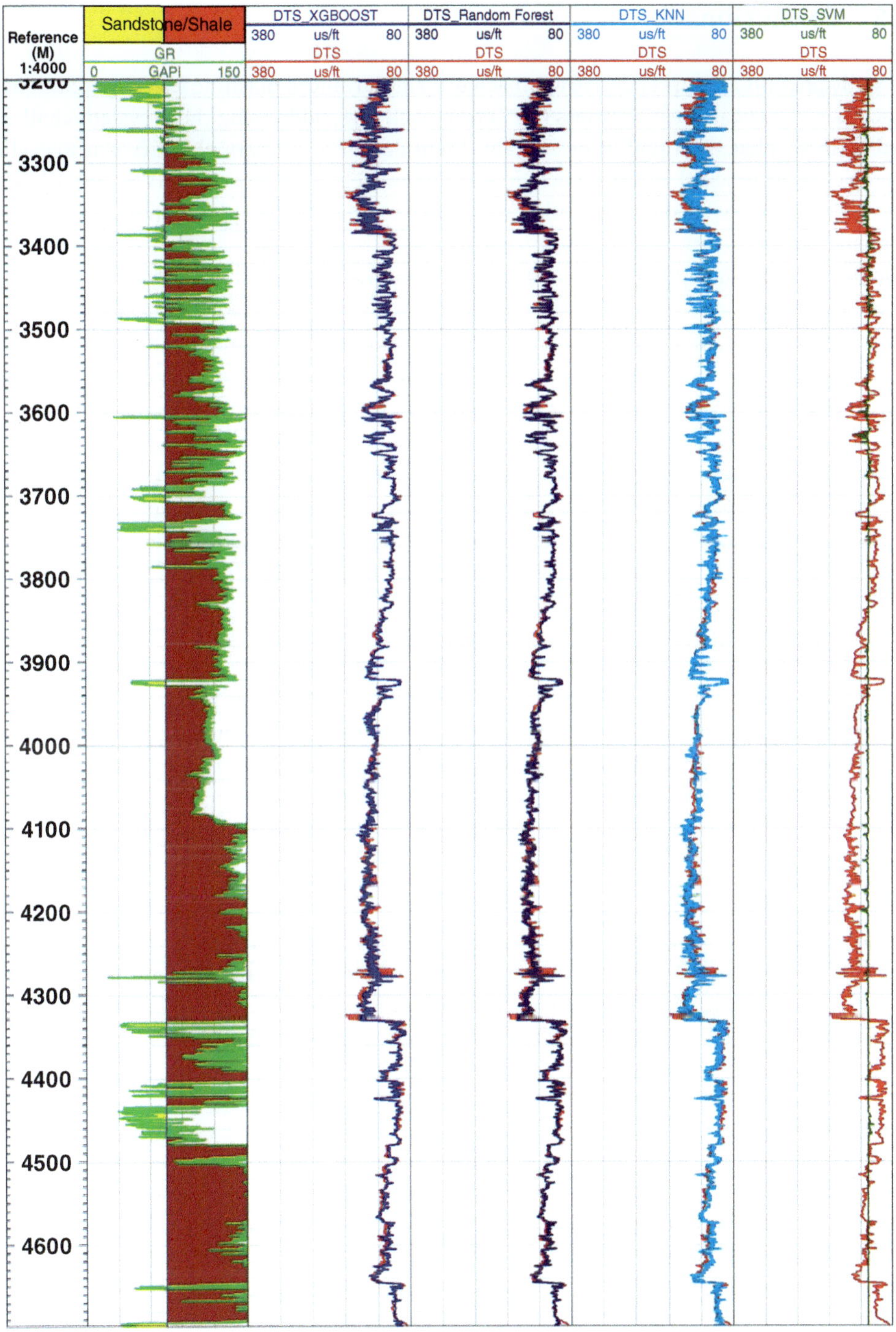

Figure 16.6 Well logs describe subsurface properties along a 1498 m open hole borehole, illustrated through multiple tracks. These tracks include lithology classification using gamma-ray (GR) logs, GR measurements, real DTS data logs, and DTS predictions from various machine learning models (RF, XGBOOST, KNN, and SVM), developed as part of the WP-2 modeling task.

16.6 Conclusions

This study presents a comprehensive approach utilizing advanced machine learning techniques to predict dynamic shear wave slowness (DTS) from well logs, addressing the prevalent challenges posed by incomplete, low-quality, or entirely absent data. The primary objective was to evaluate the predictive accuracy of various machine learning models – XGBoost, RF, SVM, and KNN – in predicting DTS from well logs. The work was executed in three distinct phases: training, validation, and prediction.

The methodology incorporated rigorous data preprocessing, including stringent quality control and assurance measures, followed by the application of the machine learning models. Performance metrics such as MAE, MSE, and the coefficient of determination (R^2) were employed to assess the models' performance. During the training and validation phase, RF demonstrated the highest R^2 value of 0.961 and the lowest MAE of 2.482, underscoring its robust predictive capability. Similarly, the XGBoost model exhibited strong performance with an R^2 value of 0.959 and an MAE of 2.780. The application of these models to wells WP-1 and WP-2 further validated their effectiveness, showcasing significant improvements in the accuracy of shear wave slowness predictions. Specifically, for WP-1, RF achieved an R^2 value of 0.93 and an MAE of 4.52, while XGBoost achieved an R^2 value of 0.91 and an MAE of 4.92. For WP-2, XGBoost recorded an R^2 value of 0.90 and an MAE of 4.93, while RF achieved an R^2 value of 0.89 and an MAE of 4.94.

This work demonstrates that machine learning techniques, particularly XGBoost and RF, can significantly enhance the quality and completeness of well log data. By addressing the limitations of traditional methods, these advanced models pave the way for more precise and cost-effective exploration and exploitation of subsurface resources. Future research should focus on integrating additional data types and exploring ensemble learning approaches to further improve prediction accuracy and applicability across different geological settings.

Acknowledgment

We would like to acknowledge the use of Python and its essential libraries, which were instrumental in our machine learning model development and data analysis. Additionally, we extend our gratitude to SONATRACH for providing the TechLog software, which was crucial for well log data processing and analysis. These tools significantly contributed to the success and robustness of our study.

References

Anemangely, M., Ramezanzadeh, A., Amiri, H., et al. (2019). Machine learning technique for the prediction of shear wave velocity using petrophysical logs. Journal of Petroleum Science and Engineering 174, pp. 306–327. https://doi.org/10.1016/j.petrol.2018.11.032.

Asoodeh, M. & Bagheripour, P. (2014). ACE stimulated neural network for shear wave velocity determination from well logs. Journal of Applied Geophysics 107, pp. 102–107. ISSN 0926-9851. https://doi.org/10.1016/j.jappgeo.2014.05.014.

Bastos, A. C., Dillon, L. D., Vasquez, G. F., et al. (1998). Core-derived acoustic, porosity & permeability correlations for computation pseudo-logs. Geological Society, London, Special Publications 136 (1), pp. 141–146. https://doi.org/10.1144/gsl.sp.1998.136.01.12.

Benali, L., Notton, G., Fouilloy, A., et al. (2019). Solar radiation forecasting using artificial neural network and random forest methods: application to normal beam, horizontal diffuse and global components. Renewable Energy 132, pp. 871–884. https://doi.org/10.1016/j.renene.2018.08.044.

Bishop, C. M. & Nasrabadi, N. M. (2006). Pattern recognition and machine learning. New York, NY: Springer, google schola 2, pp. 1122–1128. ISBN 978-0-387-31073-2.

Breiman, L. (1996). Bagging predictors. Machine learning 24, pp. 123–140. https://doi.org/10.1007/BF00058655.

Breiman, L. (2001). Random forests. Machine learning 45, pp. 5–32. https://doi.org/10.1023/A:1010933404324.

Castagna J. P, Batzle M. L., & Eastwood R. L. (1985). Relationships between compressional-wave and shear-wave velocities in clastic silicate rocks. Geophysics 50, pp. 571–581. https://doi.org/10.1190/1.1441933.

Chang, C., Zoback, M. D., & Khaksar, A. (2006). Empirical relations between rock strength and physical properties in sedimentary rocks. Journal of Petroleum Science and Engineering 51, pp. 223–237. ISSN 0920-4105. https://doi.org/10.1016/j.petrol.2006.01.003.

Chen, F., Deng, P., Wan, J., et al. (2015). Data mining for the internet of things: literature review and challenges. International Journal of Distributed Sensor Networks 2015, pp. 1–15.

Chen, T. & Guestrin, C. (2016). Xgboost: a scalable tree boosting system. In: Proceedings of the 22nd acm sigkdd international conference on knowledge discovery and data mining. pp. 785–794. https://doi.org/10.1145/2939672.2939785.

Cortes, C. & Vapnik, V. (1995). Support-vector networks. Machine Learning 20, pp. 273–297. https://doi.org/10.1007/BF00994018.

Cregger, D. M. & Lamb, T. J. (1984). Poisson's Ratio as a parameter for determining dynamic elastic modulus. Paper presented at the 25th U.S. Symposium on Rock Mechanics (USRMS), Evanston, Illinois, June 1984.

Cristianini, N. & Shawe-Taylor, J. (2000). An Introduction to Support Vector Machines. New York, NY: Cambridge University Press. https://doi.org/10.1017/CBO9780511801389.

Diaz, P., Salas, J. C., Cipriano, A., et al. (2021). Random forest model predictive control for paste thickening. Minerals Engineering 163, p. 106760, https://doi.org/10.1016/j.mineng.2020.106760.

Eaton, B. A. (1976). Graphical method predicts geopressures worldwide. World Oil: (United States) 182, pp. 51-56.

Eberli, G. P., Baechle, G. T., Anselmetti, F. S., et al. (2003). Factors controlling elastic properties in carbonate sediments and rocks. The Leading Edge 22, pp. 654–660. https://doi.org/10.1190/1.1599691.

Fan, R. S., Ya, L. I., & Tao, T. (2019). Research and application of project settlement overdue prediction based on xgboost intelligent algorithm. In: 2019 IEEE Sustainable Power and Energy Conference (iSPEC). pp. 1213–1216. Beijing, China: IEEE. https://doi.org/10.1109/iSPEC48194.2019.8975056.

Gislason, P. O., Benediktsson, J. A., & Sveinsson, J. R. (2006). Random forests for land cover classifi cation. Pattern Recognition Letters 27 (4), pp. 294–300. https://doi.org/10.1016/j.patrec.2005.08.011.

Grandi, S., Rao, R. V., & Toksöz, M. N. (2002). Geomechanical Modelling of In-situ Stresses Around a Borehole. Massachusetts Institute of Technology, Earth Resources Laboratory Cambridge, MA, USA. https://dspace.mit.edu/bitstream/handle/1721.1/67848/GRANDI.PDF?sequence=1&isAllowed=y.

Greenberg, M. L. & Castagna, J. P. (1992). Shear-wave velocity estimation in porous rocks: theoretical formulation, preliminary verification and applications1. Geophysical Prospecting 40 (2), pp. 195–209. https://doi.org/10.1111/j.1365-2478.1992.tb00371.x.

Hefner, J. T. & Ousley, S. D. (2014). Statistical classification methods for estimating ancestry using morphoscopic traits. Journal of Forensic Sciences 59 (4), pp. 883–890. https://doi: 10.1111/1556-4029.12421.

Hefner, J. T., Pilloud, M. A., Black, C. J., et al. (2015). Morphoscopic trait expression in “Hispanic” populations. Journal of Forensic Sciences 60 (5), pp. 1135–1139. https://doi: 10.1111/1556-4029.12826.

Hefner, J. T., Spradley, M. K., & Anderson, B. (2014). Ancestry assessment using random forest modeling. Journal of Forensic Sciences 59 (3), pp. 583–589, https://doi.org/10.1111/1556-4029.12402.

Joachims, T. (1998). Text categorization with support vector machines: Learning with many relevant features. In: Machine Learning: European Conference on Machine Learning (in LNCS 1398), (ed. D. Hutchison, T. Kanade, J. Kittler., et al.), pp. 137–142. Berlin: Springer. https://www.cs.cornell.edu/people/tj/publications/joachims_98a.pdf.

Khatibi, S. & Aghajanpour, A. (2020). Machine Learning: A Useful Tool in Geomechanical Studies, a Case Study from an Offshore Gas Field. Energies 13, p. 3528. https://doi.org/10.3390/en13143528.

Kiangala, S. K. & Wang, Z. (2021). An effective adaptive customization framework for small manufacturing plants using extreme gradient boosting-XGBoost and random forest ensemble learning algorithms in an Industry 4.0 environment. Machine Learning with Applications 4, p. 100024. ISSN 2666-8270. https://doi.org/10.1016/j.mlwa.2021.100024.

Krief, M., Garat, J., Stellingwerff, J., et al. (1990). A petrophysical interpretation using the velocities of P and S waves (full-waveform sonic). The Log Analyst 31 (6), pp. 355–369.

Kumar, J. (1976). The Effect of Poisson’s Ratio on Rock Properties. Paper presented at the SPE Annual Fall Technical Conference and Exhibition, New Orleans, Louisiana, October 1976. https://doi.org/10.2118/6094-MS.

Li, Y. & Ruan, X. (2005). Feature selection for cancer classification based on support vector machine. Journal of Computer Research and Development 42 (10), pp. 1796–1801. https://api.semanticscholar.org/CorpusID:13301855.

Liaw, A. & Wiener, M. (2002). Classification and regression by randomForest. R news 2 (3), pp. 18–22. ISSN 1609-3631. https://journal.r-project.org/articles/RN-2002-022/RN-2002-022.pdf.

Mo, H., Sun, H., Liu, J., et al. (2019). Developing window behavior models for residential buildings using XGBoost algorithm. Energy and Buildings 205, p. 109564. https://doi.org/10.1016/j.enbuild.2019.109564.

Naghibi, S. A. & Dashtpagerdi, M. M. (2017). Evaluation of four supervised learning methods for groundwater spring potential mapping in Khalkhal region (Iran) using GIS-based features. Hydrogeology journal 25 (1), p. 169. https://doi.org/ 10.1007/s10040-016-1466-z.

Nourafkan, A. & Kadkhodaie-Ilkhchi, A. (2015). Shear wave velocity estimation from conventional well log data by using a hybrid ant colony–fuzzy inference system: a case study from Cheshmeh-Khosh oilfield. Journal of Petroleum Science and Engineering 127, pp. 459–468. https://doi.org/10.1016/j.petrol.2015.02.001.

Olayiwola, T. & Sanuade, O. A. (2021). A data-driven approach to predict compressional and shear wave velocities in reservoir rocks, Petroleum 7 (2), pp. 199–208. ISSN 2405-6561, https://doi.org/10.1016/j.petlm.2020.07.008.

Phadke, S., Bhardwaj, D., & Yerneni, S. (2000). Marine synthetic seismograms using elastic wave equation. In: SEG International Exposition and Annual Meeting, pp. 2385–2388. Houston: SEG. https://doi.org/10.1190/1.1815941.

Pickett, G. R. (1963). Acoustic character logs and their applications in formation evaluation. Journal of Petroleum Technology 15, pp. 659–667. https://doi.org/10.2118/452-PA.

Platt, J. (1998). Fast training of support vector machines using sequential minimal optimization. In: (ed. Christopher J.C.B., Bernhard, S., Alexander, J.S) Advances in Kernel Methods: Support Vector Learning,. USA: MIT Press, pp. 185–208. ISBN 9780262283199, https://doi.org/10.7551/mitpress/1130.003.0016.

Qu, Y., Lin, Z., Li, H., et al. (2019). Feature recognition of urban road traffic accidents based on GA-XGBoost in the context of big data. IEEE Access 7, pp. 170106–170115. https://doi.org/10.1109/ACCESS.2019.2952655.

Rajabi, M., Bohloli, B., & Ahangar, E. G. (2010). Intelligent approaches for prediction of compressional, shear and Stoneley wave velocities from conventional well log data: a case study from the Sarvak carbonate reservoir in the Abadan Plain (Southwestern Iran), Computers & Geosciences 36 (5), pp. 647–664. ISSN 0098-3004. https://doi.org/10.1016/j.cageo.2009.09.008.

Ramjohn, R., Gan, T., & Sarfare, M. (2018). 3D Geomechanical Modeling for Wellbore Stability Analysis: Starfish, ECMA, Trinidad and Tobago. SPE Trinidad and Tobago Section Energy Resources Conference. https://doi.org/10.2118/191242-MS.

Reinstein, I. (2017). Xgboost, a top machine learning method on kaggle, explained. KDnuggets (October 3), https://www. kdnuggets. com/xgboost-a-top-machine-learning-method-on-kaggleexplained.html.

Rezaee, M. R., Kadkhodaie Ilkhchi, A., & Barabadi, A. (2007). Prediction of shear wave velocity from petrophysical data utilizing intelligent systems: an example from a sandstone reservoir of Carnarvon Basin, Australia. Journal of Petroleum Science and Engineering 55 (3–4), pp. 201–212. https://doi.org/10.1016/j.petrol.2006.08.008.

Scholkopf, B., Burges, C., & Vapnik, V. (1995). Extracting support data for a given task. In: Proceedings of the First International Conference on Knowledge discovery and Data Mining, (ed. U. M. Fayyad., R. Uthurusamy), pp. 252–257. Menlo Park, CA: AAAI Press. https://cdn.aaai.org/KDD/1995/KDD95-030.pdf.

Sumayli, A. (2023). Development of advanced machine learning models for optimization of methyl ester biofuel production from papaya oil: Gaussian process regression (GPR), multilayer perceptron (MLP), and K-nearest neighbor (KNN) regression models. Arabian Journal of Chemistry 16 (7), p. 104833. https://doi.org/10.1016/j.arabjc.2023.104833.

Suo, G., Song, L., Dou, Y., et al. (2019). Multi-dimensional short-term load forecasting based on XGBoost and fireworks algorithm. In: 2019 18th International Symposium on Distributed Computing and Applications for Business Engineering and Science (DCABES). pp. 245–248. IEEE. https://doi.org/ 10.1109/DCABES48411.2019.00068.

Suppachoknirun, T. & Tutuncu, A. N. (2017). Hydraulic fracturing and production optimization in eagle ford shale using coupled geomechanics and fluid flow model. Rock Mechanics and Rock Engineering 50, pp. 3361–3378. https://doi.org/10.1007/s00603-017-1357-1.

Tyralis, H., Papacharalampous, G., & Langousis, A. (2019). A brief review of random forests for water scientists and practitioners and their recent history in water resources. Water 11 (5), p. 910. https://doi.org/10.3390/w11050910.

Urso, A., Fiannaca, A., La Rosa, M., et al. (2019). Data Mining: Classification and Prediction, In: (ed. Shoba, R., Michael, G., Kenta, N., Christian, S) Encyclopedia of Bioinformatics and Computational Biology, Academic Press, pp. 384–402. ISBN 9780128114322, https://doi.org/10.1016/B978-0-12-809633-8.20461-5.

Vapnik, V. (1995). The Nature of Statistical Learning Theory. New York, NY: Springer. https://doi.org/10.1007/978-1-4757-3264-1.

Vapnik, V. (1998). Statistical Learning Theory. New York, NY: Wiley. ISBN: 978-0-471-03003-4.

Waluyo, W., Uren, N. F., & McDonald, J. A. (1995). Poisson's ratio in transversely isotropic media and its effects on amplitude response: an investigation through physical modeling experiments. In: SEG Technical Program Expanded Abstracts 1995, 585–588. Houston: Society of Exploration Geophysicists. https://doi.org/10.1190/1.1887420.

Wang, C., Deng, C., & Wang, S. (2020). Imbalance-XGBoost: leveraging weighted and focal losses for binary label-imbalanced classification with XGBoost. Pattern Recognition Letters 136, pp. 190–197. https://doi.org/10.1016/j.patrec.2020.05.035.

Wang, Z. & Xue, X. (2014). Multi-class support vector machine. In: Support Vector Machines Applications, 1e (ed. Y. Ma. & G. Guo), pp. 23–48. Cham: Springer International Publishing. https://doi.org/10.1007/978-3-319-02300-7_2.

Wu, X., Kumar, V., Quinlan, R. J., et al. (2008). Top 10 algorithms in data mining. Knowledge and Information Systems 14, pp. 1–37. https://doi.org/10.1007/s10115-007-0114-2.

Xu, R. & Luo, F. (2021). Risk prediction and early warning for air traffic controllers' unsafe acts using association rule mining and random forest. Safety science 135, p. 105125. https://doi.org/10.1016/j.ssci.2020.105125.

Xu, Y. & McDonald, J. A. (1988). Effects of velocity variation on AVO and AVA calculations. In: SEG Technical Program Expanded Abstracts 1988, 1221–1223. California: Society of Exploration Geophysicists. https://doi.org/10.1190/1.1892495.

Yang, X. D., Wang, J. M., & Zhang, L. N. (2017). Application of XGBoost in ultra-short term load forecasting. Electrical Drive Automation 39, pp. 21–25.

Yue, L., Yi, Z., Pan, J., et al. (2021). Identify M subdwarfs from M-type Spectra using XGBoost. Optik 225, p. 165535. https://doi.org/10.1016/j.ijleo.2020.165535.

Zhang, P., Liang, X., Xian, C., et al. (2022). Geomechanics simulation of stress regime change in hydraulic fracturing: a case study. Geomechanics and Geophysics for Geo-Energy and Geo-Resources 8, p. 83. https://doi.org/10.1007/s40948-022-00391-5.

Zhao, T., Zheng, S., Li, W., et al. (2018). Research on credit risk analysis based on XGBoost. Software engineering 21 (6), pp. 29–32.

Zhu, S. & Zhu, F. (2019). Cycling comfort evaluation with instrumented probe bicycle. Transportation research part A: policy and practice 129, pp. 217–231. https://doi.org/10.1016/j.tra.2019.08.009.

Section IV

Energy transition: New Perspectives on Geothermal Energy Exploration and Development and CO_2 Sequestration

17

Energy Transition and the Role of AI: Statistics, Trends, and Implications

Said Gaci[1], *Hasna Yazid*[2], *Aziz Khelalef*[3], *Mohammed Farfour*[4]

[1] *Sonatrach–Central Directorate of Research & Development (DC R&D), Boumerdes, Algeria*
[2] *Department of Computer Science, University of M'Hamed Bougerra (UMBB), Boumerdes, Algeria*
[3] *Sonatrach–Division Exploration, Boumerdes, Algeria*
[4] *Department of Earth Sciences, College of Science, Sultan Qaboos University, Muscat, Oman*

17.1 Introduction

The energy transition, which is the world's movement away from fossil fuel-based energy production and consumption systems and toward renewable energy sources, is essential to combating climate change and reaching sustainability objectives. It is the response to the greatest challenge facing humanity—climate change—and the shift to a sustainable energy system (Loncharich, 2023).

The goal of the energy transition is to encourage the replacement of massive fossil fuel facilities with clean, renewable energy sources like solar and wind power, as well as distributed generation (Harichandan et al., 2022). In contrast to centralized generation from power plants in conventional power grids, the latter permits producing electricity from sources—typically renewable energy sources—close to the point of use. Reducing energy losses and wastes is another objective of the energy transition: incorporating more efficient technologies, practices, and services. Due to their intermittency and heavy reliance on weather, renewable energy sources are being more and more integrated into the grid (Sayed-Mouchaweh, 2020).

17.2 Objectives for the Energy Transition

Adopted at the Paris Climate Conference (COP21) in December 2015, the Paris Agreement is the first-ever universal, legally enforceable global climate change agreement. It aims to limit the world average temperature increase to 1.5 °C above pre-industrial levels. In order to achieve net-zero emissions within the energy sector by 2050, it is necessary to reduce carbon dioxide emissions by about 37 gigatonnes (Gt) from 2022 levels (IRENA, 2023a). In addition, under the 1.5 °C Scenario, power generation must more than triple by 2050 (IRENA, 2023a).

Concerns about climate change and current geopolitical events are driving the global energy transition, which is a move away from fossil fuels and toward renewable sources

Geophysical Exploration for Hydrocarbon Reservoirs, Geothermal Energy, and Carbon Storage: New Technologies and AI-based Approaches, First Edition. Edited by Said Gaci.

(Harichandan et al., 2022). This transition entails substituting renewable energy sources for traditional energy systems, necessitating frameworks for policy and technological advancements (Vigoya et al., 2020). Many nations are investing in renewable technology, reducing emissions, and improving energy efficiency (Vigoya et al., 2020). While European countries concentrate on decarbonization and expanding renewable energy, the United States seeks to lessen its reliance on oil (Vigoya et al., 2020). Despite being a significant oil importer, China leads the world in investment in renewable energy (Vigoya et al., 2020). Nonetheless, the socioeconomic effects of the shift, such as changes in tax income and employment losses, call for careful consideration of equitable problems (Henry et al., 2020). To enable a fair and timely global energy transition, future research should focus on policy formulation, energy infrastructure, and economic analysis (Harichandan et al., 2022).

The urgency of the energy transition has increased in 2022, and the effects of climate change, rising oil and gas costs, the conflict in Ukraine, COVID-19, and other factors make the global energy transition critically necessary. Concerns are being raised about the need for a more sustainable energy system by the COVID-19 pandemic, the affordability of energy bills, and the Ukraine issue (IRENA, 2022).

Global capital investments in all energy transition technologies reached a record high of USD 1.3 trillion in 2022, while the amount invested in fossil fuels was nearly twice that of renewable energy (IRENA, 2023b). In the 1.5 °C scenario projected by IRENA, the proportion of renewable energy in the world's energy mix would rise from 16% in 2020 to 77% by 2050. All end-use sectors will see an increase in renewable energy, but a high rate of electrification in industries like transportation and construction would need a twelve-fold increase in renewable electricity capacity by 2050 compared to 2020 levels (IRENA, 2023b). In order to meet the 1.5 °C scenario, yearly additions to renewable power generation would need to reach an average of 1 066 GW globally between 2023 and 2050. By 2050, in the 1.5 °C scenario, electricity would take the lead as the primary energy carrier, making up more than half of all final energy use, and 94% of hydrogen produced would come from renewable sources (IRENA, 2023b).

17.3 Emerging Trends of Energy Transition and AI

17.3.1 Optimization of Energy Generation and Consumption

Artificial intelligence (AI) is becoming increasingly powerful instruments for streamlining the switch to renewable energy sources and maximizing energy output. The dependability and lifetime of renewable energy infrastructure can be increased by using AI techniques to evaluate enormous volumes of data in order to forecast and avoid equipment breakdowns (Dawodu et al., 2024). Meena et al. (2021) show that AI helps in process parameter optimization and energy output prediction from biomass in bioenergy production. The efficiency of AI algorithms in energy optimization is demonstrated by the Collaborative Energy Optimization Platform (CEOP) model (Stecyk and Miciuła, 2023). It is also expected that combining AI with other cutting-edge technologies like blockchain, quantum computing, and next-generation communication technologies would improve smart energy systems even more and ease the shift to renewable energy sources (Zhao et al., 2023). In addition, the role of AI is of high importance for renewable energy systems like solar, wind, and hydropower to adjust to shifting environmental circumstances and optimize resource allocation.

By precisely anticipating trends and enabling improved decision-making for energy management, AI tools are revolutionizing the analysis of energy consumption (Ohalete et al., 2023). They have a significant impact on the efficiency and sustainability of renewable energy systems through its applications in predictive maintenance and energy optimization across solar, wind, and hydro sources (Hamdan et al., 2024), and show a tremendous potential to achieve considerable energy savings, lower prices, and contribute to a more sustainable energy future, despite obstacles such data privacy and integration complexity (Olatunde et al., 2024; Ukoba et al., 2024).

17.3.2 Integration of Renewable Energy into Grids

An increasingly important trend for grid integration and energy transition optimization is the incorporation of AI models into renewable energy systems. Resource assessment, energy forecasting, system monitoring, and control strategies are all being improved by the application of AI techniques (Ukoba et al., 2024).

Forecasting for renewable energy, grid operations, and demand-side management coordination are all improving as a result of these technologies (Márquez, 2023). AI plays a particularly important role in predictive maintenance and energy optimization for solar, wind, and hydro sources, which increases the sustainability and efficiency of renewable energy systems (Hamdan et al., 2024). With the help of AI, smart grids can solve problems with integrating renewable energy sources and deliver benefits like lower energy costs and emissions (Yadav et al., 2021). AI's influence on the renewable energy environment is anticipated to grow as it develops, hastening the shift to a more sustainable energy future (Hamdan et al., 2024).

17.3.3 Speeding Up the Decarbonization Process

The integration of sophisticated technology with AI has the potential to expedite the energy transition towards carbon neutrality, which is a key problem. To optimize carbon capture operations, the Scalable Modeling, Artificial Intelligence, and Rapid Theoretical computations (SMART) technique integrates data-driven modeling and high-throughput calculations (Lei et al., 2023). Carbon capture and storage (CCS) is a crucial mitigation strategy that involves capturing, transporting, and storing CO_2 emissions from industrial and power generation sources in order to stabilize atmospheric greenhouse gas concentrations (English and English, 2022). These developments in CCS and AI technology present viable paths toward fulfilling climate targets and speeding up decarbonization initiatives.

The integration of electric vehicles (EVs) and the use of distributed resources are two challenges in the power system decarbonization that the technology is especially useful for tackling (Dobrić, 2023). A three-layered strategy involving economics, technology, and policy can maximize AI's potential impact on the transition of electric energy systems to a carbon-neutral state (Xie et al., 2022). One important area of concentration for decarbonization is the transportation sector, which accounts for more than 20% of greenhouse gas emissions. Renewable energy-powered electric vehicles (EVs) are seen to be a practical way to fight climate change and advance the energy transition (Bharathidasan et al., 2022). However, more industry innovation, adoption, and cooperation are needed to realize AI's full potential in this field (Márquez, 2023).

17.3.4 Advanced Business Strategies and Policy Development

By offering insights from data analysis and simulations of various scenarios, AI can help policymakers create and implement successful energy policies. Researchers suggest AI-driven policy frameworks that are in accordance with the concepts of the circular economy in order to ease this transition and establish energy systems that are efficient, fair, and sustainable (Danish and Senjyu, 2023). These frameworks address issues with data science integration into energy systems, policy planning, and implementation (Danish and Senjyu, 2023). Furthermore, new business models that address obstacles to the integration of sustainable energy technology into current structures or to their independence are made possible by digitalization (Loock, 2020). These developments present new opportunities for value creation and extraction in the dynamic energy industry, and they may be incremental or transformative (Loock, 2020).

17.4 Implications of Leveraging AI in Energy Transition

Researches show that efficiency, optimization, reliability, and innovation are AI's key benefits, and the energy sector is already taking advantage of reaping substantial financial gains from the implementation of AI, such as lower costs, better operational efficiency, and the creation of jobs (Gusc et al., 2022; Loncharich, 2023; Miskat et al., 2023; Peng et al., 2024, Wang et al., 2024).

Even if AI greatly outperforms humans in the sector, some tasks still require the knowledgeable advice of an energy specialist. Similarly, even if this technology can save money and optimize some processes, it is crucial to perform a cost-benefit analysis to ascertain whether the benefits result from using it outweigh the costs associated with using it. Another benefit of AI is reliability that depends strongly on data quality, since the best data is crucial to achieve the most consistent outcomes. Additionally, while AI fosters innovation in businesses, there are a few potential issues to take into account, like ethical and regulatory issues (Miller, 2014; Loncharich, 2023; Noorman et al., 2023).

17.5 Challenges to Apply AI in Renewable Energy Sector

In the current energy sector's transition to a digital one, AI-based technology has the potential to be a potent tool. However, there are challenges facing the energy sector and obstacles in the way of its wider application (Sayed-Mouchaweh, 2020, Hatti, 2021, Loncharich, 2023).

17.5.1 Quality and Accessibility of Data

Due to the fact that AI technologies demand large amounts of data for testing and training, high quality standards and high data availability are essential for integrating them into current energy systems. In fact, the lack of this data, such as that related to power networks and consumer, makes the implementation of AI difficult. There are three different categories of data that should be identified: data that is readily available and usable, data that exists but requires modification before use, and data that must be newly collected.

Regarding the availability of data, three obstacles could be encountered: (i) There is choice of the appropriate data type before using AI algorithms. (ii) Advanced analytics need real-time data, and data interchange in the energy sector needs to be seamless and compliant with data standards. Data sharing within the energy sector also poses a significant challenge that must be easily overcome, and (iii) Bad quality data might reduce the performance of AI algorithms which mostly rely on high-quality data, while AI decision-making and training require accurate data on variables like weather, energy production, and grid efficiency (Loncharich, 2023). Finding this information, though, can be challenging because it may contain mistakes or inconsistencies that compromise the results of AI systems.

17.5.2 Computing Capacity and Environmental Impact

The increasing demand for computing power in machine learning is causing a significant increase in power consumption. And the need for processing capacity rises in tandem with the exponential growth of complex algorithms and the large size of input data (Ahmad et al., 2022). As a result, prospective GPUs, considered supercomputers, are requested for complex machine learning tasks, and involve substantial expenses (Sayed-Mouchaweh, 2020; Višković et al., 2022).

Simultaneously, there has been a notable rise in power consumption associated with processing power, and it is anticipated to persist due to AI applications. The digital sector's energy consumption is expected to rise significantly, and the environment is threatened by the indirect carbon emissions produced during this procedure. However, the use of AI in the energy transition presents a trade-off with the pressing need for a balance (Chatterjee and Dethlefs, 2022). The energy transition could be accelerated by AI, but only if it doesn't end up being a source of excessive energy consumption in and of itself. This is where the term "green AI" first appeared, and policymakers ought to take note of it in order to protect the environment (Zamponi and Barbierato, 2022). In general, the acquisition and storage of data entail a costly expenditure owing to computational power and an ecological concern arising from energy usage.

17.5.3 Integration and Interoperability with Existing Systems

Diverse renewable energy products and systems are made by various businesses using various techniques. For AI to properly and efficiently manage energy resources, the link between these many processes must be preserved.

In this context, the energy sector presents integration issues for AI technology because of a lack of knowledge and training. Integration is made more difficult by the differences between AI technology and existing equipment. This necessitates modifications to infrastructure, workflow, and processes, such as standardizing data preservation and documentation procedures. Enhancing AI's flexibility is essential for the energy industry's highly intelligent development, whereas it can be costly and challenging to integrate (Ahmad et al., 2022).

17.5.4 Skilled Professionals' Shortage and Employees Training

Data science, machine learning, and electrical engineering specialists are needed to develop and oversee AI for green energy. A lack of these abilities may prevent advancement in the field. Therefore, to guarantee that users comprehend the outputs of AI technologies, a medium-level of human-AI cooperation is needed. The workforce should undergo extensive training sessions to get ready to manage and operate AI-powered systems and energy system assets (Višković et al., 2022). Employee contributions to the energy industry's future depend heavily on trust. Set goals and verify AI performance with the help of product managers and user researchers. Identification of possible harm by AI should be done by an ethics team. Recruiting IT experts, educating current staff members, and contracting with outside service providers are some ways that businesses might acquire AI expertise.

17.5.5 Economic Considerations

Applying AI technologies in the renewable energy sector can have significant up-front expenditures due to the initial investments in sensors, computational costs, training data, and software development. For the AI project to be financially viable, these expenses should be weighed against the anticipated outcomes. Companies can either invest in their own data centers or pay for cloud computing. AI models require continuous training and maintenance, increasing computational expenses (Sayed-Mouchaweh, 2020). The production, improvement, and management of AI software also require significant capital and finance (Ahmad et al., 2022). The long-term benefits of AI applications may be a barrier for companies.

17.5.6 Data Security and Privacy

AI is seen as a weakness in the energy sector, where privacy and security are becoming more and more crucial (Vijayalakshmi et al., 2023). Robust data and security procedures are needed to process important information regarding energy production, consumption, and performance. Ensuring compliance and safeguarding this data from cyber threats is a challenging task.

In fact, concerns regarding data security and protection arise when personal data is used in big data research. Data security pertains to safeguarding information while it is being gathered, transferred, processed, stored, and utilized. Energy networks, particularly smart grids, may be more vulnerable to cyberattacks if AI is overused (Priyadarshi et al., 2022).

Distributed denial of service attacks or fake data injection attacks are the usual sources of cyberattacks. To gain public trust in AI technology, businesses must guarantee that they are both technically rigorous and compliant with applicable laws and ethical guidelines (Danish, 2023). Although they are still in their infancy, AI technologies such as machine learning have shown to be successful at identifying Trojans (Vijayalakshmi et al., 2023). Appropriate security measures are anticipated to be created in order to safeguard consumer data and safeguard systems against these vulnerabilities as AI-powered applications continue to proliferate.

17.5.7 Model Complexity

One important factor to take into account while switching to sustainable energy sources is the complexity of energy system models (Ridha et al., 2020). Accuracy requires a certain amount of complexity, but this complexity must be matched with computational resources (Priesmann et al., 2019). This complexity may be decreased by AI, especially deep learning, which offers effective modeling techniques (Köhnen et al., 2022). To be accurate and useful, these models must also incorporate societal developments such as human behavior and societal dynamics (Trutnevyte et al., 2019).

It can be difficult to create precise and useful intelligence models for the management of renewable energy. Energy production, demand, and storage can be complexly modelled, and doing so frequently calls for specialist knowledge.

17.5.8 Difficulties with Regulation and Policy

The renewable energy industry is confronted with numerous legal and regulatory frameworks that could impact the utilization of intellectual property. Regulatory such as: licenses, taxes, and standards, impact AI's potential to increase energy efficiency

The energy industry must ensure a cheap and steady supply of energy, deal with security concerns, and an easier integration of AI- algorithms (Višković et al., 2022). These are just a few of the obstacles the industry has in regulating data analysis and AI.

It is necessary to create foundational frameworks and structures in order to expand the use cases of AI applications and uncover value propositions for society and the economy. Industry standards should be developed to assist entrepreneurs in developing new business models, and regulations should be implemented to define uniform standards for Internet of Things devices in the grid.

The government should enact laws that clarify legal issues, establish industry standards, and provide stakeholders with sustainable and equitable policy options. This will help emerging business models and create a more innovative regulatory environment (Manshreck, 2022).

17.5.9 Ethical Issues

As AI systems are managed more and more, ethical issues also become relevant. In fact, the energy transition has complex and varied ethical aspects that include concerns about justice, human well-being, and the reorganization of institutions and values (Miller, 2014). The integration of AI technologies, which offer advantages as well as ethical questions about machine learning, and larger socio-technical systems, exacerbates these difficulties even further (Stahl, 2021). Concerns about energy justice, transparency, privacy, and discrimination are particularly raised by the application of AI in the context of smart power networks (Noorman et al., 2023). But AI also has a lot of promise for the energy transition, especially in Latin America, where it may help with the adoption of new information and communication technologies (ICTs) and the integration of sustainable energy (Jimenez and Gonzalez, 2022).

17.5.10 Opposition to Change

Decision-makers and professionals accustomed to using conventional energy management techniques may be resistant to the adoption of AI-powered solutions. It can be challenging to overcome this resistance and gain acceptance for AI technology.

There is strong resistance to the shift to low-carbon energy, especially from established players in the nuclear and fossil fuel industries (Biresselioglu et al., 2024; Stefes, 2020). This resistance, which is motivated by a desire to preserve the status quo, frequently takes the form of political, economic, and technological actions. Different locations and energy types employ different oppositional strategies, such as protest and lawsuit (Sovacool et al., 2022). In fact, opposition isn't always logical, as demonstrated by the instance of wind power, where opposition is frequently founded in false beliefs (Pasqualetti, 2012). This implies that dealing with resistance to the energy transition necessitates a sophisticated comprehension of the strategies and motivations of the opposing parties.

17.5.11 Scalability

AI solutions need to grow in tandem with the ongoing improvements in energy efficiency. It is difficult to make sure AI models and algorithms can manage growing data volumes and intricate procedures.

The integration of big data and AI is facilitating the energy transition towards renewable sources and efficiency, as noted by Doukas (2022) and Márquez (2023). Through enhanced demand-side management, grid operations, and forecasting, these technologies are speeding up the shift and facilitating the creation of consumer-centric energy services. Gusc et al. (2022) underscore the capacity of blockchain, AI, and big data to facilitate sustainable energy production decision-making. Jimenez and Gonzalez (2022) emphasize how crucial it is to comprehend AI's workings and their effects on the energy value chain, especially in Latin America.

17.6 Conclusions

This chapter discusses the relationship between AI and the energy transition. It highlights the key role that AI is playing in accelerating the global transition to sustainable and renewable energy sources by examining data, trends, and challenges. In order to take full advantage of AI's potential in the energy sector to build more robust, sustainable, and efficient energy systems, increased industry innovation, adoption, and cooperation are required.

References

Ahmad, T., Zhu, H., Zhang, D., et al. (2022). Energetics systems and artificial intelligence: Applications of industry 4.0. Energy Reports 8, 334–361. https://doi.org/10.1016/j.egyr.2021.11.256.

Bharathidasan, M., Indragandhi, V., Suresh, V., et al. (2022). A review on electric vehicle: Technologies, energy trading, and cyber security. Energy Reports 8, pp. 9662–9685. ISSN 2352-4847. https://doi.org/10.1016/j.egyr.2022.07.145.

Biresselioglu, M. E., Solak, B., & Savaş, Z. F. (2024). Unveiling resistance and opposition against low-carbon energy transitions: A comprehensive review. Energy Research & Social Science 107, pp. 103354. ISSN 2214-6296. https://doi.org/10.1016/j.erss.2023.103354.

Chatterjee, J. & Dethlefs, N. (2022). Facilitating a smoother transition to renewable energy with AI. Patterns (New York, N.Y.) 3 (6), pp. 100528. https://doi.org/10.1016/j.patter.2022.100528.

Danish, M. S. S. (2023). AI in energy: Overcoming unforeseen obstacles. AI 4 (2), pp. 406–425. https://doi.org/10.3390/ai4020022.

Danish, M. S. & Senjyu, T. (2023). AI-enabled energy policy for a sustainable future. Sustainability 15, pp. 7643. https://doi.org/10.3390/su15097643.

Dawodu, S. O., Onwusinkwue, S., Osasona, F., et al. (2024). Artificial intelligence (AI) in renewable energy: A review of predictive maintenance and energy optimization. World Journal of Advanced Research and Reviews 21 (1), pp. 2487–2499.

Dobrić, G. (2023). Power system decarbonization and electrical vehicles. 2023 22nd International Symposium INFOTEH-JAHORINA (INFOTEH). IEEE, East Sarajevo, Bosnia and Herzegovina, 2023, pp. 1–5. https://doi.org/10.1109/INFOTEH57020.2023.10094047.

Doukas, H. (2022). Energy transitions, intelligence and big data: Towards a prosumer concept with energy autonomy. 2022 13th International Conference on Information, Intelligence, Systems & Applications (IISA). Corfu, Greece, 2022, pp. 1–5. https://doi.org/10.1109/IISA56318.2022.9904415.

English, J. M. & English, K. L. (2022). An overview of carbon capture and storage and its potential role in the energy transition. First Break 40 (4), pp. 35–40. https://doi.org/10.3997/1365-2397.fb2022028.

Gusc, J., Bosma, P., Jarka, S., et al. (2022). The big data, artificial intelligence, and blockchain in true cost accounting for energy transition in Europe. Energies 15 (3), p. 1089. https://doi.org/10.3390/en15031089.

Hamdan, A., Ibekwe, K. I., Ilojianya, V. I., et al. (2024). AI in renewable energy: A review of predictive maintenance and energy optimization. International Journal of Science and Research Archive 11 (01), pp. 718–729. https://doi.org/10.30574/ijsra.2024.11.1.0112.

Harichandan, S., Kar, S. K., Bansal, R., et al. (2022). Energy transition research: A bibliometric mapping of current findings and direction for future research. Cleaner Production Letters 3, 100026. https://doi.org/10.1016/j.clpl.2022.100026.

Hatti, M. (2021). Artificial Intelligence and Renewables Towards an Energy Transition- Series Title Lecture Notes in Networks and Systems. 990 pages, Springer Cham, Switzerland. https://doi.org/10.1007/978-3-030-63846-7.

Henry, M. S., Bazilian, M. D., & Markuson, C. (2020). Just transitions: Histories and futures in a post-COVID world. Energy Research & Social Science 68, p. 101668. https://doi.org/10.1016/j.erss.2020.101668.

IRENA (2023a). World Energy Transitions Outlook 2023: 1.5 °C Pathway, 1. 978-92-9260-527-8. Available at: https://www.irena.org/Publications/2023/Jun/World-Energy-Transitions-Outlook-2023.

IRENA (2023b). World Energy Transitions Outlook 2023: Executive Summary and Introduction. Available at: https://www.irena.org/Digital-Report/World-Energy-Transitions-Outlook-2023.

IRENA (2022). World Energy Transitions Outlook 2022- Executive Summary and Introduction. Available at: https://www.irena.org/Digital-Report/World-Energy-Transitions-Outlook-2022.
Jimenez, V. M. M. & Gonzalez, E. P. (2022). The Role of artificial intelligence in Latin Americas energy transition. In: IEEE Latin America Transactions 20 (11), pp. 2404–2412. https://doi.org/10.1109/TLA.2022.9904766.
Köhnen, C. S., Priesmann, J., Nolting, L., et al. (2022). The potential of deep learning to reduce complexity in energy system modeling. International Journal of Energy Research 46 (4), pp. 4550–4571. https://doi.org/10.1002/er.7448.
Lei, Q., Li, L., Chen, H., et al. (2023). Emerging directions for carbon capture technologies: A synergy of high-throughput theoretical calculations and machine learning. Environmental Science & Technology 57 (45), pp. 17189–17200. https://doi.org/10.1021/acs.est.3c05305.
Loncharich, Y. (2023). Unlocking the power of AI: Exploring emerging business models for driving the energy transition in Germany, October 2023. https://doi.org/10.13140/RG.2.2.19324.92808, Thesis for: MSc Management and Technology.
Loock, M. (2020). Unlocking the value of digitalization for the European energy transition: A typology of innovative business models. Energy research and social science 69, p. 101740. https://doi.org/10.1016/j.erss.2020.101740.
Márquez, F. P. (2023). Accelerating renewable energy by artificial intelligence. Proceedings of the 9th World Congress on New Technologies (NewTech'23). Brunel University, London, United Kingdom - August 09-11, 2023. https://doi.org/10.11159/icert23.001.
Manshreck, J. (2022). Transformation of the Electric Utility Business Model: From Edison to Musk. Berlin, Boston: De Gruyter. https://doi.org/10.1515/9783110714036.
Meena, M., Shubham, S., Paritosh, K., et al. (2021). Production of biofuels from biomass: Predicting the energy employing artificial intelligence modelling. Bioresource technology 340, p. 125642.
Miller, C. (2014). The ethics of energy transitions. In: 2014 IEEE International Symposium on Ethics in Science, Technology and Engineering, ETHICS 2014. Institute of Electrical and Electronics Engineers Inc. Article 6893445. https://doi.org/10.1109/ETHICS.2014.6893445.
Miskat, M. I., Sarker, P., Chowdhury, H., et al. (2023). Current scenario of solar energy applications in Bangladesh: Techno-economic perspective, policy implementation, and possibility of the integration of artificial intelligence. Energies 16, p. 1494.
Noorman, M., Espinosa Apráez, B., & Lavrijssen S. (2023). AI and Energy Justice. Energies 16 (5), p. 2110. https://doi.org/10.3390/en16052110.
Ohalete, N. C., Aderibigbe, A. O., Ani, E. C., et al. (2023). Data science in energy consumption analysis: A review of AI techniques in identifying patterns and efficiency opportunities. Engineering Science & Technology Journal 4 (6), pp. 357–380. https://doi.org/10.51594/estj.v4i6.637.
Olatunde, T. M., Okwandu, A. C., Akande, D. O., et al. (2024). Reviewing the role of artificial intelligence in energy efficiency optimization. Engineering Science & Technology Journal 5 (4), pp. 1243–1256. https://doi.org/10.51594/estj.v5i4.1015.
Pasqualetti, M. J. (2012). The Misdirected Opposition to Wind Power. In: Learning from Wind Power. Energy, Climate and the Environment Series (eds. J. Szarka, R. Cowell, G. Ellis et al.). Palgrave Macmillan: London. https://doi.org/10.1057/9781137265272_7.
Peng, X., Guan, X., Zeng, Y., et al. (2024). Artificial intelligence-driven multi-energy optimization: Promoting green transition of rural energy planning and sustainable energy economy. Sustainability 16 (10), p. 4111. https://doi.org/10.3390/su16104111.
Priesmann, J., Nolting, L. & Praktiknjo, A. (2019). Are complex energy system models more accurate? An intra-model comparison of power system optimization models. Applied Energy Elsevier 255, p. 113783.

Priyadarshi, N., Padmanaban, S., Hiran, K., et al. (2022). Artificial Intelligence and Internet of Things for Renewable Energy Systems. Berlin, Boston: De Gruyter. https://doi.org/10.1515/9783110714043.

Ridha, E., Nolting, L., & Praktiknjo, A. (2020). Complexity profiles: A large-scale review of energy system models in terms of complexity. Energy Strategy Reviews 30, p. 100515. Article 100515. https://doi.org/10.1016/J.ESR.2020.100515.

Sayed-Mouchaweh, M. (2020). Artificial Intelligence Techniques for a Scalable Energy Transition: Advanced Methods, Digital Technologies, Decision Support Tools, and Applications (1st edition 2020). Springer Cham, Switzerland. https://doi.org/10.1007/978-3-030-42726-9.

Sovacool, B. K., Hess, D. J., Cantoni, R., et al. (2022). Conflicted transitions: Exploring the actors, tactics, and outcomes of social opposition against energy infrastructure. Global environmental change 73, p.102473. Article 102473. https://doi.org/10.1016/j.gloenvcha.2022.102473.

Stahl, B. C. (2021). Ethical issues of AI. In: Artificial Intelligence for a Better Future. SpringerBriefs in Research and Innovation Governance. Springer, Cham. https://doi.org/10.1007/978-3-030-69978-9_4.

Stecyk, A. & Miciuła, I. (2023). Harnessing the power of artificial intelligence for collaborative energy optimization platforms. Energies 16, p. 5210. https://doi.org/10.3390/en16135210.

Stefes, C. H. (2020). Opposing Energy transitions: modeling the contested nature of energy transitions in the electricity sector. Review of Policy Research 37, pp. 292–312. https://doi.org/10.1111/ropr.12381.

Trutnevyte, E., Hirt, L. F., Bauer, N., et al. (2019). Societal transformations in models for energy and climate policy: The ambitious next step. One Earth 1 (4), pp. 423–433. https://doi.org/10.1016/j.oneear.2019.12.002.

Ukoba, K., Olatunji, K. O., Adeoye, E., et al. (2024). Optimizing renewable energy systems through artificial intelligence: Review and future prospects. Energy & Environment 35 (7), pp. 3833–3879. https://doi.org/10.1177/0958305X241256293.

Vigoya, M. F., Mendoza, J. G., & Abril, S. O. (2020). International energy transition: A review of its status on several continents. International Journal of Energy Economics and Policy 10 (6), 216–224.

Vijayalakshmi, S., Balusamy, B., & Dhanaraj, R. K. (Eds.). (2023). AI-Powered IoT in the Energy Industry: Digital Technology and Sustainable Energy Systems. Springer Cham, Switzerland. https://doi.org/10.1007/978-3-031-15044-9.

Višković, A., Franki, V., & Jevtić, D. (2022). Artificial intelligence as a facilitator of the energy transition. 2022 45th Jubilee International Convention on Information, Communication and Electronic Technology (MIPRO). pp. 494–499. Opatija, Croatia, IEEE. https://doi.org/10.23919/MIPRO55190.2022.9803700.

Wang, X., Ali, D. A., & Zhao, W. (2024). Urban energy transition and quality of life: A comprehensive statistical study on the nexus of wind clean energy adoption, job creation, and livability. Sustainable Cities and Society 108, p. 105436. https://doi.org/10.1016/j.scs.2024.105436.

Xie, L., Huang, T., Zheng, X., et al. (2022). Energy system digitization in the era of AI: A three-layered approach toward carbon neutrality. Patterns, 3 (12), 100640. https://doi.org/10.1016/j.patter.2022.100640.

Yadav, P., Chaudhary, P., & Yadav, H. (2021). Evolving Trends for Smart Grid Using Artificial Intelligent Techniques. In a book: Introduction to AI Techniques for Renewable Energy Systems, 1st Edition (eds. S. L. Tripathi, M. K. Dubey, V. Rishiwal et al.), CRC Press, Boca Raton, FL, USA. https://doi.org/10.1201/9781003104445.

Zamponi, M. E. & Barbierato, E. (2022). The dual role of artificial intelligence in developing smart cities. Smart Cities 5 (2), pp. 728–755. https://doi.org/10.3390/smartcities5020038.

Zhao, N., Zhang, H., Yang, X., et al. (2023). Emerging information and communication technologies for smart energy systems and renewable transition. Advances in Applied Energy 9, p. 100125.

18

On the Importance of Integrating Geomodeling in Geothermal Studies

Mohamed Amrouche

Senior Geophysicist (SLB)

18.1 Introduction

With the rising interest of exploring alternative energy sources to hydrocarbon, the geothermal energy constitute a milestone in the decarbonization efforts and the future energy mix. Its renewable and low-carbon natures, as well as its widely under explored potential, triggered great scientific interest of exploration from major energy companies around the world.

Understanding the subsurface of geothermal provinces can help in exploration and development of geothermal reservoirs, by providing major information about its energy potential. In this perspective, geomodeling can help reproducing numerical models of the geothermal reservoirs and their properties, by integrating geological and geophysical methods that helps to identify and characterize geothermally active provinces. These models are three-dimensional representations of the subsurface, typically divided into cells or voxels, that helps to capture the spatial distribution of various geological and physical properties that control geothermal reservoir behavior.

In fact, geomodeling is already a well-established practice in oil and gas reservoir exploration and production, where it is helping to maximize production of the target reservoirs. Applying the know-how acquired in hydrocarbon geomodeling to geothermal reservoirs can help to maximize their exploration potential. Geocellular grid modeling is a powerful and widely used technique in subsurface exploration. It involves creating a 3D representation of the subsurface geological formation using a grid composed of cells or blocks, where each cell represents a small volume of the subsurface.

In this chapter, we will discuss fundamentals of geothermal exploration and how geocellular subsurface modeling enables the creation of detailed, 3D representation of the subsurface geology, which can be essential for understanding and exploiting geothermal resource. By integrating diverse datasets such as geological, geophysical, geomechanical, and petrophysical data, geocellular models offer precise simulations of the temperature, pressure, and fluid flow within a geothermal reservoir.

Geophysical Exploration for Hydrocarbon Reservoirs, Geothermal Energy, and Carbon Storage: New Technologies and AI-based Approaches, First Edition. Edited by Said Gaci.

18.2 Geology of Geothermal Provinces

It is well-established that the heat from Earth's interior can be used as an energy source, known as the geothermal energy. The geothermal gradient issued from the earth's interior has been used for heating, bathing, as well as other agricultural and medicinal purposes since ancient Roman empire times. Nowadays, geothermal energy plays significant role in the global energy transition towards a more sustainable and low-carbon emission future. The widespread adoption of geothermal energy can significantly contribute to the future energy mix and help with the global climate goals. Therefore, a growing need to understand and develop geothermal provinces is witnessed globally, with an increasing interest on geothermal subsurface studies from both academia and industry.

The geology of geothermal reservoirs encompasses a diverse array of geological features and processes crucial for the occurrence and behavior of subsurface heat and fluid systems. Geothermal reservoirs are primarily hosted within the Earth's crust, where various geological formations serve as reservoir rocks capable of storing and transmitting heat. A schematic representation of a geothermal reservoir is illustrated in Figure 18.1, showing the interaction between the different elements constituting its subsurface geology and that we can discuss as follow.

The heat source is probably the first element defining geothermal reservoirs, and it can originate from various geological processes within the Earth's interior. The importance of the heat source in geothermal reservoirs lies in its role as a primary energy input for geothermal energy production by harnessing the natural heat stored within the Earth. The heat source can be radiogenic heat produced by the decay of radioactive isotopes, such as uranium, thorium, and potassium, within the Earth's crust and mantle, constituting a significant portion of the heat budget. It can also be generated by the cooling and crystallization of the magmatic intrusions beneath volcanic regions and tectonically active zones, creating thermal gradients in the crust, in this case, we talk about convective heat. Or simply be conductive heat, caused by the transmission of the heat from the Earth's mantle to the crust over geological times.

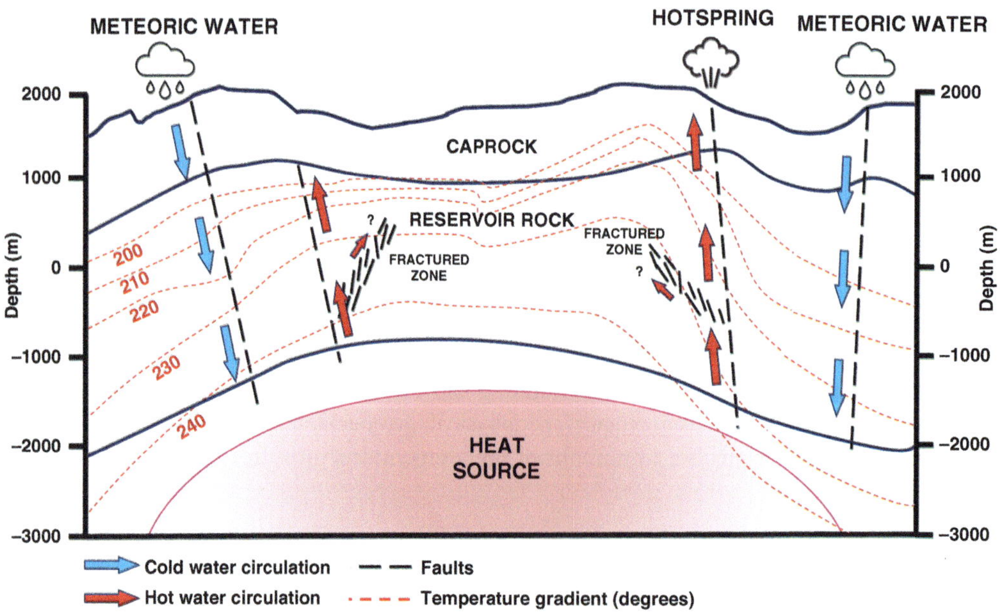

Figure 18.1 Schematic representation of the subsurface geology of a geothermal province.

The reservoir rocks serve as the host medium for the geothermal fluid and play a crucial role in the heat transfer and fluid storage. They can include igneous formations like granite and basalt, sedimentary deposits such as sandstone and limestone, or metamorphic rocks like schist and gneiss. The fluids within the reservoir rocks are typically a mixture of water and dissolved minerals, heated by the subsurface heat sources.

Geothermal fluids can exhibit varying characteristics depending on their origin, temperature, and chemical composition. They often manifest as hot water in liquid state, often reaching temperatures exceeding 50 °C (300 °F) or even higher. In shallow reservoir depths, geothermal fluids may also manifest as predominantly steam or vapor at the surface, constituting natural bathing spot and geysers.

These fluids circulations are conditioned by natural geological faults, fractures, and permeable layers ensuring the heat transfer within the reservoir. The faults and fractures in geothermal reservoirs originate from tectonic processes and stress distribution, creating zones of weakness within the reservoir formations, facilitating the migration of hot fluids from deeper reservoirs to shallower depths, where they can be accessed for energy extraction. Moreover, faults and fractures can compartmentalize the reservoir, creating distinct fluid compartments with varying temperature and pressure conditions, connecting geothermal reservoirs with larger hydrothermal systems that includes recharge area where cold water infiltrates the Earth's crust and discharge area heated fluids emerging at the surface that forms natural thermal hot springs or extracted by drilling wells from subsurface fractured zones. Understanding the distribution, orientation, and connectivity of faults and fractures is therefore critical for the characterization and modeling of geothermal reservoirs.

Similarly to hydrocarbon reservoirs, a good geothermal reservoir needs a good sealing cap rock at the top of the reservoir to ensure a sustainable geothermal system. Positioned above the reservoir, and composed of impermeable materials such as shales, anhydrite or dense volcanic rocks, the caprock acts as a barrier that prevents the upward migration of geothermal fluids towards the surface and maintaining the pressure and integrity of the geothermal system. By confining the geothermal fluids within the reservoir, the caprock ensures their long-term containment and pressurization, essential for sustaining fluid circulation and heat extraction. Also, the caprock prevents the escape of geothermal fluids to the surface, minimizing the risk of environmental contamination and ensuring the efficient utilization of the geothermal resource.

Geothermal provinces are regions characterized by their significant geothermal activity and potential for energy extraction, they are distributed across various tectonic settings around the world. These provinces are commonly found along tectonic plate boundaries, where volcanic and seismic activities are prevalent, providing the necessary heat sources for geothermal systems. Major geothermal provinces include the Pacific Ring of Fire, which encompasses countries like the United States (especially California and Nevada), Indonesia, Japan, the Philippines, and New Zealand. This region is known for its high volcanic activity, resulting in abundant geothermal resources.

In Europe, the Icelandic geothermal province is notable for its utilization of geothermal energy for heating and electricity, thanks to the island's unique geological setting on the Mid-Atlantic Ridge. The East African Rift System is another significant geothermal province, with countries like Kenya and Ethiopia harnessing geothermal energy from the tectonic rift activity. Additionally, Turkey and Italy in the Mediterranean region are known for their geothermal fields due to complex tectonic interactions. These geothermal provinces leverage the heat from the Earth's interior, making them critical areas for renewable energy development and contributing to sustainable energy solutions globally.

Given the complexity of the different geological and hydrological elements constituting geothermal reservoirs, subsurface modeling studies can be very challenging. One significant burden lies in accurately characterizing the geological structure and properties of the reservoir, which can exhibit considerable heterogeneity and anisotropy in lithology, porosity, permeability, and fracture networks. Obtaining high-resolution data to capture these features is often difficult and costly, leading to the use of cheaper and lower resolution geophysical methods, which can, in turn, leads to lot of uncertainties in reservoir characterization.

Moreover, managing reservoir pressure and geomechanics stress changes induced by fluid production poses challenges in ensuring reservoir stability and integrity over time. Modeling the fluid flow dynamics within geothermal reservoirs involves intricate multiphase and multiphysics phenomena, including fluid-rock interactions, phase changes, and heat transfer mechanisms. Predicting these processes accurately requires advanced numerical techniques and high-performance computing resources.

18.3 Exploration of Geothermal Reservoirs

Exploring the presence of a geothermal reservoir typically involves a combination of geological, geochemical, geophysical, and drilling methods to characterize subsurface structures, properties, and fluid dynamics. With the understanding on geothermal reservoir behaviors and the advent on subsurface modeling techniques, geomechanical methods and geostatistical methods are being adapted to help a better characterization of the reservoir. By integrating these diverse disciplines, geoscientists can develop comprehensive models of geothermal systems.

18.3.1 Geological Methods

Understanding the regional geology and structural context of the region helps in targeting areas with favorable geological conditions for reservoir development (Moeck, 2014; Siler et al., 2019). Geological mapping and analysis of surface features, such as hot springs, fumaroles, and altered mineral assemblages, provide initial indications of geothermal activity and potential reservoir locations.

Surface imagery enables the mapping and interpretation of geological structures, such as faults, outcropping formations, and volcanic features, which control the occurrence and behavior of geothermal reservoirs. By analyzing surface expressions of these structures and combining them with subsurface knowledge, geologists can delineate favorable exploration targets and assess the structural control on geothermal systems (Calvin et al., 2002; Littlefield and Calvin, 2014).

Remote sensing techniques, such as thermal infrared imaging (TIR) and hyperspectral analysis, can detect land surface temperature anomalies and mineralogical alterations associated with geothermal activity (Van Der Meer et al., 2014; Nishar et al., 2016; Weldeyohannes et al., 2022; Li et al., 2023). These methods are particularly useful for large-scale reconnaissance and initial stage screening of geothermal prospects by identifying surface expression of geothermal systems based on mineral assemblages that result from hydrothermal alteration.

18.3.2 Geochemical Methods

In situ geochemical measurements involve analyzing the chemical composition of surface water, gases, and soil samples to detect signatures of geothermal fluids and associated mineral alteration. They can provide insights about the composition, origin, and behavior of fluids and rocks within the Earth's crust (Belhai et al., 2022; Wang et al., 2022; Anees et al., 2024).

Through geochemical analysis, scientists can identify potential geothermal reservoirs by studying the distribution and anomalous concentration of elements and isotopes in surface such as silica, chloride, lithium, boron, and helium manifestations. These signatures offer clues about the temperature, depth, and permeability of subsurface reservoirs. Geochemical modeling also enables researchers to predict the chemical composition of geothermal fluids, aiding in resource assessment and reservoir management.

18.3.3 Geophysical Methods

Geophysical methods remain very valuable in geothermal exploration projects by providing subsurface structure, proprieties, and fluid dynamics essential for identifying and characterizing geothermal resources. Several geophysical methods are commonly employed in geothermal studies, we can cite:

18.3.3.1 Seismic Methods

Seismic reflection, refraction, and tomography can be used to image subsurface geological features, including faults, fractures, and stratigraphic variations that control the occurrence and behavior of geothermal reservoirs. Seismic methods can be divided into two groups: Active seismic methods where a seismic source is shot and reflection data gathered, processed, and interpreted to obtain details of the subsurface structure (Krawczyk et al., 2019; Bredesen et al., 2020; Gao et al., 2023), and passive seismic methods that uses natural seismic ambient noise to obtain physical parameters from surface waves dispersion measurement and analysis (Tian et al., 2022; Baoqing et al., 2022).

While seismic methods can be essential for subsurface mapping of the geothermal provinces, they can be limited by resolution and challenge interpretation, while being expensive and time-consuming specially in urban and mountainous areas, further hindering the acquisition and processing.

18.3.3.2 Electrical Methods

Electrical resistivity of rocks in geothermal surroundings is a parameter which reflects the properties of the geothermal system, or its history. This relates to the fact that the resistivity of rocks is chiefly controlled by parameters that correlate to the geothermal activity such as: Porosity, fracture intensity, alteration of the rocks, salinity of the fluids, temperature or amount of water.

Electrical methods include many different types of measurements and varying setups or configurations for the different types. The most important types are: Magnetotellurics (MT) (Di Paolo et al., 2020) and Controlled-Source Electromagnetic (CSEM) (Eltayieb et al., 2023) surveys that measures the electrical conductivity distribution in the subsurface, helping to delineate reservoir boundaries and fluid pathways. These methods are sensitive to variations in fluid salinity and temperature, providing insights into reservoir properties.

18.3.3.3 Gravity and Magnetic Methods

Potential fields methods are commonly used in geothermal exploration because of their affordability of acquisition and their wide areal coverage (Saibi et al., 2022; Feng et al., 2024). These methods aim to measure subtle fluctuation of the gravity variations caused by geological density contrasts or magnetic susceptibility variations of the subsurface geology from surface measurements. These methods aid in delineating fault systems, identifying potential fluid pathways, and mapping reservoir extents.

18.3.3.4 Distributed Acoustic Sensing (DAS) Methods

DAS is a cutting-edge technology increasingly used in geothermal reservoirs for monitoring and managing geothermal energy extraction (Azzola et al., 2023). DAS involves the use of fiber optic cables as sensors to detect acoustic signals along their length, providing high-resolution spatial and temporal data about subsurface conditions. DAS also facilitates thermal profiling by inferring temperatures changes along wellbores, thereby providing crucial insights into the thermal gradients and heat distribution within the reservoir.

18.3.4 Well Logging Methods

Wireline logging tools, including gamma-ray, resistivity, and sonic tools, are deployed in boreholes to provide direct measurements of lithology, porosity, permeability, and fluid properties within the reservoir. Gamma-ray logging measures natural gamma radiation emitted by subsurface formations, aiding in lithological identification and correlation. Resistivity logging evaluates the electrical resistivity of rocks, helping delineate permeable zones and identify fluid types within the reservoir. Sonic logging measures the travel time of sound waves through the formation, providing information on rock mechanical properties and porosity. Density logging assesses bulk density variations in the formation, aiding in lithology identification and reservoir characterization. Additionally, temperature logging measures downhole temperatures, helping delineate geothermal gradients and identify potential reservoir zones.

18.3.5 Geomechanical Methods

Geomechanical studies tackle the mechanical behavior of the subsurface rock formations, which can be essential for understanding reservoir stability, wellbore integrity and flow dynamics (Jeanne et al., 2015; Levi et al., 2024). Mechanical Earth Models (MEMs) are estimated through a combination of empirical correlations, laboratory testing, geophysical methods and numerical simulations to provide the mechanical behavior of the subsurface. They involve stress analysis by analyzing the stress state within the reservoir rock to determine how it affects the formation's mechanical behavior and the distribution and magnitude of stresses. Geomechanical engineers can predict the likelihood of rock failure, fault reactivation, and induced seismicity. This can be critical for identifying regions of high stress concentration where faults are more likely to occur and for predicting the orientation of fractures relative to stress fields which are important considerations for reservoir management and wellbore stability.

Geomechanics can also help to assess fault stability under varying stress conditions and fluid pressures, considering factors such as fault geometry, mechanical properties of rocks, pore pressure distribution, or assess the likelihood of fault reactivation and potential seismic hazard.

These studies also help identify geohazards and assess associated risks, such as induced seismicity, caprock integrity, or fault reactivation. By quantifying these risks, reservoir engineers can develop mitigation measures and contingency plans to ensure safe and sustainable reservoir operations.

18.3.6 Geostatistical Methods

Geostatistical methods are often used in geothermal modeling (Patterson et al., 2020; Deb et al., 2020; Freienstein et al., 2023). Data analytics and statistical tools such as variogram analysis, cross correlation, and spatial interpolation methods (e.g. Kriging), helps to quantity spatial variability of the reservoir properties, such as temperature, permeability and porosity. By analyzing spatial correlations and patterns in well data and geophysical measurements, geostatistics helps create realistic models of reservoir heterogeneity and continuity.

Other geostatistical simulations, such as multiple-point statistics (MPS) and stochastic algorithms (e.g. Sequential Gaussian Simulation), are employed to generate multiple realizations of the reservoir properties that honor observed data and spatial patterns.

These simulations account for uncertainties in reservoir parameters and provide probabilistic assessments of reservoir behavior, aiding in risk analysis and decision-making. In addition to spatial variability analysis and uncertainty quantification, geostatistical analysis can help on well placement design and optimization, providing insights into reservoir connectivity, preferential flow pathways, and reservoir sweep efficiency. By incorporating spatial variability and uncertainty into reservoir models, geostatistics enables engineers to identify optimal well locations, control injection and production rates, and maximize energy recovery while minimizing operational risks.

In addition to the methods described above, geoscientists often rely on conceptual models, geological descriptions, maps, and interpretative cross-sections reported in the different literature works of the area of interest. Published studies provides a wealth of empirical data that helps to refine and benchmark the geothermal models by digitizing well logs, seismic interpretations and integrate them to geothermal numerical studies. Furthermore, using peer-reviewed studies ensures that the models are based on validated and accepted scientific principles, which enhances their credibility.

Integrating these different exploration methods can help to develop a comprehensive understanding of the reservoir's structure and behavior, improve the accuracy of predictions, optimize resource exploitation, and minimize risks associated with geothermal energy production. This holistic approach aims to model and simulate the most representative subsurface model that can help to optimize drilling and production strategies, evaluate reservoir performance, and assess the long-term sustainability of geothermal projects (Figure 18.2). By simulating different scenarios and predicting reservoir responses, these models assist in maximizing energy extraction while minimizing environmental impacts and operational risks.

However, geothermal reservoirs often present complex conditions, not typically found in petroleum reservoirs. In fact, they are known to be more "dynamic" and "unpredictable" than conventional oil and gas reservoirs, as fluid circulation, heat, steam, and micro-seismicity can constantly change the dynamic of the subsurface.

This includes, but not limited to: convection cells due to local heat flux, variations and a dynamic equilibrium of energy and mass in their natural state, reservoir structure, fault locations, dual-porosity and permeability, unsealed reservoir boundaries, heat loss to caprock or the atmosphere, fluid recharge and discharge locations, etc. Geothermal modeler need to take in consideration all these specifications to build the most representative reservoir model.

18.4 Modeling the Subsurface of Geothermal Reservoirs

Geothermal subsurface modeling studies have evolved over several decades, reflecting advancements in computational techniques, geological understanding and improvement of exploration and geophysical methods. The earliest geothermal subsurface modeling efforts between 1960s and 1970s, focused on simplistic analytical models to estimate reservoir properties and heat flow (Witherspoon and Neuman, 1967; Gringarten et al., 1975). The development of computational capabilities and the advent of numerical techniques such as the finite difference method and the finite element method allowed a more detailed simulation of the geothermal reservoirs (Mercer and Faust, 1975; Coats, 1977) considering complex geometries and boundary conditions.

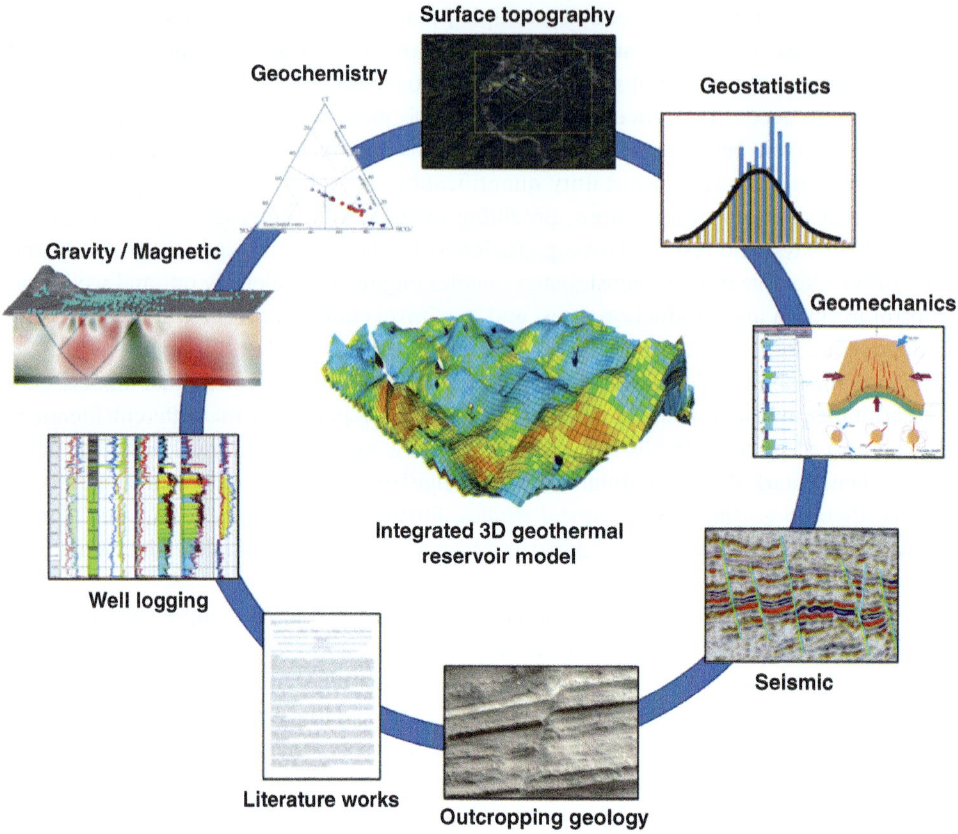

Figure 18.2 Integrating different disciplines can help to build a more reliable interpretative geothermal reservoir model.

It was not until the 1980 Stanford's special geothermal model intercomparison study that numerical models for reservoir management were generally accepted (O'Sullivan et al., 2001).

During the 1990s, there was a significant increase in geological understanding of subsurface structures and processes, driven by advancements in seismic imaging, well logging technologies, and geological mapping. This led to more sophisticated geological models that incorporated realistic representations of rock formations, fault systems, and fluid pathways. In particular, fluid/phase modeling with equations of state available to treat mixtures of water, CO_2, and dissolved solids (Anderson et al., 1992; Battistelli et al., 1997), multicomponent chemical models (Wolery, 1992), and other inverse modeling and uncertainty analysis for history matching (Finsterle and Pruess, 1995).

Pioneer works on 3D geocellular modeling for geothermal studies started emerging at the early 2000s (Pruess and Garcia, 2000; Pruess, 2002; O'Sullivan et al., 2009). As interest in geothermal energy increased and computational tools became more affordable, researchers and industry professionals adapted geocellular modeling techniques to characterize and simulate geothermal reservoirs to improve reservoir understanding and optimize geothermal energy extraction. The development of geocellular modeling techniques for reservoir characterization in the oil and gas industry laid the foundation for its application to geothermal reservoirs, as these models allowed engineers and geoscientists to create three-dimensional representations of subsurface reservoirs, incorporating geological, geophysical, and petrophysical data to simulate fluid flow and predict reservoir behavior, and can easily be replicated for geothermal settings.

Since then, several notable software packages have been developed and commercialized, helping with the democratization of 3D geocellular modeling for subsurface geothermal studies. These software provide users with advanced computational techniques to simulate complex behavior of geothermal reservoirs; they have user-friendly interfaces and comprehensive tutorials that lower the entry barrier for new users. This accessibility allows researchers from various backgrounds, including academia, industry, and government agencies, to engage in geothermal studies without requiring extensive expertise in geophysics or geology, which greatly contributed on the surge of geothermal modeling studies all around the world.

Notable examples of geothermal modeling software include TOUGH2/TMVOC, a versatile simulator for multiphase flow and heat transfer in porous and fractured media, widely used for geothermal reservoir modeling. Another prominent software package is COMSOL Multiphysics, which offers a comprehensive platform for simulating coupled multiphysics phenomena, making it suitable for studying heat transfer, fluid flow, and geomechanical processes in geothermal reservoirs. Other popular tools include PetraSim, FEHM, and GMT, each offering unique capabilities for specific aspects of geothermal modeling. In addition, conventional subsurface modeling software used in the oil and gas industry offers outstanding collaborative and integrative alternative environments for geothermal studies such as Petrel, Leap Frog, and Geosoft.

18.5 Concepts of 3D Geocellular Modeling

Geocellular modeling is a numerical method used in geosciences to realize interpretative conceptual models of the subsurface geology and petrophysical parameters, particularly in fields like petroleum engineering, hydrology, and environmental science. It aims to represent the subsurface structure of the Earth in a 3D grid or mesh of cells, with each cell of the grid representing a given volume of the subsurface, with corresponding properties such as lithology, porosity, permeability, and fluid saturation assigned to each of one of these cells making the grid.

Generating a 3D geocellular model representative of the subsurface involves several key steps from the integration of seismic interpretation results, well data and correlations, remote sensing data, as well as satellite imagery. Throughout the entire process, collaboration between geoscientists, engineers, and stakeholders is essential to ensure the accuracy, reliability, and relevance of the model for its intended application. After building the most "representative model," continuous data collection, model updating, and refinement are often necessary to incorporate new information and improve the model's performance over time.

Rectangular and Voronoi meshes are often used for geocellular modeling. Rectangular grid, also known as a Cartesian grid, is a type of grid where the spatial domain is divided into a series of uniform or non-uniform rectangular (or cuboidal in 3D) cells. Whereas, Voronoi grid is a type of spatial partitioning where the domain is divided into regions based on a set of seed points. Each region (or cell) contains all points that are closer to its seed point than to any other seed point. The main difference between using a rectangular grid and a Voronoi grid lies in their cell structures and adaptability to complex geometries. Rectangular grids, composed of uniform or non-uniform rectangular cells, offer simplicity and ease of implementation, making them suitable for problems with relatively simple and regular geometries. In contrast, Voronoi grids consist of irregularly shaped cells that adapt to the spatial distribution of seed points, providing a more accurate representation of heterogeneous and complex domains.

It is very important for the modeler to keep in mind that all geocellular models are about simplifications. Since the purpose of the model is always to help us understand, describe, or predict how a reservoir would behave in the real world without representing the real geology at the subsurface. Hence, there is no "correct" or "wrong" models of the subsurface, but rather "most representative" or "less representative" models. In fact, the statement "All models are wrong, but some are useful" from the British statistician George Box became a very common aphorism among geo-modelers, acknowledging the fact that no model can perfectly represents reality, as they are always simplified representations of very complex geological systems and phenomena. They involve assumptions, approximations, and simplifications that necessarily deviate from the true nature of the systems being modeled.

To give an idea about the simplifications geomodelers are taking to represent subsurface reality, let's assume the 3D geocellular model represented in Figure 18.3. The model represents an area of 1300 m × 800 m × 150 m made of 2496 cells, with a cellular size of 50 m × 50 m × 25 m each (Figure 18.3a). To put in perspective each cell size in the real world, we can imagine it as a big empty living room (a bed and closet are illustrated to scale dimensions in Figure 18.3b), now think about a well bore drilling which is usually of few dozens of centimeters of diameter crossing this cell from ceiling to the floor with continuous variable measurements along the well vertically or laterally. Each single cell will only have a unique value representative of the continuous log measurement inside of it. Now let us consider other surrounding cells, with no penetrating wells. Each cell will need to be populated with specific value that relates to its proximity to the well location using specific geostatistical methods.

Therefore, before engaging in any geocellular modeling workflow, it is very important for the modeler to know the objectives behind the model and ask the right questions about the purposes of this process: What do we want to achieve from this geological model? Framing the modeling study in this way should guide all decision making used for the overall construction of any subsurface model.

Advances in seismic data acquisition and processing means that we are now able to image the subsurface in far greater detail. As a result, geological interpretations are far closer to actual subsurface features. This means that geological models are being constructed with the aim of capturing as much of the heterogeneity of reservoir features as possible. This objective is often at odds with the goals of fluid simulation, where very dense and complicated geological models are often avoided by reservoir engineers as these models do not perform well in simulators; largely due to very small cell sizes required to capture this heterogeneity.

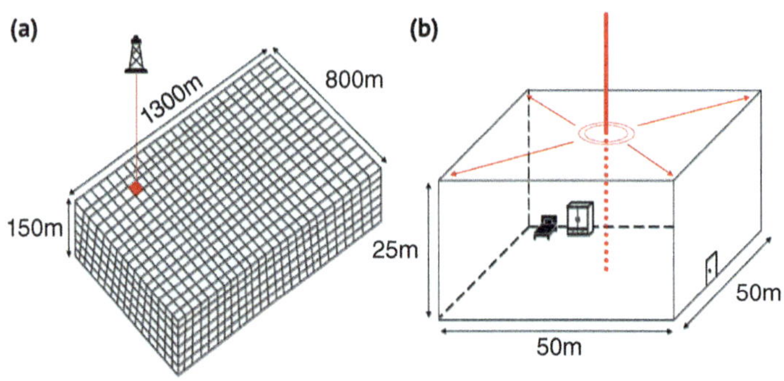

Figure 18.3 (Left a) Schematic representation of a 3D geo-cellular rectangular grid made of 2496 cells of 50 × 50 × 25 m. (Right b) Closeup of a single cell crossed by the well bore, each cell will have a single value representing the whole log data across the cell.

Reservoir engineers often prefer models that are designed to be more agile in terms of simulation performance, this typically includes only the most important features that affect fluid flow that can be rapidly manipulated in fluid flow simulators to understand the uncertainty of hydrocarbon production.

Often models are constructed without knowing the goals of all stakeholders. As a result, a model is frequently constructed for one use and then re-purposed for another workflow for which is not suitable for, leading to compromises in the model quality, delivery schedule or the scope of the later workflows. Integration of subsurface workflows has evolved significantly since the inception of three-dimensional geocellular modeling; however, we are not yet at a point where a single subsurface representation can be used for everything. Therefore, informed choices must be made. Knowing the multiple purposes for the model to be constructed will facilitate the population and detail needed for a model.

Commercial geocellular modeling software uses different algorithms and methods to build a representative geocellular model of the subsurface, but despite their differences, the technical overall concept remains more or less the same: After defining the objectives of the geocellular model and the aim to achieve with the model, modelers need to define its scope, including geographical area, and geological formations and integrate their data to build the subsurface representation.

Based on the size and the complexity of the model, user will then select the adequate type of the structural gridding process to be used and set specific gridding size (i.e. cell size). Here, it is important to consider resolution of the modeling versus the time of computation: the smaller the cell size is the higher the resolution and detail we will reproduce; however, the more hardware resources and computation time it will take to model and simulate. An example is illustrated in Figure 18.4a–c, where we used different gridding size to structure the same faulted geological model using cell sizes of = 10 m, 25 m and 50 m. Note the resolution meshing of our model with smaller cells compared to coarser cells: the bordure and the contact at faults are smoother and better reproduced at the smallest cell size gridding, while the coarser the grids, the less resolution we obtain.

The main drawback with modeling with finer grid is the large number of cells generated, affecting the calculation and the performance of the numerical model. Therefor a trade-off should be respected, given the objective to model and the computation hardware at hands.

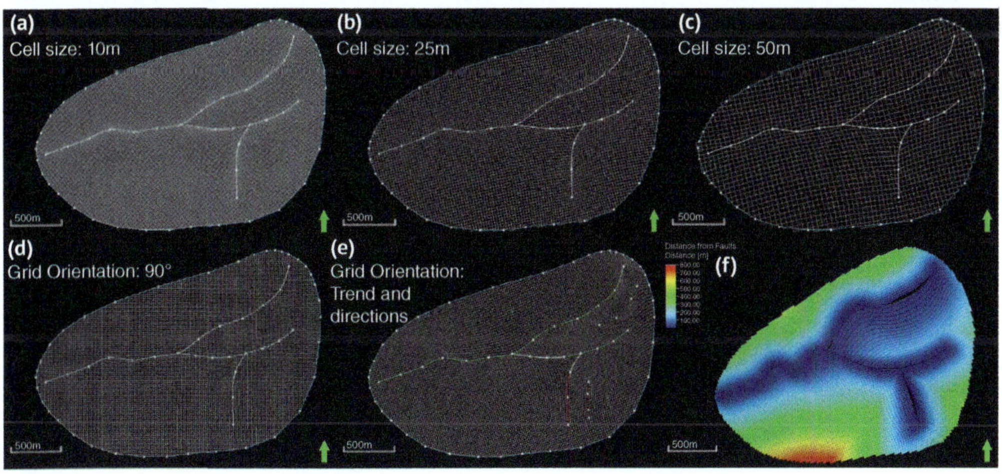

Figure 18.4 Different settings for the same 3D geocellular grid skeleton: (a) Cell size gridding of 10 m; (b) 25 m; (c) 50 m; (d) vertical cell orientation of 90°; (e) personalized grid orientation using fault network trends and cells orientations; (f) populated grid parameters with distance from faults modeling.

Another parameter to be considered while gridding the model is the structural orientation and trends of the grid. This can be controlled by setting specific orientation and directionality of the grid cells and the main fault network.

The grid orientation can impact the distribution of cells in a given direction and influence the resolution around faults and boundaries. The same example previously discussed is illustrated in Figure 18.4d–e, where we used two different orientations of the grid geometry; a 90° cells orientation in Figure 18.4d, and a manually trended gridding direction, following the direction of the faults in Figure 18.4e. We can observe the shape and distribution of the cells between the two models and how cells distribution matches better the direction of the faulting when trends and directions are implemented.

Once the gridding and directions of the grid structure are set, seismic interpretation results are then compiled to build an initial geological framework (skeleton) of the subsurface, incorporating subsurface structural features such as stratigraphic layers and other geological structures mapped from the seismic data. Vertical layering of the grid is then defined to reflect the geological deposition at each geological interval.

The structural grid geometry is generated, the model can then be populated with geometrical proprieties using pre-defined system variables, such as the volume of the cells, the layering index, etc. Each cell gets a numerical value corresponding to the selected system variable. For example, we can select the distance to object operation, to generate a geometrical property with cells values calculated from the distance to the faults as illustrated in Figure 18.4f; this can help to bias the property distribution by proximity to the fault network.

Well logs and other data are then upscaled and integrated into the grid to populate the remaining cells between the wells, using interpolation techniques based on geostatistical analysis and modeling algorithms such as the Kriging algorithm or stochastic methods. The model will then be calibrated with the observed data and uncertainty analysis are performed to minimize errors associated with model predictions, while sensitivity analysis helps to identify the most influential parameters and assess their impact on model outcome.

Finally, in the case of hydrocarbon production engineering, numerical simulation techniques are used to model physical dynamic behavior of the reservoir such as fluid flow, production history, heat transfer or contaminant transport. Similar simulation schemes can be used to fluid circulation and behavior in geothermal systems, based on the geocellular models we developed.

Geocellular modeling has a rich history that spans several decades, reflecting the evolution of computational techniques and the growing demand for accurate representations of subsurface systems. Originating in the 1960s with early reservoir simulation methods, geocellular modeling initially focused on optimizing hydrocarbon recovery in the petroleum industry. Over time, advancements in computer technology and geological data acquisition led to the development of more sophisticated models capable of capturing the complexities of subsurface geology. The 1980s saw the integration of geological data into reservoir models, enabling the representation of features such as faults, stratigraphic layers, and facies variations. By the 1990s, grid generation techniques evolved to accommodate unstructured grids, facilitating the modeling of the geological structures. The 2000s witnessed the incorporation of advanced simulation techniques, including Discrete Fracture Network (DFN) modeling and coupled multiphysics simulations.

Today, geocellular modeling finds applications beyond petroleum engineering, encompassing fields such as hydrology, environmental science, geothermal energy, and carbon capture and storage. As technology continues to advance, geocellular modeling remains at the forefront of efforts to understand and manage subsurface systems for a big range of scientific and engineering endeavors. This evolution coincides with the evolution of geothermal modeling, making the geocellular method a premium choice of modeling complex geothermal reservoirs.

18.6 Geophysical Modeling with the 3D Geocellular Grid

Each geophysical method provides its own set of information about the subsurface properties. Therefore combining results from the different geophysical surveys into the 3D geocellular model helps to update and sharpen the subsurface structure. These subsurface properties can be upscaled into the geocellular grid that divides the subsurface into a 3D matrix of cells, each representing a discrete volume with specific geological and petrophysical properties (Figure 18.5):

- Three-dimensional geophysical inversion methods involve transforming geophysical data, such as seismic, magnetic, gravity, and electrical measurements, into detailed 3D subsurface models that depict the physical properties of the Earth's interior. Recent developments on geophysical inversion methods helped with the standardization of geophysical inversion modeling within the geothermal exploration community.
- Velocity cubes are 3D representations of the seismic velocity distribution at the subsurface essentially built from well check-shots data or seismic surveys. They can be very helpful for depth conversion workflows and help to understand the general velocity trends of the subsurface. Interval velocities can be excellent indicators of subsurface irregularities and complex geological structures.
- The integration of seismic attributes into geothermal subsurface studies enhances accuracy of interpretation. Seismic attributes, which include measures like amplitude, frequency, phase, and continuity of seismic reflections can provide valuable information about the subsurface geological features. They can be combined with other geological and geophysical data, such as temperature gradient or well logs (Yasin et al., 2023).

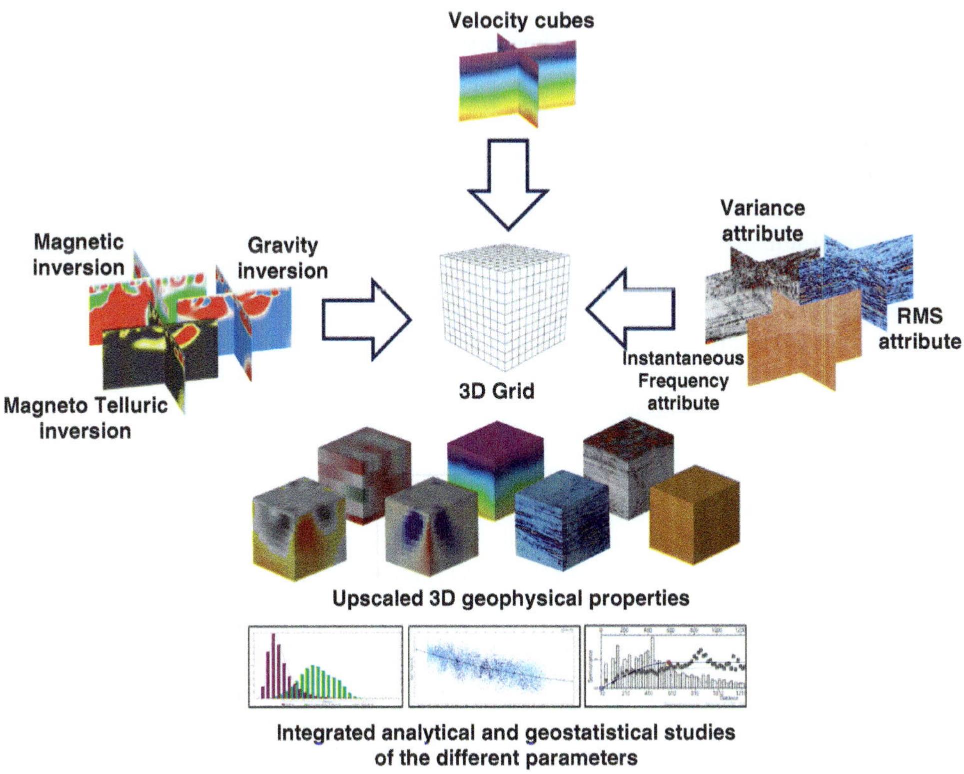

Figure 18.5 Integration of different geophysical data into a 3D geocellular grid.

By integrating different inversion results into the geocellular grid, each cell of the grid can be assigned accurate and data-driven property values through geostatistical and analytical studies that can correlate and draft convergences between the various physical properties. Combining different results is essential for geothermal studies as they provide quantitative and relatively accurate insights into the distribution and characteristics of geothermal reservoirs.

By converting raw geophysical data into meaningful geological information, these methods enables the identification and characterization of geothermal reservoirs, helping to estimate their size, structure, and thermal properties. This will improve the resolution and accuracy of the subsurface model, allowing for more precise simulations of the geothermal activity.

Potential fields methods, such as gravity and magnetic surveys, are widely used in geothermal exploration. They can help to map the physical properties such as gravity contrast and magnetic susceptibility of the, to infer geological structures and proprieties.

Gravity surveys measure the gravitational attraction of subsurface density variations, providing information on the distribution of rock types, faults, and sedimentary basins, while magnetic surveys detect variations in the Earth's magnetic field caused by differences in the magnetic properties of subsurface rocks, such as magnetization due to the presence of ferromagnetic minerals. These variations are then processed using inversion techniques to create a three-dimensional model of the subsurface. For instance, if an area has undergone hydrothermal alteration, it may exhibit reduced magnetization due to the conversion of magnetic minerals like magnetite to non-magnetic minerals such as hematite. By identifying these low-magnetization zones, geologists can infer the presence of geothermal reservoirs, as these areas often coincide with high heat flow and fluid circulation. Same applies to regions with lower density, that may indicate the presence of hot, less dense rock or fluid-filled fractures and voids, which are key indicators of geothermal activity.

To demonstrate the concept of geophysical inversion for potential fields methods, we conducted a numerical experiment summarized in Figure 18.6. The model represents a half space subsurface model with a constant density representing the background geology, we included three different density anomalies of different shapes: one square geological body with higher density value, and two other rectangular shapes with lower density values shown in Figure 18.6b. Surface gravity measurements can be simulated by forward modeling of the density contrast subsurface model; the gravity surface estimated in Figure 18.6a shows a peak positive response of the gravity above the higher density anomaly, while the lower anomalies shows negative responses at the surface gravity measurements.

Inversion model of the surface gravity alone is illustrated in Figure 18.6c. The density contrast issued from inversion shows how it could successfully spot the position of the density anomalies, without any prior knowledge of the subsurface.

This particularity can be very useful in geothermal exploration projects, where fractured zones and fluid circulations cause remarkable density contrast with the background reservoir, that can be measured at the surface using gravity measurements methods. Recent advancements in potential fields inversion studies have significantly enhanced the exploration and characterization of geothermal resources. For instance, Carrillo et al. (2022) proposed a 3D joint inversion algorithm of gravity and magnetic data application and applied it to Los Humeros geothermal field in Mexico. Saibi et al. (2022) successfully delineated the subsurface structure of two geothermal fields in the UAE, by analyzing inversion results of gravity and magnetic data acquired in the field. Melouah et al. (2023) combined structural observations with gravity and magnetic data to provide new insights and interpretations to the crustal architecture of the West Arabian Shield and their effect on the geothermal systems of the area.

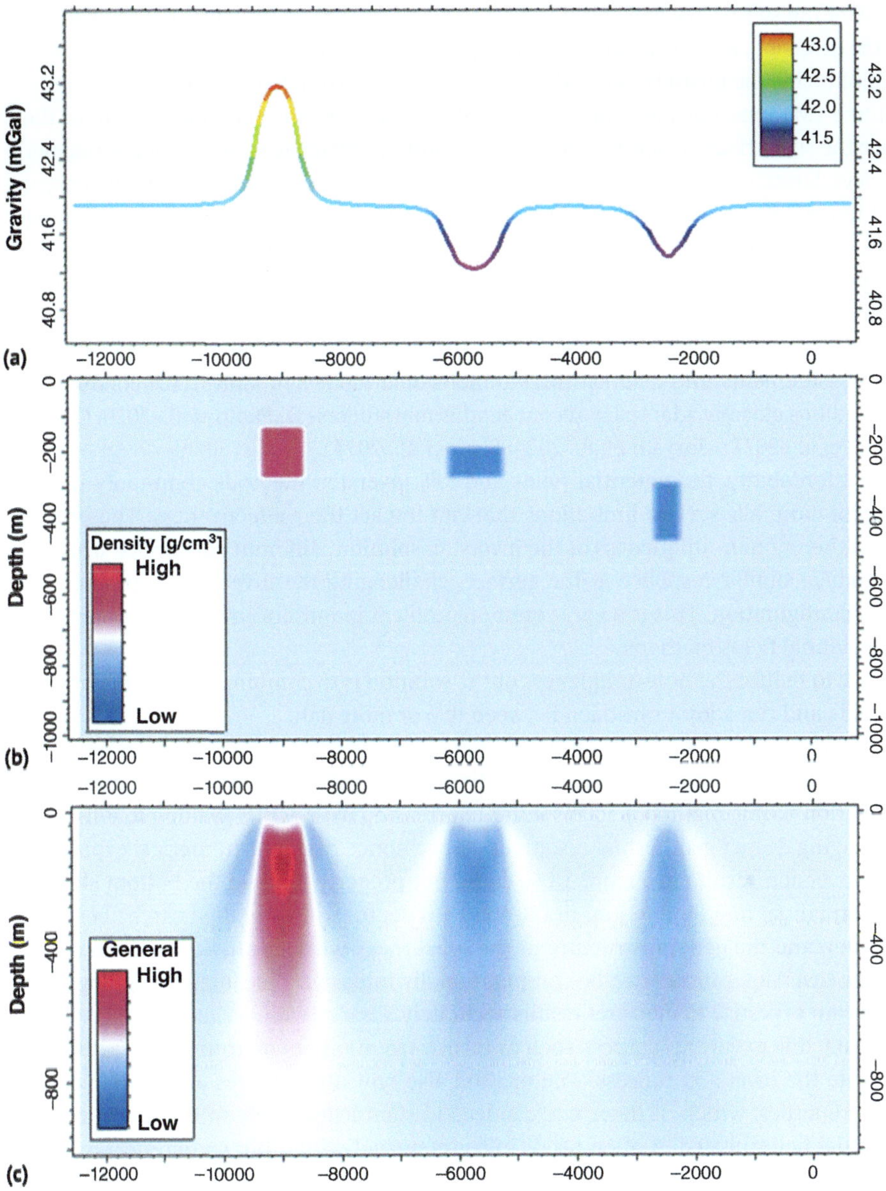

Figure 18.6 Transvers cross-section of a numerical experiment validating the feasibility of potential fields inversion. (a) Forward gravity surface estimated from the subsurface density contrast numerical model in Figure 18.5b; (b) half space numerical subsurface model with density anomalies; (c) density contrast inversion results from the gravity surface in Figure 18.5a.

Gravity and magnetic inversion results can be directly upscaled into the 3D geocellular grid and help to constrain or trend the model. It can be a density contrast or magnetic susceptibility contrast results directly obtain from inversion of the surface measurements, without any prior information on the subsurface (similar to our numerical experiment discussed earlier); in this case, we talk about geological contrasts of the physical parameters and not density and magnetic values, so they are more "highs" and "lows" values contrasts and not proper petrophysical values. It is also possible to build an inversion model of the subsurface geological layers. This will require an initial model set from well data to guide the inversion, and the inversion will then improve the subsurface structure in an iterative fashion, minimizing the misfit between the real data and the estimated models.

Magnetotelluric (MT) methods are increasingly integrated to geothermal studies due to their ability to image the subsurface structures with good resolution. By measuring the electromagnetic fields, MT surveys detect variations in electrical conductivity which are influenced by the presence of geothermal fluids, temperature and mineralogy. These surveys cover a large frequency range, with frequencies penetrating shallow depths and lower frequencies reaching deeper layers.

Similarly to potential fields methods, the non-invasive nature of MT methods makes them particularly advantageous in environmentally sensitive areas and rugged terrains where traditional seismic methods might be challenging or disruptive. Additionally, advanced inversion techniques convert these measurements into detailed three-dimensional models of subsurface conductivity, making MT inversions globally adapted in recent geothermal studies (Di Paolo et al., 2020; Quiroga et al., 2023; Saibi et al., 2021; Maryadi et al., 2024; Shah et al., 2024).

Despite their affordability, the potential fields and MT inversion methods commonly used in geothermal exploration, has several limitations that can impact their effectiveness. The primary limitation is its inherent non-uniqueness of the inversion solution, different subsurface structural settings can produce similar response at the surface, challenging the inversion to pinpoint the exact geological configuration. This is a very common problem in most of inversion problems and not limited to potential fields methods.

One alternative to reduce the non-uniqueness of the solution is to combine the inversion with a secondary variable and run a joint inversion between two or more data.

We can also reduce the non-uniqueness of the inversion by constraining it with an initial model (also known as low frequency model) built from wells data and other geophysical measurements.

Another limitation of this method is its resolution decreasing with depth, leading to difficulties in accurately imaging deeper geothermal resources. As we notice from our numerical experiment in Figure 18.6, inversion results in Figure 18.6c could not properly retrieve the bottom shape of the density abnormalities and tend to lose resolution with depth. So the method can be beneficial to spot the position and the general structure of the subsurface, without providing much details. Furthermore, the inversion process can be computationally intensive and time-consuming, often requiring significant expertise to interpret results accurately.

Noise in the data, due to various sources such as instrumentation or environmental factors, can further complicate the inversion process. The method also typically requires assumptions about the subsurface properties, which, if incorrect, can lead to significant errors in the interpretation. Moreover, potential fields inversion often needs to be integrated with other geophysical methods and geological data to produce reliable and comprehensive subsurface models, increasing the complexity and cost of geothermal exploration projects.

18.7 Faults and Fracture Network Modeling with the 3D Geocellular Grid

Faults and fractures at geothermal reservoirs are of prominent importance as they control the fluid flow pathways, reservoir connectivity and overall reservoir performance. They can have different sizes or distribution and can be very common at outcropping formations (Figure 18.7a). Fractures are mostly associated with fault and fracture networks that are conditioned within the Earth's crust by a variety of parameters interacting with geological, tectonic, and mechanical

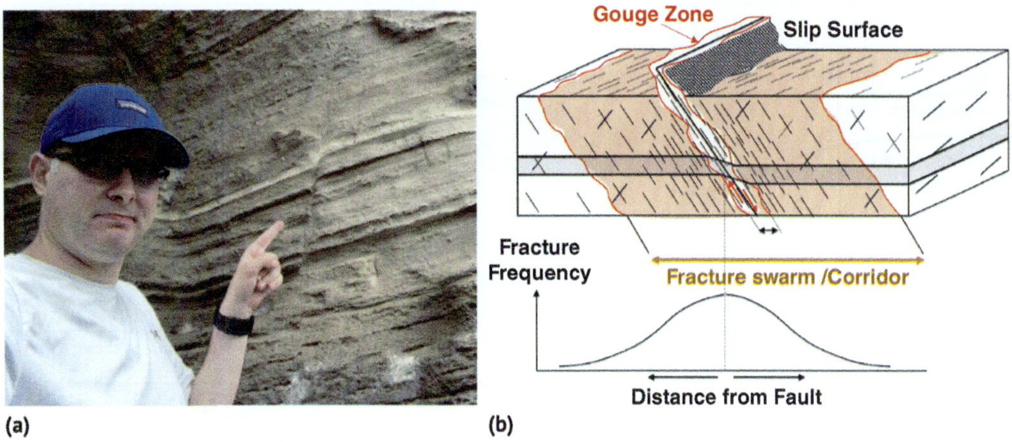

Figure 18.7 (a) Normal movement fracture of decametric scale at the Miura group formation outcrop in Japan; (b) schematic representation of the fault zone associated with fractures.

proprieties of the area. These parameters include lithology, stress regime, tectonic history, fluid pressure, temperature, and structural heterogeneities.

The purpose of modeling fractures network in a 3D geocellular grid is to simulate properties with the power to predict the reservoir geomechanical behavior. It is a multi-step process, involving geological concepts such as interpretation of thin beddings, faults, and fractures from image log data, seismic attributes, as well as outcropping studies as analogs conceptual models for fracture drivers.

These data are then compiled within the framework of geomechanical studies to provide details about the distribution, orientation, and connectivity of fractures within reservoir rocks. The obtained model describes the fracture intensity, which will be scaled in a 3D geocellular model. Depending on the analysis of the fracture data, multiple sets of fractures can be identified. Each fracture set can be the result of several tectonic events, such as over-thrusts or extensional faults, conjugate fractures related to bending or flexure of geological layers, or simple joints related to difference in lithology.

We often tend to simplify faults representation as a continuous line materializing the fault plane, but fault zones in nature are more complex than a simple geological discontinuity. Fault zones generally consist of a fault core and a damage zone that have permeabilities distinctly different from the original rock in which they form. The damage zone can include gauge zones, small subsidiary faults, fractures, and veins with a frequency and intensity reducing with the distance from the fault plane as illustrated in Figure 18.7b.

However, representations of fault as 3D planes or continuous lines in the model are fairly accepted in oil and gas reservoirs, where we focus on the reservoir structure and mainly consider the sealing and transmissibility of the modeled faults.

Simplifications of faults representation omitting fractures in geothermal reservoirs can be misleading, as the fault network developed at the subsurface controls the fluid flow pathways and reservoir connectivity.

To illustrate this particularity, let's assume a schematic subsurface representation of a given geothermal reservoir illustrated in Figure 18.8a: a first vertical well is drilled in the reservoir, where a highly fractured zone and two major faults are reported in the area, connecting the reservoir to the surface. Building a 3D geocellular model based on the conventional fault interpretation where faults are simply represented as 3D planes in Figure 18.8b, fracture network distribution inside

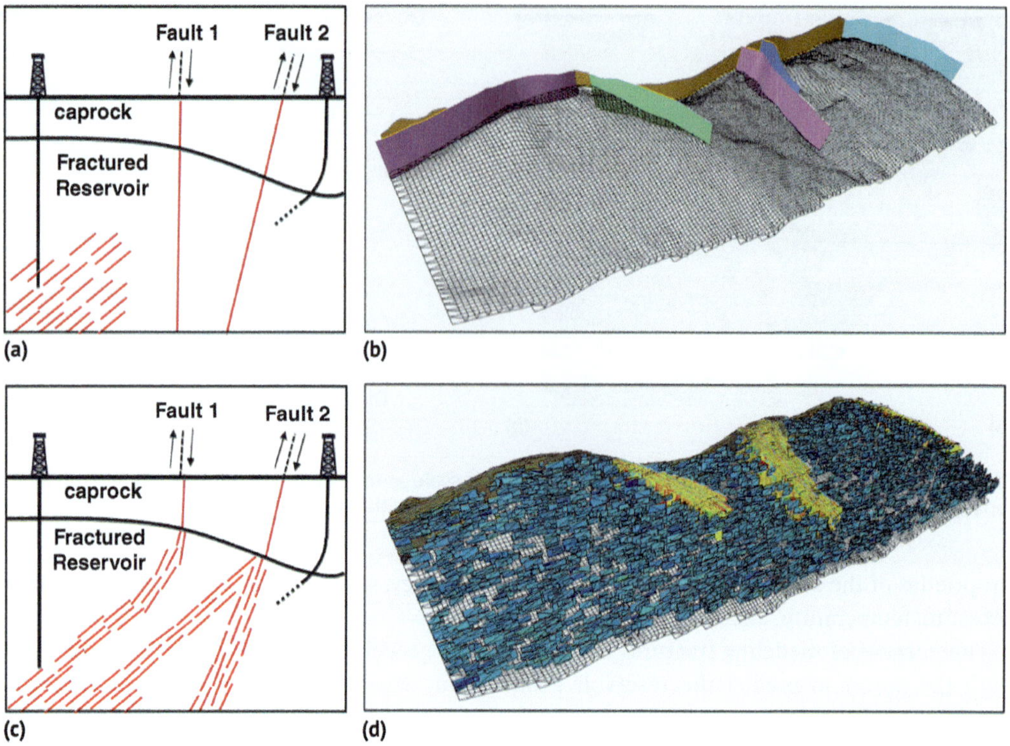

Figure 18.8 Schematic representation and 3D geocellular modeling of fracture network. (a) Schematic representation of a simplified faulted and fractured geothermal reservoir; (b) 3D geocellular model of reservoir considering simple fault planes only; (c) schematic representation of a faulted and fractured geothermal reservoir including geomechanical studies predicting the fault network distribution; (d) 3D geocellular model of reservoir including faults and fracture network modeling.

the reservoir cannot be represented properly causing serious concerns on the subsurface accuracy when we try to plan a second deviate well through the reservoir.

However, integrating geomechanical studies considering relationship between lithology, stress regime, tectonic history, fluid pressure, and temperature. We can predict the fracture and fault network patterns as illustrated in Figure 18.8c, where we can understand for example that the major faults of the area are more complex than simple linear fault plane representation, but are rather part of a complex fracture zones that follow specific trends driven by the geomechanical proprieties of the reservoir. The 3D modeled fracture network can therefore be corrected to a more representative reservoir fracture model, similar to the representation in in Figure 18.8c. The 3D fracture model built in the geocellular grid is shown in Figure 18.8d; we can see that the model includes two generations of faults: yellow faults group surrounding the major faults, representing the fracture swarm zone associated with the major faults and fragilizing the surrounding areas, and a second family of faults with a green color that follows the general tectonic folding of the structure.

3D geocellular models with fracture network are more representative of geothermal reservoirs, providing far more details on the size, the shape, and the extent of the reservoir fractures. This can be more helpful to understand the fluid behavior in the reservoir through simulation. Predicting how fluids will flow through the geothermal reservoir and how heat will be transferred from the rock to the fluid helps on engineered fractures, by predicting how induced fractures will propagate and interact with existing natural faults on Enhanced Geothermal Systems (EGSs) where natural permeability is low.

Moreover, geothermal operations can be associated with risks such as induced seismicity, where the injection or extraction of fluids can cause earthquakes. Fracture modeling helps in predicting and mitigating these risks by providing a detailed understanding of the stress state and mechanical behavior of the reservoir.

One of the most used methods for 3D fracture modeling is the DFN that helps to represent geometry and connectivity of fractures within a geological formation. In a DFN model, fractures are treated as discrete entities rather than continuous features, allowing for the representation of individual fractures and their interactions within the rock mass. However, one of the difficulties with the traditional discrete fracture modeling workflow is that the number of fractures to be modeled in the field can be extremely large. Trying to represent all of them explicitly in the model is often hampered by hardware memory and performance limitations.

Even if this is achievable, calibration of fracture modeling parameters to flow simulation results demand iterative steps in a workflow, and computational performance becomes a limiting factor.

Therefore, some software such as Petrel are offering hybrid methods where the bigger and more important fractures are modeled explicitly as discrete patches DFN, and the residual part of the distribution (smaller fractures) is statistically represented as grid properties (implicit fracture model - IFM).

Several studies have been conducted on fracture network modeling for geothermal exploration: Wang et al. (2020) proposed an embedded discrete fracture modeling method, incorporating arbitrary 3D discrete fractures into modeling fracture-dominated fluid flow and heat transfer in fractured. Janiga et al. (2022) used DFN in modeling and simulation of a horizontal well-doublet Enhanced Geothermal System (EGS). Medici et al. (2023) shared an extensive literature review of the DFN characterization for geothermal reservoirs. Barcelona et al. (2020) studied the superficial leakages of the shallow steam zone using a DFN and 3D forward train-stress at the Copahue geothermal field in Argentina.

One major challenge in fracture modeling is the difficulty in accurately characterizing the subsurface fracture network, as it often relies on sparse and indirect data, such as seismic surveys and well logs, which may not fully capture the complexity and heterogeneity of the fractures.

The resolution of these data sources can be insufficient, leading to oversimplified models that fail to represent the true nature of the fractures. The assumptions made during the modeling process, such as fracture orientation, density, and connectivity, can introduce significant uncertainties if they do not reflect the actual conditions. The dynamic nature of geothermal reservoirs, where fractures can evolve over time due to thermal and hydraulic stresses, further complicates the modeling process. Furthermore, fracture modeling often requires integration with other geophysical and geological data, which can be challenging to reconcile and may introduce additional errors. These limitations can impact the reliability of predictions regarding fluid flow and heat transfer within the geothermal reservoir, ultimately affecting the assessment and development of geothermal resources.

18.8 Updating the Property Models with Integrated Workflows

Using the previously described geophysical and fracture modeling studies to correct property models built from wells in a geocellular grid can significantly improve their reliability. It provides a more comprehensive framework that captures spatial heterogeneity and complex interactions within the geothermal reservoirs.

Combination of properties, such as porosity, permeability, thermal conductivity, densities measured from wells with geophysical and structural data into the geocellular grid, offers a more accurate simulation of fluid flow and heat transfer processes. This holistic view is crucial for optimizing the placement of wells, enhancing reservoir management strategies, and improving the prediction of geothermal energy output.

Property models built from wells relies heavily on geostatistical analysis and modeling algorithms that interpolates data between the different wells. For example:

Kriging algorithms provides an estimation of the property values at unsampled locations based on spatial autocorrelation, while Co-Kriging incorporates secondary variables to improve the prediction accuracy.

Stochastic methods such as the Sequential Gaussian Simulation (SGS) generates multiple equally probable realizations of the reservoir property, capturing the uncertainty and spatial variability; they can be particularly useful for modeling heterogeneous properties and assessing the impact of spatial variability on geothermal production.

Multi-point geostatistcs (MPS) uses training image to capture complex geological patterns and structures in the reservoir. It can be beneficial for modeling the intricate networks of fractures and faults commonly found in geothermal reservoirs.

Thermo-hydro-mechanical (THM) coupling algorithms simulate the interaction between thermal, hydraulic, and mechanical processes in geothermal reservoirs. By coupling these processes, THM algorithms provide a more comprehensive understanding of how changes in temperature, pressure, and stress fields impact reservoir properties and performance.

However, these algorithms alone can produce misleading results, especially when the wells are located at very far distances. One significant limitation is the inherent uncertainty due to sparse and unevenly distributed data, which can lead to inaccuracies in the model, simplifying assumptions and approximations are often necessary, potentially overlooking small-scale heterogeneities and dynamic processes. Therefore, combining secondary data from geophysical and structural studies can further constraint the model and reduce the range of uncertainties.

Recent examples using an integrated geocellular approach for geothermal studies we can mention: Pollack et al. (2021) introduced algorithms and open source code for stochastic inversion to stochastically invert five typical datasets in order to recreate geologically realistic structural models. Yalcin et al. (2024) presented a detailed 3D reservoir modeling of Al-Wajj basin in western Saudi Arabia using advanced modeling method based on multiple-point statistics with a training image and a fluid simulator. Huang et al., (2022) assessed the geothermal potential of an the Jidong oil field, northern China, using a 3D integrated geological model with a hydro-thermal coupled simulation.

18.9 Conclusions

3D geocellular modeling offers significant advantages on integrated geothermal subsurface studies, it provides reliable and quantitative representation of the geological formations, enabling acceptable predictions of fluid flow and heat transfer. They can facilitate the integration of multidisciplinary methods, enhancing the reliability of the reservoir characterization and aiding in the optimization of well placement and extraction strategies. This approach also allows dynamic simulations to assess the long-term sustainability of geothermal resources and helps in mitigating environmental impacts and risks associated with geothermal operations. They can also help to assess uncertainties and run multiple realizations, minimizing the error tolerance in the realized models.

However, the accuracy of geocellular models is often constrained by the availability and quality of the subsurface data, leading to significant uncertainties. Simplifying assumptions and computational constraints can limit the resolution and complexity of the models, potentially overlooking important small-scale heterogeneities and dynamic processes. The integration of diverse data types can be challenging and may require sophisticated techniques to reconcile conflicting information. Despite these limitations, geocellular modeling remains a powerful tool in geothermal exploration, offering valuable insights and aiding in the efficient and sustainable development of geothermal resources.

References

Anees, M., Sosa, G., Van den Kerkhof, A., et al. (2024). Granitoids of the western Himalaya and Karakoram as potential geothermal reservoirs – A petrological, geochemical and petrophysical study. Geothermics 121, pp. 103040. https://doi.org/10.1016/j.geothermics.2024.103040.

Anderson, G., Probst, A., Murray, L., Butler, S. (1992). An accurate PVT model for geothermal fluids as represented by H_2O-CO_2-NaCl mixtures. Proc., Seventeeth Workshop on Geothermal Reservoir Engineering, Stanford University, Stanford, California, 29–31. https://pangea.stanford.edu/ERE/pdf/IGAstandard/SGW/1992/Andersen.pdf.

Azzola, J., Thiemann, K., & Gaucher, E. (2023). Integration of distributed acoustic sensing for real-time seismic monitoring of a geothermal field. Geothermal Energy 11 (1), p. 30. https://doi.org/10.1186/s40517-023-00272-4.

Baoqing, T., Zhifeng, D., Liming, Y., et al. (2022). Microtremor survey method: a new approach for geothermal exploration. Frontiers in Earth Science 10, p. 817411. https://doi.org/10.3389/feart.2022.817411.

Barcelona, H., Maffucci, R., Yagupsky, D., et al. (2020). Discrete fracture network model of the vapor zone leakages at the Copahue geothermal field. Journal of Structural Geology (140), p. 104155.

Battistelli, A., Calore, C., & Pruess, K. (1997). The simulator TOUGH2/EWASG for modeling geothermal reservoirs with brines and noncondensible gas. Geothermics 26 (4), p. 437.

Belhai, M., Fujimitsu, Y., Barragan-Reyes, R. M., et al. (2022). Insights into geochemical characteristics of the geothermal fluids of northwestern algeria: An updated conceptual model. Geothermics 99, p. 102304. https://doi.org/10.1016/j.geothermics.2021.102304.

Bredesen, K., Dalgaard, E., Mathiesen, A., et al. (2020). Seismic characterization of geothermal sedimentary reservoirs: A field example from the Copenhagen area, Denmark. Interpretation 8 (2), pp. T275–T291. https://doi.org/10.1190/INT-2019-0184.1.

Carrillo, J., Perez-Flores, M. A., Gallardo, L. A., et al. (2022). Joint inversion of gravity and magnetic data using correspondence maps with application to geothermal fields. Geophysical Journal International 228, pp. 1621–1636. https://doi.org/10.1093/gji/ggab416.

Calvin, W., Coolbaugh, M., & Vaughan R. G. (2002). Geothermal site characterization using multi- and hyperspectral imagery. Geothermal Resources Council Transactions 26, pp. 483–484. https://www.geothermal-library.org/index.php?mode=pubs&action=view&record=1019646.

Coats, K. H. (1977). Geothermal Reservoir Modelling. Presented at the SPE Annual Fall Technical Conference and Exhibition, Denver, Colorado, 9–12 October 1977. SPE-6892-MS. https://doi.org/10.2118/6892-MS.

Deb, P., Knapp, D., Marquart, G., et al. (2020). Stochastic workflows for the evaluation of Enhanced Geothermal System (EGS) potential in geothermal greenfields with sparse data: the case study of Acoculco, Mexico. Geothermics 88, p. 101879. https://doi.org/10.1016/j.geothermics.2020.101879.

Di Paolo, F., Ledo, J., Ślęzak, K., et al. (2020). Geothermal system revealed by 3D magnetotelluric data inversion. Scientific Reports 10, p. 18181. https://doi.org/10.1038/s41598-020-75001-z.

Eltayieb, M., Werthmüller, D., Drijkoningen, G., et al. (2023). Feasibility study of controlled-source electromagnetic method for monitoring low-enthalpy geothermal reservoirs. Applied Sciences 13, p. 9399. https://doi.org/10.3390/app13169399.

Feng, X., Ma, J., Guo, R., et al. (2024). 3D gravity and magnetic inversion reveal geothermal structure dominated by radioactive heat production of granites in the Weihe basin. Tectonophysics 874, p. 230237. https://doi.org/10.1016/j.tecto.2024.230237.

Finsterle, S. & Pruess, K. (1995). Automatic History Matching of Geothermal Field Performance. Proc., Seventeenth New Zealand Geothermal Workshop, Auckland, New Zealand, 193.

Freienstein, J., Szwillus, W., Wansing, A., et al. (2023). Statistical Appraisal of Geothermal Heat Flow Observations in the Arctic. Solid Earth 15, pp. 513–533. https://doi.org/10.5194/se-15-513-2024.

Gao, K., Huang, L., & Cladouhos, T. (2023). Seismic characterization of the blue mountain geothermal field. Energies 15, p. 5822. https://doi.org/10.3390/en16155822.

Gringarten, A. C., Ramey, H. J., & Raghavan, R. (1975). Applied pressure analysis for fractured wells. Journal of Petroleum Technology 27, pp. 887–892. https://doi.org/10.2118/5496-PA.

Huang, Y., Cheng, Y., Ren, L., et al. (2022). Assessing the geothermal resource potential of an active oil field by integrating a 3D geological model with the hydro-thermal coupled simulation. Frontiers in Earth Science 9, p. 787057. https://doi.org/10.3389/feart.2021.787057.

Janiga, D., Kwa´snik, J., & Wojnarowski, P. (2022). Utilization of Discrete Fracture Network (DFN) in modelling and simulation of a horizontal well-doublet enhanced geothermal system (EGS) with sensitivity analysis of Key production parameters. Energies 15, p. 9020. https://doi.org/10.3390/en15239020.

Jeanne, P., Rutqvist, P., Dobson, F., et al. (2015). Geomechanical simulation of the stress tensor rotation caused by injection of cold water in a deep geothermal reservoir. JGR Solid Earth 120 (12), pp. 8422–8438. https://doi.org/10.1002/2015JB012414.

Krawczyk, C. M., Stiller, M., Bauer, K., et al. (2019). 3-D seismic exploration across the deep geothermal research platform Groß Schönebeck north of Berlin/Germany. Geothermal Energy 7 (15).

Levi, N., Weissl, M., & Decker, K. (2024). Assessing the hazard of fault triggering by deep geothermal energy production in an active fault system via a 1D stress model and 3D fault mapping. International Journal of Earth Sciences (Geol Rundsch) 113, pp. 583–609. https://doi.org/10.1007/s00531-023-02383-6.

Li, X., Jiang G., Tang, X., et al. (2023). Detecting Geothermal Anomalies Using Multi-Temporal Thermal Infrared Remote Sensing Data in the Damxung–Yangbajain Basin, Qinghai–Tibet Plateau. Remote Sensing 2023, 15, p. 4473. https://doi.org/10.3390/rs15184473.

Littlefield, E. F. & Calvin, M. (2014). Geothermal exploration using imaging spectrometer data over Fish Lake Valley, Nevada. Remote Sensing of Environment 140, pp. 509–518. https://doi.org/10.1016/j.rse.2013.09.007.

Maryadi, M., Sari, E. K., Zarkasyi, A., et al. (2024). Three-dimensional magnetotelluric and gravimetric imaging of Mount Lawu geothermal prospect area, Indonesia. Geothermics 118, p. 102917. https://doi.org/10.1016/j.geothermics.2024.102917.

Medici, G., Ling, F., & Shang, J. (2023). Review of discrete fracture network characterization for geothermal energy extraction. Frontiers in Earth Science 11, p. 1328397. https://doi.org/10.3389/feart.2023.1328397.

Mercer, J. & Faust, C. R. (1975). Simulation of water and vapor-dominated hydrothermal systems. Paper SPE5520, Presented at the 50th Annual Fall Meeting of the Society of Petroleum Engineers, Dallas, TX, USA, 16 pp.

Melouah, O., Ebong, E. D., Abdelrahman, K., et al. (2023). Lithospheric structural dynamics and geothermal modeling of the Western Arabian Shield. Scientific Reports 13, p. 11764. https://doi.org/10.1038/s41598-023-38321-4.

Moeck, I. S. (2014). Catalog of geothermal play types based on geologic controls. Renewable and Sustainable Energy Reviews 37, pp. 867–882. https://doi.org/10.1016/j.rser.2014.05.032.

Nishar, A., Richards, S., Breen, D., et al. (2016). Thermal infrared imaging of geothermal environments and by an unmanned aerial vehicle (UAV): A case study of the Wairakei – Tauhara geothermal field, Taupo, New Zealand. Renewable Energy 86, pp. 1256–1264. https://doi.org/10.1016/j.renene.2015.09.042.

O'Sullivan, M. J., Pruess, K., & Lippmann, M. J. (2001). State of the art of geothermal reservoir simulation. Geothermics 30, pp. 395–429. https://doi.org/10.1016/S0375-6505(01)00005-0.

O'Sullivan, M. J., Yeh, A., & Mannington, W. I. (2009). A history of numerical modelling of the Wairakei geothermal field. Geothermics 38 (1), pp. 155–168. https://doi.org/10.1016/j.geothermics.2008.12.001.

Patterson, J. R., Cardiff, M., & Feigl, K. L. (2020). Optimizing geothermal production in fractured rock reservoirs under uncertainty. Geothermics 88, p. 101906. https://doi.org/10.1016/j.geothermics.2020.101906.

Pollack, A., Cladouhos, T. T., Swyer, M. W., et al. (2021). Stochastic inversion of gravity, magnetic, tracer, lithology, and fault data for geologically realistic structural models: Patua Geothermal Field case study. Geothermics 95, p. 102129.

Pruess, K. (2002). Numerical simulation of `multiphase tracer transport in fractured geothermal reservoirs. Geothermics 1 (4), pp. 475–499. https://doi.org/10.1016/S0375-6505(02)00007 X.

Pruess, K. & Garcia, J. (2000). A systematic approach to local grid refinement in geothermal reservoir simulation. In: Proceedings of the 2000 World Geothermal Congress. Kyushu-Tohuku, Japan, pp. 2809–2814.

Quiroga, B. P., Rioseco, E. M., Kapinos, G., et al. (2023). Three-dimensional magnetotelluric inversion for the characterization of the Sol de Mañana high-enthalpy geothermal field, Bolivia. Geothermics 113, p. 102748. https://doi.org/10.1016/j.geothermics.2023.102748.

Saibi, H., Amrouche, M., Bâtir, J., et al. (2022). Magnetic and gravity modeling and subsurface structure of two geothermal fields in the UAE. Geothermal Energy 10, p. 28. https://doi.org/10.1186/s40517-022-00240-4.

Saibi, H., Khosravi, S., Cherkose, B. A., et al. (2021). Magnetotelluric data analysis using 2D inversion: A case study from Al-Mubazzarah geothermal area (AMGA), Al-Ain, United Arab Emirates. Heliyon 7 (6), p. e07440. https://doi.org/10.1016/j.heliyon.2021.e07440.

Shah, M., Yadav, K., & Sircar, A. (2024). Shallow and deep geothermal water sources identification in Unai geothermal field, Gujarat, India with applications of magnetotelluric (MT). Unconventional Resources 4, p. 100086. https://doi.org/10.1016/j.uncres.2024.100086.

Siler, D. L., Faulds, J. E., Hinz, N. H., et al. (2019). Three-dimensional geologic mapping to assess geothermal potential: examples from Nevada and Oregon. Geothermal Energy 7, p. 2. https://doi.org/10.1186/s40517-018-0117-0.

Tian, B., You, Z., Wang, G., et al. (2022). Building a geothermal formation model using microtremor array measurement. Geophysics 87 (1), pp. B23–B30. https://doi.org/10.1190/geo2021-0014.1.

Van Der Meer, F., Hecker, C., Van Ruitenbeek, F., et al. (2014). Geologic remote sensing for geothermal exploration: A review. International Journal of Applied Earth Observation and Geoinformation 33, pp. 255–269. https://doi.org/10.1016/j.jag.2014.05.007.

Wang, C., Winterfield, P., Johnston, B., et al. (2020). An embedded 3D fracture modeling approach for simulating fracture-dominated fluid flow and heat transfer in geothermal reservoirs. Geothermics 86, p. 101831. https://doi.org/10.1016/j.geothermics.2020.101831.

Wang, K., Hua, C., Ren, L., et al. (2022). Geochemical and isotopic characteristics of two geothermal systems at the Nanpu Sag, Northern Bohai Bay Basin. Frontiers in Earth Science 10, p. 844605. https://doi.org/10.3389/feart.2022.844605.

Weldeyohannes, T. T., Hailu, B. T., Muluneh, A. A., et al. (2022). Detection of geothermal anomalies in the Northern Lake Abaya geothermal field, Main Ethiopian Rift. Journal of Volcanology and Geothermal Research 430, p. 107638. https://doi.org/10.1016/j.jvolgeores.2022.107638.

Witherspoon, P. A. & Neuman, S. P. (1967). Evaluating a slightly permeable caprock in aquifer gas storage: 1. caprock of infinite thickness. Journal of Petroleum Technology 19, pp. 949–955. https://doi.org/10.2118/1634-PA.

Wolery, T. (1992). EQ3/6, A Software Package for Geochemical Modeling of Aqueous Systems: Package Overview and Installation Guide (Version 7.0). Report UCRL-MA-110662 PT1, Lawrence Livermore Natl. Laboratory, Livermore, California.

Yasin, Q., Gholami, A., Majdański, M., et al. (2023). Seismic characterization of geologically complex geothermal reservoirs by combining structure-oriented filtering and attributes analysis. Geothermics 112, p. 102749. https://doi.org/10.1016/j.geothermics.2023.102749.

Yalcin, B., Ezekiel, U., & Mai, P. M. (2024). Potential for CO_2 plume geothermal and CO_2 storage in an onshore Red Sea Rift basin, Al-Wajj, Saudi Arabia: 3D reservoir modeling and simulations. Geothermics 119, p. 102966. https://doi.org/10.1016/j.geothermics.2024.102966.

19

Advancements, Challenges, and Outlook of Geothermal Reservoir Operations

Ghoulem Ifrene[1], Singh Kuldeep[2], William Gosnold[1]

[1] *University of North Dakota, Grand Forks, North Dakota, USA*
[2] *Earth Science Department, Kent State University, Ohio, USA*

19.1 Introduction

Geothermal energy (GE) stands as a promising non-carbon source of renewable energy, harnessing the heat flux from the Earth's core (Figure 19.1). This reliable and abundant energy source has great potential for mitigating climate change (Fridleifsson et al., 2008). However, the development and exploitation of geothermal reservoirs comes with a range of environmental, economic, and social challenges that need to be addressed to ensure sustainable development (Soltani et al., 2021).

Enhanced Geothermal Systems (EGSs), also known as human-made geothermal energy (Figure 19.2), are advanced geothermal technologies designed to harness the Earth's heat for electricity production (Louie and Solomon, 2017). Unlike naturally occurring hydrothermal systems, which require heat, fluid, and natural permeability, EGS can be implemented in regions where underground rock is hot but lacks sufficient permeability or fluids (Olasolo et al., 2016).

In an EGS, fluid is injected deep into the Earth's crust under controlled conditions to create new fractures and reopen existing ones, thereby enhancing permeability. This allows the injected fluid to circulate through the hot rock, absorbing heat as it moves. The heated fluid is then pumped to the surface, where it is used to generate electricity. EGS has the potential to significantly expand geothermal energy production beyond traditional hydrothermal regions, offering a renewable energy solution capable of powering millions of homes and businesses.

EGS face significant challenges in data collection due to the nature of the geothermal environment and the technological limitations associated with it. The geological conditions where geothermal resources are found often present harsh environments that are challenging for data acquisition. For instance, drilling in deep geothermal reservoirs requires penetrating hard rock formations, which generates intense vibrations that can damage sensitive equipment. Additionally, the high temperatures (often exceeding 250°C) typical of geothermal reservoirs can exceed the operational limits of many instruments, leading to potential failures and data loss (Weers and Huggins, 2019).

Geophysical Exploration for Hydrocarbon Reservoirs, Geothermal Energy, and Carbon Storage: New Technologies and AI-based Approaches, First Edition. Edited by Said Gaci.

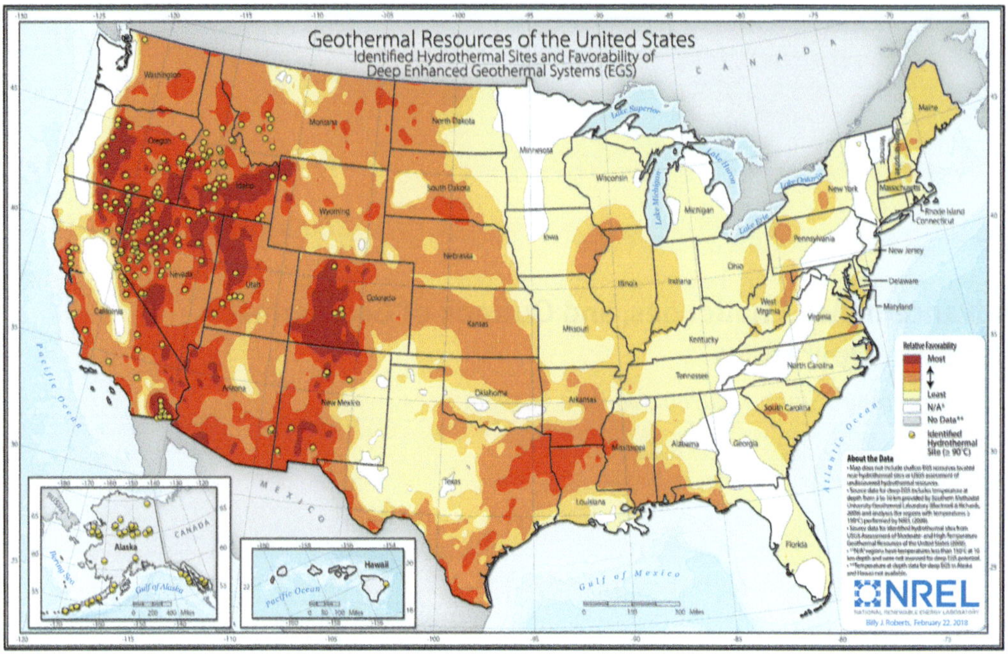

Figure 19.1 Geothermal Resources of the USA (Billy, 2018 / U.S. Department of Energy / Public Domain).

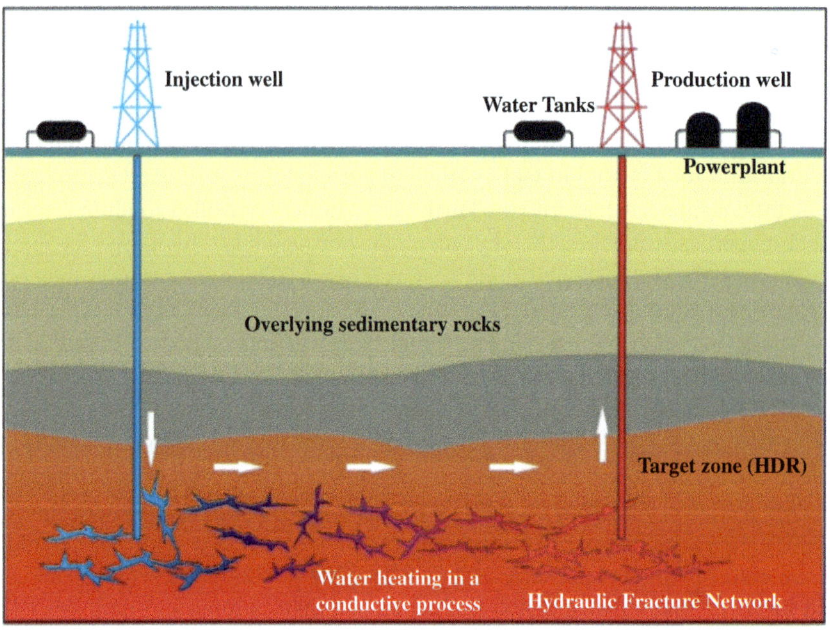

Figure 19.2 Schematic of an enhanced geothermal system (Yasin et al., 2022 / MDPI / CC BY 4.0).

Modern geothermal projects generate vast amounts of data due to advancements in sensor technology and data acquisition methods. For example, high-resolution sensors and Distributed Temperature Sensing (DTS) systems can produce detailed measurements at very small intervals. However, managing and processing this "big data" is challenging. The data sets can be too large and complex to be effectively handled using traditional storage and processing techniques (Anderson and Ussher, 1992; Weers and Huggins, 2019).

Effective geothermal exploration and development require the integration of various types of data, including geological, geochemical, geophysical, and reservoir engineering data. These data types often come in different formats and from different disciplines, making integration a complex task. Moreover, data collected during different project phases can become siloed within individual departments or teams, leading to inefficiencies and duplicated efforts. A robust data management system that facilitates data integration and accessibility is crucial but often lacking in geothermal projects (Benegar and Kennedy, 2020; Hanton et al., 2019; Weers and Huggins, 2019).

Another major challenge is the accessibility and sharing of existing data. Many valuable datasets from past geothermal projects are stored in inaccessible formats, such as physical reports, or are kept private due to proprietary concerns. This lack of access hinders the ability of current projects to build on previous work, leading to repeated mistakes and missed opportunities for innovation. Initiatives like the DOE's Geothermal Data Repository (GDR) aim to mitigate this issue by making data from funded projects publicly available, although there can be delays due to proprietary data protection periods (Weers et al., 2022).

Storing the large volumes of data generated by geothermal projects is also problematic. High data generation rates can quickly exceed the storage capacities of conventional hard drives, and certain file systems can encounter limits on the number of files or directories they can handle. These storage limitations can cause unanticipated downtime and data loss if not properly managed.

One of the primary environmental challenges in geothermal reservoir development is fluid management (Mott et al., 2022). Geothermal systems often involve the extraction of hot water and steam from deep within the Earth, which must be carefully managed to prevent environmental degradation. Direct environmental impacts include land use, atmospheric emissions, water consumption, solid waste production, and potential threats to biodiversity (Arnórsson, 2004; Bošnjaković et al., 2019; Dhar et al., 2020; Paulillo et al., 2019). For instance, land subsidence and seismic activities can occur due to the extraction and reinjection of geothermal fluids, posing significant risks to the surrounding environment and infrastructure (Buijze et al., 2019).

Geothermal projects are capital-intensive, requiring substantial upfront investment for drilling and infrastructure development. The high costs associated with locating and exploiting geothermal resources at different depths, coupled with the variability in resource quality, present significant economic challenges. Financial feasibility must be established through detailed studies and analysis to ensure that the benefits outweigh the costs. Additionally, geothermal projects often face financial risks related to the uncertainty of resource availability and productivity, which can affect the overall economic viability.

Social acceptance is crucial for the successful deployment of geothermal projects. Local communities may oppose geothermal developments due to concerns over environmental impacts, noise pollution, and the disruption of local ecosystems (Dowd et al., 2011). Ensuring community involvement and addressing their concerns through transparent and inclusive processes is essential. Social acceptance is also influenced by the perceived benefits of the project, such as job creation, economic development, and the provision of local energy sources (Kurek et al., 2021).

To overcome these challenges, it is necessary to implement a comprehensive framework that addresses technical, economic, environmental, and social aspects. This includes:

Technical Innovations: Developing advanced technologies for efficient resource extraction and management.

Economic Incentives: Providing financial incentives and support mechanisms to reduce the upfront costs and economic risks associated with geothermal projects.

Environmental Governance: Establishing robust environmental regulations and monitoring systems to mitigate the negative impacts of geothermal activities.

Community Engagement: Promoting active participation and collaboration with local communities to ensure social acceptance and support for geothermal projects.

By addressing these challenges through coordinated efforts, geothermal energy can play a significant role in achieving global renewable energy targets and reducing greenhouse gas emissions. The sustainable development of geothermal reservoirs requires a multidisciplinary approach that balances technical feasibility, economic viability, environmental protection, and social responsibility.

19.2 Geomechanical Considerations of Geothermal Reservoirs

19.2.1 Geothermal Reservoir Characterization

Geothermal reservoir characterization is a necessary process that integrates geological, geophysical, and wellbore data to understand the subsurface environment. This comprehensive approach is essential for successful and sustainable geothermal energy development, optimizing production, reducing exploration risks, and ensuring sustainable resource development. Characterizing naturally fractured reservoirs is complex due to their heterogeneous nature, requiring a multi-method approach to enhance accuracy and reliability (Aydin and Temizel, 2022; Bredesen et al., 2019; Oh et al., 2023; Wei et al., 2021).

Key parameters like porosity and permeability are used in reservoir simulations to evaluate heat extraction, storage, and production lifetime. Characterization provides a detailed understanding of the subsurface setting, rock properties, and fluid dynamics, crucial for optimizing exploration and production strategies. It helps identify favorable areas for geothermal energy extraction, reduces exploration risks, and optimizes well placement. Reservoir modeling predicts fluid flow paths and identifies optimal well locations, leading to efficient heat extraction and cost-effective operations.

19.2.1.1 Geological Characterization

Geological characterization is the foundational step in evaluating geothermal resource potential. This involves analyzing regional geological features and historical data to identify areas with favorable conditions for geothermal energy development (Aydin and Temizel, 2022). This characterization involves:

Regional Geology: Analyzing regional geological maps and well logs helps identify potential geothermal systems, focusing on areas with favorable heat sources, such as young volcanic rocks and sedimentary basins with high heat flow (Aydin and Temizel, 2022). This regional analysis is crucial for understanding the broader geological context and guiding subsequent exploration efforts.

Geothermal Surface Features: Surface expressions of geothermal activity, such as hot springs, fumaroles, and geysers, provide valuable clues about the location and potential temperature of the geothermal reservoir (Oh et al., 2023). Investigating these features can guide the placement of exploration wells and help prioritize areas for further study.

Stratigraphic Maps: Mapping geological layers helps identify favorable zones for geothermal wells. Stratigraphic data guide well placement and target areas for drilling, ensuring efficient resource exploitation (Aydin and Temizel, 2022). This detailed geological mapping is essential for understanding the vertical distribution of reservoir properties and planning effective drilling strategies.

Outcrop Analysis: Studying exposed rock formations provides insights into the geological history and characteristics of the reservoir rocks. While not directly representative of subsurface conditions, outcrop analysis offers a tangible understanding of potential reservoir properties and structures (Aydin and Temizel, 2022). This field-based approach complements subsurface data and enhances the overall geological model.

19.2.1.2 Geophysical Techniques

Geophysical surveys are essential tools for examining the subsurface, providing crucial information about the hidden structure and properties of a geothermal reservoir. These non-invasive methods offer invaluable insights that complement geological data, creating a comprehensive picture of the reservoir's characteristics. Common methods include:

Seismic Reflection Surveys: Seismic reflection surveys utilize sound waves to map subsurface rock formations and identify potential reservoir boundaries, fractures, and faults. This technique is particularly useful in delineating large-scale geological features and optimizing well placement (Bredesen et al., 2019).

Seismic Tomography: Seismic tomography creates a 3D image of the subsurface, providing detailed information about the distribution of rock types and potential fluid pathways. This advanced technique enhances the resolution of subsurface imaging and improves the accuracy of reservoir models.

Gravity Surveys: Gravity surveys measure variations in Earth's gravitational field caused by differences in subsurface rock densities, potentially revealing the location of fault zones and fluid-filled reservoirs. This method is useful for identifying large-scale density anomalies associated with geothermal systems.

Electrical Resistivity: Electrical geophysical surveys offer valuable tools for comprehensive geothermal reservoir characterization (Garg et al., 2007). These techniques provide insights into the subsurface electrical resistivity structure, which can be correlated with crucial reservoir properties such as the presence of high-temperature fluids and permeable zones favorable for geothermal energy extraction (Garg et al., 2007).

Magnetotelluric (MT) Surveys: MT surveys measure electrical conductivity variations in the subsurface, which can indicate the presence of hot fluids and geothermal systems. This technique is particularly effective in identifying deep-seated geothermal resources and mapping conductive anomalies associated with geothermal activity (Newman et al., 2008; Patro, 2017).

19.2.1.3 Wellbore Data Acquisition

Drilling and well completion activities offer a unique opportunity to gather crucial data directly from the geothermal reservoir. The information obtained during this stage serves as a vital link between the initial geological and geophysical characterization and the development of a comprehensive reservoir model.

Lithology Logs: Analyzing wellbore cuttings and core samples provides information about rock types, porosity, and permeability within the reservoir. This direct sampling is essential for validating geological models and refining reservoir properties.

Geochemical Analysis: Studying the chemical composition of formation fluids and gases can reveal information about reservoir temperature, pressure, and fluid flow dynamics (Oh et al., 2023). Geochemical data provide insights into the thermodynamic conditions within the reservoir and help predict production performance.

Well Logging: Downhole logging tools measure various physical properties of the formation surrounding the wellbore, including porosity, permeability, and temperature variations. This data is crucial for refining the reservoir model and understanding the distribution of these properties.

19.2.1.4 Reservoir Engineering Analysis

Combining geological, geophysical, and wellbore data allows for reservoir engineering analysis. This integration forms the basis for developing a comprehensive understanding of the reservoir's physical properties, fluid dynamics, and overall geothermal potential (Aydin and Temizel, 2022). This involves:

Production Monitoring: Continuous monitoring of production data, including pressure, temperature, and non-condensable gases, helps evaluate reservoir performance over time. This long-term data is crucial for assessing changes in reservoir behavior and optimizing production strategies.

Tracer Tests: By injecting a tracer and monitoring its arrival at production wells, connectivity between wells can be assessed, and crucial data like flow velocity and reservoir volume can be gathered. This information is vital for optimizing injection and production strategies.

Reservoir Modeling: Creating computer models of the geothermal reservoir to simulate fluid flow and heat transfer allows for predicting production performance under different operating conditions. Reservoir models integrate various data sources and provide a framework for scenario analysis and decision-making.

Pressure Transient Tests (PTA): PTA involves analyzing pressure changes over time to determine reservoir properties such as permeability, porosity, boundaries, and skin factor. These tests are particularly useful in understanding reservoir behavior during production and injection (Wei et al., 2021).

Temperature Transient Analysis (TTA): TTA leverages the temperature responses to changes in flow rate and pressure, providing valuable information about the reservoir. This method complements PTA and helps characterize the damaged zone near the wellbore, determine reservoir types, and address non-uniqueness issues (Kocabas, 2005).

19.2.1.5 Limitations and Mitigating Strategies for Geothermal Reservoir Characterization

Geothermal reservoir characterization techniques have inherent limitations that affect data interpretation. To overcome these limitations, a multi-method approach is crucial.

Geophysical techniques like seismic reflection surveys have resolution limitations, making it challenging to detect smaller-scale features (Bredesen et al., 2019). Geological characterization through outcrop analysis may not accurately represent subsurface conditions due to geological processes (Aydin and Temizel, 2022). Wellbore data acquisition has limited sampling and requires integration with other methods.

A multi-method approach integrates diverse datasets from geological studies, geophysical surveys, wellbore information, and downhole logging data. This integrated workflow enhances accuracy and combines data from various techniques, such as PTA and TTA (Wei et al., 2021), seismic data, well logs, and regional geological data (Bredesen et al., 2019), and electrical resistivity surveys (Garg et al., 2007). This integration provides a comprehensive understanding of the geothermal reservoir, facilitating informed development decisions.

19.2.2 Rock Mechanics Issues in Geothermal Development

Geothermal energy, derived from the Earth's subsurface heat, presents a significant renewable energy source. With geothermal resources being exploited in 32 countries, the global installed capacity stands at 16,318 MW (Gutiérrez-Negrín, 2024). The USA alone generates nearly 3900 MWe from geothermal energy in 2023, with an estimated potential exceeding 100,000 MWe for 30 years at depths of up to 3 km. Additionally, geothermal heat is extensively used for direct applications such as heating buildings (Soltani et al., 2019), agricultural (Van Nguyen et al., 2015), and oil recovery from tar sands (Settari et al., 1992).

Most commercial geothermal power generation utilizes hydrothermal systems, which have sufficient permeability and fluid content to allow hot water and steam extraction through porous structures. However, these hydrothermal resources represent only a small portion of available geothermal energy. The majority of geothermal resources consist of hot dry rock with low permeability and water content. EGSs are designed to exploit these resources by creating engineered fracture networks through which water can circulate, absorbing heat from the rock.

Understanding the in-situ stress state is crucial for geothermal reservoir development. It impacts well planning, drilling safety, reservoir stimulation, and permeability evolution (Sun et al., 2021). Various methods such as mini-frac tests, borehole breakouts, and hydraulic testing of pre-existing fractures are employed to determine the stress state. However, high temperatures and the presence of heterogeneous rock properties complicate these measurements in geothermal settings.

Reservoir stimulation in geothermal systems involves initiating and propagating fractures to enhance permeability (Sun et al., 2021). Unlike conventional hydraulic fracturing in petroleum reservoirs, geothermal systems require a nuanced approach due to the high temperatures and different rock properties. Numerical models and laboratory experiments help understand fracture behavior and optimize stimulation techniques, balancing fracture propagation and interaction with natural fractures.

Natural fractures significantly influence the success of EGS by affecting fluid flow, heat extraction, and induced seismicity. Shear slips on natural fractures, caused by pore pressure and thermal stress changes, can enhance fracture permeability through dilation. Laboratory studies and field observations provide empirical models for understanding joint closure, shear deformation, and the impact of natural fractures on hydraulic fracture propagation.

Geothermal reservoirs are affected by coupled hydro-thermo-chemo-mechanical processes that alter fracture permeability and reservoir dynamics (Rühaak et al., 2017). For instance, temperature changes due to fluid injection/extraction induce thermal stresses, which can lead to fracture propagation or closure. Chemical reactions between the rock and fluid, such as mineral dissolution and precipitation, further modify fracture aperture and permeability (Figure 19.3). Numerical models incorporating these processes are essential for predicting reservoir behavior and optimizing geothermal energy extraction.

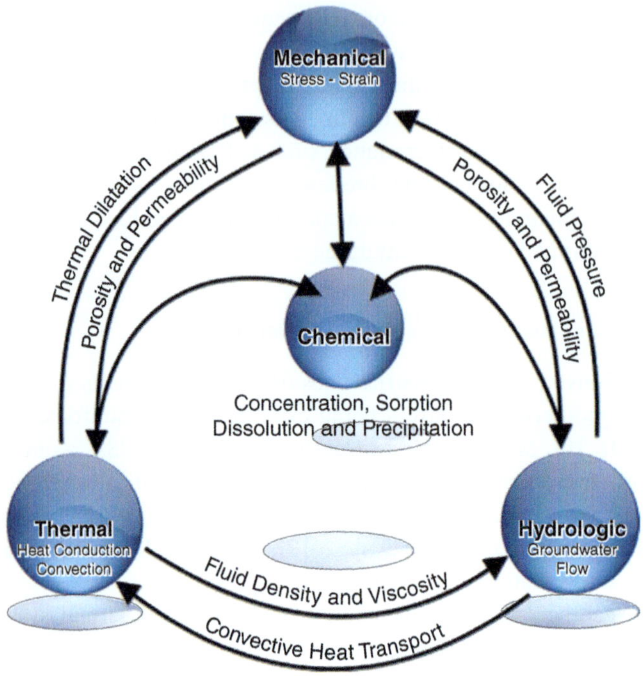

Figure 19.3 Thermo-hydro-mechanical coupling process (Rühaak et al., 2017 / with permission of Elsevier).

Geothermal reservoir stimulation often induces micro-seismic events, which provide valuable information about reservoir dynamics. Monitoring and analyzing these events help estimate the stimulated volume, fracture growth, and in-situ stress changes. However, the mechanisms of induced seismicity and its relationship with fluid flow and pressure changes remain complex and require further investigation.

Advancing geothermal energy development necessitates a thorough understanding of rock mechanics and geomechanical processes. Research in in-situ stress characterization, reservoir stimulation, natural fracture behavior, coupled processes, and induced seismicity is critical for optimizing geothermal reservoir design and ensuring sustainable energy extraction. By integrating experimental, numerical, and theoretical studies, the geothermal industry can overcome current challenges and harness the full potential of geothermal resources.

19.2.3 Fracture Characterization and Its Role in Heat Extraction

Fracture characterization is essential for understanding and enhancing the performance of EGSs (Jia et al., 2022). The efficiency of geothermal reservoirs in extracting heat largely depends on the nature, distribution, and properties of fractures within the rock (Lu, 2018). This section synthesizes insights from multiple studies to elucidate the role of fracture characterization in heat extraction from geothermal systems.

Fractures serve as primary pathways for fluid flow and heat transfer in geothermal reservoirs. The geometry, orientation, length, and connectivity of these fractures critically influence heat extraction efficiency (Xiong et al., 2022). Discrete Fracture Network (DFN) models are often employed to represent the complex nature of fracture networks. Numerical simulations indicate that well-connected fracture networks oriented optimally with respect to fluid flow can significantly enhance heat transfer efficiency (Jia et al., 2022; Lu, 2018; Xiong et al., 2022).

The aperture, or opening width of fractures, directly impacts the flow rate of the working fluid and the surface area available for heat exchange. Variations in fracture aperture can lead to substantial differences in heat extraction performance. Rougher fractures tend to enhance heat transfer due to increased turbulence and surface area for heat exchange (Chen et al., 2018).

The interaction between thermal, hydraulic, and mechanical processes in a geothermal reservoir is complex and highly interdependent (Figure 19.4). THM coupling models consider these interactions to provide more accurate predictions of reservoir behavior. For example, as cold fluid is injected into a hot reservoir, thermal contraction of the rock can alter fracture apertures, affecting both permeability and heat transfer efficiency. Studies have demonstrated that incorporating THM effects into models leads to better predictions of reservoir performance over time.

Fracture permeability is another crucial parameter influencing fluid movement through the reservoir. Higher permeability fractures allow for greater fluid flow, enhancing heat extraction. However, excessively high permeability can lead to rapid cooling and reduced thermal efficiency (Caulk et al., 2016). Porosity, primarily a matrix property, interacts with fractures to influence overall fluid storage and flow characteristics of the reservoir.

Both experimental and numerical approaches have been employed to study the heat extraction characteristics of fractured geothermal reservoirs. Experimental studies often involve creating artificial fractures in rock samples and subjecting them to heat flow experiments to measure the effects of different fracture properties (Caulk et al., 2016; Kumari et al., 2018). Numerical simulations complement these experiments by allowing the exploration of various scenarios and parameters that are difficult to replicate in a laboratory setting (Pandey et al., 2017; Shi et al., 2019; Xiong et al., 2022).

Effective characterization of fractures is essential for optimizing the heat extraction performance of geothermal reservoirs. By understanding the geometry, aperture, roughness, and THM interactions of fractures, engineers can design better stimulation and management strategies for EGS. Future research should continue to integrate experimental data with advanced numerical models to enhance the predictive capabilities and operational efficiency of geothermal energy extraction systems.

19.2.4 Managing Induced Seismicity in Geothermal Operations

Induced seismicity refers to the occurrence of earthquakes triggered by human activities such as fluid injection or extraction (Table 19.1), which can alter the stress and pressure conditions in the Earth's crust. In geothermal operations, managing this phenomenon involves understanding the mechanisms behind induced seismicity, implementing monitoring systems, and applying mitigation strategies to minimize risks.

Induced seismicity in geothermal operations often results from fluid injection, which increases pore pressure and decreases the effective stress on pre-existing faults or fractures, potentially triggering slip events. For instance, the Basel EGS project in Switzerland experienced significant induced seismicity due to fluid injection at high pressures, leading to a series of seismic events that raised public concerns and ultimately halted the project (Gaucher et al., 2015). Similarly, in the Cooper Basin, Australia, over 27 000 microseismic events were detected during the hydraulic stimulation of a geothermal reservoir, highlighting the direct relationship between fluid injection and seismic activity (Baisch et al., 2006).

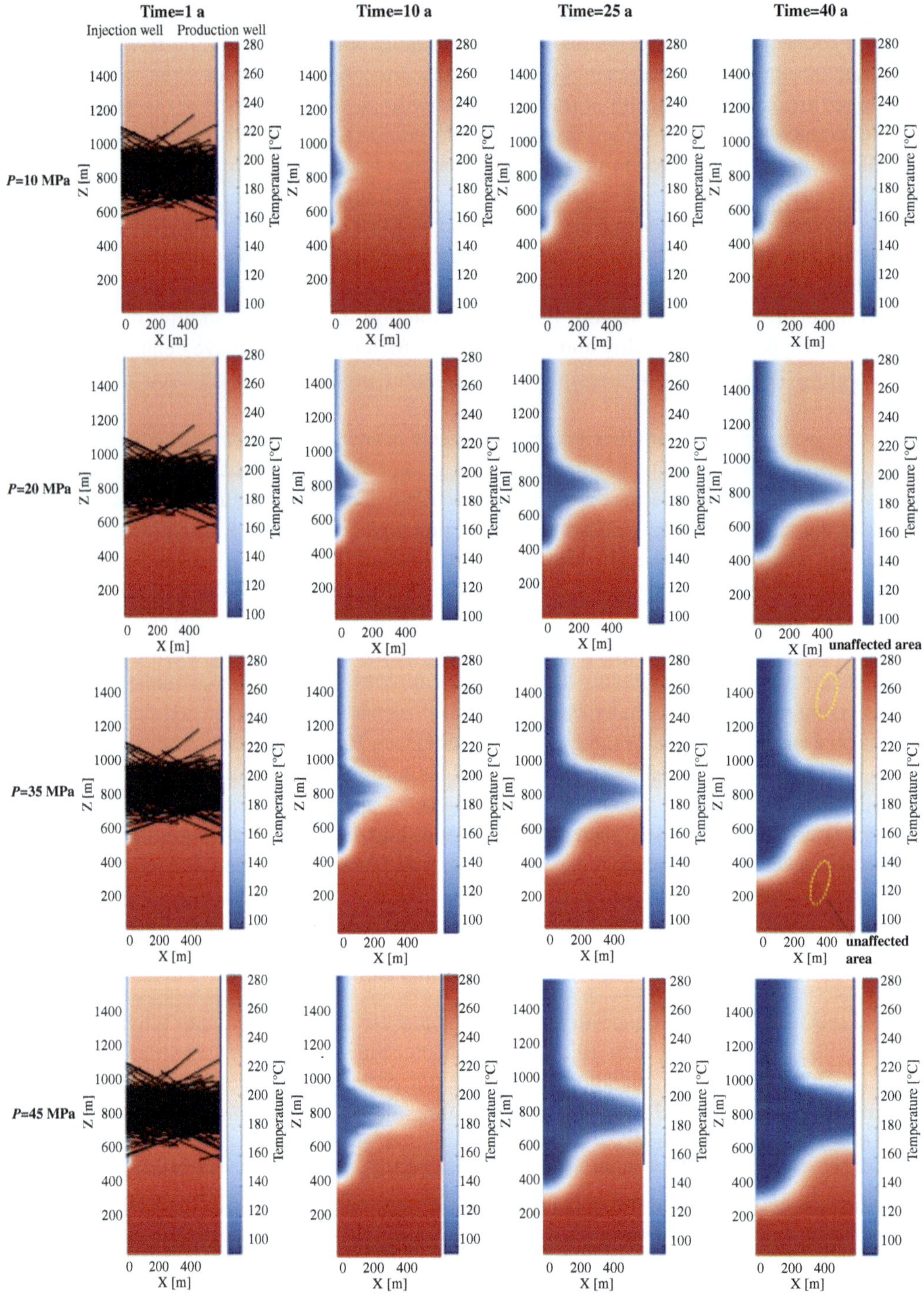

Figure 19.4 Temperature variations in the Habanero geothermal reservoir for different injection pressures over time (Xiong et al., 2022b / with permission of Elsevier).

Table 19.1 Summary of induced seismicity magnitudes at major geothermal sites.

Location	Event Date	Magnitude	References
Paralana, Australia	July 2011	Between 2.4 and 2.5	Albaric et al. (2014)
Soultz-sous-Forêts, France	Between 2000 and 2005	Between 2.5 and 2.9	Dorbath et al. (2009); Evans et al. (2012)
Basel, Switzerland	2006	Between 0.7 and 3.4	Deichmann et al. (2014)
Berlin, El Salvador	2003	3.6	Kwiatek et al. (2014)
Cooper Basin, Australia	2003	3.7	Baisch et al. (2006)

Effective management of induced seismicity relies heavily on robust monitoring systems that can detect and analyze seismic events in real-time. Seismic monitoring networks, comprising geophones and accelerometers, are deployed around geothermal sites to record seismic activity and provide valuable data on the location, magnitude, and frequency of induced earthquakes. These networks help identify the development of seismicity patterns and potential risk zones.

In the Basel EGS project, a dense seismic monitoring network was established to track induced seismicity, enabling the identification of seismic hazard zones and the development of response strategies (Gaucher et al., 2015). Similarly, the monitoring system at the Cooper Basin geothermal field utilized a local eight-station seismic network to map the hypocenter locations of induced events, aiding in the precise characterization of the stimulated reservoir (Baisch et al., 2006).

Mitigating the risks associated with induced seismicity involves several strategies, including adjusting injection rates, managing fluid volumes, and employing real-time traffic light systems (TLS) to regulate operations based on seismic activity levels.

Injection Rate Adjustments: Reducing the rate of fluid injection can lower the likelihood of triggering large seismic events. Gradual and controlled injection helps manage pore pressure increases and reduce the stress perturbation on faults and fractures.

Fluid Volume Management: Limiting the total volume of injected fluids can mitigate the extent of pressure buildup and associated seismic risks. By carefully planning the injection schedule and volumes, operators can minimize the impact on the subsurface stress regime.

Traffic Light Systems (TLSs): TLSs are operational protocols that prescribe specific actions based on the detected levels of seismic activity. These systems classify seismicity into different levels (green, yellow, red) and define corresponding operational responses, such as continuing operations, reducing injection rates, or halting activities. The Basel project implemented a TLS to manage seismic risks, with predefined thresholds triggering operational adjustments (Gaucher et al., 2015).

The Basel EGS project serves as a valuable case study for managing induced seismicity. The project faced several challenges due to significant induced seismicity, leading to the implementation of comprehensive monitoring and mitigation measures. Despite these efforts, a magnitude 3.4 event caused widespread concern, resulting in the project's suspension and highlighting the importance of proactive risk management and community engagement.

Similarly, the Cooper Basin geothermal project demonstrated the importance of detailed seismic monitoring and the need for continuous assessment of injection-induced seismicity. The project's success in accurately predicting the intersection of the stimulated fracture zone with a subsequent well underscores the value of high-resolution seismic data in guiding geothermal operations (Baisch et al., 2006).

Managing induced seismicity in geothermal operations requires a multifaceted approach involving thorough understanding of seismic mechanisms, deployment of advanced monitoring systems, and implementation of effective mitigation strategies. By learning from past experiences and continuously improving risk management practices, the geothermal industry can enhance the safety and viability of geothermal energy projects.

19.3 Drilling and Well Completion Technologies

19.3.1 Innovations in Drilling: Embracing Fishbone Techniques for Enhanced Geothermal Systems

Fishbone drilling is a technique used in the oil and gas industry to enhance the extraction of hydrocarbons from reservoirs (Magomadov et al., 2022). It involves creating multiple small lateral branches (Figure 19.5), or "fishbones," from the main wellbore. These branches extend out into the reservoir, increasing the surface area of the well and improving the drainage and flow of hydrocarbons (Atsalos, 2022).

Fishbone drilling technology represents a significant advancement in the field of geothermal energy extraction, particularly within EGSs. This innovative approach, characterized by its multi-branch well architecture, enhances the efficiency and productivity of geothermal reservoirs (Aihar et al., 2023). The Guanzhong Basin case study provides a comprehensive analysis of fishbone drilling's application (Figure 19.6), highlighting its potential to revolutionize geothermal energy exploitation (P. Wang et al., 2024). The study compared the recovery power of fishbone with non-fishbone wells, fishbone branch wells exhibit a 26.6% and 32.4% increase in heat recovery power during 20-year and 100-year operating periods, respectively, and a 51.1% increase during the 100-year period. However, fishbone branch wells experience a faster temperature drop.

Traditional geothermal energy extraction methods have relied predominantly on vertical wells, which often result in low productivity and suboptimal utilization of geothermal resources. As the demand for renewable energy sources increases and the pressure to reduce carbon emissions intensifies, there is a pressing need to develop more efficient and effective extraction techniques. Fishbone drilling, with its lateral horizontal branches, presents a promising solution by significantly enhancing heat recovery and fluid recharge rates.

19.3.2 Comparing Fishbone and Conventional Drilling Methods

Fishbone drilling represents an advanced technique in geothermal energy extraction, characterized by multiple lateral branches extending from the main wellbore. This configuration enhances the contact area between the wellbore and the geothermal reservoir, significantly improving the efficiency of heat extraction and fluid circulation (Wang et al., 2024). Studies have shown that fishbone drilling increases heat recovery power by about 60% compared to traditional vertical wells. The larger surface area for heat exchange provided by the lateral branches contributes to this improvement, making fishbone drilling a more efficient method for geothermal energy extraction.

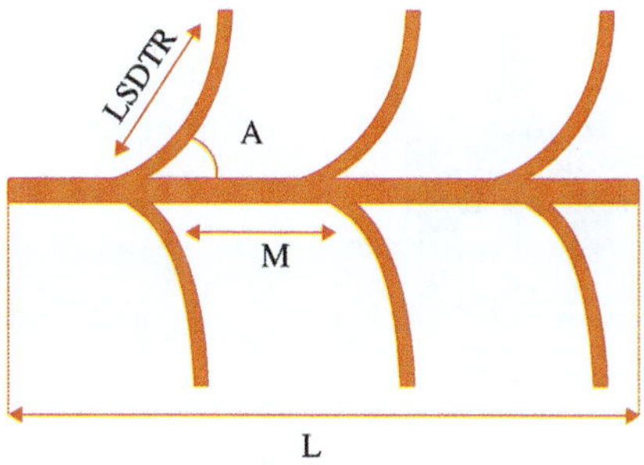

Figure 19.5 Schematic illustration of fishbone well (Manshad et al., 2019 / Springer Nature / CC BY 4.0).

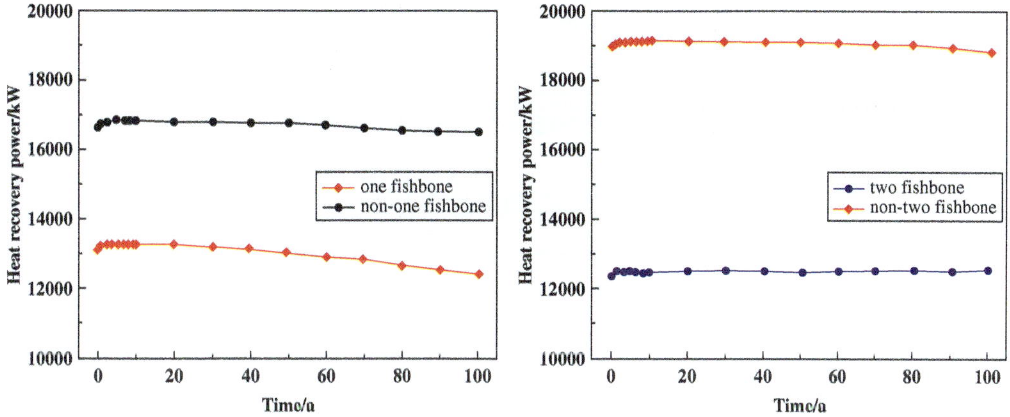

Figure 19.6 Heat extraction power simulation results of fishbone and non-fishbone wells with different structures Wang et al., 2024 / with permission of Elsevier.

The productivity of fishbone wells is substantially higher than that of conventional methods. For instance, a three-branch fishbone structure can deliver heat recovery power that is 1.5 times greater than a single-branch structure (P. Wang et al., 2024). This higher productivity is particularly beneficial in geothermal reservoirs with high thermal gradients. Additionally, the multi-branch configuration of fishbone drilling allows for more effective thermal diffusion within the reservoir. This setup ensures a more uniform spread of the cold front, preventing early thermal breakthroughs and maintaining higher operational temperatures for extended periods.

Fishbone drilling also offers flexibility in design, allowing for various configurations such as different branch angles, lengths, and depths. This adaptability makes it suitable for a wide range of geothermal reservoirs with varying geological characteristics. The ability to customize the well design to the specific conditions of the reservoir enhances the overall efficiency and effectiveness of geothermal energy extraction using fishbone drilling.

In contrast, conventional drilling methods, which typically involve vertical wells, have been the standard approach in geothermal energy extraction. However, these methods come with certain limitations when compared to fishbone drilling. Vertical wells generally provide lower heat recovery rates due to their limited contact area with the geothermal reservoir. This results in less efficient utilization of the geothermal resource, making conventional methods less effective in maximizing heat extraction.

Figure 19.7 Off-shore fishbone drilling (Oil and Gas Stimulation | Fishbones).

Conventional wells are also more prone to early thermal breakthroughs, where the cold front from the re-injected fluid reaches the production well sooner than desired. This issue reduces the efficiency and lifespan of the geothermal well. Additionally, the design of vertical wells is less flexible compared to fishbone drilling (Figure 19.7). This rigidity can be a disadvantage in heterogeneous geothermal reservoirs where adaptive well designs are necessary to optimize heat recovery. Operational challenges such as managing the balance between recharge and production and maintaining stable temperatures in the production well over time further complicate the use of conventional drilling methods.

Fishbone drilling represents a significant advancement over conventional drilling methods in the context of EGSs. The multi-branch design not only enhances heat recovery and productivity but also provides greater flexibility and scalability in well design. While conventional methods remain useful and widely applied, adopting fishbone drilling can lead to more efficient and sustainable geothermal energy extraction, addressing some of the critical limitations of traditional approaches. The choice between these methods should consider the specific geological and thermal characteristics of the geothermal reservoir, as well as the long-term operational goals of the geothermal project.

19.4 Production and Injection Optimization

19.4.1 Optimizing Geothermal Operations with Fishbone Drilling

Fishbone drilling has emerged as a highly effective method for optimizing geothermal operations, particularly within EGSs (Figure 19.8). This technique significantly enhances the contact area between the wellbore and the geothermal reservoir. The enhanced contact area results in improved efficiency of heat extraction and fluid circulation, making fishbone drilling a promising solution for maximizing the productivity of geothermal wells.

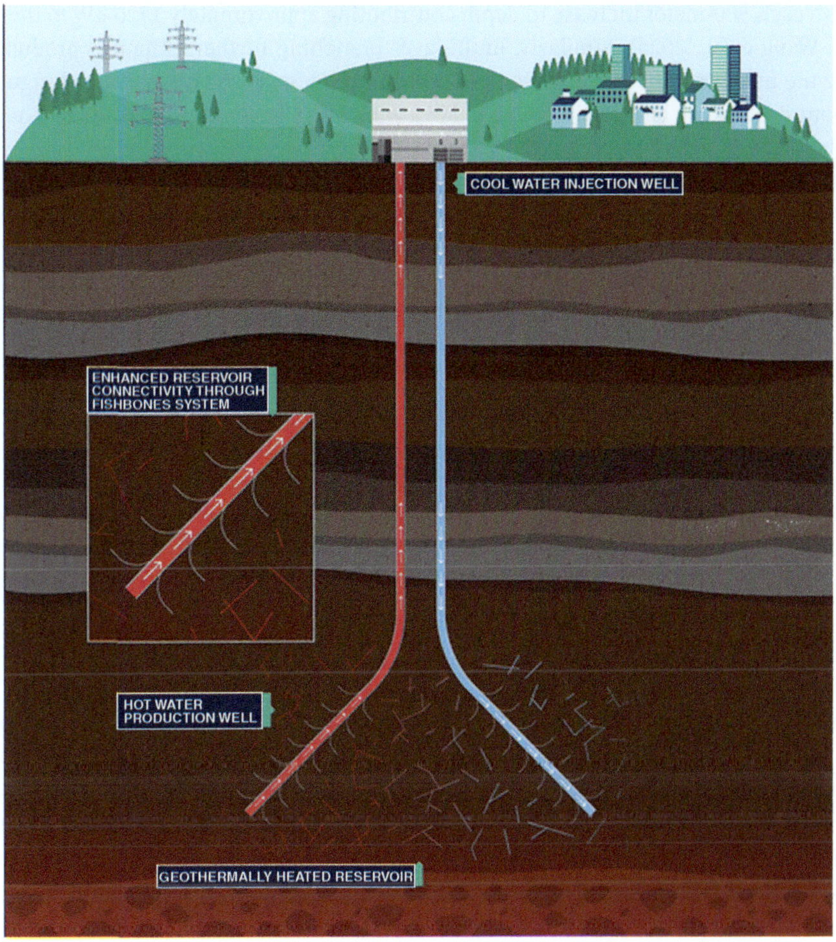

Figure 19.8 Fishbone wells fore geothermal extraction (Geothermal Stimulation | Fishbones).

One of the key advantages of fishbone drilling is its ability to enhance heat recovery power. Studies have shown that fishbone drilling can increase heat recovery power by approximately 60% compared to conventional vertical wells. This significant improvement is primarily due to the increased surface area for heat exchange provided by the multiple lateral branches. These branches allow for more effective thermal diffusion within the reservoir, preventing early thermal breakthroughs and maintaining higher operational temperatures for extended periods.

The flexibility and adaptability of fishbone drilling also contribute to its effectiveness in optimizing geothermal operations. Various parameters, such as branch number, direction, length, and angle, can be optimized to suit the specific geological characteristics of a geothermal reservoir. For instance, increasing the number of branches or adjusting their angles can significantly impact the overall heat recovery power and the thermal diffusion within the reservoir. Numerical simulations have shown that optimizing these parameters can lead to a substantial increase in heat recovery power and efficiency.

Moreover, fishbone drilling offers scalability in well design, which is crucial for adapting to different reservoir conditions. For example, deeper branch wells generally yield higher recovery power, with each 500-meter increase in depth contributing approximately 1720 kW to the thermal output (P. Wang et al., 2024). Similarly, multi-layer branching further enhances productivity by increasing the number of heat exchange points within the reservoir. This scalable design ensures that fishbone drilling can be effectively applied in a wide range of geothermal reservoirs, regardless of their depth or geological complexity.

The optimization of fishbone drilling also involves careful consideration of the spacing between the branches and the main wellbore. Closer well spacing can lead to rapid thermal breakthroughs, while larger spacing allows for more uniform temperature distribution and sustained productivity over extended periods (P. Wang et al., 2024). Adjusting the spacing between the recharge and production wells can significantly impact the system's thermal efficiency and overall productivity. This adaptability in well design is a significant advantage of fishbone drilling over conventional methods.

Fishbone drilling represents a significant advancement in optimizing geothermal operations. Its multi-branch architecture enhances heat recovery, increases productivity, and offers flexibility in well design. By optimizing various parameters, such as branch number, depth, and spacing, fishbone drilling can be tailored to suit the specific conditions of a geothermal reservoir, leading to more efficient and sustainable energy extraction. This method addresses many of the limitations of conventional drilling techniques, making it a valuable tool for maximizing the potential of geothermal resources.

19.4.2 Advanced Modeling and Simulation Techniques

Optimizing geothermal reservoirs is crucial for efficient energy production, and advanced modeling and simulation techniques play a vital role in this process. These techniques involve creating computational models that simulate the behavior of geothermal reservoirs under various conditions. By using these models, researchers and engineers can evaluate different operational scenarios, understand the complex interplay of physical processes, and optimize reservoir management strategies (Gan and Elsworth, 2016; K. Wang et al., 2022; Yin et al., 2024).

19.4.2.1 Advanced Modeling Considerations

A key aspect of advanced modeling is the representation of the fractured nature of geothermal reservoirs. There are two primary approaches to modeling fractures: the DFN model and the equivalent continuum model.

Discrete Fracture Network Models: These models explicitly represent the network of fractures within the reservoir, allowing for a more accurate representation of fluid flow pathways, particularly in naturally fractured reservoirs. DFN models incorporate fracture characteristics like size, orientation, and aperture to simulate fluid flow through these preferential pathways. This enhanced representation leads to more realistic predictions of wellbore performance and overall reservoir behavior (Gan and Elsworth, 2016; K. Wang et al., 2022; Yin et al., 2024).

Equivalent Continuum Models: Conversely, the equivalent continuum model represents the fractured rock mass as a continuum, which simplifies the model and reduces computational cost while capturing the overall behavior of the reservoir (Gan and Elsworth, 2016; K. Wang et al., 2022; Yin et al., 2024).

These models are often used in conjunction with Coupled Thermal-Hydrological-Mechanical (THM) Models, which simulate the interactions between thermal, hydrological, and mechanical processes in the reservoir. THM models account for changes in physical properties and induced effects like stress, deformation, and permeability changes (Pandey et al., 2017; Salimzadeh et al., 2018). Numerical simulations, such as finite element method (FEM) software like COMSOL Multiphysics, solve the partial differential equations governing these processes (Asai et al., 2019; Song et al., 2021; Yin et al., 2024).

Combining DFN, equivalent continuum, and coupled thermal-hydrological-mechanical (THM) models is essential for reliable simulations to accurately model geothermal reservoirs. This integrated approach accounts for the complex interactions between thermal, hydrological, and mechanical processes, which is crucial because changes in fluid pressure and temperature can induce stress and strain in the rock matrix, affecting fracture aperture and permeability, and influencing fluid flow and heat transfer.

19.4.2.2 Advanced Modeling Applications

Production Optimization: Advanced modeling and simulation are essential for optimizing the performance of geothermal reservoirs. By simulating different operational scenarios, engineers can identify optimal strategies to maximize energy extraction. For instance, researchers have used numerical simulations to evaluate the effect of various well layouts, such as doublet, triplet, and multilateral well systems. One study examined the impact of different flow schemes on heat recovery from EGSs. The study concluded that an exponential flow scheme, where the flow rate gradually increases over time, would be optimal for meeting increasing energy demands (Asai et al., 2019).

Stimulation Optimization: Simulation techniques are also crucial for optimizing reservoir stimulation strategies, which aim to enhance permeability and improve heat extraction. Researchers have used models to investigate the effects of different stimulation methods, such as hydraulic fracturing and cyclic injection (Anyim and Gan, 2020; Gan and Elsworth, 2016; Song et al., 2021; Yin et al., 2024). One study explored a stimulation strategy involving the creation of two parallel, high-permeability manifolds, one for injection and one for production (Gan and Elsworth, 2016). This approach aimed to create a uniform flow regime analogous to drilling horizontal wells. The study found that this method significantly enhanced heat sweep efficiency.

Parameter Sensitivity Analysis: Advanced modeling allows for sensitivity analysis, which helps identify the key parameters influencing geothermal reservoir performance. By systematically varying input parameters and analyzing their impact on output variables, such as production temperature, thermal breakthrough time, and energy recovery, researchers can rank the importance of various factors (Anyim and Gan, 2020; K. Wang et al., 2022; Yin et al., 2024). For example, the sensitivity of heat extraction performance to injection mass flow rate, injection temperature, rock matrix porosity, fracture permeability, and rock matrix permeability was investigated by Wang et al. (2022). Results showed that injection mass flow rate was the most significant factor. Another study found that for a given DFN, the fracture aperture, the rock matrix permeability, and the wellbore radius are the most influential parameters controlling the thermal breakthrough time, production well mass flux, and overall energy recovery (Mahmoodpour et al., 2022).

19.4.2.3 Future Trends in Modeling and Simulation

19.4.2.3.1 Development of High-Fidelity Models

High-fidelity models incorporate detailed geological, geochemical, and geomechanical data, capturing the heterogeneity and complexity of geothermal reservoirs more realistically. Traditional reservoir simulators often focus primarily on fluid flow and heat transfer. In contrast, high-fidelity models include additional physical processes for a more holistic representation. This may involve incorporating geomechanics to account for stress changes and potential rock deformation within the reservoir or incorporating chemical reactions to model the behavior of dissolved minerals in geothermal fluids. Advanced discretization techniques, such as unstructured grids, can conform to irregular shapes within the reservoir or utilize specialized methods for handling fractures and other discontinuities. High-performance computing (HPC) advancements facilitate the practical application of these models, allowing for efficient simulations even for large and intricate geothermal reservoir models.

19.4.2.3.2 Integration of Machine Learning and Artificial Intelligence Algorithms

The integration of machine learning (ML) and artificial intelligence (AI) algorithms will enhance model accuracy, improve forecasting capabilities, and enable real-time optimization of reservoir operations (Song et al., 2021; K. Wang et al., 2022; Yin et al., 2024). The geothermal industry is experiencing a surge in the use of ML and AI, particularly in areas like reservoir engineering and characterization. This trend is expected to continue, with potential for expansion into exploration and drilling, areas with high associated costs and risks (Okoroafor et al., 2022). Successful applications of ML in these areas could make geothermal projects more attractive to private investors.

Transfer learning, where AI frameworks trained on extensive datasets are applied to similar domains with limited data, presents a promising opportunity. For instance, models from oil and gas could be adapted for seismic interpretation, fault detection, and well log analysis in geothermal contexts (Okoroafor et al., 2022). However, widespread adoption of these techniques in geothermal energy production is hindered by challenges such as limited access to comprehensive datasets, the need for effective data curation from various sources, and a shortage of professionals skilled in both geothermal science and AI/ML (Okoroafor et al., 2022).

Physics-informed deep neural networks provide another innovative approach. These networks incorporate physical constraints to improve the accuracy of ML methods, especially when learning from sparse data. This approach offers more accurate parameter estimation compared to traditional methods (Tartakovsky et al., 2020).

19.4.2.3.3 Advances in Numerical Methods

Emerging numerical techniques and algorithms will enhance the simulation of complex geothermal processes.

Coupled Thermal-Hydrological-Mechanical-Chemical (THMC) Processes: Advances in simulating coupled thermal, hydraulic, mechanical, and chemical processes provide a comprehensive understanding of reservoir behavior and long-term performance. THMC models simulate interactions between thermal (heat transfer), hydrological (fluid flow), mechanical (stress and deformation), and chemical aspects of the reservoir (Pandey et al., 2017; Salimzadeh et al., 2018).

Integration of Multi-physics Processes: Traditional reservoir simulators may focus primarily on fluid flow and heat transfer. High-fidelity models, however, incorporate additional physical processes for a more holistic representation. This involves including geomechanics to account for stress changes and potential rock deformation within the reservoir and incorporating chemical reactions to model the behavior of dissolved minerals in geothermal fluids.

Advanced Discretization Techniques: Representing complex geometries and integrating multi-physics processes often require advanced discretization techniques. These techniques may involve unstructured grids, which can conform to irregular shapes within the reservoir, or utilize specialized methods for handling fractures and other discontinuities.

High-Performance Computing (HPC): The increased complexity of high-fidelity models demands significant computational resources. Advancements in HPC have facilitated the practical application of these models. Utilizing powerful computing clusters allows for efficient simulations, even for large and intricate geothermal reservoir models.

In summary, the integration of ML and AI, the development of high-fidelity models, and the simulation of coupled THMC processes are among the promising advancements that will drive future progress in geothermal reservoir operations. These advancements will enable a more detailed and accurate understanding of geothermal reservoirs, leading to more efficient and sustainable energy production.

19.4.3 Optimization of Production and Injection Strategies

19.4.3.1 Optimization of Injection and Production Parameters

Optimizing operational parameters, particularly injection and production parameters, is crucial for efficient geothermal development. These parameters directly influence production power, resource utilization, and overall extraction performance (Asai et al., 2019; Song et al., 2021).

Injection flow rate significantly impacts heat extraction efficiency. Research suggests that an exponential flow scheme, where the injection rate changes exponentially over time, might be optimal for meeting increasing energy demands (Asai et al., 2019).

Production pressure is another key parameter that influences the pressure difference between injection and production wells. This pressure differential is crucial for controlling fluid flow and heat extraction in a geothermal system (Yin et al., 2024). Existing research highlights the need to analyze pressure and temperature evolutions in fractured areas alongside key performance indicators like production temperature, thermal breakthrough time, total heat extracted, production thermal power, and injection pressure (K. Wang et al., 2022).

Fracture permeability is a critical factor influencing the effectiveness of production and injection strategies (Anyim and Gan, 2020). The permeability of fractures in geothermal reservoirs plays a crucial role in determining the dominant heat transfer mechanism and permeability evolution. High fracture permeability leads to hydro-mechanical processes dominating permeability evolution, resulting in faster fluid flow and advection-driven thermal sweep. In contrast, low fracture permeability leads to heat conduction becoming the dominant heat transfer mechanism, potentially delaying the timing of slip along fault lines. Intermediate fracture permeability, however, can trigger shear failure earlier in areas with lower stress states due to the interplay of induced thermal unloading and fluid injection, emphasizing the importance of understanding the stress state at injection and production well locations (Anyim and Gan, 2020).

19.4.3.2 Optimization of Stimulation Strategies

Optimization of stimulation strategies is essential for maximizing the productivity of fractured geothermal reservoirs. The development of high-permeability manifolds and stimulation along principal stress directions can significantly enhance fluid sweep efficiency and heat recovery, leading to improved electricity generation. However, the efficacy of these strategies is contingent upon a comprehensive understanding of key parameters, including fracture orientation and reservoir stress state.

Development of High-Permeability Manifolds: One effective strategy involves developing two parallel and high-permeability manifolds, which act as separate injection and production zones. This method aims to replicate the success of parallel wells in unconventional oil and gas reservoirs (Gan and Elsworth, 2016).

Stimulation Along Principal Stress Direction: Simulations indicate that stimulating a reservoir along the major principal stress direction, after creating parallel manifolds along the minor principal stress direction, yields the highest power generation. This strategy improves heat sweep efficiency and increases fluid flux rates to production wells, thereby contributing to higher power output (Gan and Elsworth, 2016).

Fracture Orientation and Permeability Enhancement: The orientation of fractures within the reservoir significantly impacts stimulation outcomes. Fractures oriented between 45 and 120 degrees relative to the stimulation direction result in the most significant permeability enhancement, leading to more permeable and conductive manifolds. Effective stimulation strategies consider these orientations to maximize permeability and enhance reservoir performance.

Influence of Reservoir Stress State: The success of stimulation is heavily influenced by the reservoir's stress state. Lower stress states can lead to earlier shear failure along fault or fracture planes, which affects the overall efficiency of the stimulation process (Anyim and Gan, 2020). Understanding the stress state is essential for designing effective stimulation strategies that maximize heat recovery and fluid flow.

19.4.3.3 Cyclic Injection

Cyclic injection is a technique for optimizing production and enhancing heat extraction from geothermal reservoirs (Hofmann et al., 2018; Li et al., 2022). This method involves alternating periods of fluid injection and production from the same well or well pair, promoting heat transfer from the rock matrix to the injected fluid. The benefits of cyclic injection include reduced breakdown pressure, enhanced fracture network complexity and permeability, and mitigated induced seismicity. Additionally, cyclic injection improves heat extraction performance by managing cold fronts and increasing cumulative thermal energy extraction (Yin et al., 2024).

However, production temperature decreases with increasing injection pressure and frequency, and the choice of working fluid significantly impacts reservoir temperature and decline rate. Parameter sensitivity analysis highlights the critical role of maximum injection pressure and frequency in optimizing production temperature, reservoir permeability, and cumulative thermal energy extraction (Yin et al., 2024).

While cyclic injection presents a promising approach for geothermal reservoir optimization, further research is needed to investigate its long-term impacts on reservoir modification. Specifically, the effects of maximum injection pressure, injection frequency, and working fluid type require further investigation to ensure the effectiveness and sustainability of this technique (Hofmann et al., 2018; Yin et al., 2024).

19.5 Future Directions and Research Needs

19.5.1 The Potential Usage of CO_2 as Injection Fluid for EGS

Traditionally, water has been the working fluid of choice for EGS. However, water-based systems face several challenges, including chemical reactivity with reservoir rocks leading to scaling and corrosion, environmental concerns related to water usage, and inefficiencies in heat extraction.

In light of these issues, researchers have been exploring alternative working fluids, with carbon dioxide (CO_2) emerging as a promising candidate.

The use of CO_2 in EGS has garnered attention due to its unique thermophysical properties and potential environmental benefits. CO_2, particularly in its supercritical state, offers several advantages over water, including higher heat capacity and thermal expansivity, lower viscosity, and reduced chemical reactivity. Additionally, using CO_2 in geothermal systems presents an opportunity for carbon sequestration, aligning with global efforts to mitigate climate change by capturing and storing CO_2 emissions.

- **Overview of CO_2-Based EGS Technology**
  CO_2-based geothermal systems have gained attention due to their dual benefits of heat extraction and carbon sequestration. The key types of geothermal resources include hydrothermal systems, geopressured reservoirs, sedimentary basins, and hot dry rock (HDR). EGSs are typically deployed to extract heat from HDR, which has low permeability and porosity.
  Using CO_2 as a working fluid in EGS offers several advantages over traditional water-based systems. CO_2 has a lower viscosity and higher mobility, which allows for more efficient heat transfer and extraction. Additionally, CO_2 can be stored in the subsurface, providing an opportunity for carbon sequestration. However, CO_2's lower heat capacity compared to water means that it requires a higher mass flow rate to achieve the same energy output, potentially leading to faster thermal breakthrough and shorter project lifetimes.
- **Benefits and Challenges of Using CO_2 as an Injection Fluid**
  The use of CO_2 as an injection fluid in geothermal systems presents both significant benefits and notable challenges. The unique thermophysical properties of CO_2 make it an attractive alternative to water in EGSs.

Higher Heat Extraction Efficiency: One of the primary benefits of using CO_2 as a working fluid is its higher heat extraction efficiency compared to water (F. Sun et al., 2018; C. -L. Wang et al., 2018). Numerical simulations have shown that CO_2 can extract significantly more heat from geothermal reservoirs (Pan et al., 2015; Shi et al., 2018). This is due to the favorable thermophysical properties of CO_2, such as its lower viscosity and higher expansivity, which enhance its ability to transport heat (Esteves et al., 2019; Pruess, 2008). Additionally, CO_2's mobility, which is the ratio of density to viscosity, is less sensitive to temperature changes compared to water, allowing for more consistent heat extraction rates even as the reservoir cools (Pruess, 2006).

Reduction in Water Losses: Water losses are a significant obstacle to the commercialization of water-based EGS. CO_2, being a non-ionic solvent, does not suffer from the same degree of fluid loss due to mineral dissolution and precipitation that plagues water-based systems (Taron and Elsworth, 2009). This can help in maintaining the integrity of the reservoir and the efficiency of the system over longer periods (Brown, 2000).

CO_2 Storage and Environmental Benefits: Utilizing CO_2 as a geothermal fluid offers the ancillary benefit of geological CO_2 sequestration. This dual-purpose approach not only aids in geothermal energy extraction but also contributes to reducing atmospheric CO_2 levels, thereby addressing climate change concerns (Wu and Li, 2020). The sequestration process involves trapping CO_2 in underground formations, which can potentially store vast amounts of this greenhouse gas (Fouillac et al., 2004).

Enhanced Flow Characteristics: CO_2 exhibits lower viscosity and higher density under geothermal conditions, which translates to lower power requirements for pumping and maintaining circulation in the wellbores (Pan et al., 2015). This can lead to significant operational cost savings and more efficient system performance. The high compressibility of CO_2 further enhances its flow characteristics, making it easier to circulate through the fractured rock formations typical of geothermal reservoirs (Pruess, 2008).

19.5.1.1 Challenges of Using CO_2 in Geothermal Systems

Thermal Breakthrough and Gravity Effects: One of the significant challenges of using CO_2 is the risk of premature thermal breakthrough (Pruess, 2008). The large density contrast between cold injected CO_2 and hot reservoir CO_2 can lead to preferential flow paths (Figure 19.9), particularly along the bottom of the reservoir (Cui et al., 2018). This can result in rapid cooling of the production wells and reduced overall efficiency of heat extraction (Pruess, 2008). Managing these gravity effects requires careful design of the injection and production well configurations.

Buoyancy and Recovery Issues: CO_2's buoyant nature poses challenges for its use in geothermal systems. During injection, CO_2 tends to rise due to its lower density compared to water, potentially leading to uneven distribution within the reservoir. Furthermore, during the production phase, not all injected CO_2 can be recovered, with simulations showing that at least 10% of the CO_2 remains trapped in the reservoir. This necessitates careful consideration of optimal injection and production strategies to maximize CO_2 recovery and maintain system efficiency.

Chemical Interactions and Scaling: While CO_2 does not cause the same level of mineral scaling as water, chemical interactions between CO_2, reservoir rocks, and existing aqueous phases can

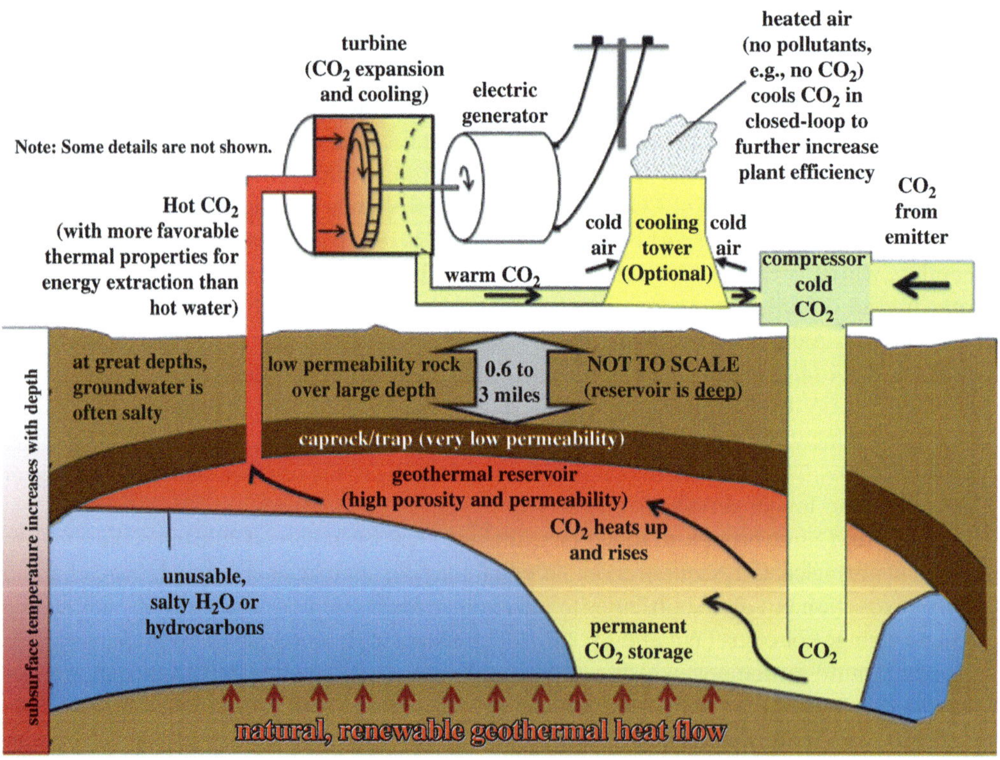

Figure 19.9 CO_2-plume geothermal (Garapati et al., 2014 / with permission of Elsevier).

still pose challenges. These interactions can alter the permeability and porosity of the reservoir, potentially leading to issues with flow rates and heat transfer efficiency over time (Xu and Pruess, 2004). Understanding and mitigating these interactions is crucial for the long-term viability of CO_2-based geothermal systems.

Salt Precipitation and Clogging: A significant challenge associated with using CO_2 as a working fluid is the precipitation of salts, particularly halite, within the reservoir fractures. As CO_2 is injected and the reservoir dries out, salt can precipitate and clog the fractures, reducing permeability and system efficiency. To mitigate this effect, researchers have explored strategies such as co-injecting brine with CO_2. However, this approach can reduce the overall heat extraction efficiency, necessitating a balance between mitigating clogging and optimizing heat recovery (Borgia et al., 2013).

Operational and Infrastructure Adjustments: Transitioning from water to CO_2 as a working fluid necessitates significant changes in the operational procedures and infrastructure of geothermal plants. This includes modifications to well designs, surface facilities, and monitoring systems to handle the different physical and chemical properties of CO_2. These adjustments can be costly and require substantial upfront investment (Brown, 2000).

Regulatory and Safety Concerns: The use of CO_2 in geothermal systems introduces new regulatory and safety considerations. Ensuring the safe storage of CO_2 and preventing leaks is paramount to both environmental protection and public safety. This requires robust monitoring and management practices to detect and address any potential issues promptly (Fouillac et al., 2004).

The utilization of CO_2 as an injection fluid in geothermal systems offers promising benefits, including higher heat extraction efficiency, reduced water losses, and potential environmental advantages through CO_2 sequestration. However, these benefits are accompanied by challenges such as thermal breakthrough risks, chemical interactions, necessary infrastructure adjustments, and regulatory concerns. Addressing these challenges through innovative engineering solutions and rigorous research is essential to fully harness the potential of CO_2-based geothermal systems.

19.5.2 Emerging Technologies

- **Closed Loop Systems**
  Closed-loop geothermal systems have emerged as a promising solution for sustainable energy production. Unlike open-loop systems, which rely on extracting groundwater, closed-loop systems circulate a working fluid through a sealed loop of pipes, eliminating the risk of environmental contamination. These systems offer numerous advantages, including reduced environmental impact and increased durability, making them an attractive option for harnessing geothermal energy (Alagoz et al., 2023).
- **Types of Closed-Loop Systems**
  Closed-loop systems can be broadly categorized into U-tube and coaxial wellbore heat exchangers (WBHEs) (Figure 19.10). U-tube heat exchangers circulate fluid through a loop of pipes that absorbs heat from the surrounding rock. In contrast, coaxial WBHEs consist of two concentric pipes, where the fluid flows down the outer pipe, absorbs heat, and returns through the inner pipe. The coaxial design offers a larger surface area for heat exchange and lower pumping energy requirements (Lo Russo et al., 2020).

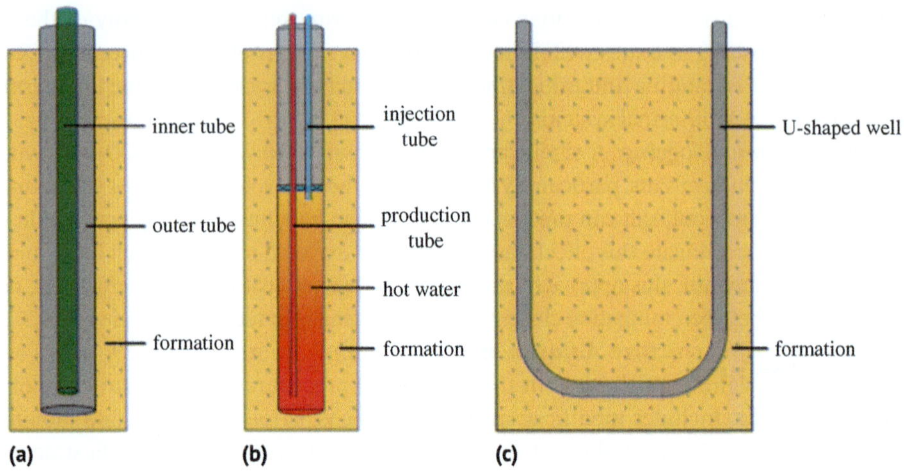

Figure 19.10 Conceptual layouts for closed-loop geothermal systems include: (a) a co-axial design, (b) a multiple string configuration, and (c) a U-shaped design Hu et al., 2020 / with permission of Elsevier.

- **Heat Transfer and System Efficiency**
  Heat transfer efficiency is critical in closed-loop geothermal systems. Efficiency is influenced by several factors, including the thermal conductivity of the surrounding rock, the flow rate of the working fluid, and the design of the heat exchanger. Studies have shown that coaxial WBHEs are more efficient than U-tube systems due to their higher surface area and better fluid dynamics (Nian and Cheng, 2018; Russo, 2020).

  The thermal conductivity of the geological formations surrounding the wellbore is a significant factor. High thermal conductivity enhances heat transfer, improving the system's overall efficiency. However, the variability of geological formations can complicate the accurate estimation of heat exchanger performance. Advanced numerical simulations are often required to optimize the design and operation of these systems.
- **Economic and Environmental Considerations**
  The economic viability of closed-loop geothermal systems depends on various factors, including the initial drilling costs, the efficiency of heat extraction, and the lifespan of the system. While the initial costs can be high, especially for deep wells, the long-term benefits, such as reduced maintenance and operational costs, can make these systems economically attractive in the long run (Beckers et al., 2022; Fowler and McClure, 2021).

  Environmentally, closed-loop systems offer significant advantages over open-loop systems. Since they do not extract groundwater, there is no risk of depleting or contaminating local water sources. This makes closed-loop systems a more sustainable and environmentally friendly option for geothermal energy production (Alagoz et al., 2023).
- **Challenges and Future Directions**
  Despite their advantages, closed-loop geothermal systems face several challenges. One major issue is the low thermal energy recovery rate, particularly in deep wells with low thermal gradients. Additionally, the long-term stability of wellbores and the potential for thermal interference between adjacent wells need to be addressed (Irani et al., 2022).

  Future research should focus on improving the heat transfer efficiency of these systems and developing more accurate simulation models to predict their performance under various geological conditions. Innovations in drilling technology and materials science could also help reduce costs and enhance the durability of closed-loop systems (Livescu and Dindoruk, 2022).

Closed-loop geothermal systems represent a promising avenue for sustainable energy production. They offer numerous advantages, including reduced environmental impact and increased durability. However, challenges such as low thermal energy recovery and high initial costs must be addressed through continued research and technological innovation. By optimizing the design and operation of these systems, we can harness geothermal energy more effectively and contribute to a more sustainable future.

19.6 Environmental and Social Considerations for Geothermal Energy Development

19.6.1 Challenges and Opportunities in Sustainable Energy Practices

- **Addressing Environmental Concerns**
  Unlike fossil fuel power plants, which can release greenhouse gases, geothermal energy systems have a relatively minimal environmental impact (Figure 19.11). Advances in technology and heightened awareness of the importance of environmental protection have further minimized these impacts. This section examines the environmental implications of geothermal energy systems.
- **Surface Disturbances and Land Use**
  The development of geothermal energy often involves significant surface disturbances. These include the construction of infrastructure such as drilling rigs, pipelines, and power plants.

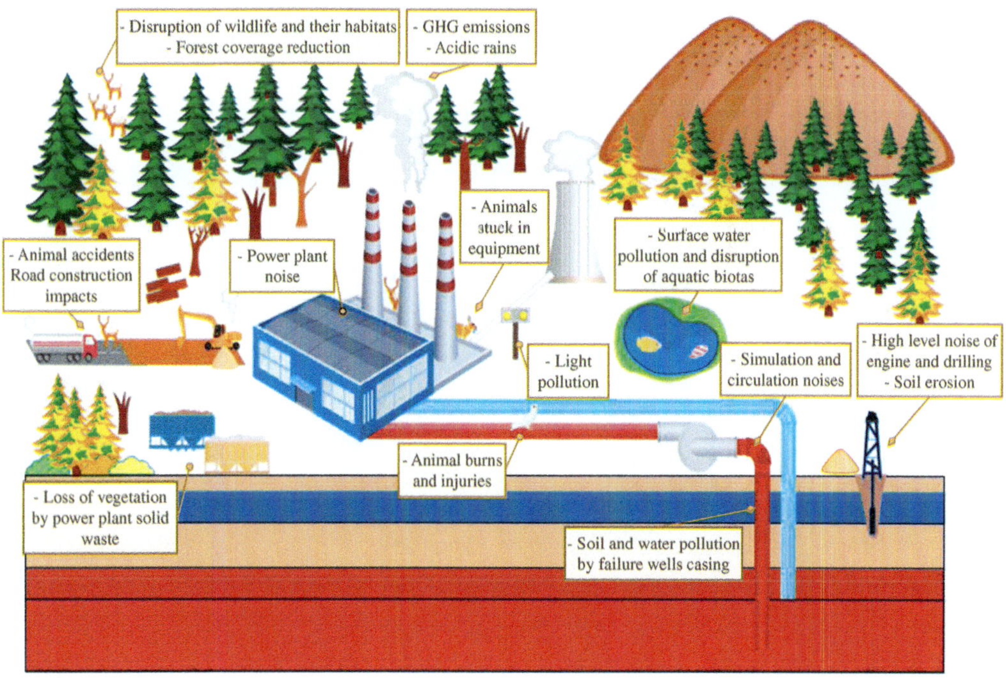

Figure 19.11 Geothermal impact on biodiversity (Soltani et al., 2021b / with permission of Elsevier).

Surface disturbances can lead to habitat disruption, soil erosion, and changes in land use patterns. For instance, in Iceland, the development of geothermal fields has caused surface disturbances that have necessitated extensive monitoring and mitigation measures (Kristmannsdóttir and Ármannsson, 2003).

- **Fluid Withdrawal and Subsidence**
  One of the primary environmental impacts of geothermal energy extraction is the physical effect of fluid withdrawal. This process can lead to land subsidence, where the ground surface sinks due to the removal of large volumes of geothermal fluids. Subsidence can damage infrastructure, alter water tables, and affect local ecosystems. Notable cases of subsidence have been observed in geothermal fields in New Zealand, Iceland, and Italy, with varying degrees of severity (Allis and Zhan, 2000; Eysteinsson, 2000).
- **Water Contamination**
  Geothermal fluids often contain high concentrations of hazardous chemicals, including heavy metals and metalloids like arsenic, boron, and mercury. The improper management of geothermal fluids can lead to the contamination of surface and groundwater resources (Figure 19.12). For example, boron is a significant contaminant in geothermal fluids, and its high concentrations can be toxic to plants, animals, and humans (Mott et al., 2022). Effective treatment and management strategies are crucial to mitigate the environmental risks posed by these contaminants.
- **Air Emissions**
  Geothermal plants emit gases such as CO_2, hydrogen sulfide (H_2S), and methane (CH_4), which can contribute to air pollution and have adverse effects on human health and the environment (Armannsson, 2003; Fridriksson et al., 2017; Nagl, 1999). While the greenhouse gas emissions from geothermal plants are relatively low compared to fossil fuel-based power

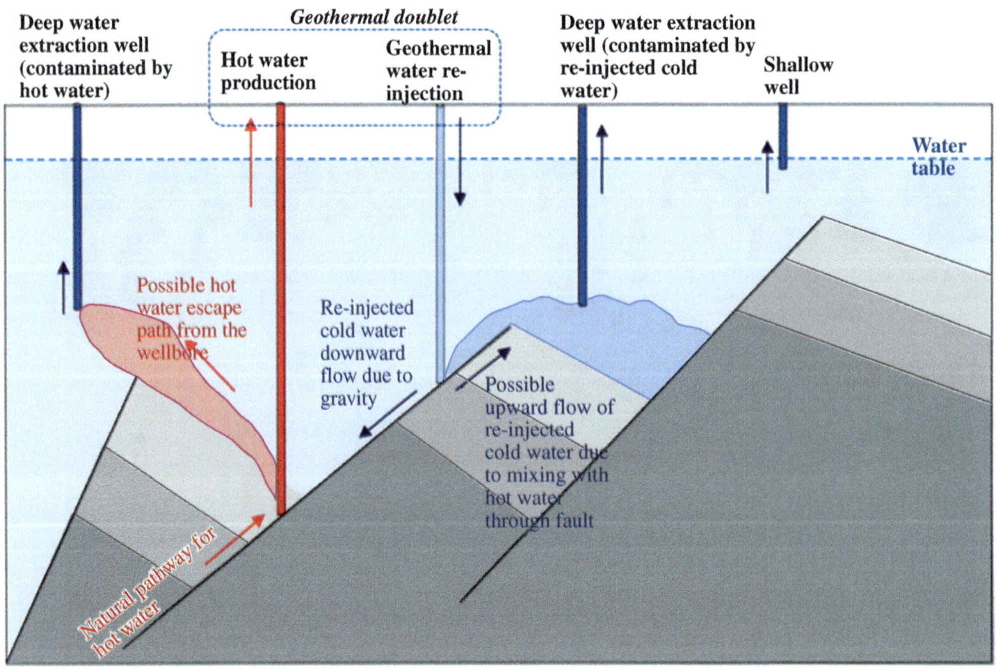

Figure 19.12 The impact of geothermal energy production on groundwater contamination (Mott et al., 2022b / with permission of Elsevier).

plants, they still need to be managed to minimize their environmental impact (Bargagli et al., 1997). Research has shown that geothermal development can lead to increased emissions of these gases, necessitating stringent monitoring and control measures (Ármannsson, 2017; Kristmannsdóttir et al., 2000; Kristmannsdóttir and Ármannsson, 2003; Layman, 2017).

- **Noise Pollution**
  Noise pollution is another concern associated with geothermal energy development. The drilling and operation of geothermal wells and power plants generate significant noise, which can disturb local communities and wildlife. Temporary noise from drilling can exceed 90 decibels, while the noise from operating plants can reach levels that require mitigation to comply with environmental regulations (Arnórsson, 2004; Kubo, 2003; Viesca, 1995; Windrem and Marr, 1981).
- **Thermal Pollution**
  Thermal pollution occurs when the waste heat from geothermal plants is discharged into the environment, affecting local ecosystems (Brock, 1975). The discharge of hot fluids into rivers, lakes, or the atmosphere can alter the thermal balance of these ecosystems, potentially harming aquatic life and changing local weather patterns (Huang and Tian, 2006). Cooling methods such as ponds or reinjection of spent fluids into the reservoir can help mitigate thermal pollution (Lund and Freeston, 2001).
- **Induced Seismicity**
  The extraction and reinjection of geothermal fluids can induce seismic activity, known as induced seismicity (Zang et al., 2014). This phenomenon occurs due to changes in the pressure and stress within the Earth's crust caused by fluid movements. In some cases, induced seismicity has been significant enough to cause concern among local populations and regulatory authorities. Studies have documented instances of induced seismicity in geothermal fields, highlighting the need for comprehensive monitoring and risk assessment (Brandsdóttir et al., 2002).
- **Biological Impacts**
  Geothermal development can also have biological impacts, affecting both terrestrial and aquatic ecosystems. The introduction of chemicals and changes in temperature can disrupt the habitat of various species, leading to changes in biodiversity. In some geothermal areas, the unique thermophilic vegetation and microorganisms can be threatened by the alterations in their environment (Kristmannsdóttir and Ármannsson, 2003; Webster and Timperley, 1995).

Effective mitigation strategies are essential to address the environmental concerns associated with geothermal energy development. These strategies include:

Comprehensive Environmental Impact Assessments (EIA): Conducting thorough EIAs before the development of geothermal projects can help identify potential environmental risks and develop mitigation measures. These assessments should consider all aspects of the environment, including physical, chemical, biological, and socio-economic factors (Kristmannsdóttir and Ármannsson, 1995).

Reinjection of Spent Fluids: Reinjecting geothermal fluids back into the reservoir can help maintain pressure levels, reduce subsidence, and minimize the risk of contamination and thermal pollution. This practice also helps in the sustainable management of geothermal resources (Hunt, 2001).

Emission Control Technologies: Implementing technologies to capture and treat emissions of CO_2, H2S, and other gases can significantly reduce the environmental impact of geothermal plants.

For example, H2S can be removed from the gas stream using scrubbers or other chemical treatments (Kristmannsdóttir and Ármannsson, 1995).

Noise Mitigation Measures: Using noise barriers, mufflers, and other soundproofing techniques can help reduce noise pollution from geothermal operations. Monitoring noise levels and implementing mitigation measures can ensure compliance with environmental regulations (Ármannsson and Kristmannsdóttir, 1992).

Monitoring and Research: Continuous monitoring of geothermal fields for changes in land elevation, groundwater quality, and seismic activity is crucial for identifying and addressing environmental impacts. Research into new technologies and methods for mitigating environmental risks can further enhance the sustainability of geothermal energy (Kristmannsdóttir et al., 2000).

Public Involvement and Education: Engaging local communities and stakeholders in the planning and development process of geothermal projects can help address their concerns and improve project outcomes. Public education on the benefits and potential impacts of geothermal energy can foster greater acceptance and support for renewable energy projects (Hongying, 2000).

While geothermal energy offers significant benefits as a renewable energy source, its development must be carefully managed to mitigate environmental concerns. By addressing surface disturbances, fluid withdrawal effects, water and air pollution, noise, thermal impacts, induced seismicity, and biological effects, geothermal energy can be harnessed sustainably. Implementing effective mitigation strategies and continuous monitoring will ensure that geothermal energy contributes positively to our energy needs without compromising environmental integrity.

- **Addressing Social Concerns**

  It faces significant social challenges that can impede its development and acceptance. Addressing these social concerns is crucial for the successful implementation of geothermal projects. Key areas of focus include environmental impacts, economic benefits, cultural considerations, and effective communication strategies (Arevalo, 2006; Dowd et al., 2011; Kubota, 2015; Soltani et al., 2021).

  The economic benefits of geothermal projects can be a significant factor in gaining social acceptance. These benefits include job creation, infrastructure development, and local economic stimulation. For instance, in the Ahuachapan and Berlin geothermal fields in El Salvador, local communities have benefited from temporary jobs created by project execution, such as pipeline network setups and drilling activities (Arevalo, 2006).

  Moreover, companies often engage in social investment plans that support local development through educational programs, health initiatives, and infrastructure improvements. These initiatives not only improve the quality of life for local residents but also foster a positive relationship between the geothermal company and the community.

  Cultural considerations are another critical aspect of social acceptance for geothermal projects. In Japan, for example, the development of geothermal power has sometimes faced opposition from hot spring inn managers and local governments who fear that geothermal activities might negatively impact the traditional hot spring (onsen) culture, which is deeply embedded in Japanese society (Kubota, 2015).

  To address these concerns, developers must engage in meaningful dialogue with local stakeholders, respecting and preserving cultural heritage while demonstrating the minimal impact of geothermal projects on hot spring resources. By providing clear, factual information about the benefits and risks of geothermal energy, developers can help alleviate fears and build trust within the community (Arevalo, 2006; Dowd et al., 2011; Kubota, 2015; Soltani et al., 2021).

Effective communication is vital for addressing social concerns and fostering acceptance of geothermal projects. Transparent, honest, and continuous communication with all stakeholders including local communities, governments, and environmental groups is essential. In Australia, understanding public perceptions and concerns about geothermal energy through surveys and public consultations has been crucial in shaping policies and gaining social acceptance.

Educational campaigns that inform the public about the environmental and economic benefits of geothermal energy, as well as the safety measures in place, can help demystify the technology and reduce opposition. Additionally, involving the community in decision-making processes and addressing their concerns promptly and respectfully can significantly enhance the social acceptance of geothermal projects.

Addressing social concerns in geothermal energy projects requires a multifaceted approach that encompasses environmental stewardship, economic development, cultural sensitivity, and effective communication. By prioritizing these areas, geothermal developers can foster social acceptance, ensuring the successful and sustainable implementation of geothermal energy projects.

- **Integrating Geothermal with Other Renewables**
  The global energy landscape is undergoing a significant transformation as the urgency to shift from fossil fuels to renewable energy sources becomes more apparent (Dincer, 2000). Hybrid energy systems, which integrate two or more renewable energy sources (Figure 19.13), offer a viable solution to enhance the efficiency, reliability, and sustainability of energy production. Among the various combinations, the integration of geothermal energy with other renewable sources such as solar and wind has shown promising potential (Olabi et al., 2020). This section delves into the various configurations, advantages, and applications of hybrid energy systems, with a particular focus on geothermal energy.

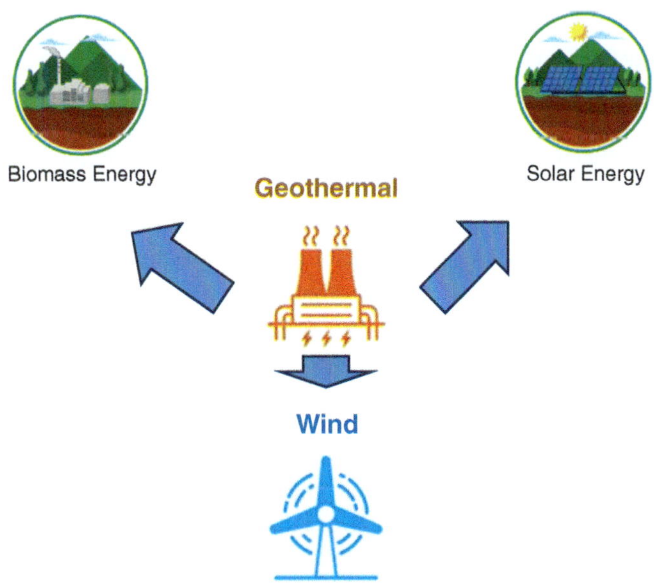

Figure 19.13 Most common geothermal based hybrid systems.

Hybrid energy systems represent a crucial step towards a sustainable and resilient energy future. By integrating geothermal energy with other renewable sources, these systems can provide reliable, efficient, and eco-friendly energy solutions. Continued research and development in hybrid system optimization, new energy combinations, and advanced control strategies will further enhance the potential of these innovative systems. As the demand for clean energy grows, hybrid geothermal systems will play a vital role in meeting global energy needs while minimizing environmental impact.

Hybrid geothermal systems have a wide range of applications, primarily in heating, cooling, and power generation:

Heating and Cooling: Hybrid geothermal systems are particularly effective for heating and cooling applications. Ground source heat pumps (GSHP) combined with solar thermal systems can efficiently provide space heating and cooling (Hackel and Pertzborn, 2011). For instance, solar-assisted ground source heat pumps (SAGSHP) use solar collectors to supplement geothermal energy, reducing the need for extensive borehole heat exchangers and lowering installation costs (Duarte et al., 2021).

These systems are also employed in district heating and cooling networks. For example, a hybrid GSHP system can distribute heating and cooling through a common water loop, serving multiple buildings efficiently. The integration of cooling towers into geothermal systems can further enhance performance by maintaining soil thermal balance, thereby improving the coefficient of performance (COP).

Power Generation: Although less common than heating and cooling applications, hybrid geothermal systems can also be used for power generation. By combining geothermal energy with solar or wind power, these systems can achieve higher efficiency and more reliable electricity production. For instance, a solar-geothermal hybrid power plant can use concentrated solar power (CSP) to boost geothermal steam production, increasing the overall power output (McTigue et al., 2018). Such systems can be particularly effective in regions with high solar insolation and geothermal potential.

- **Hybrid Energy Systems**

  Geothermal energy, derived from the Earth's internal heat, provides a stable and continuous energy supply, making it an ideal candidate for integration into hybrid systems. However, the standalone efficiency of geothermal energy can be relatively low. Hybrid systems can overcome this limitation by combining geothermal energy with other renewable sources to enhance overall performance and output.

  Hybrid geothermal systems offer numerous advantages, including enhanced efficiency, reduced environmental impact, and improved reliability. By combining multiple energy sources, these systems can optimize the use of available resources, reduce operational costs, and lower greenhouse gas emissions. Additionally, the integration of renewable energy sources can mitigate the intermittency issues associated with solar and wind power.

  However, hybrid systems also present several challenges. The complexity of integrating multiple energy sources requires advanced control strategies to ensure optimal performance (El Haj Assad et al., 2021; Sharmin et al., 2023). Issues such as thermal imbalance, system compatibility, and initial capital costs must be carefully managed. Advanced control methods, such as fuzzy logic controllers and model predictive control (MPC), have been developed to address these challenges.

- **Geothermal-Solar Hybrid Systems**
  The integration of geothermal and solar energy is one of the most common hybrid configurations. Solar energy, despite its abundance, is intermittent and subject to weather and daylight variations. Combining it with geothermal energy, which is available continuously, can provide a more stable and reliable energy supply.
  A typical geothermal-solar hybrid system might use solar thermal collectors to generate additional heat, which can supplement the geothermal energy. This integration can reduce annual operating costs and CO_2 emissions. For instance, solar collectors can assist in ground heat recovery, thereby stabilizing the geothermal system. Conversely, geothermal heat exchangers can support solar thermal plants by providing a secondary heat source. Such systems have been shown to increase efficiency by up to 3.6% compared to standalone systems.
  Another effective configuration involves the use of photovoltaic thermal (PVT) systems integrated with ground source heat pumps (GSHP). This setup allows for the simultaneous production of electrical and thermal energy, leveraging both solar and geothermal resources. The design considerations for these systems often focus on optimizing the working fluid to ensure maximum efficiency.
- **Geothermal-Wind Hybrid Systems**
  Geothermal-wind hybrid systems, although less common, offer significant potential, especially in regions with substantial wind resources. Wind energy can complement geothermal energy by providing high energy output during windy periods, thus balancing the continuous but lower-grade geothermal energy. These systems can enhance overall energy security and reduce dependence on any single energy source.
- **Geothermal-Biomass Hybrid Systems**
  Integrating geothermal energy with biomass energy can further enhance system efficiency. Biomass energy, derived from organic materials, can provide a steady energy supply that complements the geothermal base load. In a dual-fluid biomass-geothermal power plant, geothermal energy can preheat the biomass feedstock, improving the overall efficiency of the system. This configuration can also reduce greenhouse gas emissions and reliance on fossil fuels (Beckers et al., 2015; Briola et al., 2019; Thain and DiPippo, 2015).

References

Aihar, A., Bouabdallah, N., Ifrene, G., et al. (2023). Comparing fishbone drilling and hydraulic fracturing in Ultra-low permeability geothermal reservoirs. 2023 AAPG Rocky Mountain Section Meeting, Bismarck, North Dakota, June 4–6, 2023. https://doi.org/10.1306/80737Aihar2023.

Alagoz, E., AlNasser, F., Ozkan, Y., et al. (2023). Overview of Closed–Loop Enhanced Geothermal Systems. International Journal of Earth Sciences Knowledge and Applications 5 (1), pp. 158–164.

Albaric, J., Oye, V., Langet, N., et al. (2014). Monitoring of induced seismicity during the first geothermal reservoir stimulation at Paralana, Australia. Geothermics 52, pp. 120–131.

Allis, R. G. & Zhan, X. (2000). Predicting subsidence at Wairakei and Ohaaki geothermal fields, New Zealand. Geothermics 29 (4–5), pp. 479–497.

Anderson, E. B. & Ussher, G. N. H. (1992). GDManager–An Integrated Geothermal Data Management System. Proceedings of 13th NZ Geothermal Workshop, University of Auckland. pp. 121–126.

Anyim, K. & Gan, Q. (2020). Fault zone exploitation in geothermal reservoirs: Production optimization, permeability evolution and induced seismicity. Advances in Geo-Energy Research 4 (1), pp. 1–12. https://doi.org/10.26804/ager.2020.01.01.

Arevalo, A. S. (2006). Environmental and social issues in geothermal in El Salvador. Workshop for Decision Makers on Geothermal Projects in Central America, Organized by UNU-GTP and LaGeo in San Salvador, El Salvador, 26.

Armannsson, H. (2003). CO_2 emission from geothermal plants. International Geothermal Conference 103, pp. 56–62.

Ármannsson, H. (2017). Carbon dioxide emissions from Icelandic geothermal areas. Procedia Earth and Planetary Science 17, pp. 104–107.

Ármannsson, H. & Kristmannsdóttir, H. (1992). Geothermal environmental impact. Geothermics 21 (5–6), pp. 869–880.

Arnórsson, S. (2004). Environmental impact of geothermal energy utilization. Geological Society, London, Special Publications 236 (1), pp. 297–336.

Asai, P., Panja, P., McLennan, J., et al. (2019). Effect of different flow schemes on heat recovery from Enhanced Geothermal Systems (EGS). Energy 175, pp. 667–676. https://doi.org/10.1016/j.energy.2019.03.124.

Atsalos, E. F. (2022). Fishbone wells: Drilling, stimulation and productivity [Master's thesis, Aristotle University of Thessaloniki]. IKEE Institutional Repository. https://doi.org/10.26262/heal.auth.ir.339870.

Aydin, H. & Temizel, C. (2022). Characterization of Geothermal Reservoirs. 47th Workshop on Geothermal Reservoir Engineering, Stanford, California, USA, pp. 7–9.

Baisch, S., Weidler, R., Vörös, R., et al. (2006). Induced seismicity during the stimulation of a geothermal HFR reservoir in the Cooper Basin, Australia. Bulletin of the Seismological Society of America 96 (6), pp. 2242–2256.

Bargagli, R., Cateni, D., Nelli, L., et al. (1997). Environmental impact of trace element emissions from geothermal power plants. Archives of Environmental Contamination and Toxicology 33, pp. 172–181.

Beckers, K. F., Lukawski, M. Z., Aguirre, G. A., et al. (2015). Hybrid low-grade geothermal-biomass systems for direct-use and co-generation: from campus demonstration to nationwide energy player. Proceedings, 40th Workshop on Geothermal Reservoir Engineering, Stanford University, Stanford, California.

Beckers, K. F., Rangel-Jurado, N., Chandrasekar, H., et al. (2022). Techno-economic performance of closed-loop geothermal systems for heat production and electricity generation. Geothermics 100, p. 102318.

Benegar, J. & Kennedy, R. (2020). Geothermal Data Management and Best Practices. Proceedings, 8th African Rift Geothermal Conference (ARGeo-C8) Nairobi, Kenya.

Billy J. R. (2018). Geothermal Resource of the United States: Identified Hydrothermal Sites and Favorability of Deep Enhanced Geothermal Systems (EGS). National Renewable Energy Laboratory https://www.nrel.gov/gis/assets/images/geothermal-identified-hydrothermal-and-egs.jpg.

Borgia, A., Pruess, K., Kneafsey, T. J., et al. (2013). Simulation of CO_2-EGS in a fractured reservoir with salt precipitation. Energy Procedia 37, pp. 6617–6624.

Bošnjaković, M., Stojkov, M., & Jurjević, M. (2019). Environmental impact of geothermal power plants. Tehnički Vjesnik 26 (5), pp. 1515–1522.

Brandsdóttir, B., Franzson, H., Einarsson, P., et al. (2002). Seismic monitoring during an injection experiment in the Svartsengi geothermal field, Iceland. Jökull 51, pp. 43–52.

Bredesen, K., Dalgaard, E., Mathiesen, A., et al. (2019). Seismic characterization of geothermal sedimentary reservoirs: A field example from the Copenhagen area, Denmark. Interpretation 8 (2), pp. T275–T291. https://doi.org/10.1190/INT-2019-0184.1.

Briola, S., Gabbrielli, R., & Bischi, A. (2019). Off-design performance analysis of a novel hybrid binary geothermal-biomass power plant in extreme environmental conditions. Energy Conversion and Management 195, pp. 210–225.

Brock, T. D. (1975). Predicting the ecological consequences of thermal pollution from observations on geothermal habitats. Environmental Effects of Cooling Systems at Nuclear Power Plants, pp. 599–621.

Brown, D. W. (2000). A hot dry rock geothermal energy concept utilizing supercritical CO_2 instead of water. Proceedings of the Twenty-Fifth Workshop on Geothermal Reservoir Engineering, Stanford University, pp. 233–238.

Buijze, L., van Bijsterveldt, L., Cremer, H., et al. (2019). Review of induced seismicity in geothermal systems worldwide and implications for geothermal systems in the Netherlands. Netherlands Journal of Geosciences 98, p. e13.

Caulk, R. A., Ghazanfari, E., Perdrial, J. N., et al. (2016). Experimental investigation of fracture aperture and permeability change within Enhanced Geothermal Systems. Geothermics 62, pp. 12–21.

Chen, Y., Ma, G., & Wang, H. (2018). Heat extraction mechanism in a geothermal reservoir with rough-walled fracture networks. International Journal of Heat and Mass Transfer 126, pp. 1083–1093.

Cui, G., Ren, S., Rui, Z., et al. (2018). The influence of complicated fluid-rock interactions on the geothermal exploitation in the CO_2 plume geothermal system. Applied Energy 227, pp. 49–63.

Deichmann, N., Kraft, T., & Evans, K. F. (2014). Identification of faults activated during the stimulation of the Basel geothermal project from cluster analysis and focal mechanisms of the larger magnitude events. Geothermics 52, pp. 84–97.

Dhar, A., Naeth, M. A., Jennings, P. D., et al. (2020). Geothermal energy resources: Potential environmental impact and land reclamation. Environmental Reviews 28 (4), pp. 415–427.

Dincer, I. (2000). Renewable energy and sustainable development: a crucial review. Renewable and Sustainable Energy Reviews 4 (2), pp. 157–175.

Dorbath, L., Cuenot, N., Genter, A., et al. (2009). Seismic response of the fractured and faulted granite of Soultz-sous-Forêts (France) to 5 km deep massive water injections. Geophysical Journal International 177 (2), pp. 653–675.

Dowd, A. -M., Boughen, N., Ashworth, P., et al. (2011). Geothermal technology in Australia: Investigating social acceptance. Energy Policy 39 (10), pp. 6301–6307.

Duarte, W. M., Paulino, T. F., Tavares, S. G., et al. (2021). Feasibility of solar-geothermal hybrid source heat pump for producing domestic hot water in hot climates. International Journal of Refrigeration 124, pp. 184–196.

El Haj Assad, M., Ahmadi, M. H., Sadeghzadeh, M., et al. (2021). Renewable hybrid energy systems using geothermal energy: hybrid solar thermal–geothermal power plant. International Journal of Low-Carbon Technologies 16 (2), pp. 518–530.

Esteves, A. F., Santos, F. M., & Pires, J. C. M. (2019). Carbon dioxide as geothermal working fluid: An overview. Renewable and Sustainable Energy Reviews 114, pp. 109331.

Evans, K. F., Zappone, A., Kraft, T., et al. (2012). A survey of the induced seismic responses to fluid injection in geothermal and CO_2 reservoirs in Europe. Geothermics 41, pp. 30–54.

Eysteinsson, H. (2000). Elevation and gravity changes at geothermal fields on the Reykjanes Peninsula, SW Iceland. Proceedings World Geothermal Congress, pp. 559–564.

Fouillac, C., Sanjuan, B., Gentier, S., et al. (2004). Could sequestration of CO_2 be combined with the development of enhanced geothermal systems. Third Annual Conference on Carbon Capture and Sequestration, Alexandria, VA, Vol 4.

Fowler, G. & McClure, M. (2021). A Feasibility study on three geothermal designs: deep closed-loop (with and without conductive fractures) and open-loop circulation between multifractured laterals. Transactions-Geothermal Resources Council 45 (c), pp. 635–653.

Fridleifsson, I. B., Bertani, R., Huenges, E., et al. (2008). The possible role and contribution of geothermal energy to the mitigation of climate change. IPCC Scoping Meeting on Renewable Energy Sources, Proceedings, Luebeck, Germany 20 (25), pp. 59–80.

Fridriksson, T., Merino, A. M., Orucu, A. Y., et al. (2017). Greenhouse gas emissions from geothermal power production. Proceedings, 42nd Workshop on Geothermal Reservoir Engineering. California: Stanford University, Stanford.

Gan, Q. & Elsworth, D. (2016). Production optimization in fractured geothermal reservoirs by coupled discrete fracture network modeling. Geothermics 62, pp. 131–142. https://doi.org/10.1016/j.geothermics.2016.04.009.

Garapati, N., Randolph, J. B., Valencia, J. L., et al. (2014). CO_2-plume geothermal (CPG) heat extraction in multi-layered geologic reservoirs. Energy Procedia 63, pp. 7631–7643.

Garg, S. K., Pritchett, J. W., Wannamaker, P. E., et al. (2007). Characterization of geothermal reservoirs with electrical surveys: Beowawe geothermal field. Geothermics 36 (6), pp. 487–517. https://doi.org/10.1016/j.geothermics.2007.07.005.

Gaucher, E., Schoenball, M., Heidbach, O., et al. (2015). Induced seismicity in geothermal reservoirs: A review of forecasting approaches. Renewable and Sustainable Energy Reviews 52, pp. 1473–1490.

Geothermal Stimulation | Fishbones. (n.d.). Retrieved July 7, 2024, from https://www.fishbones.as/geothermal.

Gutiérrez-Negrín, L. C. A. (2024). Evolution of worldwide geothermal power 2020–2023. Geothermal Energy 12 (1), pp. 14.

Hackel, S. & Pertzborn, A. (2011). Effective design and operation of hybrid ground-source heat pumps: three case studies. Energy and Buildings 43 (12), pp. 3497–3504.

Hanton, C., Edwards, A., Richardson, I., et al. (2019). The Integration of Data Management and Geological Modelling in a Geothermal SubsurfaceTeam. Transaction-Geothermal Resource Council 43, pp. 1043–1055.

Hofmann, H., Zimmermann, G., Zang, A., et al. (2018). Cyclic soft stimulation (CSS): a new fluid injection protocol and traffic light system to mitigate seismic risks of hydraulic stimulation treatments. Geothermal Energy 6 (1), pp. 1–33. https://doi.org/10.1186/s40517-018-0114-3.

Hongying, L. (2000). Geothermal environmental impact assessment studies in Hebei Province. China. Report 12, pp. 231–266.

Hu, Z., Xu, T., Feng, B., et al. (2020). Thermal and fluid processes in a closed-loop geothermal system using CO_2 as a working fluid. Renewable Energy 154, pp. 351–367.

Huang, S. & Tian, T. (2006). Study of environmental impact in geothermal development and utilization. Proceedings of the 7th Asian Geothermal Symposium.

Hunt, T. M. (2001). Five lectures on environmental effects of geothermal utilization. United Nations University Geothermal Training Programme 2000, Report 1, Reykjavı́k, Iceland, pp. 109Irani, M., Ghannadi, S., Nick Daprocida, N. D., et al. (2022). On Numerical Modelling of the Hydraulic-Fractured Closed Loop Systems: Single Producer. SPE Canadian Energy Technology Conference, pp. D021S018R001.

Jia, Y., Tsang, C. -F., Hammar, A., et al. (2022). Hydraulic stimulation strategies in enhanced geothermal systems (EGS): a review. Geomechanics and Geophysics for Geo-Energy and Geo-Resources 8 (6), pp. 211.

Kocabas, I. (2005). Geothermal reservoir characterization via thermal injection backflow and interwell tracer testing. Geothermics 34 (1), pp. 27–46. https://doi.org/10.1016/j.geothermics.2004.09.003.

Kristmannsdóttir, H. & Ármannsson, H. (1995). Environmental impact of geothermal utilization in Iceland. Proceedings of the World Geothermal Congress 1995, Florence, Italy, pp. 2731–2734.

Kristmannsdóttir, H. & Ármannsson, H. (2003). Environmental aspects of geothermal energy utilization. Geothermics 32 (4–6), pp. 451–461.

Kristmannsdóttir, H., Sigurgeirsson, M., Ármannsson, H., et al. (2000). Sulfur gas emissions from geothermal power plants in Iceland. Geothermics 29 (4–5), pp. 525–538.

Kubo, B. M. (2003). Environmental management at Olkaria geothermal project, Kenya. Proceedings from the International Geothermal Conference, pp. 72–80.

Kubota, H. (2015). Social acceptance of geothermal power generation in Japan. Proceedings World Geothermal Cogress, pp. 19–25.

Kumari, W. G. P., Ranjith, P. G., Perera, M. S. A., et al. (2018). Experimental investigation of quenching effect on mechanical, microstructural and flow characteristics of reservoir rocks: Thermal stimulation method for geothermal energy extraction. Journal of Petroleum Science and Engineering 162, pp. 419–433.

Kurek, K. A., Heijman, W., van Ophem, J., et al. (2021). The contribution of the geothermal resources to local employment: Case study from Poland. Energy Reports 7, pp. 1190–1202.

Kwiatek, G., Bulut, F., Bohnhoff, M., et al. (2014). High-resolution analysis of seismicity induced at Berlín geothermal field, El Salvador. Geothermics 52, pp. 98–111.

Layman, E. B. (2017). Geothermal projects in Turkey: extreme greenhouse gas emission rates comparable to or exceeding those from coal-fired plants. Proceedings of the 42nd Workshop on Geothermal Reservoir Engineering, Stanford, CA, USA, pp. 13–15.

Li, N., Xie, H. P., Hu, J. J., et al. (2022). A critical review of the experimental and theoretical research on cyclic hydraulic fracturing for geothermal reservoir stimulation. Geomechanics and Geophysics for Geo-Energy and Geo-Resources 8 (1). p. 7. https://doi.org/10.1007/s40948-021-00309-7.

Livescu, S. & Dindoruk, B. (2022). Coupled Well-Reservoir Heat Modelling for Closed-Loop Geothermal Wells-A Feasibility Study. SPE Improved Oil Recovery Conference?, pp. D031S016R002.

Louie, E. P. & Solomon, B. D. (2017). Geography of geothermal energy technologies. In Handbook on the Geographies of Energy, pp. 163–176. Edward Elgar Publishing eBooks, Cheltenham, UK. https://doi.org/10.4337/9781785365621.00021.

Lu, S. -M. (2018). A global review of enhanced geothermal system (EGS). Renewable and Sustainable Energy Reviews 81, pp. 2902–2921.

Lund, J. W. & Freeston, D. H. (2001). World-wide direct uses of geothermal energy 2000. Geothermics 30 (1), pp. 29–68.

Magomadov, I., Balhasan, S. A., Elsayed, M., et al. (2022). Multilateral Fishbone Wells Utilization to Unlock the Production Potential and Analyze the Interference Phenomena of Marginal Field with Depletion Drive Mechanism. Abu Dhabi International Petroleum Exhibition and Conference, pp. D011S015R004.

Mahmoodpour, S., Singh, M., Turan, A., et al. (2022). Simulations and global sensitivity analysis of the thermo-hydraulic-mechanical processes in a fractured geothermal reservoir. Energy 247, pp. 123511. https://doi.org/10.1016/j.energy.2022.123511.

Manshad, A. K., Dastgerdi, M. E., Ali, J. A., et al. (2019). Economic and productivity evaluation of different horizontal drilling scenarios: Middle East oil fields as case study. Journal of Petroleum Exploration and Production Technology 9, pp. 2449–2460.

McTigue, J. D., Castro, J., Mungas, G., et al. (2018). Hybridizing a geothermal power plant with concentrating solar power and thermal storage to increase power generation and dispatchability. Applied Energy 228, pp. 1837–1852.

Mott, A., Baba, A., Mosleh, M. H., et al. (2022). Boron in geothermal energy: Sources, environmental impacts, and management in geothermal fluid. Renewable and Sustainable Energy Reviews 167, p. 112825.

Nagl, G. J. (1999). Controlling H2S emissions in geothermal power plants. Bulletin d'Hydrogeologie 17, pp. 393–402.

Newman, G. A., Gasperikova, E., Hoversten, G. M., et al. (2008). Three-dimensional magnetotelluric characterization of the Coso geothermal field. Geothermics 37 (4), pp. 369–399.

Nian, Y. -L. & Cheng, W. -L. (2018). Evaluation of geothermal heating from abandoned oil wells. Energy 142, pp. 592–607.

Oh, H., Akar, S., Davalos Elizondo, E., et al. (2023). Subsurface Characterization for Evaluating Geothermal Resource Potential from Existing Oil and Gas Wells in Tuttle, Oklahoma. National Renewable Energy Laboratory (NREL), Golden, CO (United States).

Fishbones AS. (n.d.). *Geothermal stimulation* Retrieved July 7, 2024, from https://www.fishbones.as/oil-and-gas.

Okoroafor, E. R., Smith, C. M., Ochie, K. I., et al. (2022). Machine learning in subsurface geothermal energy: Two decades in review. Geothermics 102, 102401. https://doi.org/10.1016/j.geothermics.2022.102401.

Olabi, A. G., Mahmoud, M., Soudan, B., et al. (2020). Geothermal based hybrid energy systems, toward eco-friendly energy approaches. Renewable Energy 147, 2003–2012.

Olasolo, P., Juárez, M. C., Morales, M. P., et al. (2016). Enhanced geothermal systems (EGS): A review. Renewable and Sustainable Energy Reviews 56, pp. 133–144.

Pan, L., Freifeld, B., Doughty, C., et al. (2015). Fully coupled wellbore-reservoir modeling of geothermal heat extraction using CO_2 as the working fluid. Geothermics 53, pp. 100–113.

Pandey, S. N., Chaudhuri, A., & Kelkar, S. (2017). A coupled thermo-hydro-mechanical modeling of fracture aperture alteration and reservoir deformation during heat extraction from a geothermal reservoir. Geothermics 65, pp. 17–31.

Patro, P. K. (2017). Magnetotelluric studies for hydrocarbon and geothermal resources: Examples from the Asian region. Surveys in Geophysics 38 (5), pp. 1005–1041.

Paulillo, A., Striolo, A., & Lettieri, P. (2019). The environmental impacts and the carbon intensity of geothermal energy: A case study on the Hellisheiði plant. Environment International 133, pp. 105226.

Pruess, K. (2006). Enhanced geothermal systems (EGS) using CO_2 as working fluid—A novel approach for generating renewable energy with simultaneous sequestration of carbon. Geothermics 35 (4), pp. 351–367.

Pruess, K. (2008). On production behavior of enhanced geothermal systems with CO_2 as working fluid. Energy Conversion and Management 49 (6), pp. 1446–1454.

Rühaak, W., Heldmann, C. -D., Pei, L., et al. (2017). Thermo-hydro-mechanical-chemical coupled modeling of a geothermally used fractured limestone. International Journal of Rock Mechanics and Mining Sciences 100, pp. 40–47.

Russo, S. Lo. (2020). Abandoned oil and gas wells exploitation by means of closed-loop geothermal systems: A review. *Geoingegneria Ambientale e Mineraria*, 160 (3), pp. 3–11.

Salimzadeh, S., Paluszny, A., Nick, H. M., et al. (2018). A three-dimensional coupled thermo-hydro-mechanical model for deformable fractured geothermal systems. Geothermics 71, pp. 212–224. https://doi.org/10.1016/j.geothermics.2017.09.012.

Settari, A., Ito, Y., & Jha, K. N. N. (1992). Coupling of a fracture mechanics model and a thermal reservoir simulator for tar sands. Journal of Canadian Petroleum Technology 31 (9), pp. 9–20. https://doi.org/10.2118/92-09-03.

Sharmin, T., Khan, N. R., Akram, M. S., et al. (2023). A state-of-the-art review on geothermal energy extraction, utilization, and improvement strategies: conventional, hybridized, and enhanced geothermal systems. International Journal of Thermofluids 18, p. 100323.

Shi, Y., Song, X., Li, J., et al. (2019). Numerical investigation on heat extraction performance of a multilateral-well enhanced geothermal system with a discrete fracture network. Fuel 244, pp. 207–226.

Shi, Y., Song, X., Shen, Z., et al. (2018). Numerical investigation on heat extraction performance of a CO_2 enhanced geothermal system with multilateral wells. Energy 163, pp. 38–51.

Soltani, M., Kashkooli, F. M., Dehghani-Sanij, A. R., et al. (2019). A comprehensive study of geothermal heating and cooling systems. Sustainable Cities and Society 44, pp. 793–818.

Soltani, M., Kashkooli, F. M., Souri, M., et al. (2021). Environmental, economic, and social impacts of geothermal energy systems. Renewable and Sustainable Energy Reviews 140, p. 110750.

Song, G., Song, X., Li, G., et al. (2021). An integrated multi-objective optimization method to improve the performance of multilateral-well geothermal system. Renewable Energy 172, pp. 1233–1249. https://doi.org/10.1016/j.renene.2021.03.073.

Sun, F., Yao, Y., Li, G., et al. (2018). Performance of geothermal energy extraction in a horizontal well by using CO_2 as the working fluid. Energy Conversion and Management 171, pp. 1529–1539.

Sun, Z., Jiang, C., Wang, X., et al. (2021). Combined effects of thermal perturbation and in-situ stress on heat transfer in fractured geothermal reservoirs. Rock Mechanics and Rock Engineering 54, pp. 2165–2181.

Taron, J. & Elsworth, D. (2009). Thermal–hydrologic–mechanical chemical processes in the evolution of engineered geothermal reservoirs. International Journal of Rock Mechanics and Mining Sciences 46 (5), pp. 855–864.

Tartakovsky, A. M., Marrero, C. O., Perdikaris, P., et al. (2020). Physics-Informed Deep Neural Networks for Learning Parameters and Constitutive Relationships in Subsurface Flow Problems. Water Resources Research 56 (5), p. e2019WR026731. https://doi.org/10.1029/2019WR026731.

Thain, I. & DiPippo, R. (2015). Hybrid geothermal-biomass power plants: applications, designs and performance analysis. Proceedings World Geothermal Congress, Melbourne, Australia, pp. 19–25.

Van Nguyen, M., Arason, S., Gissurarson, M., et al. (2015). Uses of geothermal energy in food and agriculture – Opportunities for Developing Countries. FAO, Rome., Italy.

Viesca, A. L. (1995). Environmental Impacts Associated with Geothermal Exploration, Development, and Power Generation. In: Energy and Mineral Potential of the Central American-Caribbean Region, Springer, Berlin, Heidelberg, pp. 283–286.

Wang, C. -L., Cheng, W. -L., Nian, Y. -L., et al. (2018). Simulation of heat extraction from CO_2-based enhanced geothermal systems considering CO_2 sequestration. Energy 142, pp. 157–167.

Wang, K., Liu, Z., Zeng, T., et al. (2022). Performance of enhanced geothermal system with varying injection-production parameters and reservoir properties. Applied Thermal Engineering 207, p. 118160. https://doi.org/10.1016/j.applthermaleng.2022.118160.

Wang, P., Wang, S., Wang, G., et al. (2024). Exploitation technology and productivity analysis of hydrothermal geothermal multi-branch wells. Geothermics 117, p. 102883.

Webster, J. G. & Timperley, M. H. (1995). Biological impacts of geothermal development. Brown, KL (Convenor), Environmental Aspects of Geothermal Development. World Geothermal Congress, pp. 97–117.

Weers, J., Anderson, A., & Taverna, N. (2022). The geothermal data repository: Ten years of supporting the geothermal industry with open access to geothermal data. National Renewable Energy Lab. (NREL), Golden, CO (United States).

Weers, J. & Huggins, J. (2019). Getting data out of the ground: modern challenges facing EGS collab, the DOE geothermal data repository, and the geothermal industry. 44th Workshop on Geothermal Reservoir Engineering.

Wei, C., Cheng, S., She, J., et al. (2021). Numerical study on the heat transfer behavior in naturally fractured reservoirs and applications for reservoir characterization and geothermal energy development. Journal of Petroleum Science and Engineering 202, pp. 108560. https://doi.org/10.1016/j.petrol.2021.108560.

Windrem, P. F. & Marr, G. L. (1981). Environmental problems and geothermal permitting. Natural Resources Law 14, p. 675.

Wu, Y. & Li, P. (2020). The potential of coupled carbon storage and geothermal extraction in a CO_2-enhanced geothermal system: a review. Geothermal Energy 8 (1), p. 19.

Xiong, F., Sun, H., Ye, Z., et al. (2022). Heat extraction analysis for nonlinear heat flow in fractured geothermal reservoirs. Computers and Geotechnics 144, p. 104641.

Xu, T. & Pruess, K. (2004). Numerical simulation of injectivity effects of mineral scaling and clay swelling in a fractured geothermal reservoir. (No. LBNL-55113). Lawrence Berkeley National Lab. (LBNL), Berkeley, CA (United State).

Yasin, Q., Majdański, M., Awan, R. S., et al. (2022). An analytical hierarchy-based method for quantifying hydraulic fracturing stimulation to improve geothermal well productivity. Energies 15 (19), p. 7368.

Yin, T., Dai, H., Zhuang, D., et al. (2024). Numerical simulation study of enhanced geothermal system modification and heat mining performance under cyclic injection. Applied Thermal Engineering 240, p. 122259. https://doi.org/10.1016/j.applthermaleng.2023.122259.

Zang, A., Oye, V., Jousset, P., et al. (2014). Analysis of induced seismicity in geothermal reservoirs–An overview. Geothermics 52, pp. 6–21.

20

Multiscale Reservoir Characterization of a CO_2 Storage Aquifer: Mineralogical, Geomechanical, and Petrophysical Analyses for a CCS Project in North Dakota?

Aimen Laalam[1], Ahmed Merzoug[2], Hichem Aymen Katib Chellal[3]

[1] *Colorado School of Mines, Golden, Colorado, USA*
[2] *University of Texas at Austin, Austin, Texas, USA*
[3] *University of North Dakota, Grand Forks, USA*

20.1 Introduction

In the global efforts to combat climate change and achieve the ambitious targets set by the Paris Agreement, Carbon Capture, Utilization, and Storage (CCUS) technology has emerged as a crucial tool. The technology plays a vital role in capturing carbon dioxide (CO_2) emissions from various sources, including industrial processes and the atmosphere, thereby preventing the release of CO_2 into the atmosphere and contributing to the establishment of a circular carbon economy (Friedlingstein et al., 2020). Industries such as steel, cement, and chemical manufacturing, known for their significant CO_2 emissions, can greatly benefit from the implementation of CCUS technologies (Fuss et al., 2014). The Global Carbon Capture and Storage Institute (GCCSI) emphasizes the importance of targeted interventions in these sectors to effectively reduce greenhouse gas emissions (Wang et al., 2022). The Intergovernmental Panel on Climate Change (IPCC) and the International Energy Agency (IEA) stress the urgency of limiting global warming to below 2 °C to avoid severe climate impacts, underscoring the role of CCUS in achieving net-zero emissions by 2050 (Salvia et al., 2021). Projections suggest that the deployment of CCUS technologies can significantly reduce atmospheric CO_2 concentrations and help mitigate the effects of global warming by 2100 (Ren et al., 2022). This technology not only aids in combating climate change but also contributes to safeguarding ecosystems and ensuring a sustainable future (Jung et al., 2020). Countries like China have set ambitious goals of "Carbon Peak" and "Carbon Neutrality," highlighting the importance of CCUS in achieving these targets (Saraji et al., 2013). Further research by Fuss et al. (2014) delves into the concept of negative emissions and the potential of technologies like CCUS in achieving climate change mitigation goals (Fuss et al., 2014). Wang et al. (2022) identify potential CCUS clusters and optimize pipeline networks in China, emphasizing the importance of strategic planning for effective carbon capture and storage (CCS) (Wang et al., 2022). Salvia et al. (2021) analyze local-level plans in the EU to achieve carbon neutrality by 2050, emphasizing the need for comprehensive strategies at various governance levels (Salvia et al., 2021). Ren et al. (2022) provide insights into the research

Geophysical Exploration for Hydrocarbon Reservoirs, Geothermal Energy, and Carbon Storage: New Technologies and AI-based Approaches, First Edition. Edited by Said Gaci.

landscape of CCUS in China and the United States, highlighting the importance of technological innovation in achieving emission reduction targets (Ren et al., 2022). Jung et al. (2020) analyze international policies and projects on CCU for climate change mitigation, showcasing the global efforts towards sustainable development (Jung et al., 2020). These studies collectively underscore the significance of CCUS technologies in the transition towards a low-carbon future and the achievement of carbon neutrality goals (Ampomah et al., 2016). This book chapter delves into the characterization of a saline aquifer for its potential use in CO_2 storage. It begins by providing a comprehensive overview of CCS technology, detailing the various phases of CCS projects, highlighting major global projects, and discussing the use of saline aquifers for CO_2 storage. Additionally, it explores the role of machine learning in enhancing CCS technology. Following the introductory overview, the chapter presents a detailed case study that focuses on the storage aspects of CCS. This case study thoroughly examines the storage system, including an in-depth characterization of the saline aquifer and the geological layers above and below it that serve as confining zones. This analysis aims to assess the aquifer's suitability for effectively and safely storing CO_2.

20.2 CCS Overview

From 1870 to 2015, approximately 232,000 tons of CO_2 were emitted by human activities, with current estimates suggesting around 40 gigatons of CO_2 are released annually (IPCC, 2018). To achieve the ambitious 1.5°C goal set in the Paris Agreement, the IPCC underscores the necessity for negative emission technologies to remove approximately 10 Gt of CO_2 each year. In terms of potential storage capabilities, about 800 sedimentary basins have been identified as suitable for underground CO_2 storage, which could play a critical role in achieving these targets (Bachu, 2015). According to the Intergovernmental Panel on Climate Change (IPCC) and the International Energy Agency (IEA), to avoid the most catastrophic impacts of climate change, it is critical to limit warming to below 2°C. Achieving net-zero emissions by 2050 will require a massive increase in renewable energy adoption, complemented by technologies like CCUS. The IPCC underscores that if deployed effectively, CCUS could be instrumental in reducing atmospheric CO_2 concentrations to 450 ppm by 2100, highlighting its potential to significantly mitigate global warming effects. This strategy is essential in our toolkit for combating climate change, safeguarding ecosystems, and ensuring a sustainable future for the planet.

20.2.1 CCS Project Phases

20.2.1.1 Carbon Capture

Capturing CO_2 entails navigating a series of technical and regulatory hurdles, particularly given the diverse flue gas emissions from sectors like iron and steel, cement production, and metal smelting. These industries face additional challenges in aligning with local power plant regulations, which dictate operational frequency, maintenance schedules, and managing decreases in power output. The adoption of large-scale post-combustion CO_2 capture technology marks a pivotal advancement, offering extensive benefits and transforming industry practices (Ma et al., 2022).

Different CO_2 capture techniques significantly impact the overall costs of CCS projects, with potential cost increases ranging from 70% to 80%. Among the prevalent methods are pre-combustion, post-combustion, and oxy-combustion. While post-combustion is tailored for exhaust gases with low CO_2 concentrations, pre-combustion techniques efficiently separate and store both CO_2 and hydrogen, achieving up to 90% capture efficiency but potentially reducing plant efficiency (Madejski et al., 2022). Oxyfuel combustion, known for superior energy efficiency, incurs an energy cost increase of 4% to 8% compared to post-combustion methods (Wienchol et al., 2020). Raza et al. (2017) assessed recent successes of advanced CO_2 capture technologies. For instance, the Port Arthur facility, employing a dual pressure swing adsorption (PSA) approach, successfully captured about 1 million tons of CO_2. The SaskPower BD3 facility has sequestered over 4.14 million metric tons since 2014, meeting 75% of its annual target. Petra Nova, in Texas, is recognized as the largest post-combustion CO_2 capture plant globally, with an annual capacity of approximately 1.4 million metric tons. The Oxy Low Carbon DAC project aims to extract 1 megaton of CO_2 per year by 2025, intending to use the captured CO_2 for enhanced oil recovery. Another notable project is the Illinois Industrial Carbon Capture and Storage Plant, which started injecting 1 million metric tons of CO_2 per year in 2017, showcasing the application of CCS technology in biofuel production (Raza et al., 2017).

20.2.1.2 CO_2 Transport

The behavior of CO_2 storage is significantly influenced by its solubility in water and the interfacial tension between CO_2 and water. These properties are both affected by environmental conditions; they increase under higher pressures and decrease with rising temperatures, as noted by Matter et al. (2014). For large-scale transportation across significant distances, pipelines are predominantly employed. CO_2 can be transported in three physical states: gaseous, liquid, and solid, with the gaseous form being the most utilized for practical purposes. The movement of carbon on a grand scale typically involves pipelines, maritime shipping, and storage tanks. Specifically, four types of vessels are dedicated to transporting liquefied carbon from large-scale sources. Reports from the Intergovernmental Panel on Climate Change (IPCC) between 2001 and 2005 (IPCC, 2001) indicate that these maritime vessels are essential for moving liquefied carbon to coastal distribution centers. The costs associated with this transportation method depend on various factors including the geographical location of the sites (both onshore and offshore) and the specifications of the pipeline infrastructure (Bennaceur, 2008). When CO_2 is under supercritical conditions, it is compressed to a density of approximately 900 kg/m^3. This state allows for more efficient transportation due to its lower density and reduced pressure drop per unit length, as detailed by Pires et al. (2011). Before beginning the CO_2 injection into storage sites, it's crucial to assess the operational costs thoroughly. The IPCC estimated in 2001 that constructing a 250 km pipeline could cost between \$1.80 and \$22.50 per metric ton of CO_2. Moreover, an analysis by the IPCC suggested that transportation costs for CCS are manageable if the distance from the source to potential storage basins is under 300 km (IPCC, 2001).

20.2.1.3 CO_2 Storage

Global sedimentary basins have the potential to store between 8200 and 35 000 gigatons of CO_2. Selecting sites for CO_2 storage requires careful evaluation of physical properties and how CO_2 behaves under different pressures and temperatures (de Coninck and Benson, 2014). Supercritical CO_2 storage in deep geological formations can handle 50 to 100 million tons per project, suitable for commercial rates (Kelemen et al., 2019). Optimal rocks for storage include sandstone, limestone, dolomite, and basalt, with shale and low-permeability carbonate rocks serving as effective caprocks due to their impermeability (Raza et al., 2016). CO_2 storage capacity in sedimentary reservoirs generally ranges from 0.5 to 5.0 metric tons per square meter, influenced by design and physical parameters. CO_2 sequestration is feasible in oil reservoirs, sandstone aquifers, coal beds, and salt beds (Bachu, 2003), with deep saline aquifers and depleted reservoirs showing promise. Experimental methods like carbon mineralization and storage in marine gas hydrates are also being explored. Table 20.1 summarizes the main storage methods currently available for CO_2 sequestration.

Table 20.1 Main storage options for CO_2 storage.

Storage option	Description	Pros	Cons
Depleted Hydrocarbon Reservoirs (Zheng et al., 2020)	Former oil and gas fields repurposed for CO_2 storage, leveraging existing geology known for its capability to trap fluids.	Established geology with proven cap integrity and available infrastructure.	Limited capacity relative to global emissions and potential risks of leakage over long-term.
Deep Saline Aquifers (Ma et al., 2022)	Deep underground formations saturated with brine that can be injected with CO_2. These are widespread and can be found in many regions around the world.	High storage capacity: less impacted by economic fluctuations compared to oil reservoirs.	Risk of inducing seismic activity due to increased pore pressure; management of over pressurization and caprock integrity are crucial.
Unmineable Coal Seams (National Academies of Sciences Engineering and Medicine, 2019)	Coal deposits that are not economically viable for mining can be used for CO_2 storage, with the added benefit of enhancing methane recovery.	Provides an opportunity to enhance coalbed methane production.	Storage capacity is generally lower than other geological formations; regional feasibility varies.
Carbon Mineralization (Zhang and DePaolo, 2017)	A process where CO_2 reacts with abundantly available minerals like basalt to form stable carbonates.	Permanently converts CO_2 into rock, preventing any future leak. Environmental and health benefits.	Higher costs involved in the process and the need for specific types of rock formations which may not be available everywhere.
Marine Gas Hydrates (Sloan and Koh, 2008)	CO_2 is injected into oceanic methane hydrate formations where it replaces methane, which can then be collected.	Potential to recover methane for use as a natural resource, providing an energy source.	Experimental technology with significant technical challenges and environmental concerns related to disturbing ocean sediments.

20.2.2 Major Successful CCUS Projects

According to Ma et al. (2022), the construction of large-scale CCS projects in developed countries can span 5–10 years, beginning with meticulous site selection processes. However, these projects are not without their risks, which include potential leaks, accidents, environmental impacts, and public health issues, thereby escalating the social risks associated with their deployment in local communities (Ma et al., 2022). Cherepovitsyn et al. (2020) argue that successful CCS implementations necessitate a combination of persistent CO_2 emission sources and proximate underground storage options such as depleted oil fields or deep saline aquifers, along with robust financial backing and a favorable regulatory framework to support project infrastructure development (Cherepovitsyn et al., 2020). Documented examples include Norway's Sleipner project, which has been operational since 1996 and has successfully stored about 20 million tonnes of CO_2 annually, a feat supported by Norway's carbon tax policy (Chadwick et al., 2010). Similarly, the Weyburn Project, starting in 2000, has become the most extensive interdisciplinary CCS project to date, capturing about 35 million tonnes of CO_2 at a rate of 1.8 Mt/year with governmental collaboration (Hitchon, 2012; Brown et al., 2017). Furthermore,

Table 20.2 Major CCUS projects worldwide.

Name	Year	CO_2 storage capacity
Sleipner, Norway	1996	0.9 Mt/year, stored about 20 Mt so far
Weyburn, Canada	2000	1.8 Mt/year, stored about 35 Mt so far
In Salah, Algeria	2004–2011	1.2 Mt/year; ~ Stored 3.8 million tonnes of CO_2
Tomakomai, Japan	2016–2019	Cumulative 300,000 t of CO_2
Snøhvit Norway	2008	0.7 Mt/year
Ketzin, Germany	2008–2013	Stored 67,271 t of CO_2
Aquistore, Canada	2015	Stored 350,000 t of CO_2
CO_2CRC Otway, Australia	2009	Stored 80,000 t of CO_2

Source: Adopted from Nath et al., 2024 / with permission of Elsevier.

the In-Salah project in Algeria focuses on CO_2 reinjection into aquifers, contributing significantly to the global CCS dialogue (Herzog, 2016). Collectively, Norway's Sleipner and Snøhvit facilities have sequestered over 22 million tons of CO_2 since their respective commencements in 1996 and 2008 (Martin-Roberts et al., 2021). Projections by Martin-Roberts et al. (2021) estimate that by 2050, the annual storage capacity should range between 3.0 and 6.5 gigatons (Martin-Roberts et al., 2021). Currently, the Global CCS Institute reports 65 major projects worldwide, with 26 operational, highlighting the critical need for expansive CCUS deployment to combat climate change (Ma et al., 2022). Table 20.2 summarizes the main CCS projects worldwide.

20.2.3 Carbon Storage in Saline Aquifers

The effective sequestration of CO_2 in deep saline aquifers is a critical component of CCS strategies, crucial for mitigating climate change. Extensive research has been conducted to enhance our understanding of the processes involved and to optimize the storage capacity and safety of these methods. Riaz et al. (2006) performed a linear stability analysis of density-driven miscible flow in porous media, providing foundational insights into the dynamics of CO_2 storage in saline aquifers (Riaz et al., 2006). This was followed by Xu et al. (2006) who investigated the influence of permeability on buoyancy-driven flow (Xu et al., 2006), and Bachu et al. (2007) who discussed methodologies for estimating CO_2 storage capacity, emphasizing the importance of accurate capacity assessments for large-scale CCS projects (Bachu et al., 2007). Rutqvist (2012) reviewed the geomechanics of CO_2 storage in deep sedimentary formations, particularly saline aquifers, contributing critical insights into the geomechanical aspects essential for safe and efficient CO_2 sequestration (Rutqvist, 2012). In 2013, Saraji et al. highlighted the viability of deep saline aquifers for carbon sequestration due to their large capacity and widespread availability, focusing on safe trapping mechanisms. They also investigated the wettability of supercritical CO_2/water/quartz systems, measuring contact angles and interfacial tensions at reservoir conditions, which aid in predicting and enhancing sequestration efficiency (Saraji et al., 2013). Cheng et al. (2012) explored the role of permeability anisotropy in enhancing CO_2 sequestration, furthering our understanding of geomechanical influences on CO_2 storage effectiveness (Cheng et al., 2012). Research continued with Bahadur et al. (2016), who utilized small-angle neutron scattering to examine porosity and flow dynamics during CO_2 sequestration, providing insights into the structural characteristics and fluid behaviors within storage reservoirs (Bahadur et al., 2016). More recent studies include Ampomah et al. (2016), who evaluated CO_2 storage mechanisms in CO_2-enhanced oil recovery sites, underscoring the significance of CCUS in reducing atmospheric CO_2 levels (Ampomah et al., 2016). Zhang et al. (2020) focused on modeling interfacial tension to predict CO_2 storage capacity in saline aquifers, utilizing supervised learning approaches and sensitivity analyses to refine sequestration processes (Zhang et al., 2020). This was complemented by He et al. (2022), who also used modeling techniques to optimize sequestration efficiency and capacity (He et al., 2022). Lastly, Luu et al. (2022) investigated the geomechanical implications of CO_2 injection, particularly induced seismicity in the Illinois Basin, contributing to the ongoing assessment of the safety and feasibility of geological carbon sequestration (Luu et al., 2022).

20.2.4 Machine Learning for CCS

Integrating machine learning to solve complex CCS problems was recognized as a useful tool in four main aspects. First, predicting CO_2 PVT (Pressure-Volume-Temperature) properties and phase behavior; second, CO_2 monitoring and leakage detection; third, operational constraint optimization; and finally, surrogate models to reduce the computational time of numerical simulation. Several authors leverage machine learning techniques to predict CO_2 PVT properties and trapping mechanisms. Ahmadi et al. (2016) proposed using a particle swarm-optimized feed-forward neural network and genetic algorithm-optimized least squares support vector machines to predict the viscosity and thermal conductivity of CO_2 in a sequestration environment. They trained their models using data collected from the literature (Ahmadi et al., 2016). Mohammadian et al. (2016) applied neural networks and genetic programming to predict CO_2 solubility in water aquifers given temperature, pressure, NaCl salinity, and stream purity. Their model was trained using sensitivity measurements from laboratory experiments (Mohammadian et al., 2016). Najafi-Marghmaleki et al. (2017) used feed-forward neural networks, radial basis function neural networks, least square support vector machine optimized with particle swarm optimization, and adaptive neuro-fuzzy inference system with hybrid optimization to predict the impact of gas impurities on CO_2 mole fraction in liquid and gas phases. They compare their model to the Peng-Robinson equation of state results. Their models take temperature, pressure, and critical properties of H_2, N_2, and O_2 to output the mole fraction in the liquid and the vapor phases (Najafi-Marghmaleki et al., 2017). Kim et al. (2017) trained a neural network to predict the storage efficiency of CO_2 storage in deep saline aquifers. The training data is generated using sensitivity analysis of a synthetic numerical reservoir simulation. Their neural network predicts the indexes of residual gas trapping and solubility gas trapping given reservoir porosity, thickness, permeability depth, residual gas saturation, and sequestration time. The neural network is then tested on a real-case numerical simulation, which shows high accuracy (Kim et al., 2017). Vo Thanh et al. (2020) applied neural network to predict oil recovery and CO_2 storage in enhanced oil recovery in residual oil zones. The training data was generated using synthetic numerical reservoir simulation. The model takes reservoir thickness, porosity, permeability, residual oil saturation, residual water saturation, gas injection rate, and production bottom hole pressure. The model is tested on five residual oil zones in the Permian basin, showing good agreement with history-matched reservoir model results (Vo Thanh et al., 2020). Safaei-Farouji et al. (2022) used Gaussian process regression, radial basis function, and random forest techniques to predict the interfacial tension of the CO_2-brine system from data from laboratory experimental work. Their models take the pressure, temperature, monovalent molality, bivalent molality, and mole fraction of CH_4 and N_2 to predict interfacial tension. They showed that random forest outperformed the other models and applied on Cuu Long Basin offshore in Vietnam showing good agreement with field data (Safaei-Farouji et al., 2022). Vo Thanh and Lee (2022) applied Gaussian process regression, support vector machines, and random forest to predict CO_2 residual and solubility trapping indexes in saline aquifers. The model permeabilities, porosity, residual gas saturation, salinity, injection rate, and permeability anisotropy ratio were used as input to predict the indexes. They applied the model on Cuu Long Basin offshore in Vietnam and compared it to results from numerical simulation, showing good prediction accuracy (Vo Thanh and Lee, 2022). Machine learning techniques have also assisted with CO_2 monitoring and leak detection. Zhong et al. (2019) used convolutional short-term memory to build a 2D spatiotemporal anomaly detection tool for CO_2 leakage. The model takes porosity, permeability, and injection rate. The pressure at the monitoring well is then predicted to detect potential anomalies in the monitoring pressure. They successfully applied their anomaly detection tool to Cranfield, Mississippi, CO_2 EOR project data (Zhong et al., 2019). Sinha et al. (2020) compared neural networks, convolutional neural networks, long short-term memory, and convolutional long short-term memory in anomaly detection for harmonic pulse testing techniques. Their models take pressure at the monitoring well, injection, and cumulative injection rates to predict the future pressure behavior at the monitoring well (Sinha et al., 2020). Bakhshian and Romanak (2021) developed DeepSense. They report that LSTM outperformed other models when applied to the Cranfield, Mississippi, CO_2 EOR project data (Bakhshian and Romanak, 2021). Park et al. (2023) introduced DeepNRMS, an autoencoder architecture that assists in monitoring CO_2 storage behavior using seismic data. The model cleans seismic images from the noise for better interpretability. DeepNRMs were applied to synthetic data and Aquistore CO_2 storage sites, showing promising results (Park et al., 2023). Hernandez Mejia et al. (2024) proposed using a long short-term memory autoencoder and isolation forest for anomaly detection at monitoring wells to detect CO_2 leakage. Their model analyzes the bottom hole pressure data at the monitoring well to flag any pressure anomalies compared with autoencoder results using the Tukey method (Hernandez Mejia et al., 2024). Liu et al. (2024) applied a neural network, gradient boosting, and multivariate adaptive regression spline to predict leakage risks of methane and CO_2 into shallow aquifers using reservoir depth, permeability, net-to-gross ratio, reservoir pressure, water saturation, oil composition, and time. They report that using several machine learning models for different input ranges adds flexibility to the model and increases prediction (Liu et al., 2024). He et al. (2024) proposed a workflow for CO_2 leak detection using a long short-term memory algorithm and Bayesian framework. The LSTM algorithm is used as a proxy to reservoir simulation, assuming cases with up to 3 leakage locations, and the affine invariant ensemble algorithm Bayesian framework inverts the CO_2 leakage location from the observed data and the LSTM algorithm (He et al., 2024). Harati et al. (2024) compared support vector machines, k-nearest neighbor regression, decision tree regression, random forest, and artificial neural network for predicting CO_2 leakage location and size given wellhead and bottom hole pressures and temperatures. They found that K-nearest neighbor regression outperformed the other models in detecting leaks. They generated the data using a steady-state multiphase flow simulator with 100% CO_2 fluid injection (Harati et al., 2024). CO_2 storage wells operating constrained has been another area of interest to several researchers. Cameron and Durlofsky (2012) proposed a workflow for optimizing CO_2 trapping in aquifers. They use the Hooke-Jeeves direct search algorithm to optimize well location, injection rates, and brine cycling timing and volumes to minimize CO_2 mobility and maximize trapping. They showed that injecting brine significantly decreases CO_2 mobility (Cameron and Durlofsky, 2012). Cihan et al. (2015) proposed a methodology using the constrained differential evolution algorithm to optimize well location and injection rate and minimize brine extraction. Their work maximizes CO_2 trapping while constraining pressure changes to ensure cap rock integrity and prevent fault reactivation (Cihan et al., 2015). Sun et al. (2021) introduced a general workflow to optimize CO_2 storage well location and control to maximize CO_2 storage while preventing fault reactivation using a safety factor. A flow-geomechanics coupled reservoir model was used to build an artificial neural network-based proxy model to be used as a part of the net present environmental value function. The workflow uses the steepest ascent algorithm to find the optimal well location and injection rate while accounting for fault reactivation safety (Sun et al., 2021). Jahandideh et al. (2021) proposed a data assimilation workflow to track CO_2 plumes using microseismic events. They used two numerical models for flow and microseismic data prediction. The microseismic response is then assimilated using an ensemble-based stochastic point process filter, and the ensemble smoother with multiple data assimilation. The workflow was applied to the Farnsworth field in Texas (Jahandideh et al., 2021). Zou and Durlofsky (2023) developed a general framework to optimize well locations and injection constraints of horizontal/deviated CO_2 sequestration wells. They aimed to minimize CO_2 mobility (i.e. plume) and maximize storage efficiency under operational constraints. They use particle swarm optimization and differential evolution for a multi-fidelity optimization process. The reservoir grid resolution increases during the course of optimization to reduce the computational complexity related to running a high-resolution optimization process (Zou and Durlofsky, 2023). Zheng et al. (2023) proposed a general optimization workflow for CO_2 geological storage to maximize CO_2 storage and minimize the risks associated with cap rock integrity with the change of injection rates. CO_2 leakage and stress changes were incorporated into the optimization function. Their models allowed for storing larger CO_2 quantities under geomechanical and leakage constraints (Zheng et al., 2023). Chen et al. (2024a) proposed a data assimilation and forecasting workflow to incorporate different data types of CO_2 storage sites. The workflow assimilates data at well location and spatial time-lapse CO_2 plume evolution (inferred from seismic) using an advanced ensemble smother multiple data assimilation method called ES-MDA-GEO. They were able to constrain their model prediction better and showed that their approach is robust to noise of 1–10% (Chen et al., 2024a). Other researchers leverage machine learning for building reservoir surrogate models to predict CO_2 plume and migration. Zhong et al. (2019) applied a conditional deep convolutional generative adversarial network to build a surrogate model for 2D CO_2 plume prediction. The CO_2 plume is tracked by predicting saturation distribution at time '*t*' from the initial permeability distribution (Zhong et al., 2019). Wen et al. (2021) developed CCSNet, a 2D machine learning surrogate model to reduce the computational expense of numerical simulation. A spatial-temporal model based on a modified convolutional neural network-based autoencoder architecture. They use their model to estimate the CO_2 plume, pressure distributions, mass balance, solubility trapping, and sweep efficiency given reservoir pressure, temperature, permeability, thickness, relative permeability, capillary pressure, and completion and operational constraints. Their model shows a 10^3 to 10^4 reduction in computational time with high accuracy (Wen et al., 2021). Tang et al. (2021) proposed using U-Net as a 2D surrogate model for reservoir simulation of CO_2 plume and pressure. Their models take permeability distribution and injection rates to predict CO_2 plume. They use an ensemble smother multiple data assimilation framework with the surrogate model to match the history of reservoir permeability, assuming the same well location. They used dimensionality reduction techniques to reduce the parameters and data spaces (Tang et al., 2021). Tang et al. (2022) used recurrent U-Net to build a 3D surrogate model for CO_2 fluid flow and geotechnical surface displacement. The model predicts 2D surface displacement and 3D pressure and saturation distributions given permeability and porosity distributions with fixed well locations. They also applied the surrogate model to history to match the geological model, including permeability and porosity, using rejection sampling from principal component analysis-based geomodelling (Tang et al., 2022). Yan et al. (2022) proposed a 3D deep learning workflow to predict the pressure changes due to CO_2 injection in a heterogeneous porous media from permeability, porosity, rate, and time. The deep learning model uses the Fourier Neural Operator and efficient coarsening techniques to reduce the computational time and preserve accuracy. They reported lower training time and memory requirements. Their model has a 10^3 speedup in predicting the deep learning model compared to the numerical simulation (Yan et al., 2022). Chu et al. (2022) developed a CO_2 plume and pressure radial proxy model using residual U-Net Fourier neural operator in Shaly-sandstone. They showed that moderate shale layer length, with lower permeability, restricts CO_2 migration and improves storage efficiency (Chu et al., 2022). Witte et al. (2023) utilized the Fourier neural operator and 3D wavelet transform to build a CO_2 storage surrogate model. They showed that their proposed model performs better than a simple Fourier neural operator in predicting CO_2 front and calculation speed by applying it to Norway's Sleipner field scale CO_2 storage project (Witte et al., 2023). Jiang and Durlofsky (2024) combine adversarial autoencoder and convolutional long short-term memory to build a spatiotemporal surrogate model for pressure and saturation fields. They apply the surrogate model as a data-space inversion technique to leverage the ensemble smoother with multiple data assimilation for history matching the reservoir model to the observed data (Jiang and Durlofsky, 2024). Chen et al. (2024b) proposed using ResNetUNet, long-term memory, and transformers to predict static and dynamic earth surface displacement, given that the reservoir structure assumes a constant injection rate and volume (Chen et al., 2024b). The application of machine learning assisted in solving several CO_2 sequestration problems. Machine learning methods can be used as an alternative for fluid properties calculations, support optimization and decision-making under different constraints, and aid in the monitoring and detection of geological CO_2 leaks. They also support numerical simulation efforts by reducing the computational time. These models can be useful for local application to areas where the models have been developed. However, these applications are limited to training data and selected input design. For instance, no models account for relative permeability changes in the reservoir or residual saturation hysteresis in the 3D proxy models. Thus, extending these methods beyond the training data and selected input to production mode is very challenging. The practitioner needs to be aware of this limitation.

20.3 Case Study: Carbon Storage in the Broom Creek Saline Aquifer, Williston Basin, North Dakota

Developing a geological storage site for CO_2 can take approximately 10 years and involves extensive site assessment and data gathering (Nath et al., 2024). There are three main risks associated with underground CO_2 storage. First, there are potential direct leakage paths within geological formations that could include channels through underground layers (Heuberger et al., 2017). Second, injecting CO_2 into subterranean structures might trigger seismic events, potentially creating or widening fractures in the caprocks that could lead to CO_2 escape (Azar et al., 2010). Additionally, natural seismic activity can damage the structural integrity of CO_2 storage facilities, including borehole walls, increasing the chances of leaks (Heuberger et al., 2016). During the injection and extraction phases, wellbores are subjected to various structural stresses such as changes in diameter, ruptures, or collapses. Over time, the casings and cement linings of older wells can degrade, posing further risks to the stability and safety of the geological storage system.

20.3.1 Overview of the Study Area

This project site is on the central portion of the Williston Basin. The Williston Basin is an intracratonic sedimentary basin covering approximately 150,000 square miles, with its depocenter near Watford City, North Dakota. Overall, the stratigraphy of the Williston Basin has been well studied, particularly the numerous oil-bearing formations. The Williston Basin has been identified as an excellent candidate for long-term CO_2 storage because of, in part, the thick sequence of clastic and carbonate sedimentary rocks and the basin's subtle structure character and tectonic stability (Peck et al., 2014). The target CO_2 storage reservoir is the Broom Creek Formation, a predominantly sandstone horizon lying about 5900 ft below DGC's Great Plains Synfuels Plant, a good source for CO_2. Mudstones, siltstones, and interbedded evaporites of the Opeche Formation unconformably overly the Broom Creek and serve as the primary confining zone (Figure 20.1). The Amsden Formation (dolostone, limestone,

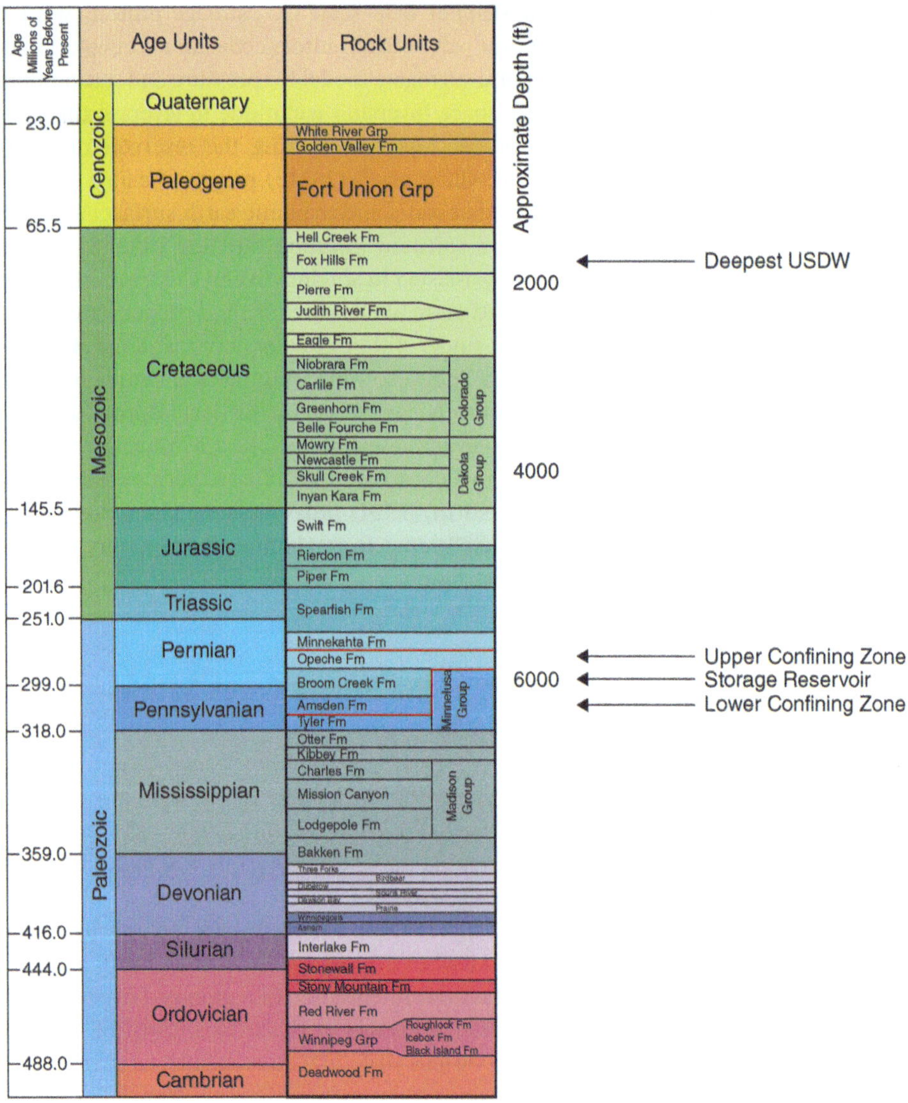

Figure 20.1 The stratigraphic column identifying the storage reservoir, confining zones, and lowest USDW addressed in this permit application for the CO_2 Sequestration Project. *Source*: (North Dakota Department of Mineral Resources, n.d.) / State of North Dakota.

and anhydrite) unconformably underlies the Broom Creek Formation and serves as the lower confining zone (Figure 20.1). Together, the Opeche, Broom Creek, and Amsden comprise the CO_2 storage complex for the CO_2 Sequestration Project. Including the Opeche Formation, there is ~1100 ft of impermeable formations between the Broom Creek Formation and the next overlying porous zone, the Inyan Kara Formation. An additional ~2700 ft of impermeable intervals separates the Inyan Kara and the lowest USDW, the Fox Hills Formation (Figure 20.1). In 2023, Laalam and Tomomewo conducted a comprehensive study of various geological storage methods for hydrogen in North Dakota, with a focus on saline aquifers and salt caverns. They identified three significant aquifers and five principal salt formations suitable for hydrogen storage. Among these, the Broom Creek Formation was noted for its potential viability (Laalam et al., 2023). This formation, which also holds promise for CO_2 storage, is a component of the upper unit of the Permian Minnelusa Group located predominantly within the Williston Basin. It features a distinctive geological makeup, consisting primarily of sandstone with interspersed layers of dolostone and anhydrite, as detailed by Sorensen et al. (2009) (Sorensen et al., 2009). A visual representation of the Broom Creek Formation's location within the Williston Basin is provided in Figure 20.1. The Broom Creek Formation is capped by the Permian Opeche Formation, predominantly composed of continuous layers of red shale and silt with interspersed anhydrite and halite, as illustrated in Figure 20.2. Additionally, a substantial Cretaceous-age shale layer, spanning several thousand meters, is positioned above the lowest drinking water aquifer, providing an extra protective layer above the Broom Creek Formation. Conversely, the formation's lower boundary is sealed by the Amsden Formation, noted for its low permeability and its makeup of anhydrite, dolomite, and shale. The base of the Broom Creek Formation lies approximately 1615 m (about 5300 ft) above the interface with the Precambrian basement rock, a depth that mitigates concerns about seismic activity potentially induced by injection processes (Peck et al., 2019). Research into the Broom Creek Formation has primarily focused on its potential for CO_2 sequestration, particularly aiming to mitigate greenhouse gas emissions from fossil fuel-based power generation. Comprehensive studies of the formation's lithology, petrophysical attributes, mineral composition, and geomechanical properties have been undertaken to evaluate its suitability for CO_2 storage and the robustness of its sealing capabilities. These studies include detailed analyses of its permeability, porosity, structural resilience, and the prevailing in-situ stress conditions (Leroux et al., 2020).

20.3.2 Data Sources

To evaluate the suitability of the injection and confining zones for CO_2 storage and containment at the CO_2 Sequestration site, multiple data sets were utilized. These included both pre-existing data and new data specifically collected for this project. The pre-existing data, sourced from the NDIC online database, comprised publicly accessible well logs and formation top depths for 120 wellbores within a 5472 mi^2 area centered on the proposed storage site. This information was instrumental in determining the depth, thickness, and spatial distribution of the underground geological formations. Additionally, laboratory measurements from core samples of the Broom Creek Formation, obtained from five wells, were analyzed. These analyses helped establish correlations between measured petrophysical properties and those estimated from well-log data, which were then integrated with the new, site-specific data. The site-specific data collection effort involved drilling several wells to acquire subsurface geological data essential for the development of a CO_2 storage facility permit. This included geophysical well logs, fluid analyses, and 2D seismic data, focusing on the Opeche, Broom Creek, and Amsden Formations (Figure 20.2). These site-specific data sets were critical in assessing the viability of the storage complex for the safe and permanent sequestration of CO_2. They also provided key inputs for constructing geological models, conducting numerical simulations of CO_2 injection, performing geochemical and geomechanical analyses, and developing monitoring

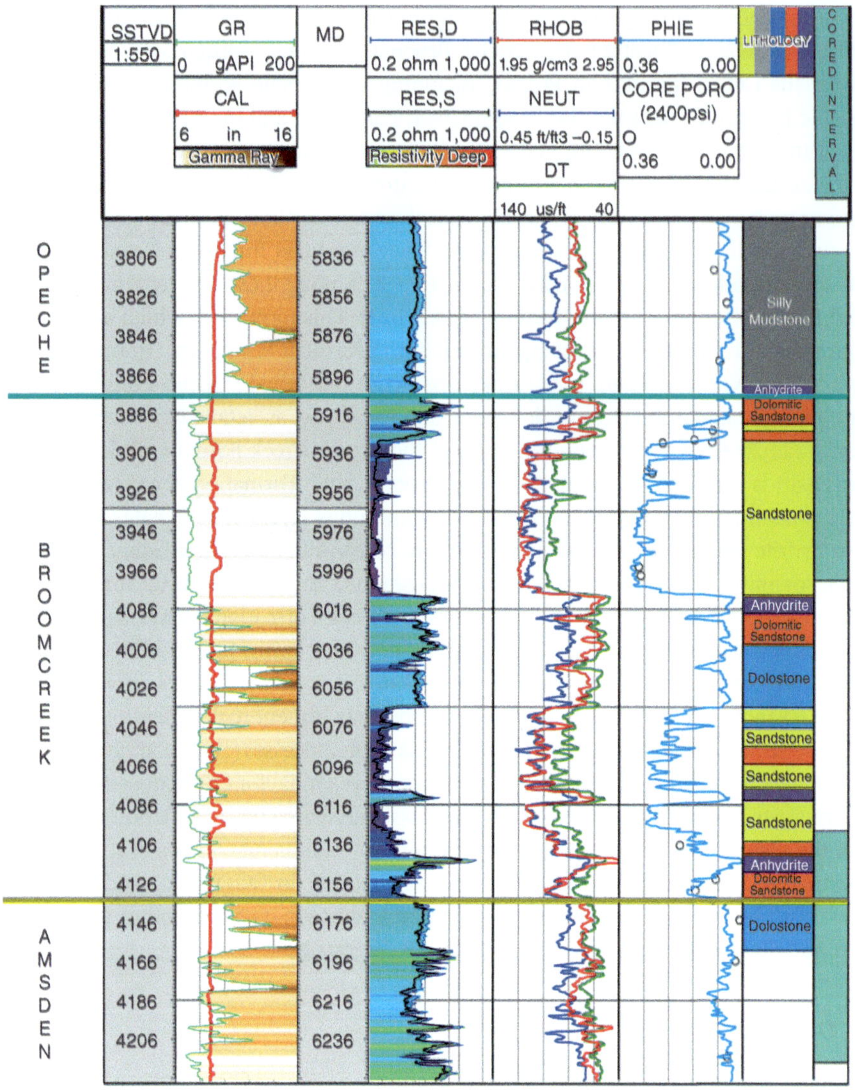

Figure 20.2 Schematic Showing Vertical relationship of coring (rightmost track) and core plug porosity (third track from right) intervals in the Opeche, Broom Creek, and Amsden Formations in the study well. *Source*: (North Dakota Department of Mineral Resources, n.d.) / State of North Dakota.

technologies. This comprehensive dataset enhanced understanding of the subsurface conditions and informed the development of monitoring strategies and influenced the design and operational plans of site infrastructure and equipment (North Dakota Department of Mineral Resources, n.d.).

20.3.3 Geophysical Well Logs Interpretation

Geophysical well-log interpretation involved gathering openhole wireline geophysical logs across the entire exposed section of the wellbore. The suite of logs collected included caliper, gamma ray (GR), density, porosity, dipole sonic, resistivity, combinable magnetic resonance (CMR), spectral GR, and specialized logs such as fracture finder or image logs. These logs were instrumental in identifying formation top depths, interpreting lithology and petrophysical properties, and generating synthetic seismic traces to correlate depth with seismic time. Formation tops were identified from the Pierre Formation to the Amsden Formation, enhancing the existing dataset for 120 wellbores over a 5472 mi^2 area. This integration helped delineate the geological scope, depth, and thickness of subsurface layers and aided in the creation of structural surfaces for geological modeling. Within the specific study well, the Broom Creek Formation was documented at a depth of 5906 feet with a thickness of 258 feet, composed of 134 feet of sandstone, 35 ft of dolostone, 24 ft of anhydrite, and 65 ft of dolomitic sandstone. Throughout the study area, the thickness of the Broom Creek Formation ranged from 163 to 322 ft, averaging 249 ft. Data from offset wells indicated that the net sandstone thickness varied from 24 to 205 ft, averaging 99 ft across the simulation model area.

The upper boundary of the Broom Creek Formation was delineated by a shift from the high GR signatures of the mudstones and siltstones of the Opeche Formation to the lower GR signatures of the Broom Creek Formation's sandstone and dolostone (Figure 20.3). The base of the

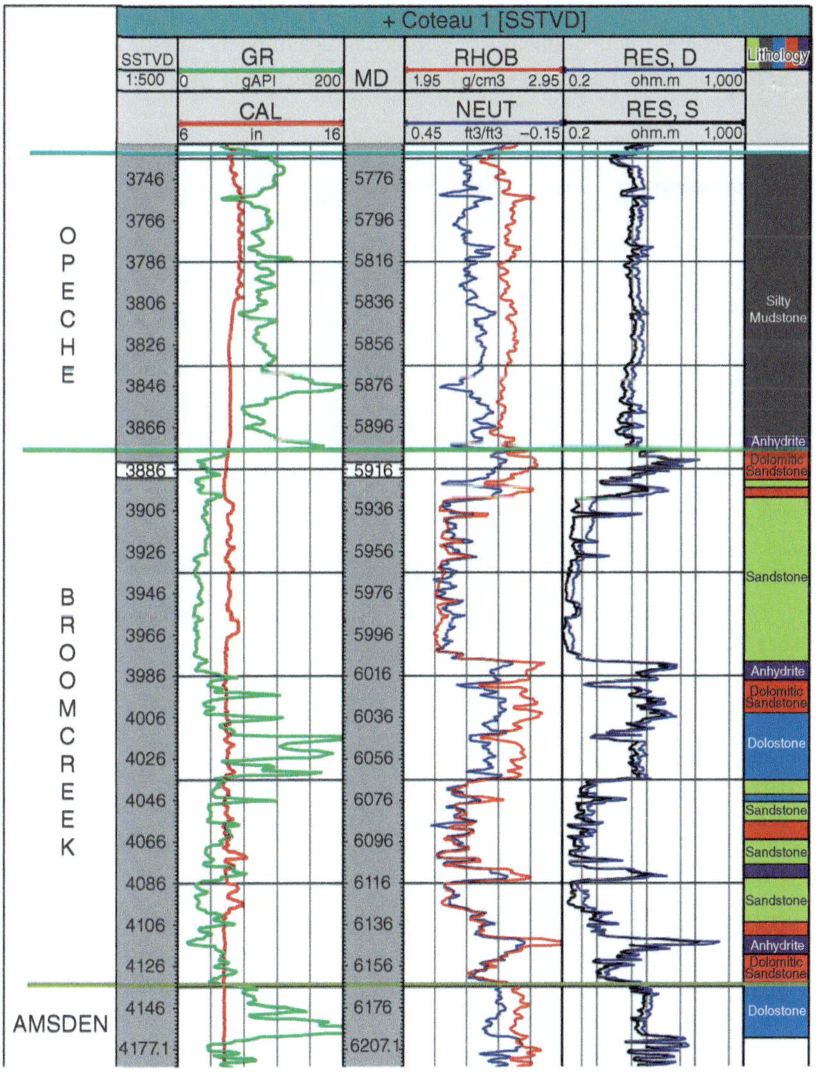

Figure 20.3 Well log display of the interpreted lithologies of the Opeche, Broom Creek, and upper Amsden formations. *Source*: (North Dakota Department of Mineral Resources, n.d.) / State of North Dakota.

Amsden Formation was marked by a high GR signature of argillaceous dolostone, which could be consistently traced across the CO_2 Sequestration Project area. Additionally, 2D seismic data collected for site characterization supported the structural correlation and provided more precise estimations of reservoir thickness. The structural analysis suggested minimal major stratigraphic discontinuities near the study well, as shown in Figures 20.4 and 20.5.

20.3.4 Core Analysis

A 283 ft core sample was extracted from the Broom Creek storage complex in the study well. Detailed analyses were performed on this core to characterize the lithologies of the Broom Creek, Opeche, and Amsden Formations, aligning these findings with well log data. The core analysis encompassed a variety of tests, including measurements of porosity and permeability, X-ray diffraction (XRD), X-ray fluorescence (XRF), relative permeability testing, thin-section analysis, capillary entry pressure measurements, and triaxial geomechanics testing. These analyses were crucial for refining geologic, geochemical, and geomechanical models and for inputting predictive simulations (North Dakota Department of Mineral Resources, n.d.).

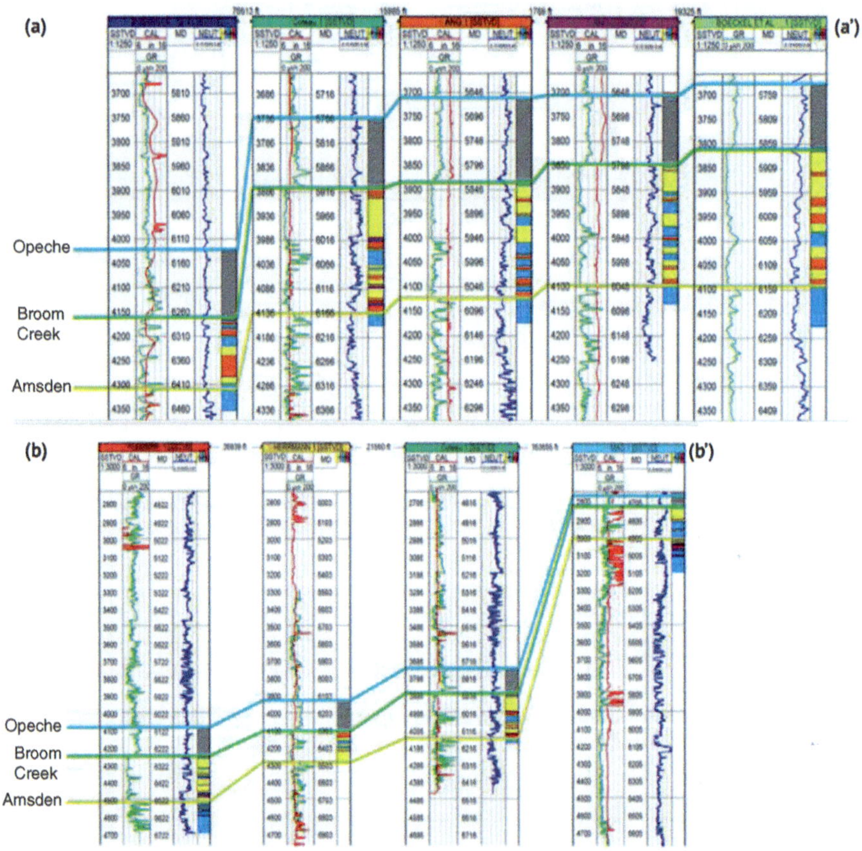

Figure 20.4 Regional well log stratigraphic cross sections of the Opeche and Broom Creek Formations flattened on the top of the Amsden Formation. The logs displayed in tracks from left to right are GR (green) and caliper (red); neutron porosity (blue); and interpreted lithology log. *Source*: (North Dakota Department of Mineral Resources, n.d.) / State of North Dakota.

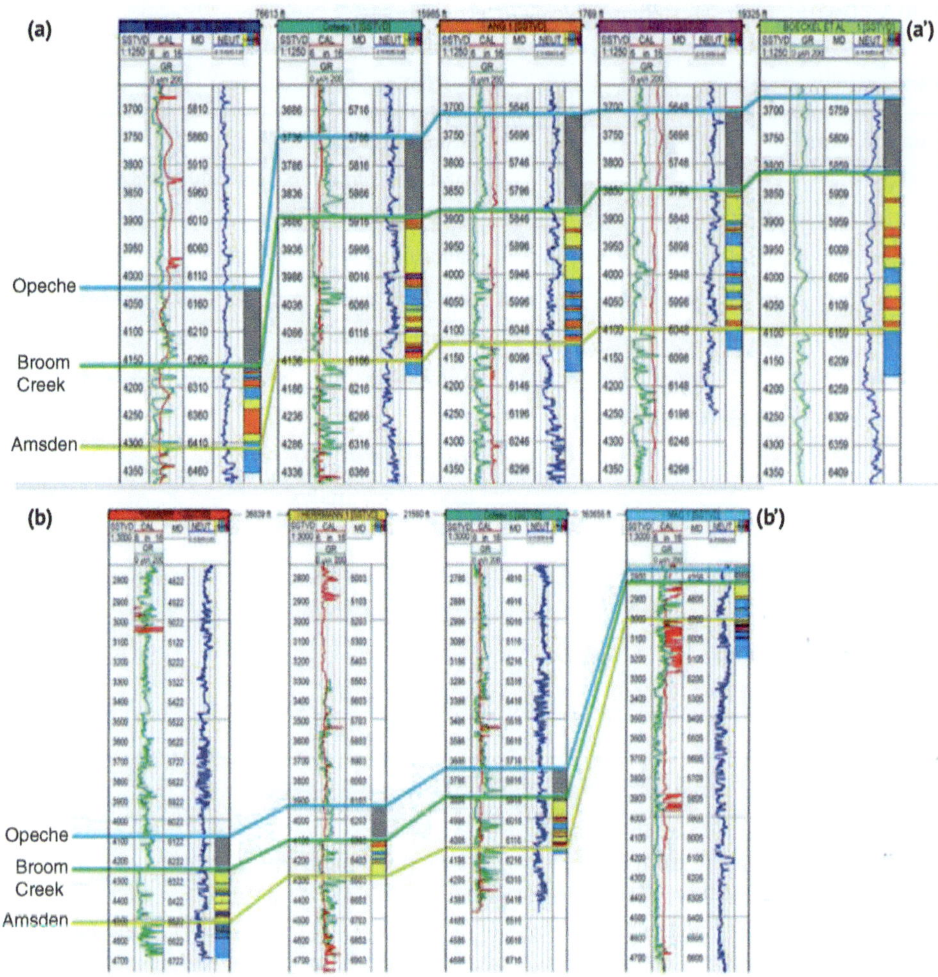

Figure 20.5 Regional well log cross sections showing the structure of the Opeche, Broom Creek, and Amsden Formations. The logs displayed in tracks from left to right are GR (green) and caliper (red); neutron porosity (blue); and interpreted lithology log. *Source*: (North Dakota Department of Mineral Resources, n.d.) / State of North Dakota.

From this core, twenty-two 1-inch-diameter core plug samples were collected from the sandstone and dolostone layers. Three of these samples, taken from depths of 5941.9′, 5969.9′, and 5994.4′, were duplicated and reoriented 90 degrees from the original orientation to assess potential orientation-dependent permeability variations within the reservoir. The other 19 samples were analyzed to map out porosity and permeability distributions across the formation. The Broom Creek Formation core samples demonstrated a porosity range of 1.41–34.39% at 800 psi and 7.88–30.34% at 2400 psi. Permeability values varied widely from 0.13 to 12 300 millidarcies (mD) at 800 psi and from 0.118 to 3990 mD at 2400 psi, reflecting the heterogeneity between the sandstone and dolostone lithofacies in the Broom Creek Formation (Table 20.3). In particular, the sandstone portion of the core showed porosity values between 8.73 and 34.39% at 800 psi, and between 7.88 and 30.34% at 2400 psi, with averages of 25.10 and 21.28%, respectively. The permeability of these sandstone samples ranged from 3.22 to 9660 mD at 800 psi and from 2.92 to 3990 mD at 2400 psi, with geometric averages of 728.35 mD and 221.84 mD, respectively. For the dolostone samples, porosity ranged from

Table 20.3 Description of CO_2 storage reservoir (injection zone).

Property	Description
Formation name	**Broom creek**
Lithology	Sandstone, dolostone, dolomitic sandstone, anhydrite
Formation top depth, ft	5906
Thickness, ft	Sandstone 134
	Dolostone 35
	Dolomitic Sandstone 65
	Anhydrite 24
Capillary entry pressure (CO_2/brine), psi	0.72
Geologic properties	
Formation/property	**Laboratory analysis**
Broom creek (sandstone)	
Porosity, %	21.28 (7.88–30.34)
Permeability, mD	221.84 (2.92–3990)
Broom creek (dolostone)	
Porosity, %	8.79 (8.66–8.94)
Permeability, mD	0.180 (0.118–0.361)

Source: (North Dakota Department of Mineral Resources, n.d.).

1.41 to 12.31% at 800 psi and from 8.66 to 8.94% at 2400 psi, with averages of 6.64 and 8.79%, respectively. Dolostone permeability ranged from a minimal 0.001 to 1.62 mD at 800 psi and from 0.118 to 0.361 mD at 2400 psi, with geometric averages of 0.109 and 0.180 mD (Table 20.3 and Figure 20.6). These results highlight the varied structural and compositional properties within the Broom Creek Formation.

Sandstone intervals in the Broom Creek Formation are associated with low GR, low density, high porosity (neutron, density, and sonic), low resistivity due to high porosity and brine salinity, and high sonic velocity measurements. The dolostone intervals in the formation are associated with an increase in GR measurements compared to the sandstone intervals, in addition to high density, low porosity (neutron, density, and sonic), high resistivity, and low sonic velocity measurements.

20.3.5 Mineralogy

The Broom Creek Formation is primarily characterized by fine- to medium-grained sandstone, with smaller quantities of carbonates and anhydrites. For detailed mineralogical analysis, 131.25 ft of the formation were explored, from which 22 depth intervals were sampled for X-ray Diffraction (XRD) and X-ray Fluorescence (XRF) chemical analysis. Eighteen of these samples were further used to create thin sections, providing a robust confirmation of the formation's mineral constituents. In the upper sections of Broom Creek, thin-section analysis revealed that

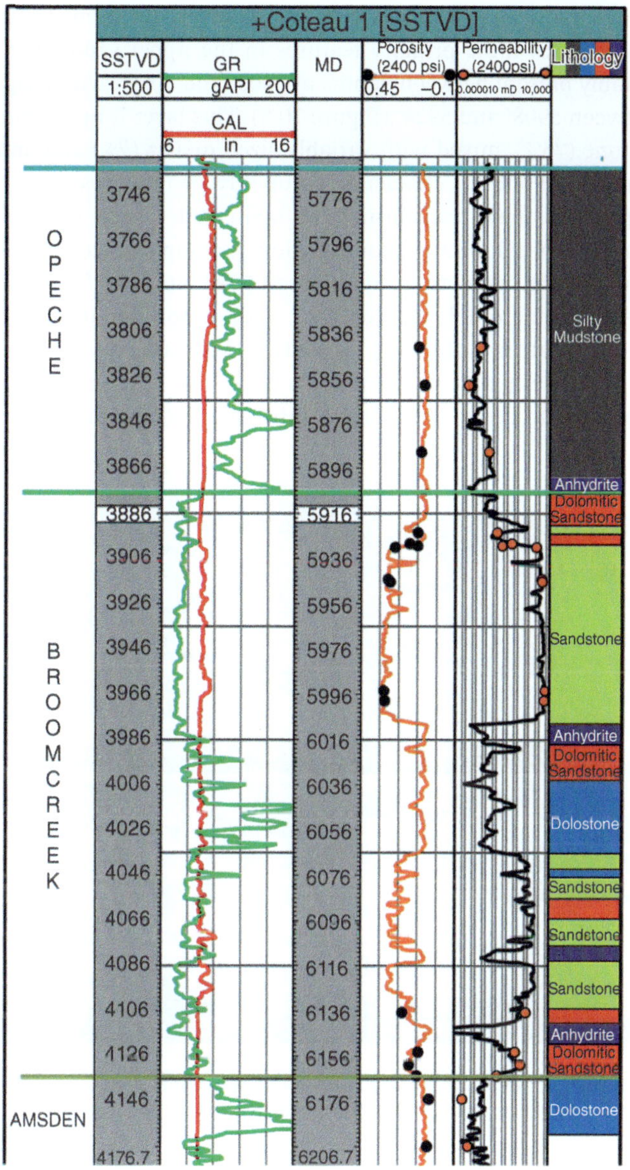

Figure 20.6 Vertical distribution of core-derived porosity and permeability values in the CO_2 sequestration project storage complex. *Source*: (North Dakota Department of Mineral Resources, n.d.) / State of North Dakota.

quartz is the predominant mineral, constituting 84% of the composition. This is accompanied by smaller proportions of feldspar (6%), dolomite (5%), and anhydrite (5%). Anhydrite typically crystallizes between quartz grains, reducing intercrystalline porosity. The quartz often displays overgrowth and occasionally, signs of dissolution. The grains generally exhibit long (straight) to tangential contact and are frequently rimmed by thin layers of red-brown to dark-red iron oxides. Porosity within these intervals varies from 15 to 34%, although one sample at a depth of 6146 feet showed significantly reduced porosity (9%) due to extensive anhydrite cementation. Figure 20.7 illustrates these primary features in the upper sand layer of the formation. Carbonates are notably present between depths of 5903′ and 6001′, with a significant dolostone layer identified between 5908′ and 5924′ (Figure 20.8). This layer is predominantly very fine- to fine-grained dolostone (75%), mixed with variably sized quartz (7%) and anhydrite (18%). The average porosity here is about 8%, characterized mainly by intercrystalline and moldic structures. Diagenetic processes have transformed original calcite grains into dolomite, with some bivalve shell fragments also exhibiting dolomitization. A compact dolomite section was encountered just before ending the first coring run between 5999′ and 6001′. This section is primarily composed of dolosparite/micro-dolosparite (78%), with minor amounts of clay (11%) and iron oxides within the rock matrix. Anhydrite, present as clasts and veins, accounts for 7% of the composition, and very fine quartz grains make up 4%. The observed porosity in this section averages 7%, largely due to the dissolution of anhydrite and the presence of open fractures.

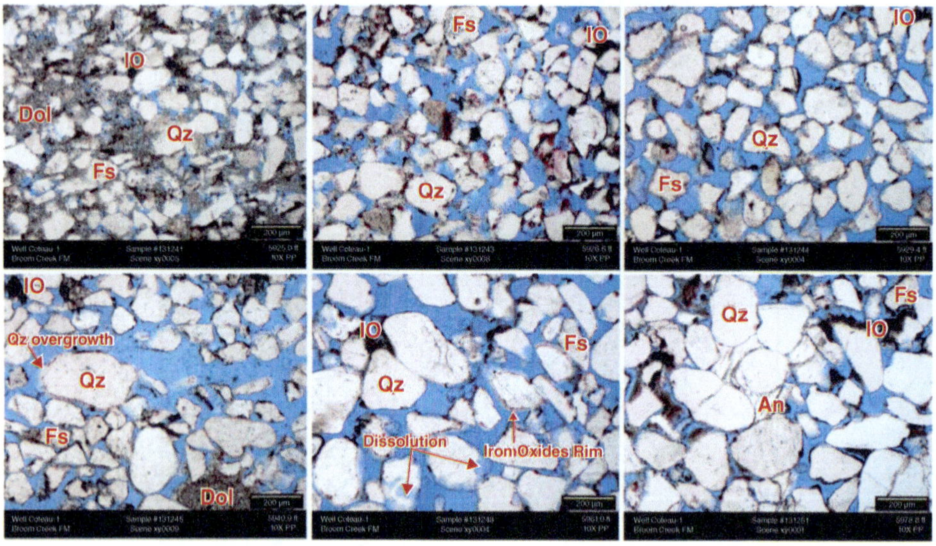

Figure 20.7 Thin sections from the upper sand interval of the Broom Creek formation. *Source*: (North Dakota Department of Mineral Resources, n.d.) / The Department of Mineral Resources.

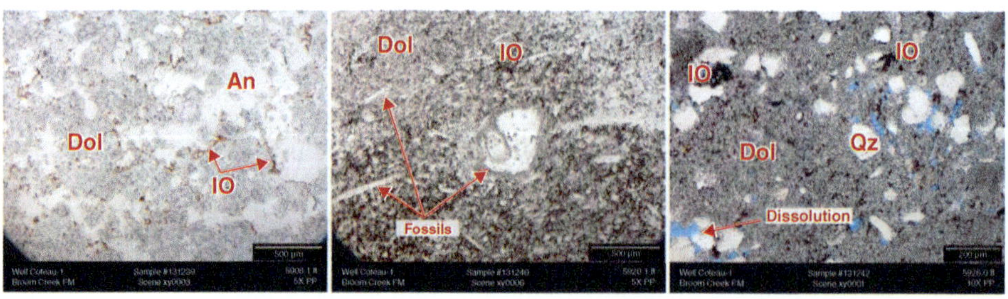

Figure 20.8 Thin sections from the three carbonate depth intervals of the upper Broom Creek formation. *Source*: (North Dakota Department of Mineral Resources, n.d.) / The Department of Mineral Resources.

It's important to note that the fractures observed at this scale, particularly in thin-section samples, may be artifacts introduced during the sample preparation process, evident in Figure 20.9. The deepest carbonate occurrence within the Broom Creek Formation is observed between the depths of 6130′ and 6163′. Particularly from 6160′ to 6163.25′, this segment features quartz-rich dolomite, often referred to as sandy dolomite, which primarily consists of micro-dolomite (54%), quartz (35%), feldspar (10%), and a small amount of clay (1%). The presence of iron oxides is also notable in this layer. In this interval, quartz minerals exhibit some signs of dissolution, and the grains are generally separated by a dolomitic matrix and occasionally by iron oxide cements, presenting tangential contacts between them. The observed porosity, averaging 9%, primarily results from the dissolution of feldspar. These specific characteristics of the carbonate segment are depicted in Figure 20.10, illustrating the distinct mineral composition and structural properties at these depths.

X-ray diffraction (XRD) data from the samples corroborated the facies interpretations derived from core descriptions and thin section analyses. The core from the Broom Creek Formation is predominantly made up of quartz, feldspar, carbonates, anhydrite, and clay, along with other minor minerals. These findings are detailed in Figure 20.11.

X-ray fluorescence (XRF) data for the Broom Creek Formation, illustrated in Figure 20.12, confirm the presence of sandstone and dolomite intervals, indicated by high percentages of SiO_2 (71–98%), CaO (19–36%), and MgO (13–21%). Additionally, elevated levels of CaO and SO_3 at depths of 5908, 6141, and 6154 feet suggest the presence of anhydrite beds within the formation. The data also show minimal clay content across all samples, with percentages ranging from as low as 0.04% up to 10.54%.

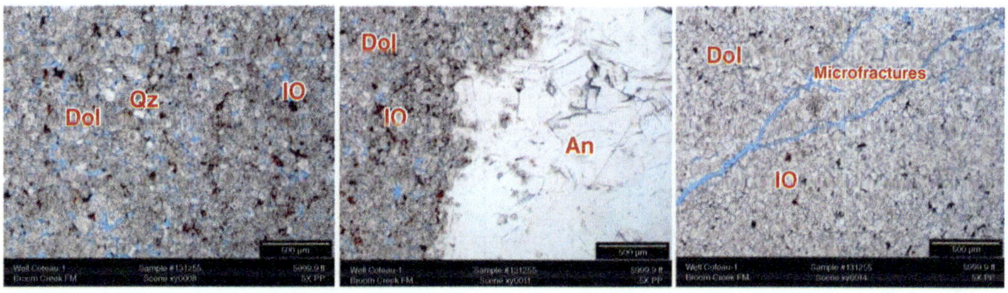

Figure 20.9 Thin section from the carbonate depth interval of the middle Broom Creek formation. *Source:* (North Dakota Department of Mineral Resources, n.d.) / The Department of Mineral Resources.

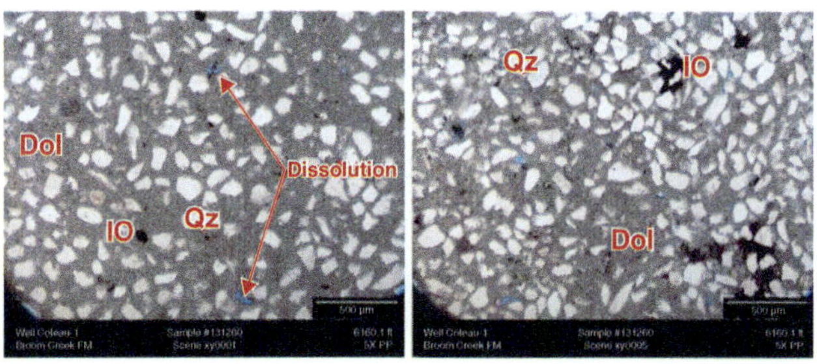

Figure 20.10 Thin section from a carbonate depth interval of the lower Broom Creek Formation. *Source:* (North Dakota Department of Mineral Resources, n.d.) / The Department of Mineral Resources.

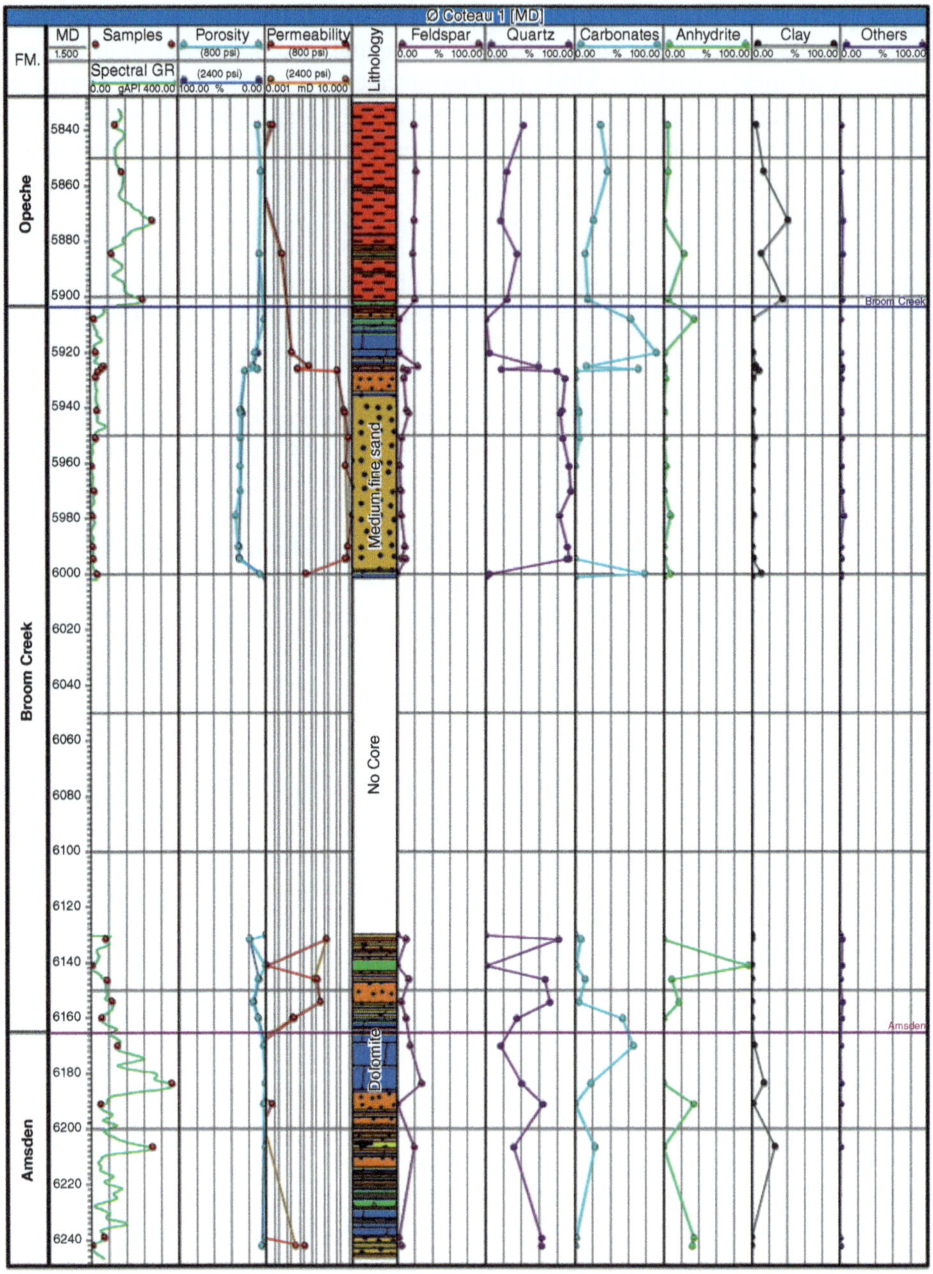

Figure 20.11 Described core and laboratory-derived mineralogic characteristics of the Opeche, Broom Creek, and Amsden formations. *Source*: (North Dakota Department of Mineral Resources, n.d.) / State of North Dakota.

20.3.6 Geomechanical Evaluation

During the drilling of the testing well, challenging hole conditions prevented the performance of an openhole MDT microfracture in situ stress test, which would have measured formation breakdown pressure, fracture closure pressure, fracture propagation pressure, and minimum

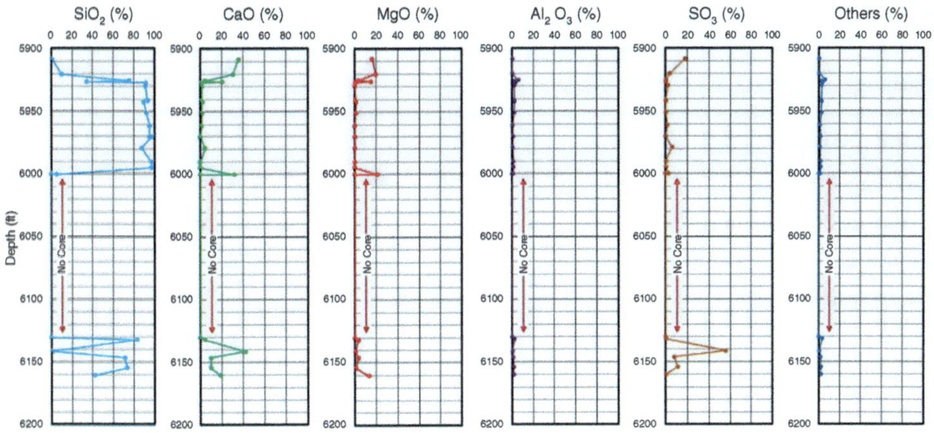

Figure 20.12 XRF data from the Broom Creek formation from the Study Well. *Source*: (North Dakota Department of Mineral Resources, n.d.) / State of North Dakota.

horizontal stress. To compensate for this data gap, a 1D Mechanical Earth Model (MEM) for the Opeche, Broom Creek, and Amsden Formations was developed using core data from laboratory tests and well log data.

The 1D MEM was instrumental in calculating the formation breakdown pressure, fracture closure pressure, and fracture propagation pressure for the Broom Creek Formation. Breakdown pressure was calculated by setting the minimum tangential stress around the wellbore circumference to zero and employing the Kirsch (1898), Aadnoy (2008), and Grandi et al. (2002) equations. It was assumed that the fracture propagation pressure is equivalent to the fracture pressure, allowing for the estimation of a maximum threshold where connected flow might be sustained. This estimated fracture pressure was also taken as the fracture closure pressure.

The fracture closure pressure was defined using the minimum horizontal stress (Shmin), which can typically be estimated using a modified Eaton calculation method. This method is seen as providing a lower boundary for the reservoir's fracture closure pressure or the maximum stress before a breakdown in the system's competency occurs. The modified Eaton formula, which has been validated against field test data and is widely accepted in the industry, is detailed in Eq. (20.1):

$$P = \frac{\nu}{1-\nu}\left((S_v - \alpha_v)P_p\right) + \alpha_H P_p \tag{20.1}$$

where:

- P is pressure.
- ν is Poisson's ration.
- S_v is the vertical stress.
- α_v is the vertical Biot's constant.
- α_H is the horizonal Biot's constant.
- P_p is pore pressure.

The results of the calculations are shown in Table 20.4.

The average estimated fracture propagation pressure gradient of 0.71 psi/ft agrees with the average fracture propagation values determined from microfracture in situ stress tests in other regional wells. So, there is confidence in the calculated value for fracture propagation pressure gradient and the predicted maximum BHP.

Table 20.4 Broom Creek microfracture interpreted results from the study well.

Parameter	Value
Pressure (psi)	
Breakdown	5193 psi
Avg. Fracture Propagation	4263 psi
Avg. Closure	4014 psi
Gradient (psi/ft)	
Breakdown	0.85
Avg. Fracture Propagation	0.71
Avg. Closure	0.71

Source: (North Dakota Department of Mineral Resources, n.d.) / State of North Dakota.

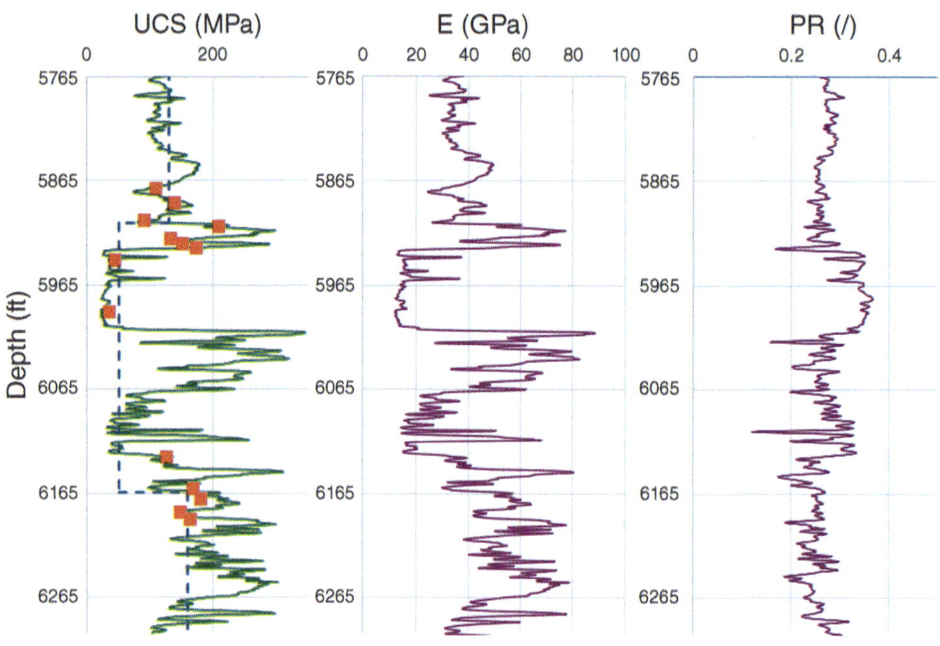

Figure 20.13 Elastic and mechanical properties of the study region.

The necessary elastic properties, such as Young's Modulus (E) and Poisson's Ratio (PR), along with the mechanical property of Unconfined Compressive Strength (UCS), were calculated and calibrated using laboratory experiments documented in well reports. Figure 20.13 displays the calibrated elastic and mechanical properties for the study region that are ready for application. Due to a limited number of calibration points, the dashed line in the first track illustrates the average UCS values utilized for constructing the simulation model in this study.

Similarly, average values derived from log data for Young's Modulus and Poisson's Ratio were employed to develop a three-zone simulation model. This model aims to assess the mechanical integrity of the rock within the storage system. The reliance on average values was primarily due

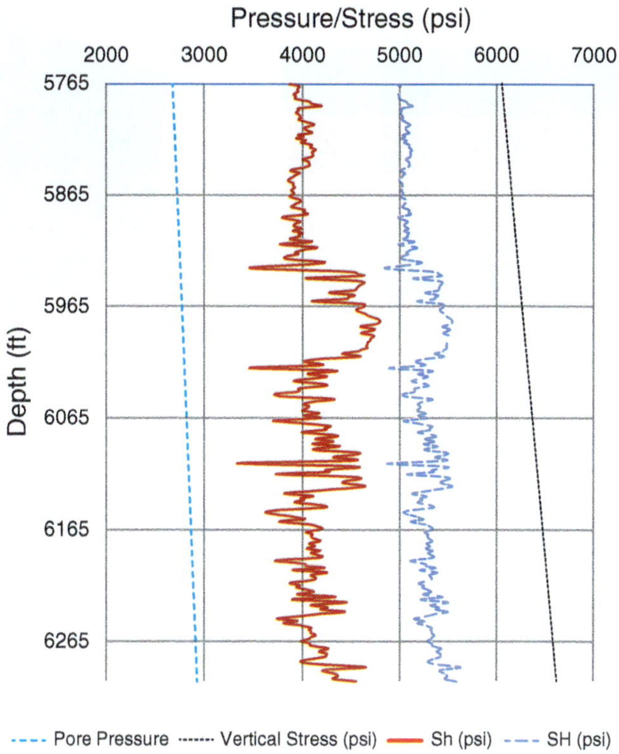

Figure 20.14 *In situ* stress field in the study region.

to the extensive computational resources required to run a flow-geomechanics coupled simulation model that incorporates the detailed variability observable in the logs.

Figure 20.14 displays the pore pressure and in situ stress, calculated using Eaton's stress estimation model, alongside reported values from formation test data obtained in the study area's offset wells.

20.3.6.1 Geomechanics-Flow Coupled Simulation Model Description

A shoebox model was developed to simulate the storage complex, utilizing average values from the previously mentioned logs for petrophysical and geomechanical properties. This three-zone model assigns the average value for each formation across its entire thickness. The model spans a horizontal area of 20,000 × 20,000 ft and vertically from the top of the Opeche formation at 5765 ft to the bottom of the Amsden formation at 6300 ft. The model is divided into 30 layers, each with a thickness of 18 ft. In the horizontal plane, the model comprises 50 by 50 grid cells, with grid cell sizes gradually increasing from 50 ft near the well to 1000 ft at the model's outer edges.

The model features a single well centrally located (see figure 20.15), designed to inject CO_2 at a rate of 1 Mt/year. The maximum operating bottomhole pressure is restricted to 90% of the minimum horizontal stress observed in the reservoir formation. To simulate this, the CMG-GEM compositional simulator was employed, utilizing a sequential one-way geomechanics coupling scheme. This setup was used to model a 30-year injection phase followed by a 100-year monitoring period. Since no relative permeability or capillary pressure curves were available in the well report, the curves used in this study were sourced from recent studies by Chellal et al. (2023).

Figure 20.15 Simulation model.

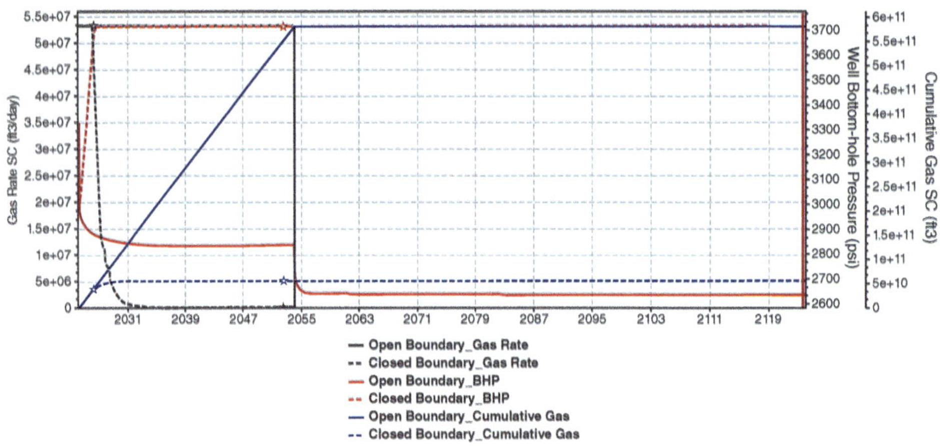

Figure 20.16 Injection well performance metrics in a closed boundary vs. open boundary model.

The model was tested under both open and closed boundary conditions, as illustrated in Figure 20.16. In the open boundary scenario, the model successfully managed 30 years of CO_2 injection at a rate of 1 MT/year, resulting in a total of 30 MT of CO_2 injected by the end of the period. This scenario saw a moderate increase in the injector bottom hole pressure, rising by 300 psi above the initial formation pore pressure. Conversely, the closed boundary system could only accommodate 10 MT of CO_2, with the maximum bottom hole pressure surging by 1120 psi, reaching the upper limit of injection pressure set at 90% of the fracture pressure. Consequently, the closed boundary system appears more prone to rock failure. To further assess this risk, the potential for rock failure in both systems will be analyzed using Mohr circles. This analysis will detail the stress states at the beginning, the end of the injection period, and following the monitoring period, providing deeper insights into the structural integrity under different boundary conditions.

20.3.6.2 Open Boundary Model

Figure 20.17 displays the variation in stress within the cap rock of the open boundary model, particularly above the injector where the influence of injection is most pronounced. The changes in effective stress in this region are minimal. Given the high Unconfined Compressive Strength

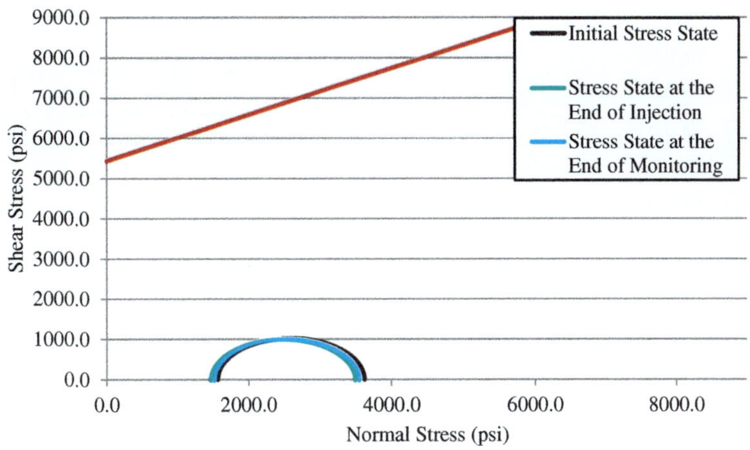

Figure 20.17 Changes in the effective stress of the cap rock in open boundary model.

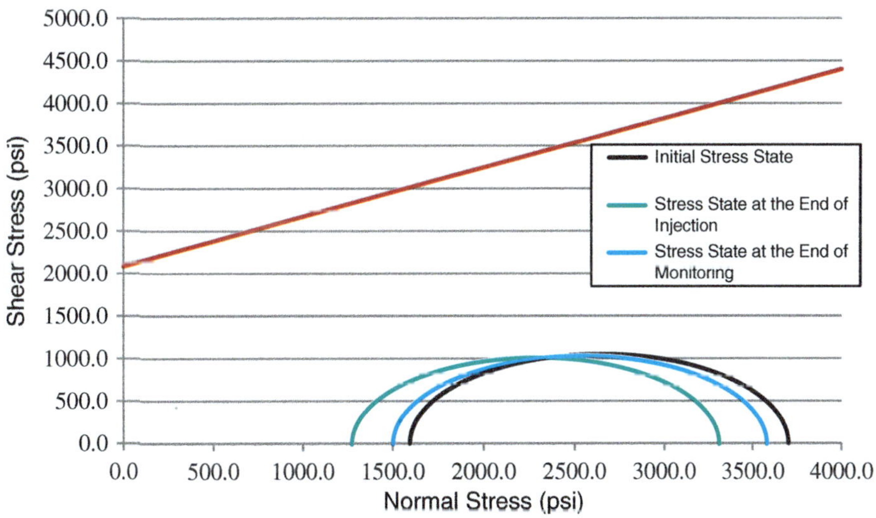

Figure 20.18 Changes in the effective stress of the reservoir rock in open boundary model.

(UCS) of the rock, which indicates its robustness, the likelihood of intact rock failure in the cap rock is considered very low.

Conversely, as depicted in Figure 20.18, the reservoir underwent significant changes in effective stress due to the injection process. However, by the end of the monitoring period, the stress state nearly returned to its initial condition. Despite the lower strength of the reservoir rock compared to the caprock, which makes it more susceptible to failure, the observed changes in stress relative to the mechanical strength of the rock suggest that mechanical failure is unlikely to occur.

In Figure 20.19, it is evident that the bedrock experienced only a minor change in stress state compared to the reservoir rock. The high mechanical strength of the bedrock renders these stress variations insignificant in terms of rock integrity. The Mohr circles, when compared to the rock's cohesion, are very small, indicating that these stress changes are not sufficient to cause compressional failure of the rock.

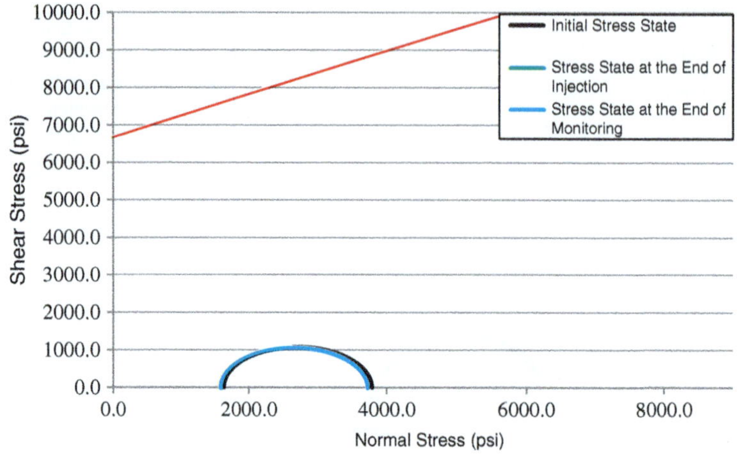

Figure 20.19 Changes in the effective stress of the bedrock in open boundary model.

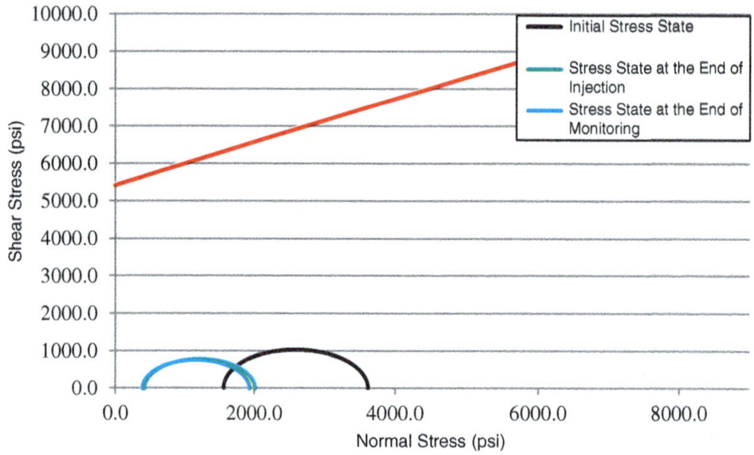

Figure 20.20 Changes in the effective stress of the caprock in closed boundary model.

20.3.6.3 Closed Boundary Model

In the closed boundary model, despite a smaller amount of CO_2 being injected into the storage system, there was a more significant and considerable change in the stress state of the cap rock compared to the open boundary model, where the changes were negligible. This contrast can be observed when comparing results in the earlier figures with those presented in Figure 20.20.

Similarly, as observed in the cap rock, a considerable change in the stress state was also noted in the reservoir rock, as illustrated by Figure 20.21. This change suggests that tensile failure is more probable than compressional failure, given that the Mohr circles approach zero confining stress. Furthermore, the size of the circles is smaller than the rock's cohesion, which effectively rules out the possibility of compressional failure.

The bedrock experienced a significant change in its in-situ stress state; however, due to the formation's very high strength, compressional failure of the rock is highly unlikely (Figure 20.22).

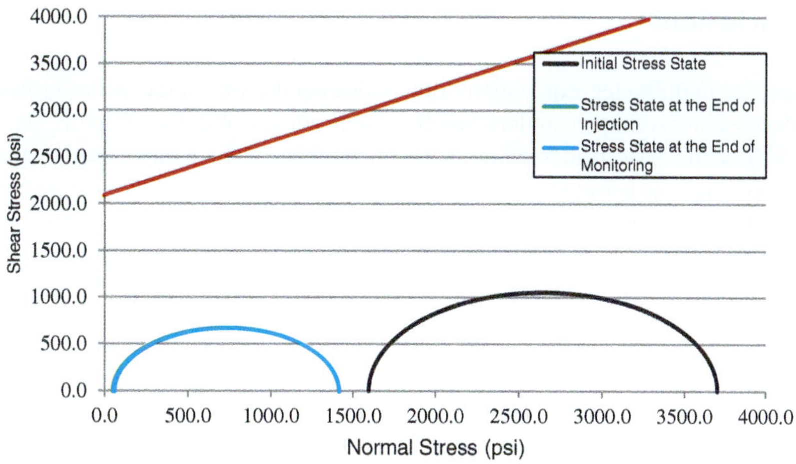

Figure 20.21 Changes in the effective stress of the reservoir rock in closed boundary model.

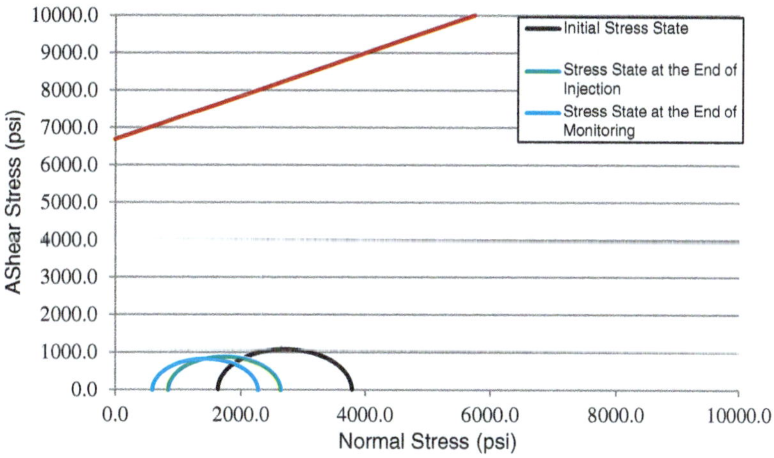

Figure 20.22 Changes in the effective stress of the bedrock in open boundary model.

20.3.6.4 Discussion

Geomechanical characterization plays a crucial role in CO_2 storage projects. Our study highlights the challenges associated with closed boundary conditions in CO_2 storage, primarily due to the limited volume of CO_2 that can be stored and the higher risk of compromising the mechanical integrity of the storage system. In such settings, the changes in the in-situ stress state are significantly more pronounced. If operational limits, such as maintaining the bottom hole pressure (BHP) within 90% of the fracture gradient, are not strictly adhered to, there is a high likelihood of mechanical failure occurring within the storage formation, as well as in the cap rock and bedrock.

Conversely, open boundary storage systems demonstrate smaller changes in in-situ stresses, thereby reducing the likelihood of mechanical failure across various regions of the storage system, including the reservoir, cap rock, and bedrock. This makes open boundary conditions more favorable for the long-term stability and safety of CO_2 storage projects.

20.4 Conclusions

In conclusion, the multifaceted exploration of the saline aquifer within the Williston Basin for CO_2 storage, as detailed in this chapter, underscores the significant role of Carbon Capture Utilization and Storage (CCUS) technologies in addressing global climate change. This comprehensive study not only outlines the geological and petrophysical characteristics of the Broom Creek Formation but also highlights its capacity and integrity for long-term CO_2 sequestration. Through the integration of advanced mineralogical, geomechanical, and petrophysical analyses, the research supports the feasibility of the aquifer as a robust carbon storage solution. Further, the application of machine learning techniques in enhancing the accuracy of CO_2 storage models exemplifies the potential of technological advancements in optimizing CCUS processes. These innovations are pivotal in refining storage efficiency, monitoring, and leakage detection, which are critical to ensuring the safety and effectiveness of CCUS projects. However, the successful deployment of CCUS technologies requires more than just technological advancement; it demands comprehensive policy support, regulatory frameworks, and public acceptance. Future research should focus on reducing the costs and improving the efficiency of carbon capture technologies, exploring economically viable carbon utilization pathways, and rigorously assessing the environmental impacts of CCUS technologies. Moreover, integrating CCUS with renewable energy systems could further enhance the sustainability and viability of these technologies as a climate change mitigation strategy. This holistic approach will not only help in mitigating the adverse effects of climate change but also support the transition towards a low-carbon future, ensuring the environmental integrity and sustainability of global ecosystems.

References

Aadnøy, B. S., & Belayneh, M. (2004). Elasto-plastic fracturing model for wellbore stability using non-penetrating fluids. Journal of Petroleum Science and Engineering, 45 (3–4), pp. 179–192. https://doi.org/10.1016/j.petrol.2004.07.006.

Ahmadi, M. A., Kashiwao, T., Rozyn, J., et al. (2016). Accurate prediction of properties of carbon dioxide for carbon capture and sequestration operations. Petroleum Science and Technology 34 (1), pp. 97–103. https://doi.org/10.1080/10916466.2015.1107847.

Ampomah, W., Balch, R., Cather, M., et al. (2016). Evaluation of CO_2 storage mechanisms in CO_2 enhanced oil recovery sites: Application to morrow sandstone reservoir. Energy & Fuels 30 (10), pp. 8545–8555. https://doi.org/10.1021/acs.energyfuels.6b01888.

Azar, C., Lindgren, K., Obersteiner, M. et al. (2010). The feasibility of low CO_2 concentration targets and the role of bio-energy with carbon capture and storage (BECCS). Climatic Change 100 (1), pp. 195–202.

Bachu, S. (2003). Screening and ranking of sedimentary basins for sequestration of CO_2 in geological media in response to climate change. Environmental Geology 44 (3), pp. 277–289.

Bachu, S. (2015). Review of CO_2 storage efficiency in deep saline aquifers. International Journal of Greenhouse Gas Control 40, pp. 188–202. https://doi.org/10.1016/j.ijggc.2015.01.007.

Bachu, S., Bonijoly, D., Bradshaw, J., et al. (2007). CO_2 storage capacity estimation: Methodology and gaps. International Journal of Greenhouse Gas Control 1 (4), pp. 430–443. https://doi.org/10.1016/s1750-5836(07)00086-2.

Bahadur, J., Medina, C., He, L., et al. (2016). Determination of closed porosity in rocks by small-angle neutron scattering. Journal of Applied Crystallography 49 (6), pp. 2021–2030. https://doi.org/10.1107/s1600576716014904.

Bakhshian, S. & Romanak, K. (2021). DeepSense: A physics-guided deep learning paradigm for anomaly detection in soil gas data at geologic CO_2 storage sites. Environmental Science & Technology 55 (22), pp. 15531–15541. https://doi.org/10.1021/acs.est.1c04048.

Bennaceur, K. (2008). CO_2 Capture and Storage: a Key Carbon Abatement Option. International Energy Agency, pp. 81–107. Paris: OECD Publishing. https://doi.org/ 10.1787/9789264041417-en.

Brown, K., Whittaker, S., Wilson, M., et al. (2017). The history and development of the IEA GHG Weyburn-Midale CO_2 Monitoring and Storage Project in Saskatchewan, Canada (the world's largest CO_2 for EOR and CCS program). Petroleum 3 (1), pp. 3–9.

Cameron, D. A. & Durlofsky, L. J. (2012). Optimization of well placement, CO_2 injection rates, and brine cycling for geological carbon sequestration. International Journal of Greenhouse Gas Control 10, pp. 100–112. https://doi.org/10.1016/j.ijggc.2012.06.003.

Chadwick, A., Williams, G., Delepine, N., et al. (2010). Quantitative analysis of time-lapse seismic monitoring data at the Sleipner CO_2 storage operation. The Leading Edge 29 (2), pp. 170–177.

Chellal, H. A. K., Malki, M. L., & Egenhoff, S. (2023). Assessing the feasibility of above injection zone pressure monitoring of CO_2 geologic storage in the Williston Basin, North Dakota, USA. Paper presented at the International Geomechanics Symposium, Al Khobar, Saudi Arabia, October 2023. https://doi.org/10.56952/IGS-2023-0359.

Chen, B., Morales, M. M., Ma, Z., et al. (2024a). Assimilation of geophysics-derived spatial data for model calibration in geologic CO_2 sequestration. SPE Journal pp. 1–10. https://doi.org/10.2118/212975-PA.

Chen, W., Li, B., & Tian, Y. (2024b). CarbonNet: How computer vision plays a role in climate change? Application: Learning geomechanics from subsurface geometry of CCS to mitigate global warming. https://doi.org/10.48550/ARXIV.2403.06025.

Cheng, P., Bestehorn, M., & Firoozabadi, A. (2012). Effect of permeability anisotropy on buoyancy-driven flow for CO_2 sequestration in saline aquifers. Water Resources Research 48 (9). https://doi.org/10.1029/2012wr011939.

Cherepovitsyn, A., Chvileva, T., & Fedoseev, S. (2020). Popularization of carbon capture and storage technology in society: Principles and methods. International Journal of Environmental Research and Public Health 17 (22), pp. 8368.

Chu, A. K., Benson, S. M., & Wen, G. (2022). Deep-learning-based flow prediction for CO_2 storage in shale–sandstone formations. Energies 16 (1), pp. 246. https://doi.org/10.3390/en16010246.

Cihan, A., Birkholzer, J. T., & Bianchi, M. (2015). Optimal well placement and brine extraction for pressure management during CO_2 sequestration. International Journal of Greenhouse Gas Control 42, pp. 175–187. https://doi.org/10.1016/j.ijggc.2015.07.025.

de Coninck, H. & Benson, S. M. (2014). Carbon dioxide capture and storage: Issues and prospects. Annual Review of Environment and Resources 39 (1), pp. 243–270. https://doi.org/10.1146/annurev-environ-032112-095222.

Friedlingstein, P., O'Sullivan, M., Jones, M., et al. (2020). Global carbon budget 2020. Earth System Science Data 12 (4), pp. 3269–3340. https://doi.org/10.5194/essd-12-3269-2020.

Fuss, S., Canadell, J., Peters, G., et al. (2014). Betting on negative emissions. Nature Climate Change 4 (10), pp. 850–853. https://doi.org/10.1038/nclimate2392.

Grandi, S. R., & Nafi, T. MGeomechanical Modeling of In-Situ Stresses Around a Borehole. (2002). http://hdl.handle.net/1721.1/67848.

Harati, S., Rezaei Gomari, S., Rahman, M. A. et al. (2024). Performance analysis of various machine learning algorithms for CO_2 leak prediction and characterization in geo-sequestration injection wells. Process Safety and Environmental Protection 183, pp. 99–110. https://doi.org/10.1016/j.psep.2024.01.007.

He, X., Zhu, W., Alsinan, M. et al. (2022). CO_2 storage capacity prediction in deep saline aquifers: Uncertainty and global sensitivity analysis. https://doi.org/10.2523/iptc-22463-ms.

He, X., Zhu, W., Kwak, H. et al. (2024). Deep learning-assisted Bayesian framework for real-time CO_2 leakage locating at geologic sequestration sites. Journal of Cleaner Production 448, pp. 141484. https://doi.org/10.1016/j.jclepro.2024.141484.

Hernandez Mejia, J. L., Imhof, M., & Pyrcz, M. J. (2024). Anomaly detection for geological carbon sequestration monitoring. SSRN Preprint. https://doi.org/10.2139/ssrn.4701351.

Herzog, H. (2016). Lessons learned from CCS demonstration and large pilot projects. An MIT Energy Initiative Working Paper.

Heuberger, C. F., Staffell, I., Shah, N., et al. (2016). Quantifying the value of CCS for the future electricity system. Energy & Environmental Science 9 (8), pp. 2497–2510.

Heuberger, C. F., Staffell, I., Shah, N. et al. (2017). A systems approach to quantifying the value of power generation and energy storage technologies in future electricity networks. Computers & Chemical Engineering 107, pp. 247–256.

Hitchon, B. (ed.) (2012). Best Practices for Validating CO_2 Geological Storage: Observations And Guidance from the IEAGHG Weyburn-Midale CO_2 Monitoring and Storage Project. Geoscience Publishing.

Intergovernmental Panel on Climate: IPCC (2018). Global Warming of 1.5°C. An IPCC Special Report on the impacts of global warming of 1.5 °C above pre-industrial levels and related global greenhouse gas emission pathways. In: The context of strengthening the global response to the threat of climate change (ed. V. Masson-Delmotte, P. Zhai, H. -O. Pörtner, et al.), p. 24. World Meteorological Organization, Geneva, Switzerland.

Intergovernmental Panel on Climate: IPCC, T. I. (2001). Third Assessment Report (TAR). IPCC. Geneva, Switzerland, p. 995.

Jahandideh, A., Hakim-Elahi, S., & Jafarpour, B. (2021). Inference of rock flow and mechanical properties from injection-induced microseismic events during geologic CO_2 storage. International Journal of Greenhouse Gas Control 105, pp. 103206. https://doi.org/10.1016/j.ijggc.2020.103206.

Jiang, S., & Durlofsky, L. J. (2024) History matching for geological carbon storage using data-space inversion with spatio-temporal data parameterization. International Journal of Greenhouse Gas Control, 134, p. 104124. https://doi.org/10.1016/j.ijggc.2024.104124.

Jung, S., Lee, S., Min, J., et al. (2020). Analysis of the state of the art of international policies and projects on CCU for climate change mitigation with a focus on the cases in Korea. Sustainability 13 (1), pp. 19. https://doi.org/10.3390/su13010019.

Kelemen, P., Benson, S. M., Pilorgé, H., et al. (2019). An overview of the status and challenges of CO_2 storage in minerals and geological formations. Frontiers in Climate 1, p. 9.

Kim, Y., Jang, H., Kim, J., et al. (2017). Prediction of storage efficiency on CO_2 sequestration in deep saline aquifers using artificial neural network. Applied Energy 185, pp. 916–928. https://doi.org/10.1016/j.apenergy.2016.10.012.

Kirsch, G. (1898). Die Theorie der Elastizitat und die Bedurfnisse der Festigkeitslehre. Zantralblatt Verlin Deutscher Ingenieure, 42, pp. 797–807.

Laalam, A., Tomomewo, O. S., & Chellal, H. A. K. (2023). A preliminary geomechanical study of underground hydrogen storage in the Inyan Kara saline aquifer in the Williston Basin. Paper presented at the International Geomechanics Symposium, Al Khobar, Saudi Arabia, October 2023. https://doi.org/10.56952/IGS-2023-0045.

Leroux, K., Klapperich, R., Ayash, S., et al. (2020). Subtask 1.3 – Integrated carbon capture and storage for North Dakota ethanol production. United States Department of Energy (DOE). https://www.osti.gov/biblio/1631156.

Liu, L., Mehana, M., Chen, B., et al. (2024). Reduced-order models for the greenhouse gas leakage prediction from depleted hydrocarbon reservoirs using machine learning methods. International Journal of Greenhouse Gas Control 132, pp. 104072. https://doi.org/10.1016/j.ijggc.2024.104072.

Luu, K., Schoenball, M., Oldenburg, C., et al. (2022). Coupled hydromechanical modeling of induced seismicity from CO_2 injection in the Illinois basin. Journal of Geophysical Research Solid Earth 127 (5), pp. e2021JB023496. https://doi.org/10.1029/2021jb023496.

Ma, J., Li, L., Wang, H., et al. (2022). Carbon capture and storage: History and the road ahead. Engineering 14 (7), pp. 33–43. https://doi.org/10.1016/j.eng.2021.11.024.

Madejski, P., Chmiel, K., Subramanian, N., et al. (2022). Methods and techniques for CO_2 capture: Review of potential solutions and applications in modern energy technologies. Energies 15 (3), pp. 887.

Martin-Roberts, E., Scott, V., Flude, S., et al. (2021). Carbon capture and storage at the end of a lost decade. One Earth 4 (11), pp. 1569–1584.

Matter, J. M., Stute, M., Hall, J., et al. (2014). Monitoring permanent CO_2 storage by in situ mineral carbonation using a reactive tracer technique. Energy Procedia 63, pp. 4180–4185.

Mohammadian, E., Motamedi, S., Shamshirband, S., et al. (2016). Application of extreme learning machine for prediction of aqueous solubility of carbon dioxide. Environmental Earth Sciences, 75 (3), pp. 215. https://doi.org/10.1007/s12665-015-4798-4.

Najafi-Marghmaleki, A., Barati-Harooni, A., & Mohammadi, A. H. (2017). Impact of gas impurities on CO_2 mole fraction: Application in carbon capture and storage (CCS) processes. International Journal of Greenhouse Gas Control 57, pp. 173–184. https://doi.org/10.1016/j.ijggc.2016.12.008.

Nath, F., Mahmood, M. N., & Yousuf, N. (2024). Recent advances in CCUS: A critical review on technologies, regulatory aspects and economics. Geoenergy Science and Engineering 238, pp. 212726. https://doi.org/10.1016/j.geoen.2024.212726.

National Academies of Sciences Engineering and Medicine. (2019). Negative Emissions Technologies and Reliable Sequestration: A Research Agenda. Washington, DC: The National Academies Press. https://www.nap.edu/catalog/25259/negative-emissions-technologies-and-reliablesequestration-a-research-agenda.

North Dakota Department of Mineral Resources (2023). Case 29450 - Certificate, draft permit, fact sheet, and storage facility permit application. https://www.dmr.nd.gov/dmr/sites/www/files/documents/Oil%20and%20Gas/Class%20VI/DGC/C29450.pdf.

Park, M. J., Frigerio, J., Clapp, B., et al. (2023). DeepNRMS: Unsupervised deep learning for noise-robust CO_2 monitoring in time-lapse seismic images. https://doi.org/10.48550/ARXIV.2310.07897.

Peck, W. D., Ayash, S. C., Klapperich, R. J., et al. (2019). The North Dakota integrated carbon storage complex feasibility study. International Journal of Greenhouse Gas Control, 84, pp. 47–53. https://doi.org/10.1016/j.ijggc.2019.03.001.

Peck, W. D., Liu, G., Klenner, R. C. L., et al. (2014). Storage capacity and regional implications for large-scale storage in the basal Cambrian system: Plains CO_2 Reduction (PCOR) Partnership Phase III Task 16 Deliverable D92 for U.S. Department of Energy National Energy Technology Laboratory Cooperative Agreement No. DE-FC26-05NT42592, EERC Publication 2014-EERC-05-12, Grand Forks, North Dakota, Energy & Environmental Research Center, March. https://edx.netl.doe.gov/dataset/storage-capacity-and-regional-implications-for-large-scale-storage-in-the-basalcambrian-system.

Pires, J. C. M., Martins, F. G., & Alvim-Ferraz, M. C. M. (2011). Recent developments on carbon capture and storage: An overview. Chemical Engineering Research and Design 89 (9), pp. 1446–1460.

Raza, A., Gholami, R., Rezaee, R., et al. (2017). Assessment of CO_2 residual trapping in depleted reservoirs used for geosequestration. Journal of Natural Gas Science and Engineering 43, pp. 137–155.

Raza, A., Gholami, R., Sarmadivaleh, M., et al. (2016). Integrity analysis of CO_2 storage sites concerning geochemical-geomechanical interactions in saline aquifers. Journal of Natural Gas Science and Engineering 36, pp. 224–240.

Ren, Q., Wei, S., Du, J., et al. (2022). What do we know about research on carbon capture, utilization and storage (CCUS) in China and the United States? Insights from the systematic review. Preprint on Research Square. https://doi.org/10.21203/rs.3.rs-2108142/v1.

Riaz, A., Hesse, M., Tchelepi, H., et al. (2006). Onset of convection in a gravitationally unstable diffusive boundary layer in porous media. Journal of Fluid Mechanics 548, pp. 87–111. https://doi.org/10.1017/s0022112005007494.

Rutqvist, J. (2012). The geomechanics of CO_2 storage in deep sedimentary formations. Geotechnical and Geological Engineering 30 (3), pp. 525–551. https://doi.org/10.1007/s10706-011-9491-0.

Safaei-Farouji, M., Vo Thanh, H., Sheini Dashtgoli, D., et al. (2022). Application of robust intelligent schemes for accurate modelling interfacial tension of CO_2 brine systems: Implications for structural CO_2 trapping. Fuel 319, pp. 123821. https://doi.org/10.1016/j.fuel.2022.123821.

Salvia, M., Reckien, D., Pietrapertosa, F., et al. (2021). Will climate mitigation ambitions lead to carbon neutrality? An analysis of the local-level plans of 327 cities in the EU. Renewable and Sustainable Energy Reviews 135, pp. 110253. https://doi.org/10.1016/j.rser.2020.110253.

Saraji, S., Goual, L., Piri, M., et al. (2013). Wettability of supercritical carbon dioxide/water/quartz systems: Simultaneous measurement of contact angle and interfacial tension at reservoir conditions. Langmuir 29 (23), pp. 6856–6866. https://doi.org/10.1021/la3050863.

Sinha, S., De Lima, R. P., Lin, Y. et al. (2020). Normal or abnormal? Machine learning for the leakage detection in carbon sequestration projects using pressure field data. International Journal of Greenhouse Gas Control 103, pp. 103189. https://doi.org/10.1016/j.ijggc.2020.103189.

Sloan, E. D. & Koh, C. A. (2008). Clathrate Hydrates of Natural Gases. Boca Raton, FL: CRC Press.

Sorensen, J., Bailey, T., Dobroskok, A., et al. (2009). Characterization and modeling of the Broom Creek Formation for potential storage of CO_2 from coal-fired power plants in North Dakota. Search and Discovery 1, pp. 80046.

Sun, Z., Xu, J., Espinoza, D. N., et al. (2021). Optimization of subsurface CO_2 injection based on neural network surrogate modeling. Computational Geosciences 25 (6), pp. 1887–1898. https://doi.org/10.1007/s10596-021-10092-9.

Tang, H., Fu, P., Sherman, C. S., et al. (2021). A deep learning-accelerated data assimilation and forecasting workflow for commercial-scale geologic carbon storage. International Journal of Greenhouse Gas Control 112, pp. 103488. https://doi.org/10.1016/j.ijggc.2021.103488.

Tang, M., Ju, X., & Durlofsky, L. J. (2022). Deep-learning-based coupled flow-geomechanics surrogate model for CO_2 sequestration. International Journal of Greenhouse Gas Control, 118, p. 103692. https://doi.org/10.1016/j.ijggc.2022.103692.

Vo Thanh, H. & Lee, K. -K. (2022). Application of machine learning to predict CO_2 trapping performance in deep saline aquifers. Energy 239, pp. 122457. https://doi.org/10.1016/j.energy.2021.122457.

Vo Thanh, H., Sugai, Y., & Sasaki, K. (2020). Application of artificial neural network for predicting the performance of CO_2 enhanced oil recovery and storage in residual oil zones. Scientific Reports 10 (1), pp. 18204. https://doi.org/10.1038/s41598-020-73931-2.

Wang, P., Wu, X., Ge, G., et al. (2022). Identification of potential CCUS clusters and its pipeline network optimization in China. E3S Web of Conferences, 358, pp. 02015. https://doi.org/10.1051/e3sconf/202235802015.

Wen, G., Hay, C., & Benson, S. M. (2021). CCSNet: A deep learning modeling suite for CO_2 storage. Advances in Water Resources 155, pp. 104009. https://doi.org/10.1016/j.advwatres.2021.104009.

Wienchol, P., Szlęk, A., & Ditaranto, M. (2020). Waste-to-energy technology integrated with carbon capture– Challenges and opportunities. Energy 198, pp. 117352.
Witte, P. A., Konuk, T., Skjetne, E., et al. (2023). Fast CO_2 saturation simulations on large-scale geomodels with artificial intelligence-based Wavelet Neural Operators. International Journal of Greenhouse Gas Control 126, pp. 103880. https://doi.org/10.1016/j.ijggc.2023.103880.
Xu, X., Chen, S., & Zhang, D. (2006). Convective stability analysis of the long-term storage of carbon dioxide in deep saline aquifers. Advances in Water Resources 29 (3), pp. 397–407. https://doi.org/10.1016/j.advwatres.2005.05.008.
Yan, B., Robert Harp, D., Chen, B., et al. (2022) Improving deep learning performance for predicting large-scale geological CO_2 sequestration modeling through feature coarsening. Scientific Reports, 12 (1). https://doi.org/10.1038/s41598-022-24774-6.
Zhang, J., Feng, Q., Zhang, X., et al. (2020). A supervised learning approach for accurate modeling of CO_2–brine interfacial tension with application in identifying the optimum sequestration depth in saline aquifers. Energy & Fuels 34 (6), pp. 7353–7362. https://doi.org/10.1021/acs.energyfuels.0c00846.
Zhang, S. & DePaolo, D. J. (2017). Rates of CO_2 mineralization in geological carbon storage. Accounts of Chemical Research 50 (9), pp. 2075–2084. https://doi.org/10.1021/acs. accounts.7b00334.
Zheng, F., Jha, B., & Jafarpour, B. (2023). Controlled CO_2 injection into storage reservoirs to minimize geomechanical risks under geologic uncertainty. SPE Annual Technical Conference and Exhibition, Day 2 Tue, October 17, 2023, D021S020R005. https://doi.org/10.2118/215155-MS.
Zheng, J., Chong, Z. R., Qureshi, M. F., et al. (2020). Carbon dioxide sequestration via gas hydrates: A potential pathway toward decarbonization. Energy Fuels 34 (9), pp. 10529–10546.
Zhong, Z., Sun, A. Y., & Jeong, H. (2019). Predicting CO_2 plume migration in heterogeneous formations using conditional deep convolutional generative adversarial network. Water Resources Research 55 (7), pp. 5830–5851. https://doi.org/10.1029/2018WR024592.
Zou, A. & Durlofsky, L. J. (2023). Integrated framework for constrained optimization of horizontal/deviated well placement and control for geological CO_2 storage. SPE Journal 28 (05), pp. 2462–2481. https://doi.org/10.2118/212228-PA.

21

Anthropogenic Carbon Sequestration into the Subsurface: Caveats and Pitfalls

Steven A. Tedesco

Running Foxes Energy Inc., Centennial, Colorado, USA

21.1 Introduction

Carbon dioxide (CO_2) sequestration has become critical according to several governments and parts of society to reduce human impact on Earth. A variety of CO_2 sequestration methods have been proposed to reduce climate change. A number of these methods are in their early stages such as using oceans, soils, combining with cement, etc. A proposed method is to extract CO_2 from the atmosphere and sequester in either underground such as depleted oil and gas fields or saline aquifers or within an inert material such as cement or bicarbonate soda (Service, 2018; McDonnell, 2021). The method utilizing cement or inert materials at present is uneconomic and the number of facilities needed to reduce CO_2 is potential beyond the capabilities presently for the world economies. Cost predictions of sequestering CO_2 vary from as little $92 a ton to over $600 depending upon the source (Service, 2018; McDonnell, 2021). Paddison (2024) recently indicated the carbon capture and sequestration (CCS) facility in Iceland is presently closer to $1000 a ton for sequestering CO_2 but hopes to be at $300–$350 a ton by 2030 and $100 a ton by 2050.

Presently, there two general methodologies for disposal of CO_2: (i) depleted oil, gas, and CO_2 reservoirs, and (ii) reservoirs that are saline aquifers. The utilization of subsurface reservoirs requires either (i) a regular Class II for CO_2 use in EOR (Jie et al., 2019) or (ii) a Class VI well specifically for permanently entombing the CO_2 in a saline aquifer. Potential other reservoirs for using for CO_2 sequestration are reservoirs used for geothermal and/or those used to extract lithium and rare earths. The permitting and approval of Class VI wells are managed for the most part by the Environmental Protection Agency (EPA) except in North Dakota, Wyoming and Louisiana where those states have primacy. Several other states are in the process of obtaining primacy for Class VI wells from the EPA. Despite the urgency by the present and previous political administrations, the EPA has only approved two Class VI wells in its history as of July 2024. Over 100 are in process but it is expected to take up to three or four years to approve many of them.

Geophysical Exploration for Hydrocarbon Reservoirs, Geothermal Energy, and Carbon Storage: New Technologies and AI-based Approaches, First Edition. Edited by Said Gaci.

This article focuses on, in a general sense, sequestering CO_2 in depleted oil, gas, and CO_2 reservoirs or brine filled aquifers. The reservoir and project criteria between the two types of processes are generally similar with the difference in petroleum or CO_2 present or not. Within depleted reservoirs, there is typically significant amount of data known about the reservoir, whereas with virgin, saline aquifers, reservoirs, their characteristics can only be speculated about from adjacent wells and 2D/3D seismic surveys, before actual development.

21.2 CO_2 Incentives

Carbon sequestration (CO_2) is becoming an industry that is driven by tax credits. The incentives for taking CO_2 out of the atmosphere or from industrial sources is based most recently on the Inflation Reduction Act (IRA) by the US government (U.S. Government, 2022). Canada, Australia and the European Union have similar programs that are further along and, in some cases, mandatory. The present US is as follows:

1. IRA allows for up to $130 per metric ton of CO_2 into subsurface reservoirs. For direct air capture technology, the tax credit is $180 per metric ton. If a project is financed with tax-exempt bonds, the credit is decreased by no more than 15%. The carbon credits require the disposal operation to complying with various Federal and State requirements in terms of wages, benefits and apprenticeship requirements.
2. IRA significantly reduced the yearly threshold for the amount of CO_2, as present in previous laws, that must be captured for a facility to qualify for the credit. For direct air capture facilities, the threshold for obtaining the CO_2 is reduced from 100,000 metric tons to 1000 metric tons; for certain electricity-generating facilities, from 500,000 metric tons to 18,750 metric tons; and for other facilities, from 100,000 metric tons to 12,500 metric tons.
3. IRA delays the deadline for starting construction of facilities and carbon capture equipment from December 31, 2025, to December 31, 2032. A facility that starts after that date cannot receive credits.
4. IRA allows tax-exempt entities, including state and local governments, to claim credits as cash refunds (rather than offsets against taxes otherwise payable). The election is also available to other taxpayers, but only for a maximum of five years.
5. IRA allows taxpayers to sell credits tax-free by transferring the credits to other taxpayers in exchange for cash.

It is likely as with most government programs that they will be modified and extended. Wind and solar have been modified and extended several times. At least the European Union and Australia allow the use of these credits that are generated in the United States to be applied to their credit needs.

21.3 Chemistry

CO_2 is a chemical compound that is made up of one carbon atom covalently double bonded to two oxygen atoms. At room temperature CO_2 is found in a gas state. Atmospheric CO_2 is the primary carbon source for life on Earth. In the atmosphere, CO_2 is transparent to visible light but absorbs infrared radiation, thus acting as a greenhouse gas to the planet. CO_2 is soluble in water and is found in subsurface brines and gases, groundwater, lakes, ice caps, and seawater. When CO_2 dissolves in water, it forms carbonate and mainly bicarbonate (HCO^{-3}), which causes ocean acidification as atmospheric CO_2 levels increase (National Research Council, 2010).

CO_2 is colorless and at low concentrations, the gas is odorless sufficiently high concentrations, it has a sharp, acidic odor. At standard temperature and pressure, the density of CO_2 is around 1.98 kg/m^3, about 1.53 times that of air. CO_2 has no liquid state at pressures below 0.51795(10) MPa (5.11177(99) atm). At a pressure of 1 atm (0.101325 MPa), the gas deposits directly to a solid at temperatures below −194.6855(30) K (−78.4645(30) °C) and the solid transforms directly to a gas above this temperature. In its solid state, CO_2 is commonly called dry ice. CO_2 is denser than dry air by 53% and is retained in the atmosphere for long periods of time.

21.4 Carbon Dioxide

Natural sources of CO_2 in the atmosphere are (Wikipedia, 2024a):

1. Animals exhale CO_2 as a waste product;
2. Plant decay release CO_2;
3. Volcanic eruptions release erratic volumes of CO_2. These volumes can be significant enough to equal all of the CO_2 generated in a year by human and/or biologic activities;
4. Natural CO_2 seeps from hot springs and geysers;
5. Thawing of permafrost; Dissolution of carbonate rocks;
6. Forest fires.

CO_2 concentration in the atmosphere since late in the Precambrian to Earth's pre-industrial era was regulated by organisms and geological phenomena (Wikipedia, 2024a). Plants, algae and cyanobacteria use energy from sunlight to synthesize carbohydrates from CO_2 and H_2O in a process called photosynthesis, which produces oxygen as a waste product. In turn, oxygen is consumed and CO_2 is released as waste by all aerobic organisms when they metabolize organic compounds to produce energy by respiration. CO_2 is released from organic materials when they decay or combust, such as in forest fires. Since plants require CO_2 for photosynthesis, and humans and animals depend on plants for food, CO_2 is necessary for the survival of life on earth. Anthropogenic CO_2 that is causing concern is released into the atmosphere through:

1. Burning of fossil fuels;
2. Releasing of CO_2 due to building construction and road building;
3. Forest clearing for farming, logging or building of structures, agriculture activities, etc.

It is estimated that approximately half of the CO_2 emissions into the atmosphere are absorbed by soil, certain rock types, and oceans. CO_2 can eventually be sequestered permanently saline acquires, depleted oil and gas fields, salt domes and depleted CO_2 reservoirs. Other forms of CO_2 sequestration such as in oceans, soils, and minerals are in early stages of investigation to determine if they could viable (Wikipedia, 2024a).

CO_2 is a versatile industrial material used for:

1. Inert gas in welding and fire extinguishers,
2. As a pressurizing gas in air guns,
3. Used in enhanced oil recovery (EOR) in petroleum reservoirs,
4. As a supercritical fluid solvent in decaffeination and supercritical drying,
5. Added to carbonated beverages like seltzer and beer for effervescence.

CO_2 is a trace gas in Earth's atmosphere which presently approximately 421 parts per million (ppm), or about 0.04% (as of May 2022) having risen from pre-industrial levels of 280 ppm or about 0.025% Figure 21.1) (European Environment Agency, 2024). Human civilization is the

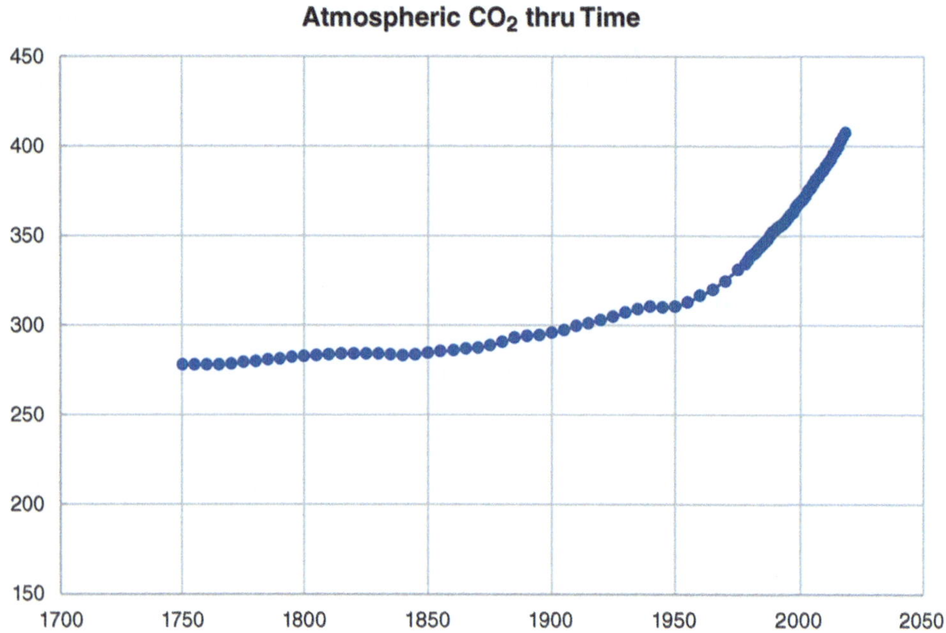

Figure 21.1 Atmospheric CO_2 through time (European Environment Agency, 2024; Gapminder Foundation, 2024; Macrotrends, 2024; Wikipedia, 2024b).

primary cause of the increase in CO_2 concentrations and part of the cause of climate change. The Earth is presently in a warming period between glacial ages. Normal biologic activity releases CO_2 critical for the normal carbon cycle for life on earth. Volcanic or venting of the earth can out do any human release of CO_2. However, volcanic events are generally unpredictable and cannot be controlled.

CO_2 is critical in the carbon cycle. CO_2 levels 100 ppm or lower will cause all photosynthesis to stop thus ending most of life, specifically on land, on the planet (Manitoba, 2014). At 100–150 ppm CO_2, the plants photosynthesis stops. At this low level a plant will not be able to extract CO_2 and photosynthesis is restricted and plants will eventually die (Manitoba, 2014). Above 150 and optimally above 250 ppm CO_2, the rate of photosynthesis will be consistent with growing conditions outside of a controlled environment (Manitoba, 2014). With no other limiting factors such as heat, light, and nutrients, the plants will photosynthesize at a rate consistent with ambient conditions in the natural environment. The optimal level for maximum plant photosynthesis is 1200 ppm CO_2, assuming no other limiting factors (Manitoba, 2014). At 10,000 ppm of CO_2 the photosynthetic rate in the plants will become very low due to the closing of the plant stomata and the exclusion of air into the leaf interior (Manitoba, 2014). This level of CO_2 is sufficient to cause toxic effect on the plants and cause damage and eventually death of a plant (Manitoba, 2014).

CO_2 at low concentration has little toxicological effects on biologic or human life (Permentier et al., 2017). At concentrations in the atmosphere of more than 5% CO_2 causes the development of hypercapnia and respiratory acidosis (Permentier et al., 2017). Severe acidosis increases the effects of parasympathetic nervous activity, possibly by interfering with the hydrolysis of acetylcholine by acetylcholinesterase, resulting in a reduction of the respiration and circulation (Permentier et al., 2017). Concentrations greater than 10% CO_2 in the atmosphere may cause convulsions, coma, and death (Permentier et al., 2017). CO_2 levels of more than 30% in the atmosphere rapidly cause consciousness in seconds (Permentier et al., 2017).

The approach in the media and government at present is CO_2 at net zero implying no CO_2 in the atmosphere. To achieve sustainable levels of CO_2 for plant growth and less dangerous levels to humans, the amount of anthropogenic CO_2 in the atmosphere is somewhere around 270 ppm. This amount of CO_2 is generally equivalent to where the molecule was in the atmosphere in the second half of the 1700s (Figure 21.1). The downside is that plant life will slow, reduction in crop yields, and not be as robust worldwide as present at 421 ppm.

21.5 Potential Sequestration Locations

CCS projects sequester CO_2 into a saline aquifer and depleted natural and CO_2 gas reservoirs, by either structural and/or solubility trapping, which is the dissolution of the gas/liquid into the formation reducing its viscosity and allowing it to move more easily. Trapping of CO_2 into a saline aquifer is not dependent upon structure but more on solubility trapping. Structural tapping of CO_2 is the primary form of sequestration followed by solubility trapping (Zhao et al., 2014; Leslie et al., 2021). Residual, mineral and capillary pinning trapping also occurs when the CO_2 reaches it maximum distance (Zhao et al., 2014). Figure 21.2 is an illustration of the trapping mechanisms from Arif (2017). In addition, Figure 21.3 is the trapping efficiency over time scale.

Determining the efficiency of the reservoir and type of trapping that occurs is dependent on a multiple of geologic, geophysical, and geochemical characteristics. In some cases, the collection of data is significantly easy (oil, gas, and CO_2 depleted reservoirs) to limited and non-existent (saline aquifers, coal-bed, soil, and oceans) and requires extensive collection. The approach to any CCS project is reducing risk to near zero of the CO_2 breaching the seal or migrating away from area of the reservoir designated for storage.

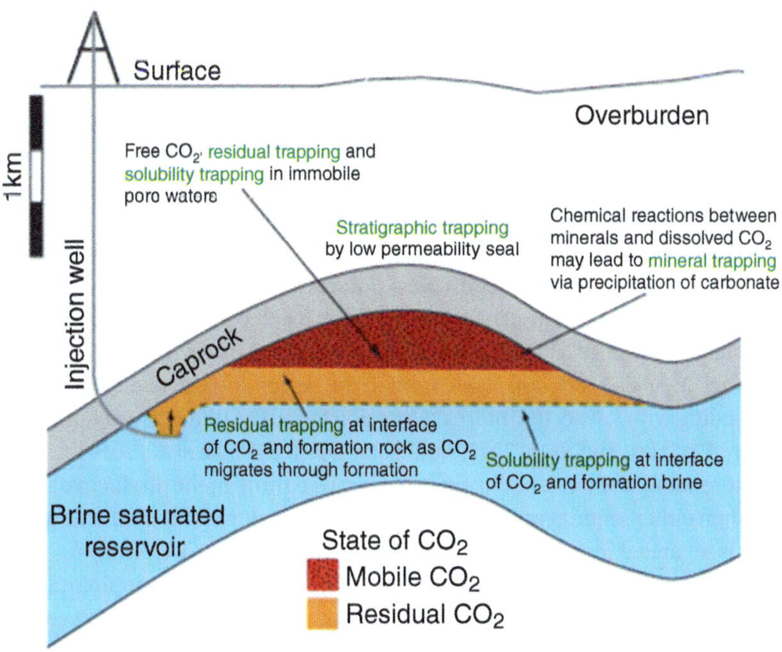

Figure 21.2 Conceptualization of the in-situ mechanisms (Arif, 2017).

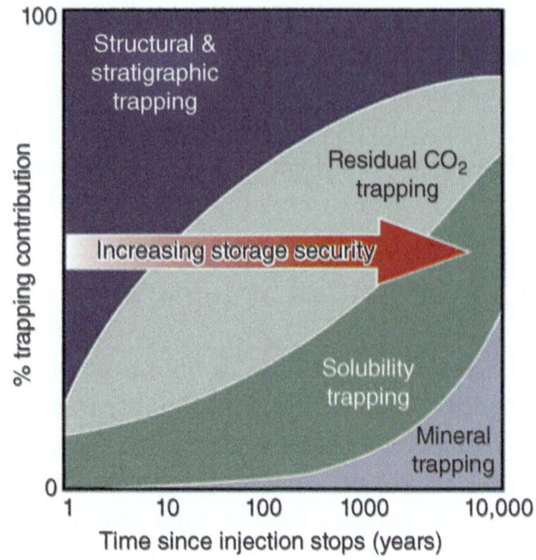

Figure 21.3 Trapping mechanism efficiency over timescale (Arif, 2017).

The potential places in the subsurface and near surface for sequestration are as follows:

21.6 Sequestration in Hydrocarbon and Carbon Dioxide Reservoirs

CO_2 is used in Enhance Oil Recovery (EOR) systems in existing oil reservoirs to increase production by being injected in a dense or supercritical phase at typical reservoir pressures and temperatures that helps move unmovable oil to the surface (Mungan, 1983; Lee and Kam, 2013; Jia et al., 2019). The primary goal of using CO_2 is to reduce viscosity and increase mobility of the oil so that it can be extracted. CO_2 flooding of oil reservoirs is considered more efficient than water, natural gas, polymers and surfactants in recovery and typically, CO_2 is used typically after waterflooding (Mungan, 1983; Lee and Kam, 2013; Ettehadtavakkol, 2016). The process of using CO_2 can be generally categorized into miscible and immiscible CO_2 flooding. Miscible CO_2 flooding mixes completely with the oil, creating a single-phase fluid (Figure 21.4). This occurs when the reservoir pressure is above the minimum miscibility pressure, allowing CO_2 to dissolve in the oil, causing it to swell, and reduces its viscosity. CO_2 is miscible in petroleum under certain reservoir pressure, temperature, oil composition, and typically depths below 2000 ft (606 m). The oil and CO_2 enter an oleic phase. In general, CO_2 with oil works pressures greater than 1500 psi, temperatures less than 140°F (60°C), high concentration C_4 through C_6 petroleum compounds and 30^0 or higher API (Ettehadtavakkol, 2016). CO_2 is used over other gases because it forms a high density, high viscosity, and stable front in the reservoirs (Lee and Kam, 2013). The CO_2 creates miscible flooding with lowers the interfacial tension to zero resulting in no oil trapped by capillary forces ((Mungan, 1983; Lee and Kam, 2013; Ettehadtavakkol, 2016; Jia et al., 2019). CO_2 in flooding oil reservoirs will store some of the gas/liquid but some is cycled through the productive zone, produced with the oil, and then either sequestered in the reservoir, cycled out with oil production and re-injected or vented after it is separated from the oil at the surface facility. The CO_2 is typically injected into the reservoir with alternating cycles of water or surfactant in what is either a water-alternating-gas (WAG) system (Figure 21.4) or surfactant-gas (SAG) system, respectively (Mungan, 1983; Lee and Kam, 2013).

Immiscible CO_2 flooding occurs when the reservoir pressure is below the minimum miscibility pressure. CO_2 in this method does not fully mix with the oil, is less effective than miscible

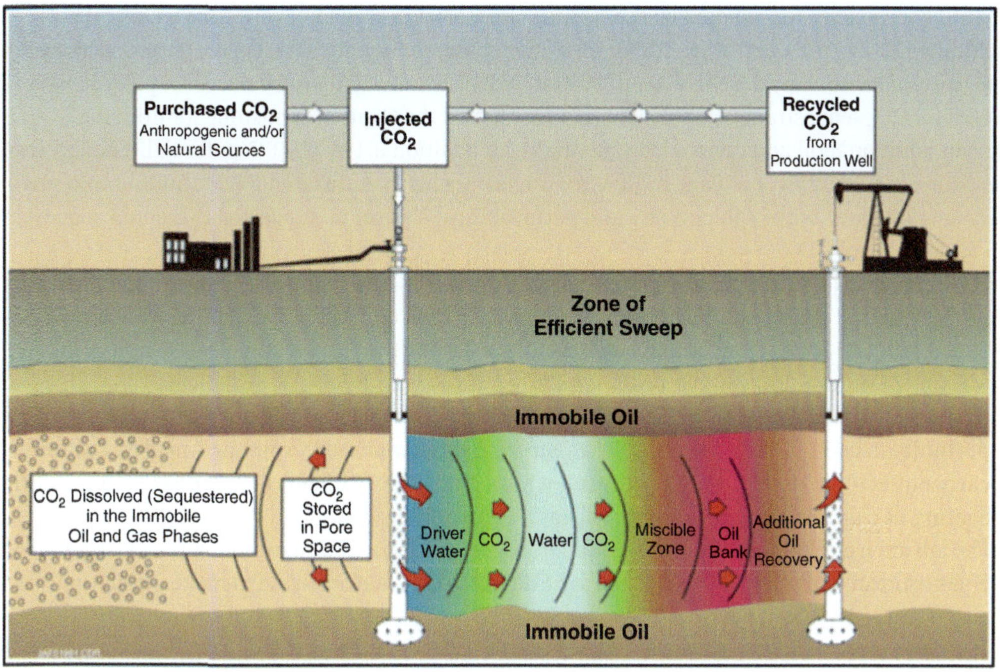

Figure 21.4 Diagram of CO_2 flooding of an oil reservoir using anthropogenic CO_2.

flooding, but instead pushes the oil toward the production wells like a waterflood would do through a process called immiscible displacement.

The CO_2 effects on the oil can be summarized as follows:

1. CO_2 dissolves in the oil, causing it to swell and expand which increases the oil's volume and reduces its viscosity, enhancing its mobility;
2. CO_2 reduces the interfacial tension between oil and water, increasing the flow of oil from the reservoir;
3. CO_2 injection into the reservoir helps maintain reservoir pressure, temporarily counteracting the natural decline in pressure as oil is extracted.
4. CO_2 can change the wettability of the reservoir rock, making it more water-wet, which improves oil displacement efficiency.

The Big Piney-La Barge Field operated by Exxon in western Wyoming produces predominantly CO_2 specifically for EOR flooding in several oil fields in eastern Wyoming. Similar CO_2 fields are producing the gas in southwestern Colorado, such as the Doe Canyon and McElmo fields, that is shipped to oil fields in Permian Basin, Texas, for EOR. The Sheep Mountain Field on the northwest flank of the Raton Basin in Colorado is a depleted CO_2 field that was used for EOR and could be repurposed to store CO_2 (Ning et al., 2024). The Sheep Mountain Field has produced CO_2 since 1983 and has produced 1.7 TCF from approximately 4000 ft (1212 m) from two reservoirs, the Dakota (Cretaceous) and Entrada (Jurassic) sandstones. The EPA (2024a) estimated that the USA generated 6.3 billion metric tons of CO_2 in 2022. Ning et al. 2024 estimated that it would take 30 years to put back 80% of the original CO_2 in place back into the Sheep Mountain Field. There are several other depleted CO_2 fields that could be used to store anthropogenic CO_2.

There is presently no limit in terms of potential depleted oil and gas fields for CO_2 storage around the world. The Wellington Field in Sumner County, Kansas, produces oil from Mississippian rocks and has been extensively studied by the Kansas Geological Survey for a variety of technical aspects including CO_2 EOR (Lueck et al., 2016). Extensive cores, wells, and a 3D seismic survey were available to evaluate not only the Mississippian strata but also underlying Arbuckle Group (Ordovician) as well for CO_2 sequestration. Webb et al. 2019 evaluated the Cypress Sandstone for utilizing CO_2 to recover additional oil and concluded potentially an additional 196 MMBO potential. The Cypress sandstone is a Pennsylvanian age multistory fluvial system that are laterally continuous and vertically amalgamated with high porosity and permeability. Webb et al. suggested this could sequester 7.6 billion tons of CO_2.

Carter et al. 2017, evaluated parts of the Appalachian Basin to determine what depleted oil and gas fields, natural gas storage facilities, mined-rock caverns and salt caverns were available. These depleted fields provide in an initial assessment of parts of the Appalachian Basin found 113 gas fields and 12 natural gas storage units for CO_2 storage. Trentham (2011) indicated that there is an additional 10 BBO of additional oil to be recovered using CO_2. Childers (2012) presented the opportunity to use 1 million tons of CO_2 a year from industrial sources in Los Angeles, California to carry out EOR on the Wilmington Graben which is offshore. Childers (2012) discussed the critical nature of using wireline logs, cores, seismic, and reservoir data.

The advantage to depleted oil and gas fields, especially those that have had a major or large oil and gas company involved, is there tends to be a plethora of data available to make evaluation simpler and less risky than a new area.

21.7 Risk Assessment Analysis and Characterization of a Reservoir for CO_2 Sequestration

The amount of data collected to characterize the reservoir and associated seal are critical for containing anthropogenic CO_2. In the case of injecting CO_2 into a depleted oil, gas, or CO_2 field, there is a plethora amount of data available generally.

The categories of data can be summarized as follows:

1. Subsurface geology of the reservoir – In defining an area/reservoir for CO_2, an optimal site has to be determined. The optimal site is dependent upon identifying a reservoir, either sandstone or carbonate that has moderate to high and continuous porosity, permeability, and storage capacity. Databases such as reported by Lewis et al. (2019) and Nuttall et al. (2019) in the Appalachian and Michigan basins are a good start to look for suitable CCS projects.
2. Seal – The optimal seal or caprock for CO_2 sequestration is composed of clay minerals that are inert with no discontinuities, unconformities, fractures, and faults. Integrity of the seal is defined by the caprock lithology further defined as mineralogical composition, wettability, porosity, and permeability. Seal capacity can be defined as the column height of the stored CO_2 that the seal can hold back before the capillary entry pressure is exceeded (Daniel and Kaldi, 2008; McLaughlin et al., 2014; Petersen et al., 2022). There are three mechanisms of leakage by any gas or fluid through the seal: diffusion, capillary breakthrough and migration along faults and fractures (Daniel and Kaldi, 2008; Petersen et al., 2022). Diffusion is considered negligible and if faults or extensive fracture systems are present, the seal may not be effective. If faults and fractures are not an issue then primary concern is what is the capillary pressure in which breakthrough will occur. Knowing what the breakthrough pressure is will help determine the column height of the CO_2 plume that can be sequestered in the underlying reservoir. The thickness of the strata, pore throat size, and capillary pressures are critical factors in determining the viability of any CO_2 sequestration project. The thicker the seal the better.

Medina et al. (2020) analyzed using mercury injection methods for capillary pressure on the Upper Maquoketa carbonaceous mudstones and shale from the Indiana, Kentucky, Ohio, and Pennsylvania, with an assumed 20% pore brine displacement, determined the rocks could contain a CO_2 column of 5000 ft (1500 m).

3. Aeromagnetic/Gravity – Detailed airborne or ground aeromagnetic/gravity data can be critical to identifying basement faults whose presence may affect the integrity of the seal and reservoir. Basement faulting appearance may not be clearly defined by subsequent 2D/3D seismic due to the lack of usable data in the basement rocks. Integrating aeromagnetic/gravity data with geophysical and geologic data will potentially identify fault and fracture issues especially with the seal.
4. Seismic – 2D or 3D seismic is critical to defining the reservoir extent, thickness and integrity of the seal, any heterogeneity in the reservoir, faulting, discontinuities, etc. Seismic is critical to defining the topography, reservoir and seal continuity, any potential fault or discontinuities, and any internal changes (lithologic) in reservoir or seal. One of the main concerns is the presence of faults or fracture systems that are initially missed because of insufficient data acquisition, poor processing, poor interpretation, or most likely they do not appear until the CO_2 causes reactivation (McBride et al., 2013).
5. Test or Exploration Well(s) – In areas where there is no geologic data, it becomes critical to obtain logs, cores, and cuttings from test or exploration wells. Meneguolo et al. 2022 discussed a test well in the EL001 license on the Horda Platform targeting potential reservoirs of the Early Jurassic Dunlin Group as part of the Norwegian Longship CCS Project. The offshore exploitation license EL001 is the first area on the Norwegian continental shelf to be formally administered solely as a CO_2 storage site (Meneguolo et al., 2022). The companies Equinor, Shell and Total Energies were granted permission to develop the CCS prospect in 2019, and in early 2021 they formed the Northern Lights joint venture. "Aurora," within EL001 (Meneguolo et al., 2022). The license EL001 is the Norwegian full-scale CCS designated geological storage complex project (Meneguolo et al., 2022). The result of the test wells at the EL001 is several sandstone intervals were identified in the Dunlin Group that had excellent potential to store CO_2.
6. Mudlogs –While drilling, mudlogs can be critical to determine the effectiveness of the seal caprock. During drilling for oil and gas, mudlogs are used to help evaluate the potential of a reservoir. In some hydrocarbon traps the mudlogs will indicate the presence of hydrocarbons in the overlying seal prior to reaching the petroleum productive reservoir. Some reservoirs with heavy oil or oil with a high pourpoint do not leak significantly because of the presence of large hydrocarbon molecules that have difficulty in migrating. Oils that contain significant amount of wax also have problems of migration migrating through the seal. Seals that yield hydrocarbons while drilling, in some cases, can be indicating the seal would be ineffective for storing CO_2 in the underlying reservoir. Petersen et al. (2022) used a similar approach in evaluating the Nini West oil field in the North Sea (Mohammadkhani et al., 2024).
7. Drill Stem Tests (DST) – These can be used to give an idea of the permeability of the reservoir and therefore its ability to take fluids and gases. DSTs allow collection of fluids for analysis and determination of any issues in relation to CO_2.
8. Fluid Analysis – Samples of the fluid from the reservoir are necessary to determine any potential issue with the CO_2 being injected. Keister et al. (2019) reported on the effects on reservoir mineralogy from the injection of 1,555,417 metric tons of CO_2 into Dover 33 oil productive Niagaran (Silurian) reef field as Dover 33 for EOR of an oil recovery. Keister et al. (2019) used isotopic carbon from the CO_2, which has heavy δ^{13} signature, to determine if there were carbonates deposited during the EOR process. Generally, it was

found that there was no conclusive evidence that CO_2 caused dissolution, worm-holes in rock and deposition of carbonate minerals or what was present was related to diagenesis. This is an example of where the CO_2 had no or minimal effect on the existing reservoir. Geochemical reactions between injected fluids and minerals can alter rock mechanical properties and strength parameters, influencing geomechanical behavior and potentially impacting storage integrity.

9. Cores – Analysis of the reservoir using cuttings yields limited data in terms of geomechanical, porosity, permeability, potentially fluid analysis, isotope analysis, geomechanical testing, stimulate CO_2, and flooding or sequestration.
10. PetrologyA – Dissolution and/or precipitation of minerals during CO_2 injection is believed to be the main reason for flow and injectivity changes in reservoir properties, particularly within carbonate rock. Cores with a wide range of characteristics undergo flooding experiments to identify the effects of CO_2 on rock mineralogy. XRD-XRF, thin-section CT scanning, porosity, and permeability analysis prior to and post core flooding are critical to understand the effects in terms of benefits and issues caused by CO_2 injection. Zareei et al. (2022) evaluated changes in carbonate rock and found there was a 400-fold in the permeability and 50% during CO_2 flooding.
11. Micro-seismicity – One of the negatives of injecting fluids or gases into the ground is that it can, under the right pressures and temperatures, induce seismicity. This is well-documented in unconventional petroleum plays in relation to disposal of waste water from oil and gas wells into injection wells (Hincks et al., 2018; Van der Baan, 2021). In the cases of the induced seismicity in the unconventional petroleum plays, seismicity can be reduced by reducing the amount of fluid and lower pressures put into the disposal wells daily, (Monies, 2015; Van der Baan, 2021). Some of this seismicity can be imperceptible to humans but still register on sensitive instruments. De la Muela et al. (2024) at the Hontomin CO_2 sequestration site facility in Spain found the intensity of the induced seismicity was related to length of injection and pressure. This highlights the potential need for micro-seismicity analysis of the potential CCS reservoir and caprock.
12. Fault and Fracture Analysis – The Earth is highly faulted and fractured and must be analyzed for each CCS facility in to determine if there will be issues with leakage. Initial assessment is by using aeromagnetic, gravity and 2D/3D data. Faults and fracture systems can act as baffles in plume movement. Blanco-Martin et al. (2022) discussed modeling for subseismic faults and fractures that may reactivate during injection. CO_2, unlike hydrocarbons, can react and dissolve the mineralization along fault and fractures allowing them to reactivate. 4D seismic monitoring though the life of CCS project can work effectively to identify faults and fractures that may reactivate or act as baffles (Nazarian and Furre, 2022). In Nguyen et al.'s (2019) study of 300,000 tons of CO_2 injected into the fluvial sandstones of the Liujiagou Fm on the Shenhua, CCS demonstration project reported cyclic pressure build-up implying that fractures and boundaries were impacting the process. The pressure dropped temporarily when a boundary was breached, probably along fractures and then the plume moved to the next compartment. Nguyen et al. (2019) suggested the breach occurred along unknown fracture systems that connected the various fluvial compartments.
13. Geomechanics – Heterogeneity in rock properties can introduce geomechanical challenges for CO_2 injection. Variations in strength, stress, and deformation characteristics may lead to localized failures or compromise containment integrity under CO_2 injection. Favorable or unfavorable stress and strain variations, and volumetric responses during testing need to be determined. The geomechanical stability caprock or seal is critical and weaknesses or failures during laboratory testing on cores may indicate its inability to provide long-term containment. This also can help if increased fracturing occurs during CO_2 injection causing seismicity.

Anisotropic rock behavior may not be apparent in the assessments. Neglecting transversely isotropic properties, where rocks behave differently in various directions, can lead to inaccurate prediction on response of the rock to stress and deformation and accordingly potential failure zones.

In carbonaceous mudstone or shale reservoirs, there is more likely to be reactivation of faulting due to the low porosity and permeability resulting in pressure build up (Ozotta and Ostadhassan, 2019). Elevated pore pressure resulting from injection can influence geomechanical stability, potentially leading to issues like failure/fracture initiation, surface uplift, fault reactivation, and inducing seismicity if not carefully managed.

14. Hydrogeology – In any CCS project, the groundwater and brine movement in the subsurface needs to be studied based on mechanical, chemical and thermal interaction. It is critical to evaluate any freshwater zones above the reservoir used for CCS. Evaluating existing groundwater wells to establish baseline data in hydrochemistry, flowrates, porosity, conductivity (permeability), hydrodynamic dispersion, pressure, temperature, lithology, any constraining barriers, and constant monitoring ensures that CO_2 plume is not contaminating the groundwater aquifer(s).
15. Temperature – Consideration of reservoir temperature may be part of initial assessments especially for coalbed reservoirs. The temperature of the CO_2 being injected is different from reservoir temperature. This can potentially cause local cooling or thermal shock and induce thermal stress; thermal cycling potentially cause failure of the gas/liquid to entire the rest of the zone.
16. Modeling – Modeling is critical in the permitting process to determine the viability of a reservoir to store CO_2. Unlike hydrocarbons, which are relatively inert, CO_2 is reactive with various minerals that could be present in the reservoir rock, especially carbonates, making modelling difficult. Unlike hydrocarbons, CO_2, because it can modify porosity and permeability chemically, can migrate rapidly to potentially areas in the reservoir or other formations that present containment and breaching issues. Modeling is only as good as the data that is put into the model. In many cases, especially with saline aquifers, there typically is limited geological data available and this can cast significant doubt on the validity of any model.

    Modeling programs divide the reservoir into cells of equal proportion. If the cells are too small or too large, this may impact the analysis of the reservoir. Neural networks have been proven useful to evaluate a location for a potential CO_2 storage (Cyphers et al., 2013).

    Dressel (2015) presented a workflow in Figure 21.5 for modeling a project and applied it nine projects funding the National Energy Technology Laboratory. One of the conclusions of Dressel (2015) work is that higher risk areas are located outside of petroleum producing regions. Data tends to be more plentiful in terms of geologic, geophysical and geochemical in petroleum regions. Higher risk, even after a test well, tends to remain for potential targets in non-petroleum areas.

21.8 Sequestration in Saline Aquifers

Saline reservoirs are considered the primary focus of CO_2 injection. The concept requires capturing CO_2 from industrial or air capture sources, liquefying it, and then injecting it into these deep geological formations. CO_2 is injected in a supercritical state, which allows it to displace the denser brine in the reservoir. CO_2 is sequestered in the reservoir and is thought to react with dissolved salts to form carbonates. The issue with these types of reservoirs may already be at maximum capacity in terms of filled pore space and adding to much CO_2 may cause fracturing stimulation of the overlying seal and cause breaching (Gorecki et al., 2009). However, putting CO_2 into these

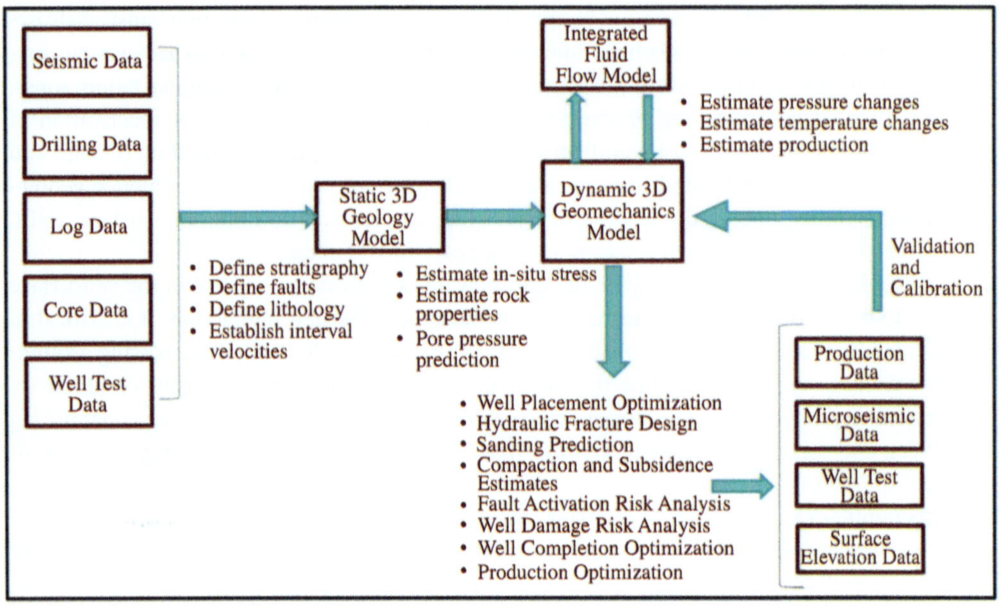

Figure 21.5 Workflow model for evaluating a projects favorability for CO_2 storage (Dressel, 2015/U.S. Department of Energy/Public Domain).

type of reservoirs leads to permanent sequestration and no recycling occurs as in oil fields due to EOR techniques. This also results in receiving the maximum tax credit per ton of CO_2 in the USA.

Saline aquifers used for CO_2 storage must be mapped and modeled in detail in order to determine heterogenous issues in the CCS reservoir (Gorecki et al., 2009; Callis et al., 2023). Medina and Rupp (2012) used this approach to evaluate the potential of the Mt. Simon sandstone (Cambrian) in the northeastern USA on a regional basis. The Mt. Simon sandstone is composed of three distinct lithologic units with different clay content (Medina and Rupp, 2012):

1. The upper unit has relatively high gamma-ray values because of a mixture of argillaceous strata;
2. The middle unit is a clean sandstone based on low gamma response, cuttings, and cores;
3. The lower most unit has increased clay unit with depth based on gamma-ray response. In evaluating the reservoir, the storage capacity of each unit has to be modeled to determine flow characteristics of the reservoir based on mechanical logs and cores.

The Tricarb Consortium for Carbon Sequestration (TCCS) drilled a stratigraphic test well, the NYSTA-Tandem Lot #1, located in Rockland County, New York, to determine if the Triassic age Passaic Formation in the Newark Rift Basin in 2011 (Olsen et al., 1996; Slater et al., 2013; Zakharova and Goldberg, 2014; Olsen et al., 2016; Zakharova et al., 2016; Zakharova et al., 2020) (Figure 21.6). Only two other wells have penetrated the Newark Basin Triassic section to basement and they were on the flanks of the structure. The goal of the NYSTA-Tandem Lot #1 well was to characterize the northern Newark basin for potential geologic carbon storage. The Newark Basin lies in close proximity to several major cities and therefore a potential easy disposal of large volumes of CO_2. Prior to drilling of the well, a 2D survey was shot to determine potential reservoir targets (Zakharova and Goldberg, 2014; Zakharova et al., 2020). The NYSTA-Tandem Lot #1 was drilled to about 6885 ft (2086 m) deep was drilled intersecting the Passaic Formation and the Palisade dolerite sill. Cores were taken from 4570 to 4880 ft (1384–1479 m) in the strata above the Palisade sill

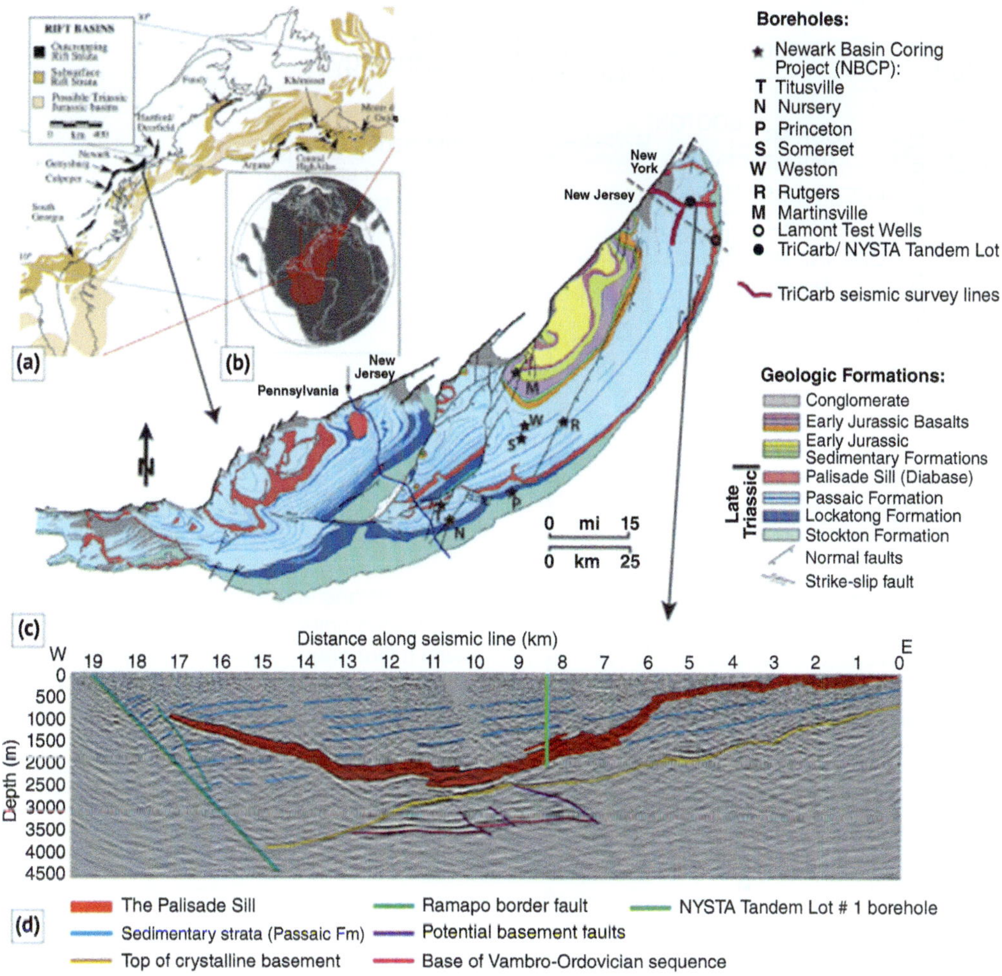

Figure 21.6 Geologic map and setting of the Newark rift basin. (a) Mesozoic rift system of eastern North America; (b) the extent of the rift zone and associated with the Central Atlantic magmatic province within the Pangaean subcontinent; (c) geologic map of the Newark Basin showing major formations and borehole discussed here; (d) Migrated reflection seismic profile across the northern basin acquired by Tricarb with geologic interpretation by Zakharova et al., 2020/Society of Exploration Geophysicists.

(Zakharova and Goldberg, 2014; Zakharova et al., 2020). The well upon completion of drilling and logging of the well was plugged and abandoned.

Based on the seismic section in Figure 21.6 the NYSTA-Tandem Lot #1 was not drilled on any definable structure but toward the basin center. Structure would provide better containment as it would create closure for trapping the CO_2. However, the sole purpose of the NYSTA-Tandem Lot #1 well was to quantify the reservoir/seal characteristics for potential CO_2 permanent storage.

The Late Triassic Stockton Fm is the basal strata of the Triassic section in the Newark Basin (Figure 21.7). The Stockton Fm is composed of predominantly coarse-grained fluvial sediments and was thought to be the primary reservoir for carbon sequestration (Zakharova et al., 2020). The formation has multiple mudstones with excellent sealing capacity and multiple reservoirs with 8–18% porosity (Zakharova et al., 2020). Collins (2017) estimated that Stockton is potentially a good host reservoir for two to three gigatons of CO_2. This estimate was based on limited data and

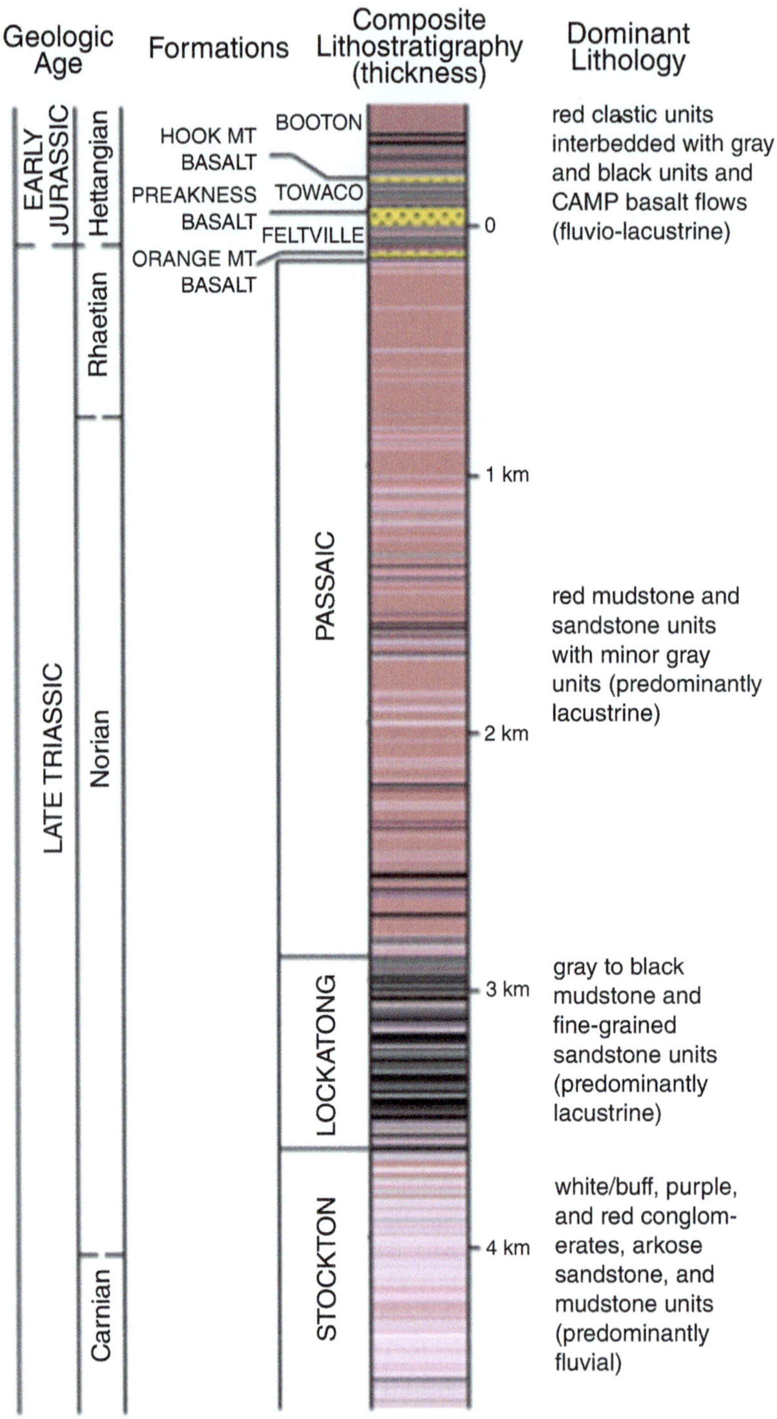

Figure 21.7 Simplified lithostratigraphy of the Newark Basin. The depth and thickness of lithologic units vary across the basin. The figure indicates a composite lithostratigraphy of the drill cores from the NYSTA-Tandem Lot #1 plus two offsetting shallower wells. The stratigraphy column also integrated outcrop studies and other geotechnical cores for wells penetrating the Jurassic geologic section.

thus reflecting the wide range of potential CO_2 to be stored. The negatives for the Stockton Fm based on the data from the NYSTA-Tandem Lot #1 are (Zakharova et al., 2020):

1. Lateral heterogeneity which makes continuity of reservoir unclear and potentially limits storage capacity;
2. Limited data on injectivity, potentially low permeability. It was unfortunate the well was plugged and abandoned before any testing could be done;
3. Presence of basement faulting.

Stockton Fm is overlain by the lacustrine sediments of the Lockatong Fm (Figure 21.7). The Lockatang Fm has limited potential due to limited thickness of reservoirs, unknown continuity of reservoirs and fracturing identifying in overlying seals. Overlying the Lockatang Fm is the Passaic Fm which is composed of red-beds and lacustrine deposits (Figure 21.7). Zakharova et al. (2020) define the Passaic Fm has potential reservoirs with up to 15% porosity, minerals of the formation is highly reactive to CO_2, potential massive basalts as seals and potential storage of 1 gigaton of CO_2 (Goldberg et al., 2010). The intrusive Palisade sill is present at different depths, stratigraphic intervals, and in different areas across the basin as seen in Figure 21.7 (Zakharova et al., 2020). The sill is more contiguous and discordant across the northern part of the basin and could act as a seal.

Zakharova et al. (2020) summarized the difficulty of evaluating an entire basin in that the single borehole could not answer numerous questions such as reservoir heterogeneity needed to be defined, areal extent of the reservoirs, and seal homogeneity and extent of any fractures. The nature of rift basins is that they are subject to ongoing tectonic activity that results in rapid deposition of fan type deposits that generally are arkosic in nature. There may be clean or well-sorted fine-grained that could act as reservoirs but with only one well, it is difficult to target them with significant additional drilling or acquisition of 3D seismic surveys.

Howat et al. (2017) evaluated the Precambrian, Cambrian, and Ordovician rocks in parts of Ohio, USA, to determine where potential zones within these strata may be optimal for CO_2 sequestration. The area has an extensive database as a result of oil and gas drilling, which makes initial assessment having lower risk.

21.9 Sequestration in Coal Seams

Coal is formed from decaying plant material modified through millions of years by temperature and pressure. Large scale coal deposits are found from Pennsylvanian to Tertiary ages. Coal is generally mined for (i) fuel in power plants or (ii) specific metallogenic process where by certain rank of coal, medium-volatile bituminous, that are used in the coking processes for steel production. The optimal coal for power generation has a very low ash content and high organic matter to maximize BTU content. Burning of coal has contributed significantly to CO_2 into the atmosphere and in some countries are being phased out as fuel for power generation. However, in some countries, coal is expanding in China, Japan, and India.

Coal is composed of 50% or more organic matter that is further defined by minerals or macerals that represent various components of the plant. The four most important macerals are (i) vitrinite, (ii) liptinite, (iii) inertinite, and (iv) fusinite. Inertinite and fusinite are composed mainly of marginally organic minerals that contain clays, carbonates, sulfides, and other non-organic molecules and compounds. Inertinite and fusinite are generally composed of the non-organic materials of the plant and thus contribute very little to providing energy in power production and inhibit coals use in the coking process. These minerals can be removed during the floatation process where the coal is crushed and floated in a fluid heavier than the coals. The coal, being lighter than the fluid floats across and is recaptured minus the heavier clays, carbonates, cherts, sulfides, or other heavy compounds. This optimizes the coal for burning in a power plant. Coals used for power plant generation range in rank from peat to low volatile coals. Typically, in the USA and Canada, the coals used for power generation is lignite to high-volatile bituminous in rank. Metallogenic coal tends to be very low in ash and are typically higher rank coals, such as medium volatile in rank.

The rank of coal seams is based on BTU, fixed carbon, volatile matter, and ash content (Figure 21.8). The higher the rank, the less volatile matter, water, and ash and the higher the fixed carbon. Reflectance, use of incident light on vitrinite macerals, is used to also determine the rank of coal. The reflectance value, R_0, increases with rank. Both methods have positives and negatives to determining coal rank, sometimes are not in agreement but their use together does result in a clearer picture of the rank of a coal.

Another characteristic of increase in coal rank is the reduction in permeability and porosity in a seam. The higher the coal rank the lower the permeability which is generally present as fractures or cleats. The cleats in the lower rank coals typically are continuous pathways, have numerous

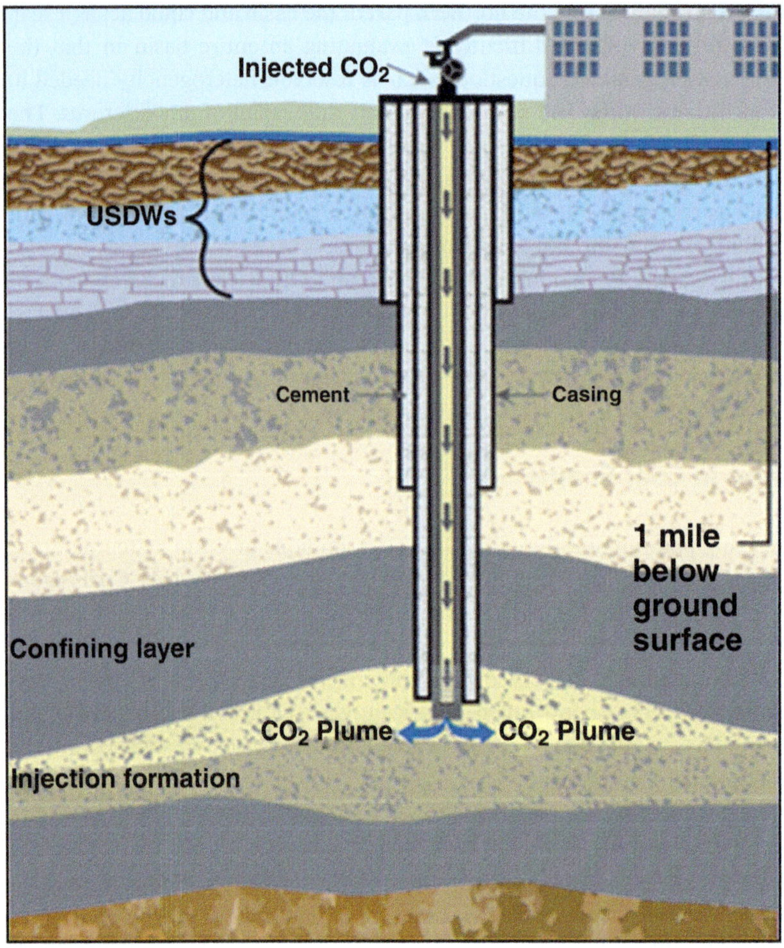

Figure 21.8 Stages of coalification relative to their approximate coal rank and R_0 (Adapted from Knetucky Geological Survey, 2024).

mineral deposits, water and gases present. As a coal becomes higher rank, the water and gases are expelled. The minerals deposited along the cleats maybe dissolved and expelled with the water and gases or preserved within the cleats.

Coals produce gas by depressurizing and dewatering the reservoir to allow methane to desorb from the coal matrix. N_2 and CO_2 will also desorb through time and usually later in the production cycle. The desorption process of CH_4 from coals during coal-bed methane production along with the removal of water increases porosity and permeability. The source of the CH_4, N_2, and CO_2 in the coal is either from: (i) thermogenic processes during the maturation of the coal and increasing rank and/or (ii) through biologic processes when microbes eat the coal and expel CH_4, N_2 and CO_2. Because of the ability of coal adsorb CH_4, N_2, and CO_2, it has been proposed that the seams can be used to sequester CO_2. The focus in sequestering CO_2 into deep unmineable coals, based on present economics, for potential storage of anthropogenic CO_2. Most of these coal seams in the eastern part of the USA and Canada tend to be less than 10 ft (3 m) thick, which results in a thin reservoir. Coal seams in the western part of the USA tend to be younger, Mesozoic to Tertiary, thicker as they are typically related the fluvial deposition then vast deltas as in the Pennsylvanian. Vishal et al. (2016) and Hosseinzadeh et al. (2024) put together a detailed summary the response of coal to CO_2 sequestration:

1. The affect of thermos-hydro-mechanical-chemical processes on coal for CO_2 storage need to be evaluated under actual conditions and in relation to the seal integrity, formation damage effects, and fault reactivation;
2. Detailed SEM analysis provides images and elemental composition data on a microscopic scale, is crucial for evaluating the fundamental geological and geomechanical aspects of CO_2 sequestration in coal seams. The microstructure of the coal is related to the formation of micro-cracks, altered surface roughness, and any mineral dissolution, which impacts CO_2 and water saturation;
3. CT scanning is invaluable for examining the microstructural characteristics of coal samples that help determine potential applications in CCUS;
4. XRD analysis provides insights into the spatial arrangement and dimensions of aromatic rings, which provides an understanding of the chemical structure of coal and the behavior of the in-place coalbed methane;
5. Analysis of silicate and carbonate minerals present within the coal seam and their potential reaction with the CO_2 being sequestered;
6. The adsorption capacity of coal. This is optimal in coal rank from high volatile bituminous to anthracite;
7. The optimal depth for CO_2 sequestration in coals is 1000–4300 ft (300–1300 m);
8. The strength of the coal is diminished by water saturation so that coals that are dry, like in Alberta or low water saturated coals such as in the Appalachian, Cherokee, Forest City, and Warrior basins are optimal for CCS.
9. When CO_2 is injected permeability declines as pressure increases and further decreases when it is in a supercritical form by 60%. This partly due to swelling of the coal as it absorbs CO_2;
10. When pore pressure is constant and as temperature increases the mechanical properties of the coal affect CO_2 sequestration.

There have been several attempts to inject CO_2 into coal seams such as in the Pennsylvanian coals in Illinois as reported by Greenberg et al. (2008). The coals in the Illinois Basin are significantly undersaturated with CH_4 and several CBM projects were never economic or viable, (Demir et al., 2004; Tedesco, 2004). There were several CO_2 sequestration pilot projects in the Illinois Basin that were being funded by states and Federal governments. The most recent report on CO_2 sequestration in the Illinois Basin coal seams was in 2008. No results after 2008 may be due funding was no longer available or there were technical issues with CO_2 injection. Coals when injected with CO_2 typically swell reducing permeability and porosity through adsorption into the coal matrix. This generally impedes injectivity and reduces the amount of CO_2 that can be injected. Coals potentially are CCS sites but their use seems to be limited.

21.10 Sequestration in Carbonaceous Mudstones

Sequestration of CO_2 into carbonaceous mudstones or shale is either to recover additional CH_4 and oil and/or sequester CO_2 into the reservoir. Khosrokhavar et al. (2014) and Lyu et al. (2021) discussed the advantage of using global shale projects both as EOR for recovering additional hydrocarbons and sequestering gigatons of CO_2 (Liu et al., 2019). As an example of regional analysis of the suitability of carbonaceous mudstones as seals and reservoirs, Lohr and Hackley (2018) investigated the Tuscaloosa shale of Mississippi and Alabama USA and its ability to sequester CO_2 and concluded that it was thick enough and laterally continuous to retain a CO_2 column 75–837 ft (23–255 m) in the Lower Tuscaloosa when the seal is near 100% water saturation.

Investigations have been done by various authors in using CO_2 for EOR. CO_2 increases recovery of oil and gas from carbonaceous mudstones or shale reservoirs (Jia et al., 2019; Lei et al., 2019). Initially, the focus of CO_2 use in carbonaceous mudstone is in the form of a "Huff and Puff" process in which the CO_2 would be injected, remain in the reservoir for a set period in which it interacts, and changes the characteristics of the petroleum present, then the well is flowed back and hopefully would recover more oil or gas then would be available by primary methods. CO_2 "Huff and Puffs" have been successful in conventional oil reservoirs so logically they should be successful in carbonaceous mudstones (Jia et al., 2019; Lei et al., 2019). The recovery of oil and gas from mudstones is low when compared to conventional reservoirs and the primary recovery from these reservoirs is focused on long horizontal laterals with large fracture stimulations. Carbonaceous mudstone reservoirs typically have nano-level permeability and porosity. Lee et al. (2020) found injecting CH_4/CO_2 into a carbonated rich shale gas reservoir did not significantly impact the recovery of CH_4 and sequestered 6.1% of the CO_2 injected. Lee et al. (2020) found reactions of the reservoir brine with the CO_2 reduced the potential of sequestration from 14.6 to 6.1%.

Liu et al. (2019) carried out experimental CO_2 and nitrogen (N_2) EOR on shale cores. In both gas cases, the highest oil recovered was in the first "Huff and Puff" cycle and thereafter begins decreasing with every cycle. Oil recovery decreased more rapidly with N_2. Liu et al. (2019) concluded the following:

1. There are three drive modes that occur during CO_2 injection, saturation and flowback which are gas cap drive, dissolved gas drive, and miscible drive.
2. CO_2 could reduce the interfacial tension from 22.4dyn/cm to 0, and decrease the oil viscosity by 40–60%.
3. N_2 could not reach miscible condition as the N_2 had very high minimum miscible pressure.
4. N_2 only changed the interfacial tension from 22.4 dyn/cm to 13.8 dyn/cm and reduced the oil viscosity by 10%.
5. The high CO_2 solubility and miscibility resulted in a higher dissolved gas drive speed and miscible drive, leading to a faster oil production rate in the following huff period.

6. The injectivity of CO_2 was better in the ultra-low permeability shale core than N_2 due to the solubility differences in crude oil (Liu et al. 2019).
7. Injected gas extracted the light oil components, and the majority of light oil components were produced in the gas phase. The heavy oil components were produced in the oil phase by dissolve gas drive. Molecular diffusion had a significant effect on oil recovery during CO_2.
8. CO_2 had better performance for all the oil components than N_2 (Liu et al. 2019).

The development of hydraulic fracturing and horizontal drilling techniques has promoted the exploitation of shale gas resources (Lyu et al., 2021). Supercritical CO_2 has shown potential to enhance shale gas recovery replacing water as the stimulation fluid. CO_2 has a better stimulation and rock-breaking capability than water. CO_2 fracturing creates rougher and more complex fracture networks than hydraulic fracturing, causing an increase in permeabilities (Lyu et al., 2021). Some of the injected CO_2 can be safely sequestered in shale reservoirs, thereby lowering carbon emissions and accessing CO_2 tax credits. Problems with CO_2 and carbonaceous mudstone reservoirs is what is the in action of the gas/liquid with the lithology and in situ brine (Lyu et al., 2021). Lyu et al. (2021) suggested using CO_2 for:

1. Drilling the wells as the threshold pressure and cutting energy required for CO_2 in the system are lower than water. However, the capacity of CO_2 to carry cuttings still needs to be improved;
2. Fracture stimulation with CO_2 can create more irregular multiple fractures, rougher and more complex fracture surfaces than liquid CO_2 than water. The proppant carrying ability of CO_2 is poor, which can lead to screening out earlier or misplacement of the sand. The pressures needed to move CO_2 is very high, and there is safety concerns at the surface facilities that have to be addressed;
3. The processes of CO_2 interaction with the carbonaceous mudstones and associated brines are not well understood;
4. Sequestration of some of the CO_2 will be by solubility, adsorption, and mineral trapping.

Murugesu et al. (2023) determined from using low pressure CO_2 and N_2 sorption on pure clay minerals and shale samples from seven different formations was able to characterize nano-porosity (< 200 nm) of shale from ultra-micro to mesopores. Murugesu et al. (2023) concluded:

1. That organic-rich shale has nanopores that are dominated by microporosity, whereas clay-rich shale has a significant fraction of meso- to macropores within the total pore volume;
2. CO_2 adsorption capacity of shales increases with the total organic content (TOC), clay and their maturity level;
3. Carbonate, quartz, K-feldspar, plagioclase, and pyrite s by themselves show insignificant effect on CO_2-adsorbed volume. Calcite may trap some CO_2 molecules;
4. Dissolution of feldspar minerals and simultaneous precipitation of secondary aluminum minerals such as kaolinite;
5. The dissolution of feldspar, where K+ and Si+ ions are released from feldspar surface, depends on several collective factors such as brine pH and composition, temperature, pressure, and chemical affinity. Similar to K-feldspar, plagioclase and aluminosilicate also dissolve in CO_2 -saturated brine, releasing calcium cations. Calcium cations could permanently trap CO_2 by forming calcium carbonate;
6. Pyrite dissolves in CO_2 rich environment and in the presence of oxygen and then simultaneously precipitate siderite;
7. There is a positive correlation between clay, expandable clay, thermal maturity and CO_2 adsorbed specific surface area;
8. There is a positive correlation between hydrogen index, TOC and CO_2 -adsorbed volume.

Mahmoud et al. (2020) evaluated carbonaceous mudstones with different lithology, different thermal maturity levels, and different assemblages of carbonates and clays. The amount of CO_2 in shale formations is related to the adsorption capacity of these rocks, different temperatures varying from 50 to 150 °C and different percentages of CO_2 in the gas mix. Mahmoud et al. (2020) study adsorption behavior of CH_4, CO_2, and 10/90 CO_2/ CH_4 mixture on shales samples was investigated at 50, 100, and 150 °C and a maximum pressure of 45 bars pressure. The results of the study were:

1. Variations in TOC and mineralogy resulted in varying amounts of CO_2 adsorption in shale samples at different temperatures and pressures;
2. The analysis indicated the shale has a high affinity to CO_2, while adsorption entropy and enthalpy were less compared to CH_4;
3. All three shales showed high adsorption uptake with the increase in CO_2 percentage in the gas. This is explained by the high affinity of some of the shale constituents to CO_2;
4. The temperature has a very strong effect on shale adsorption. Some of the shale samples showed endothermic behavior at 50–100 °C leading to high adsorption uptake at 100 °C. However, the adsorption was exothermic at 150 °C with a high decline in adsorption. Another shale sample exhibited endothermic adsorption behavior leading to an increase in the adsorption with increasing temperature. This behavior is attributed to the changes in the crystallinity of clay mineral in shale and the degradation of shale organic matter with temperature;
5. Shales with high TOC showed significantly higher CH_4 uptake compared to low TOC shales (SH1 and SH3) at a temperature of 150 °C because of clay mineral in shale and the alteration of shale organic;
6. The presence of clays hinders the adsorption of CO_2 because of the swelling effect on the clay surface.

Overall, shale mineralogy, maturity and total organic carbon strongly influence the uptake capacity and consequently gas reserve estimations in shale suggesting the importance of characterization of rock lithology and organic matter (Mahmoud et al., 2020).

Merey and Sinayuc (2016) analyzed samples from the Dadas shale (Silurian) from Turkey samples were conducted at 25, 50, and 75 °C and up to approximately 2000 psia by using pure CO_2 and pure CH_4. They used the Langmuir isotherm and Ono-Kondo lattice models (three-layer and monolayer) and the data concluded the latter model fits the resulting data the best. It was concluded that gas concentration changes the adsorption potential of CO_2 from 39.2% to 71.8% between 5000 psia and 500 psia.

Storing CO_2 directly into a carbonaceous mudstone or shale is in the early stages of investigation leading to exploitation. The difficulties in using these lithologies to store CO_2 is in defining the clay and organic materials present, the composition of the brine water, mapping internal lithological architecture in order to define any geochemical, geological, and geomechanical issues, fracturing and faulting possibilities, and the likely amount of CO_2 that will be retained. The typical lateral well in a shale play is from a mile to three miles in length, which provides significant variability in lithology and clay/organic material content. Thus, it is critical to collect a sufficient amount of data to determine as many factors and parameters as possible when initiating CO_2 EOR and sequestration.

21.11 Mineral Sequestration

Mineral sequestration or carbonization of CO_2 has been proposed as a possible process for reducing climate change. CO_2 trapping in minerals, specifically in the form of solid carbonate salts, is a slow process. CO_2 mineral sequestration of some kind is likely to occur during injection into depleted oil and gas fields and saline aquifers. CO_2 is injected into the groundwater/brine forming carbonic acid. The carbonic acid reacts with silicates and slowly dissolves calcium, magnesium, alkalis, and silica thus leaving residue of clay minerals.

Mineral sequestration occurs when CO_2 is injected into hydrated serpentinite or olivine (dunite) deposits as the reaction will result in magnesium carbonate being deposited which is stable and non-toxic. The more iron that is present, the less efficient the reaction will be. It has been proposed to inject CO_2 in basalt rich areas in the deep ocean. The CO_2 would mix with sea water then react with the basalt over long periods of time depositing minerals such as magnesium carbonate.

Zeolitic imidazolate frameworks (ZIF) are metal-organic frameworks similar to zeolites and have good porosity, chemical stability and thermal resistance could potentially store CO_2. ZIF could be used as filters at industrial sites to keep emissions of CO_2 out of the atmosphere. Mineral sequestration outside of that occurs in depleted oil, gas, and CO_2, and saline aquifers is in the early stages of evaluation as an effective CCS project.

21.12 Sequestration in Oceans

Annually, the ocean is thought to actively sequester up one-third of anthropogenic CO_2 emissions (Jain et al., 2012). The ocean's ability to sequester carbon from the atmosphere exerts an important control on global climate (Boyd et al., 2019). The carbon cycle drives carbon storage in the deep ocean and is thought to function via gravitational settling of organic particles from surface waters (Boyd et al., 2019). Jain et al. (2012) and Boyd et al. (2019) proposed ocean sequestration strategies that will speed up this process to reduce atmospheric CO_2 concentrations. The long-term effectiveness and potential side effects to the environment of using the oceans to sequester anthropogenic CO_2 are unknown (Jain et al., 2012). To date, there has been no significant attempt commercially to inject anthropogenic CO_2 into the oceans.

21.13 Sequestration in Soils

CO_2 sequestration in soils has been proposed on the basis they have organic matter, generally decaying plant matter, of varying percentages around the world (Melillo and Gribkoff, 2015; Paustain et al., 2019). It has been estimated with the 10% of Earth's land is used for raising crops and this could potential sequester over 1 billion tons a year in CO_2. Doing this would require a significant change in farming methods and may not be politically possible. Decaying plant material holds a significant amount of CO_2 that the vegetation took in while alive. This CO_2 is released back into the atmosphere if not held in place by the organic matter (Melillo and Gribkoff, 2015). If this CO_2 was to be returned to the atmosphere, it would increase greenhouse gas (Melillo and Gribkoff, 2015). Over the last 12,000 years, the growth of farmland has been estimated to have released 110 billion metric tons of carbon which is equivalent to 80 years of US emissions (Sanderman et al., 2017). The upside to that release is in the increase in plant growth that has occurred with its release.

21.14 Class VI Wells

Class VI wells (Figure 21.9) in the US for CO_2 disposal are typically managed by the Environmental Protection Agency (EPA). Presently, there are 48 projects and 141 well applications under review. Three states presently have primacy over the EPA for Class VI wells are Louisiana, North Dakota,

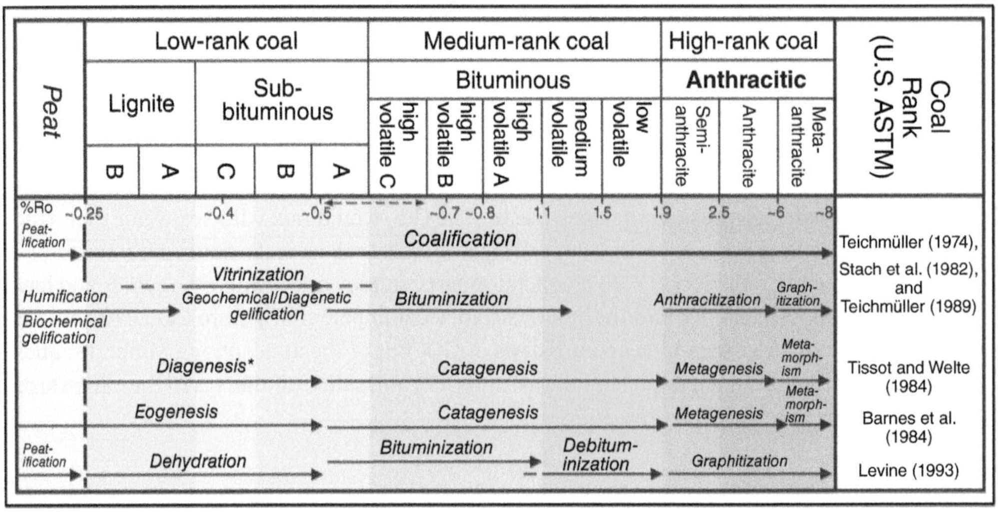

Figure 21.9 Ideal diagram of injection of CO_2 into the subsurface. Multiple casing strings to protect and isolate the groundwater from CO_2 injection. Figure 21.8 is not too scale (Adapted from EPA, 2024).

and Wyoming. Presently, other states are in the process of applying to the EPA to obtain primacy. The main reasons for taking over primacy is 1) political in that individual states want to limit Federal interference in managing subsurface operations in their state, and 2) the EPA takes inordinate amount of time to approve Class VI wells which is delaying disposing of CO_2 under several Federal programs, and 3) the EPA typically lacks the personnel with experience in subsurface operations critical to understanding and evaluating data.

Requirements for a permitting a Class IV well (Figure 21.8) are (EPA, 2024b):

1. Site characterization requirements to ensure the geology in the project area can receive and contain the CO_2 within the zone where it will be injected, including that the area is free of faults and fractures and that induced seismicity is not a concern.
2. Requirements to predict the extent of the injected CO_2 plume and associated pressure front for the project using computational modeling, and to identify and address any deficiencies of existing wells within the area of review through corrective action. The area of review includes the area where the injected plume and its associated pressure front may impact pore fluids.
3. Well construction requirements to ensure the Class VI injection well is constructed in a manner that will prevent any CO_2 from leaking outside of the injection zone (Figure 21.9). Basically, casing specifically designed to handle CO_2 so that it does not corrode the injection string and containment groundwater sources.
4. Outline testing and monitoring of the injection well, groundwater quality, and the movement of the CO_2 plume and pressure front throughout the life of the project, including after CO_2 injection has ended, until the permitting authority determines no additional monitoring is needed to ensure that the geologic sequestration (GS) project does not pose an endangerment to Underground Sources of Drinking Water (USDWs).
5. Operating requirements to ensure the injection activity is appropriate to the well's construction and geologic characteristics so that it will not endanger USDWs or human health.
6. Plug parameters for the injection well in a manner that will entomb the CO_2 and not allow its movement that would endanger USDWs.

7. Requirements for the operator to establish and maintain financial instruments sufficient to cover the cost of corrective action, plugging the injection well, post-injection site care, and emergency and remedial response for the GS project (i.e. financial responsibility). Financial responsibility can be in the amounts of several tens of millions of US dollars.
8. Develop and maintain a site-specific emergency and remedial response plan.
9. The EPA requires all owner(s) or operator(s) of a Class VI well to report all testing and monitoring results to the permitting authority to ensure the project is operating in compliance with all permit and regulatory requirements.

There is a more detail to each one of these items that needs to be expanded upon in any application for a Class VI well. The EPA requires extensive detail for a Class VI and is focused on preventing breach of the seal, any contamination of the shallow groundwater and keeping the CO_2 plume in the reservoir and not migrate any distance to where it is no longer contained.

21.15 Case Histories/Models

21.15.1 Farnsworth Unit

The Farnsworth Unit (Figure 21.10) is located in the Anadarko Basin, Ochiltree County, Texas, and produces oil from the Pennsylvanian age Morrow sandstone reservoir and has been subject to CO_2-EOR flooding starting in 2010 (Rose-Coss et al., 2015; Balch et al., 2020). The field was discovery in 1955, over 100 wells were completed by 1960 and waterflooding started in 1964. For the CO_2-EOR flooding the mineralogy, provenance, diagenetic history, depositional environment and porosity types was analyzed from a 3D seismic survey, cores (from 48 wells), oil and water production, thin sections, and logs (Rose-Coss et al., 2015; Balch et al., 2020). Figure 21.11 is from Rose-Coss et al. 2015 and shows the continuity of the Morrow sandstone fluvial or marine sandstone located in an incised valley. This data was used to create a detailed scale lithofacies geologic model for the field. The source of the anthropogenic CO_2-CCS was from an ethanol plant in in Liberal, Kansas, and a fertilizer plant in Borger, Texas (Rose-Coss et al., 2015) and 500,000–600,000 tons of CO_2 are injected per year (Balch et al., 2020). The EOR project has been highly successful and one of the reasons is the significant amount of data available from the field producing oil for several years. Figure 21.12 is the production profile for the field and includes simulated oil, gas, and pressure rates from 1955 to mid-2018 (Balch et al., 2020). Figure 21.11 indicates the impact of water and CO_2 flooding.

It is unknown if a 4D seismic survey is present and has been used to track movement of the plume. Many EOR projects for oil production have multiple injector wells, which can help track the CO_2 plume. CO_2-CCS specific projects in non-petroleum areas focused on saline aquifers tend to have only a single injection and monitor well which precludes having the ability to watch the plume seeing the plume migrate in real time and physically measure it. The advantage of having multiple wells is the plume in EOR project provides real-time data collection of its movement. Changes then can be made quickly in real time adjusting injection rates.

21.15.2 Aquistore CO_2 Sequestration Site

The Aquistore CO_2 sequestration site (Figure 21.13) is located on the northwestern flank of the Williston Basin near the town of Estevan, Saskatchewan, Canada, and was chosen to test the viability of storing liquid CO_2 in an underground reservoir (Rostron et al., 2014; Worth et al., 2014).

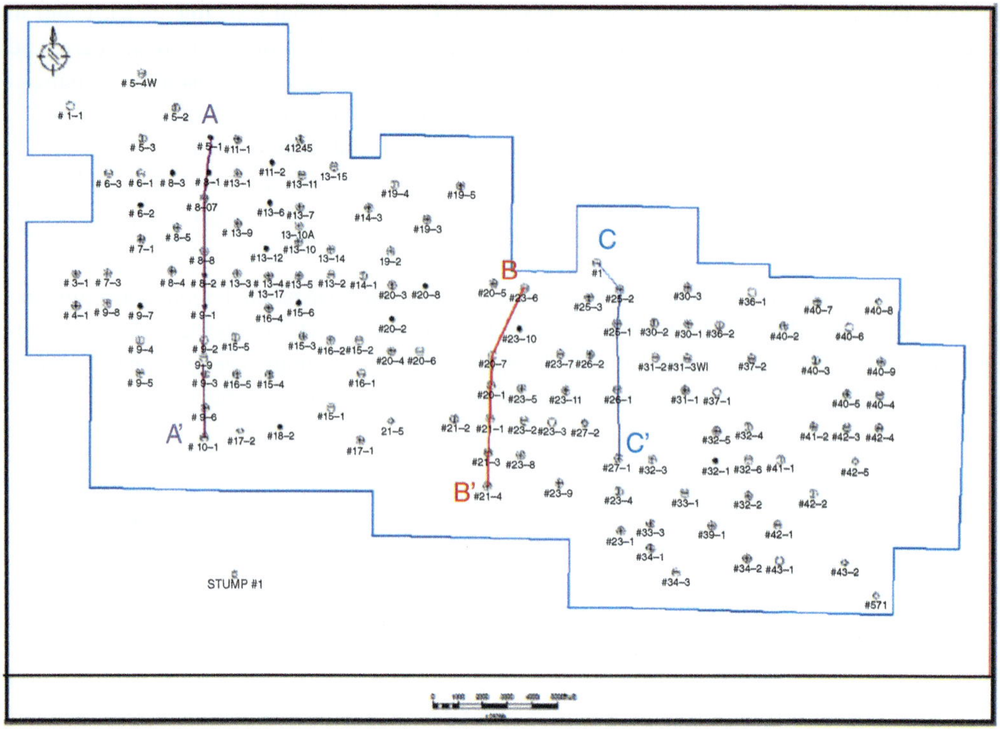

Figure 21.10 Farnsworth Unit located in the Anadarko Basin, Ochiltree County, Texas and produces from the Pennsylvanian age Morrow sandstone and was subject to CO_2 injection for an EOR project (Rose-Cross et al. 2015/U.S. Department of Energy/Public Domain). (a), (b), and (c) are cross-sections from Rose-Coss et al., 2015 but only (b) is present in Figure 21.11.

The Cambrian Deadwood Fm and Winnipeg sandstone as CCS reservoirs were initially investigated utilizing a 3 × 3 mi (5 × 5 km) 3D seismic and determined there was no faulting in the area (Rostron et al., 2014). Two wells were drilled, one an injector well that reached total depth 11,206 ft (3396 m) and one an observation well that reached total depth of 11,220 ft (3400 m) in the Precambrian (Figure 21.14). The two wells are 500 ft (150 m) apart from each other (Roston et al., 2014). The reservoir has several highly permeable zones, approximately 500 ft (150 m) thick and is overlain by the Ice Box shale which acts as the primary seal (Rostron et al., 2014). The secondary seal is the Prairie Evaporite (Devonian) and is a thick evaporite sequence that is continuous across most of the Williston Basin.

Monitoring of the Aquistore project was extensive in terms putting in a:

1. 4D seismic geophone array to track the CO_2 plume in the reservoir through time (Worth et al., 2016; White et al., 2024);
2. 50 shallow semi-permanent soil gas probes at less than 7 ft (2 m) were placed at 46 locations to test for gaseous hydrocarbon compounds and isotopic data (Rostron et al., 2014; Worth et al., 2016);
3. Forty shallow monitoring wells, of which several drilled for the project and others were local groundwater supply wells. These 40 wells were used in the area to establish a baseline for hydrogeological and hydrochemical data (Klappstein and Rostron, 2014; Rostron, et al., 2014). Figure 21.13 shows the location map of the Aquistore CO_2 Facility near Estevan, Saskatchewan, Canada (Klappstein and Rostron, 2014). Note the source of the CO_2 is a local power plant.

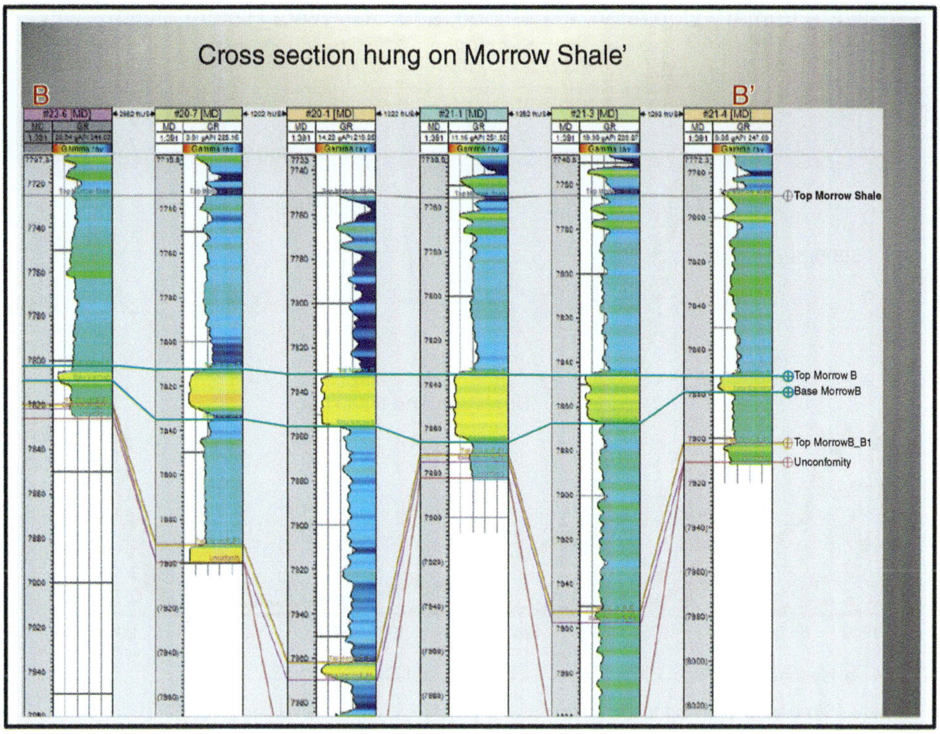

Figure 21.11 Cross-section B through the Farnsworth Unit indicating the Pennsylvanian Morrow sandstone (yellow), which is oil productive and presently undergoing a CO_2 EOR flood (Rose-Cross et al. 2015/U.S. Department of Energy/Public Domain).

Figure 21.15 is a cross-section through the 3D survey indicating the extensive nature of the Deadwood Fm/Winnipeg sandstone, Ice Box shale, which acts as the primary seal and the Prairie Evaporite which acts the secondary seal (Rostron et al., 2014). The seismic data indicates there is no Precambrian topographic highs, which can create different issues in that the reservoir or seal may not be present or partially present.

Overtime, the approximately 600,000 tons of CO_2 been disposed in the Deadwood Fm/Winnipeg sandstone through the injection well. The CO_2 plume that developed has been tracked over time by the 4D seismic array that was installed at the start of the project. Figure 21.16 is the some of the 4D analysis at 36,000, 102,000, 150,000, and 272,000 tons. White et al. (2024) interpretation the presence of an unknown NW-SE trending fault that seems to act as a baffle to the CO_2 plume and redirected its migration.

The Aquistore CO_2 storage facility is an example of what needs to be done in terms of evaluating before CO_2 injection and then continue monitoring after injection has started. The primary and secondary seals ensure, hopefully, there is no leakage into the shallow groundwater or to the surface.

21.15.3 Sleipner Field

The Sleipner Field was the world's first offshore CCS facility (Figure 21.17). The Sleipner Field was discovered in 1974 and produces natural gas and condensate from approximately 8200 ft

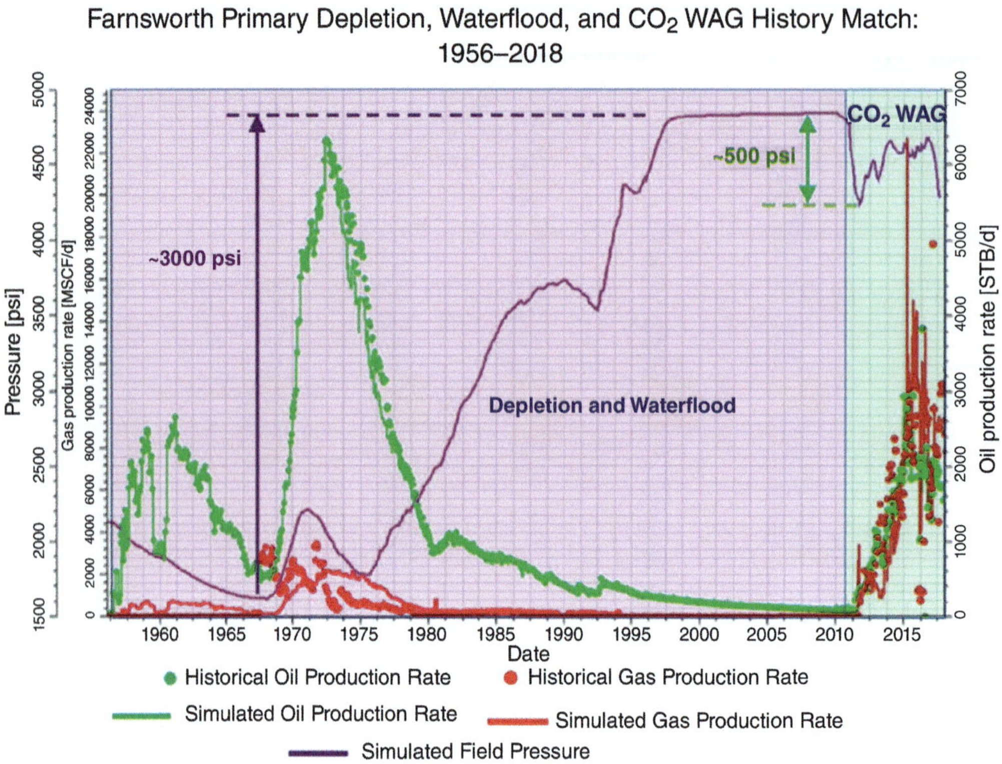

Figure 21.12 Oil and gas production history of the Farnsworth Unit with simulated oil, gas, and pressure match (Balch et al., 2020/U.S. Department of Energy/Public Domain).

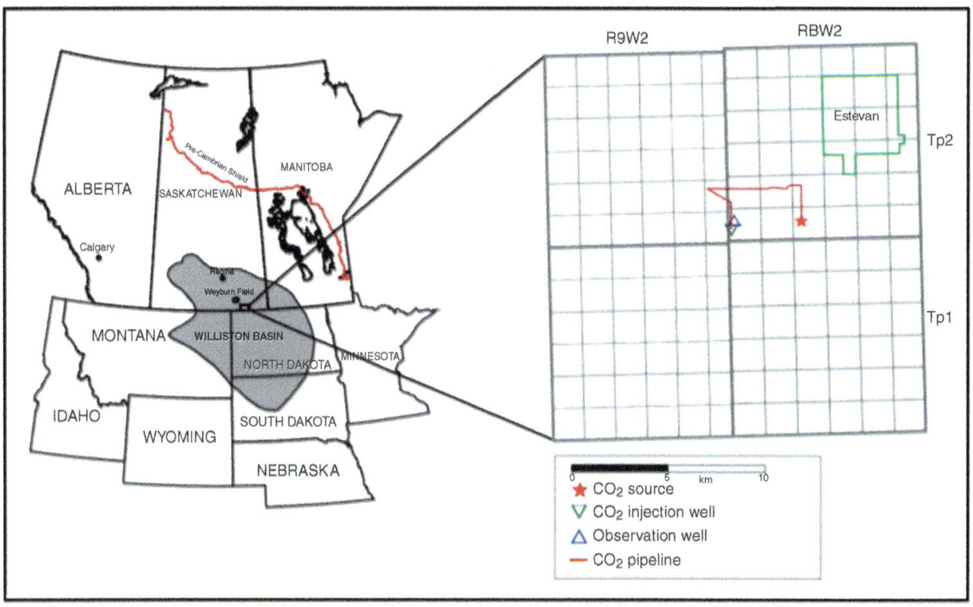

Figure 21.13 Location of the Aquistore CO_2 sequestration site (Rostron et al., 2014; Worth et al., 2014).

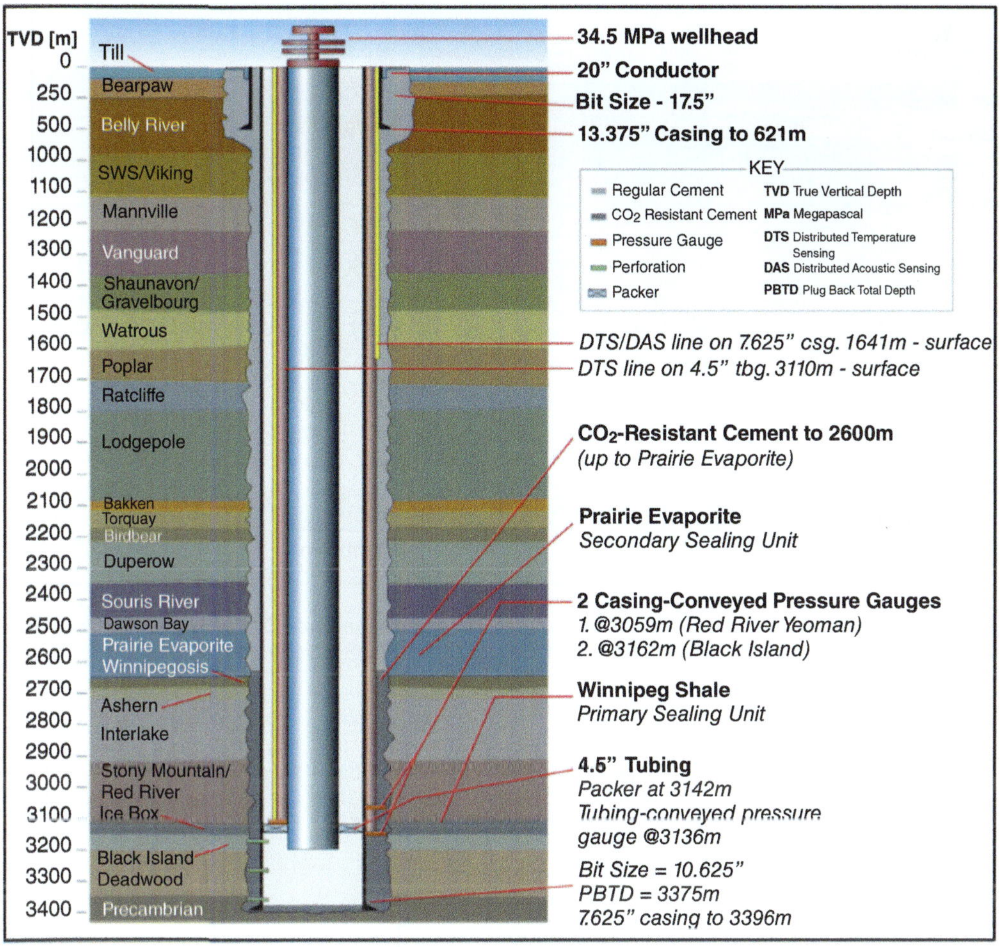

Figure 21.14 Wellbore diagram and stratigraphy of the injection well, 01/5-6-2-8W2, at the Aquistore CO_2 storage facility (Rostron, et al., 2014/with permission of Elsevier).

(2500 m) and is divided into two areas, designated east and west (Figure 21.17). The natural gas and condensate extracted at Sleipner contains approximately 9% CO_2, which is separated and then re-injected into Utsira Fm (Miocene) at 2640 ft (800 m) below the top of the seabed. The Utsira Fm is 500 ft (150 m) thick, a large aquifer sandstone but with thin layers of shale throughout, 27–40% porosity, several Darcy permeability, and is overlain by 2640 ft (800 m) of clay sediment of Miocene through Pleistocene age that is an effective seal (Nazarian and Furre, 2022). Approximately, 2800 tons of CO_2 are separated and injected daily in to the Utsira Fm. Based on seismic data, nine sandstone reservoir layers have been defined in the Utsira Fm that showed changes in the CO_2 plume through time (Nazarian and Furre, 2022). There is an additional sandstone described as the "wedge" above the Utsira Fm that the CO_2 plume has migrated into. Sleipner passed its twentieth anniversary in 2016 with and as of 2017 over 17 million tons of CO_2 have been injected and stored. The operator of the field set up seismic and gravity monitoring of the CO_2 plume. CO_2 from the adjacent Alfa Nord and Gudrun fields are also being injected into the Utsira Fm.

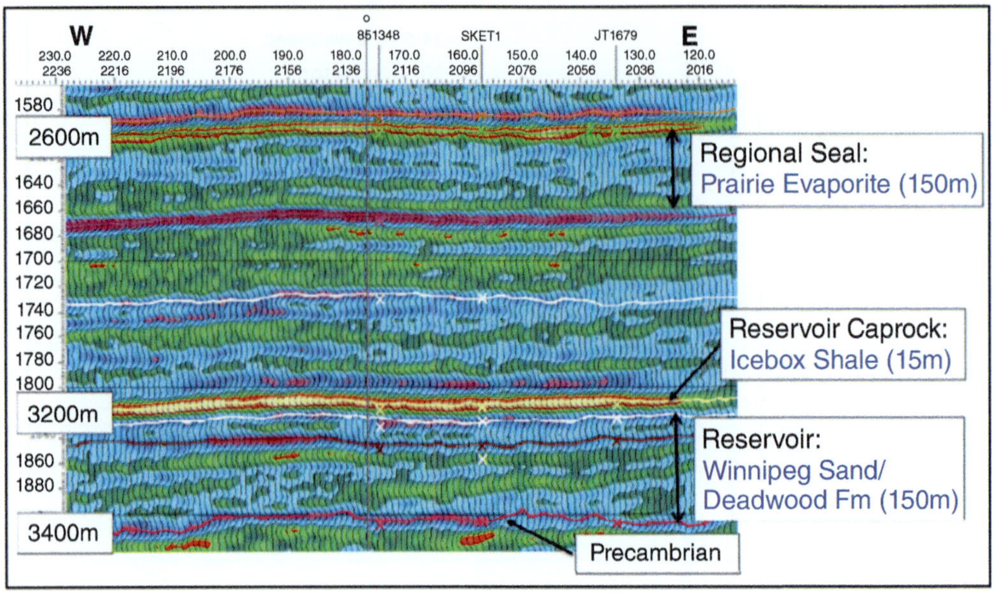

Figure 21.15 A slice of the 3D seismic survey that was shot at the Aquistore CO_2 storage facility (Rostron, et al., 2014/with permission of Elsevier). Note the contiguous nature of the reservoirs, primary and secondary seals.

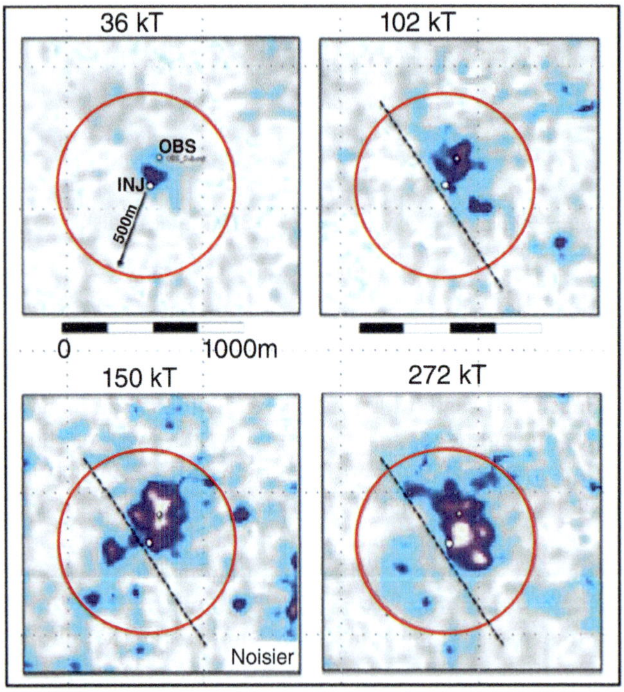

Figure 21.16 Time-lapse amplitude differences for the upper Deadwood interval (White et al., 2024). Observed amplitude differences in the 4D seismic volumes are interpreted as zones of CO_2 saturation.

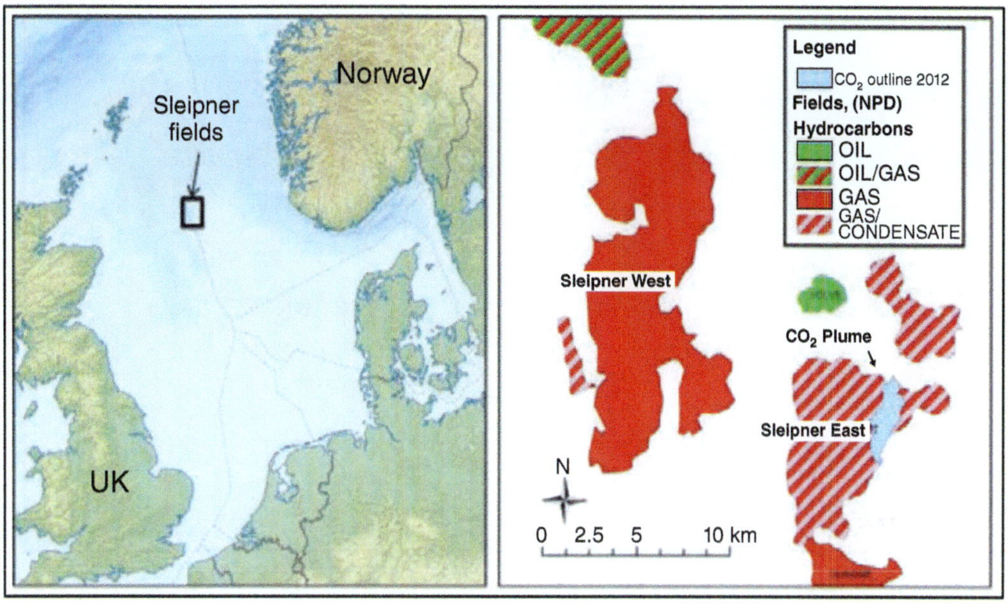

Figure 21.17 Location of the Sleipner East and West fields and the CO_2 plume (Furre et al., 2017/with permission of Elsevier).

The European Union in 2014 mandated that gas movement, shell stability, and effective remedies had to be assessed and defined in case of breach of the seal or reservoir. It was discovered that the CO_2 plume had moved up close to the caprock but has not caused any leaking (Furre et al., 2017). The CO_2 is in a solubility trap and based on modeling once injection ceases will become stable. Seismic data presented by Furre et al. (2017) showed how the seismic section had changed from being mappable to being difficult to map through area where the plume presently is located.

The CO_2 in the Utsira Fm is very close to critical point due to the shallow depth of the storage reservoir (Nazarian and Furre, 2022). Nazarian and Furre, 2022 modelled the 4D data that was available and had to introduce several factors in order to match the modeling to the actual results. Nazarian and Furre (2022) work attempted to stimulate the CO_2 plume to actual data. They had to include a thermal history and an increase in salinity of the brine in the from sea water to 150,000 TDS. The interpretation of the data from 2010 onward indicated that the CO_2 had migrated into a sand wedge overlying the Utsira Fm and underlying the seal. Nazarian and Furre (2022) identified from the seismic data several chimneys or vertical channels that allowed migration of the gas/liquid towards the caprock.

Previous work according to Nazarian and Furre (2022) had focused on the sand wedge whereas their study had focused on the nine layers of the Utsira Fm and the sand wedge. Their study which was more definitive still demonstrates the problem of not having data on numerous variables, known and unknown. Modeling unto itself is only as good as the data available and used.

21.15.4 Stenlille Gas Storage

In Denmark, a present-day operating gas storage facility (30 years) called Stenlille (Figure 21.18a and b) is being converted to partially store CO_2 on the flank of the structure (LÜthie et al., 2023).

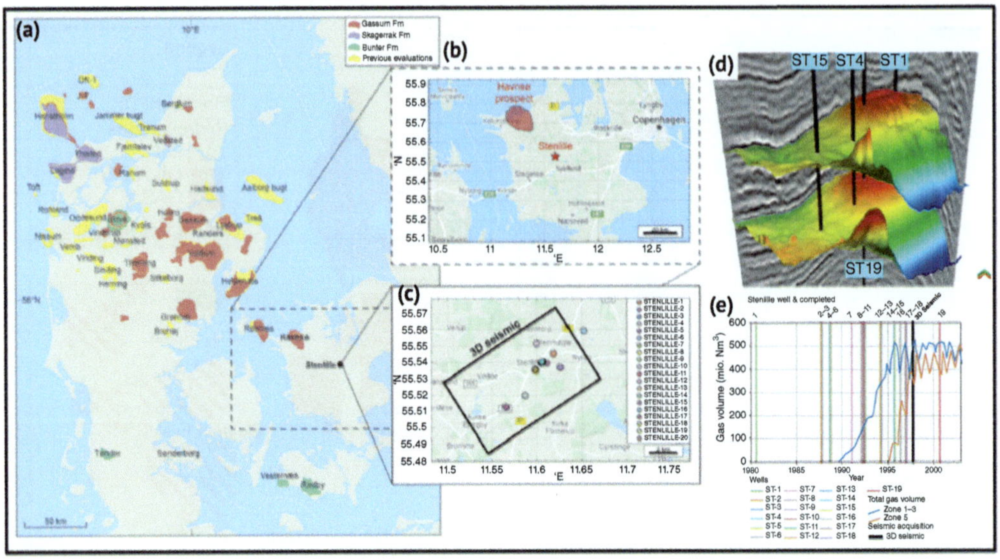

Figure 21.18 Location maps showing: (a) identified CO_2 structures in the Danish area relevant for CO_2 storage within different formations; (b) the Sjælland island with the Havnsø prospect indicated; (c) the Stenlille area with well locations and the 3D seismic survey indicated; (d) the Stenlille structure with two horizon surfaces bounding the Gassum reservoir target plotted together with seismic sections and four wells in 3D view; and (e) net gas volume history in Zone 1–3 and Zone 5 and the 3D seismic survey (Bredesen et al., 2024/The Geological Society of London/CC BY 4.0).

The field presently stores natural gas in Gassum Fm (U. Triassic to L. Jurassic) sandstone reservoirs at 500 ft (150 m) thick, approximately 5000 ft (1450 m) from surface and porosities are typically 25% with up to 7 darcies permeability (Figure 21.19). The overlying Jurassic Fjerritslev Fm is approximately 500 ft (150 m) thick tight seal rock composed of mudstones (Figure 21.19). The four-way structural closure at the facility is caused by a salt pillow at depth (Gregersen et al., 2023). The field has extensive 2D and 3D data and logs from 20 wells. A deeper zone, the Bunter Sandstone Fm (Lower to Middle Upper Triassic) has also been considered as an additional CCS zone, but it has lower porosity and permeability. Both the Bunter and Gassum Fm are also considered potential geothermal reservoirs (Kristensen et al., 2016).

Berendsen et al. (2024) in Figure 21.18 discussed other projects such as the Havnsø Field, but the present-day data set is not as extensive as Stenlille. If one looks forward to a hydrocarbon free world, then natural gas storage facilities, because of the large data sets generally available, make feasibility analysis of these sites for CO_2 easier. Figure 21.18c indicates the layout of a significantly large 3D survey (Berendsen et al., 2024). Figure 21.18d indicates an interpretation of the 3D survey by Berendsen et al. (2024) shows the structural closure for Zones 1 and 3 of the Gassum Fm. Figure 21.18e is the total volume of natural gas placed in Zone 1 and 3 into the Gassum Fm since 1989.

Extensive analysis of the 3D has been done by Berendsen et al. (2024) and others have defined the potential faults that may impact CO_2 sequestration (Figure 21.20). Whether these faults are open, closed, and potentially filled with mineralization is unknown. The faults if present within the reservoir may act as baffles and deflect the CO_2 plume as it moves through the reservoir. Injection of CO_2 is supposed to begin in 2026 (Lüthie et al., 2023).

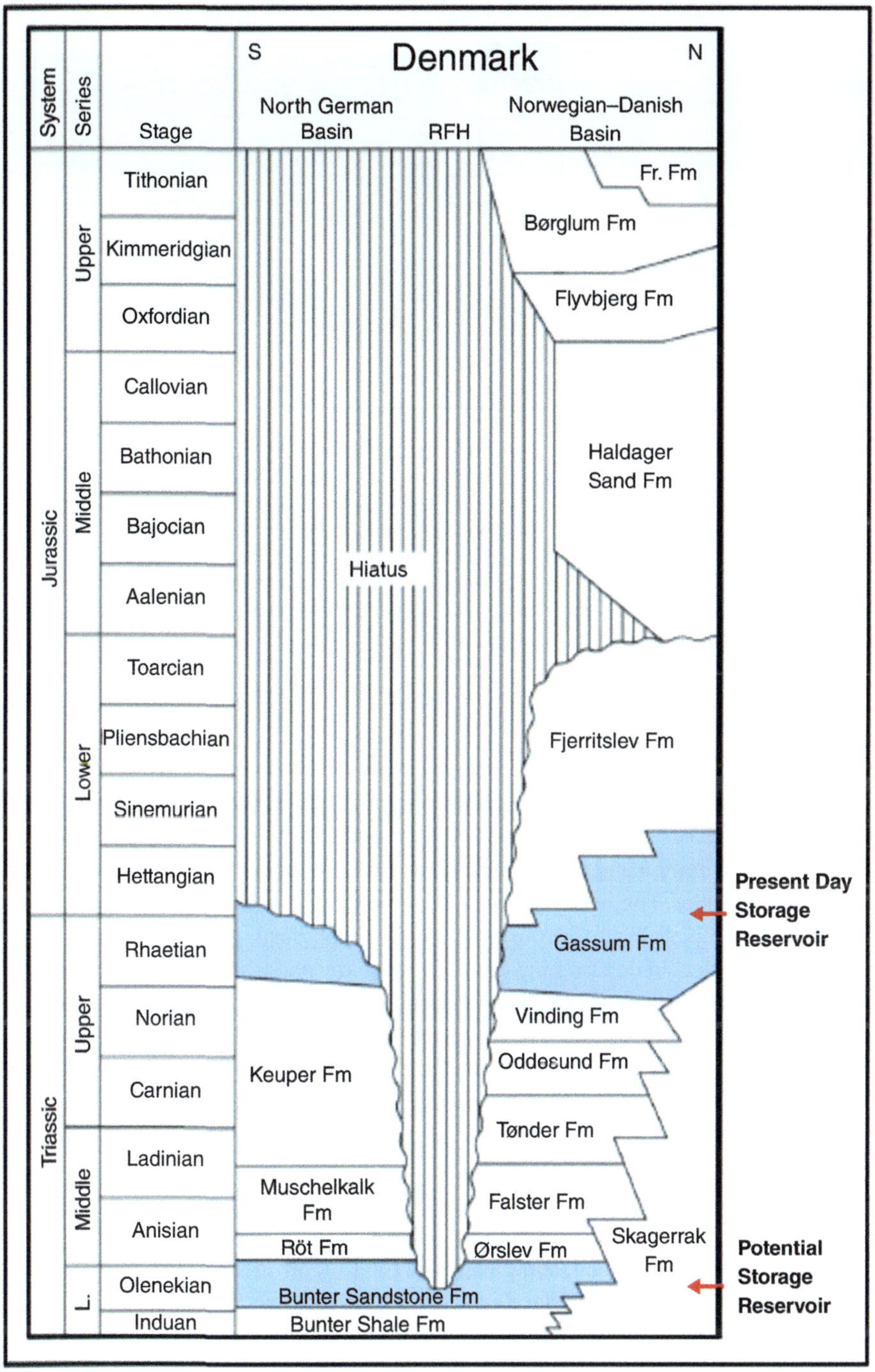

Figure 21.19 Stratigraphic column for the Norwegian-Danish Basin and the North German Basin (Kristensen et al, 2016/Springer Nature/CC BY 4.0). The present-day storage reservoir (Upper Triassic–Lower Jurassic Gassum formations (highlighted in blue) and the potential Lower to Middle Triassic Bunter Sandstone (highlighted in blue).

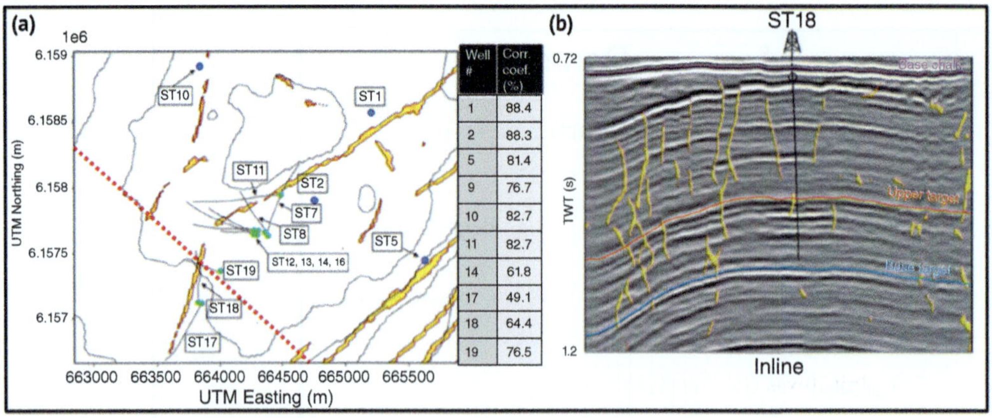

Figure 21.20 (a) A fault probability attribute map derived from a machine learning approach from the upper part of the Gassum interval with the well-path projections plotted on top (Bredesen et al., 2023). Blue dots are the locations of vertical wells, whereas the deviated well paths are shown by the black curves extending out from the well heads (green dots). Elevation iso-contours represent the upper target horizon (Berendsen et al., 2024). (b) A cross-section through the 3D volume intersecting the Stenlille-18 (ST18) well path (intersecting red dotted line on the map). The yellow vertical lines are potential faults through the geologic section. The red line represents the upper target and blue represents the lower target zones as described by Bredesen et al., 2024/The Geological Society of London/CC BY 4.0.

21.16 Summary

There are potentially numerous places to dispose of anthropogenic CO_2 in the subsurface just based on data from the petroleum industry. Utilizing depleted oil, gas and CO_2 reservoirs to store or recover additional hydrocarbons is a common and economic practice. In addition, tax credits are available for using CO_2 to recover hydrocarbons. These projects in the USA use Class II wells that are easy to permit and maintain. However, a complete evaluation using geologic, geophysical, geochemical, and engineering data for any site is critical in determining the viability of any one reservoir. The lack of data for a depleted field may impede using it as EOR. Carbonaceous mudstone reservoirs provide an even more unique opportunity to store hydrocarbons as they typically are using a "Huff and Puff" method of oil or gas recovery using CO_2. The injection of CO_2 in the horizontal lateral frees additional oil and sequesters some of the CO_2 used. CO_2 used in carbonaceous mudstones optimally fracture stimulate these reservoirs providing new fracture pathways for oil to be produced. Tax credits also make these projects very economic and can weather volatile oil prices.

Storage of anthropogenic CO_2 in a saline aquifer is more difficult because of the cost and expense of collecting datasets. Typically, this limits analysis and may not be sufficient to reduce risk and implement a project. Or a project is drilled, completed and the CO_2 plume migrates in the wrong direction, or breaches the seal/caprock. Surface issues, such as local resistance to putting in a CO_2 facility also may increase costs or prevent a project from going forward. The justification for the economics on these types of projects is solely for the tax credits. If a project in the USA is not initiated prior to 2032, it will not be eligible for tax credits. However, like most government tax programs, Congress is notorious for extending programs that economically beneficial to their constituents despite which political party it is.

The EU and other countries have taken a more urgent approach to getting CO_2 sequestration projects implemented. In the USA, despite the Inflation Reduction Act (IRA), a variety of Federal and State agencies are not moving quickly to approve projects that are in saline reservoirs. This delays utilization of funds and tax credits from IRA and consequently this causes an economic impact. CO_2 sequestration industry is relatively new and rapidly growing. Over the next 10 years, there will be several projects implemented and the results will be interesting to see.

References

Arif, M. (2017). Experimental investigation of wettability of Rock-CO_2-Brine for improved reservoir characterization, PhD Dissertation, Curtin University. p. 195.

Balch, R., Will, R., & McMillan, M. (2020). Recent advancements in mechanical earth modeling at Farnsworth Unit. Texas USA, Search and Discovery. Article 80722, AAPG, p. 46.

Blanco-Martin, L., Jahangir, E., Rinaldi, A. P., et al. (2022). Evaluation of possible reactivation of undetected faults during CO_2 injection. International Journal of Greenhouse Gas Control 121, p. 103794. https://doi.org/10.1016/j.ijggc.2022.103794.

Boyd, P. W., Claustre, H., Levy, M., et al. (2019). Multi-facet particle pumps drive carbon sequestration in the ocean. Nature 568, pp. 327–335. https://doi.org/10.1038/s41586-019-1098-2.

Bredesen, K., Smit, F., Lorentzen M., et al. (2023). Improved delineation of the Gassum Formation reservoir zones using seismic impedance inversions: implications for exploiting the Stenlille aquifer gas storage facility as a CO_2 storage demonstration site, onshore Denmark, Geological Survey of Denmark and Greenland (GEUS). Øster Voldgade 10, 1350. Copenhagen K, Denmark. https://www.lyellcollection.org.

Callis, C., Kovscek, A. R. & Benson, S. M. (2023). Assessing the impact of dip angle on carbon storage in saline reservoirs to aid site selection. International Journal of Greenhouse Gas Control 129, p. 103966. https://doi.org/10.1016/j.ijggc.2023.103966.

Carter, K., Patchen, D. Moore, J., et al. (2017). Mapping and reservoir characterization of geologic intervals for NGL storage applications. Search and Discovery Article 80617, AAPG. p. 51.

Childers, W. E. (2012). Characterization of Pliocene and Miocene formations in the Wilmington Graben, Offshore Los Angeles, for large-scale geologic storage of CO_2. Search and Discovery Article 80215, AAPG. p. 46

Collins, D. J. (2017). Characteristic of the Triassic Newark Basin of New York and New Jersey for geologic storage of carbon dioxide. Washington D.C.: U.S. Department of Energy Office of Fossil Energy. https://doi.org/10.2172/1368193.

Cyphers, S. R., Jonsson, H., & Kperna, G. J. Jr. (2013). Geologic characterization for the U.S. SECARB anthropogenic test: Combining modern and vintage well data to predict reservoir properties. Search and Discovery Article 41156, AAPG, p. 2.

Daniel, R. & Kaldi, J. (2008). Evaluating seal capacity of caprocks and intraformational barriers for the geosequestration of CO_2. Sydney: PESA Eastern Australian Basins Symposium III, September 14–17, pp. 473–484.

De la Muela, A. S., Ramos, A., Perez-Lopez, R., et al. (2024). Micro-seismicity associated with CO_2 geological storage activities in Hontomin, Spain. International Journal of Greenhouse Gas Control 135, p. 104169. https://doi.org/10.1016/j.ijggc.2024.104169.

Demir, I., Morse, D. G., Elrick, S. D., et al. (2004). Delineation of the coalbed methane resources of Illinois, Illinois Geological Survey. Circular 564, p. 74.

Dressel, B. (2015). Recent site characterization studies being investigated by the Department of Energy (DOE)/National Energy Technology Laboratory (NETL) for potential geologic storage of CO_2. Search and Discovery Article 80496, AAPG, p. 46.

EPA. (2024a). Climate Change Indications: US Greenhouse Gas Emissions. www.epa.gov/climate-indicators/climate-change-indicators-us-greenhouse-gas-emissions.

EPA. (2024b). Class VI - Wells used for Geologic Sequestration of Carbon Dioxide. www.epa.gov/uic/class-vi-wells-used-geologic-sequestration-carbon-dioxide#ClassVIWell.

Ettehadtavakkol, A, (2016). Storage of CO_2 in depleted/producing oil reservoirs, Part V. In: Geologic Carbon Sequestration (ed. V. Vishal & T. N. Singh), pp. 185–212. New York: Springer Nature.

European Environment Agency. (2024). Trends in atmospheric concentrations of CO_2 (ppm), CH_4 (ppb) and N_2O (ppb) between 1800 and 2017. European Environment Agency. www.eea.europa.eu/data-and-maps/daviz/atmospheric-concentration-of-carbon-dioxide-5.

Furre, A., Eriken, O., Alnes, H., et al. (2017). 20 years of monitoring CO_2-injection at Sleipner. Energy Procedia 114, pp. 3016–3926.

Gapminder Foundation. (2024). World population. www.gapminder.org.

Greenberg, S. E., Finley, R. J., Frailey, S. M., et al. (2008). Carbon sequestration in the Illinios Basin, Abstract, GSA. North-Central Section, 42nd Annual Meeting 40, p. 82.

Gregersen, U., Smit, F. W. H., & Vosgerau, H. (2023). Maturation for potential CO_2 storage in Denmark – the Stenlille and HavnsØ structures. Geologic Survey of Denmark and Greenland. www.ngwm2024.se/wp-content/uploads/2023/12/3_Gregersen_U.pdf.

Goldberg, D. S., Kent, D. V., & Olsen, P. E. (2010). Potential on-shore and off-shore reservoirs for CO_2 sequestration in Central Atlantic magmatic province basalt. Proceeding of the National Academy of Sciences 107 (4), pp. 1327–1332. https://doi.org/10.10173/pnas.0913721107.

Gorecki, C., Sorenson, J. A., Brenner, J. M., et al. (2009). Development of storage coefficients for carbon dioxide storage in deep saline formation. In: SPE International Conference on CO_2 Capture, Storage, and Utilization. San Diego, California: Volume SPE Paper 126444.

Hincks, T., Aspinall, W., Cooke, R., et al. (2018). Oklahoma's induced seismicity strongly linked to wastewater injection depth. Science 359 (6381), pp. 1251–1255.

Hosseinzadeh, S., Haghighi, M., Salmachi, A., et al. (2024). Carbon dioxide storage within coal reservoirs: A comprehensive review. Geoenergy Science and Engineering 241, p. 18.

Howat, E., Gupta, N., Kelley, M., et al. (2017). Integrated sub-basin scale exploration for carbon storage targets: Advanced characterization of geologic reservoirs and caprocks in the Upper Ohio River Valley. Search and Discovery Article 30492, AAPG, p. 53.

Jain, R., Urban, L., Balbach, H., et al. (2012). Chapter Thirteen – Contemporary issues in environmental assessment. In: Handbook of Environmental Engineering Assessment (eds. R. Jain, L. Urban, H. Balbach et al.), pp. 361–447. https://doi.org/10.1016/B978-0-12-388444-2.00013-0.

Jia, W., McPherson, B., Pan, F., et al. (2019). A study of multiphase flow in CO_2-EOR: Impacts of three-phase relative permeability and hysteresis models. Search and Discovery Article 80669, AAPG, p. 30.

Keister, L, Place, M., Conner, A., et al. (2019). Investigation of potential geochemical reactions in large-scale carbon-dioxide-enhanced oil recovery (CO_2-EOR) carbonate reservoirs. AAPG, Search and Discovery Article 80680, p. 3.

Kentucky Geological Survey. (2024). Stages of coalification, Kentucky Geological Survey. www.uky.edu/KGS/coal/coal-anthracite.php.

Khosrokhavar, R., Griffiths, S., & Wolf, K. (2014). Shale gas formations and their potential for carbon storage: Opportunities and Outlook, Environ. Process 1, pp. 595–611.

Klappstein, G. & Rostron, B. (2014). Shallow hydrogeological and hydrochemical characterization of the Aquistore CO_2 sequestration site in Estevan, Saskatchewan, Canada. Energy Procedia 63, pp. 4971–4976.

Kristensen, L., Hjuler, M. L., Frykman, P., et al. (2016). Pre-drilling assessments of average porosity and permeability in the geothermal reservoirs of the Danish area. Geotherm Energy 4, (6), p. 27.

Lee, J. H., Cho, V. J., & Lee, K. S. (2020). Effects of aqueous solubility and geochemistry on CO_2 injection for shale gas reservoirs. Scientific Reports 10 (1), p. 2071.

Lee, S. & Kam, S. I. (2013). Enhanced oil recovery by using CO_2 foams: Fundamental and field applications. In: Enhanced Oil Recovery: Field Cas Studies (ed. J. J. Sheng), pp. 23–61. New York: Elsevier.

Lei, L., Sua, Y., Haoa, Y., et al. (2019). A comparative study of CO_2 and N_2 huff-n-puff EOR performance in shale oil production. Journal of Science and Petroleum Engineering 181, p. 10.

Leslie, R., Cavanaugh, A. J., Haszeldine, R. S., et al. (2021). Quantification of solubility trapping in natural and engineered CO_2 reservoirs. Petroleum Geoscience. https://doi.org/10.1144/petgeo2020-120.

Lewis, J. E., Carter, K., Dinterman, P., et al. (2019). An overview of the MRCSP petroleum fields 2019 database release. AAPG, Search and Discovery Article 80686, AAPG, p. 2.

Lohr, C. D. & Hackley, P. C. (2018). Using mercury injection pressure analyses to estimate sealing capacity of the Tuscaloosa marine shale in Mississippi, USA: Implications for carbon dioxide sequestration. International Journal of Greenhouse Gas Journal 78, pp. 375–387.

Lueck, A., Raef, A., & Watney, W. L. (2016). Characterizing a Mississippian carbonate reservoir for CO_2-EOR and carbon sequestration: Applicability of existing rock physics models and implications to feasibility of time lapse monitoring programs in the Wellington Oil Field, Sumner County, Kansas. Search and Discovery Article 80530, AAPG, p. 3.

Lüthie, M., Haspang, M. P., & Nielsen, C. M. (2023). Europe's first full-scale onshore CO_2 storage project. GeoExPro October 6th, p. 5. www.geoexpro.com/europes-first-full-scale-onshore-co_2-storage-project/.

Lyu, Q., Tan, J., Li, L., et al. (2021). The role of supercritical carbon dioxide for recovery of shale gas and sequestration in gas shale reservoirs. Energy Environ Science 14, pp. 4203–4227.

Macrotrends. (2024). World Population 1950–2024. www.macrotrends.net/global-metrics/countries/WLD/world/population.

Mahmoud, M., Hamza, A., Hussein, I., et al. (2020). Carbon dioxide EGR and sequestration mature and immature shale: Adsorption study. Journal of Petroleum Science and Engineering 188, p. 10.

Manitoba. (2014). Greenhouse CO_2 Supplement. https://www.gov.mb.ca/agriculture/crops/crop-management/co2-supplement.html.

McBride, J. H., Keach II, R. W., Chandler, C. K., et al. (2013). Testing 3D seismic attribute strategies for subtle fault mapping. Search and Discovery Article 41232, AAPG, p. 3.

McDonnell, T. (2021). The world's biggest carbon-sucking machine is switching on in Iceland. World Economic Forum, September 13. www.weforum.org/agenda/2021/09/worlds-biggest-carbon-machine-iceland/.

McLaughlin, J. F., Ganshin, Y., Quillman, S., et al. (2014). Mitigating risks associated with long-term CCS: Characterizing the geologic history and heterogeneity of sealing strata. Energy Procedia 63, pp. 4999–5009.

Medina, C. & Rupp, J. A. (2012). Lithostratigraphic division of the Mt. Simon sandstone (Cambrian) along Cincinnati Arch: Implications for reduction of uncertainty in estimations of geologic sequestration storage capacity. Search and Discovery Article 41041, AAPG, p. 4.

Medina, C. R., Mastalerz, M, Lahann, R. W., et al. (2020). Seal capacity of Upper Ordovician units in the Midwest region: A quantitative approach using mercury injection capillary pressure. Search and Discovery Article 80725, AAPG, p. 3.

Melillo, J. & Gribkoff, E. (2015). Soil-based carbon sequestration. Climate Portal, Massachusetts Institute of Technology. www.climate.mit.edu/explainers/soil-based-carbon-sequestration.

Meneguolo, R., Sundal, A., Martinius, A. W., et al. (2022). Impact of the lower Jurassic Dunlin Group depositional elements on the Aurora CO_2 storage site, EL001, northern North Sea, Norway. International Journal of Greenhouse Gas Control 119, p. 103723. https://doi.org/10.1016/j.ijggc.2022.103723.

Merey, S. & Sinayuc, C. (2016). Analysis of carbon dioxide sequestration in shale gas reservoirs by using experimental adsorption data and adsorption models. Journal of Natural Gas Science and Engineering 36 (Part A), pp. 1087–1105.

Mohammadkhani, S., Fogden A., & Olsen, D. (2024). Fluid-rock interactions during intermittent injection of supercritical CO_2: An investigation for CO_2 storage in the depleted Nini West oil field, Danish North Sea. International Journal of Greenhouse Gas Control 135, p. 104141. https://doi.org/10.1016/j.ijggc.2024.104141.

Monies, P. (2015). More than 635 Oklahoma disposal wells directed to reduce depths, volumes. The Oklahoman, November 29th. www.oklahoman.com/story/business/energy-resource/2015/11/29/more-than-635-oklahoma-disposal-wells-directed-to-reduce-depths-volumes/60706666007/.

Mungan, N. (1983). Chapter IV: Carbon dioxide flooding. In: Improved Oil Recovery (ed. W. T. Dowd), pp. 116–163. Washington D.C.: Interstate Oil Compact Commission.

Murugesu, M. P., Joewondo, N., & Prasad, M. (2023). Carbon storage capacity of shale formations: Mineral control on CO_2 adsorption. International Journal of greenhouse Gas Control 124, p. 17.

National Research Council. (2010). Ocean Acidification: A National Strategy to Meet the Challenges of a Changing Ocean. Washington, DC: The National Academies Press. https://doi.org/10.17226/12904.

Nazarian, B. & Furre, A. (2022). Simulation study of Sleipner plume on the entire Utsira using a multi-physics modelling approach. In: 16th International Conference on Greenhouse Gas Control Technologies. GHGT-16, October 23rd–27th, p. 14. Lyon, France. www.papers.ssrn.com/sol3/papers.cfm?abstract_id=4274191.

Nguyen, M., Zhang, Y., Zhang, X., et al. (2019). Fluvial channels boundary conditions in relation to the success of Shenhua CCS demonstration project. Ordos Basin, Search and Discovery Article 80651, AAPG, p. 27.

Ning, Y., Tura, A., & Guiltinan, E. (2024). CO_2 sequestration into a depleted natural CO_2 Dome: Sheep Mountain. Colorado, AAPG, Abstract, National Convention, August.

Nuttall, B. C., Sparks, T. N., & Greb, S. F. (2019). Mapping storage and enhanced gas recovery for organic shale in the Midwest Regional Carbon Sequestration Partnership. Search and Discovery Article 80667, AAPG, p. 4.

Olsen, P. E., Kent, D. V., Cornet, B., et al. (1996). High-resolution stratigraphy of the Newark rift basin (early Mesozoic, eastern North America). GSA Bulletin 108 (1), pp. 40–77.

Olsen, P. E., Kinney, S. T., Zakharova, N. V., et al. (2016). New insights on rift basin development and geological carbon cycle, mass extinction, and carbon sequestration from outcrops, and new core, drill holes and seismic lines from the northern Newark Basin (New York and New Jersey). www.researchgate.net/publication/310804939_New_insights_of_rift_basin_development_and_the_geological_carbon_cycle_mass_extinction_and_carbon_from_outcrops_and_new_core_drill_holes_and_seismic_lines_from_the_northern_Newark_Basin_New_York_and_N/citation/download.

Ozotta, O. & Ostadhassan, M. (2019). Geomechanical analysis of CO_2 sequestration in the Bakken Formation. Search and Discovery Article 80691, AAPG, p. 2.

Paddison, L. (2024). The 'world's largest' vacuum to suck climate pollution out of the air just opened. Here's how it works, CNN, May 8th. www.cnn.com/2024/05/08/climate/direct-air-capture-plant-iceland-climate-intl/index.html.

Paustain, K., Larson, E., Kent, J., et al. (2019). Soil C sequestration as a biological negative emission strategy. Frontiers in Climate 1, p. 482133. www.frontiersin.org/journals/climate/articles/10.3389/fclim.2019.00008/full.

Permentier, K., Vercammen, S., Soetaert, S., et al. (2017). Carbon dioxide poisoning: A literature review of an often-forgotten cause of intoxication in the emergency department. International Journal of Medicine Emergency Medicine 10 (14), p. 4. www.ncbi.nlm.nih.gov/pmc/articles/PMC5380556/pdf/12245_2017_Article_142.pdf.

Petersen, H., Springer, N. Weibel R., et al. (2022). Sealing capability of the Eocene-Miocene Horda and Lark formations of the Nini West depleted oil field – implications for safe CO_2 storage in the North Sea. Greenhouse Gas Control 118, p. 103675. https://doi.org/10.1016/j.ijggc.2022.103675.

Rose-Coss, D., Wampomah, W., Hutton, A. C., et al. (2015). Geologic characterization for CO_2-EOR simulation: A case study of the Farnsworth Unit, Anadarko Basin, Texas. Search and Discovery Article 80484, AAPG, p. 29.

Rostron, B., Whiter, D., Hawkes, C, et al. (2014). Characterization of the Aquistore CO_2 project storage site, Saskatchewan, Canada, Energy Procedia, 63, p. 2977–2984.

Sanderman, J., Hengl, T., & Fiske, G. J. (2017). Soil carbon debt of 12,000 years of human land use. PNAS 114, (36). pp. 9575–9580. https://doi.org/10.1073/pnas.1706103114.

Service, R. F. (2018). Cost plunges for capturing carbon dioxide from the air: Approach could make it affordable to produce gasoline from air, water, and renewable electricity. Science, June 7th. www.science.org/content/article/cost-plunges-capturing-carbon-dioxide-air.

Slater, B. E., Smith, L., Tymchak, M. P., et al. (2013). Preliminary results from the Tricarb deep stratigraphic well drilled into the Newark Rift Basin, Rockland County, New York. Search and Discovery Article 80303, AAPG, p. 4.

Tedesco, S. (2004). Coalbed methane potential and activity in the Illinois Basin. AAPG, Search and Discovery Article #10058, p. 8.

Trentham, B. (2011). Residual oil zones: The long-term future of enhanced oil recovery in the Permian Basin and Elsewhere. Search and Discovery Article 40787, AAPG, p. 41.

U.S. Government. (2022). Inflation Reduction Act. www.home.treasury.gov/policy-issues/inflation-reduction-act and www.congress.gov/bill/117th-congress/house-bill/5376.

Van der Baan, M. (2021). Earthquakes triggered by underground fluid injection modelled for a tectonically active oil field. Nature, News and Views, July. www.nature.com/articles/d41586-021-01997-7.

Vishal, V., Sudhakaran, A., Tiwari, A. K., et al. (2016). A review summary on multiple aspects of coal seam sequestration, Part IV. In: Geologic Carbon Sequestration (ed. V. Vishal & T. N. Singh), pp. 161–182. New York: Springer Nature.

Webb, N., Grigsby, N., & Frailey, S. (2019). CO_2 storage and EOR resources assessment of the Cypress sandstone residual oil zone play in the Illinois Basin. Search and Discovery Article 11263, AAPG, p. 5.

White, D., Movahedzadeh, Z., & Klimovych, M. (2024). CO_2 plume migration over 8-1/2 years of injection at the Aquistore CO_2 storage site. AAPG International Convention, August 27th to 30th, Houston, TX: AAPG.

Wikipedia. (2024a). Carbon Dioxide. www.en.wikipedia.org/wiki/Carbon_dioxide.

Wikipedia. (2024b). Estimates of Historical World Population. www.en.wikipedia.org/wiki/Estimates_of_historical_world_population.

Worth, K., White, D., Chalaurnyk Sorenson, J., et al. (2014). Aquistore project measurement, monitoring, and verification: from concept to CO2 injection. Energy Procedia, 63, pp. 3202–3208.

Worth, K., White, D., Chalaturnyk, R. et al. (2016). Aquistore: Tear one – injection, data, results. Energy Procedia 114, pp. 5624–5635.

Zakharova, N. V. & Goldberg, D. S. (2014). In situ stress analysis in the north Newark Basin: Implications for induced seismicity from CO_2 injection, AGU. Journal of Geophysical Research: Solid Earth 119, pp. 2362–2374.

Zakharova, N. V., Goldberg, D. S., Olsen, P. E., et al. (2016). New insights into lithology and hydrogeology of the northern Newark Rift Basin, AGU. Geochemistry, Geophysics, Geosystems 17 (6), pp. 2070–2094.

Zakharova, N. V., Goldberg, D. S., Olsen, P. E., et al. (2020). Reservoir and sealing properties of the Newark rift basin formations: Implications for carbon sequestration, SEG. The Leading Edge, Special Section: CO_2 39 (1), pp. 38–46.

Zareei, D., Rostami, B. & Kostarelos, K. (2022). Petrophysical changes of carbonate rocks related to CO_2 injection and sequestration. International Journal of Greenhouse Gas Control 117, p. 103648. https://doi.org/10.1016/j.ijggc.2022.103648.

Zhao, B., MacMinn, C. W., & James, R. (2014). Residual trapping, solubility trapping and capillary pinning complement each other to limit CO_2 migration in deep saline aquifers. Energy Procedia 63, pp. 3833–3839.

Index

Note: Page numbers in *italics* and **bold** refers to figures and tables respectively.

Geophysical Exploration for Hydrocarbon Reservoirs, Geothermal Energy, and Carbon Storage: New Technologies and AI-based Approaches, First Edition. Edited by Said Gaci.